INTERNATIONAL GLOSSARY OF TECHNICAL TERMS FOR THE PULP AND PAPER INDUSTRY

From the Publishers of
PULP & PAPER, PULP & PAPER INTERNATIONAL
POST'S PULP & PAPER DIRECTORY
INTERNATIONAL PULP & PAPER DIRECTORY
ATLAS OF THE WESTERN EUROPEAN PULP AND PAPER INDUSTRY

Miller Freeman Publications, Inc.

OTHER BOOKS, DIRECTORIES, AND MARKETING AIDS FOR THE PULP AND PAPER INDUSTRY

Fiber Conservation and Utilization
Proceedings of the seminar sponsored by
Pulp & Paper, Chicago, Illinois, May 1974.

Transport and Handling in the Pulp and Paper Industry
Proceedings of the first international symposium
sponsored by Pulp & Paper International,
Rotterdam, Holland, April 1974.

An Introduction to Paper Industry Instrumentation
by John R. Lavigne

Metrication for the Pulp and Paper Industry
by Kenneth E. Lowe

Synthetic Polymers and the Paper Industry
by Vladimir M. Wolpert

Practical Computer Applications for the Pulp and Paper Industry
edited by Kenneth E. Lowe

Post's Pulp and Paper Directory
Complete annual directory of the pulp, paper,
paperboard, and converting industries of
North America, updated annually and including
the mill buyer's guide.

The International Pulp & Paper Directory
Biennial directory of pulp, paper, and
paperboard mills of the world containing
detailed information on all mills with over
10,000 tons annual capacity.

The Pulp and Paper Mill Map
Full color map showing the exact location of
every pulp, paper, and paperboard mill in the
USA and Canada. Also identifies the six major
forest areas of North America with a legend
describing species found in each.

Atlas of the Western European Pulp and Paper Industry
Showing all pulp, paper, and paperboard mills
in Western Europe with over 5,000 tons capacity.

Copyright ©1976 by
Miller Freeman Publications, Inc.,
500 Howard Street, San Francisco,
California 94105, USA.
Printed in the United States of America.

Library of Congress Catalog Card Number: 74-20168
International Standard Book Number: 0-87930-037-X

First edition, March 1976

INTERNATIONAL GLOSSARY OF TECHNICAL TERMS FOR THE PULP AND PAPER INDUSTRY

Terms most commonly used for
Forestry, Wood Handling, Pulping, Papermaking,
Finishing, Converting, Testing, and
Research & Development

Paul D. Van Derveer and Leonard E. Haas, Editors

ENGLISH SVENSKA DEUTSCH FRANÇAIS ESPAÑOL

Translated by
Fernand Porcile, Paris, France
Helga Rothe, Düsseldorf, Federal Republic Germany
Olle Andersson, Solna, Sweden
Luis and Eva Gonzalez-Agudo, Madrid, Spain

US $30
First Edition
1976

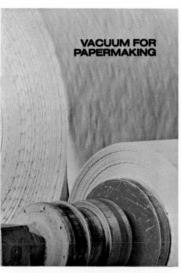
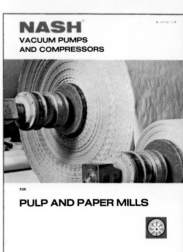

CONTENTS

Preface   7

1   Glossary of English Terms   9

2   Language Indexes
   Swedish   145
   German   165
   French   195
   Spanish   217

3   Index to Advertisers   234

Svenska

Deutsch

Français

Español

HOW TO USE THE GLOSSARY

The glossary uses the letter-by-letter system of alphabetizing in which word breaks and punctuation are disregarded. Additionally, hyphens are ignored; abbreviations are alphabetized as they appear, not as if they were spelled out; phrases are alphabetized as they appear, not by inverting to put an arbitrary key word first; numbers are spelled out.

When the English term is known
Turn to Section 1. The English terms are numbered and listed alphabetically. The Swedish, German, French, and Spanish equivalents will be found opposite each English term. Within each language, synonyms for the same term are separated by commas. A dash (—) is printed in a column where there is no equivalent term in that particular language. Insofar as the idioms of each language allow, equivalents are given in the same part of speech, syntax, number, tense, person, or mood as the English term.

When the Swedish, German, French, or Spanish term is known
Turn to the corresponding index in Section 2. All the equivalent terms in a particular language that appear in Section 1 are indexed alphabetically. Nonspeakers of Swedish should note that the letters Å, Ä, and Ö are alphabetized after the letter Z. Nonspeakers of Spanish should note that CH, LL, and Ñ are alphabetized as separate letters, following C, L, and N respectively. Following each term is a code number which refers back to the numbered English terms in Section 1. Some common terms have several English equivalents and are therefore followed by several code numbers.

ANVISNINGAR FÖR ORDBOKENS BEGAGNANDE

Ordboken begagnar ord-för-ord systemet för bokstaveringen, så att man bortsett från mellanrummet vid sammansatta ord. Bindestreck bortses också från; förkortningar är alfabetiska såsom de framgår, dvs inte som om de vore utskrivna; uttryck är alfabetiska i den ordning orden är placerade i själva uttrycket, dvs icke genom att man placerat ett godtyckligt nyckelord först; nummer är utskrivna.

När Det Engelska Uttrycket Är Känt

Gå till Del 1. De engelska uttrycken är numrerade och i alfabetisk ordning. De svenska, tyska, franska och spanska motsvarigheterna återfinnes mitt emot det engelska uttrycket. Synonymer inom varje språk för ett speciellt uttryck är åtskilda medelst kommatecken. Saknas motsvarighet för ett uttryck i något språk anges detta med tankstreck (-). I den mån som de idiomatiska uttrycken i varje språk tillåter, är motsvarigheterna angivna i samma ordklass, syntax, antal, tempus, person eller modus som det engelska uttrycket.

När Det Svenska, Tyska, Franska eller Spanska Uttrycket Är Känt

Gå till motsvarande index i Del 2. Alla synonyma uttryck i ett visst språk som förekommer i Del 1 är angivna i alfabetisk ordning. För ickesvensktalande bör noteras att bokstäverna Å, Ä och Ö följer efter Z. Icke-spansktalande bör notera att CH, LL och Ñ förekommer såsom enskilda bokstäver, respektive efter C, L och N. Efter varje uttryck följer ett kodnummer, vilket hänvisar till de numrerade engelska uttrycken i Del 1. Vissa vanliga uttryck ha flera engelska motsvarigheter och är därför följda av flera kodnummer.

UTILISATION DU LEXIQUE

Le lexique utilise le système alphabétique lettre par lettre dans lequel il est fait abstraction des intervalles entre les mots et de la ponctuation. De plus, il n'est pas tenu compte des traits d'union; les abbréviations sont reportées alphabétiquement telles qu'elles apparaissent, non pas comme si elles étaient orthographiées; les expressions sont reportées alphabétiquement telles qu'elles apparaissent, non pas en faisant une inversion de mots pour placer, en tête, un mot clé arbitraire; les nombres sont orthographiés.

Quand le terme anglais est connu

Se reporter à la Section 1. Les termes anglais sont numérotés et inscrits alphabétiquement. Les équivalents suédois, allemand, français et espagnol seront trouvés en face de chaque terme anglais. A l'intérieur de chaque langue, les synonymes d'un même terme sont séparés par une virgule. Un tiret est imprimé dans la colonne d'une langue pour laquelle il n'existe pas de terme équivalent. Autant que les idiotismes de chaque langue le permettent, les équivalents sont donnés dans la même classe de mot, la même syntaxe, le même nombre, le même temps, la même personne et le même mode que le terme anglais.

Quand le terme suédois, allemand, français ou espagnol est connu

Se reporter à l'index correspondant dans la Section 2. Tous les termes équivalents dans une langue particulière, qui apparaissent à la Section 1, sont inscrits alphabétiquement. Les personnes ne parlant pas le suédois noteront que les lettres Å, Ä, et Ö sont inscrites alphabétiquement après la lettre Z. Les personnes ne parlant pas l'espagnol noteront que les lettres CH, LL, et Ñ sont inscrites alphabétiquement et séparément à la suite des lettres C, L, et N respectivement. Chaque terme est suivi d'un code chiffré qui se réfère au terme anglais numéroté à la Section 1. Quelques termes courants possèdent plusieurs équivalents anglais et sont donc suivis de plusieurs codes chiffrés.

ANWEISUNG ZUM GEBRAUCH DES WÖRTERBUCHES

Dieses Wörterbuch ist nach Buchstaben alphabetisch angeordnet; Trennungen, Satzzeichen und Bindestriche bleiben unbeachtet. Abkürzungen werden alphabetisch aufgeführt, wie sie gelesen werden, nicht wie sie geschrieben erscheinen. Ausdrücke, die aus mehreren Worten bestehen, sind aufgeführt, wie sie gelesen werden, nicht unter Voransetzung eines beliebig gewählten Schlüsselwortes; Zahlen werden ausgeschrieben.

Wenn der englische Ausdruck bekannt ist
Sektion I aufschlagen. Die englischen Ausdrücke sind numeriert und alphabetisch aufgeführt. Die entsprechenden schwedischen, deutschen, französischen und spanischen Ausdrücke sind dem englischen Ausdruck gegenübergestellt. In jeder Sprache werden die verschiedenen Synonyme für den gleichen Ausdruck durch ein Komma getrennt. Wenn es in einer der Sprachen keinen gleichwertigen Ausdruck gibt, wird ein Strich (-) in die entsprechende Sparte eingesetzt.

Soweit es die idiomatischen Redewendungen jeder Sprache erlauben, werden die Übersetzungen dem englischen Ausdruck in Anordnung, Satzbau, Nummer, Zeitform, Person oder Aussageweise angeglichen.

Wenn der schwedische, deutsche, französische oder spanische Ausdruck bekannt ist
Das entsprechende Sprachverzeichnis in Sektion II aufschlagen. Alle Ausdrücke einer bestimmten Sprache, die in Sektion I vorkommen, sind hier alphabetisch aufgeführt. Im Schwedischen ist zu beachten, dass die Buchstaben Å, Ä und Ö nach dem Buchstaben Z kommen. Im Spanischen werden CH, LL und N als getrennte Buchstaben je nach C, L und N aufgeführt. Hinter jedem Ausdruck steht eine Schlüsselnummer, die sich auf die numerierten englischen Ausdrücke in Sektion I bezieht. Allgemeinen Ausdrücken, für die es im Englischen mehrere Übersetzungen gibt, werden daher mehrere Schlüsselnummern beigeordnet.

COMO USAR EL GLOSARIO

El glosario usa el sistema de alfabetización de letra-por-letra en el que no se consideran ni la separación de palabras por sílabas ni la puntuación. Además: se ignoran los guiones; las abreviaciones se alfabetizan como aparecen, no como si fuesen deletreadas; las frases se alfabetizan como aparecen, no invirtiéndolas para colocar una palabra clave arbitratia primero; los números se deletrean.

Cuando se conoce el término inglés
Pase a la Sección I. Los términos ingleses están numerados y puestos en lista alfabeticamente. Los equivalentes suecos, alemanes, franceses y españoles se encuentran frente a cada término inglés. Dentro de cada idioma, los sinónimos para el mismo término están separados por comas. En una columna en la que no hay término equivalente en ese respectivo idioma aparece un guion (-). Hasta donde las expresiones idiomáticas de cada idioma lo permitan,

los equivalentes se dan en la misma parte de la oración, sintaxis, número, tiempo, persona o modo que en el término inglés.

Cuando se conoce el término sueco, alemán, francés o español
Pase al índice correspondiente en la Sección II. Todos los términos equivalentes en un respectivo idioma que aparecen en la Sección I están en el índice alfabeticamente. Quienes no hablen sueco deben notar que las letras Å, Ä y Ö aparecen alfabeticamente después de la letra Z. Quienes no hablen español deben notar que CH, LL y Ñ aparecen alfabeticamente como letras separadas siguiendo a C, L y N respectivamente. En seguida de cada término hay un número clave que se refiere a los anteriores términos ingleses numerados en la Sección I. Algunos términos comunes tienen varios equivalentes ingleses y así están seguidos de varios números clave.

PREFACE

Within the covers of this glossary the reader will find an extensive compendium of terms in five languages — all relating to the pulp and paper industry, from the forests to the marketplace. The paper industry includes a number of diverse operations — wood processing, material handling, pulping, papermaking, converting, warehousing, shipping, marketing, et cetera. And it utilizes complex processes involving many disciplines — civil, mechanical, electrical, and chemical engineering; biology, botany, physics, chemistry, and others. Each of these operations and disciplines has contributed terms peculiar to the paper industry — terms interwoven through the years with terms that grew up with the industry when papermaking was more art than science.

To some extent, the people involved in the industry have their own jargon — the words sound familiar, but they have a specific meaning to the workers, supervisors, and managers who make pulp and paper. For instance, "secondary fiber" describes the fiber obtained from recycled wastepaper; "freeness" is a property of the pulp relating to the drainage of water through a mat of fiber; the "calender" does not show days and months as does a calendar (note the slight difference in spelling), but is an item of equipment on the paper machine which compresses and smooths the paper.

These technical terms, words, and phrases have been compiled through years of research by the editors of PULP & PAPER and PULP & PAPER INTERNATIONAL. The initial version was in English — that is, as spoken in the North American pulp and paper mills, which comprise a large portion of the world's paper industry. The glossary was then translated into the four languages the editors felt would provide the broadest coverage of the present-day worldwide paper industry. Unquestionably, it would have been desirable to include as well several other languages, but the size of such a glossary would have exceeded practical limits.

We feel that the five languages included in this glossary — English, Swedish, German, French, and Spanish — will serve well those who wish to communicate with people in other countries concerning some aspect of the pulp and paper industry. Moreover, the need for a glossary of this nature is growing rapidly. Each year we see a worldwide increase in the trade of pulp and paper products and raw materials, as well as of machinery, equipment, and chemicals for pulp and paper manufacturers.

In order to publish a volume that would have industry-wide use, we sought the advice of industry specialists all over the world. Knowing the importance of keeping the volume within manageable size, while alphabetically cross-referencing all entries, we were obliged to cut the final list to approximately 6,200 terms.

We believe this volume covers the most important and the most frequently used terms for the industry, and is completely up-to-date. Its handy size and clear organization will be of great help to both specialists and nonspecialists faced with a need for translation in one or more of the five languages.

We would like to thank the translators, Fernand Porcile for the French, Dr. Olle Andersson for the Swedish, Helga Rothe for the German, and Luis and Eva Gonzalez for the Spanish. France's Centre Technique also gave assistance for some terms as did the Swedish Pulp & Paper Association for certain difficult Swedish equivalents.

Paul D. Van Derveer
Leonard E. Haas
Brussels
December 1975

ENGLISH

ENGLISH	SVENSKA	DEUTSCH	FRANÇAIS	ESPAÑOL

A

	ENGLISH	SVENSKA	DEUTSCH	FRANÇAIS	ESPAÑOL
100	a.c. (alternating current)	växelström	Wechselstrom	courant alternatif	C.A. (corriente alterna)
101	a.d. (air dry)	lufttorr	lufttrocken	sec à l'air	seco al aire
102	A-flute	A-pipa	A-Welle (der Wellpappe), grobe Riffelung	grande cannelure	canaladuras A
103	abaca	abaca	Manilahanf, Abaka	chanvre de Manille	cáñamo de Manila
104	abaca pulp	abaca-massa	Abakazellstoff	pâte de chanvre de Manille	pasta de abacá
105	Abies (Lat.)	Abies	Tanne, Tannenholz	sapin	abeto
106	abietic acids	abietinsyra	Abietinsäuren	acides abiétiques	ácidos abiéticos
107	abrade	nöta	abschaben, abreiben, abschleifen	raser, user en frottant	desgastar
108	abrasion	nötning	Abrieb, Abnutzung, Abscheuerung	abrasion	abrasión
109	abrasion resistance	nötningsmotstånd	Abriebfestigkeit, Scheuerfestigkeit	résistance à l'abrasion	resistencia a la abrasión
110	abrasiveness	nötningsförmåga	Abriebgrad, Abriebeffekt	pouvoir abrasif	poder abrasivo
111	abrasive papers	slippapper	Schleifpapiere, Schmirgelpapiere	papiers abrasifs (emeri)	papeles abrasivos
112	abrasives	slippulver	Schleifmittel	abrasifs	abrasivos
113	absolute humidity	absolut fuktighet	absolute Feuchtigkeit	humidité absolue	humedad absoluta
114	absorb	absorbera	absorbieren, aufsaugen, aufnehmen	absorber	absorber
115	absorbency	absorbtionsförmåga	Absorptionsvermögen, Saugfähigkeit	pouvoir absorbant, ascension capillaire	poder absorbente
116	absorbent	absorberande	saugfähig	absorbant	absorbente
117	absorbent papers	absorberande papper	saugfähiges Papier	papiers absorbants	papeles absorbentes
118	absorber	absorbator	Absorptionsgerät, Absorptionsmittel	absorbant	absorbedor
119	absorbing pad	absorbtionskudde	Saugpolster	tampon absorbant	almohadilla absorbente
120	absorption	absorption	Absorption	absorption	absorción
121	absorption coefficient	absorbtionskoefficient	Absorptionskoeffizient	coéfficient d'absorption	coeficiente de absorción
122	absorption measurement	absortionsmätning	Absorptionsmessung	mesure de l'absorption	medida de la absorción
123	absorption tester	absorbtionsprovare	Saughöhenmessgerät	appareil d'essais d'absorption	probador de absorción
124	absorption tower	absorbtionstorn	Absorptionsturm	tour d'absorption	torre de absorción
125	absorptive	absorberande	aufnahmefähig, saugähig	absorbant	absorbente
126	absorptive capacity	absorbtionsförmåga	Absorptionsvermögen	pouvoir absorbant	porosidad
127	absorptivity	absorbtionsförmåga	Saugfähigkeit, Aufnahmefähigkeit	absorption	absorbencia
128	acacia	akasia	Akazie	acacia	acacia
129	acacia gum, acacine	akasiagum	Akaziengummi, Akazin	gomme arabique	goma de acacia, acacina
130	accelerated aging test	prov genom accelererad åldring	Schnellalterungstest	essai d'oxydation accélérée	prueba de envejecimiento acelerado
131	accelerating agent	accelereringsmedel	Beschleunigungsmittel	produit accélérant	agente acelerador
132	acceptability	accepterbarhet, antaglighet	Eignung, Akzeptierbarkeit	admissibilité	admisibilidad
133	acceptance sampling system	kontrollprovtagning	Vorsortiersystem	–	sistema de clasificación previa
134	acceptance test	godkännandeprov	Eignungstest, Abnahmeprüfung	essai de réception	prueba de aceptación
135	accepted chips	accepterad flis	Gut-Chips, Hackschnitzel-Gutstoff	copeaux classés, copeaux acceptés	virutas aceptadas

	ENGLISH	SVENSKA	DEUTSCH	FRANÇAIS	ESPAÑOL
136	accepted stock	accepterad mäld	Gutstoff	pâte épurée	pasta depurada
137	accepts	accepterat material	Gutstoff	pâte acceptée, pâte épurée	pasta depurada
138	accessibility	accessibilitet	Zugänglichkeit	accessibilité	accesibilidad
139	accessory	tillbehör	zusätzlich, Zubehör, Zusatzeinrichtung	accessoire, secondaire	accesorio
140	accordian fold	dragspelsbälg	Zickzackfalz, Zickzackfalzung	plié en accordéon	plegado en acordeón
141	accumulation	ackumulering	Anhäufung, Ansammlung, Speicherung	accumulation	acumulación
142	accumulator	ackumulator	Akku, Kraftspeicher, Sammelbehälter	accumulateur	acumulador
143	accuracy	noggrannhet	Genauigkeit, Präzision, aufgewandte Sorgfalt	exactitude, précision	precisión
144	Acer (Lat.)	Acer	Ahorn	érable	puntiagudo
145	acetaldehyde	acetaldehyd	Acetaldehyd	acétaldéhyde	aldehida acética
146	acetals	acetaler	Acetal	acétyl	acetales
147	acetate fiber	acetatfiber	Acetatfaser	fibre d'acétate	fibra de acetato
148	acetate pulp	acetatmassa	Acetatzellstoff	pâte à l'acétate	pasta de acetato
149	acetate rayon	acetatsilke	Acetatseide	rayonne à l'acétate	rayón de acetato
150	acetates	acetater	Acetate	acétate	acetatos
151	acetic acid	ättiksyra	Essigsäure	acide acétique	ácido acético
152	acetic anhydride	ättiksyreanhydrig	Essigsäureanhydrid	anhydride acétique	anhídrido acético
153	acetone	aceton	Azeton	acétone	acetona
154	acetonitrile	acetonitril	Azetitril, Acetonitril	acétonitrile	acetonitrilo
155	acetylation	acetylering	Acetylierung	acétylation	acetilación
156	acetyl groups	acetylgrupper	Acetylgruppen	groupement acétyl	grupos acetilos
157	acid	syra	sauer, Säure	acide, lessive	ácido
158	acid chlorides	sura klorider	Säurechloride	chlorures acides	cloruros ácidos
159	acid dyes	sura färgämnen	saure Farbstoffe, saure Farben, lichtbeständige Farbstoffe	colorants acides	colorantes ácidos
160	acid dyestuff	surt fägêmne	saure Farbstoffe	colorant acide	colorante ácido
161	acid free	syrafri	säurefrei	sans acide	sin ácido
162	acid-free paper	syrafritt papper	säurefreies Papier	papier exempt d'acide	papel sin ácido
163	acid groups	syragrupper	saure Gruppen, Karboxylgruppe	groupement acides	grupos ácidos
164	acid halides	sura halider	Säurehalogenid	hydracides	haluros ácidos
165	acidic	sur (syra-)	säurehaltig	acidique	acidificador
166	acidification	surgörning	Ansäuerung, Ansäuern	acidification	acidulación
167	acidity	surhet	Acidität, Säuregrad, Säurehaltigkeit	acidité	acidez
168	acid making	syratillverkning	Säureherstellung	fabrication d'acide	acidificar
169	acid plant	syrafabrik	Sulfitanlage	salle de préparation de la lessive au bisulfite	sala de preparación de lejía
170	acidproof	syrafast	säurefest, säurebeständig	résistant à l'acide	resistente a los ácidos
171	acidproof brick	syrafast tegel	säurebeständiger Mauerstein, säurebeständiger Backstein, säurebeständiger Ziegelstein	brique résistant aux acides	ladrillo resistente a los ácidos
172	acidproof lining	syrafast murning, syrafast fodring	säurefeste Auskleidung	revêtement résistant aux acides	forro resistente a los ácidos
173	acidproof paper	syrahärdigt papper	säurefestes Papier	papier résistant aux acides	papel resistente a los ácidos
174	acid resistance	syramotstånd	Säurebestäandigkeit	résistance aux acides	resistencia a los ácidos
175	acid resistance tests	syramotståndsprov	Säurebeständigkeitsprüfung	essais de résistance aux acides	pruebas de resistencia a los ácidos
176	acid resistant	syrahärdig	säurefest, säurebeständig	résistant aux acides	antiácido
177	acid-resistant paper	syrahärdigt papper	säurefestes Papier	papier résistant aux acides	papel resistente a los ácidos
178	acid size	surt lim	Freiharzleim	collage à la résine	cola ácida
179	acid stable size	syrafast lim	säurebeständiger Leim	colle stable en milieu acide	cola resistente a los ácidos
180	acid sulfite	sur sulfit	saures Sulfit, Bisulfit	bisulfite	bisulfito
181	acid treatment	syrabehandling	Behandlung mit Säure	traitement à l'acide	tratamiento ácido
182	acid wash	syratvätt	Säurewäsche, Reinigung durch Säure	lavage à l'eau acidulée	lavado con agua acidulada
183	acoustical board	akustikplatta	Schalldämpfungsplatte, Akustikplatte	carton pour isolation acoustique	cartón acústico
184	acoustic insulation	ljudisolering	Schallisolierung	isolation acoustique	aislamiento acústico
185	acoustic properties	akustiska egenskaper	akustische Eigenschaften	propriétés acoustiques	propiedades acústicas
186	across the grain	tvêrs fibrerna	quer zur Faser	sens travers	a través de las fibras
187	acrylamide	akrylamid	Acrylamid	acrylamide	acrilamida
188	acrylates	akrylater	Acrylat, Acrylsäureester	acrylates	acrilatos
189	acrylic acid	akrylsyra	Acrylsäure	acide acrylique	ácido acrílico
190	acrylic compounds	akrylföreningar	Acrylverbindungen	composés acryliques	compuestos acrílicos
191	acrylic fiber	akrylfiber	Acrylfaser	fibre acrylique	fibra acrílica
192	acrylonitrile	akrylnitril	Acrylnitril	acrylonitrile	acrilonitrilo
193	activated carbon	aktivt kol	aktiviertes Kohlepapier, Aktivkohle, Absorptionskohle	carbone activé	carbón activado

	ENGLISH	SVENSKA	DEUTSCH	FRANÇAIS	ESPAÑOL
194	activated sludge	aktiverat slam	Belebtschlamm	boue activée	lodo activado
195	activated sludge process	aktiverat slam-process	Belebtschlammverfahren	procédé d'épuration par boues activées	tratamiento del lodo activado
196	active alkali	aktivt alkali	aktives Alkali	alcali actif	álcali activo
197	active chlorine	aktivt klor	Aktives Chlorgas	chlore actif	cloro activo
198	active sulfur	aktivt svavel	aktiver Schwefel	soufre actif	azufre activo
199	actual weight	aktuell vikt	tatsächliches Gewicht	poids réel	peso real
200	adding-machine paper	adderingsmaskinpapper	Additionsmaschinenpapier	papier pour caisses enregistreuses	papel para máquinas sumadoras
201	addition polymers	additionspolymer	Additionspolymer	polymères de complément	polimenoros de adición
202	additives	tillsatser	Papierhilfsstoffe, Zusätze	additifs	aditivos
203	address-label paper	adresslappapper	Adressenaufkleberpapier	papier pour étiquettes	papel para etiquetas de direcciones
204	adhesion	adhesion	Adhäsion, Haftfähigkeit, Haftfestigkeit	adhérence	adhesión
205	adhesive papers	klisterpapper	Adhäsivpapier, Haftpapier	papiers adhésifs	papeles adhesivos
206	adhesives	tillsatsmedel	Klebstoffe	adhésifs	adhesivos
207	adhesive migration	bindemedelsvandring	Bindemittelwanderung	migration de la colle	migración del adhesivo
208	adhesive tape	klistertejp	Klebestreifen, Klebeband	bande adhésive, bande gommée	cinta adhesiva
209	adiabatic conditions	adiabatiska förhållanden	adiabatische Zustände	conditions adiabatiques	condiciones adiabáticas
210	adjustable speed	reglerbar hastighet	regulierbare Geschwindigkeit, einstellbare Geschwindigkeit	vitesse variable	velocidad ajustable
211	adjusting color	justeringsfärg	Einstellen der Farbe	corriger la teinte	corrección de coloración
212	adjusting screw	justerskruv	Einstellspindel	vis réglable	tornillo de ajuste
213	admixture	blandning	Beimischung, Zusatz	mixture	aditivo
214	adsorb	adsorbera	adsorbieren, aufnehmen, ansaugen	adsorber	adsorber
215	adsorbate	adsorbat	Adsorbat, adsorbierter Stoff, adsorbierte Substanz	produit adsorbé	adsorbido
216	adsorbent	adsorbent	Adsorbens (der aufnehmende Stoff)	adsorbant	adsorbente
217	adsorption	adsorption	Adsorption	adsorption	adsorción
218	adsorptive capacity	adsorptionsfömåga	Adsorptionsfähigkeit, Adsorptionsvermögen	pouvoir adsorbant	capacidad de adsorción
219	adsorptivity	adsorptivitet	Adsorptionsfähigkeit	adsorption	capacidad de adsorción
220	aerate	lufta	lüften, auflockern	aérer	ventilar
221	aeration	luftning	Lüftung, Belüftung	aération	ventilación
222	aeration tank	luftningsgarvning	Belüftungstank	fosse d'aération	cuba de aireación
223	aerator	luftningsmedel	Abwasser-Belüfter	aérateur	aireador
224	aerobic process	aerob process	aerobes Verfahren	procédé aérobique	proceso aeróbico
225	affinity	affinitet	Affinität	affinité	afinidad
226	after-dryer	eftertork	Nachtrockner	post-sécheur	post-secador
227	agalite	agalit	Agalit, faseriges Mg-Silikat, Faserkalk	agalite (poudre d'amiante)	polvo de amianto
228	agar	agar	Agar	agar	agar
229	agate marble paper	agatpapper	Achatimitationspapier	papier marbré à l'agate	papel ágata jaspeado
230	age	ålder	altern, reifen, aushärten, ablagern, Alter	vieillesse	período
231	ageing	åldring	Altern, Alterung, Reifen	vieillissement, jaunissement, oxydation	envejecimiento (del papel)
232	agent	medel	Agens, Reaktionsmittel, Mittel	agent, produit	agente
233	agglomerate	agglomerat	agglomerieren, zusammenbacken, zusammenballen, sich klumpen	aggloméré	aglomerar
234	agglomeration	agglomeration	Agglomerat, Anhäufung, Ballung	agglomération	aglomeración
235	aggregate	aggregat	anhäufen, Anhäufung, Aggregat, Zuschlagstoff	assemblage	agregar
236	aging	åldring	Altern, Alterung, Reifen	vieillissement	envejecimiento
237	aging resistance	åldringsmotstånd	Alterungsbeständigkeit	résistance au jaunissement	resistencia al envejecimiento
238	aging test	åldringsprov	Alterungsprüfung	essai de jaunissement	prueba de envejecimiento
239	agitate	omröra	rühren, durchrühren, bewegen	agiter	agitar
240	agitation	omrörning	Bewegung, Erregung, Rühren	agitation	agitación
241	agitator	omrörare	Rührwerk, Rührer	agitateur	agitador
242	air	luft	Luft, lüften, der Luft aussetzen	air	aire
243	air bell	luftklocka	Luftblase, Schnalle (Papierfehler)	soufflette, ampoule	burbuja
244	air blade, air brush	luftblad, luftborste	Luftmesser,Luftbürste, Luftrakel	lame d'air	capa de aire
245	air blower	luftblåsare	Kühldüsen (am Glättwerk)	ventilateur	ventilador soplador
246	airborne	luftburen	durch Luft befördert, mittels Luft befördert	aéroporté	transportado por aire
247	air-brush coater	luftborstebestrykare	Luftbürstenstreicher	coucheuse à lame d'air	estucadora de labio soplador
248	air-brush coating	luftborstebestrykning	Luftbürstenstrichauftrag	couchage à la lame d'air	estucado por labio soplador
249	air bubbles	luftbubblor	Luftblasen	bulles d'air	burbujas de aire
250	air cap	luftgap	Hochleistungshaube	capsule d'air	cápsula de aire

	ENGLISH	SVENSKA	DEUTSCH	FRANÇAIS	ESPAÑOL
251	air chamber	luftkammare, luftrum	Luftkammer, Windkessel	chambre à air	cámara de aire
252	air content	lufthalt	Luftgehalt	teneur en air	contenido de aire
253	air cooled	luftkyld	luftgekühlt	refroidi à l'air	enfriado por aire
254	air cooling	luftkylning	Luftkühlung	refroidissement à l'air	enfriamiento por aire
255	air curtain	luftridå	Luftschleier	–	cortina de aire
256	air deckle	luftdäckel	Luftstrahl zur Begrenzung der Papierbahn auf der Siebpartie	bords barbés obtenus sur machine par jet d'air	marco de aire
257	air dried	lufttorkad	luftgetrocknet	séché à l'air	secado al aire
258	air-dried paper	lufttorkat papper	luftgetrocknetes Papier	papier séché à l'air	papel secado al aire
259	air dry (a.d.)	lufttorr	lufttrocken	sec à l'air	seco al aire
260	air dryer	luttorkare	Lufttrockner	sécheur à air	secador al aire
261	air drying	lufttorkning	an der Luft trocknen	séchage à l'air	secado al aire
262	air entrainment	luftindragning	Lufteinzug	entraînement par air	arrastre por aire
263	air filter	luftfilter	Luftfilter	filtre à air	filtro de aire
264	air-filtration paper	luftfiltreringspapper	Luftfilterpapier	papier pour filtres à air	papel para filtración de aire
265	air-float dryer	luftkuddetorkare	Schwebetrockner	sécheur à feuille aéroportée	sequería de hoja llevada por aire
266	airflow	luftflöde	Luftstrom, Luftbewegung	courant d'air	corriente de aire
267	airfoil	airfoil	Tragfläche, Tragflügel	système de séchage par feuille aéroportée	superficie aerodinámica
268	air heater	luftvärmare	Lufterhitzer	réchauffeur d'air	recalentador de aire
269	air impacter	luftblåsare	Luftpinsel (zum Verteilen der überschüssigen Streichfarbe)	lame d'air pour égaliser la couche sur une fonceuse	impactadora de aire
270	air impingement	luftinblåsning	Aufprall, Auftreffen der Luft	épuration par l'air	impacto de aire
271	air intake	luftintag	Lufteinlass, Luftansaugstutzen	prise d'air	toma de aire
272	air jet	luftstråle	Luftdüse, Luftstrahl	jet d'air	chorro de aire
273	air knife	luftkniv	Luftmesser,Luftrakel	lame d'air	capa de aire
274	air-knife coater	luftknivbestrykare	Luftrakelstreicher	coucheuse à lame d'air	estucadora de labio soplador
275	air-knife coating	luftknivbestrykning	Luftmesser, Luftschaberstreichen	couchage à lame d'air	estucado de labio soplador
276	air-knife mark	luftknivsmärke	Markierung durch Luftmesser verursacht	marque de la lame d'air	marca del labio soplador
277	airlaid	luftlagd	wirrverlegt, durch Luft verlegt (Fasern)	système pneumatique de formation de la feuille	apilado por aire
278	air-loaded tension device	pneumatisk spänningsanordning	luftbetätigte Spannungskontolle	dispositif de tension pneumatique	tensor cargado de aire
279	airmail paper	luftpostpapper	Luftpostpapier	papier avion	papel avión
280	air nozzle	luftmunstycke	Luftdüse	conduite d'air	tobera de aire
281	air permeability	luftpermeabilitet	Luftdurchlässigkeit	perméabilité à l'air	permeabilidad al aire
282	air pollution	luftförorening	Luftverschmutzung, Luftverunreinigung	pollution de l'air	contaminación atmosférica
283	air quality	luftkvalitet	Reinheitsgehalt der Luft	qualité de l'air	calidad del aire
284	air roll	luftrulle	Luftwalze, Blasenwalze (zur Beseitigung von Luftblasen im Papier)	rouleau souffleur	rodillo soplador
285	air velocity pressure	stagnationstryck	Staudruck	vitesse de pression de l'air	presión dinámica
286	albumins	albuminer	Albumin, Eiweissstoffe	albumines	albúminas
287	album paper	albumpapper	Albumpapier	papier pour albums photo	papel de álbum
288	alcohol groups	alkoholgrupper	Alkoholgruppen	groupes alcooliques	grupos de alcohol
289	alcohol lignins	alkoholligniner	Holzfaserstoff	lignines alcooliques	ligninas de alcohol
290	alcohols	alkoholer	Alkohole	alcools	alcoholes
291	aldehyde groups	aldehydgrupper	Aldehydgruppen	groupes aldéhydiques	grupos aldehidos
292	aldehydes	aldehyder	Aldehyde	aldéhydes	aldehidos
293	alder	al	Erle	aune	aliso
294	algae	alger	Algen, Seetang	varech	algas
295	algicides	algicider	Algenbekämpfungsmittel	algicides	algicidas
296	alginates	alginater	Alginat	alginates	alginatos
297	alignment	inriktning	Ausrichten, Ausrichtung, Ausfluchtung, Fluchtlinie	alignement	alineación
298	aliphatic compounds	alifatiska föreningar	aliphatische Verbindungen	composés aliphatiques	compuestos alifáticos
299	alkali	alkali	Alkali	alcali	álcali
300	alkali cellulose	alkalicellulosa	Alkalicellulose	alcali cellulose	álcali celulosa
301	alkali extraction	alkaliextraktion	Alkaligewinnung	extraction de l'alcali	extracción de álcali
302	alkali lignins	alkaliligniner	Alkalilignine	lignines alcalines	álcali ligninas
303	alkali resistance	alkaliresistans	Alkalibeständigkeit	résistance à l'alcali	resistencia a los álcalis
304	alkali-resistant cellulose	alkalihärdig cellulosa	alkalifeste Cellulose, laugenbeständige Cellulose	cellulose résistant à l'alcali	celulosa resistente a los álcalis
305	alkali-soluble	alkalilöslig	alkalilöslich	soluble dans l'alcali	álcali soluble
306	alkali-staining resistance	motstånd mot alkalifärgning	resistent gegen alkalische Verfärbung	résistance aux taches d'alcali	resistencia a la coloración por álcali
307	alkali treatment	alkalibehandling	alkalische Behandlung	traitement à l'alcali	tratamiento al álcali
308	alkaline filler	alkaliskt fyllmedel	alkalischer Füllstoff	charge alcaline	carga alcalina

	ENGLISH	SVENSKA	DEUTSCH	FRANÇAIS	ESPAÑOL
309	alkaline process	alkalisk process	alkalischer Prozess, alkalisches Verfahren	procédé alcalin	proceso alcalino
310	alkaline pulp	alkalisk massa	Alkalizellstoff	pâte alcaline	pasta alcalina
311	alkaline pulping	alkalisk uppslutning	alkalischer Zellstoffaufschluss	procédé alcalin de fabrication des pâtes	elaboración de pasta alcalina
312	alkaline sizing	alkalisk limning	alkalische Leimung	collage à l'alcali	encolado al álcali
313	alkalinity	alkalitet	Alkalinität, Alkalität, Alkaleszenz	alcalinité	alcalinidad
314	alkaliproof	alkalifast	alkalifest	résistant à l'alcali	resistente a los álcalis
315	alkaliproof paper	alkalifast papper	laugenbeständiges Papier	papier résistant à l'alcali	papel resistente a los álcalis
316	alkaliproof soap-box board	alkalifast tvålförpackningkarrtong	alkalifester Seifenschachtelkarton	carton pour caisses à savon résistant à l'alcali	cartón para embalaje del jabón resistente al álcali
317	alkylaryl sulfonates	alkylarylsulfonater	Alkylarylsulfonat	sulfonates d'alkylaryle	alquilaril sulfonatos
318	alkylation	alkylering	Alkylierung	alkylation	alquilización
319	alkyl groups	alkylgrupper	Alkylgruppen	groupes alkyl	grupos alcohilo
320	alloy	legering	Legierung	alliage	aleación
321	Alnus (Lat.)	Alnus	Erle	aulne	aliso
322	alpha cellulose	alfacellulosa	Alphacellulose	alpha cellulose	alfa celulosa
323	alpha gauge	alfastrålningsmätare	Alphamessgerät	jauge alpha	indicador alfa
324	alpha-protein	alfaprotein	Alphaprotein	alpha protéine	alfa proteína
325	alpha pulp	alfamassa	Alphazellstoff	pâte d'alfa	pasta alfa
326	alternating current (a.c.)	växelström	Wechselstrom	courant alternatif	coriente alterna (C.A.)
327	alum	alun	Alaun	alun	alumbre
328	aluminates	aluminater	Aluminat	aluminates	aluminatos
329	aluminium (Brit.)	aluminium	Aluminium	aluminum	aluminio
330	aluminum	aluminium	Aluminium	aluminum	aluminio
331	aluminum compounds	aluminiumföreningar	Aluminiumverbindungen	composés d'alumine	compuestos de alumino
332	aluminum foil	aluminiumfolie	Alufolie	feuille mince d'aluminum	laminilla de aluminio
333	aluminum paper	aluminiumpapper	Aluminiumpapier, Silberpapier	papier d'aluminum	papel aluminio
334	aluminum stearate	aluminiumstearat	Aluminiumstearat	stéarate d'alumine	estearato de aluminio
335	aluminum sulfate	aluminiumsulfat	Aluminiumsulfat, schwefelsaure Tonerde	sulfate d'alumine	sulfato de aluminio
336	alum spots	alunfläckar	Alaunflecken	taches d'alun	manchas de alumbre
337	amides	amider	Amide	amides	amidas
338	amine polymers	aminpolymerer	Aminpolymere	polymères amines	aminas polimeras
339	amines	aminer	Aminosäuren	amines	aminas
340	amino acids	aminosyror		acides aminés	aminoácidos
341	aminoethyl cellulose	aminoetylcellulosa	Aminoäthylcellulose	aminoéthyl cellulose	celulosa aminoetílica
342	amino groups	aminogrupper	Aminogruppen	groupes aminés	grupos amino
343	aminopropyl cellulose	aminopropylcellulosa	Aminopropylcellulose	aminopropyl cellulose	celulosa aminopropilo
344	ammeter	ampermeter	Amperemeter, Strommesser	ampèremètre	amperímetro
345	ammonia	ammoniak	Ammoniak, Ammon	ammoniaque	amoniaco
346	ammonium base liquor	ammoniumlut	Ammonium-Bisulfitlauge, Salmiakgeist	lessive à l'ammoniaque	lejía de base amónica
347	ammonium bisulfite	ammoniumbisulfat	Ammonium-Bisulfit	bisulfite d'ammoniaque	bisulfito de amonio
348	ammonium compounds	ammoniumföreningar	Ammoniumverbindungen	composés d'ammoniaque	compuestos de amonio
349	ammonium hydroxide	ammoniumhydroxid	Ammoniumhydroxid	hydroxyde d'ammoniaque	hidróxido de amonio
350	ammonium stearate	ammoniumstearat	Ammoniumstearat	stéarate d'ammoniaque	estearato de amonio
351	ammonium sulfate	ammoniumsulfat	Ammoniumsulfat,schwefelsaures Ammoniak	sulfate d'ammoniaque	sulfato de amonio
352	ammunition paper	ammunitionspapper	Kartuschenpapier, Patronenhülsenpapier	papier pour cartouches	papel de municiones
353	amorphous	amorf	amorph, nicht kristallin	amorphe	amorfo
354	amorphous cellulose	amorf cellulosa	amorphe Cellulose	cellulose amorphe	celulosa amorfa
355	amorphous material	amorft material	amorphes Material	produits amorphes	material amorfo
356	amphoteric	amfoter	amphoter	amphotère	anfótero
357	amylase	amalys	Amylase	amylase	amilasa
358	amylose	amylos	Amylose	amylose	amilosa
359	anaerobic process	anaerobisk process	anaerobes Verfahren	procédé anaérobique	proceso aneróbico
360	analog	analog	analog, sinngemäss	analogue	análogo
361	analog computer	analogdator	Analogrechner	calculateur analogique	calculadora analógica
362	analog system	analogsystem	Analogsystem	ordinateur analogique	sistema analógico
363	analog to digital converter	analog-digitalomvandlare	Analog-Digital-Umsetzer	convertisseur analogique à numérique	convertidor de sistema analógico en numérico
364	analysis	analys	Analyse, Bestimmung, Zerlegung	analyse	análisis
365	analysis of variance	variansanalys	Varianzanalyse, Streuungsanalyse	analyse de variation	análisis de variación
366	analytical filter paper	analytiskt filterpapper	Analysen-Filterpapier	papier filtre analytique	papel filtro analítico
367	anatase	anatas	Anatas (Titandioxid)	anatase	anatasa
368	anchor bolt	ankarbult	Ankerbolzen, Fundamentanker, Ankerschraube	boulon de scellement	perno de anclaje

13

	ENGLISH	SVENSKA	DEUTSCH	FRANÇAIS	ESPAÑOL
369	anemometer	anemometer	Windmessgerät	anémomètre	anemómetro
370	angiosperms	angiospermer	bedecktsamige Pflanzen	angiospermes	angioesperma
371	anhydrides	anhydrider	Anhydride	anhydres	anhidridos
372	anhydrite	anhydrit	Anhydrite	anhydrite	anhidrita
373	aniline	anilin	Anilin	aniline	anilina
374	aniline dyes	anilinfärgämnen	Anilinfarbstoffe	colorants à l'aniline	colorantes de anilina
375	aniline printing	anilintryckning	Anilindruck, Flexodruck	impression à l'aniline	impresión de anilina
376	animal glue	djurlim	Tierleim	colle animale, gélatine	cola animal
377	animal size	djurlom	Tierleim	collage en cuve à la gélatine	cola animal
378	anion exchange	anjonbyte	Anionenaustausch	échange d'anions	intercambio aniónico
379	anion exchanger	anjonbytare	Sauerstoffionenaustauscher	échangeur d'anions	cambiador aniónico
380	anionic	anjonisk	anionisch, anionenaktiv	anionique	aniónico
381	anionic compounds	anjoniska föreningar	anionische Verbindungen	composés anioniques	compuestos aniónicos
382	anions	anjoner	Anionen	anions	aniones
383	annual ring	årsring	Jahresring	anneau d'accroissement annuel	anillo de crecimiento
384	annular	ringformig	ringförmig	annulaire	anular
385	annuli	ringar	Ringe, Kreisringe	parties annulaires	anillos
386	anode	anod	Anode	anode	ánodo
387	antacid	antisyra	Säuren entgegenwirkend, neutralisierend, särewidrig	anti-acide, neutre	antiácido
388	antacid manila paper	syraskyddande manillapapper	Kabelisolierpapier	papier mince neutre	papel de manila antiácido
389	antiblocking agent	blockningshindrande medel	Antiblockmittel (das ein unerwünschtes Kleben oder Zusammenbacken verhindert)	produit empêchant l'adhérence en surface	agente antiadherente
390	antichlors	antifärger	Antichlor	antichlores	anticloros
391	anticrawl agent	krypningshindrande medel	Anticrawlmittel, Mittel zur Verhinderung der Wanderung (z.B. von Lösungsmitteln)	produit anti-dérapant	agente antiarrastre
392	antideflection	deflektionshindrande	Biegeschutz	antiflexion	antideflexión
393	antideflection roll	stabiliseringsvals	durchbiegungsfreie Walze	rouleau anti-déflecteur	rodillo antideflector
394	antifalsification paper	förfalskningssäkert papper	Sicherheitspapier, Fälschungsschutzpapier	papier de sécurité	papel infalsificable
395	antifoam	skumsläckande medel	Schaum entgegenwirkend, Schaumverhütungsmittel	anti-mousse	antiespumante
396	antifogging agent	dimhindranade medel	Nebelschutzmittel	produit anti-buée	agente antinebuloso
397	antioxidants	antioxderingsmedel	Antioxydationsmittel	antioxydants	antioxidantes
398	antique book paper	antikbokpapper	Antikbuchpapier	papier impression non apprêté	papel litos antiguo
399	antique bristol	antikbristol	Antikbristol	bristol non apprêté	Brístol antiguo
400	antique eggshell paper	halvmatt antikpapper	Antikpapier mit rauher Oberfläche (Eierschaleneffekt)	papier non apprêté couleur coquille d'oeuf	papel de China antiguo
401	antique finish	antikfinish	Antikausrüstung (rauhe Oberfläche)	fini antique	acabado antiguo
402	antique glazed paper	antikglättat papper	einseitig hochsatiniertes Antikpapier	papier satiné fini antique	papel satinado antiguo
403	antique laid bond	antikfinpapper	gerippte Antik-Feinpost, feingeripptes Dockumentenpapier	papier vergé fini antique	títulos verjurados antiguos
404	antique paper	antikpapper	Antikpapier	papier grainé	papel antiguo
405	antirust	rostskyddande	rostfrei, nichtrostend	anti-rouille	antioxidante
406	antirust papers	rostskyddande papper	Rostschutzpapier	papiers anti-rouille	papeles antioxidantes
407	antiskid coating	glidningsfri bestrykning	Gleitschutzbeschichtung	couchage anti-dérapant	estucado antideslizante
408	antiskid treatment	friktionsbehandling	Gleitschutzpräparation, rutschfeste Ausrüstung	traitement anti-dérapant	tratamiento antideslizante
409	antistatic agent	medel mot statisk elektricitet	Antistatikmittel, Mittel zur Verhinderung elektrostatischer Aufladung	produit antistatique	agente antiestático
410	antitarnish	missfärgningskyddande	Anlaufschutz	anti-ternissure	antideslustrante
411	antitarnish board	missfärgningskartong	Anlaufschutzkarton, Rostschutzkarton	carton pour coutellerie	cartón antideslustrante
412	antitarnish paper	missfärgningskyddande papper	Silbereinpackpapier	papier pour coutellerie	papel anticorrosivo
413	antitarnish tissue	missfärgningskyddande tunnpapper	Silberschutzseidenpapier	mousseline pour coutellerie	muselina anticorrosiva
414	apparent density	skenbar densitet	Schüttgewicht,scheinbare Dichte	grammage	densidad aparente
415	apparent specific volume	skenbar bulk	spezifisches Volumen	poids spécifique apparent	volumen específico aparente
416	apparent viscosity	apparent viskositet	Viskosität (bei bestimmtem Geschwindigkeitsgefälle)	viscosité apparente	viscosidad aparente
417	appearance	anblick	Aussehen, Erscheinen	aspect	aspecto
418	application	tillämpning	Auftrag, Einsatz, Anwendung	usage, emploi	impregnación
419	applicator	anbringare, applikator	Auftragsvorrichtung, Streichvorrichtung, Elektrode	enducteur	estucador

14

	ENGLISH	SVENSKA	DEUTSCH	FRANÇAIS	ESPAÑOL
420	applicator roll	påläggningsvals	Auftragswalze	rouleau enducteur	rodillo estucador
421	approach flow	inflöde	Zufluss	courant de pâte en avant de la machine à papier	corriente de pasta antes de la máquina
422	approach-flow system	inflödessystem	Zufliessystem, Stoffverteiler, Verteiler	dispositif d'amenée du courant de pâte	sistema de corriente de pasta antes de la máquina
423	appropriation	bevillning	Zuteilung, Zuweisung	appropriation	apropiación
424	apron	bröstläder	Schürze, Siebleder	tablier de machine à papier	delantal
425	apron board	virabord	Siebleiste (zur Regulierung der Entwässerung am Sieb)	traverse de fixation du tablier	cartón acolchado
426	aqueous	vatten-	wässerig, wasserhaltig	aqueux	acuoso
427	arabic gum	gummi arabicum	Gummi arabicum	gomme arabique	goma arábiga
428	arabinose	arabinos	Arabinose	arabinose	arabinosa
429	arching	valvbildning	sich wölbend, krümmend	arqué	curvatura
430	armature paper	armeringspapper	Isolierpapier	papier pour usages électriques	papel para inducidos
431	aromatic acids	aromatiska syror	aromatische Säuren	acides aromatiques	ácidos odoríferos
432	aromatic compounds	aromatiska föreningar	aromatische Verbindungen	composés aromatiques	compuestos aromáticos
433	aromatic groups	aromatiska grupper	aromatische Gruppen	groupes aromatiques	grupos aromáticos
434	arrowroot starch	arrowroot-stärkelse	Pfeilwurzmehl, Marantastärke	fécule d'arrow root	almidón de arrurruz
435	arsenical paper	arsenikpapper	arsensaures Papier, arsenhaltiges Papier	papier d'arsenic	papel arsenical
436	art cover	konsttrycksomslag	Glückwunschkartenkarton, Kunstdruckkarton	couverture couchée	encuadernación estucada
437	articulating paper	artikulationspapper	Zahnabdruckpapier	papier à articuler (pour dentistes)	papel para articular
438	artificial leather	konstläder	Kunstleder	simili cuir	cuero artificial
439	artificial leather paper	konstläderpapper	Kunstlederpapier	papier simili cuir	papel simil cuero
440	artificial parchment	konstgjort pergament	Pergamentersatz	simili sulfurisé	pergamino artificial
441	art paper	konsttryckpapper	Kunstdruckpapier	papier couché	papel estucado
442	art parchment	konsttryckpergament	Dokumentenpergament	parchemin couché	estucado pergamino
443	art poster board	konsttryckaffichpapper	Plakatkarton (häafig ein- oder beidseitig verstärkt)	carton couché pour affiches	cartulina estucada para carteles
444	art vegetable parchment	vegetabiliskt pergament för konsttryck	Echtdokumentenpergament	parchemin végétal couché	pergamino vegetal estucado
445	asbestine	asbestartad	asbestartig Magnesiumsilikat, faseriges Magnesiumsilikat, fast reines Magnesiumsilikat	poudre d'amiante	asbestina
446	asbestos	asbest	Asbest	amiante	amianto
447	asbestos board	asbestkartong	Asbestpappe	carton d'amiante	cartón asbestos
448	asbestos cement board	asbestcementpapper	Asbestzementpappe	fibro-ciment	cartón cemento de asbesto
449	asbestos felt	asbestfilt	Asbestfilz	feutre d'amiante	filtro de asbesto
450	asbestos fiber	asbestfiber	Asbestfaser	fibre d'amiante	fibra de amianto
451	asbestos paper	asbestpapper	Asbestpapier	papier d'amiante	papel asbestino
452	asbestos roofing felt	asbesttakpapp	Asbest-Dachpappe	feutre d'amiante pour toitures	fieltro asbestino para tejados
453	asbestos waterproofing felt	asbesttätningspapper	asphalt-imprägnierter Asbestfilz (zum Abdichten)	feutre d'amiante imperméable	fieltro asbestino impermeabilizante
454	ascorbic acid	askorbinsyra	Ascorbinsäure	acide ascorbique	ácido ascórbico
455	asexual reproduction	asexuell reproduktion	ungeschlectliche Fortpflanzung	reproduction cryptogame	reproducción asexual
456	ash	aska	Esche, Asche	cendre	ceniza
457	ash content	askhalt	Aschegehalt	teneur en cendres	contenido en cenizas
458	ashless	askfri	aschefrei	sans cendres	sin cenizas
459	ashless filter paper	askfritt filterpapper	aschefreies Filterpapier	papier filtre sans cendres	papel filtro sin cenizas
460	ash tree	ask	Eschenbaum	frêne	fresno
461	aspen	asp	Espe	tremble	álamo tremblón
462	asphalt	asfalt	Asphalt	asphalte, bitume	asfalto
463	asphalt emulsion	asfaltemulsion	Asphaltemulsion	émulsion d'asphalte	asfalto emulsificado
464	asphalt felt	asfaltfilt	Asphaltfilz, mit Asphalt imprägnierte Dachpappe	feutre bituminé	fieltro asfaltado
465	asphalt-laminated paper	asfaltlaminerat papper	Asphaltpapier, mit Asphalt beschichtetes Papier	papier goudron stratifié	papel laminado al asfalto
466	asphalt papers	asfaltpapper	Asphaltpapier	papiers goudronnés	papeles asfálticos
467	asphalt roofing	asfalttakpapp	Asphalt-Dachabdeckung	carton goudronné pour toitures	techado asfáltico
468	asphalt sheathing paper	asfaltisoleringspapper	Asphalt-Isolierpappe	papier goudronné pour construction	papel asfáltico para puesta en hojas
469	asphalted board	asfaltkartong	Asphaltpappe	carton goudronné	cartón asfaltado
470	asphalting board	asfaltkartong	Asphaltrohpappe, mit Asphalt zu imprägnierende Pappe	carton support pour goudronnage	cartón para asfaltar
471	asphalting paper	asfaltpapper	Asphalt-Rohpapier	papier support pour goudronnage	papel para asfaltar
472	aspiration	luftning	Ansaugung	aspiration	aspiración

	ENGLISH	SVENSKA	DEUTSCH	FRANÇAIS	ESPAÑOL
473	assimilation	assimilering	Assimilation, Anpassung, Angleichung	assimilation	asimilación
474	assistant superintendent	biträdande förman	Betriebsassistent, stellvertretender Betriebsleiter	sous-directeur	superintendente auxiliar
475	asymmetry	asymmetri	Asymmetrie	asymétrie	asimetría
476	atmosphere	atmosfär	Atmosphäre	atmosphère	atmósfera
477	atmospheric pressure	atmosfäriskt tryck	Atmosphärendruck, Luftdruck	pression atmosphérique	presión atmosférica
478	atomic bond	atombindning	Atombindung	liaison atomique	enlace atómico
479	atomization	finfördelning	Feinzerstäubung, Zerstäubung, Verneblung	atomisation	atomización
480	attapulgite	attapulgit	entwässertes Aluminiumsilikatmineral	attapulgite	atapulguita
481	attenuation	försvagning	Abschwächung, Verminderung, Verdünnung	atténuation	atenuación
482	autoclave	autoklav	Autoklav, Schnellkocher	autoclave	autoclave
483	autoclaving	behandling i autoklav	im Autoklav dämpfen, im Autoklav kochen	stériliser, porter à haute température	autoclave
484	automatic control	automatisk kontroll	Selbtsteuerung, Selbstregelung	contrôle automatique	control automático
485	automatic guide roll	automatisk ledvals	automatische Regulierwalze	rouleau-guide automatique	rollo-guía automático
486	automatic wire guide roll	automatisk viraledvals	Sieblaufregler	rouleau guide-toile automatique	rollo-guía de tela automático
487	automation	automation	Automation, Automatisierung	automatisation	automatización
488	automobile board	bilklädselkartong	Karosseriepappe	celloderme pour carrosserie	cartón para automóviles
489	autoslice	automatiskt inställbar läpp	selbsttätiger Schaber	règle automatique	auto-rebanador
490	auto tire wrap	däcksomslag	Autoreifeneinschlag, Autoreifenverpackung	papier pour emballer les pneus	embalaje de ruedas de coche
491	autotype paper	autotypipapper	Autotypiepapier (zur Vervielfältigung)	papier pour liasses et formulaires en continu	papel autotipia
492	available chlorine	tillgänglikt klor	aktives Chlor, nutzbares Chlor	teneur en chlore actif	cloro activo
493	axial	axiell	axial, achsial	axial	axial
494	axial flow pump	axialflödeslump	Axialpumpe	pompe à hélice	bomba axial
495	axis	axel	Achse, Richtlinie	ligne d'axe	eje
496	azo compounds	azpföreningar	Azoverbindungen	composés d'azote	compuestos azoicos
497	azure laid writing paper	blått ribbat skrivpapper	geripptes Schreibpapier, bläuliches Schreibpapier	papier écriture vergé bleuté	papel de escritura azulado

B

498	b.d. (bone dry)	absolut torr	absolut trocken	sec absolu	peso seco absoluto a cien grados
499	BOD (biochemical oxygen demand)	BS (biokemisk syreförbrukning)	biochemischer Sauerstoffbedarf	B.O.D. (consommation d'oxygène)	consumo de oxígeno
500	Btu (british thermal unit)	BTU	BTU (0,252 kcal)	unité de chaleur anglaise	unidad térmica británica
501	B-flute	B-flute	B-Welle (der Wellpappe), Feinwelle	petite cannelure	canaladuras B
502	babbitt metal, babbitt	babbittmetall	Babbittmetall, Lagermetall	anti-friction (métal blanc)	metal antifricción
503	back	bak-	unterstützen, auf der Rückseite verstärken, Rücken, Buchrücken, Rückseite, Rückwand	côté transmission	folio verso
504	back flow	återflöde	Rücklauf, Rückfluss	courant de retour	contraflujo
505	backing	bakgrund	Rückschicht, Unterlage	doublage, soutien	apoyo
506	backing board	bakgrundspapp	Aufklebekarton, Beklebekarton	carton à doubler	cartón de apoyo
507	backing paper	bakgrundspapper	Unterklebepapier, Matrizenpapier	papier pour matrices	papel de apoyo
508	backing roll	stödvals	Gegendruckwalze	rouleau de soutien	rodillo de sostén
509	backing-roll mark	stödvalsmarkering	Walzenmarkierung	marque du rouleau de soutien, bulle	marca del rodillo de sostén
510	backing sheet	bakgrundspapper	Unterlegpapier	feuille de doublage	hoja de apoyo
511	backing wire	stödvira	Stützsieb, Bodendrähte	toile de soutien	tela soporte
512	back liner	stöd-liner	Kartonrückschicht	doublure arrière	forro interior
513	back-lining paper	stöd-liner	Buchrückenpapier	papier pour doublure arrière	papel para forrado interior
514	back pressure	mottryck	Gegendruck	contrepression	contrapresión
515	backside	baksida	Antriebsseite (der Papiermaschine)	côté commande, transmission	lado de transmisión
516	backstand	understöd	Abrollständer, Rollengestell	dévidoir pour bobines	soporte posterior
517	back tender	passare	Papiermaschinengehilfe	aide-conducteur	ayudante de conductor
518	bacteria	bakterie	Bakterien	bactéries	bacteria
519	bacterial count	bakterieräkning	Bakterienzahl (in einem bestimmten Papier oder Karton)	nombre de bactéries sur une surface donnée	número de bacterias

	ENGLISH	SVENSKA	DEUTSCH	FRANÇAIS	ESPAÑOL
520	bactericides	baktericider	bakterientötende Mittel, Bakterizide	bactéricides	bactericidas
521	baffle	baffel	Staublech, Leitblech, Staukörper, Zwischenwand	chicane, déflecteur	deflector
522	baffle board	baffelbräda	Schutzbrett	planche de garde	cartón deflector
523	baffle plate	baffelplatta	Ablenkplatte, Schaumlatte	chicane	placa deflectora
524	baffle slice	linjal	Blende	règle de caisse d'arrivée de pâte	regla deflectora
525	baffled perforated roll	hålvals	Lamellenlochwalze	rouleau perforé avec déflecteur	rodillo perforado de deflección
526	bag	påse, säck	Beutel, Sack, Tüte, einsacken, in Beutel füllen	sac	saco
527	bagasse	bagasse	Bagasse (Zuckerrohrrückstände)	bagasse	bagazo
528	bagasse cutter	bagassskärmaskin	Bagasseschneider	coupeuse à bagasse	cortadora de bagazo
529	bagasse-depithing machine	märgupplösare för bagass	Bagasse-Entmarkungsmaschine	machine pour ôter la moelle de la bagasse	máquina desmeduladora de bagazo
530	bagasse fiber	bagassfiber	Bagassefaser	fibre de bagasse	fibra de bagazo
531	bagasse paper	bagasspapper	Bagassepapier	papier à base de bagasse	papel de bagazo
532	bagasse pulp	bagassmassa	Bagassezellstoff	pâte de bagasse	pasta de bagazo
533	bagginess	pösighet	Beuteligkeit, Durchhängen der Bahn	défaut de la bobine provenant d'une tension non uniforme de la feuille et provoquant la formation de plis	formación de bolsas
534	bagging	säcktillverkning	Sackwand, Einsacken, in Beutel verpacken	vieux sacs de jute classés	arpillera
535	bagging machine	säckmaskin	Einsackmaschine	machine à ensacher	máquina para hacer sacos
536	baggy	pösig	wellig, beutelig, bauchig (aufgrund zu loser Wicklung)	défectueux	forma de saco
537	bag liners	säck-liner	Tütenfutter, innerste Schicht mehrlagiger Tüten oder Säcke	doublures de sacs	forros para sacos
538	bag machine	säckmaskin	Sackherstellmaschine, Beutelherstellmaschine	machine à sacs	máquina de sacos
539	bag-making machine	säckmaskin	Sackherstellmaschine, Beutelherstellmaschine	machine à fabriquer les sacs	máquina para fabricar sacos
540	bag paper	säckpapper	Tütenpapier, Sackpapier, Beutelpapier	papier pour sacs	papel para sacos
541	bag-sealing machine	säckslutningsmaskin	Beutelmaschine, Sackschliessmaschine, Siegelmaschine	machine à fabriquer les sacs	máquina selladora de sacos
542	bakers' wrap	brödomslag	leichtgewichtiges Kekspapier, leichtgewichtiges Broteinwickelpapier	papier pour boulangerie	embalajes de tahona
543	balance	balans	Gleichgewicht, Gegengewicht, Balance	balance, équilibre, solde	equilibrio
544	bald cypress	bald cypress	Sumpfzypresse	cyprès chauve	ciprés pelado
545	bale	bal	Ballen, Bündel	balle	fardo
546	baled	balad	gebündelt, in Ballen verpackt	mis en balles	empacado
547	bale pulper	balupplösare	Ballenauflöser, Ballenzerreisser	triturateur (pour balles entières)	hidrodesfibrador
548	baler	balningsmaskin	Ballenpresse, Packpresse, Ballenpacker	presse â balles	empacadora
549	baling	baltillverkning	in Ballen verpacken	mise en balles	empacado
550	baling paper	balpapper	Balleneinpackpapier	macule pour balles	papel para empacado
551	baling press	balpress	Ballenpresse	presse à emballer	prensa de enfardar
552	baling wire	baltråd	Binderdraht, Packdraht	fil de fer de cerclage	tela de enfardar ·
553	ball bearing	kullager	Kugellager	roulement à billes	rodamiento a bolas
554	ball mill	kulkvarn	Kugelmühle	raffineur à boulets	molino de bolas
555	balloon logging	ballongtimring	Holztransport mittels Fesselballon, Holzeinbringung mittels Fesselballon	transport des bois par ballon	arrastre de troncos por globo aerostático
556	ball valve	kulventil	Kugelventil	clapet à bille, soupape à boulet	válvula de bola
557	balsam fir	balsamgran	Balsamtanne, Fichtenholz	sapin baumier	abeto balsámico
558	balsam poplar	balsampoppel	Balsampappel	peuplier baumier	chopo balsámico
559	bamboo	bambu	Bambus	bambou	bambú
560	bamboo paper	bambupapper	Bambuspapier	papier à base de bambou	papel bambú
561	bamboo pulp	bambumassa	Bambuszellstoff	pâte de bambou	pasta de bambú
562	banding	bandning	mit Banderole versehen	cercler avec du feuillard	guardacantar
563	banding machine	bandningsmaskin	Banderoliermaschine	machine à cercler avec du feuillard	máquina cercadora
564	band stock	bandmassa	Banderolenstoff, Banderolenpapier	feuillard	caja de cerco
565	bank-note paper	sedelpapper	Banknotenpapier	papier pour billets de banque	papel moneda

	ENGLISH	SVENSKA	DEUTSCH	FRANÇAIS	ESPAÑOL
566	bank paper	finpapper	Bankpostpapier, hochwertiges Schreibpapier	papier pour machine à écrire	papel bancario
567	bank stock	bankpappersmassa	Wertzeichenpapier, Banknotenpapier	stock de papier M.A.E.	reservas bancarias
568	bar	bom	Ausstreichleiste, Stab, Stange	barre, lame (de pile)	barra
569	bar coater	stavbestrykare	Stabstreicher	coucheuse à lame	estucadora de cuchillas
570	barge	pråm	Kahn, Lastkahn	péniche, chaland	barcaza
571	barium	barium	Barium	baryum	bario
572	barium carbonate	bariumkarbonat	Bariumkarbonat, kohlensaures Barium, Witherit	carbonate de baryum	carbonato de bario
573	barium sulfate	bariumsulfat	Bariumsulfat	sulfate de baryum	sulfato de bario
574	bark	bark	Rinde, Borke, entrinden, schälen	écorce	corteza
575	bark dryer	barktorkare	Rindentrockner	sechoir à écorcés	secadora de corteza
576	barker	barkningsmaskin	Schälmaschine, Rindenschälmaschine	écorceuse	descortezadora
577	barking	barkning	Schälen, Entrinden	écorçage	descortezado
578	barking drum	barktrumma	Entrindungstrommel	tambour écorceur	tambor descortezador
579	bark peeling machine	avbarkningsmaskin	Rindenschälmaschine	machine à écorcer	máquina descortezadora
580	bark press	barkpress	Rindenpresse	presse à écorcer	prensa de corteza
581	bark specks	barkfläckar	Rindenflecken (im Papier)	taches d'écorce	motas de corteza
582	barometric leg	barometerben	Fallwasserrohr, Flüssigkeitssäule	trompe, siphon à eau	sifón
583	barometric pressure	barometertryck	Aussenluftdruck, Luftdruck	pression barométrique	presión barométrica
584	barrel	fat	Fass, Tonne, Trommel	fût, tonneau, tambour	barril
585	barrier	spärr, barriär	Schranke, Sperre, Barriere	matière de protection	materia de protección
586	barrier material	spärrmaterial	Sperrschichtmaterial, Schutzmaterial, Zwischenschichtmaterial	matériaux de protection	material de protección
587	barrier paper	tätningspapper	Isolierpapier, Sperrschichtpapier	papier protecteur	papel de protección
588	barrier sheet	spärrskikt	Zwischenpapier, Sperrschichtpapier, Dichtungspapier	feuille de protection	hoja de protección
589	barring	spärrning	Querstreifigkeit im Papier	barrer	obstrucción
590	baryta	baryt	Baryt, Bariumoxid	baryte, blanc fixe	barita
591	baryta board	barytkartong	Barytkarton, Barytpappe	carton baryté	cartón con suspensión de barita
592	baryta paper	barytpapper	Barytpapier	papier baryté	papel con suspensión de barita
593	barytes	baryter	Bariumsulfat, Schwerspat	barytes	baritina
594	basalt tackle	basaltgarnityr	Basaltzuwinde, Basalteinsatz (im Refiner)	cylindre en lave	revestimiento de basalto
595	base	bas	Base (chem.), Basis, Grundfläche, basieren auf, beruhen auf	base, support	base
596	base paper	baspapper	Rohpapier	papier support	papel soporte
597	base plate	grundplatta	Grundplatte, Fundamentplatte, Sockel	plaque de fondation	placa de asiento
598	base stock	basmassa	Rohstoff	papier support (enduction, couchage etc.)	pasta soporte
599	basewad paper	vaddbaspapper	Schrotpatronenpapier	papier ouaté	papel con asiento de guata
600	basic alum	råalun	echtes Alaun (im Vergleich zum Papiermacheralaun)	alun basique	alumbre básico
601	basic dyes	grundfärgämnen	basischer Farbstoff	colorants basiques	colorantes básicos
602	basicity	alkalitet	chemische Basizität, Basität	basicité	basicidad
603	basic size	grundformat	Normalformat, Handelsformat	format de base	cola básica
604	basis weight	ytvikt	Flächengewicht	grammage	peso de la resma de papel
605	basswood	bastträ	Linde, Bast, amerikanischer Schwarzlindenholz	tilleul d'Amérique	tilo de América
606	bast fiber	bastfiber	Bastfaser	fibre libérienne	fibra de liber
607	bast paper	bastpapper	Bastpapier	papier de liber	papel de liber
608	batch process	satsvis process	Chargenbetrieb	procédé discontinu	proceso discontinuo
609	batch processing	behandling satsvis	diskontinuierlicher Betrieb, diskontinuierliche Arbeitsweise, partieweise Herstellung	fabrication en discontinu	procedimiento discontinuo
610	batch pulper	satspulper	Chargenpulper	triturateur en discontinu	desfibrador discontinuo
611	bathroom tissues	badrumspapper	Kosmetik-Tissue, Hygiene-Tissue	papier pour salle de bains	muselinas para el baño
612	battery board	batteripanel	Trockenbatteriekarton	carton pour accumulateurs	cartón para pilas
613	battery paper	batteripapper	Akkumulatorenpapier	papier pour accumulateurs	papel para pilas
614	Bauer-McNett classifier	Bauer-McNett fraktioneringsapparat	Bauer-McNett Fraktioniergerät, Bauer-McNett-Sortierer	appareil pour l'étude des fibres par fractionnement	clasificador Bauer-McNett
615	bead coater	pärlbestrykare	Tropfstreicher	coucheuse à bourrelet	estucadora bordeadora
616	beading	omfalsning, vulstning	Verzierung, Bördeln, Umbiegen, Wulst	enroulement très serré des bords (transformation)	moldear
617	beaker	bägare	Becherglas	bécher	cubilete
618	bearing	lager	Lager, Auflager	palier, roulement	cojinete
619	bearing housing	lagerhus	Lagergehäuse	logement de palier	soporte de cojinete

	ENGLISH	SVENSKA	DEUTSCH	FRANÇAIS	ESPAÑOL
620	beat	mala	schlagen, mahlen, zerfasern, rühren	raffiner	refinar
621	beatability	malbarhet	Mahlbarkeit, Mahlfähigkeit	aptitude au raffinage	aptitud al refino
622	beaten	mald	gemahlen	raffiné	refinado
623	beaten stuff tester (Brit.)	malningsprovare	Mahlgradprüfer	essayeur de degré de raffinage	refinómetro
624	beater	kvarn	Holländer	pile raffineuse	pila refinadora
625	beater additive	mäldtillsatsmedel	Holländerzusatz	tout produit non fibreux introduit dans la composition de fabrication	aditivo de refino
626	beater bar	kvarnkniv	Holländermesser	lame de pile	cuchilla de pila
627	beater chest	malningskar	Holländertrog, Holländerbütte	cuvier de pile	caja de pila
628	beater colored	mäldfärgad	massegefärbt	coloré à la pile	coloreado en pila
629	beater dyeing	mäldfärgning	Massefärbung	coloration dans la pile	coloración en pila
630	beater filling	mäldfyllning	Holländereinsatz, Holländerbemesserung,	remplissage de la pile	guarnición de pila
631	beater furnish	massa för malning	Stoffeintrag im Holländer	quantité chargée dans la pile	composición
632	beater loading	fyllmedeltillsats vid malning	Füllstoffzugabe im Holländer, Holländerbeschickung	chargement de la pile	carga de pila
633	beater man	holländraförare	Holländermüller, Holländermahler	gouverneur de pile	encargado de refinos
634	beater plate	motskär	Holländergrundwerk	platine de pile	platina de holandesa
635	beater roll	kvarnkubb	Holländerwalze	cylindre de la pile	cilindro de holandesa
636	beater room	mälderi	Holländersaal	salle des piles, salle de fabrication	cuarto de pila
637	beater-sized	mäldlimnad	in der Masse geleimt, massegeleimt	collé en pile	encolado en pila
638	beater sizing	mäldlimning	Masseleimung	collage en pile	encolado en pila
639	beater tackle	malningsgarnityr	Holländerbemesserung	lamage d'une pile	armadura de pila
640	beater tests	malningsprov	Mahlgradprüfung	essais de pile	pruebas en pila
641	beating	malning	Schlagen, Mahlen, Mahlung	raffinage	refinado
642	beating degree·	malningstillstånd	Stärke der Mahlung, Mahlgrad	degré de raffinage	grado de refinado
643	beating rate	malningshastighet	Mahlgeschwindigkeit	taux de raffinage	coeficiente de refinado
644	beating time	malningstid	Mahlzeit	temps de raffinage	tiempo de refinado
645	beaverboard	bäverpapp	Hartfaserplatte, Baupappe	carton pour visières	cartón de fibra para paredes
646	bed knife	motskärsbom	Untermesser eines Querschneiders	couteau fixe	cuchilla fija
647	bed load	malningstryck	Grundbelastung, Geschiebe	charge de base	carga en la pila
648	bedplate	motskär	Grundwerk, Mahlfläche, Messerblock, Geriffelte Platte (im Holländer)	platine de pile, plaque d'assise	platina de holandesa
649	beech	bok	Buche	hêtre	haya
650	beech pulp	bokmassa	Buchenzellstoff	pâte de hêtre	pasta de haya
651	beer filter paper	ölfilterpapper	Bierfiltrierpapier	papier filtre pour brasseries	papel filtro para cerveza
652	beer mat board	ölglaspapp	Bierdeckelkarton	carton pour dessous de bocks	cartón para posavasos
653	belt	remband	Riemen, Gurt, Gürtel, Band	courroie	correa
654	belt conveyor	remtransportör	Förderband, Bandtransport	transporteur à courroie	transportador de correa
655	belt drive	remdrift	Riemenantrieb	commande par courroie	transmisión por correa
656	belt shifter	rembytare	Ausrückhebel, Riemenausrücker	fourchette de débrayage	horquilla de correa
657	Ben Day screen	Ben Day-sil	Ben Day Sortierer	procédé Ben Day (écran)	depuradora Ben Day
658	bender	böjare	falzfähiger Karton, biegungsfähiger Karton	fléchisseur	dobladora
659	bending	böjning	Biegen, Krümmen, Durchbiegen	courbure, flexion	flexión
660	bending chip	falsbar gråpapp	Faltschachtelkarton aus Schrenzpappe	carton ordinaire pour boîtes pliantes	viruta de dobladura
661	bending fatigue tester	böjutmattningsprovare	Dauerbiegeprüfgerät	flexiomètre	probador de fatiga de flexión
662	bending machine	böjningsmaskin	Biegemaschine, Langfalzmaschine	machine à courber	máquina dobladora
663	bending strength	böjhållfasthet	Biegefestigkeit, Querfestigkeit	résistance à la flexion	resistencia a la flexión
664	bend quality	böjkvalitet	Biegequalität, Biegeeigenschaft	aptitude au pliage	calidad de curvatura
665	bend strength	böjhållfasthet	Biegefestigkeit, Querfestigkeit	résistance au pliage	fuerza de flexión
666	bend tests	böjprov	Biegeversuche	essais de flexion	pruebas de flexión
667	beneficiation	fördel	Aufbereitung	amélioration	beneficio
668	bentonite	bentonit	Betonit	bentonite	bentonita
669	benzene	bensin	Benzol	benzène	benceno
670	benzoates	bensoater	Benzoate	benzoates	benzoatos
671	benzoic acid	bensoesyra	Benzoesäure	acide benzoïque	ácido benzoico
672	benzyl cellulose	bensylcellulosa	Benzylcellulose	benzyl cellulose	celulosa al benzilo
673	benzyl groups	bensylgrupper	Benzylgruppen	groupes benzyl	grupos benzilo
674	beryllium	beryllium	Beryllium	béril	berilio
675	beta cellulose	betacellulosa	Betacellulose	bêta cellulose	betacelulosa
676	beta gauge	betastrålningsmätare	Dickenmessgerät (für Flächengewichtsmessung)	jauge bêta	indicador beta

	ENGLISH	SVENSKA	DEUTSCH	FRANÇAIS	ESPAÑOL
677	beta-ray gauge	betastrålningsmätare	Betastrahlenmessgerät	appareil pour mesurer le rayonnement bêta	indicador de rayos beta
678	beta rays	betastrålar	Betastrahlen	rayons bêta	rayos beta
679	Betula (Lat.)	Betula	Birke	bétula	Betula
680	bevelled gear, bevel gear	knoväxel	Kegelradantrieb, Kegelradgetriebe	engrenage cônique, engrenage d'angle	engranaje cónico
681	bias	snedbelastning, obaland	schräg, schief, diagonal, einseitige Wirkung, beeinflussen	biais, tendance, but	bias
682	bible paper	bibelpapper	Bibelpapier	papier bible	papel biblia
683	bibulous paper	läskpapper	absorbierfähiges Papier, Löschpapier	papier absorbant	papel absorbente
684	bicarbonates	bikarbonater	Bikarbonate	bicarbonates	bicarbonatos
685	bicomponent fiber	tvåkompnentfiber	Zweikomponentenfaser	fibre à deux composants	fibra bicomponente
686	billboard paper	affichpapper	Plakatpapier, lichtechtes Reklamepapier	papier pour affiches	papel para tableros de avisos
687	billing-machine paper	fakturapapper	Rechenmaschinenpapier	papier pour machines comptables	papel para máquinas facturadoras
688	bill paper	fakturapapper	Wechselpapier, Scheckpapier	papier pour effets de commerce	papel para letras de cambio
689	bin	binge, behållare	Silo, Behälter, Bunker, Vorratsbunker	coffre, silo	arca
690	binary system	binärt system	binäres System, aus 2 Elementen bestehendes System	système binaire	sistema binario
691	binder	bindemedel	Bindemittel	gangue, liant	ligante
692	binders' board	bokbindarkartong	Buchbinderpappe	carton pour reliure	cartón para encuadernaciones
693	binders' waste	bokbindaravfall	Buchbinderbeschnitt, Buchbinderchrenz	rognures de relieur	desperdicios de encuadernación
694	binding	bindning	Binden, Bindung, Einband	reliure	encuadernación
695	binding power	bindningsstyrka	Bindevermögen, Klebkraft, Bindekraft	pouvoir liant	poder ligante
696	binding tape	bindningsremsa	Klebestreifen	bande adhésive	ribete de encuadernación
697	biochemical oxygen demand (BOD)	biokemisk syreförbrukning (BS)	biochemischer Sauerstoffbedarf	B.O.D. (consommation d'oxygène)	consumo de oxígeno
698	biocides	biocider	Biozide (toxische Stoffe zur mikrobiologischen Kontrolle)	biocides	biócidos
699	biodegradable	biodegraderbar	biologisch abbaubar	biodégradable	biodegradable
700	biodegradation	biosönderfall	biologischer Abbau	biodégradation	biodegradación
701	biological control	biologisk kontroll	biologische Kontrolle	contrôle biologique	control biológico
702	biological effluent treatment	biologisk avfallsbehandling	biologische Abwasserreinigung	traitement biologique des effluents	tratamiento biológico
703	biological oxygen demand	biologisk syreförbrukning	biologischer Sauerstoffbedarf	B.O.D.	consumo de oxígeno
704	biological tests	biologiska prov	biologische Prüfung	essais biologiques	pruebas biológicas
705	biological treatment	biologisk behandling	biologische Aufbereitung	traitement biologique	tratamiento biológico
706	biphenyls	bifenyl	Biphenyle, zweiwertige Phenyle	biphényles	bifenilos
707	birch	björk	Birke	bouleau	abedul
708	birch pulp	björkmassa	Birkenzellstoff	pâte de bouleau	pasta de abedul
709	bisulfates	bisulfater	Bisulfate	bisulfates	bisulfatos
710	bisulfite liquor	bisulfitlut	Bisulfitlauge	lessive au bisulfite	lejía al bisulfito
711	bisulfite pulp	bisulfitmassa	Bisulfitzellstoff	pâte au bisulfite	pasta al bisulfito
712	bisulfite pulping	bisulfituppslutning	Bisulfitaufschlussverfahren, Bisulfitaufschluss	fabrication de pâtes au bisulfite	fabricación de pasta al bisulfito
713	bisulfites	bisulfiter	Bisulfite	bisulfites	bisulfitos
714	bitumen	bitumen	Bitumen	bitume	betún
715	bituminous board	bitumenpapp	Bitumenpappe	carton bituminé	cartón bituminoso
716	bituminous emulsion	bitumenemulsion	bitumige Emulsion, Bitumenemulsion	émulsion de bitume	emulsión bituminosa
717	black album paper	svartalbumpapper	schwarzes Albenpapier	papier noir pour albums	papel negro para álbum
718	black ash	svart aska	Schwarzesche, unreines Soda	résidu de la combustion de la lessive noire	ceniza negra
719	blackening	svartning	Grauverdrücken am Kalander, Schwärzung durch zu hohe Feuchtigkeit beim Kalandrieren	noircissement, plombage	ennegrecimiento
720	black liquor	svartlut	Schwarzlauge, Ablauge	liqueur (lessive) noire	licor negro
721	black poplar	svartpoppel	Schwarzpappel	peuplier noir	álamo negro
722	black spruce	svartgran	Schwarzfichte	épicéa noir	epícea negro
723	blade	blad	Schaber, Messer, Klinge	lame	cuchilla
724	blade coater	bladbestrykare	Messerstreicher, Rakelstreicher	coucheuse à lame	estucadora de cuchilla
725	blade coating	bladbestrykning	Rakelstreichung	couchage par lame	estucado por cuchilla
726	blanc fixe	blanc fix	Blanc fixe, Bariumsulfat	blanc fixe	blanco fijo
727	blank-book paper	–	unbedrucktes Buchdruckpapier	papier registre	papel para libros en blanco
728	blanket mark	filtmarkering	Filztuchmarkierung, Gummituchmarkierung	marque du feutre	señal de mantilla

	ENGLISH	SVENSKA	DEUTSCH	FRANÇAIS	ESPAÑOL
729	blanking paper	filtpapper	ungestrichenes Plakatpapier	papier non collé pour affiches	papel a troquelar
730	blank news	otryckt tidningspapper	aus Sekundärstoff unbedrucktes Zeitungsdruckpapier zur Herstellung von Faltschachtelkarton	rognures blanches	papel periódico en blanco
731	blanks	tomrum	Stanzlinge, jede Art von steifer Pappe mit glatter und bedruckbarer Oberfläche	carton contrecollé	impresos en blanco
732	blasting paper	dynamitpatronpapper	Sprengpulverpapier	papier pour munitions	papel de voladura
733	bleach	blekning, bleka	bleichen, Bleiche	blanchiment	agente blanqueante
734	bleachability	blekbarhet	Bleichfähigkeit	aptitude au blanchiment	blanqueabilidad
735	bleachable	blekbar	bleichfähig, bleichbar	blanchissable	blanqueable
736	bleachable pulp	blekbar massa	bleichfähiger Zellstoff	pâte blanchissable	pasta blanqueable
737	bleach demand	blekningsefterfrågan	Bleichmittelbedarf	capacité de blanchiment	necesidad de blanqueo
738	bleached pulp	blekt massa	gebleichter Zellstoff	pâte blanchie	pasta blanqueada
739	bleacher	blekare	Bleicher	pile blanchisseuse	blanqueador
740	bleachery	blekeri	Bleicherei, Bleichanlage	blanchisserie	blanquería
741	bleaching	blekning	Bleichen, Bleiche	blanchiment	blanqueamiento
742	bleaching effluent	blekeriavlopp	Bleichereiabwasser	effluent de blanchiment	efluente de blanqueo
743	bleaching liquor	blekvätska	Bleichlösung	solution de blanchiment	lejía de blanqueo
744	bleaching powder	blekpulver	Bleichpulver, Kalziumhypochlorit	chlorure de chaux	cloruro de cal
745	bleaching tower	blektorn	Bleichturm	tour de blanchiment	torre de blanqueo
746	bleach liquor	blekvätska	Bleichlösung, Chlorlauge	solution de blanchiment	lejía de blanqueo
747	bleach plant	blekeri	Bleichanlage	installation de blanchiment	planta de blanqueo
748	bleach requirement	blekningsbehov	Bleichmittelbedarf (zur Erreichung eines bestimmten Weissgehalts im Zellstoff)	aptitude au blanchiment	requisitos de blanqueo
749	bleach-resistant papers	blekningsresistenta papper	bleichfähiges Papier	papiers résistant au blanchiment	papeles resistentes al blanqueo
750	bleach scale	blekningsskala	Flecken durch das Bleichmittel hervorgerufen	taches provoquées par des résidus de blanchiment	escala de blanqueo
751	bleed	blöda, läcka a.o.	verlaufen, verwaschen, Auslaufen der Druckfarbe	rogner, soutirer	sangrar
752	bleeding	blödning, läckning a.o.	Oberflächenverfärbung, Ablassen, Durchschlagen der Farbe	soutirage	sangría
753	bleeding resistance	blödningsmotstånd, läckningsmotstånd	Durchschlagfestigkeit	résistance d'un papier goudronné à la transudation du goudron	resistencia a sangrar
754	bleed trim	fållande rems	Buchbeschnitt	largeur rognée	corte de sangría
755	blemish	fläck	Fehler, Makel	ternir, tacher	imperfección
756	blend	blandning	mischen, vermengen	mélange	mezcla
757	blender	blandare	Mischer, Mischgerät	mélangeur du milieu, brosses du milieu (couchage)	mezclador
758	blister	blåsa	Blase, blasig werden	cloque, ampoule	ampolla
759	blister, coating	blisterbestrykning	Bläschenstreicher	cloque de couchage	ampolla (estucado)
760	blister cut	blisterutskott	Bahnriss in der Falzlinie	coupure d'ampoule	corte de ampollas
761	blistering	blåsbildning	Blasenpackung	cloquage	ampollamiento
762	blister pack	blisterförpackning	Blisterpackung	ampoule de protection	ampolla
763	blister, ply separation	blisterskiktseparation	Schichttrennung im Karton infolge von Blasen	cloque à la séparation des couches	ampolla (separación de capas)
764	blocking	spärrning	Zusammenbacken von Papierbogen, Zusammenkleben von Papierbogen	adhérence en surface	adherencia en superficie
765	blocking, calender roll	glättvalsspärrning	Blockierung der Kalanderwalze	adhérence au rouleau de calandre	adherencia al rodillo de calandra
766	blocking resistance	spärrningsmotstånd	Blockfestigkeit, Gleitfähigkeit	résistance à l'adhérence	resistencia a la adherencia
767	bloodproof paper	blodtätt papper	blutundurchlässiges Papier, Fleischeinwickelpapier	papier résistant aux taches de sang	papel resistente a la sangre
768	blood resistance	blödningsmotstånd	Blutundurchlässigkeit	résistance au sang	resistencia a la sangre
769	blotting board	läskkartong	Löschpapier (schwergewichtiges)	carton pour buvards	cartulina secante
770	blotting paper	läskpapper	Löschpapier	papier buvard	papel secante
771	blow	blåsa	Schlag, Stoss, blasen	soufflage, dégazage	evacuar por soplado
772	blower	blåsare	Gebläse, Lüfter, Ventilator	ventilateur	ventilador soplador
773	blowing	blåsning	Abblasen (der Papierbahn vom Filz), Spalten der Kartonbahn	ventilation	sopladura
774	blow line	blåsledning	Spaltspur (Fehler im Karton)	installation de soufflage	línea de soplado
775	blow-line refining	blåsledningsrening	Druckmahlung nach der Ausblasstation	installation de raffinage par soufflage	refinado en línea de soplado
776	blow liquor	blåsningslut	Schwarzlauge, Ablauge	liqueur de soufflage	lejía sopladora
777	blow off	blåsa bort	abblasen, ausstossen, auslassen (Dampf)	évacuer par soufflage	descargar a presión

	ENGLISH	SVENSKA	DEUTSCH	FRANÇAIS	ESPAÑOL
778	blow pit	flisgrop	Diffuseur, Kochergrube, Waschgrube	fosse de décharge par soufflage	tina de descarga por soplado
779	blow roll	blåsvals	Blaswalze	rouleau souffleur	rodillo soplador
780	blow tank	blåsningstank	Abblasetank	fosse de décharge	tanque de soplado
781	blow tests	blåsningsprover	Spalttest, Spaltprüfung	essais de soufflage	pruebas de soplado
782	blow valve	blåsningsvärde	Ausblaseschieber (Kocher), Gebläseventil	vanne de décharge	válvula de soplado
783	blue	blå	blau, bläuen, blau färben, blau werden	bleu, azuré	azular
784	blue-glass method, blue-glass test	blåglasmetod, blåglasprov	Blauglasprüfung	essai de pâtes au verre bleu	método del cristal azul, prueba del cristal azul
785	blueprint paper	blåkopiepapper	Blaupauspapier, Lichtpauspapier	papier au ferrocyanure	papel ferroprusiato
786	blue tracing paper	blåkopiepapper	Pauspapier (für Zeichnungen)	papier calque au ferrocyanure	papel calco para planos
787	blush coating	blush coating	–	couchage rougeâtre	estucado soflama
788	blushing	blushing	Anlaufen, Trübung	rougeur	soflamado
789	board	kartong, papp	Pappe, Karton, Brett, Tafel	carton	cartón
790	board and box lining	kartongfoder	Kartonkaschierung (hauptsächlich aus Schrenzpappe)	doublage de cartons et caisses	forro de cartón y cajas
791	board liner	kartongfoder	Kaschieranlage, Kaschierlage	machine à doubler le carton	segunda capa (cartón)
792	board machine	kartongmaskin	Kartonmaschine	machine à carton	máquina de cartón
793	board mill	kartongbruk	Kartonfabrik, Pappenfabrik	cartonnerie	cartonería
794	body stock	basmäld	Trägerstoff, Rohstoff	papier support	soporte
795	bogus	falsk	falsch, nachgemacht, unecht	à base de matières de récupération	simili
796	bogus back lining	imiterad fodring	imitierte Rückseitenverstärkung	doublure à base de vieux papiers	forro posterior simili
797	bogus board	imitationskartong	Karton aus Sekundärfaserstoff	carton à base de vieux papiers	cartón simili
798	bogus bristol	imitationsbristol	Bristolkarton, Bristolimitation	bristol à base de vieux papiers	bristol simili
799	bogus corrugating medium	imitationsvell	Wellpappenwelle aus Sekundärfaserstoff	cannelure vieux papiers	cartón ondulado simili
800	bogus drawing paper	imitationsritpapper	geleimtes Zeichenpapier aus Sekundärfaserstoff	papier à dessin à base de vieux papier	papel de dibujo simili
801	bogus duplex	imitationsduplex	einseitig gefärbter Duplex-Karton	carton duplex à base de vieux papiers	duplex simili
802	bogus kraft	imitationskraftpapper	Kraftimitation	simili kraft	kraft simili
803	bogus lining paper	imitationsfodring	Kaschierpapier aus Sekundärstoff	papier couverture à base de vieux papiers	papel forro simili
804	bogus manila	imitationsmanilla	Manilaimitation	papier d'emballage à base de vieux papiers	manila simili
805	bogus mill wraps	imitationsförpackningspapper	dickes Einschlagpapier, voluminöses Einschlagpapier	macule à base de vieux papiers	–
806	bogus papers	imitationspapper	Rückschichtenpapier geringer Festigkeit (aus Sekundäfaserstoff)	papiers à base de matières de récupération	papeles simili
807	bogus pasting papers	imitationsklisterpapper	Beklebepapier aus Sekundärfaserstoff, Beklebekarton aus Sekundärfaserstoff	papiers à doubler à base de matières de récupération	papeles simili para contracolado
808	bogus saturating paper	imitationsläskpapper	Trägerpapier für Asphaltimprägnierung	papiers à imprégner à base de matières de récupération	papeles simili de impregnación
809	bogus screenings	imitationskopiepapper	Papier aus Sortierstoff, Papier aus aufbereitetem Altpapier	déchets	rechazos simili de depuración
810	bogus tag	imitationsetikett	Imitationsanhänger, Imitations-Etikettenpapier	étiquettes à base de vieux papiers	etiqueta simili para equipaje
811	bogus wrapping	imitationsförpackningspapper	Packpapier aus Sekundärfaserstoff, Verpackungspapier aus Sekundärfaserstoff	emballages à base de vieux papiers	embalaje simili
812	boiler	kokare	Kessel, Kocher	chaudière, lessiveur	caldera
813	boiler feedwater	kokarmatarvatten	Kesselspeisewasser	eau d'alimentation de la chaudière	agua de alimentación para caldera
814	boiler house	kokeri	Kesselhaus	chaufferie	cuarto de calderas
815	boiling point	kokpunkt	Siedepunkt	point d'ébullition	temperatura de ebullición
816	bole	trädstam	Baumstamm	tronc d'arbe	tronco
817	bond	bindning	binden,verbinden, verkleben, Bindung, Verklebung, Band, Valenz	liaison, adhérence	unión
818	bonded area	bindningsyta	Faseroberflächenbindung	zone d'adhérence	área unida
819	bonding	bindning	Binden, Verbinden, Verkleben	liaison entre deux feuilles de multijets	adherencia
820	bonding agent	bindemedel	Bindemittel	liant	ligante
821	bonding strength	bindningsstyrka	Haftfestigkeit, Spaltfestigkeit, Bindekraft, Klebefestigkeit	résistance à l'arrachage	fuerza de adherencia

	ENGLISH	SVENSKA	DEUTSCH	FRANÇAIS	ESPAÑOL
822	bond paper	finpapper	Bankpostpapier, Banknotenpapier, Wertpapier	papier écriture coquille, M.A.E.	papel bond
823	bond strength	bindningsstyrka	Bindefestigkeit, Haftfestigkeit	résistance à l'arrachage	fuerza de unión
824	bone dry (b.d.)	absolut torr	absolut trocken	sec absolu	seco absoluto
825	bone-dry weight	absolut torr vikt	absolutes Trockengewicht	poids sec absolu	peso seco absoluto a cien grados
826	book	bok	Buch, verbuchen, aufschreiben, notieren, eintragen	livre	libro
827	book-backing paper	bokbaspapper	Buchrückenpapier	papier pour doublure de livres	papel soporte para libros
828	book back liner	bokfoderpapper	Buchrückenbeklebepapier	papier à doubler les livres	contraforro para libros
829	book basis	bokbaspapper	Standard-Buchseitenformat	papier édition	soporte para libros
830	bookbinders' paper	bokbindarpapper	Vorsatzpapier	papier pour reliure	papel para encuadernadores
831	bookboard	bokkartong	Buchbinderpappe	carton pour livres	cartón libro
832	book-cover paper	bokpärmpapper	Schutzumschlagpapier	papier pour couvrir les livres	papel para cubiertas de libro
833	book jacket	bokomslag	Buchhülle, Schutzumschlag	jaquette	cubre libro
834	bookkeeping-machine paper	bokföringsmaskinpapper	Geschäftsbücherpapier, Registerpapier	papier pour machines comptables	papel para máquinas contables
835	book lining	försättspapper	stark geleimtes Vorsatzpapier	doublage de livres	forro para libros
836	book-match board	tändsticksplånkartong	Rundsiebkarton für Streichholzbriefe	carton pour pochettes d'allumettes	cartón para fósforos
837	book paper	bokpapper	Buchpapier, Buchdruckpapier, Werkdruckpapier	papier d'édition	papel para libros
838	book shavings	bokpapperspån	Buchbeschnitt	rognures de livres	recortes de libros
839	book stock	bokpapper	de-inkter Stoff, entfärbter Stoff	stock de livres	reserva de libros
840	book wrapper	bokomslag	Bucheinbindepapier, Schutzumschlagpapier	papier pour emballage de livres	envoltura de libros
841	boom	bom	Ladebaum, Hafensperre, Hochkonjunktur, Aufschwung	boom	cadena de troncos
842	borates	borater	Borate	borates	boratos
843	borax	borax	Borax	borax	bórax
844	bordered pit	kantad por	behöfter Tüpfel	ponctuation aréolée	tina orillada
845	boric acid	borsyra	Borsäure	acide borique	ácido bórico
846	boring	urborrning	Bohrung, Bohren	prélèvement de rondelles	toma de muestras circulares
847	borohydrides	borhydrider	Borhydrid	hydrures de bore	ácidos borhídricos
848	boron	bor	Bor	bore	boro
849	boron compounds	borföreningar	Borverbindungen	composés de bore	compuestos de boro
850	bottle-cap board	kapsylkartong	Karton für Flaschenkapseln	carton pour capsules de bouteilles	cartón para cápsulas de botella
851	bottom couch roll	guskvals	untere Gautschwalze	rouleau inférieur de presse humide	rodillo de prensa húmeda inferior
852	bottom felt	underfilt	Unterfilz, Abnahmefilz	feutre inférieur	fieltro inferior
853	bottom felt press	underfiltpress	Unterfilzpresse	feutre de presse inférieure	prensa de fieltro inferior
854	bottom press	underpress	untere Presse	presse inférieure	prensa inferior
855	bottom press roll	under pressvals	untere Presswalze	cylindre de presse inférieure	rodillo de prensa inferior
856	bottom roll	undervals	untere Walze	rouleau inférieur	rodillo inferior
857	bound water	bundet vatten	gebundenes Wasser	eau liée	agua fijada
858	bowed roll	böjd vals	bombierte Walze, gekrümmte Walze	rouleau cintré	rodillo inclinado
859	bowl	tank	Schale, Napf, Bassin, Kalanderwalze	rouleau de calandre	rodillo de calandra
860	bowl glazing	våtglättning	kalandrieren, satinieren, Friktionsglättung	calandrage	abrillantado por calandra
861	bowl paper	–	Kalanderwalzenpapier	papier pour rouleaux de calandres	papel parabólico
862	box	låda	Schachtel, Kiste, Kasten, in Schachtel verpacken	boîte, caisse	caja
863	boxboard	lådkartong	Kartonagenkarton, Schachtelpappe	carton pour boîtes	cartón para cartonajes
864	boxboard container	papplåda	Behälter aus Karton	caisse carton	caja de cartón
865	box clippings	–	Schachtelbeschnitt	rognures de caisses	recortes de cajas
866	box compression	lådkomprimering	Druckwiderstand eines Kartons, Kompressionswiderstand eines Kartons	résistance d'une caisse à la compression	compresión de caja
867	box-cover paper	lådomslagspapper	Schachtelabdeckpapier	papier à couvrir les boîtes en carton	papel para tapas de caja
868	boxed writings	lådförpackat skrivpapper	in Karton verpacktes Schreibpapier	papier écriture emballé en caisse	letras en cajetín
869	box enamel paper	lådemaljpapper	einseitig gestrichenes und hochsatiniertes Kartonagenpapier (für Kaschierzwecke)	papier couché à couvrir les boîtes	papel esmaltado para cajas
870	box liners	lådfodring	Schachtelfutter	revêtement de boîtes	forros de caja

23

	ENGLISH	SVENSKA	DEUTSCH	FRANÇAIS	ESPAÑOL
871	box machine	lådmaskin	Schachtelmaschine	machine à fabriquer les caisses	máquina para cajas
872	box partitions	låddelar	Schachteltrennwand (Steg)	cloisons de caisses	división de caja
873	brace	klämma	versteifen, absteifen, verstärken, Klammer, Verbindungsklammer	tirant	puntal
874	bracket trimmer	kantsprits	Konsolenstützbalken	massicot	guillotina de cartela
875	braille paper	blindskriftpapper	Blindenschriftpapier	papier braille (pour aveugles)	papel braille
876	brakes	bromsar	Bremsen, Bremsvorrichtung, bremsen, hemmen, brechen (Hanf)	freins	frenos
877	braking	bromsning	Bremsen,Bremsung	freinage	frenaje
878	brass	mässing	Messing	laiton	latón
879	bread-bag paper	brödpåspapper	Brotbeutelpapier	papier pour sacs à pain	papel para bolsas de pan
880	bread wrap	brödomslag	Broteinschlagpapier	papier pour emballage du pain	embalaje de pan
881	bread wrappers	brödomslag	Broteinschlagpapier	papier d'emballage pour boulangeries	envolturas para pan
882	break	sönderdela	unterbrechen, reissen, zerschlagen, Bruch, Riss, Lücke, Knick	cassure, rupture	ruptura
883	breaker	uppslagningsmaskin	Halbzeugholländer	pile défileuse	pila desfibradora
884	breaker bars	upplösarknivar, pulperknivar	Brecherrippen	lames de pile	barras del molino
885	breaker beater	uppslagningsholländare	Auflöseholländer, Stetigholländer	pile sans platine	pila refinadora holandesa
886	breaker stack	staplad dubbelpress	Feuchtglättwerk	lisse	lisa
887	breaker trap	partikelfälla	–	purgeur de pile	purgador de pilas desfibradoras
888	break-even point	jämviktspunkt	Rentabilitätsschwelle	seuil de rentabilité	punto de rotura uniforme
889	breaking	upplösning	Brechen, Zerreissen	défilage, broyage	trituración
890	breaking length	avslitningslängd	Reisslänge	longueur de rupture	longitud de ruptura
891	breaking strength	avslitningshållfasthet	Reissfestigkeit, Bruchfestigkeit	résistance à la rupture	resistencia a la rotura
892	break resistance	avslitningsmotstånd	Reisswiderstand, Bruchwiderstand	résistance à la rupture	resistencia a la rotura
893	breast box (Brit.)	bröstvalslåda	Stoffauflaufkasten	caisse d'arrivée de pâtes	caja de entrada
894	breast roll	bröstvals	Brustwalze	rouleau de tête	rodillo cabecero
895	brightener	blekmedel	Aufheller	éclaircissant	abrillantador
896	brightness	ljushet	Helligkeit, Weissgrad	éclat, brillant, blancheur	blancura
897	brightness reversion	eftergulning	Weissgradabfall, Nachdunklung von gebleichten Stoffen	inversion de l'éclat	inversión de blancura
898	Brinell hardness	brinellhårdhet	Brinell-Härte	dureté brinell	dureza de Brinell
899	Brinell tester	Brinellprovare	Brinell-Härteprüfer	duromètre brinell	ensayo por máquina Brinell
900	bristol board	bristolkartong	Bristolkarton	carton bristol	brístol
901	bristols	bristol	Karton (0.15 mm und stärker)	bristols	cartulinas brístol
902	brittleness	sprödhet	Brüchigkeit, Sprödigkeit	fragilité	fragilidad
903	broadleaf	bredblad	Laubbaum	feuillu	de hoja ancha
904	brocade paper	brokadpapper	Brokatpapier	papier brocart	papel brochado
905	broke	utskott	Ausschuss, Kollerstoff	cassé de fabrication	recortes de fabricación
906	broke bundle	utskottsknippe	angebrochener Pack, angebrochener Ballen	balle de cassés	bala de recortes de fabricación
907	broken edge	bruten kant	abgestossene Rollenränder, eingerissene Rollenränder	bord irrégulier	borde quebrado
908	broken ream	brutet ris	unvollständiges Ries, angebrochenes Ries	rame incomplète	resma incompleta
909	broke pulper	utskottspulper	Ausschuss-Pulper	broyeur de cassés	desfibrador de recortes
910	bromine	brom	Brom	brome	bromo
911	bromine compounds	bromföreningar	Bromverbindungen	composés de brome	compuestos de bromo
912	bronze	brons	Bronze, Rotguss, bronzieren	bronze	bronce
913	bronzing	bronsering	Bronzieren	durcir, bronzer	bronceado
914	bruising	krossning	zerquetschen, zerreiben, verschroten	écrasement des fibres	machacadura
915	brush	borste	Bürste, Stromabnehmer, bürsten, abbürsten	brosse, balai	cepillo
916	brush coater	borstbestrykare	Bürstenstreicher	fonceuse à brosses	estucadora de cepillos
917	brush coating	borstbestrykning	Bürstenstrich, Bürstenstreichen	couchage par brosses	estucado por escobillas
918	brush dampener	borstfuktare	Bürstenfeuchter	mouilleuse à brosses	humectadora de escobillas
919	brush enamel paper	borstemaljpapper	einseitig oder beidseitig gestrichenes und vor dem Satinieren gebürstetes Papier	papier couché une face à la brosse	papel charol
920	brush finish	borstfinish	gebürstet	fini brosse	acabado al cepillo
921	brush-finish coating	borstfinishbestrykning	Bürstenglättung	couchage à la brosse	estucado con acabado al cepillo
922	brush glazing	borstglättning	Bürstenglättung	glaçage sur brosseuse	lustrado con cepillo
923	brushing	borstning	Bürsten, Ausstreichen	affleurage, fibrillation	refinado ligero
924	brushing out	fibrillering	Fibrillierung	affleurer	afloramiento
925	brush marks	borstmarkering	Bürstenmarkierung, Borstenmarkierung	marques de brosse	señales del cepillo

	ENGLISH	SVENSKA	DEUTSCH	FRANÇAIS	ESPAÑOL
926	brush polishing	borstpolering	Bürstenglättung	lustrage sur brosseuse	lustrado con cepillo
927	brush polishing machine	borstpoleringsmaskin	Glanzbürstmaschine	fonceuse à brosses	máquina de satinar al cepillo
928	brush roller	borstvals	Bürstenwalze	cylindre à brosses	cepillo rotativo
929	bubble	bubbla	sprudeln, Blasen bilden, perlen, Luftblase, Gasblase, Bläschen, Tröpfchen	cloque, bulle	burbuja
930	bubble coating	bubbelbestrykning	Bläschenstreichen	couchage cloqué	estucado con ampollas
931	bubbling	bubbling	Durchblasen, Durchperlen	cloquage	ampollamiento
932	bucket conveyor	skoptransportör	Becherwerk	transporteur à godets	transportador de cangilones
933	buckeye	bocköga	Rosskastanie	marronier d'Inde	abertura de enlejiado
934	buckling	knäckning	Schnallenbildung, Wölbung, sich verziehen	gondolage	abarquillamiento
935	bud	knopp	okulieren, Knospe	bouton	brote
936	buffer	buffert	dämpfen, puffern, Puffer, Stossdämpfung	tampon	tampón
937	buffering agent	buffer	Pufferlösung	agent tampon	agente amortiguador
938	buffing paper	sämskskinnimitation	Schmirgelpapier (für Lederindustrie)	papier bulle	papel para pulir
939	building board	fiberskiva	Baupappe	carton de construction	cartón para paneles de construcción
940	building paper	byggnadspapper	Konstruktionspapier	papier de construction	papel para paneles de construcción
941	built-in strain	inbyggd töjning	in das Papier eingearbeitete Spannung, zurückgebliebene Spannung während der Trocknung	déformation incorporée	tensión integrada
942	built-in stress	inbyggd spänning	eingearbeitete Druckbelastung	tension incorporée	carga integrada
943	bulge resistance	utbuktningsmotstånd	Ausbeulfestigkeit	résistance au bombement	resistencia al combado
944	bulging	utbuktning	Ausbauchen, Ausbeulen, sich bauschen, Ausbauchung, Verbeulung	renflement	protuberante
945	bulk	bulk	Volumen (von Papier oder Karton), dicke Griffigkeit	main, bouffant, vrac, volume, masse	mano
946	bulk density	skrymdensitet	Schüttdichte, Schüttgewicht	densité de masse	densidad en masa
947	bulk handling	bulkhantering	Massenabwicklung	manutention en vrac	manipulación de los bultos
948	bulkhead	skott	Schott, Zwischenwand	fronteau, cloison	compuerta
949	bulk index	bulkindex	spezifisches Volumen	indice de volume	índice de mano
950	bulking board	bulkkartong	voluminöse Pappe, voluminöser Karton, Volumenkarton (nicht kalandriert)	carton léger	cartón esponjoso
951	bulking book paper	bulkbokpapper	Werkdruckpapier, hochvoluminöses Papier	papier impression bouffant	papel esponjoso para libros
952	bulking paper	bulkpapper	voluminöses Papier, Dickdruckpapier	papier bouffant	papel esponjoso
953	bulking pressure	bulktryck	Tasterdruck bei der Dickenmessung	pression sous laquelle l'épaisseur d'un rame est déterminée	presión de cuerpo
954	bulking thickness	bulktjocklek	Dicke eines Blattes oder eines Papierstapels von bestimmter Bogenzahl	épaisseur d'une rame sous une pression donnée	espesor de cuerpo
955	bulk modules	bulkmoduler	Kompressionsmodul, Elastizitätsmodul	compressibilité	módulos de cuerpo
956	bulk storage	bulklagring	en gros Lagerung, lose Aufbewahrung	stockage en vrac	almacenamiento de bultos
957	bulky	bulkig, voluminös	auftragend, voluminös	bouffant	corpulento
958	bull screen	grovsil	Splitterfänger	tamis à buchettes	criba de troncos
959	bundle	knippe, bunt	Bündel, Bund, zusammenbündeln	faisceau, balle	fardo
960	bundling	buntning	Bündelung	mise en balles	empaquetado
961	buoyancy	flytförmåga	Schwimmkraft, Trägfähigkeit	élasticité	flotación
962	buret	byrett	Bürette	burette	bureta
963	burlap finish	grov finish	Sackleinensatinage	fini entoilé	acabado tela
964	burner	brännare	Brenner, Bunsenbrenner	bruleur	quemador
965	burnished finish	blankpolering	Hochglanzsatinage	lissé à la calandre de friction, lissé à la pierre	acabado bruñido
966	burnisher	polerare	Hochglanzpolierkalander, Hochglanzsatinierkalander	brunisseur	bruñidor
967	burnishing	blankpolering	Glätten, Polieren, Kalandern	action de brunir	bruñido
968	burnt	bränd	verbrannt, übertrocknetes Papier, brüchiges Papier, überhitzter Zellstoff	brulé	quemado
969	burr	skärpa	abgraten, rauhe Kante (Metall), Grat, Naht, Werkzeug zur Oberflächenbehandlung von Schleifsteinen	décrasse-meule	herramienta para avivar la muela

	ENGLISH	SVENSKA	DEUTSCH	FRANÇAIS	ESPAÑOL
970	burring	skärpning	Schärfen (des Schleifsteins)	rhabillage des meules d'un défibreur	avivamiento de muela
971	burr number	skärptal	Nahtzahl	caractéristique de la molette de rhabillage des meules de défibreur	número de rodana
972	burst	sprängtryck	bersten, platzen, zerspringen, Bahnriss	éclater	reventamiento
973	burst factor	sprängfaktor	Berstfaktor	indice d'éclatement	índice de reventamiento
974	bursting	sprängning	Bersten, Platzen	éclatement	reventamiento
975	bursting strength	spränghållfasthet	Berstfestigkeit	résistance à l'éclatement	fuerza de reventamiento
976	bursting strength tester	sprängtrycksprovare	Berstdruckprüfer	appareil d'essai de résistance à l'éclatement	probador de resistencia al reventamiento
977	burst ratio	sprängindex	Berstindex	indice d'éclatement	índice de reventamiento
978	burst strength	spränghållfasthet	Berstfestigkeit	résistance à l'éclatement	resistencia al reventamiento
979	burst tester	sprängtrycksprovare	Berstdruckprüfer	éclatomètre	probador de reventamiento
980	burst tests	sprängtrycksprov	Berstdruckprüfung	essais d'éclatement	pruebas de reventamiento
981	business forms	affärsformulär	Geschäftsformular	formulaires commerciaux	impresos de negocios
982	butadiene	butadien	Butadien	butadiène	butadieno
983	butanes	butaner	Butan	butanes	butanos
984	butchers' manila	köttomslag	manilafarbenes Fleischeinwickelpapier	papier mince pour boucherie	manila para carnicerías
985	butchers' paper	köttomslag	geleimtes Fleischeinwickelpapier	papiers pour boucherie	papel para carnicerías
986	butchers' wrap	köttomslag	Fleischeinwickelpapier	emballages pour boucherie	embalaje de carnicerías
987	butt	tjockända	anstossen, stumpf aneinanderfügen, unteres Stammende (Baum)	souche (d'arbre), billot, gros bout	tope
988	butted splice	stumskarv	stumpf gegeneinanderstossende Anklebestelle ohne Überlappung	ajouture	pegadura asteada
989	butt end	tjockända	dickes Endstück (Baum), Rollenrest	gros bout	tope
990	butterfly valve	spjällventil	Drosselklappe, Drosselventil	vanne à papillon	válvula de mariposa
991	butter wrap	smöromslag	Buttereinwickelpapier	papier à beurre	embalaje para mantequilla
992	butt splicer	skarvdon	Bahnanklebevorrichtung	système de collage bout-à-bout de deux feuilles de papier	empalmador
993	button specks	knappfläckar	Knopfflecken (von Hadern herrührend)	boutóns (dans le papier)	granos de arena
994	butyl acetate	butylacetat	Butylacetat	acétate de butyl	acetato de butilo
995	butyl groups	butylgrupper	Butylgruppen	groupes butyl	grupos de butilo
996	butyl rubber	butylgummi	Butylkautschuk	butylcaoutchouc	caucho butilo
997	butyrates	butyrater	Butyrat	butyrates	butiratos
998	butyric acid	smörsyra	Buttersäure	acide butyrique	ácido butírico
999	butyryl groups	butyrylgrupper	Butyrylgruppen	groupes butyryl	grupos butiril
1000	by-product	biprodukt	Nebenprodukt	sous-produit	subproducto

C

1001	C 1 S (coated one side)	ensidig bestrykning	einseitg gestrichen	couché une face	estucado una cara
1002	C 2 S (coated two sides)	dubbelsidig bestrykning	beidseitig gestrichen	couché deux faces	estucado dos caras
1003	COD (chemical oxygen demand)	kemisk syreförbrukning	chemischer Sauerstoffbedarf	C.O.D.	demanda química de oxígeno
1004	C-flute	C-flute	C-Welle (der Wellpappe), mittlere Riffelung	cannelure moyenne	canaladuras C
1005	cable	kabel	Kabel, Seil, Tau	câble	cable
1006	cable logging	kabelloggning	Holztransport mittels Kabelseil	transport des arbres abattus par câble	arrastre por cable
1007	cable paper	kabelpapper	Kabelpapier	papier pour câbles	papel para cables
1008	cadmium	kadmium	Kadmium	cadmium	cadmio
1009	cadmium compounds	kadmiumföreningar	Kadmium-Verbindungen	composés de cadmium	compuestos de cadmio
1010	calcination	kalcinering	Brennen, Rösten, Kalzinierung	calcination	calcinación
1011	calcium·	kalcium	Kalzium, Calcium	calcium	calcio
1012	calcium-base liquor	kokvätska på kalciumbas	Calciumlauge	liqueur à base de calcium	lejía a base de calcio
1013	calcium-base sulfite	kalciumsulfit	Calciumsulfit	bisulfite de calcium	sulfito a base de calcio
1014	calcium bisulfite	kalciumbisulfit	Calciumbisulfit	bisulfite de calcium	bisulfito a base de calcio
1015	calcium carbonate	kalciumkarbonat	Calciumcarbonat	carbonate de calcium	carbonato de calcio
1016	calcium chloride	kalciumklorid	Calciumchlorid	chlorure de calcium	cloruro de calcio
1017	calcium compounds	kalciumföreningar	Calciumverbindungen	composés de calcium	compuestos de calcio
1018	calcium hydroxide	kalciumhydroxid	Calciumhydroxid	hydrate de calcium	hidróxido de calcio

	ENGLISH	SVENSKA	DEUTSCH	FRANÇAIS	ESPAÑOL
1019	calcium hypochlorite	kalciumhypoklorit	Calciumhypochlorit	hypochlorite de calcium	hipoclorito de calcio
1020	calcium lignosulfonate	kalciumlignosulfonat	Calciumlignosulfonat	lignosulfonate de calcium	lignosulfonato de calcio
1021	calcium oxide	kalciumoxid	Calciumoxid, gebrannter Kalk	chaux vive	cal viva
1022	calcium sulfate	kalciumsulfat	Calciumsulfat, Gips	sulfate de calcium	sulfato de calcio
1023	calcium sulfite	kalciumsulfit	Calciumsulfit	sulfite de calcium	sulfito de calcio
1024	calculator	räknemaskin	Rechner, Berechner	calculatrice	calculadora
1025	calender	glätt	Kalander, Walzwerk, Glättwerk	calandre	calandra
1026	calender blackened	glättningssvärtad	grauverdrückt (Satinierfehler)	plombé (à la calandre)	ennegrecido en calandra
1027	calender blackening	fläckbildning i glätt	Grauverdrücken am Kalander	plombage (à la calandre)	ennegrecimiento en calandra
1028	calender box	glättlåda	Kalanderkasten	boîte à eau de calandre	caja de calandra
1029	calender broke	glättningsutskott	Kalanderausschuss	cassé à la calandre	recortes de fabricación en calandra
1030	calender colored	glättningsfärgad	Kalandergefärbt, oberflächengefärbt	coloré en surface sur calandre	coloreado en calandra
1031	calender crushed	glättningskrossad	im Kalander verdrückt	écrasé au passage dans la calandre	desfibrado en calandra
1032	calender-crush finish	glättfinish	feuchtgeglättetes Papier, zerdrücktes Papier	écrasé au passage dans la calandre	acabado de desfibrado en calandra
1033	calender cuts	glättrems	Kalanderfalten, Satinierfalten	plis de calandre	cortadas
1034	calender dyed	glättfärgad	im Kalander gefärbt	coloré sur calandre	coloración en calandra
1035	calendered	glättad	kalandriert	calandré	satinado
1036	calendered paper	glättat papper	kalandriertes Papier, santiniertes Papier	papier calandré	papel satinado
1037	calender finished	kalandrering	im Kalander ausgerüstet, im Kalander oberflächensatiniert	fini calandré	acabado en calandra
1038	calendering	kalandrering	Kalandrieren, Satinieren	calandrage, lustrage	calandrado
1039	calenderman	kalanderförare	Kalanderführer	calandreur	conductor de calandra
1040	calender marked	kalandermarkerad	Eindrücke im Papier infolge— defekter Kalanderwalzen oder darauf haftender Fremdstoffe	marqué sur la calandre	marcado en calandra
1041	calender marking	kalandermarkering	Kalandermarkierung infolge defekter oder unsauberer Kalanderwalzen	marquage sur la calandre	marcado de calandra
1042	calender operator	kalanderförare	Kalanderführer	calandreur	conductor de calandra
1043	calender pick	kalandernappning	Das Rupfen der klebenden Kalanderwalzen Fasern einer Papierbahn an den Kalanderwalzen	arrachage causé par la calandre	arranque de calandra
1044	calender roll	kalandervals	Kalanderwalze	rouleau de calandre	rodillo de calandra
1045	calender-roll paper	pappersvals	Kalanderwalzenpapier	papier pour rouleaux de calandre	papel de rodillo de calandra
1046	calender scabs	glättflagor	Markierung von Papier oder Pappe durch an den Walzen haftende Pigmentteilchen	marques de calandre	cubrejuntas de calandra
1047	calender scales	glättvalsbelägg	Kalanderstaub, Kalanderspäne	écailles de calandre	escamas de calandra
1048	calender section	glättparti	Kalanderpartie	section des calandres	sección de calandra
1049	calender-sized paper	glättpapper, kalanderlimmat papper	kalandergeleimtes Papier	papier paraffiné sur calandre	papel encolado en calandra
1050	calender sizing	glättlimning, kalanderlimning	Kalanderleimung	paraffinage sur calandre	encolado de calandra
1051	calender spots	glättningsfläckar	Kalanderflecken, Satinierflecken, Kalanderschäden	plaquages de calandre	manchas de calandra
1052	calender stack	glättstapel	Trockenglättwerk	calandre de bout de machine	lisa
1053	calender-stack crumbs	glättsmulor	Satinagestaub	croûte de calandre	migas de lisa
1054	calender staining	glättfläckning	Satinierfärbung	coloration sur calandre	manchado de calandra
1055	calender streaks	glättveck	Satinierstreifen	raies en long sur papier calandré	estrías de calandra
1056	calender vellum finish	velängglättning	Velinausrüstung mit besonders glatter Oberfläche	fini vélin sur calandre	acabado vitela en calandra
1057	calender water box	fuktlåda	Kalanderwasserkasten	caisse à eau de rouleau de calandre	caja de agua para calandra
1058	calibrate	kalibrera	kalibrieren, eichen	calibrer	calibrar
1059	calibration	kalibrering	Kalibrierung, Eichung	étalonnage	contraste
1060	caliper	tjocklek	die unter spezifische Bedingungen gemessene Dicke einer Bahn	épaisseur d'un papier, épaisseur d'un carton	espesor
1061	caliper-gauge	tjockleksmätare	Dickenmessgerät, Dickenmesser	pied à coulisse, calibre	calibre
1062	calorimeter	kalorimeter	Kalorimeter, Wärmemesser	calorimètre	calorímetro
1063	calorimetry	kalorimetri	Wärmemessung, Kalorimetrie	calorimétrie	calorimetría
1064	cam	kam	Nocke, Nockenscheibe	came	leva
1065	cambered roll	bomberad vals	bombierte Walze	rouleau bouché	rodillo bombeado
1066	cambium	kambium	Cambium	cambium	cambium
1067	cambric finish	batistfinish	leinengeprägte Oberfläche	fini entoilé	acabado tela
1068	can	fat, tunna	zylinderiger Behälter (aus Karton-Faserplatte-Papier-oder Metallfolie)	bidon, boîte à conserves	recipiente

	ENGLISH	SVENSKA	DEUTSCH	FRANÇAIS	ESPAÑOL
1069	Canadian Standard freeness	Candian Standard freeness	CSF (Mahlgrad)	indice d'égouttge Canadien Standard	desgote standard canadiense
1070	can board	vätskeförpackningskartong	Dosenkarton, Büchsenkarton	récipient en cärton	cartón para recipientes
1071	canister	kanister	Kanister, Behälter (aus Pappe-Papier-Film-oder Folie)	boîte en fer-blanc	bote
1072	can stock	–	Dosenkartonstoff, Büchsenkartonstoff	carton pour la partie cylindrique de récipients	stock de recipientes
1073	cantilever	kantilever	freitragend, Auskragung	montage en porte-à-faux	voladizo
1074	cantilevered design	kantileverkonstruktion	freitragende Ausführung	montage en porte-à-faux	diseño de voladizo
1075	cantilever type fourdrinier	kantileverviraparti	Langsiebmaschine mit Aufahrvorrichtung	table plate type Cantilever	mesa plana de tipo voladizo
1076	capacitance	kapacitans	elektrostatische Belastbarkeit, elektrostatische Kapazität	capacité	capacitancia
1077	capacitor paper	kondensatorpapper	Kondensatorpapier	papier pour condensateurs	papel para condensadores
1078	capacitor tissue	kondensatorpapper	Kondensatortissue, Kondensseidenpapier	papier mince pour condensateurs	muselina para condensadores
1079	capacitor	kondensator	Kondensator	condensateur	condensador
1080	capacity	kapacitans	Kapazität, Leistungsfähigkeit, Aufnah, mefähigkeit, tato-Ausstoss einer in Betrieb befindlichen maschine	capacité	capacidad
1081	capillarity	kapilläritet	Kapillarität, Kapillarwirkung	capillarité	capilaridad
1082	capillary	kapillär	kapillar, haarförmig, haarfein	capillaire	capilar
1083	capital	kapital	haupt-, Kapital, Fond, grosser Anfangsbuchstabe	capital	capital
1084	cap paper	kapsylpapper	Flaschenverschlusspapier, Pflanzenschutzpapier	mousseline	papel Cap para envolver
1085	capping paper	kapsylpapper	Flaschenkapsel, Pflanzenschutzpapier	papier pour capsules de bouteilles	papel de coronamiento
1086	caprolactam	kaprolaktan	Kaprolactam	caprolactam	caprolactam
1087	capsule	kapsel	Kapsel, Druckdose	capsule	cápsula
1088	captive pulp	utskottsmassa	durch Eigenverbrauch gebundene Zellstoffmenge	pâte vendue à une filiale	pasta cautiva
1089	carbamates	karbamater	Karbamat	carbamates	carbamatos
1090	carbamides	karbamider	Karbamid, Harnstoff	carbamides	carbamidas
1091	carbides	karbider	Karbid	carbures	carburos
1092	carbohydrates	kolhydrater	Kohlenhydrat	hydrates de carbone	hidratos de carbono
1093	carbon	kol	Kohle, Kohlenstoff, Kohlepapier	carbone	carbono
1094	carbonates	karbonater	mit Kohlendioxid saturieren, kohlensaures Salz, Karbonat	carbonates	carbonatos
1095	carbonation	karbonering	Verbindung mit Kohlensäure, Kohlensäuresättigung	carbonisation	carbonatación
1096	carbon black	kimrök	Russ, Karbonschwärze, Druckerschwärze	noir de fumée	negro de humo
1097	carbon coating	karbonisering	Kohlebeschichtung für Kohlepapier	enduction du papier carbone	estucado de carbono
1098	carbon disulfide	koldisulfid	Kohlenstoffdisulfid, Schwefelkohlenstoff	disulfure de carbone	bisulfuro de carbono
1099	carbon fiber	kolfiber	Kohlenstoffaser	fibre de carbone	fibra de carbono
1100	carbonization	förkolning, karbonisering	Verkohlung, Entgasung, Trockendestillation	carbonisation	carbonización
1101	carbonized roll	kolvals	Kohlenstoff beschichtete Walze	bobine de papier carbone	rodillo carbonizado
1102	carbonizing paper	karbonråpapper	Kohlerohpapier	papier support carbone	papel cebolla
1103	carbon paper	karbonpapper	Kohlepapier	papier carbone	papel carbón
1104	carbon spots	karbonfläckar	Kohleflecken	taches de carbone	manchas de carbón
1105	carbon tetrachloride	karbontetraklorid	Tetrachlorkohlenstoff	tétrachlorure de carbone	tetracloruro de carbón
1106	carbonyl groups	karbonylgrupper	Karbonylgruppen	groupes carbonyl	grupos carbonilo
1107	carbonyls	karbonyler	Karbonyle	carbonyls	carbonilos
1108	carbonyl sulfide	karbonylsulfid	Karbonylsulfid	sulfure de carbonyl	sulfuro carbonilo
1109	carborundum paper	karborundumpapper	Karborundpapier, Schmirgelpapier	papier abrasif	papel carborundo
1110	carboxyalkylation	karboxy-alkylering	Karboxalkylierung	carboxyalkylation	carboxialquilación
1111	carboxyalkyl cellulose	karboxyalkylcellulosa	Karboxyalkylcellulose	carboxyalkyl cellulose	celulosa de carboxialquilo
1112	carboxyalkyl groups	karboxyalkylgrupper	Karboxyalkylgruppen	groupes carboxyalkyl	grupos carboxialquilo
1113	carboxyethylation	karboxyetylering	Karboxyäthylierung	carboxyéthylation	carboxietilación
1114	carboxyethyl cellulose	karboxyetylcellulosa	Karboxyäthylcellulose	carboxyéthyl cellulose	celulosa de carboxietilo
1115	carboxyethyl groups	karboxyetylgrupper	Karboxyäthylgruppen	groupes carboxyéthyl	grupos carboxietilo
1116	carboxylation	karboxylering	Karboxylierung	carboxylation	carboxilación
1117	carboxyl groups	karboxylgrupper	Karboxylgruppen	groupes carboxyl	grupos carboxilo
1118	carboxylic acids	karboxylsyror	Karbonsäuren	acides carboxyliques	ácidos carboxílicos
1119	carboxymethylation	karboxymetylering	Karboxymethylierung	carboxyméthylation	carboximetilación
1120	carboxymethyl cellulose	karboxymetylcellulosa	Karboxymethylcellulose	carboxyméthyl cellulose	celulosa de carboximetilo
1121	carboxymethyl groups	karboxymetylgrupper	Karboxymethylgruppen	groupes carboxymérhyl	grupos carboximetilo
1122	card	kort	krempeln, karden, Krempel, Karde, Karte	carte mince	tarjeta

	ENGLISH	SVENSKA	DEUTSCH	FRANÇAIS	ESPAÑOL
1123	cardboard	kartong	Karton, Pappe, Vollpappe	carton assez fort	cartulina
1124	cardboard finish	kartongfinish	Kartonveredlung, Kartonausrüstung, Kartonoberfläche	fini obtenu par pression entre deux cartons épais	acabado cartón
1125	car liner	bilklädsel	Wagenplane, Kraftpapier-Schutzdecke	celloderme pour carrosserie	forro para coches
1126	car lining	bilklädsel	Wagenplanierung	protection du plancher d'un camion avec du celloderme	forrado para coches
1127	carload	billast	Wagenladung, Waggonladung	wagon complet	vagonada
1128	carload lot	bilplats	Mindestfrachtmenge an Papier	tonnage habituellement livré dans un wagon complet	carro completo
1129	carnauba wax	karnubavax	Karnaubawachs	cire de carnauba	cera de carnauba
1130	carpet tissue	mattväv	Teppich-Tissue, Sulfatzellstoffpapier von hoher Zugfestigkeit für Webzwecke	papier mince pour mettre sous les tapis	tejido de alfombras
1131	Carpinus (Lat.)	Carpinus	Carpinus	charme	Carpinus
1132	carrageenan	carragen-pektin	Seetang	extrait gélatineux de Canagheen	musgo perlado
1133	carrying roll	bärvals	Laufrolle	rouleau porteur	rodillo portador
1134	carton	papplåda	Karton, Kasten, Versandschachtel	boîte pliante en carton	cartonaje
1135	cartoning machine	lådmaskin	Kartonherstellmaschine	machine à fabriquer les boîtes pliantes	máquina para cartonajes
1136	carton labels	lådetiketter	Kartonaufkleber, Schachtelaufkleber, Paketaufkleber	étiquettes pour boîtes pliantes	etiquetas para cartonajes
1137	carton-liner paper	lådfoderpapper	fettdichtes Papier zur Kartoninnenauskleidung, gewachstes Papier zur Kartoninnenauskleidung	papier à doubler les boîtes pliantes	papel forro para cajas de cartón
1138	carton-sealing paper	lådförseglingspapper	heissgesiegeltes Papier für Lebensmittelkarton, mit Isolierlack beschichtetes Papier für Lebensmittelkarton	papier imperméable pour emballer les caisses carton contenant des produits alimentaires	papel para sellado de cajas de cartón
1139	cartridge paper	karduspapper	Kartuschenpapier	papier pour cartouches	papel para cartuchos
1140	case	låda	Behälter, Gehäuse, Etui, Schachtel, Kiste, Hülle	caisse, boîte, écrin	caja
1141	case hardening	ythärdning	Einsatzhärten	cémentation	cementación
1142	casein	kasein	Kasein	caséine	caseína
1143	case labels	lådetiketter	Schachteletiketten	étiquettes pour caisses, étiquettes pour boîtes	etiquetas para cajas
1144	case-lining paper	lådfoderpapper	Kistenausschlagpapier	papier pour doubler des boîtes, papier pour doubler des caisses	papel forro para cajas
1145	case lot	lådmängd	verpackter Papierstapel	papier à plat livré emballé	partida de cajas
1146	cash flow	cash flow	cash flow, Zu- und Abgang von Barmitteln	cash flow	recursos generados
1147	cash-register paper	kassaregisterpapper	Registrierkassenpapier	papier pour rouleaux de caissses enregistreuses	papel para cajas registradoras
1148	casing	låda	Verkleidung, Auskleidung, Gehäuse, Bezug	capot, carter	envoltura
1149	cast-coated paper	gjutbestruket papper	gussgestrichenes Papier	papier couché chrome superbrillant	papel estucado por extrusión
1150	cast coater	gjutbestrykare	Gusstreicher	coucheuse au glacis	estucadora por extrusión
1151	cast coating	gjutbestrykning	Gusstreichen	couchage au glacis	estucado por extrusión
1152	casting	gjutning	Vergiessen, Giessen, Giessling, Abguss	fusion	fundición
1153	cast iron	gjutjärn	Gusseisen, gusseisern	fonte	hierro colado
1154	catalog paper	katalogpapper	Katalogpapier	papier pour catalogues	papel para catálogos
1155	Catalpa (Lat.)	Catalpa	Trompetenbaum, Antilleneiche	Catalpa	Catalpa
1156	catalyst	katalysator	Katalysator, Beschleuniger	catalyseur	catalizador
1157	cathode	katod	Kathode	cathode	cátodo
1158	cathode-ray tube	katodstrålerör	Braun'sche Röhre, Kathodenstrahlröhre, Elektronenstrahlröhre	lampe à rayons cathodiques	tubo de rayos catódicos
1159	cation	katjon	Kation	cation	catión
1160	cation exchange	katjonbyte	Kationenaustausch	échange de cations	intercambio de cationes
1161	cation exchanger	katjonbytare	Kationenaustauscher	échangeur de cations	intercambiador de cationes
1162	cationic	katjonisk	kationisch	cationique	catiónico
1163	cationic compounds	katjonföreningar	kationische Verbindungen	composés cationiques	compuestos catiónicos
1164	cationic starch	katjonstärkelse	kationische Stärke	amidon cationique	almidón catiónico
1165	catwalk	tvärbrygga	Laufsteg, Laufgang	passerelle	pasarela
1166	caustic	kaustik	kaustisch, ätzend, beissend, Ätzmittel, Beize	caustique	cáustico
1167	causticity	kausticitet	Kaustizität	causticité	causticidad
1168	causticize	kausticitera	kaustifizieren, in Laugenform überführen	caustifier	causticar

	ENGLISH	SVENSKA	DEUTSCH	FRANÇAIS	ESPAÑOL
1169	causticizer	kausticiterare	Schmelzlöser	caustificateur	causticador
1170	causticizing	kausticering	Kaustifizieren	caustification	caustificación
1171	caustic soda	kaustisk soda, lut	Ätznatron, Natriumhydroxid	soude caustique	sosa cáustica
1172	cavitation	kavitering	Kavitation, Hohlraumbildung, Blasenbildung	cavitation	cavitación
1173	cedar	ceder	Zeder	cèdre	cedro
1174	Cedrus (Lat.)	Cedrus	Zedernholz, Zedernöl	cèdre	Cedrus
1175	ceiling board	takpapp	Dämmpappe	carton pour toitures	cartón para techos
1176	celite	kelit	Celit	célite (terre de diatomée)	celita
1177	cell	cell	Zelle, Farbnäpfchen (Tiefdruck)	cellule	célula
1178	cellophane	cellofan	Cellophan, Zellglas	cellophane	celofán
1179	cell structure	cellstruktur	Zellenstruktur, Zellenaufbau	structure des cellules	estructura de la célula
1180	cellular board	cellpapp	gedeckte Wellpappe, Zellstrukturpappe	carton à alvéoles	cartón celuloso
1181	cellular structure	cellstruktur	Zellstruktur	structure cellulaire	estructura celular
1182	cellulose	cellulosa	Zellulose, Cellulose	cellulose	celulosa
1183	cellulose acetate	cellulosaacetat	Celluloseacetat, Azetylcellulose	acétate de cellulose	acetato de celulosa
1184	cellulose butyrate	cellulosabutyrat	Cellulosebutyrat	butyrate de cellulose	butirato de celulosa
1185	cellulose content	cellulosahalt	Cellulosegehalt	teneur en cellulose	contenido de celulosa
1186	cellulose copolymers	cellulosapolymerer	Cellulose-Copolymere	copolymères de cellulose	copolímeros de celulosa
1187	cellulose degradation	cellulosanedbrytning	Celluloseabbau	dégradation de la cellulose	degradación de la celulosa
1188	cellulose derivatives	cellulosaderivat	Cellulosederivate	dérivés de la cellulose	derivados de la celulosa
1189	cellulose esters	cellulosaestrar	Celluloseester	esters de cellulose	ésteres de celulosa
1190	cellulose ethers	cellulosaetrar	Celluoseäther	éthers de cellulose	éteres de celulosa
1191	cellulose fiber	cellulosafiber	Cellulosefaser	fibre de cellulose	fibra de celulosa
1192	cellulose film	cellulosafilm	Zellglas, Klarsichtfolie	cellophane	celofán
1193	cellulose nitrate	cellulosanitrat	Cellulosenitrat, Nitrocellulose	nitrate de cellulose	nitrato de celulosa
1194	cellulose plastics	cellulosaplast	Kunststoffe auf Cellulosebasis	plastiques à base de cellulose	plásticos de celulosa
1195	cellulose solvents	cellulosalösningsmedel	Zellulose-Lösungsmittel	dissolvants de la cellulose	disolventes celulósicos
1196	cellulose structure	cellulosastruktur	Cellulosestruktur	structure de la cellulose	estructura de celulosa
1197	cellulose wadding	cellulosavadd	Zellstoffwatte	ouate de cellulose	guata de celulosa
1198	cellulose xanthate	cellulosaxantat	Cellulosexanthogenat	xanthate de cellulose	xantato celulósico
1199	cell wall	cellulosacellvägg	Zellwand	paroi de cellule	pared de célula
1200	cell-wall thickness	cellväggtjocklek	Zellwandstärke	épaisseur de la paroi de cellule	espesor de la pared de célula
1201	Celsius scale	celsiusskala	Celsiusskala	thermomètre celsius	escala Celsius
1202	cement-sack paper	cementsäckpapper	Zementsackpapier	papier pour sacs ciment	papel para sacos de cemento
1203	center ply	centerskikt	Mittelschicht einer Pappe, Einlage	couche centrale	capa central
1204	center stock	centerskiktmassa	Füllung, Zwischenlage, Mittellage	matières premières pour couche centrale	núcleo de pasta
1205	center winding	centervalsrullning	Freiwicklung	bobinage sur mandrin	bobinado sobre mandril
1206	centipoise	centipoise	Zentipoise, Centipoise	centipoise	centipoise
1207	centrifugal	centrifugal	zentrifugal	centrifuge	centrífugo
1208	centrifugal blower	centrifugalblåsare	Turbogebläse, Schleudergebläse, Zentrifugalgebläse	ventilateur centrifuge	ventilador soplador centrífugo
1209	centrifugal classifier	centrifugalfraktionerare	Zentrifugalsortierer	classeur centrifuge	clasificador centrífugo
1210	centrifugal filtration	centrifugalfiltrering	Zentrifugalfiltrierung	filtration centrifuge	filtración centrífuga
1211	centrifugal screen	centrifugalsil	Zentrifugalsortierer, Schleudersortierer	épurateur centrifuge	depurador centrífugo
1212	centrifugal vacuum compressor	centrifugalkompressor	Turbogebläse	compresseur aspirant centrifuge	compresor centrífugo de vacío
1213	centrifugation	centrifugering	Schleuderung, Zentrifugierung	centrifugation	centrifugación
1214	centrifuge	centrifug	schleudern, zentrifugieren, Zentrifuge	centrifuge	centrífugo
1215	centripetal	centripetal	zentripetal	centripète	centrípeto
1216	ceramic coating	keramisk bestrykning	keramische Beschichtung	revêtement de céramique	estucado cerámico
1217	ceramic fiber	keramisk fiber	Keramikfaser	fibre de céramique	fibra cerámica
1218	ceramics	keramik	Keramik	céramique	cerámica
1219	ceramic suction box cover	keramisk suglådebeläggning	keramische Siebsaugerbeläge	revêtement en céramique de la face supérieure d'une caisse aspirante	tapa cerámica de sifón
1220	chain barker	kedjedisintegrator	Kettenentrindungsmaschine	écorceuse à chaîne	descortezador de cadena
1221	chain conveyor	kedjetransportör	Kettenförderer, Transportkette, Förderkette	transporteur à chaîne	cadena transportadora
1222	chalking	kalkning	Abkreiden, Ausschwitzen, Druckfehler (das Farbpigment wird von der Papieroberfläche weggewischt)	poudrage	desprendimiento de polvo de una tinta

ENGLISH	SVENSKA	DEUTSCH	FRANÇAIS	ESPAÑOL
1223 chalky appearance	kalkaktigt utseende	kreidiges Aussehen	aspect crayeux	aspecto gredoso
1224 Chamaecyparis (Lat.)	Chamaecyparis	Chamaecyparis	cèdre	Chamaecyparis
1225 champion coater	championbestrykare	Champion-Streicher	coucheuse type Champion	estucadora principal
1226 change of deckle	däckelbyte	Formatneueinstellung	changement de couverte	cambio de cubierta
1227 characteristic	karakteristisk	charakteristisch, Besonderheit, Eigenschaft, Merkmal	caractéristique	característico
1228 charcoal	aktivt kol	Holzkohle, Zeichenkohle, chemische Aktivkohle	charbon de bois	carbón vegetal
1229 charcoal drawing paper	kolritpapper	Kohlezeichenpapier	papier pour dessin au fusain	papel para carboncillo
1230 charcoal paper	kolpapper	Kohlezeichenpapier	papier pour dessin au fusain	papel carbón vegetal
1231 chart	diagram	Karte, Tabelle, graphische Darstellung	diagramme, graphique	ábaco
1232 chart paper	diagrampapper	Landkartenpapier, Diagrammpapier	papier pour diagrammes, pour cartes géographiques	papel para mapas
1233 chattering of rolls	valssmatter	Schlagen von Walzen, Schlagen von Rollen	vibration des rouleaux	chirrido de los rodillos
1234 checking	kontroll	Prüfung, Kontrolle, Überprüfen, Nachprüfen	contrôle, vérification	control
1235 check list	checklista	Bewertungsliste	liste de contrôle	lista de comprobación
1236 check paper	checkpapper	Scheckpapier, Sicherheitspapier	papier pour chèques	papel para cheques
1237 check valve	kontrollventil	Rückschlagventil, Absperrventil	vanne d'arrêt	válvula de retención
1238 cheese rolls	bibanor	Nebenbahnen	–	orillos
1239 chelates	kelater	Chelate	chelates	quelatos
1240 chelating agent	kelatmedel	Cheliermittel	chelateur	agente quelato
1241 chelation	kelatering	Chelatbildung, Chelierung	chelation	quelación
1242 chemical barking	kemisk barkning	chemische Holzentrindung	écorçage chimique	descortezado químico
1243 chemical bond	kemisk bindning	chemische Bindung	liaison chimque	unión química
1244 chemical consumption	kemikalieförbrukning	chemischer Verbrauch, chemischer Bedarf	consommation chimique	consumo químico
1245 chemical debarking	kemisk barkning	chemische Entrindung	écorçage chimique	descortezado químico
1246 chemical degradation	kemisk nedbrytning	chemischer Abbau	dégradation chimique	degradación química
1247 chemical fiber paper	papper av kemisk fiber	Chemiefaserpapier	papier sans bois	papel fibra química
1248 chemical filter paper	kemifilterpapper	chemisch reiner Filterpapier	papier filtre pour laboratoires	papel filtro químico
1249 chemical loss	kemisk förlust	chemischer Verlust	perte chimique	pérdida química
1250 chemical oxygen demand (COD)	kemisk syreförbrukning	chemischer Sauerstoffbedarf (COD)	C.O.D.	demanda química de oxígeno
1251 chemical properties	kemiska egenskaper	chemische Eigenschaften	propriétés chimiques	propiedades químicas
1252 chemical pulp	kemisk massa	chemischer Zellstoff	pâte chimique	pasta química
1253 chemical púlping	kemisk uppslutning	chemischer Zellstoffaufschluss	fabrication de pâtes chimiques	elaboración de pasta química
1254 chemical reaction	kemisk reaktion	chemische Reaktion	réaction chimique	reacción química
1255 chemical recovery	kemisk återvinning	Chemikalienrückgewinnung	récupération chimique	recuperación química
1256 chemical resistance	kemisk motståndskraft	chemische Beständigkeit, Chemikalienbeständigkeit	résistance chimique	resistencia química
1257 chemicals	kemikalier	Chemikalien	produits chimiques	productos químicos
1258 chemical tests	kemiska prov	chemische Prüfung	analyses chimiques	pruebas químicas
1259 chemical-to-wood ratio	kemikalie/vedkvot	Verhältnis zwischen Holz und Chemikalienzugabe (bei der Sulfatkochung)	rapport produits chimiques/bois	relación químico-maderera
1260 chemical treatment	kemisk behandling	chemische Behandlung	traitement chimique	tratamiento químico
1261 chemical wood pulp	kemisk massa	Zellstoff, Holzzellstoff	pâte de bois chimique	pasta de madera química
1262 chemigroundwood	chemigroundwood	chemischer Holzschliff	pâte chimicomécanique	mecano-química
1263 chemimechanical pulp	kemisk-mekanisk massa	Refinerschliff	pâte michimique	pasta mecano-química
1264 cheque paper (Brit.)	checkpapper	Scheckpapier	papier pour chèques	papel para cheques
1265 cherry	körsbär	Kirschbaumholz	cerisier	cerezo
1266 chest	kar	Bütte, Kasten	cuvier, réservoir, caisse	caja
1267 chestnut tree	kastanjeträd	Kastanienbaum	chataignier	castaño
1268 cheviot	cheviot	Granitimitation	papier fantaisie pour couvrir les boîtes en carton	cheviot
1269 chief chemist	chefskemist	Chefchemiker	chimiste en chef	jefe químico
1270 chief engineer	överingenjör	Oberingenieur	ingénieur en chef	ingeniero jefe
1271 chill roll	kylvals	Kühlwalze	rouleau refroidisseur	rodillo de enfriamiento
1272 chilled roll	kyld vals	geriffelte Walze, Hartgusswalze	rouleau en fonte	rodillo enfriado
1273 chimney	skorsten	Kamin	cheminée	chimenea
1274 china board	porslinskartong	China-Pappe (meist gefärbt z.B. für Theaterbillets)	carton hydrochine	cartón de China
1275 china clay	china clay, porslinslera	China Clay, Kaolin	kaolin	caolin
1276 chip	flis	Hackschnitzel, Schnitzel, Hackspan	copeau	viruta
1277 chip bin	flisbinge	Hackschnitzelsilo	silo à copeaux	arca de virutas

	ENGLISH	SVENSKA	DEUTSCH	FRANÇAIS	ESPAÑOL
1278	chipboard	chipboard	Schrenzpappe, Graupappe, Spanplatte (US)	carton gris (de vieux papiers)	cartulina de estraza
1279	chip charger	flislastare	Kocherfüllapparatur, Hackschnitzelzuteiler	chargeur de copeux	cargador de virutas
1280	chip chute	flisschakt	Spänefang, Rutsche für die Hackschnitzel	chute des copeaux	vertedor de virutas
1281	chip classification	flisklassificering	Chipklassierung, Spangutklassierung	tri des copeaux	clasificación de virutas
1282	chip classifier	flissåll	Chipklassiergerät	trieur de copeaux	clasificador de virutas
1283	chip conveyor	flistransportör	Spänetransportanlage, Chipförderanlage	transporteur de copeaux	transportador de virutas
1284	chip crusher	fliskross	Hackschnitzelmühle, Hackzerkleinerungsmaschine	broyeur de copeaux	triturador de virutas
1285	chip dimensions	flisdimensioner	Spanabmessungen	taille des copeaux	dimensión de las virutas
1286	chip feeder	flismatare	Chiptransport, Chipzugabe, Hackschnitzeltransport	rigole à copeaux	alimentador de virutas
1287	chip groundwood	flisslip	Schliff aus Chips, Schliff aus Hackschnitzel	copeaux pour pâte mécanique	pasta mecánica de virutas
1288	chip meter	flismätare	Hackschnitzeldosierer	jauge de copeaux	medidor de virutas
1289	chip packer		Kocherfüllapparatur	bourreur de copeaux	enfardador de virutas
1290	chip packing	flispackning	Einfüllen der Chips in den Kocher	répartition et bourrage des copeaux	enfardeladura de virutas
1291	chipper	flishugg	Hacker, Hackmaschine, Spanmesser	coupeuse à bois	troceadora
1292	chipper knife	flishuggkniv	Hackmesser, Zerspanungsmesser	lame de coupeuse à bois	cuchilla de troceadora
1293	chip pile	flishög	Hackschnitzelstapel	tas de copeaux	pila de virutas
1294	chipping	flishuggning	Absplittern, Hobelspan, Spanen	mise en copeaux	virutado
1295	chip screen	flissåll	Spansieb, Spänesieb	tamis à copeaux	criba para astillas
1296	chip silo	flissilo	Hackschnitzelsilo	silo à copeaux	silo de virutas
1297	chlorates	klorater	Chlorat	chlorates	cloratos
1298	chlorides	klorider	Chlorid	chlorures	cloruros
1299	chlorinated lignins	klorerade ligniner	chloriertes Lignin	chlorolignine	cloroligninas
1300	chlorinated starch	klorerad stärkelse	chlorierte Stärke	amidon chloruré	almidón clorado
1301	chlorination	klorering	Chlorierung	chloruration	cloruración
1302	chlorination stage	kloreringssteg	Chlorierungsstufe	chloruration en plusieurs stades	etapa de cloruración
1303	chlorine	klor	Chlor	chlore	cloro
1304	chlorine cell	klorcell	Chlorzelle	cellule d'électrolyse pour la production du chlore	pila de cloro
1305	chlorine compounds	klorföreningar	Chlorverbindungen	composés de chlore	compuestos de cloro
1306	chlorine dioxide	klorioxid	Chlordioxid	peroxyde de chlore	dióxido de cloro
1307	chlorine gas	klorgas	Chlorgas	gaz de chlore	gas de cloro
1308	chlorine monoxide	klormonoxid	Chlormonoxid	protoxyde de chlore	monóxido de cloro
1309	chlorine number	klortal	Chlorzahl	indice de chlore	índice de cloro
1310	chlorine process	kloreringsprocess	Chlorierverfahren	procédé au chlore	proceso de cloro
1311	chlorine requirement	klorbehov	Chlorbedarf	aptitude au blanchiment au chlore	demanda de cloro
1312	chlorites	kloriter	Chlorite	chlorites	cloritos
1313	chloroacetic acids	klorättikssyra	Chloressigsäure	acides chloroacétiques	ácidos cloracéticos
1314	chlorolignins	klorligniner	Chlorlignine	chlorolignines	cloroligninas
1315	chlorosis	kloros	Chlorose, Gelbblättrigkeit	chlorose	clorosis
1316	chromates	kromater	Chromat	chromates	cromatos
1317	chromaticity	kromaticitet	Chromatizität, Farbart	chromaticité	cromaticidad
1318	chromatic paper	kromatpapper	fleckiges Papier, scheckiges Papier	papier marbré	papel cromático
1319	chromatographic paper	kromatografipapper	chromatographisches Papier	papier pour chromatographie	papel cromatográfico
1320	chromium	krom	Chrom	chrome	cromo
1321	chromium compounds	kromföreningar	Chromverbindungen	composés de chrome	compuestos de cromo
1322	chromium oxide	kromoxid	Chromgrün, Chromoxid	oxyde de chrome	óxido de cromo
1323	chromo board	kromkartong	Chromokarton	carton couché chrome	cartón cromo
1324	chromo paper	krompapper	Chromopapier	papier couché chrome	papel cromo
1325	chuck	chuck	Rollenkonus	mandrin (de tour)	mandril
1326	cigarette paper	cigarrettpapper	Zigarettenpapier	papier à cigarettes	papel para cigarrillos
1327	cigarette tissue	cigarrettpapper	Zigarettenseidenpapier	papier à cigarettes	papel de fumar
1328	circuit breaker	kretsbrytare	Stromunterbrecher, selbsttätiger Ausschalter	coupe-circuit	disyuntor
1329	circular chart	cirkulärt diagram	Kreisblatt (für Kreisblattschreiber)	diagramme circulaire	diagrama circular
1330	circular cutter	cirkulär skärarkniv	Rundschneidemaschine	couteau circulaire	cuchilla circular
1331	circumferential register	perifert register	Umfangspasser, umfangsbedingte	–	registro circunferencial
1332	circumferential speed	periferihastighet	Umfangsgeschwindigkeit	vitesse circonférentielle	velocidad circunferencial
1333	clad digester	inklädd kokare	plattierter Kocher	lessiveur en métal plaqué	lejiadora chapada

THOSE WHO CAN, DO.

Those who can't, sit back and blame the economy.

The people at Hooker decided some time back that things would get better only if people did something to make them better.

That's why the Hooker Chemicals & Plastics Corp. is backing the pulp and paper industry every way it can through its people, its technology and its capital. Increasing its chlorate, chlor-alkali production through new plants. Applying its new technology to reduce polluting effluents in bleaching processes. Developing the latest in diaphragm and membrane cell technology.

You can see what we are up to in Taft, Louisiana where we've added 815 tons a day to our chlor-alkali capacity. You can see it in the new H Series diaphragm cells, installed not only at Taft, Louisiana and Niagara Falls, New York but in licensee plants at home and abroad.

You can see it in the increased use of our SVP® process for the effluent free production of chlorine dioxide. And, in the not too distant future, you will be seeing more of what we can do as our new MX™ membrane cells begin commercial production of caustic and chlorine.

Hooker is proud of its close working relationship to the pulp and paper industry. We like to talk about it, but more important we like to do something about it.

Actions, after all, speak louder than words.
Hooker Chemicals & Plastics Corp.
Niagara Falls, New York 14302
Telephone: (716) 285-6655

	ENGLISH	SVENSKA	DEUTSCH	FRANÇAIS	ESPAÑOL
1334	cladding	inklädning	Auskleiden, Plattieren	plaquer un métal sur un autre	chapeado
1335	clad metal	inklädningsmetall	plattiertes Metall	métal plaqué	metal chapado
1336	clad-steel plate	stålklädd platta	stahlplattiert	plaque revêtue d'acier	placa de acero chapeado
1337	clamp	klämma	Klemme, Klammer, Spannvorrichtung	crampon, collier	grapa
1338	clamp marks	klämmningmärken	Klammermarkierung	marques d'agrafe	marcas de grapa
1339	clarification	klarläggande	Klärung, Abwasserklärung	clarification	clarificación
1340	clarifier	klargörare	Klärapparat, Klärmittel	clarificateur	clarificador
1341	clarify	klara	klären, klar machen, reinigen	clarifier	clarificar
1342	classification	klassning	Klassifizierung, Einordnung, Sortierung, Einstufung	classement, fractionnement	clasificación
1343	classifier	klassare	Fraktioniergerät, Klassiergerät	appareil pour l'étude des fibres par la méthode du fractionnement	clasificadora
1344	clay	lera	Clay, Tonerde, Ton, Lehm	argile, glaise	arcilla
1345	clay-coated	lerbestruken	clay-gestrichen, pigmentgestrichen	couché au kaolin	estucado de arcilla
1346	clay-coated blanks	lerbestrukna ark	ein- oder beidseitig pigmentgestrichener Karton	papiers et cartons coucheé au kaolin	formularios estucados caolina
1347	clay-coated boxboard	lerbestruken kartong	pigmentgestrichener Faltschachtelkarton	carton couché au kaolin	cartonaje estucado caolina
1348	clay-filled paper	lerfyllt papper	mit Füllstoffpigment angereichertes Papier	papier très chargé à haute teneur en cendres	papel relleno de arcilla
1349	clay lump	lerklump	Pigmentklumpen, Pigmentstück	pâton de glaise	grumo de arcilla
1350	cleaner	renare	Rohrschleuder, Wirbelsichter	produit de nettoyage, épurateur	depurador
1351	cleaning	rening	Reinigen, Säubern	débardage, décapage, épuration	depuración
1352	clear cutting	renskärning	Kahlschlag	coupe rase	cortado limpio
1353	clipper seam	clippersöm	Clippernaht	couture rognée	costura de abrazadera
1354	clog	täppa	verstopfen, klumpig werden, sich zusammenballen, Hindernis, Verstopfung	obstruer	obstruir
1355	clone	klon	Klon	clone	clónico
1356	closed circuit	sluten krets	geschlossener Stromkreis	circuit fermé	circuito cerrado
1357	closed hood	sluten kåpa	geschlossene Trockenhaube (Trockenpartie)	hotte fermée	cubierta cerrada
1358	closed-loop control	styrning med återkoppling	geschlossener Regelkreis	régulation en boucle fermée	control de bucle cerrado
1359	closed system	slutet system	geschlossenes System	système en circuit fermé	sistema cerrado
1360	closed journal type dandy roll	egoutör med sluten lagring	Zapfenegoutteur	rouleau égoutteur à tourillon fermé	rodillo desgotador para soporte de cojinete cerrado
1361	closed type flow box	sluten inloppslåda	geschlossener Stoffauflauf	caisse d'alimentation en circuit fermé	caja de entrada cerrada
1362	close formation	sluten arkbildning	geschlossene Blattbildung	épair fondu, plein, régulier	transparencia uniforme
1363	cloth finish	klotfinish	Appretur, Webeffekt, Webstruktur im Papier	gaufrage toile	gofrado tela
1364	clothing	beklädnad	Bespannung, Bekleidung	habillage	guarniciones
1365	cloud box covering	ångkåpa	melierte Schachtelkaschierung, marmorierte Schachtelverkleidung	nuageux, chiné	–
1366	cloud effect	molnighetseffekt	wilde Formierung (in der Blattbildung)	nuageux	efecto anublado
1367	cloudiness	molnighet	Wolkigkeit (vom Papier), ungleichmässige Durchsicht	aspect nuageux	nubosidad
1368	cloudy	molnig	wolkig, trübe, unklar	nuageux	nuboso
1369	cloudy formation	flockig arkibildning	wilde Blattformierung (am Stoffauflauf)	épair nuageux	transparencia nubosa
1370	clump	klump	Klumpen	bloc, groupe	flocular
1371	clutch	koppling	kuppeln, erfassen, ergreifen, Kupplung	embrayage	embrague
1372	coagulant	koagulermedel	Koagulationsmittel, Koagulierungsmittel	coagulant	coagulante
1373	coagulate	koagulera, koagulerat material	koagulieren	coaguler, figer	coagular
1374	coagulating	koagulerande	Ausflockung, Gerinnung, Festwerden	coagulant	coagulación
1375	coagulating agent	koaguleringsmedel	Gerinnungsmittel, Koagulierungsmittel	produit coagulant	agente coagulante
1376	coagulation	koagulering	Ausflockung, Festwerden, Gerinnen	coagulation	coagulación
1377	coarse	grov, rå	grob, rauh	grossier, brut	basto
1378	coarse finish	rå yta	rauhe Oberfläche	fini brut	acabado basto
1379	coarse papers	grovt papper	Papiere mit rauher Oberfläche	papier brut (emballage)	papeles bastos
1380	coarse screen	grov sil	Grobsortierer	premier épurateur dégraisseur	depurador de desbaste

china clay

---

svenska	**kaolin**
deutsch	**Kaolin**
français	**kaolin**
español	**caolín**

quality

---

svenska	**kvalitet**
deutsch	**Qualität**
français	**qualité**
español	**calidad**

china clay = quality in any language -

	ENGLISH	SVENSKA	DEUTSCH	FRANÇAIS	ESPAÑOL
1381	coaster board	brick-kartong	Bierfilz, Bierdeckelpappe	carton pour dessous de bocks	papel absorbente para platillos
1382	coat	bestrykning	Auftrag, Strich, Überzug	couche, enduit	capa de impregnación
1383	coatability	bestrykbarhet	Auftragsfähigkeit, Streichfähigkeit	aptitude au couchage	impregnabilidad
1384	coated	bestruken	gestrichen, überzogen	couché	estucado
1385	coated art paper	konsttryckpapper	gestrichenes Kunstdruckpapier	papier couché classique	papel estucado para impresión artística
1386	coated blanks	bestrukna ark	ein- oder beidseitig gestrichener Karton	rognures de papier couché	impresos estucados
1387	coated board	bestruken kartong	gestrichener Karton	carton couché	cartón estucado
1388	coated bond paper	bestruket finpapper	gestrichenes Bankpostpapier, gestrichenes Banknotenpapier	papier écriture couché	papel bond estucado
1389	coated book paper	bestruket boktryckpapper	gestrichenes Buchdruckpapier	papier d'édition couché	papel estucado para libros
1390	coated boxboard	bestruken kartong	gestrichener Kartonagenkarton	carton couché pour boîtes	cartonajes estucados
1391	coated bristol	bestruken bristolkartong	gestrichen Bristol	bristol couché	bristol estucado
1392	coated glassine	bestruken pergamyn	gestrichenes Pergamin	papier cristal couché	papel transparente estucado
1393	coated magazine paper	bestruket journalpapper	gestrichenes Illustrationsdruckpapier	papier couché pour magazines (LWC)	papel estucado para revistas
1394	coated offset paper	bestruket offsetpapper	gestrichenes Offsetdruckpapier	papier offset couché	papel estucado offset
1395	coated papers	bestrukna papper	gestrichene Papiere	papiers couchés	papeles estucados
1396	coated printing paper	bestruket tryckpapper	gestrichene Druckpapiere	papier d'impression couché	papel estucado para imprimir
1397	coated tag	bestruket etikettpapper	gestrichen Anhängeetiketten	étiquette couchée	etiquetas estucadas
1398	coater	bestrykare	Streicher, Streichanlage, Streichaggregat	coucheuse, fonceuse	estucadora
1399	coating	bestrykning	Streichen, Strich	couchage, revêtement	estucado
1400	coating additive	bestrykningstillsats	Strichzusatz, Streichhilfsstoffe	additif de couchage	aditivo del estucado
1401	coating band	bestrykningsband	Streichfarbenstreifen	traînée de couchage	cinta estucada
1402	coating base	råpapper för bestrykning	Streichunterlage, Strichträgermaterial	support de couche	soporte del estucado
1403	coating base paper	råpapper för bestrykning	Streichrohpapier	papier support de couche	papel soporte de estucado
1404	coating clay	bestrykningslera	Streichclay, Streichpigment	kaolin	arcilla para estucar
1405	coating color	bestrykningssmet	Streichfarbe	bain de couchage	baño de estucado
1406	coating composition	bestrykningsrecept	Streichfarbenzusammensetzung	sauce de couchage	composición del estucado
1407	coating defect	bestrykningsfel	Streichdefekt	défaut de couchage	defecto del estucado
1408	coating grade clay	bestrykningslera	ein zum Streichen geeigneter Clay	kaolin	arcilla de estucado
1409	coating lump	bestrykningsklump	Streichklumpen	pâton de couchage	grumos en el estucado
1410	coating machine	bestrykningsmaskin	Streichmaschine, Streichanlage	coucheuse, fonceuse	estucadora
1411	coating mixture	bestrykningsblandning	Streichmasse	bain de couchage	baño de estucado
1412	coating pan	bestrykningskärl	Streichtrog, Streichnapf, Lackiertrommel	cuvette d'une coucheuse	bandeja de estucado
1413	coating pick	bestrykningsnappning	Strichabhebung	arrachage du couchage	arrancamiento del estucado
1414	coating raw stock	råpapper för bestrykning	Steichrohmaterial	support de couche	soporte de estucado
1415	coating skip	bestrykningsmiss	beim Streichen ausgelassene Stelle im Papier, Fehlstelle	couchage irrégulier	cargador de estucadora
1416	coating slip	bestrykningsslipp	Fehlstelle beim Streichen	couche, sauce (de couchage)	baño de estucado
1417	coating weight	bestrykningsvikt	Strichgewicht	poids de la couche	peso del estucado
1418	Cobb tests	Cobb-prov	Cobb-Test, Cobb-Prüfung des Leimungsgrads	essais Cobb	pruebas de Cobb
1419	cock	kran	Absperrvorrichtung, Lufthahn, Wasserhahn	robinet	grifo
1420	cockle	cockle	Schnalle (im Papier), Beule, Schwiele	crispage, gode, ondulation	arrugar
1421	cockle cuts	cockle-utskott	Falten	plis de calandre	agujetas
1422	cockled paper	cockle-papper	welliges Papier, faltiges Papier	papier gondolé	papel arrugado
1423	cockle finish	cockle-finish	wellige Oberfläche, Kräuseleffekt	fini gondolé	acabado con arrugas
1424	cockling	cockling	Schnallenbildung, Kräuseligwerden, Beuligwerden	gondolage, crispage	abarquillamiento
1425	COD value	bikromattal	Bichromatzahl	–	índice de bicromato
1426	coefficient of absorption	absorptionskoefficient	Absorptionskoeffizient	coefficient d'absorption	coeficiente de absorción
1427	coffer dam	kassun	Umhüllung des Abwasserrohrs	–	dique provisorio
1428	cohesion	kohesion	Kohäsion, Bindekraft	cohésion	cohesión
1429	cold caustic	kall kaustisk soda	Kaltalkali	soude à froid	cáustico frío
1430	cold caustic pulp	massa från kall kaustisk soda	kaltalkalibehandelter Zellstoff	pâte défibrée à froid	pasta cáustica fría
1431	cold grinding	kallslipning	Kaltschleifen	défibrage à froid	desfibrado en frío
1432	cold soda process	kallsodaprocess	Kalnatronverfahren	procédé à la soude à froid	proceso de sosa fría
1433	cold soda pulp	kallsodamassa	Kaltnatronzellstoff	pâte à la soude à froid	pasta a la sosa fría
1434	cold soda pulping	kallsodauppslutning	Kaltnatronzellstoffaufschluss	fabrication de pâtes par le procédé de la soude à froid	fabricación de pasta a la sosa fría
1435	collet	chuck	Hals (masch.), Haltevorrichtung	collet	collar
1436	colloidal	kolloidal	kolloidal	colloïdal	coloidal

ENGLISH	SVENSKA	DEUTSCH	FRANÇAIS	ESPAÑOL
1437 colloid mill	kolloidkvarn	Kolloidmühle	moulin colloïdal	fábrica de coloide
1438 colloids	kolloider	Kolloid	colloïdes	coloides
1439 color	färg	Farbe	couleur	color
1440 colored	färgad	gefärbt, farbig	coloré	coloreado
1441 colored papers	färgade papper	farbige Papiere, gefärbte Papiere, Buntpapiere	papiers colorés	papeles coloreados
1442 color fastness	färgäkthet	Farbechtheit, Farbbeständigkeit	solidité de la couleur à la lumière	permanencia del colorante
1443 colorimeter	kolorimeter	Kolorimeter, Farbmesser	colorimètre	colorímetro
1444 colorimetric purity	kolorimetrisk renhet	kolorimetrische Reinheit	pureté colorimètrique	pureza colorimétrica
1445 colorimetry	kolorimetri	Farbmessung, Kolorimetrie	colorimétrie	colorimetría
1446 coloring	färgning	färbend, Färbung, Farbanstrich	coloration	coloración
1447 coloring material	färgmaterial	Farbstoff	colorants	material de coloración
1448 coloring pigment	färgpigment	Farbpigment	pigment colorant	pigmento de coloración
1449 color lake	färgbad	Farbsumpf, küstliches Pigment	couleur laque	laca de color
1450 color lump	färgklump	Farbklumpen	pâton de couleur	grumo del color
1451 color matcher	färgjustering	Vergleichskolorimeter	colorimètre à comparaison	colorímetro de comparación
1452 color matching	färgjustering	Farbabstimmung	comparaison des couleurs	colorimetría de comparación
1453 color reversion	eftergulning	Farbumkehrung	inversion des couleurs	reversión de color
1454 color specification	färgspecifikation	Farbbestimmung	norme	especificación de color
1455 color spots	färgfläckar	Farbflecken	tache de couleur	manchas de color
1456 colour (Brit.)	färg	Farbe	couleur, bain de couchage	color
1457 column strength	pelarhållfasthet	Knickfestigkeit (phys.)	résistance en colonne	resistencia al pandeo
1458 combination board	kombinationskartong	Mehrlagenkarton	carton multijet	cartón combinado
1459 combination chipboard	kombinationskartong	Karton mit Schrenzlage	carton gris multijet	cartulina de estraza
1460 combination felt	kombinationsfilt	Kombinationsfilz	feutre à toile nappée, feutre combiné	fieltro de combinación
1461 combined board	kombinerad kartong	Kombikarton, Verbundkarton	carton redoublé	cartón combinado
1462 combined press	kombinationspress	Kombipresse	presse combinée	prensa combinada
1463 combined sulfur dioxide	kombinerad svaveldioxid	gebundenes Schwefeldioxid	bioxyde de soufre	dióxido combinado de azufre
1464 combining calender	kombinationsglätt	Doublierkalander	calandres combinées	calandra de combinación
1465 combustible paper	kombustibelt papper	kombustibles Papier, leichtbrennbares Papier, Zigarettenpapier	papier à cigarettes combustible	papel combustible
1466 combustion	förbränning	Verbrennung, Verbrennen	combustion, ignition	combustión
1467 combustion chamber	förbränningskammare	Brennkammer, Brennschacht, Verbrennungsraum	chambre de combustion	cámara de combustión
1468 combustion products	förbränningsprodukter	Verbrennungsprodukte	produits de combustion	productos de combustión
1469 commercial match	tändsticksaskpapper	Imitation (Zweitfertigung einer bestimmten Qualität wobei diese nicht 100%ig dem Original entsprechen muss)	papier ne correspondant pas à la commande mais néanmoins vendable	fósforo comercial
1470 commercial tissues	tunnpapper	handelsübliches Tissue	papiers minces autres que pour emballage	muselinas comerciales
1471 communication papers	kommunikationspapper	Kommunikationspapier (z.B. für Fernschreibgeräte sowie Lochstreifenpapier etc.)	papiers culturels	papeles de correspondencia
1472 compaction	packning	Verdichtung	compacité, tassement	compactación
1473 compactor	packare	Verdichter, Stauchanlage (für non-wovens)	compacteur	apisonador
1474 compatibility	kompatibilitet	Verträglichkeit, Vereinbarkeit	compatibilité	compatibilidad
1475 composite can	sammansatt burk	ein aus verschiedenem Material zusammengesetzter Behälter	récipient cylindrique en carton, complexe metal-carton	recipiente mixto
1476 compressed air	komprimerad luft	Druckluft, Pressluft	air comprimé	aire comprimido
1477 compressibility	kompressibilitet	Kompressibilität, Komprimierbarkeit	compressibilité	compresibilidad
1478 compression	kompression	Druck, Verdichtung, Druckfestigkeit	compression	compresión
1479 compression strength	kompressionshållfasthet	Druckfestigkeit	résistance à la compression	resistencia a la compresión
1480 compression tests	kompressionsprov	Druckprobe, Stauchversuch	essais de compression	pruebas de compresión
1481 compression wood	tryckved	Druckholz, Rotholz	bois de compression	madera a comprimir
1482 compressor	kompressor	Kompressor	compresseur	compresor
1483 computer	dator	Computer, Rechner	ordinateur	calculadora
1484 computer program	datorprogram	Computerprogramm	programme d'ordinateur	programa de calculadora
1485 computing-machine paper	datorpapper	Computerpapier	papier pour ordinateur	papel para máquinas computadoras
1486 concave ground roll	konkavslipad vals	konkav geschliffene Walze, Hohlschliffwalze	meule concave	rodillo cóncavo de base
1487 concentration	koncentration	Konzentration, Anreicherung	concentration	concentración
1488 concentrator	koncentrator	Eindampfer, Eindicker, Anreicherungsapparat	épaississeur, ramasse-pâte	concentrador
1489 concertina fold (Brit.)	concertina fold	Zickzackfaltung	pliure en accordéon	plegado en acordeón
1490 Concora tests	Concoraprov	Concora-Test, Concora-Versuch	essais Concora	pruebas Cóncora

	ENGLISH	SVENSKA	DEUTSCH	FRANÇAIS	ESPAÑOL
1491	condensates	kondensat	Kondensat	condensats	condensados
1492	condensation	kondensation	Kondensierung, Niederschlag	condensation	condensación
1493	condensation polymerization	kondensationspolymerisation	Kondensationspolymerisation	condensation polymérisation	polimeración de condensación
1494	condensation resin	kondensationsharts	Kondensationsharz	résine de condensation	resina de condensación
1495	condenser	kondensor, kondensator	Kondensator, Verdampfer	condensateur	condensador
1496	condenser paper	kondensatorpapper	Kondensatorpapier	papier pour condensateurs	papel de condensadores
1497	condenser tissue	kondensatorpapper	Kondensatortissue, Kondensatorseidenpapier	papier mince pour condensateurs	muselina para condensadores
1498	conditioned	konditionerad	konditioniert, klimatisiert	conditionné	acondicionado
1499	conditioned paper	konditionerat papper	konditioniertes Papier, klimatisch ausgerüstetes Papier	papier conditionné	papel acondicionado
1500	conditioner	konditioneringsapparat	Konditioniergerät, Konditionierapparat	laveur de feutre	acondicionador
1501	conditioning	konditionering	Konditionieren, Klimatisieren	conditionnement	acondicionamiento
1502	conduction	ledning	Fortleitung, Leitungsfähigkeit	transmission	conducción
1503	conductive paper	konduktivt papper	leitfähiges Papier	papier pour isolation électrique	papel conductivo
1504	conductivity	ledningsförmåga	Leitfähigkeit, Leitvermögen	conductivité	conductividad
1505	cone	kon, kotte	Kegel, Konus, Zapfen (bot.)	cône	cono
1506	cone pulley	kondrev	Stufenscheibe	poulie cônique	polea de cono
1507	conformability	formbarhet	Anpassungsfähigkeit, Anpassungsvermögen	adaptabilité	conformidad
1508	conical refiner	konisk raffinör	Kegelrefiner	raffineur cônique	refino cónico
1509	conifer	conifer, barrträd	Konifere	conifère, résineux	conífera
1510	coniferous wood	barrträ	Nadelholz	bois de conifères, bois de résineux	madera de coníferas
1511	conservation	konservering	Konservierung, Erhaltung, Schutz	conservation	conservación
1512	consistency	konsistens	Konsistenz, Stoffdichte	consistance, concentration, densité	consistencia
1513	consistency control	konsistensreglering	Stoffdichteregulierung	contrôle de consistance	control de consistencia
1514	consistency controller	konsistensregulator	Stoffdichteregler	contrôleur de consistance	consistómetro
1515	consistency recorder	konistensskrivare	Stoffdichteanzeige	enregistreur de consistance	indicador de consistencia
1516	consistency regulator	konsistensregulator	Stoffdichteregler	régulateur de densité	regulador de la consistencia
1517	constant speed	konstant hastighet	konstante Geschwindigkeit, unveränderliche Geschwindigkeit	vitesse constante	velocidad constante
1518	construction material	konstruktionsmaterial	Konstruktionsmaterial, Baumaterial, Werkstoff	matériaux de construction	materiales para la construcción
1519	construction paper	konstruktionspapper	Konstruktionspapier	papier pour construction	papel para la construcción
1520	consumption	konsumtion	Verbrauch, Konsum, Aufwand	consommation	consumo
1521	contact angle	kontaktvinkel	Berührungswinkel, Randwinkel	angle de contact	ángulo de contacto
1522	contactless measurement	kontaktfri mätning	berührungslose Messung	mesure sans contact	medida sin contacto
1523	container	behållare	Container, Behälter	récipient, conteneur	contenedor
1524	container board	containerkartong	Schachtelpappe, Vollpappe	carton compact pour caisses	cartón para cajas
1525	container bottom	behållarbotten	Behälterboden	fond d'un récipient	fondo de cajas de cartón
1526	containerization	övergång till containerhantering	Zusammenfassung einzelner Ladungen in Containereinheiten	containerisation	embalaje en cajas de cartón
1527	container liner	containerliner	Container-Reederei	doublure pour boîtes en carton	forro de cajas de cartón
1528	container testing	containerprovning	Containerprüfung	essai de résistance d'un récipient	pruebas de cajas de cartón
1529	contaminate	förorena	verschmutzen, verunreinigen	contaminer, souiller	contaminar
1530	contamination	förorening	Verschmutzung, Verunreinigung, Vergiftung	contamination	contaminación
1531	continuous	kontinuerlig	kontinuierlich, durchgehend, ununterbrochen	continu	continuo
1532	continuous digester	kontinuerlig kokare	Stetigkocher, kontinuierlicher Kocher	lessiveur en continu	lejiadora en continuo
1533	continuous forms	kontinuerliga formar	Endlosformular, Endlosformularsatz, Endlosformulardruck (EFD)	formulaires en continu (LFC)	impresos en continuo
1534	continuous freeness tester	kontinuerlig freenessprovare	kontinuierlich arbeitender Mahlgradprüfer	essayeur de raffinage en continu	refinómetro en continuo
1535	continuous process	kontinuerlig process	kontinuierliches Verfahren, Durchfahrbetrieb	procédé en continu	proceso continuo
1536	continuous pulping	kontinuerlig kokning	kontinuierliche Zellstoffkochung, Stetigaufschluss	fabrication de pâtes en continu	desfibrado en continuo
1537	contracoater	motströmsbestrykare	Streichanlage mit Kontaktauftrag im Gegenlauf, Gegenlauf-Streichanlage	coucheuse à rouleaux inversés, "contracoater"	contraestucadora
1538	contraflow	motströms	Gegenfluss, Gegenstrom	contre-courant	contraflujo
1539	contraries	främmande material	Fremdstoffe	impuretés	impureza aparente
1540	contrast ratio	kontrastkvot	Kontrastverhältnis	opacité de contraste	relación de contraste

	ENGLISH	SVENSKA	DEUTSCH	FRANÇAIS	ESPAÑOL
1541	control	kontroll, styrning, reglering	Kontrolle, Regelung, Steuerung, Betätigung	contrôle, réglage	control
1542	control equipment	reglerutrustning	Steuergerät, Regelgerät	matériel de contrôle	equipos de control
1543	control gate	reglerlucka	Regelschieber	vanne de contrôle	regulador
1544	controlled atmosphere	kontrollerad atmosfär	Schutzgas	atmosphère contrôlée	atmósfera controlada
1545	controller	reglerdon	Kontrolleur, Kontrollinstrument	contrôleur	controlador
1546	control panel	kontrollpanel	Schalttafel, Instrumentenbrett, Bedienungspult	tableau de contrôle	cuadro de mando
1547	control system	kontrollsystem	Regelsystem, Organisationsprogramm	système de contrôle	sistema de control
1548	convection	konvektion	Konvektion, Fortpflanzung, Übertragung, Wärmemitführung	convection	convección
1549	convection dryer	konvektionstork	Konvektionstrockner	sécheur par convection	secadora de convección
1550	convection heating	konvektionsvärmning	Wärmeübertragung, Strahlungsheizung	séchage par convection	calefacción de convección
1551	conversion	omvandling	Umwandlung, Umsetzung, Überführung	conversion	transformación
1552	conversion coating	konvertbestrykning	Separatstreichen, Streichen ausserhalb der Maschine	couchage hors machine	estucado de manufactura
1553	converted starch	konverterad stärkelse	umgewandelte Stärke	amidon transformé	almidón convertido
1554	converter	konverter	Konverter, Umwandler, Umformer, Transformator	transformateur, façonnier	transformador
1555	converting	konvertering	Verarbeitung, Veredlung, Umwandlung	transformation, façonnage	transformación
1556	converting industry	förädlingsindustri	verarbeitende Industrie, Veredlungsindustrie	industrie de la transformation	industria de la transformación
1557	converting machine	konverteringsmaskin	Verarbeitungsmaschine	machine de transformation	máquina transformadora
1558	converting paper	förädlat papper	verarbeitungsfähiges Papier	papier destiné à être transformé	papel para transformación
1559	converting plant	förädlingsfabrik	Verarbeitungsbetrieb, Veredlungsanlage	atelier de transformation	planta de transformación
1560	converting process	förädlingsprocess	Veredlungsmethode, Verarbeitungsmethode	transformation	proceso de transformación
1561	conveying	transport	fördern, weiterleiten, überführen, übertragen	transport	transporte
1562	conveyor	transportör	Förderer, Förderband, Transporteur, Zubringer	transporteur	transportador
1563	conveyor belt	bandtransportör	Transportband, Förderbahn	courroie de transporteur	cinta transportadora
1564	conveyor screw	transportskruv	Förderschnecke, Transportschnecke	transporteur à vis	tornillo de transportadora
1565	cook	kok	kochen, aufschliessen (Holz)	cuisson, lessivage	cocción
1566	cooker	kokare	Kocher	lessiveur	lejiadora
1567	cooking	kokning	Kochen, Aufschluss	cuisson, lessivage	cocción
1568	cooking acid	koksyra	Kochsäure	acide de cuisson	ácido de cocción
1569	cooking cycle	kokningscykel	Kochprozess, Kochdauer	durée de cuisson	ciclo de lejiación
1570	cooking degree	koktemperatur	Grad des Aufschlusses	degré de cuisson	grado de lejiación
1571	cooking liquor composition	kokvätskesammansättning	Kochlaugenzusammensetzung	composition de la lessive de cuisson	composición de la lejía de cocción
1572	cooking liquor	kokvätska	Kochlauge	lessive de cuisson	lejía de cocción
1573	cooler	kylare	Kühler, Kühlapparat	refroidisseur, réfrigérant	enfriador
1574	cooling	kylning	Kühlen, Kühlung	refroidissement	enfriamiento
1575	cooling cylinder	kylcylinder	Kühlzylinder	cylindre refroidisseur	cilindro enfriador
1576	cooling system	kylsystem	Kühlsystem	installation de refroidissement	sistema enfriador
1577	cooling water	kylvatten	Kühlwasser	eau de refroidissement	agua de enfriamiento
1578	copier	kopieringsmaskin	Kopiergerät	copieur	copiadora
1579	copolymers	kopolymerer, sampolymerer	Copolymere	copolymères	copolímeros
1580	copper	koppar	Kupfer	cuivre	cobre
1581	copper number	koppartal	Kupferzahl	indice de cuivre	índice de cobre
1582	copying paper	kopiepapper	Kopierpapier, Durchschlagpapier	papier pour copies de lettres	papel copia
1583	copying tissue	kopiepapper	Kopierseidenpapier	pelure pour doubles de lettres	papel de seda para copias
1584	cord	kord	zuschnüren, festbinden, aufschichten, klaftern, Schnur, Kordel, Klafter (Holz)	corde	cuerda
1585	core	hylsa	Hülse, Hülsenkern	mandrin, noyau	mandril, núcleo
1586	core board	hylskartong	Wickelhülsenpappe, Spulenpappe	carton pour mandrins	cartón para mandriles
1587	core paper	hylspapper	Hülsenpapier	papier pour mandrins	papel en mandril
1588	core plug	hylsplugg	Hülsenspund, Hülsenpropfen	bouchon (centre de bobine)	centro de bobina
1589	core shaft	hylsspindel, rullspindel	Hülsenstange	broche d'enroulage	barra de bobinadora
1590	core stock	hylsmassa	Hülsenpappe, Mittelschichtmaterial	matières premières pour mandrins	pasta central

	ENGLISH	SVENSKA	DEUTSCH	FRANÇAIS	ESPAÑOL
1591	core waste	hylsutskott	nach Abwicklung der Rolle an der Hülse haftendes Papier	chutes de mandrins	rechazos de núcleo
1592	core winder	hylsmaskin	Hülsenwickler, Hülsenwickelmaschine	machine à enrouler les mandrins	bobinadora de mandril
1593	corn starch	majsstärkelse	Maisstärke	amidon	almidón de maíz
1594	corona	korona	Koronaentladung, Glimmentladung	couronne	corona
1595	corona discharge	koronaurladdning	Korona	décharge corona	descarga por efecto corona
1596	corrode	korrodera	korrodieren	corroder, ronger	corroer
1597	corrosion	korrosion	Korrosion	corrosion	corrosión
1598	corrosion inhibitor	korrosionsinhibitor	Korrosionsschutzmittel	anti-corrosion	agente anticorrosivo
1599	corrosion mechanism	korrosionsmekanism	Korrosionsmechanismus	processus de corrosion	mecanismo de corrosión
1600	corrosion prevention	korrosionshinder	Korrosionsschutz, Korrosionsverhütung	neutralisation de la corrosion	prevención de corrosión
1601	corrosion products	korrosionsprodukt	Korrosionsprodukte	produits corrosifs	productos corrosivos
1602	corrosion resistance	korrosionsmotstånd	Korrosionsbeständigkeit, Rostfestigkeit	résistance à la corrosion	resistencia a la corrosión
1603	corrosion tests	korrosionsprov	Korrosionsprüfung, Korrosionsversuch	essais de corrosion	pruebas de corrosión
1604	corrosive	korrosiv	korrodierend, zerfressend, ätzend, Korrosions-, Ätzmittel	corrosif	corrosivo
1605	corrosivity	korrosionsbenägenhet	Korrosivität, Korrosionsanfälligkeit	corrosivité	corrosividad
1606	corrugated	korrugerad	gewellt, gerippt, geriffelt	ondulé	ondulado
1607	corrugated board	wellpapp	Wellpappe	carton ondulé	cartón ondulado
1608	corrugated cap	wellförslutning	ein- oder doppelseitige Wellpappe zum Auskleiden von Fässern oder Körben	couvercle en carton ondulé pour les futs en carton	cápsula corrugada
1609	corrugated container	wellpapplåda	mit Wellpappe ausgeschlagen Container mit Wellpappe ausgeschlagene Behälter	caisse en carton ondulé	caja de cartón ondulado
1610	corrugated container board	wellpapp för lådor	Vollwellpappe	carton ondulé pour caisses	cartón ondulado para cajas
1611	corrugated roll	wellpapprulle	geriffelte Walze	rouleau cannelé	rodillo ondulado
1612	corrugated sheet	wellskikt	Wellblech	feuille ondulée	lámina acanalada
1613	corrugated wrapping	wellpappemballage	Wellpackpapier (für zerbrechliches Gut)	emballage ondulé	embalaje corrugado
1614	corrugating	korrugering	Wellen	cannelure	canaladura
1615	corrugating board	wellkartong	Wellpappenrohstoff	carton ondulé	cartón a ondular
1616	corrugating machine	wellmaskin	Wellpappenmaschine	onduleuse	máquina de ondular
1617	corrugating medium	wellpapper	Mittelwellenrohpapier	papier pour ondulé	cartón a ondular
1618	corrugation	korrugering	Riefelung, Riffeln, Rillen	ondulation	ondulación
1619	corrugator	korrugeringsmaskin	Wellenmaschine	machine à onduler	máquina de ondular
1620	corrugator roll	korrugeringsvals	Riffelwalze (einer Wellpappenmaschine)	machine à onduler	rodillo de máquina de ondular
1621	cortex	bark, cortex	Cortex, primäre Rinde	cortex, écorce caryocostine	corteza
1622	cost	kostnad	kosten, Kosten, Preis, Unkosten	cout	coste
1623	cost engineering	kostnadsstrukturering	Kostenplanung	technique des couts	ingeniero de costos
1624	cotton	bomull	Baumwolle	coton	algodón
1625	cotton cellulose	bomullscellulosa	Baumwollcellulose	cellulose de coton	celulosa de algodón
1626	cotton fiber content paper	bomullshaltigt papper	baumwollfaserhaltiges Papier	papier contenant des fibres de coton	papel con fibra de algodón
1627	cotton linters	bomullslinter	Baumwollinters	linters	borra de algodón
1628	cotton pulp	bomullsmassa	Baumwollcellulose	pâte de linters	pasta de algodón
1629	cottonwood	cottonwood	kanadische Pappel	liard, peuplier du Canada	chopo del Canadá
1630	couch	gusk	gautschen, Gautsche	coucher	estucar
1631	couch felt	guskfilt	Gautschfilz	feutre coucheur	fieltro estucador
1632	couching	guskning	Gautschen	coucher	estucado
1633	couch jacket	guskfilt	Gautschwalzenbezug	manchon de presse humide	manchón de prensa
1634	couch mark	guskmarkering	Gautschwalzenmarkierung	marque du rouleau coucheur	marca del estucado
1635	couch pit	guskgrop	Gautschbruchbütte	fosse sous la presse humide	fosa bajo la prensa húmeda
1636	couch press	guskpress	Gautschpresse	presse coucheuse	prensa húmeda
1637	couch roll	guskvals	Gautschwalze	cylindre de presse coucheuse	rodillo de prensa húmeda
1638	couch squirt	guskknekt	Gautschknecht	jet de couchage	tobera de estucado
1639	coulometry	coulometri	Voltmessung, Strommessung	colorimétrie	culombimetría
1640	count	räkning	zählen, fünfhundert Count = fünfhundert Blatt .	compter	cuenta, contar
1641	counter	räknare	Zähler, Zählwerk, Schalter	compteur, compte-tours, pavillon	marca de papel
1642	counter board	räknarpanel	Schuhpappe aus Sekundärfaserstoff	carton pour contreforts de chaussures	cartón contrafuerte
1643	countercurrent	motström	Gegenstrom, Gegenlauf, gegenläufig	contre-courant	contraflujo

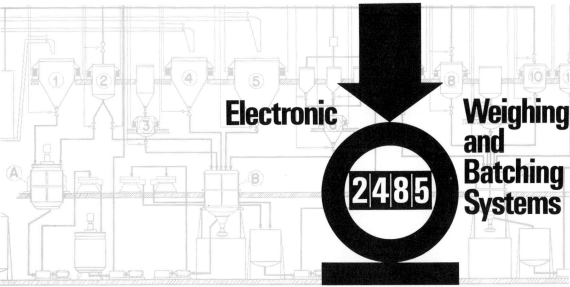

Electronic Weighing and Batching Systems

2|4|8|5

for the Pulp and Paper Industry

Experience, Specialisation, Service
From more than 10 years experience in the electronic weighing field NORDEN weighing systems are designed for successful operation under the arduous conditions met within the Pulp and Paper industry. NORDEN specializes in the Development, Design and Manufacture of packaged systems for the control of material flow.

NORDEN Vector Compensated Load Cell
Force transducer, vector compensating elements and load mounting plate are built into one compact unit, thus simplifying installation and ensuring high operational accuracy and stability.

NORDEN Digital Weighing System
Digital "step by step" null balance of measuring bridge. Modular unit

Weighing Bin on Vector Compensated Load Cells

Pulp or Paper Baleing
Weighbridge in cutting machine with electronic pre-set weight control improves production capacity and ensures accurate bale weights.

Pulp Digester Weighing
Automatic filling of all components and programmed discharging, with continous control of net content including weight-time recording.

Wood Chip Weighing
By Belt conveyor weighing equipment combined with programming system for automatic routing.

NORDEN Conveyor Weighing System
With Thyristor control for constant feed-rate or proportional mixing.

Conveyor Weighing Instrument

Tensile Stress Control
In Paper machines or Rewinders through weighing of "spring loaded roller" provides instantaneous reading and voltage output.

Tambour Weighing
Load cells built into travelling crane and digital weighing system provides accurate gross production control. Print out on adding machine or computer operation.

Coating Kitchen
Batch weighing system with automatic programming together with print out on adding machine, or computer operation ensures high production capacity with reliable quality control, at low investment costs.

Platform Weighing
Net production control through load cells built into lowering table of Rewinders or weighbridges in the ground transport system.

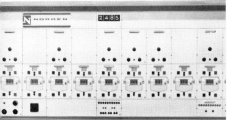

NORDEN gives operational guarantee for the complete system including mechanical handling equipment carefully chosen from reputable manufacturers.

construction provides a flexible system and simplifies servicing requirements. Provision for Print out, Adding machine and Computer operation with closed loop control are standard features.

Subsidiaries in:
Germany
NORDEN Automation Systems GmbH
Wetter
Holland / Belgium
NORDEN-TEMPO Autom. Systems N.V.
Woudenberg
Austria
NORDEN Automation Systems GmbH
Vienna

NORDEN
AUTOMATION SYSTEMS AG

CH-8002 Zürich/Switzerland, Lavaterstr. 76, Telephone 01/36 56 57, Telex 54 123

	ENGLISH	SVENSKA	DEUTSCH	FRANÇAIS	ESPAÑOL
1644	countercurrent flow	motström	gegenläufige Fliessrichtung	débit à contre-courant	contraflujo
1645	countercurrent process	motströmsprocess	Gegenstromverfahren	procédé à contre-courant	proceso de contraflujo
1646	counter flow	motström	Gegenstrom, Gegenströmung	contre-courant	contracorriente
1647	counter roll	motvals	Gegenrolle, Gegenwalze	papier d'emballage pour magasins de détail	contrarodillo
1648	counter sheet	motark	abgepackte Packpapierformate	feuille de papier d'emballage	contrahoja
1649	counting	räkning	Zählen, Zählung	comptage	contado
1650	couple	par	kuppeln, verankern, koppeln, das Paar	accoupler	acoplar
1651	coupling	koppling	Kupplung, Verbindungsstück, Kupplungsstück	accouplement, transmission	acoplamiento
1652	coupling agent	kopplingsmedel	Verbindungshilfsmittel	agent de transmission	agente de acoplamiento
1653	covalent bond	kovalent bindning	kovalente Bindung	liaison covalente	unión covalente
1654	cover	täckning	bedecken, umhüllen, ummanteln, Decke, Deckel, Umschlag, Überzug, Schutzhaube	couvercle, carter, couverture	cárter
1655	coverage	täckning	Belegung	domaine traité, connaissance, couverture	cubierta
1656	covering	täckning	Umkleidung, Umhüllung, Bezug, Abdeckung	couverture	cubrimiento
1657	covering power	täckningsfömåga	Deckkraft	pouvoir couvrant ·	capacidad de cubrimiento
1658	cover paper	täckpapper	Umschlagpapier	papier pour couverture de livres	papel para cubiertas
1659	crack	spricka	springen, knicken, reissen, platzen, Riss, Sprung	craquelure	grieta
1660	cracked edge	sprick-kant	rissige Kante einer Bahn	bord craquelé	borde agrietado
1661	crackershell board	crackershell board	Karton für Süsswarenverpackung	carton pour pâtissiers	cartón para cartuchos
1662	cracking	sprickning	Krachen, Reissen, Brechen, Springen	décollement de la couche	rotura
1663	crackle	krackelera	knarren, knistern, verpuffen	craquelure	carteo
1664	crane	kran	Kran	grue	grúa
1665	crater	krater	Krater, Trichter, Krater (im Oberflächenanstrich)	cratère	cráter
1666	cratering	kraterbildning	Bildung kleiner Eindruckstellen im gestrichen Papier durch das Aufbrechen von Luftblasen im Strich	formation de cratères	formación de cráteres
1667	crazing	bristning	rissig werden	fendillement, craquelure	cuarteamiento
1668	creasability	vikbarhet	Falzfähigkeit	résistance au gondolage	aptitud al acanalado
1669	crease	veck	Faltenbildung, falten, kniffen, umbiegen	faux-pli	falso pliegue
1670	crease, calender	kalanderveck	Kalanderfalten	faux-pli de calandre	cortadas de calandra
1671	crease proofing	vecksäkring	knitterfest machen	protection contre les faux-plis	tratamiento antipliegues
1672	creaser	veckare	Siekeneisen	machine à rainer	fileteador
1673	crease recovery	veckåterhämtning	Rückbildung von Falten (Erholzeit)	récupération des faux-plis	rehacimiento de pliegues
1674	crease retention	veckretention	Falzbeständigkeit (Verharren eines Papiers im gefalzten Zustand)	fixation des faux-plis	retención de los pliegues
1675	creasing	veckning	Verknittern, Faltenbildung	gondolage	acanalado
1676	creasing-resistance tester	veckningsmotståndsprovare	Knitterfestigkeitsprüfung	appareil pour contrôler la résistance au gondolage	probador de resistencia a los pliegues
1677	creasing strength	veckhållfasthet	Druckfalzfestigkeit	résistance au gondolage	resistencia al plegado
1678	creep	krypning	kriechen, wandern	ascension, capillarité	arrastramiento
1679	creep tests	krypprovning	Standfestigkeitsprüfung	essais de capillarité	pruebas de deslizamiento
1680	crepe	kräppning	kreppen	crêpé	crepé
1681	creped	kräppad	gekreppt	crêpé	rizado
1682	creped paper	kräppat papper	gekrepptes Papier	papier crêpé	papel rizado
1683	creped wadding	kräppad vadd	Zickzackwatte	ouate de cellulose crêpée	guata rizada
1684	crepe finish	kräppningsfinish	Kreppausrüstung, gekreppte Oberfläche	fini crêpé	acabado crepé
1685	crepe paper	kräppapper	Krepp-Papier	papier crêpé	papel crepé
1686	crepe ratio	kräppningskvot	Kreppverhältnis, Dehnungsverhältnis in Zellstoffwatte	taux de crêpage	relación de rizado
1687	creping	kräppning	Kreppen	crêpage	rizado
1688	creping doctor	kräppskaber	Kreppschaber	docteur crêpeur	rascador de rizado
1689	creping machine	kräppmaskin	Kreppmaschine	machine à crêper	máquina rizadora
1690	creping tissue	kräppat tunnpapper	Kreppseidenpapier	papier mince crêpé	muselina para rizar
1691	crescent former	crescentformare	verstellbarer Schraubenschlüssel	en forme de croissant	plantilla de lúnula
1692	cresols	kresoler	Kresol	crésols	cresoles
1693	crew	besättning	Mannschaft, Personal, Belegschaft	équipe	equipo

	ENGLISH	SVENSKA	DEUTSCH	FRANÇAIS	ESPAÑOL
1694	crill	kryll	feinster Faserabrieb während der Refinermahlung	fines	–
1695	crimp	krimp	kräuseln, umbörteln, börteln	pli	pliegue
1696	crimping	krimpning	Börteln, Kräuseln	gaufrage	plegado
1697	crinkled	veckad	gekräuselt	plissé	ondulado
1698	critical flow	kritiskt flöde	kritische Strömungsgeschwindigkeit	débit constant	caudal crítico
1699	critical path methods	critical-path metoder	Netzplan, Netzplantechnik	méthodes du chemin critique	métodos de trayectorias críticas de rutas de proceso
1700	critical pressure	kritiskt tryck	kritischer Druck	pression exacte	tensión crítica
1701	critical temperature	kritisk temperatur	kritische Temperatur	température exacte	temperatura crítica
1702	critical velocity	kritisk hastighet	kritische Geschwindigkeit, Durchsackgeschwindigkeit, Grenzgeschwindigkeit	vitesse exacte	velocidad crítica
1703	crocking	nedsotning	Abfärben	qui déteint	hollinado
1704	cross-bottom bag	kanisterpåse	Kreuzbodenbeutel	sac à fond croisé	saco de fondo transversal
1705	cross-creped	korskräppad	quergekreppt	crêpé croisé	rizado transversal
1706	crosscurrent flow	tvärsflöde	Querströmung	écoulement à courant transversal	flujo de corriente transversal
1707	cross direction	tvärsriktning	Querrichtung, quer zur Laufrichtung	sens travers	sentido transversal
1708	cross linking	tvärbindning	Vernetzung	enchaînement par liaison transversale	conexión transversal
1709	cross profile	tvärprofil	Querprofil	profil transversal	perfil transversal
1710	cross recovery	tvärs-återhämtning	verbundene Rückgewinnung (z.B. Sulfat und NSSC)	récupération transversale	recuperación transversal
1711	cross section	tvärsektion, tvärnitt	Querschnitt, Querprofil	coupe	seccionar
1712	crown	bombering	bombieren, ballig drehen, Wölbung	bombé, couronne	bombeado
1713	crowned roll	bomberad vals	bombierte Walze	rouleau bombé	bobina abombada
1714	crown filler	bomberingsinlägg	wässriges Calciumsulfat	sulfate de calcium hydraté	sulfato de calcio precipitado
1715	crowning	bombering	Bombierung	bombage	bombeado
1716	crude oil	råolja	Rohöl	pétrole brut	petróleo bruto
1717	crush	krossning	zerdrücken, zerreiben, zerkleinern	écraser, broyer	triturar
1718	crushed	krossad	verdrückt, zerrieben, zerkleinert	écrasé	triturado
1719	crushed core	kollapsad kärna	beschädigte Hülse, zerdrückte Hülse	mandrin écrasé	madera de corazón triturada
1720	crushed finish	förtryckt botten	melierte Oberfläche (hervorgerufen durch starke Druckausübung in der Nasspartie)	fini écrasé	acabado abrasivo
1721	crushed roll	deformerad rulle	eingedrückte Papierrolle, zerdrückte Papierrolle	bobine écrasée	bobina aplastada
1722	crusher	krossare	Grobzerkleinerungsmaschine, Schleudermühle	broyeur, concasseur	desfibrador
1723	crushing	krossning	Zerkleinern, Zermalmen, Stauchen (Wellpappe), Zerdrücken	écrasement	marca de bayeta
1724	crushing strength	krossningsstyrka	axiale Festigkeit (bei Rohren), Druckfestigkeit	résistance au chiffonage	resistencia al arrugado
1725	crushing test	krossprov	Druckprobe, Druckversuch, Stauchversuch	essais de résistance au chiffonage	pruebas de trituración
1726	crush resistance	krossmotstånd	Druckfestigkeit	résistance à l'écrasement	resistencia al aplastamiento
1727	cryogenics	kryogenik	Kälteerzeugung	cryogénie	criógenos
1728	crystal	kristall	Kristall	cristal	cristal
1729	crystalline cellulose	kristallincellulosa	kristalline Cellulose	cellulose cristallisée	celulosa cristalina
1730	crystalline region	kristallint område	kristalliner Bereich	région cristallinisée	región cristalina
1731	crystallinity	kristallinitet	Kristallinität	cristallinité	cristalinidad
1732	crystallization	kristallisering	Kristallbildung, Kristallisation	cristallisation	cristalización
1733	crystal structure	kristallstruktur	Kristallstruktur	structure cristalline	estructura de cristal
1734	cuam rayon	cuam-rayon	Kuprafaser, Kupraseide	rayonne cupro-ammoniacale	rayón cuam
1735	cull	kassera	aussortieren, auslesen, Papier zweiter Wahl, stehenlassen	rebut	escoger
1736	culled paper	utskott	als zweiter Wahl aussortiertes Papier	papier mis au rebut	papel entresacado
1737	cultivation	kultivering	Anbau, Zucht, Kultur	culture	cultivo
1738	cunit	cunit	Volumeneinheit für Faserholz, Masseinheit für Faserholz	mesure américaine de volume pour bois non écorcés	cien pies cúbicos de madera compacta
1739	cup board	bägarkartong	Becherkarton	carton pour gobelets	cartón para vasos
1740	cup former	bägarformare	Becherformmaschine	emboutisseuse pour gobelets	calibre para vasos
1741	cup paper	bägarpapper	Trinkbecherpapier	papier pour gobelets	papel para vasos
1742	cuprammonium hydroxide	kopparammoniumhydroxid	Cuproxam	hydrate de cuprammonium	hidróxido de cuprammonio
1743	cuprammonium viscosity	kopparammoniumviskositet	Celluloseviskositöt in Kupferoxidammoniak	viscosité du cuprammonium	viscosidad del cuprammonio

	ENGLISH	SVENSKA	DEUTSCH	FRANÇAIS	ESPAÑOL
1744	Cupressus (Lat.)	Cupressus	Zypresse	Cupressus	ciprés
1745	cupriethylenediamine hydroxide	kopparetylendiaminhydroxid	Kupferäthylendiaminhydroxid	hydrate de cuproéthylènediamine	hidróxido de cuprietilenamina
1746	cupriethylenediamine viscosity	kopparetylendiaminviskositet	Kupferäthylendiamin-Viskosität	viscosité du cuproéthylènediamine	viscosidad de la cuprietilenamina
1747	cups	bägare	Becher, Schalen	gobelets, godets	alvéolos
1748	cure	mogna	aushärten, nachbehandeln	résine synthétique	acondicionar
1749	curing	mognad	Aushärten, Nachbehandeln	conditionnement	acondicionamiento
1750	curing agent	mognadsmedel	Mittel zur Haltbarmachung, Aushärtmittel	produit correcteur	agente acondicionador
1751	curl	krökning	Rollneigung, rollen, sich rollen	gode, ondulation	abarquillamiento
1752	curlating	kurlering	Kräuseln	procédé pour améliorer la qualité des pâtes	rizado
1753	curlator	kurlator	Curlator (Kräuselmaschine)	produit permettant d'améliorer qualité des pâtes	rizador
1754	curling	curling	Rollen, Verwerfen (von Papier)	gondolage, roulage des bords du papier	arrollamiento de los bordes
1755	curling test	curl-prov	Rollneigungsprüfung	essai de roulage des bords	pruebas de arrollamiento de los bordes de una hoja de papel
1756	curly paper	buckligt papper	welliges Papier, gekräuseltes Papier	papier frisé	papel rizado
1757	currency paper	sedelpapper	Banknotenpapier	papier pour billets de banque, papier fiduciaire	papel moneda
1758	current density	rådande denistet	Belastung, Stromdichte	densité courante, normale	densidad de corriente
1759	curtain coater	ridåbestrykare	Giessbeschichtungsanlage, Steinberg-Streicher	coucheuse à nappe	estucadora en capas
1760	curtain coating	ridåbestrykning	Giessbeschichtung, Vorhangstreichen	couchage sur coucheuse à nappe	estucado en capas
1761	curve	kurva	krümmen, biegen, Kurve, krümmung	courbe	combadura
1762	curved roll	krökt vals	gekrümmte Walze, Knickwalze	rouleau cintré	rodillo curvo
1763	cushioning	dämpning	Dämpfen, Abfedern	amortir	empaque
1764	cut	snitt	schneiden, abhauen, trennen	coupure, fuite	corte
1765	cut, blade	bladsnitt	Messerschnitt	coupure de lame	corte por cuchilla
1766	cut, hair	hårfint snitt	Haarriss	coupure de poil	mondar
1767	cut-off squirt	gusk-knekt	Gautschknecht	coupe-feuille hydraulique	manguera de corte
1768	cut scored	knivrepad	Markierung der Kartonoberfläche zur Vereinfachung der Falzung	coupe par mollettes sur cylindre en acier trempé (bobineuse-trancheuse)	ranurado de corte
1769	cut size	snittstorlek	Formatschneiden	format	formato de corte
1770	cutter	skärmaskin	Querschneider	coupeuse, rogneuse	cortadora
1771	cutter broke	skärmaskinsutskott	Schnittabfall	cassée de massicot	recortes de cortadora
1772	cutter dust	skärmaskinsdamm	Schnittstaub	poussière de massicot	polvo de cortadora
1773	cutter-sorter	skärmaskin-sorterare	Sortierquerschneider	coupeuse-trieuse	cortadora clasificadora
1774	cutterman	skärmaskinförare	Formatschneider	coupeur	cortador
1775	cutting	skärning	Schneiden, Zeitungsausschnitt	coupe, découpage	corte
1776	cutting tool	skärverktyg	Schneidwerkzeug, spanabhebendes Werkzeug	outil de découpage	herramienta para cortar
1777	cyanamides	cyanamider	Cyanamide	cyanamides	cianamidas
1778	cyanates	cyanater	Cyanate	cyanates	cianatos
1779	cyanides	cyanider	Cyanide	cyanures	cianuros
1780	cyanoethylation	cyanoetylering	Cyanoäthylierung	cyanoéthylation	cianoetilación
1781	cyanoethyl cellulose	cyanoetylcellulosa	Cyanoäthylcellulose	cyanoéthyl cellulose	celulosa cianoetil
1782	cyanuric compounds	cyanuronföreningar	Tricyanverbindungen	composés de cyanure	compuestos cianúricos
1783	cycle	cykel	Zyklus, Kreislauf, Arbeitsgang	cycle	ciclo
1784	cycle beater	cirkulationshollåndare	Umlaufholländer	raffineur associé à un cuvier	pila refinadora de rotación
1785	cyclic compounds	cykliska föreningar	Ringverbindungen	composés cyliques	compuestos cíclicos
1786	cyclic loads	cyklisk belastning	Wechselbeanspruchung	charges cycliques	cargas cíclicas
1787	cyclodextrins	cyclodextriner	Zyclodextrin	cyclodextrines	ciclodextrinas
1788	cyclohexane	cyklahexan	Cyclohexan	cyclohexane	ciclohexano
1789	cyclone	cyklon	Zyklon, Staubentferner, Wirbelsichter	cyclone	ciclón
1790	cyclone separator	cyklonseparator	Staubabscheider	séparateur à cyclone	separador centrífugo
1791	cylinder	cylinder	Zylinder, Trockenzylinder	cylindre, forme ronde	cilindro
1792	cylinder board	rundvirakartong	Rundsiebkarton	carton de forme ronde	cartón para cilindros
1793	cylinder bristols	rundvirabristol	gegautschter Bristolkarton	bristols fabriqués sur forme ronde	cartulina para cilindros
1794	cylinder dried	cylindertorkad	maschinengetrocknet	séché sur cylindre	secado en cilindro
1795	cylinder kraft liner	rundvirakraftliner	Kraft-Beklebekarton	kraftliner fabriqué sur forme ronde	forro kraft del cilindro
1796	cylinder machine	rundviramaskin	Rundsiebmaschine	machine à forme ronde	máquina de formas redondas

ENGLISH	SVENSKA	DEUTSCH	FRANÇAIS	ESPAÑOL
1797 cylinder mold	rundviraformare	Rundsiebzylinder	forme ronde	forma redonda
1798 cylinder paper	rundvirapapper	mit Büttenrand versehenes aber in der Maschine hergestelltes Papier	papier fabriqué sur forme ronde	papel de cilindro
1799 cylinder press	cylinderpress	Rundsiebpresse	presse à cylindre	prensa de cilindros
1800 cylinder vat machine	rundviramaskin	Rundsiebmaschine	machine à forme ronde	tina de cilindro
1801 cymenes	cymener	Cymol	cymènes	cimenos
1802 cypress	cypress	Zypresse	cyprès	ciprés

D

1803 d.c. (direct current)	likström	Gleichstrom	courant continu	corriente continua
1804 dam	damm	Damm, Wehr, Staulatte	barrage, cloison mobile de la fosse du défibreur	presa
1805 damage	skada	beschädigen, Schaden	dégât, dommage	daño
1806 damp	fuktig	befeuchten, dämpfen, Dunst, Feuchte	humide	húmedo
1807 dampener	fuktare	Befeuchter, Dämpfer, Anfeuchter	humecteur	humectadora
1808 dampening	fuktning	Anfeuchten, Befeuchten, Dämpfen	humidification	humectación
1809 dampening machine	fuktningsmaskin	Feuchtmaschine	humecteuse, mouilleuse	máquina humectadora
1810 damping	dämpning	Abdämpfen	humectage	humectación
1811 damping roll	dämpvals	Feuchtglätte, Feuchtpresse	rouleau mouilleur	rodillo humectador
1812 damping stretch	dämpningssträckning	Feuchtdehnung	allongement	faja húmeda en una hoja
1813 damp streaks	dämpveck	Feuchtigkeitsstreifen	traînées humides dans la feuille	huellas húmedas
1814 dancer roll	reglervals	Tänzerwalze, Losrolle	rouleau danseur	rodillo igualador
1815 dancing roller	reglervals	Papierleitwalze	rouleau tendeur oscillant	rodillo igualador
1816 dandy	egotör	Egoutteur	rouleau égoutteur	desgotador
1817 dandy coater	egotörbestrykare	Egoutteur-Streicher	coucheuse sur rouleau égoutteur	estucadora con rodillo desgotador
1818 dandy mark	egotörmarkering	Egoutteur-Markierung, Egoutteur-Wasserzeichen	marque du rouleau égoutteur, piqûre au filigrane	marca del rodillo desgotador
1819 dandy pick	egotörnappning	Rupfen des Egoutteurs	arrachage dû au rouleau égoutteur	arrancamiento en el rodillo desgotador
1820 dandy roll	egotör	Egoutteur-Walze, Siebwalze	rouleau égoutteur, filigraneur, vergeur	rodillo desgotador
1821 data	data	Unterlagen, Angaben, Befunde, Mess und Versuchswerte	données, chiffres, renseignements	datos
1822 data logger	datainsamlare	Datenerfassung	enregistreur de données	tabulador automático de datos
1823 data processing	databearbetning	Datenverarbeitung	utilisation des données	proceso de datos
1824 data recorder	dataregistrerare	Datenaufzeichner, Datenaufzeichengerät	enregistreur de données	registrador de datos
1825 data recording	dataregistrering	Datenregistrierung, Datenaufzeichnung	enregistrement de données	registro de datos
1826 data retrieval	dataåtervinning	Datensuchprogramm, Datenarchiv	relèvement de données	recuperación de datos
1827 data storage	datalagring	Datenspeicherung	emmagasinage de données	almacenamiento de datos
1828 data tables	datatabeller	Datenlisten, Datentabellen	tableaux de données	índices de datos
1829 data transmission	dataöverföring	Datenübertragung	transmission de données	transmisión de datos
1830 daylight	dagsljus, etage	Tageslicht, Einbauhöhe	lumière du jour	luz diurna
1831 deadening felt	dödande filt	Dämmpappe, Teppichboden, Bodenbelagunterlage	feutre amortisseur	filtro insonorizante
1832 dead white	kritvit	neutrales Weiss	blanc neutre	blanco neutro
1833 deadwood	dedwood	Dürrholz	bois mort	leña muerta
1834 deaerate	avlufta	entlüften	elimination de l'air contenu dans la pâte	desaerear
1835 deaeration	avluftning	Entlüftung, Luftabscheider	elimination de l'air contenu dans la pâte	desaereación
1836 deaerator	avluftare	Entlüfter	appareil pour éliminer l'air contenu dans la pâte	desaereador
1837 debark	barka	entrinden	écorcer	descortezar
1838 debarker	barkningsmaskin	Entrindungstrommel	écorceuse	descortezadora
1839 debarking	barkning	Entrinden, Schälen	écorçage	descortezado
1840 decalcomania paper	decalcomaniapapper	Abziehbilderpapier	papier pour décalcomanie	papel calcomanía
1841 decay	försämras, åldras	verfaulen, vermodern, zerfallen, zersetzen, Verfall, Abbau	pourriture, altération du bois	deterioro
1842 decay resistance	åldringsmotstånd	Dauerhaftigkeit, Haltbarkeit, Alterungsbeständigkeit	résistance à la pourriture	resistencia al deterioro

	ENGLISH	SVENSKA	DEUTSCH	FRANÇAIS	ESPAÑOL
1843	decayed wood	rutten ved	faules Holz, verfaultes Holz	bois pourri	madera putrefacta
1844	deceleration	deceleration	Abnehmen der Geschwindigkeit, Bremswirkung	décélération	deceleración
1845	deciduous	löv- årligen avfallande	laubtragend, belaubt	feuillu	caduco
1846	deciduous tree	lövträd	Laubbaum (der jährlich sein Laub abwirft)	arbre feuillu, angiosperme	árbol frondoso
1847	deciduous wood	lövträ	Laubholz	bois feuillu	madera de frondosas
1848	decker	upptagningsmaskin	Eindicker, Eindickzylinder	épaississeur	espesador
1849	deckering	avvattning	Eindicken, Entwässern	épaississement, concentration	espesamiento
1850	deckle	däckel	Deckel, Auflaufrahmen, Randbegrenzung, Randsleiste	couverte, formette tiroir (de caisse aspirante)	marco
1851	deckle board	däckelbord	unbeschnittener Karton, rauhrandiger Karton	réglette, couverte fixe	cubierta fija
1852	deckle edge	däckelkant	Formatbegrenzungsleiste	bord à la cuve (barbé)	bordes con barbas
1853	deckle-edged board	däckelkantad kartong	Büttenrandkarton	carton à bords barbés	cartón con barbas
1854	deckle-edged paper	däckelkantat papper	Büttenrandpapier	papier à bords barbés	papel barba
1855	deckle stained	däckelfläckat	an den Rändern gefärbt, an den Rändern befleckt	taché sur les bords	coloreado en el marco
1856	deckle strap	däckelrem	Deckelriemen	courroie-guide	correa guía
1857	decolorization	avfärgning	Entfärbung	décoloration	decoloración
1858	decomposition	nedbrytning	Zersetzung, Auflösung, Zerfall, Abbau	décomposition	descomposición
1859	decorative papers	dekorationspapper	Dekorationspapier	papiers pour décoration	papeles decorativos
1860	deculator	dekulator	Dekulator, Entlüfter (zwischen Sortierer und Einlaufkasten)	appareil pour éliminer l'air contenu dans la pâte	deculador
1861	defect	defekt, skada	Mangel, Defekt, Fehler, Fehlstelle	défaut	defecto
1862	defiber	defibrera	mahlen, zerfasern	défibrer	desfibrar
1863	defibering	defibrering	Defibrierung, Zerfaserung	défibrage	desfibrado
1864	defiberize	defibrera	defibrieren, entstippen, in Einzelfasern zerlegen	défibrer	desfibrar
1865	defibrated pulp	defibrerad massa	aufgeschlagener Zellstoff	pâte défibrée	pasta desfibrada
1866	defibration	defibrering	Defibrierung, Zerfaserung	défibration	desfibrado
1867	defibrator	defibrator	Defibrator	triturateur	desfibradora
1868	defibrator process	debreringsprocess	Defibratorverfahren, Defibratormethode	trituration	proceso del desfibrado
1869	deflaker	knutlösare	Entstipper	dépastilleur	despastillador
1870	deflect	böja, deflektera	ablenken, abweichen	dévier	desviar
1871	deflection	deflektion	Durchbiegung	flexion, déviation	deflexión
1872	deflector	deflektor	Abstreifer, Ablenkblech, Prallblech	déflecteur	deflector
1873	deflocculating agent	antiflockningsmedel	Ausflockungsmittel	produit défloculateur	agente desfloculador
1874	deflocculation	deflockulering	Ausflockung	défloculation	desfloculación
1875	defoamer	skumbekämpningsmedel	Entschäumer	produit anti-mousse	antiespuma
1876	deformation	deformation	Deformierung, Verformung, Formveränderung	déformation	deformación
1877	degasser	avgasare	Entgaser, Entgasungsvorrichtung	dégazeur	desgasificador
1878	degassing	avgasning	Entgasen, Entgiften	dégazage	desgasificado
1879	degradation	nedbrytning, degradering	Abbau, Degradation	dégradation	degradación
1880	degree of beating	malgrad	Stärke der Mahlung, Mahlungsgrad	degré de raffinage	grado de refino
1881	degree of bleaching	blekningsgrad	Bleichgrad	degré de blanchiment	grado de blanqueo
1882	degree of orientation	orienteringsgrad	Orientierungsgrad	degré d'orientation	grado de orientación
1883	degree of polymerization	polymerisationsgrad	Polymerisationsgrad	degré de polymérisation	grado de polimerización
1884	degree of substitution	substitutionsgrad	Substitutionsgrad	degré de substitution	grado de substitución
1885	dehydrate	dehydrera	entwässern, dehydratisieren	deshydrater	deshidratar
1886	dehydration	avvattning, dehydrering	Entwässerung, Dehydratisierung	deshydratation	deshidratación
1887	dehydrator	dehydrator	wasserentziehendes Reagens, Verdampfer	deshydrateur	deshidratador
1888	dehydrogenation	dehydrering	Dehydrierung	deshydrogénation	deshidrogenación
1889	deink	avsvaärta	de-inken, bedrucktes Papier entfärben, entschwärzen	désencrer	destintar
1890	deinked pulp	avsvärtad massa	de-inkter Zellstoff	pâte de vieux papiers désencrés	pasta destintada
1891	deinking	avsvärtning	De-inking, Entfärben, Entschwärzen	désencrage	destintado
1892	deionization	avjonisering	Entionisierung	desionisation	desionización
1893	delaminate	delaminera	aufspalten (z. B. bei Karton in einzelne Schichten)	cliver	deslaminar
1894	delamination	delaminering	Aufspaltung in Schichten oder Lagen, Schichtablösung, Spaltbarkeit	séparation des plis	deslaminación
1895	delignification	uppslutning	Entlignifizierung	délignification	deslignificación
1896	delimbing	kvistning	entästen	ébranchage	poda de ramas

	ENGLISH	SVENSKA	DEUTSCH	FRANÇAIS	ESPAÑOL
1897	deliquesce	smälta bort	zerfliessen, zergehen (chem.), schmelzen	deliquescence	licuarse
1898	delivery	leverans	Abgabe, Auslass, Ablage (von Papierbogen), Lieferung, Förderung	livraison, fourniture	caudal
1899	dendrology	dendrologi	Dendrologie, Baumkunde	dendrologie	dendrología
1900	denier	denier	Denier (Garnzahl)	denier	negador
1901	Dennison wax test	Dennisons vax-prov	Dennison Wachsprüfung	essai aux cires Dennison	prueba de cera Dennison
1902	densification	packning	Verdichtung	épaississement	densificación
1903	densitometer	densitometer	Densitometer, Schwärzungsmesser	densimètre	densitómetro
1904	densitometry	densitometri	Densitometrie	densimétrie	densitometría
1905	density	densitet	Dichte, Dicke, spezifisches Gewicht, Wichte	densité, concentration	densidad
1906	density measurement	densitetsmätning	Dichtemessung	mesure de densité	medición de la densidad
1907	density meter	densitetsmätare	Dichtemesser	densimètre	densímetro
1908	densometer	densitometer	Densometer, Luftdurchlässigkeits prüfgerät	porosimètre	densímetro
1909	Denver cell	Denvercell	Denver-Zelle (Bauart einer Flotationszelle für den De-inking Prozess)	cellule de Denver (désencrage)	célula de Denver
1910	deodorization	deodorering	Geruchsbeseitigung	désodorisation	desodorización
1911	deoxygenation	deoxidering	Sauerstoffentzug	désoxygénation	desoxigenación
1912	depithing	avhartsning	Entmarkung der Bagasse, Bagassefaserreinigung	extraction de la moelle	desmedular
1913	depolarization	depolarisation	Depolarisation	dépolarisation	despolarización
1914	deposition	avsättning	Absetzen, Absatz, Abscheidung, ablagern	dépôt	deposición
1915	deposit	avsätta	Ablagerung, Bodensatz, Abscheidung	dépôt, précipitation	depósito
1916	depreciation	avskrivare	Wertverlust, Wertminderung	dépréciation	depreciación
1917	depth	djup	Tiefe	profondeur, hauteur	profundidad
1918	deragger	halvtygsholländare	Zopfwinde beim Altpapier-Pulper	effilocheur de chiffons	desfilochadora
1919	descaling	inkrustborttagning	entzundern, entschlacken, dekapieren	détartrage	desincrustación
1920	desensitization	desensibilisering	Desensibilisierung, Unempfindlichmachen	désensibilisation	desensibilización
1921	desiccant	desikkant	Trocknungsmittel	siccatif	desecante
1922	desiccator	desikkator	Exsikkator	exsiccateur	desecador
1923	design	konstruktion	Ausführung, Entwurf, Bauart, Konstruktionsart	dessin, étude, conception	diseño
1924	design speed	offererad hastighet	berechnete Geschwindigkeit, zugrundegelegte Geschwindigkeit	vitesse prévue	velocidad de régimen
1925	desorption	desorption	Desorption	désorption	desabsorción
1926	desuperheater	överhettare	Dampfentspanner	désurchauffeur	desrecalentador
1927	detection	detektering	Auffindung, Entdeckung, Nachweis	découverte, détection	detección
1928	detergent	tvättmedel	Detergens, Waschmittel, Reinigungsmittel	détersif	detergente
1929	detoxication	avgiftning	Entgiftung	désintoxication	desintoxicación
1930	dew point	daggpunkt	Taupunkt	point de rosée	punto de rocío
1931	dewater	avvattna	entwässern (besonders durch Pressen)	épaissir	desecar
1932	dewatering	avvattning	Entwässerung	épaississement, concentration, déshydration	espesamiento
1933	dewaxing	avvaxning	Paraffinieren, Entwachsen	débarrasser de cire	desparafinaje
1934	dextrin gum	dextrintillsats	Dextrinleim	gomme de dextrine	dextrinacaucho
1935	dextrins	dextriner	Dextrin	dextrines	dextrinas
1936	dextrose	dextros	Dextrose, Traubenzucker	dextrose	dextrosa
1937	diagnosis	diagnos	Diagnose	diagnostic	diagnosis
1938	diagram	diagram	Diagramm, graphische Darstellung, Schema	diagramme, épur, courbe	diagrama
1939	dial	visare	wählen, Skalenscheibe, Ziffernblatt	cadran	cuadrante
1940	dialdehyde cellulose	dialdehydcellulosa	Dialdehydcellulose	dialdéhydrate cellulose	celulosa dialdehida
1941	dialdehyde starch	dialdehydstärkelse	Dialdehydstärke	amidon dialdéhydate	almidón dialdehido
1942	dialysis	dialys	Dialyse	dialyse	diálisis
1943	diameter	diameter	Durchmesser, Durchschnitt, Stärke	diamètre	diámetro
1944	diaphragm	diafragma, membran	Diaphragma, Membran, Blende, Lochblende	diaphragme	diafragma
1945	diatomaceous earth	diatomacejordart	Diatomeenerde, Kieselgur	terre d'infusoires	tierra diatomácea
1946	diatomaceous silica	diatomacekisel	amorphes Silika (Füllstoff)	diatomite	sílice diatomácea
1947	diatoms	diatomer	Diatom, Kieselalgen	diatomes	diatomas
1948	diazo copying	ozalidkopiering	Diazokopierung	tirage sur papier diazo	copias diazo
1949	diazo papers	ozalidpapper	Diazopapiere, Lichtpauspapiere	papiers diazo	papeles diazo
1950	dichloroethane	dikloretan	Dichloräthan	dichloroéthane	dicloroetano

	ENGLISH	SVENSKA	DEUTSCH	FRANÇAIS	ESPAÑOL
1951	dichloromethane	diklormetan	Dichlormethan	dichlorométhane	diclorometano
1952	dicotyledons	dikotyledoner	dikotylen, zweikeimblättrig	dicotylédons	dicotiledones
1953	dicyandiamide	dicyanidamid	Dicyandiamid	dicyandiamide	diciandiamida
1954	die	stämpel	schwächer werden, nachlassen, Pressform, Prägestempel, Matrize, Werkzeug	filière, matrice	sacabocados
1955	die cut	fräsa	stanzen, ausstanzen	coupé à l'emporte-pièce	recorte a troquel
1956	die cutter	stämpelfräs	Stanzer, Stanzwerkzeug	emporte-pièce	recortadora a troquel
1957	die cutting	stämpelfräsning	Stanzen	découpe à l'emporte-pièce	recortar a troquel
1958	die embossing	prägling	Prägen	impression gaufrée	estampación con cuño
1959	dielectric	dielektrisk	dielektrisch, nichtleitend	diélectrique	dieléctrico
1960	dielectric constant	dielektricitetskonstant	Dielektrizitätskonstante	constante diélectrique	constante dieléctrica
1961	dielectric loss	dielektrisk förlust	dielektrischer Verlust	perte diélectrique	pérdida dieléctrica
1962	dielectric paper	dielektrikumpapper	dielektrisches Papier	papier isolant	papel dieléctrico
1963	dielectric heating	dielektrisk uppvärmning	dielektrische Erwärmung	chauffage diélectrique	caldeo dieléctrico
1964	dielectric properties	dielektriska egenskaper	dielektrische Eigenschaften	propriétés diélectriques	propiedades dieléctricas
1965	dielectric strength	dielektrisk hållfasthet	dielektrische Festigkeit, Durchschlagsfestigkeit	pouvoir diélectrique	resistencia dieléctrica
1966	diethyl ether	dietyleter	Diäthyläther	ether diéthyl	dietil éter
1967	differential drive	differentialdrift	Differentialantrieb	commande différentielle	transmisión diferencial
1968	differential gear	differentialväxel	Differentialgetriebe, Ausgleichgetriebe	engrenage différentiel	engranaje diferencial
1969	diffraction	diffraktion	Diffraktion, Beugung	diffraction	difracción
1970	diffuse porous wood	poröst trä	zerstreutporiges Holz	bois à pores diffus	madera de porosidad uniforme
1971	diffuser	diffusör	Diffuseur	diffuseur	difusor
1972	diffuser washer	diffusörtvätt	Diffuseurwaschanlage, Waschgrube, Kochergrube	laveur par diffusion	pila lavadora de difusión
1973	diffuser washing	diffusörtvättning	Waschvorgang	lavage par diffusion	lavado por difusión
1974	diffusion	diffusion	Diffusion, Streuung	diffusion	difusión
1975	digest	koka	kochen, aufschliessen, digerieren	lessiver	lejiar
1976	digester	kokare	Zellstoffkocher	lessiveur	lejiadora
1977	digester charge	kokarsats	Beschickung des Digesters, Digestercharge	charge du lessiveur	carga de lejiadora
1978	digester house (Brit.)	kokeri	Kochraum, Kochgebäude	bâtiment des lessiveurs	cuarto de lejiación
1979	digester lining	kokarinmurning	Auskleidung des Digesters	chemisage du lessiveur	revestimiento de lejiadora
1980	digesterman	kokarförare	Kocher	ouvrier des lessiveurs	operador de lejiadora
1981	digesting	kokning	Kochen, Aufschliessen, Digerieren	cuisson	lejiación
1982	digestion	kokring	Aufschluss, Kochung	lessivage, cuisson	lejiación
1983	digital computer	digitaldator	Digitalcomputer, Digitalrechner	calculateur numérique	ordenador digital
1984	digital control	digitalreglering	Zahlensteuerung	contrôle numérique	control digital
1985	digital system	digitalsystem	Digitalsystem	ensemble des programmes constructeurs pour mettre en oeuvre l'ordinateur	sistema digital
1986	diisocyanates	diisocyanater	Diisocyanat	diisocyanates	diisocianatos
1987	dilatancy	dilatans	Volumenvergrösserung, Ausdehnvermögen	dilatabilité	dilatancia
1988	dilatant	dilatant	dilatant	dilatant	dilatante
1989	diluent	förtunnare	Verdünnungsmittel	diluant	diluente
1990	dilute	späda	verdünnen, verflüssigen	diluer, délayer	diluir
1991	dilution	spädning	Verdünnung, Übergang	dilution	desleimiento
1992	dimension	dimension	Dimension, Ausdehnung, Abmessung, Grösse	dimension	dimensión
1993	dimensional stability	dimensionsstabilitet	Dimensionsstabilität, Formbeständigkeit	inertie du papier	inercia del papel
1994	dimensional stabilization	dimensionsstabilisering	formbeständig machen, dimensionsstabil machen	stabilisation de l'inertie	estabilización del papel
1995	dimer	dimer	Dimer	dimère	dímero
1996	dimethyl disulfide	dimetyldisulfid	Dimethyldisulfid	bisulfure de diméthyl	dimetil bisulfuro
1997	dimethyl formamide	dimetylformamid	Dimethylformamid	diméthyl formamide	dimetil formamida
1998	dimethyl sulfate	dimetylsulfat	Dimethylsulfat	sulfate de diméthyl	dimetil sulfato
1999	dimethyl sulfide	dimetylsulfid	Dimethylsulfid	sulfure de diméthyl	dimetil sulfuro
2000	dimethyl sulfone	dimetylsulfon	Dimethylsulfon	sulfone de diméthyl	dimetil sulfona
2001	dimethyl sulfoxide	dimetylsulfoxid	Dimethylsulfoxid	sulfoxyde de diméthyl	dimetil sulfóxido
2002	diode	diod	Diode, Gleichrichter	diode	diodo
2003	dip coater	doppbestrykare	Tauchstreicher	coucheuse au plongé	estucadora por inmersión
2004	dip coating	doppbestrykning	Heisstauchen, Tauchstreichen, Tauchauftrag	couchage au plongé	estucado por inmersión
2005	dipole	dipol	Dipol	dipôle	dipolo
2006	direct current (d.c.)	likström	Gleichstrom	courant continu	corriente continua
2007	direct dyes	direkta färgämnen	substantive Farbstoffe	colorants directs	colorantes directos
2008	direct heating	direktupphettning	direktgeheizt	chauffage direct	calefacción directa

	ENGLISH	SVENSKA	DEUTSCH	FRANÇAIS	ESPAÑOL
2009	direction	riktning	Richtung, Vorschrift, Laufrichtung	sens, direction	dirección
2010	direct steaming	direkt ångtillförsel	Direktdämpfung, direktes Dämpfen	étuvage direct	tratamiento directo en estufa de vapor
2011	directory paper	katalogpapper	Adressbuchpapier, Telefonbuchpapier	papier pour annuaires	papel para anuarios
2012	dirt	smuts	Schmutz, Verunreinigung	impureté, saleté	impureza
2013	dirt count	prickräkning	Schmutzfleckenzahl	nombre d'impuretés sur une surface donnée de papier	número de impurezas sobre el papel
2014	disc	skiva	Scheibe, Platte, Lamelle	disque	disco
2015	discharge	urladda, tömma	entladen, abladen, ausströmen lassen, ablassen, Auslass, Ausfliessen, Ablassen	décharge, vidange	descarga
2016	discharge head	tömningstryck	Druckhöhe, Förderdruck, Auslaufkopf	hauteur de refoulement	altura de descarga
2017	discharge valve	tömningsventil	Auslassventil	vanne de décharge	válvula de descarga
2018	discoloration	missfärgning	Entfärbung, Gilben, Verfärbung	décoloration, jaunissement	decoloración
2019	disc cutter	saxverk med cirkelkniv	Kreismesser (z. B. am Längsschneider)	coupeuse à disques	cortadora de discos
2020	disc filter	skivfilter	Scheibenfilter	filtre à rondelles	filtro de discos
2021	disc knife	cirkelkniv	Kreismesser	couteau circulaire	cuchilla de disco
2022	disc refiner	skivraffinör	Scheibenrefiner	raffineur à disques	refino de discos
2023	dishing	konkavisering	Randwelligkeit, Hochwölben der Ränder eines Papierstapels	arrondir	cóncavo
2024	disintegration	uppslagning	Auflösung, Zersetzung, Zerfall	désintégration, désagrégation	desintegración
2025	disintegrator	disintegrator, pulper	Zerfaserer, Auflöser, Aufschlaggerät	désintégrateur, concasseur, broyeur	desintegrador
2026	disk	skiva	Scheibe, Planscheibe, Teller, Lamelle	disque	disco
2027	dispersant	dispergeringsmedel	Dispergiermittel	dispersant	dispersador
2028	dispersing	dispergering	dispergierend	dissipation de chaleur	dispersión
2029	dispersion	dispersion	Ausbreitung, Zerstreuung, Dispersion	dispersion	dispersión
2030	disposable	umbärlig	Einwegprodukt	à jeter	disponible
2031	disposal	förfogande	Beseitigung, Vernichtung, Veräusserung, zur Verfügungstellung	disposition, arrangement	disposición
2032	dissolve	lösa, upplösa	auflösen, lösen	dissoudre	disolver
2033	dissolved oxygen	löst syre	aufgelöster Sauerstoff, eine in Wasser aufgelöste Sauerstoffmenge	oxygène dissous	oxígeno disuelto
2034	dissolver	upplösare	Auflösebehälter	dissolveur	disolvente
2035	dissolving	dissolvering, upplösning	Lösen, Auflösen	dissolvant	disolutivo
2036	dissolving pulp	dissolveringmassa	Chemiezellstoff	pâte pour usages textiles à dissoudre	pasta disoluble
2037	dissolving tank	dissolveringtank	Auflösebehälter, Auflöser	cuvier à dissoudre	tina para disolver
2038	distillate	destillat	Destillat	distillat	destilado
2039	distillation	destillation	Destillation, Destillieren	distillation	destilación
2040	distilled water	destillerat vatten	destilliertes Wasser, acqua destilliert	eau distillée	agua destilada
2041	distortion	distorsion	Verzerrung, Verformung, sich schief ziehen (vom Filz)	distorsion	deformación
2042	distribution	distribution, fördelning	Verteilung, Verbreitung, Vertrieb (von Waren)	distribution, négoce de papiers, partage	distribución
2043	distributor roll	fördelningsvals	Verteilerwalze, Ausstreichwalze	rouleau distributeur	cilindro distribuidor
2044	disulfides	disulfider	Disulfid	disulfures	bisulfuros
2045	diversification	diversifiering	Diversifizierung	diversification	diversificación
2046	divided press	delad press	geteilte Presse, Presse mit zwei Walzenpaaren	presse divisée	prensa dividida
2047	divider	delare	Teiler, Pappe oder Papier zum Abtrennen von Füllgut (z. B. in Keks- oder Pralinenschachteln)	diviseur, entre-lame, bande de papier pour séparer les rames	elementos de madera separando las cuchillas
2048	doctor	schaber	Schaber, Rakel, Abstreifer, Abstreichmesser	docteur, râcle	doctor
2049	doctor blade	schaberblad	Schaber, Streichschaber	lame de docteur	cuchilla del rascador
2050	doctor broke	schaberutskott	Pressenausschuss, an den Presschabern bei Abriss anfallender Ausschuss	cassés s'accumulant sur le docteur	recortes de rasqueta
2051	doctor dust	schaberdamm	Schaberstoff	poussière de docteur	polvo de la rasqueta
2052	doctor frame	schaberhållare	Streichschaberbalken	monture de docteur	armadura de rasqueta
2053	doctor holder	schaberhållare	Schaberhalterung, Rakelhalterung	monture de docteur	armadura de rasqueta
2054	doctor kiss coater	stavbestrykare	Rakelauftragsmaschine, Messerauftragsmaschine	râcle de couchage par léchage	estucadora de lamedura por cuchilla

	ENGLISH	SVENSKA	DEUTSCH	FRANÇAIS	ESPAÑOL
2055	doctor marks	schabermarkering	Schabermarkierung	raies de docteur	marcas de la rasqueta
2056	doctor ridges	schaberveck	Streichschabermarkierung	raies de docteur	arrugas de rasqueta
2057	doctor roll	schabervals	Dosierwalze, Abstreifwalze	rouleau égalisateur	rodillo de rasqueta
2058	doctoring	schabring	rakeln, mit dem Messer aufstreichen	corriger la teinte	limpieza con rasqueta
2059	dog hair	hundhår	falsche Haare (Textilien)	poils de chien	–
2060	dolomite	dolomit	Dolomit	dolomite	dolomita
2061	dormancy	passivitet	Knospenruhe, Samenruhe	repos	letargo
2062	dosage	dosering	Dosierung	dosage	dosificación
2063	double bond	dubbelbindning	Doppelbindung	double liaison	unión doble
2064	double calendered	dubbelglättad	doppelt satiniert	calandré deux fois	satinado dos caras
2065	double coated	dubbelbestruken	zweifach gestrichen, doppelseitig gestrichen	couché deux faces	estucado dos caras
2066	double coater	dubbelbestrykare	Doppelstreichanlage	coucheuse deux faces	estucadora dos caras
2067	double-deck dryer	tvåvåningstork	zweistöckig angeordnete Trockenzylinder	sécheur à deux rangées	secadora de cubierta doble
2068	double deckle	dubbeldäckel	Papier mit zwei Büttenrändern	avec barbes sur les deux bords	cubierta doble
2069	double-disc refiner	dubbelskivraffinör	Doppelscheibenrefiner	raffineur à double disque	refino de doble disco
2070	double divided press	dubbel delad press	Doppelfilzpresse	presse double divisée	prensa doble dividida
2071	double-drum reel	rullstol med dubbel bärcylinder	Doppeltragtrommelroller	bobineuse à double tambour	enrolladora de tambor doble
2072	double-drum rereeler	rullstol med dubbel bärcylinder	Doppeltragtrommelumroller	rebobineuse à double tambour	reenrolladora de tambor doble
2073	double-drum wind	rullstol med dubbel bärcylinder	Doppeltragwalze	bobinage sur double tambour	bobinado a tambor doble
2074	double-faced corrugated board	tvåwellpapp	kaschierte Wellpappe, doppelseitige Wellpappe	carton ondulé double-double	cartón ondulado dos caras
2075	double-faced paper	dubbelmönstrat papper	Duplex-Papier (auf der Vorderseite eine andere Farbe als auf der Rückseite)	papier double	papel dos caras
2076	double facer	dubbelmönstrare	Doppelkaschieranlage	machine à onduler double face	revés doble
2077	double felted press	tvåfiltpress	Nasspresse mit doppelter Bespannung	presse à double feutre	prensa afieltrada doble
2078	double manila-lined chipboard	dubbeltäckt gråpapp	doppelseitig kaschierte Maschinenschrenzpappe	carton gris doublé manille sur les deux faces	cartulina de estraza encartelado de manila dos caras
2079	double manila-lined newsboard	dubbeltäckt slipmassekartong	bedruckbarer doppelseitiger Karton (dessen Mittellage aus altem Zeitungsdruck besteht)	carton de vieux papiers doublé manille sur les deux faces	cartón prensa manila encartelado dos caras
2080	double sizing	tvåsidig limning	Doppelleimung	double encollage	encolado doble
2081	double vat lined	dubbeltäckt	doppelseitig in der Maschine kaschiert	carton recouvert sur les deux faces de papier fabriqué sur forme ronde	revestido doble en tina
2082	double-wall box	dubbelväggig låda	doppelwandige Kartonschachtel	caisse à doubles parois	caja de cartón con paredes dobles
2083	Douglas fir	Douglasgran	Douglastanne	sapin de Douglas	abeto Douglas
2084	downtime	stillestånd	Stillstandszeit, Unterbrechung	temps d'arrêt	período de paralización de trabajo
2085	drain	dränera, avvattna	entwässern, ablaufen, abfliessen lassen, entleeren, Abfluss, Entwässerung	filtrer, égoutter	escurrir
2086	drainage	dränage, avvattning	Drainage	égouttage	escurrimiento
2087	drainage factor	dränagefaktor	Entwässerungsfaktor, Drainagefaktor	facteur d'égouttage	índice de desgote
2088	drainage foil	dränage	Entwässerungsleiste, Foil	racle d'égouttage	lámina de desgote
2089	drainage resistance	dränagemotstånd	Entwässerungswiderstand	résistance à l'égouttage	resistencia de desgote
2090	drainage time	dräneringstid	Entwässerungszeit	temps d'égouttage	tiempo de escurrimiento
2091	drainer, draining chest	dränagelåda	Absetzkasten, Absetzbütte	caisse d'égouttage	caja de escurrido
2092	drainer stock	dräneringsmassa	Stoff im Absetzkasten	pâtes contenus dans la caisse d'égouttage, papiers contenus dans la caisse d'égouttage	pasta de desgote
2093	draining	dränering	Entwässern, Drainage	égouttage	escurrido
2094	drapability	draperbarhet	Drapierfähigkeit	aptitude au drapement	aptitud al plegado
2095	draughting paper (Brit.)	kalkerpapper	Zeichenpapier	papier à dessin	papel para dibujar
2096	draw	drag	Zugspannung, Verschiebung des geschnittenen Papiers beim Randbeschnitt	attirer, entraîner	arrastre
2097	draw down	neddragning	Abstreifen überschüssiger Streichfarbe für Testzwecke	tirage	descenso del nivel
2098	draw down tests	neddragningsprov	Oberflächenauftragstest	essais de tirage	pruebas de descenso del nivel
2099	drawing papers	ritpapper	Zeichenpapier	papiers à dessin	papeles de dibujo
2100	dregs	grums	Bodensatz, Teersatz	résidus, lies	hez

doctoring at its best

	ENGLISH	SVENSKA	DEUTSCH	FRANÇAIS	ESPAÑOL
2101	dried-in strain	uttorkad töjning	zurückgebliebene Papierspannung während der Trocknung, eingetrocknete Spannung	retrait de la feuille au cours du séchage	tensión en seco
2102	dried-in stress	uttorkad spänning	eingetrocknete Druckbeanspruchung	tension de la feuille au cours du séchage	tensión en seco
2103	drift	drift	treiben, abtreiben, triften, Abdrängung, Drift	force, tendance, portée	basuras
2104	drive	drift	antreiben, fahren, Antrieb, Getriebe	commande	mando
2105	driven roll	driven vals	angetriebe Walze, Antriebswalze	rouleau entraîné	rodillo conducido
2106	driving	drivande	Fahren, Antreiben	commande	transmisión
2107	driving side	drivsida	Führerseite, Antriebsseite	côté comande	lado de transmisión
2108	drop leg	sugben	Fallrohr	colonne barométrique	columna de agua
2109	drop-off	nedfall	Schnallen, abfallen, Abfallen der Stoffbahn vom Filz	diminuer	caducar
2110	drum	trumma, cylinder	Trommel, Seiltrommel	tambour	tambor
2111	drum barker	trumbarkare	Entrindungstrommel	écorceuse à tambour	descortezador de tambor
2112	drum chipper	cylinderflishugg	Topfscheibenzerspaner	déchiqueteur à tambour	troceadora de tambor
2113	drum dryer	cylindertork	Walzentrockner	sécheur à tambour	secadora de tambor
2114	drum filter	trumfilter	Scheibenfilter	filtre à tambour	filtro de tambor
2115	drum washer	cylindertvätt	Waschtrommel, Siebtrommel	tambour laveur	tambor lavador
2116	drum winding	cylinderrullning	Tragtrommelwickler, Tragtrommelaufwickler	bobinage sur tambour porteur	bobinado sobre tambor
2117	dry	torr	trocken, trocknen, trocken werden	sec	seco
2118	dry broke	torrutskott	Trockenausschuss	cassés secs	recortes secos
2119	dry coating	torrbestrykning	Trockenstreichen (ohne Lösungsmittel)	couchage sur par préalablement gommé	estucado en seco
2120	dry creping	torrkräppning	Trockenkreppung	crêpage à sec	rizado en seco
2121	dry end	torrände	Trockenpartie	sécherie	extremidad seca
2122	dryer	torkare, torkcylinder	Trockner, Trockenzylinder	sécheur, cylindre sécheur	secador
2123	dryer felt	torkfilt	Trockenfilz	feutre sécheur	fieltro secador
2124	dryer felt marks	torkfiltmarkering	Trockenfilzmarkierung	marques du feutre sécheur	marcas del fieltro secador
2125	dryer hood	torkhuv	Trockenhaube	hotte sécheuse	campana de la secadora
2126	dryer part	torkparti	Trockenteil, Trockenpartie	sécheur, sécherie	secadero
2127	dryer pick marks	torknappningsmarkering	Rupfstelle in der Bahn (entstanden am Trockenzylinder infolge zu hoher Bahnfeuchtigkeit)	marques d'arrachage du sécheur	marcas de arranque en la secadora
2128	dryer pocket	torkficka	Magazintrockner, Trockentasche	poche	cámara de secadero
2129	dryer section	torkparti	Trockenpartie	sécherie	sequería
2130	dryer spear	torkkäpp	Trockenstange	lance de séchoir	punta secadora
2131	dry felt	torkfilt	Trockenfilz	feutre sec	fieltro seco
2132	dry finish	torkfinish	maschinenglatt	satiné	alisado
2133	dry-forming process	torrformningsprocess	Trockenformierungsmethode	procédé de formation à sec	proceso de fabricación en seco
2134	drying	torkning	Trocknen, trocknend	séchage	secado
2135	drying agents	torkmedel	Trocknungsmittel, Trockenmittel	produits siccatifs	agentes secadores
2136	drying capacity	torkkapacitet	Trocknungsvermögen	capacité de séchage	capacidad de secado
2137	drying cracks	torksprickor	Platzen, Bersten durch Übertrocknen	craquelures de déchage	grietas del secado
2138	drying cylinder	torkcylinder	Trockenzylinder	cylindre sécheur	cilindro secador
2139	drying loft	torkvind	Trockenboden, Trockenspeicher	séchoir, étendeur	secadero
2140	drying oven	torkugn	Wärmeschrank, Trockenkammer	étuve	estufa
2141	dryness	torrhet	Trockengehalt, Trockenheit, trockener Zustand	siccité	sequedad
2142	dry rub resistance	torrgnuggmotstånd	Trockenabriebfestigkeit	résistance au frottement à sec	resistencia a la fricción en seco
2143	dry-waxed paper	torrvaxat papper	trockenes Wachspapier	papier paraffiné avec saturation	papel parafinado en seco
2144	dry strength	torrhållfasthet	Trockenfestigkeit	force à l'état sec	resistencia en seco
2145	dry strength agent	torrhållfasthetsmedel	Trockenfestigkeitsmittel	agent d'amélioration de la résistance à l'état sec	agente resistente en seco
2146	dry vat former	rundviraparti utan tråg	Trockenformierung (Rundsieb)	forme ronde non noyée	plantilla de cuba en seco
2147	dual press	dualpress	Doppelpresse	presse double	prensa doble
2148	duct	trumma	Leitung, Leitkanal, Gang, Kanal, Röhre	conduit, carneau	conducto
2149	dull coated paper	mattbestruket papper	mattgestrichenes Papier	papier couché mat	papel estucado mate
2150	dull finish	matt finish	matte Oberfläche, Mattglanz	surface mate	acabado mate
2151	dumping	dumpning	Ableeren, Auslassen, Auskippen, Abladen	vidange	descarga
2152	dunnage	stuvningsgods	Unterlegbohlen	fardage	madera para estibar
2153	duplex	duplex	Duplex	duplex	duplex
2154	duplex board	duplexkartong	Duplex-Karton	carton duplex	cartón duplex
2155	duplex bristol	duplexbristolkartong	Duplex-Bristol	bristol duplex	bristol duplex
2156	duplex coater	duplexbestrykare	Duplexstreicher, Zweifachstreicher	coucheuse duplex	estucadora duplex

Are you really acquainted with Cellier?

photos - fontana + thomasset - photos X

You mean, the coating colour specialists?
Of course, and even world leaders in the field - 80 % of the world market!
But do you also know their far-reaching range of activities?

As a matter of fact, CELLIER means Hardware and Know-how, more than thoroughly tested for the following applications :
– dispersion and storage of pigments at high consistency
– satin white preparation
– soluble binder cooking : starch, P.V.A., casein, etc.
– surface treatment by special applications :
   electrostatic Repro-paper
   hot-melt coating, P.V.C., fungicide, gumming papers
   silicone, self-releasing, self-adhesive, transfer papers...
– dyeing on size-press
– continuous dyeing of decorating papers
– preparation of stock additives for paper and board and their proportioning at machine head
– distribution of noble pigments (TiO2)

WALLPAPERS : complete production units. New methods for the production of printing inks.
– measuring lines : viscosity, PH, consistency
– control of paper, processing lines from the coating station to the converting plant

From the simplest production unit to the wholly automated plant, CELLIER's equipment is tailor-made to your specifications.

Sales and after sales service in the UK, Cellier Eng. G.B. Ltd.

ENGLISH	SVENSKA	DEUTSCH	FRANÇAIS	ESPAÑOL
2157 duplex felt	duplexfilt	Duplexfilz	feutre duplex	fieltro duplex
2158 duplex finish	duplexfinish	Duplexoberflächenausrüstung (beide Oberflächen verschieden ausgerüstet)	papier dont le fini sur chaque face est différent	acabado duplex
2159 duplex paper	duplexpapper	Duplex-Papier	papier duplex, papier bicoloré	papel duplex
2160 duplex super	duplex rullmaskin	Duplex-Super	papier surcalandré dont le fini sur chaque face est différent	super duplex
2161 duplicating paper	dupliceringspapper	Vervielfältigungspapier, Abzugspapier, Kopierpapier	papier pour duplicateurs	papel para duplicador
2162 durability	varaktighet	Haltbarkeit	longévité, permanence	permanencia
2163 durometer	durometer	Härtemesser, Härteprüfer	duromètre	durómetro
2164 dust	damm	Staub	poussière	polvo
2165 dust collector	dammuppsamlare	Staubfänger, Staubsammler, Staubabscheider	collecteur à poussières	captador de polvo
2166 dust control	damningskontroll	Staubkontrolle	contrôle des poussières	control de polvo
2167 dust filter	dammfilter	Staubfilter	filtre à poussière	filtro de polvo
2168 dusting	damning	Abstauben, Entstauben, Staubbildung, Verstäuben	dépoussiérage, peluchage	desempolvado
2169 dwell time	uppehållstid	Verweilzeit	durée de séjour	tiempo de detención
2170 dye	färg, färgämne	färben, Farbe, Farbstoff	colorant	colorante
2171 dyed	färgad	gefärbt	coloré, teinté	coloreado
2172 dyeing	färgning	Färben	coloration, teinture	coloración
2173 dyestuffs	färgämne	Farbstoffe	matières colorantes	colorantes
2174 dynamic tests	dynamiska prover	Dauerversuch, dynamische Prüfung, Fallprobe	essais de travail	pruebas dinámicas

E

ENGLISH	SVENSKA	DEUTSCH	FRANÇAIS	ESPAÑOL
2175 E-flute	E-pipa	E-Welle	microcannelure	canaladuras E
2176 early wood	vårved	Frühholz	bois de printemps	madera de primavera
2177 ease of solubility	lättlöslighet	schnelles Auflösevermögen	facilité de solubilité	facilitación de solubilidad
2178 ecology	ekologi	Ökologie	écologie	ecología
2179 economic analysis	ekonomisk analys	Wirtschaftsanalyse	analyse économique	análisis económico
2180 economizer	ekonomiser	Abgasvorwärmer, Speisevorwärmer, Betriebswasservorwärmer	économiseur	economizador
2181 ecotype	ecotyp	Ökotyp	écotype	ecotipo
2182 eddy current	virvelström	Induktionsstrom	courant de foucault, tourbillonnaire	corriente parásita
2183 edge	kant	Kante, Rand	bord	borde
2184 edge crush tests	kantkrossprov	Kantendruckfestigkeitstest	essais de résistance des bords à l'écrasement	pruebas de aplastamiento de los bordes
2185 edge curl	kantdeformation	Randwelligkeit	roulage des bords	arrollamiento del borde
2186 edge cutter	kantskärare	Randschneidmaschine, Besäummaschine	bordeuse	chorro cortador de orillo
2187 edge protector	kantskydd	Kantenschutz	carton épais pour protéger les fonds des bobines emballées	protector de orillo
2188 edge tearing resistance	inrivhållfasthet	Kanteneinreissfestigkeit	résistance des bords du papier à la déchirure	resistencia al rasgado del borde
2189 effective alkali	effektivt alkali	effektiver Alkaligehalt	alcali actif	álcali efectivo
2190 effervescence	skumbildning	Aufwallen, Schäumen	effervescence	efervescencia
2191 efficiency	effektivitet	Leistungsfähigkeit, Nutzungswert, Nutzeffekt	rendement, efficience	eficiencia
2192 efficiency of drying	torkeffektivitet	Trockenleistung, thermischer Wirkungsgrad	rendement du séchage	eficiencia del secado
2193 effluent	avlopp	Abwasser, Abfluss	effluent	efluente
2194 effluent treatment	avloppsbehandling	Abwasserreinigung, Abwasserklärung, Abwasseraufbereitung	traitement des effluents	tratamiento de los efluentes
2195 effulgence	glans	Strahlen, Glanz	lustre	brillantez
2196 egg cartons	äggkartonger	Eierschachtel, Eierkarton	cartons pour l'emballage des oeufs	cajas para huevos
2197 eggshell finish	äggskalsfinish	Eierschalenmattierung	demi-brillant	acabado China
2198 ejector	ejektor	Ausheber, Auswerfer, Ausdrückvorrichtung, Ausrückvorrichtung	éjecteur	eyector

	ENGLISH	SVENSKA	DEUTSCH	FRANÇAIS	ESPAÑOL
2199	elastic calender bowl (Brit.)	elastisk kalandervals	elastische Kalanderwalze	rouleau élastique de calandre	rodillo elástico de calandra
2200	elasticity	elasticitet	Elastizität, Federkraft, Dehnbarkeit	élasticité	elasticidad
2201	elastic strength	elastisk hållfasthet	Spannkraft	résistance élastique	resistencia elástica
2202	elastomers	elastomer	Elastomere, gummiartige Stoffe	élastomères	elastomeros
2203	elder	äldre	Holunder, Flieder	sureau	saúco
2204	electrical conductivity	elektrisk konduktivitet	elektrische Leitfähigkeit	conductivité électrique	conductividad eléctrica
2205	electrical engineering	elektroteknik	Elektrotechnik	technique de l'électricité	ingeniería eléctrica
2206	electrical insulating paper	elisoleringspapper	Isolierpapier, Kondensatorpapier	papier diélectrique	papel dieléctrico
2207	electrical paper	elektropapper	Kondensatorpapier, Papier für Isolationszwecke	papier pour usages électriques	papel dieléctrico
2208	electrical properties	elektriska egenskaper	elektrische Eigenschaften	propriétés électriques	propiedades eléctricas
2209	electric circuit	elektrisk krets	Stromkreis	circuit électrique	circuito eléctrico
2210	electric conductor	elektrisk ledare	elektrischer Leiter, Elektrizitätsleiter	conducteur électrique	conductor eléctrico
2211	electric connector	elektrisk anslutning	galvanische Verbindung, Stecker	raccordement électrique	conector eléctrico
2212	electric controller	elektrisk reglering	Servo-Regler, Regelgerät, Kontroller	appareil de contrôle électrique	combinador eléctrico
2213	electric converter	elektrisk omvandlare	Konverter, Umformer	transformateur électrique	convertidor eléctrico
2214	electric current	elektrisk ström	elektrischer Strom	courant électrique	corriente eléctrica
2215	electric discharge	elektrisk urladdning	elektrische Entladung	décharge électrique	descarga eléctrica
2216	electric drive	elektrisk drift	elektrischer Antrieb	commande électrique	accionamiento eléctrico
2217	electric fuse	elektrisk säkring	elektrische Sicherung	fusible électrique	fusible eléctrico
2218	electric impedance	elektrisk impedans	Impedanz, Scheinwiderstand	impédance électrique	impedancia eléctrico
2219	electric insulation	elektrisk isolering	elektrische Isolierung, Isolierstoff	isolation électrique	aislamiento eléctrico
2220	electric motor	elektrisk motor	elektrischer Motor, Elektromotor	moteur électrique	motor eléctrico
2221	electric power	elektrisk kraft	Elektroantrieb, Stromleistung, Kraftstrom	énergie électrique	energía eléctrica
2222	electric power distribution	elektrisk kraftdistribution	Stromverteilung, Leistungsbilanz	distribution d'énergie électrique	distribuidor da fuerza eléctrica
2223	electric power transmission	elektrisk kraftledning	Kraftübertragung	transmission d'énergie électrique	transmisión de fuerza eléctrica
2224	electric relay	elektriskt relä	elektrische Relais	relais électrique	relé eléctrico
2225	electric resistance	elektriskt motstånd	elektrischer Widerstand	résistance électrique	resistencia eléctrica
2226	electric utility	elektrisk apparat	öffentliches Kraftwerk	réseau public de distribution d'électricité	servicio público de electricidad
2227	electrode	elektrod	Elektrode	électrode	electrodo
2228	electrolysis	elektrolys	Elektrolyse	électrolyse	electrólisis
2229	electrolyte	elektrolyt	Elektrolyt	électrolyte	electrólito
2230	electrolytic cell	elektrolytisk cell	Elektrolysenbad	cellule d'électrolyseur	pila electrolítica
2231	electromagnetic field	elektromagnetiskt fält	elektromagnetisches Feld	champ électromagnétique	campo electromagnético
2232	electronic control	elektronisk reglering	Elektronensteuerung, elektronische Regelung	contrôle électronique	control electrónico
2233	electronic equipment	elektronisk utrustning	Elektronengerät	matériel électronique	material electrónico
2234	electronic instrument	elektroniskt instrument	elektronische Ausrüstung	matériel électronique	instrumento electrónico
2235	electronics	elektronik	Elektronentechnik, Elektronik, Elektronenphysik	électronique	electrónica
2236	electron	elektron	Elektron	électron	electrón
2237	electrophoresis	elektrofores	Elektrophorese	électrophorèse	electroforesis
2238	electrophotographic paper	elektrofotografiskt papper	Xerographie-Papiere	papier électrophotographique	papel electrofotográfico
2239	electrostatic charge	elektrostatisk laddning	elektrostatische Aufladung	charge électrostatique	carga electroestática
2240	electrostatic coating	elektrostatisk bestrykning	elektrostatische Beschichtung	couchage électrostatique	estucado electroestático
2241	electrostatic copying	elektrostatisk kopiering	elektrostatisches Kopierverfahren	tirage électrostatique	copiado electroestático
2242	electrostatic precipitation	elektrostatisk utfällning	Elektrofilterung	précipitation électrostatique	precipitación electroestática
2243	electrostatic precipitator	elektrostatiskt fällningsmedel	Elektrofilter zur Staubabscheidung, elektrischer Gasreiniger	précipitateur électrostatique	precipitante electroestático
2244	electrostatics	elektrostatik	Elektrostatik	électrostatique	electroestática
2245	elementary fibril	elementarfibrill	Elementarfibrille	fibrille élémentaire	fibrilla elemental
2246	elm	alm	Ulme, Ulmenholz	orme	olmo
2247	elongation	förlängning	Ausdehnung, Bruchdehnung, Dehnung	allongement de rupture	alargamiento
2248	elongation at rupture	brottförlängning	Bruchdehnung, Zerreissdehnung	allongement avant rupture	alargamiento a la tracción
2249	eluate	eluera	–	éluat	eluir
2250	elution	eluering	Elution, Extraktion	élution	elución
2251	elutriate	elutriat	elutrieren, abschlämmen, schlämmen, dekantieren	décanter	elutriar
2252	elutriation	elutriering	Auswaschung, Schlämmen	décantation	elutriación
2253	embedding	inbäddning	Einbettung	emboîter, encastrer	recubrimiento
2254	embossed	präglad	geprägt	bosselé, gaufré	gofrado
2255	embossed paper	präglat papper	geprägtes Papier, Prägepapier	papier gaufré	papel gofrado

	ENGLISH	SVENSKA	DEUTSCH	FRANÇAIS	ESPAÑOL
2256	embosser	präglare	Prägekalander	gaufreuse	gofrador
2257	embossing	prägling	Aufprägung, Prägen, Quellen des Gummituchs beim Offsetdruck	gaufrage	goframiento
2258	embossing calendar	präglingskalander	Prägekalander	calandre gaufreuse	gofradora
2259	embossing paper and board	präglingspapper och -kartong	Prägen von Papier und Karton	papiers et cartons gaufrés	papel y cartón para gofrar
2260	emission	emission	Emission, Ausströmen, Ausstrahlung, Ausstossen	émission	desprendimiento
2261	emission spectroscopy	emissionsspektroskopi	Emissionsspektroskopie	spectroscopie d'émission	espectroscopia de emisión
2262	emulsification	emulgering	Emulgierung	émulsionnement	emulsificación
2263	emulsifier	emulgator	Emulgator, Emulgierungsmittel	émulsionnant	emulgente
2264	emulsify	emulgera	emulgieren	émulsionner	emulsionar
2265	emulsion	emulsion	Emulsion	émulsion	emulsión
2266	enamel	emalj	emaillieren, Emaille, Lack, Schmelzglasur	couché	esmalte
2267	enameled	emaljerad	emailliert, mit Lack überzogen, mit Schmelz überzogen	couché, émaillé	esmaltado
2268	enameled paper	emaljerat papper	Kunstdruckpapier, Hochglanzpapier	papier glacé	papel esmaltado
2269	encapsulation	inkapsling	Einkapselung	encapsulation	encapsulación
2270	encrust	inkrustering	überkrusten	incruster	incrustar
2271	end bands	ändband	dickes Papier zum Schutz der Papierrollenenden während des Transports, Stirndeckel	papier épais pour protéger les bords des bobines emballées	cintas de remate
2272	endless wire	ändlös vira	Langsieb	toile sans fin	tela metálica sin fin
2273	endless woven felt	ändlös vävd filt	rundgewebter Filz, Filzschlauch, umlaufender Filz	feutre tissé sans fin	fieltro tejido sin fin
2274	endosperm	endosperm	Endosperm, Nährgewebe	endosperme	endospermo
2275	endothermic	endotermisk	endotherm, wärmeaufzehrend, wärmeaufnehmend	endothermique	endotérmico
2276	endothermic reaction	endotermisk reaktion	endotherme Reaktion	réaction endothermique	reación endotérmica
2277	energy	energi	Energie, Kraft	énergie, force	energía
2278	energy balance	energibalans	Energiebilanz	bilan énergétique	equilibrio de energía
2279	energy consumption	energiförbrukning	Energieverbrauch	consommation d'énergie	consumo de energía
2280	energy transfer	energiomvandling	Energieübertragung	transfert d'énergie	transferencia de energía
2281	Engelmann spruce	Engelmanngran	Engelmann-Fichte	épicéa d'Engelmann	epícea de Engelmann
2282	engineer	ingenjör	erbauen, errichten, konstruieren, Ingenieur, Techniker	ingénieur	ingeniero
2283	engineering	teknologi	Maschinenbau	ingénierie	ingeniería
2284	engine sizing	mäldlimning	Masseleimung	collage en pile	encolado en pila
2285	english finish	engelsk finish	halbglänzend, mattsatiniert	papier fortement apprêté	papel fuertemente alisado
2286	enthalpy	entalpi	Enthalpie	enthalpie	entalpía
2287	entomology	entomologi	Entomologie, Insektenkunde	entomologie	entomología
2288	entrainment	indragning	Einzug, Einziehen von Luft (z. B. in den Stoff)	entraînement	embarque en tren
2289	entropy	entropi	Entropie	entropie	entropía
2290	envelope machine	kuvertmaskin	Briefumschlagmaschine	machine à enveloppes	máquina para sobres
2291	envelope paper	kuvertpapper	Briefumschlagpapier	papier pour enveloppes	papel para sobres
2292	envelopes	kuvert	Briefumschläge, Hüllen	enveloppes	sobres
2293	environment	miljö	Umwelt, Umgebung	environnement	medio ambiente
2294	enzymes	enzymer	Enzym	enzymes	enzimas
2295	epibromohydrin	epibromohydrin	Epibromhydrin	epibromhydrine	epibromohidrin
2296	epichlorohydrin	epiklorohydrin	Epichlorhydrin	epichlorhydrine	epiclorohidrin
2297	epoxy resins	epoxihartser	Epoxyharz, Epoxydharze	résines époxy	resinas epoxi
2298	equation	ekvation	Gleichung	équation	ecuación
2299	equilibrium	jämvikt	Gleichgewicht, Ausgleich	équilibre	equilibrio
2300	equipment	utrustning	Ausrüstung, Einrichtung, Anlage	outillage, matériel, équipement	equipo
2301	equivalent weight	ekvivalent vikt	Äquivalentgewicht, Verbindungsgewicht	poids d'une rame par rapport à une autre d'un poids différent	peso equivalente
2302	erasability	raderbarhet	Radierbarkeit, Radierfestigkeit	résistance au grattage	resistencia al rozamiento
2303	erasable parchment bond	raderbart pergament	radierfestes Echtpergament	papier parcheminé grattable	pergamino bond borrable
2304	erosion	erosion	Erosion, Abtragung	érosion	erosión
2305	esparto	esparto	Esparto	alfa	esparto
2306	esparto paper	espartopapper	Espartopapier	papier d'alfa	papel de esparto
2307	esparto pulp	espartmassa	Espartozellstoff	pâte d'alfa	pasta de esparto
2308	ester groups	estergrupper	Estergruppen	groupes esters	grupos éster
2309	esterification	förestring	Veresterung	estérification	esterificación
2310	esters	estrar	Ester	esters	ésteres
2311	ethane	etan	Äthan	éthane	etano

	ENGLISH	SVENSKA	DEUTSCH	FRANÇAIS	ESPAÑOL
2312	ethanol	etanol	Äthanol	éthanol	etanol
2313	ethanolamines	etanolaminer	Äthanolamin	éthanolamines	etanolaminas
2314	ether groups	etergrupper	Äthergruppen	groupes éther	grupos éter
2315	etherification	företring	Ätherbildung	éthérification	eterificación
2316	ethers	etrar	Äther	éthers	éteres
2317	ethyl acetate	etylacetat	Äthylacetat, Essigsäureäthylester, Essigäther	acétate d'éthyl	etil acetato
2318	ethyl acrylate	etylakrylat	Äthylacrylat	acrilate d'éthyl	etil acrilato
2319	ethylamine	etylamin	Äthylamin	éthylamine	etilamina
2320	ethylation	etylering	Äthylierung	éthylation	etilación
2321	ethyl cellulose	etylcellulosa	Äthylcellulose, AT-Zellulose	éthyl cellulose	etil celulosa
2322	ethylene groups	etylengrupper	Äthylengruppen	groupes éthylène	grupos etilenos
2323	ethylenediamine	etylendiamin	Äthylendiamin	éthylènediamine	etilenodiamina
2324	ethyleneimine	etylenimin	Äthylenimin	éthylèneinine	etilenimina
2325	ethyl groups	etylgrupper	Äthylgruppen	groupes éthyl	grupos etilos
2326	ethyl mercaptan	etylmerkaptan	Äthylmercaptan	éthyl mercaptan	etil mercaptan
2327	eucalyptus	eukalyptus	Eukalyptus	eucalyptus	eucalipto
2328	eutrophication	eutrofering	in einen eutropischen Zustand versetzen	eutrophication	eutrofisación
2329	evacuate	evakuera	evakuieren, entleeren, räumen	évacuer	evacuar
2330	evaluation	utvärdering	Auswertung, Schätzung, Berechnung	évaluation, estimation, expertise	evaluación
2331	evaporate	avdunsta	verdampfen, eindampfen, verdunsten	évaporer	evaporar
2332	evaporation	avdunstning, indunstning	Verdampfung, Eindampfung	évaporation	evaporación
2333	evaporator	indunstare	Verdampfer, Einenger	évaporateur	evaporador
2334	even-aged stands	jämnåriga plantor	gleichaltriger Waldbestand, gleichaltriger Baumbestand	peuplements d'arbres de même âge	bosque de misma edad
2335	evolution	utveckling	Evolution, Entwicklung	évolution	evolución
2336	excelsior tissue	excelsiorpapper	Papierwolle	frisons de papier	muselina de rizo
2337	excess of coating	överflöd av bestrykning	Streichfarbenüberschuss	excès de couchage	exceso de estucado
2338	exciter	omrörare	Erreger, Erregermaschine	excitateur	excitatriz
2339	exhaust	tömma	absaugen, auspumpen, entweichen, auf die Faser aufziehen (von Farbstoffen und Chemikalien)	échappement	escape
2340	exhaust fan	tömningsfläkt	Absauger, Entlüfter, Saugventilator	ventilateur aspirant	ventilador de aspiración
2341	exhaust gas	avgas	Abgas, Auspuffgas	gaz d'échappement	gas de escape
2342	exhaust hood	ångkåpa	Abzughaube	hotte	campana de aspiración
2343	exothermic reaction	exoterm reaktion	exotherme Reaktion	réaction exothermique	reacción exotérmica
2344	expandable box	expanderlåda	stark dehnbare Kartonagenpappe (fur Verpackungszwecke und Kartons)	boîte extensible	caja extensible
2345	expandable mandrel	expanderspindel	ausziehbare Rollstange	mandrin extensible (compensateur)	mandril extensible
2346	expandable paper	expanderpapper	stark dehnbares Papier (für Verpackungszwecke)	papier extensible	papel extensible
2347	expander roll	expandervals	Breitstreckwalze	rouleau déplisseur	rodillo expansor
2348	expansion	expansion	Ausbreitung, Erweiterung, Ausdehnung	expansion, dilation	expansión
2349	experimental paper machine	experimentpappersmaskin	Versuchspapiermaschine	machine à papier expérimentale	máquina experimental de papel
2350	experimentation	försök	Experimentierung	expérience, essai	experimentación
2351	exploded fibers	exploderade fibrer	getrennte Fasern (durch Behandlung mit Dampf)	fibres séparées par explosion	fibras reventadas
2352	export pulp	exportmassa	Exportzellstoff	pâte exportée	pasta de exportación
2353	extender	förlängare	Leim-Streckmittel, Verschnittmittel	extenseur	dilatador
2354	extensibility	töjbarhet	Dehnungsfähigkeit, Dehnbarkeit	extension, allongement	extensibilidad
2355	extensible paper	töjbart papper	Papier mit hoher Dehnung	papier extensible	papel extensible
2356	extensometer	töjningsmätare	Dehnungsmesser	extensomètre	extensómetro
2357	extraction	extraktion	Extraktion, Auslaugung, Wurzelziehen (math.)	extraction, épuisement	extracción
2358	extraction chamber	extraktionskammare	Ableerkammer (beim Pulper), Abzugskammer	chambre d'extraction	campana de extracción
2359	extraction plate	extraktionsplatta	gelochte Abzugsplatte (beim Pulper)	plaque d'extraction	placa de extracción
2360	extractive	extraktiv	extraktiv, ausziehend, auslaugend	qui est extrait	extractivo
2361	extractor	extraktionsmedel	Extraktor, Extraktionsapparat	extracteur, boîte de vapeur	caja de vapor
2362	extruder	extruder	Extruder	extrudeuse	extrusionadora
2363	extrusion	extrudering	Extrusion, Extrudieren	extrusion	extrusión
2364	extrusion coater	strängsprutningsbestrykare	Extrusionstreicher	coucheuse par extrusion	estucadora por extrusión
2365	extrusion coating	strängsprutningsbestrykning	Extrusionsbeschichtung, Spritzbefilmen	couchage par extrusion	estucado por extrusión

	ENGLISH	SVENSKA	DEUTSCH	FRANÇAIS	ESPAÑOL
2366	extrusion die	sprutmunstycke	Strangpressform, Strangpresswerkzeug	filière d'une extrudeuse	hilera de una extrusionadora
2367	extrusion head	strängsprutningsbad	Spritzkopf	tête d'extrusion	entrada de extrusión
2368	exudation	utsvettning	Ausschwitzung, Ausscheidung	exsudation	exudación

F

2369	fob (free on board)	F.O.B.	frei an Bord	franco à bord	F.O.B. (franco a bordo)
2370	fabric	tyg	Gewebe, Tuch, Stoff	tissu	tela
2371	fabric finish paper	linnepressat papper	Leinenpapier	papier gaufré imitation toile	papel acabado tela
2372	fabric press	filtvirapress	Fabric-Presse, Siebtuchpresse, Gewebebandpresse	presse à tissu	prensa con tela de plástico
2373	face	yta	Fläche, Stirnfläche, Stirnseite, Oberfläche, Körperlänge (einer Walze)	largeur de table (machine)	cara
2374	facial tissues	ansiktsservett	Gesichtstücher	papiers à démaquiller	papel para desmaquillado
2375	facility	medel	Einrichtung, Anlage	facilité	facilidad
2376	facing paper	planskiktpapper	leichtgewichtiges Deckpapier, Überzugspapier, Beklebepapier	papier à doubler	papel de revestimiento
2377	facings	planskikt	Verblendung, Verkleidung	doublures	revestimientos
2378	fadeometer	urblekningsmätare	Fadeometer, Lichtechtheitsprüfer	appareil de mesure de la résistance de la coloration à la lumière, "fadeomètre"	fadómetro
2379	fading	urblekning	verblassen, verschiessen, Schwund	affaiblissement d'une teinte, altération	debilitación
2380	fading resistance	utblekningsmotstånd	Farbbeständigkeit, Farbechtheit	résistance à l'affaiblissement	resistencia a la debilitación
2381	fading test	urblekningsprov	Farbfestigkeitsprüfung	essais de résistance à l'affaiblissement	pruebas de debilitación
2382	Fagus (Lat.)	Fagus	Fagus	Fagus, hêtre	haya
2383	failure	brott, bristning	Versagen, Ausfall, Störung, Panne	cassure, rupture, échec	falla
2384	fan	fläkt	anblasen, ventilieren, Gebläse, Lüfter, Ventilator, Papierformat	ventilateur	ventilador
2385	fan blade	fläktblad	Ventilatorflügel, Gebläseflügel	pale de ventilateur	pala de ventilador
2386	fanning	fläktning	Zugabnahme, Anfachung, Sortier- und Prüfmethode von Papier	mise en éventail	puesta en abanico
2387	fan pump	blandningspump	Mischpumpe, Flügelradpumpe, Turbogelbläsepumpe	pompe de mélange	bomba de aletas
2388	fast color	beständig färg	lichtechte Farbe, lichtechter Farbstoff	couleur solide à la lumière	color fijo
2389	fastness	färgäkthet	Schnelligkeit, Echtheit	solidité à la lumière	resistencia de un colorante
2390	fast pulp	härdig massa	schnell entwässernder Zellstoff	pâte maigre	pasta magra
2391	fatigue	utmattning	Ermüdung	fatigue	fatiga
2392	fatigue failure	utmattningsbrott	Dauerbruch, Ermüdungsbruch	rupture dûe à la fatigue	ruptura por fatiga
2393	fatigue resistance	utmattningsmotstånd	Ermüdungsbeständigkeit	résistance à la fatigue	resistencia a la fatiga
2394	fatigue tests	utmattningsprov	Dauerschwingversuch, dynamische Untersuchung	essais de fatigue	pruebas de fatiga
2395	fatty acids	fettsyror	Fettsäuren	acides gras	ácidos grasos
2396	feasibility	rimlighet	Durchführbarkeit, Ausführbarkeit	possibilité	posibilidad
2397	featheredge	kilkant	scharfe Kante, zugespitzte Kante	pyrotechnie	canto vivo
2398	featheredged board	kilkantpapp	büttenrandiger Karton	carton pour pyrotechnie	cartón de canto vivo
2399	featheredge deckle	kilkantdäckel	rauhe Kante, abgeschrägte Kante	bords barbés amincis	cubierta en bisel
2400	featheredge paper	kilkantpapper	büttenrandiges Papier	papier pour pyrotechnie	papel de canto vivo
2401	feathering	avtunning	Auslaufen eines Tintenstrichs auf ungeleimtem Papier	bavure (encre)	biselado
2402	featherweight paper	fjäderviktpapper	federleichtes Papier	papier bouffant	papel esponjoso
2403	feed	matning	zuführen, speisen, Zufuhr, Beschickung	alimentation	alimentación
2404	feedback	återkoppling	Rückkopplung	réalimentation	realimentación
2405	feedback control	återkopplingskontroll	automatische Regelung	contrôle de la réalimentation	control de realimentación
2406	feed conveyor	matningstransportör	Transportgerät, Fördergerät	rouleau transporteur	transportador alimentador
2407	feeder	matare	Eintragsvorrichtung, Beschickungsanlage	dispositif d'alimentation, margeur	alimentador
2408	feed forward control	frammatningskontroll	Vorwärtsregelung	commande par action directe	control de reenvío de alimentación
2409	feed guide stop	stoppare för matningsregulator	Anschlag	butée d'engagement de la feuille	tope de guía de alimentación

	ENGLISH	SVENSKA	DEUTSCH	FRANÇAIS	ESPAÑOL
2410	feed hopper	matningstratt	Einlauftrichter, Beschickungstrichter, Einfülltrichter	trémie de chargement	tolva de alimentación
2411	feeding	matning	Zuführung, Speisung, Vorschub	alimentation, chargement	alimentación
2412	feed water	matningsvatten	Speisewasser	eau d'alimentation	agua de alimentación
2413	feel	känsel, känsla	Griff, Griffigkeit, Gefühl, fühlen, anfühlen, empfinden	toucher	palpar
2414	fell	fällning	fällen, hauen, schlagen (von Holz)	abattre (un arbre)	talar
2415	feller buncher	knippare	Holzfäller, Holzhauer, Waldarbeiter	abatteuse empileuse	talador agrupador de troncos
2416	feller skidder	fällarmede	Holzfäller-Skidder	abatteuse débardeuse	talador arrastrador de troncos
2417	felling	fällning	Fällen, Holzschlag	abattage	tala
2418	felt	filt	Filz	feutre	fieltro
2419	felt conditioner	filtkonditionerare	Filzinstandhalter, Filzreiniger	conditionneur de feutre, laveur de feutre	lavador de fieltros
2420	felt conditioning	filtkonditionering	Filzkonditionierung, Filzinstandhaltung	conditionnement du feutre	limpieza de fieltros
2421	felt direction mark	filtriktningsmarkering	Laufpfeil im Filz, Filzrichtungsmarkierung	flèche de direction de marche (sur le feutre)	flecha de dirección
2422	felt dryer	filttorkare	Filztrockner, Filztrockenzylinder	sécheur de feutre	secador de fieltros
2423	felt finish	filtfinish	Filzprägung (Papieroberflächenau srüstung in der Nasspresse)	vergé au feutre	acabado fieltro
2424	felt hairs	filthår	Filzhaare	poils de feutre	pelos del fieltro
2425	felting	filtning	Verfilzen	feutrage	afieltrado
2426	felt marking	filtmarkering	Filzmarkierung (in den Filz hineingewebtes Muster)	marquage au feutre	marcado del fieltro
2427	felt mark	filtmarkering	Filzmarkierung, Filzfehler	marque du feutre	marcas del fieltro
2428	felt paper	filtpapper	Filzpapier	papier feutre	papel fieltro
2429	felt roofing	lumptakpapp	Dachpappe	carton feutre pour toitures	techado de fieltro
2430	felts, deadening	filt, dödning	Dachpappe	feutre amortisseur	fieltros insonorizantes
2431	felts, saturating	filt, mättning	imprägnierter Filz	feutre bituminés	fieltros de saturación
2432	felt side	filtsida	Oberseite	côté feutre, envers	cara del fieltro
2433	felt stretcher	filtsträckare	Filzspanner	tendeur de feutre	tensor del fieltro
2434	felt stretching	filtsträckning	Filzspannung	étirage du feutre	tensado de fieltros
2435	felt-stretching roll	filtsträckvals	Filzspannwalze	rouleau tendeur de feutre	rodillo tensador de fieltros
2436	felt washing	filttvätt	Filzwäsche	lavage du feutre	lavado de fieltros
2437	ferrule	skoning	Zwinge, Buchse, Muffe	virole	férula
2438	fertilization	gödning	Fruchtbarmachung, Düngung	fécondation	fertilización
2439	fertilizer	gödningsmedel	Dünger, Düngemittel	engrais	fertilizante
2440	festoon	slinga	Girlande, Gehänge	boucle d'accrocheuse	festón
2441	festoon dried	hängtorkad	im Hängetrockner getrocknet (z. B. Streichpapier)	séché sur accrocheuse	secado en enganchadera
2442	festoon dryer	hängtork	Hängetrockner	accrocheuse sécheuse	enganchador secador
2443	festoon drying	hängtorkning	Girlandentrocknung, Hängetrocknung	séchage sur accrocheuse	secado en enganchadera
2444	festooning	hängning i slingor	Runzelbildung (Farbe)	sécher sur accrocheuse	enguirnaldar
2445	fiber	fiber	Faser, Fiber	fibre	fibra
2446	fiber analysis	fiberanalys	Faseranalyse	analyse des fibres	análisis de fibras
2447	fiberboard	fiberskiva	Hartfaserplatte, Faserstoffplatte	carton cuir (pour chaussures)	cartón duro
2448	fiber bonding	fiberbindning	Faserbindung	liaison des fibres	unión de fibras
2449	fiber bundle	fiberknippe	Faserbündel	faisceau de fibres	haz de fibras
2450	fiber can	pappburk	Fiberkanne	récipient en carton	recipiente de fibra
2451	fiber cement products	fibercementprodukter	faserige Zusatzstoffe für Zement	articles en fibrociment	fibrocementos
2452	fiber classification	fiberfraktionering	Faserfraktionierung	étude des fibres par la méthode du fractionnement	clasificación de la fibra
2453	fiber classifier	fiberfraktioneringsapparat	Faser-Fraktioniergerät	appareil pour l'étude des fibres par la méthode du fractionnement	clasificador de fibras
2454	fiber composition	fibersammansättning	Faserzusammensetzung	composition fibreuse	composición de la fibra
2455	fiber content	fiberhalt	Fasergehalt	teneur en fibres	contenido de fibra
2456	fiber cut	fibersnittning	Faserschnitt	coupure de fibre	corte de fibra
2457	fiber diameter	fiberdiameter	Faserdurchmesser	diamètre d'une fibre	diámetro de la fibra
2458	fiber dimensions	fiberdimensioner	Faserabmessungen	dimensions d'une fibre	dimensiones de la fibra
2459	fiber drum	pappbehållare	Fass aus Fasermaterial	tambour en fibre	tambor de fibra
2460	fiber entanglement	fiberflätning	Faserverflechtung	enchevêtrement des fibres	maraña de fibra
2461	fiber fraction	fiberfraktion	Faserfraktion	fractionnement des fibres	fracción fibrosa
2462	fiber furnish	mäld	Faserstoffeintrag	teneur en fibres	composición fibrosa de la pasta
2463	fiberize	defibrera	defibrieren, zerfasern	dépastiller	despastillado
2464	fiber knots	fiberknutar	Faserknoten	noeuds de fibres	nudos de la fibra
2465	fiber length	fiberlängd	Faserlänge	longueur de la fibre	longitud de la fibra
2466	fiber length distribution	fiberlängdsfördelning	Faserlängenverteilung	répartition des longueurs de fibres	distribución longitudinal de la fibra

	ENGLISH	SVENSKA	DEUTSCH	FRANÇAIS	ESPAÑOL
2467	fiber mat	fibermatta	Fasermatte, Faserfilz	matelas de fibres	capa de fibras
2468	fiber recovering	fiberåtervinning	Fasererholung	récupération des fibres	recuperación de fibras
2469	fiber recovery	fiberåtervinning	Faserrückgewinnung	récupération des fibres	recuperación de fibras
2470	fiber saturation point	fibermättnadspunkt	Fasersättigungspunkt	point de saturation des fibres	punto de saturación de la fibra
2471	fiber structure	fiberstruktur	Faserstruktur	structure des fibres	estructura fibrosa
2472	fiber tow	fibersträng	Faserwerg	étoupe	fibra-estopa
2473	fibre (brit.)	fiber	Faser	fibre	fibra
2474	fibrillae	fibriller	Fibrillen	fibrille	fibrilla
2475	fibrillate	fibrillera	zerfasern, zermahlen	fibriller	fibrilado
2476	fibrillating	fibrillering	Zermahlen, Zerlegen in Spaltfäserchen	fibrillation	formación de fibrilla
2477	fibrillation	fibrillering	Zermahlung, Zerlegung in Spaltfäserchen	fibrillation	fibrilación
2478	fibrils	fibriller	Fibrille	fibrilles	fibrillas
2479	fibrous	fiberfibriller	faserig	fibreux	fibroso
2480	fibrous filler	fiberfiller	faseriger Füllstoff	charge fibreuse	carga fibrosa
2481	filament	tråd	Faden, Staubfaden, Draht	filament	filamento
2482	filament yarn	garn	endloses Garn, Endlosfaden	fil textile	filamento textil
2483	file folder	vikmapp	Schnellhefter, Ordner	papier pour dossiers	clasificador
2484	fill	fyllning	füllen, auffüllen, Verlegen des Filzes	largeur utile maximum d'une machine à papier	rellenar
2485	filled	fylld	gefüllt, voll, beschwert, verlegt (Filz)	rempli, garni	relleno
2486	filled board	fylld kartong	Mehrlagenkarton	carton multiforme	cartón rellenado
2487	filled bristol	fylld bristolkartong	Bristolkarton (dessen Mittellage aus einer anderen Faserzusammensetzung besteht als die Decklagen)	bristol multijet fabriqué sur forme ronde	bristol rellenado
2488	filled felt	fylld filt	verlegter Filz, verstopfter Filz	feutre saturé	filtro rellenado
2489	filled newsboard	fylld slipmassekartong	mit Füllstoff gearbeitete Graupappe	carton multijet de vieux papiers pour boîtes montées	cartón prensa rellenado
2490	filled paper	fyllt papper	aschehaltiges Papier	papier chargé, papier multijet	papel rellenado
2491	filled pulpboard	fylld kartong	aschehaltige Zellstoffpappe	carton multijet, intérieur vieux papiers, fabriqué sur forme ronde	cartón de pasta rellenado
2492	filled roll	pappersglättvals	Kalanderwalze (aus Papier)	rouleau de papier d'une calandre	rodillo lleno
2493	filler	fyllmedel	Füllstoff	charge	carga
2494	filler clay	lerfyllmedel	Füllstoffclay	charge minérale	arcilla para rellenar
2495	filler paper	glättvalspapper	Einlagenpapier	papier pour intérieur de cartons	papel para rellenar
2496	filler retention	fyllmedelsretention	Füllstoffretention	rétention de charge	retención de la carga
2497	filling	fyllande	Auffüllung, Füllmasse, Bemesserung (Holländer)	remplissage, première couche d'un couchage	guarnición
2498	film	film	Film, Folie, Belag, Schicht	film, pellicule	película
2499	film former	filmbildare	Filmbildner	machine de formation de film	plantilla de película
2500	filter	filter	Filter	filtre	filtro
2501	filterability	filtrerbarhet	Filtrierbarkeit	filtrabilité	filtrabilidad
2502	filter aid	filtermedel	Filtrations-Hilfsmittel	adjuvant de filtration	apoyo filtrante
2503	filter bed	filterbädd	Filterbett	lit de filtration	lecho filtrante
2504	filter cake	filterkaka	Filterkuchen	gâteau de filtre presse	torta de filtro prensa
2505	filtering	filtrering	Filtrieren	filtration	filtrar
2506	filter mass	filtermassa	Filtermasse, Filtriermasse	masse filtrante	masa filtrante
2507	filter paper	filterpapper	Filtrierpapier	papier filtre	papel filtro
2508	filtrate	filtrat	Filtrat, filtrierte Flüssigkeit	solution filtrée, filtrant	filtrado
2509	filtration	filtrering	Filtrierung	filtration	filtración
2510	fineness	finhet	Feinheit, Feingehalt	finesse (pâte mécanique)	fineza
2511	fine papers	finpapper	Feinpapiere	papiers sans bois	papeles finos
2512	fines	finmaterial	Feinstoff	fines, parcelles	finos
2513	fine screen	finsil	Feinsortierer, Nachsortierer, Feinstoffsortierer	classeur	criba fina
2514	fine screening	finsilning	Feinsortierung, Nachsortierung	épuration des fines	cribado fino
2515	fines removal	finmaterialborttagning	Entfernung von pulverigem Material, Entfernung von kurzen Faserbestandteilen	élimination des fines	separación de finos
2516	finish	finish	beenden, aufhören, fertigstellen, Oberflächenausrüstung, Oberflächenbeschaffenheit, Oberflächenzustand	fini, apprêt	apresto

	ENGLISH	SVENSKA	DEUTSCH	FRANÇAIS	ESPAÑOL
2517	finishing	finishprocess	Ausrüstung, Endverarbeitung, Zurichtung	finissage	acabado
2518	finishing broke	finishutskott	Ausschuss bei der Ausrüstung	cassés de finissage	recortes en el acabado
2519	finishing room	finishsal	Ausrüstungssaal	atelier de finissage	sala de acabado
2520	fir	fura, tall, gran	Tanne	sapin	abeto
2521	fire	eld, brand	anzünden, entzünden, Feuer, Brand, Flamme	feu, incendie	fuego
2522	fire brick	brandtegel	feuerfester Ziegel, Schamottestein	brique réfractaire	ladrillo refractario
2523	fire detector	branddetektor	Feuermelder, Feueranzeiger	détecteur d'incendie	detector de fuego
2524	fire prevention	brandskydd	Feuerverhütung	mesures de précaution contre l'incendie	prevención de incendios
2525	fireproof crepe	eldfast kräpp	feuerfestes Krepp-Papier	papier crêpé incombustible	crepé incombustible
2526	fireproofing	eldsäkring	feuerfest machen, feuerfest ausrüsten	rendre incombustible	ignífugo
2527	fireproof paper	eldfast papper	brandsicheres Papier	papier inflammable	papel incombustible
2528	fire protection	brandskydd	Brandschutz	mesures de protection contre l'incendie	protección contra el fuego
2529	fire resistance	brandmotstånd	Feuerfestigkeit	résistance au feu	resistencia al fuego
2530	fire retardant	brandhämmande	feuerhemmend	retardeur d'incendie	retardador de incendios
2531	first dryer	första torkcylindern	erster Trockenzylinder, Vortrockner	embarqueur	embarcador
2532	first main press	första huvudpress	erste Hauptpresse, Nasspresse	presse coucheuse principale	primera prensa principal
2533	first press	första press	erste Presse, Vorpresse	première presse coucheuse	primera prensa
2534	fish eyes	fiskögon	Zellstofflecken, glasige Flecken, Schleimflecken, Fischaugen (Materialfehler)	oeil de poisson	de ciento ochenta grados
2535	fitting	passning	Installation, Montage, Apparatur, Ausstattung, Einbau	garnissage, habillage	guarnición
2536	fixed cost	fast kostnad	Fixkosten, Festkosten	frais fixes	costo fijo
2537	flaking	flagning	Abblättern, Flocken	arrachage	arrancamiento
2538	flamejet drying	flamtorkning	Flammstrahltrocknung	séchage à la flamme	secado al chorro de llama
2539	flameproof	brandsäker	flammsicher, nicht brennbar, feuerfest machen	résistant à la flamme	incombustible
2540	flameproof paper	brandsäkert papper	flammsicheres Papier	papier incombustible,ignifugé	papel incombustible
2541	flame retardant	brandhindrande	flammhemmend	retardeur de flamme	retardador de llamas
2542	flammability	brännbarhet	Entflammbarkeit, Entzündbarkeit	inflammabilité	inflamabilidad
2543	flange	fläns	flanschen, bördeln, Flansch, Bördel	bride (tuyau), joue (poulie)	brida
2544	flash-dried pulp	flingtorkad massa	schnellgetrockneter Zellstoff	pâte séchée à la vapeur	pasta de secado instantáneo
2545	flash drying	flingtorkning	Schnelltrocknung	séchage à la vapeur	secado instantáneo
2546	flashing	blixtrande	Aufblitzen, Aufflodern	séchage à la vapeur	centelleo
2547	flash point	övertändningspunkt	Entflammungspunkt, Flammpunkt	point éclair (huiles)	punto inflamador
2548	flatbed press	planpress	Flachdruckpresse	presse en blanc	prensa plana
2549	flat box	suglåda	Flachsauger	caisse aspirante	caja aspirante
2550	flat crush resistance	plankrossningsmotstånd	Flachstauchfestigkeit	résistance à l'écrasement à plat	resistencia al aplastado
2551	flat crush tests	plankrossningsprov	Flachstauchtest	essais d'écrasement à plat	pruebas de aplastamiento
2552	flat finish	plan finish	glanzlose Glätte	fini mat	acabado mate
2553	flatness	planhet	Flachliegen, Planlage (des Papiers)	aplat du papier	aplanado
2554	flat paper	plant papper	flachliegendes Papier	papier à plat	papel liso
2555	flat roll	slätvals	plattgedrückte Papierrolle, glatte Walze	bobine écrasée	rodillo aplanado
2556	flat screen	plansil	Flachsortierer	épurateur plat	criba plana
2557	flaw	fel	Fabrikationsfehler, Defekt, Riss im Papier	fente, fêlure, défaut	defecto
2558	flaw detector	feldetektor	Fehleranzeige, Fehlstellensucher	détecteur de défauts	detector de defectos
2559	flax	lin	Flachs, Lein	lin	lino
2560	flax board	linnekartong	Flachspappe	carton isolant (de paille de lin)	cartón lino
2561	flesh tanks	tank för spontan avdunstning	Entspannungsgefässe	–	tanques de destensado
2562	flexibility	flexibilitet	Flexibilität, Biegsamkeit	flexibilité	flexibilidad
2563	flexing	böjning	Biegungs-, Biegen, Beugen	courbure, déformation	flexión
2564	flexography	flexografi	Flexographie	flexographie	flexografía
2565	flexural resistance	böjmotstånd	Biegesteifigkeit	résistance à la flexion	resistencia a la flexión
2566	flexural strength	böjhållfasthet	Biegefestigkeit	rigidité	fuerza flexional
2567	flint paper	flintpapper	Flintpapier, Schmirgelpapier	papier lissé à l'agathe	papel esmeril
2568	floating dryer	luftkuddetork	Schwimmtrockner	sécheur flottant	secadora flotadora
2569	floating knife coater	floating knife-bestrykare	Streichmaschine mit Luftrakel	coucheuse avec râcle entre deux rouleaux	estucadora de cuchillas flotantes
2570	flocculant	flockningsmedel	Ausflockungsmittel	floconneux	floculante
2571	flocculate	flocka	ausflocken	floculat	flocular
2572	flocculating agent	flockningsmedel	Flockungsmittel	produit floculant	agente de floculación

	ENGLISH	SVENSKA	DEUTSCH	FRANÇAIS	ESPAÑOL
2573	flocculation	flockning	Koagulation, Ausflockung	floculation	floculación
2574	flocculator	flockningsmedel	Flockulator, Flockungsbecken	bassin de floculation	floculador
2575	flock	flock	Wollflocke, Menge, Haufen, Schar	flocon	copo
2576	flock coating	veluriserat papper	Flockenstreichen	couchage floconneux	estucado a copos
2577	flocking	flockning	in Scharen, in Haufen	floculation	floculación
2578	flock paper	flockpapper	Samttapete, Velourspapier	papier genre velours	papel rugoso para empapelar
2579	floc	flock	Flöckchen	flocon	grumo
2580	flong	matrispapp	Matrizenpapier, Druckform, Giessform, Stanzform	flan de clicherie	cartón para matrices
2581	flooded nip coater	flödat nyp-bestrykare	Streicher mit überfluteter Einlaufzone	coucheuse à pince submergée	estucadora de inundación entre rodillos
2582	flooding	flödning	überfluten, überfliessen, überschwemmen	inondation	inundación
2583	flotation	flotering	Flotation, Schwimmen, Schweben	flottation	flotación
2584	flotation agent	floteringsmedel	Flotationsmittel	produit de flottation	agente de flotación
2585	flotation deinking	floteringsavsvärtning	Flotations-De-Inken	désencrage par flottation	destintado de flotación
2586	flotation saveall	floteringsåtervinnare	Flotationsstoffänger	ramasse-pâte à flottation	recogepasta por flotación
2587	flour	mjöl	Mehlstoff (des Holzschliffs)	farine	harina
2588	flow	flöde	Fliessen, Strömen, Durchfluss, Zufluss	courant, écoulement, débit	flujo
2589	flow agent	flödningsmedel	Fliessmittel	agent d'écoulement	agente de flujo
2590	flow box	inloppslåda	Stoffauflauf	caisse d'arrivée de pâte, caisse de tête	caja de entrada
2591	flow chart	flödesdiagram	Fliessbild, Fliessdiagramm, Fliesschema, Arbeitsplan	courbe de débit	cuadro de flujo
2592	flow control	flödesreglering	Durchsatzregelung	contrôle du débit	control del flujo
2593	flow measurement	flödesmätning	Strömungsmessung	mesure du débit	medición del flujo
2594	flowmeter	flödesmätare	Strömungsmesser	débitmètre	medidor de flujo
2595	flow rate	flöde	Durchsatz, Durchfluss, Durchsatzgeschwindigkeit	vitesse d'écoulement	evaluación del flujo
2596	flow spreader	flödesspridare	Querstromverteiler, Verteiler (im Stoffauflauf)	répartiteur de pâte	repartidor de pasta
2597	flue gas	förbränningsgas	Rauchgas, Abgas	gaz de carneau	gas de combustión
2598	fluff (Brit.)	ludd, damm	flockig werden, Flaum, Federflocke, feiner Staub	duvet, bourre	pelusa
2599	fluffed pulp	dammformig massa	Flockenzellstoff	pâte en bourre	pasta con pelusa
2600	fluffing tendency	damningstendens	Staubungsneigung	tendance à se mettre en bourre	tendencia a la pelusa
2601	fluid	fluid	flüssig, gasförmig, Flüssigkeit	fluide	flúido
2602	fluid dynamics	hydrodynamik	Strömungslehre	dynamique des fluides	dinámicas de los flúidos
2603	fluid flow	vätskeflöde	Grenzfläche	coulée	flujo flúido
2604	fluid mechanics	hydromekanik	Strömungstechnik	mécaniques des fluides	mecánica de los flúidos
2605	fluid shear	skjuvströmning	Flüssigkeitsscherkraft	–	corte hidráulico
2606	fluidics	fluidik	Fluidik	liquides	flúidica
2607	fluidity	fluiditet	Dünnflüssigkeit, Flüssigkeitsgrad, Fliessfähigkeit	fluidité	fluidez
2608	fluidization	fluidisering	Wirbelschichttechnik, Wirbelschichtverfahren, Staubfliessverfahren	fluidification	fluidificación
2609	fluidized bed	fluidiserad bädd	Fliessbett, Wirbelschicht, Flüssigbett	lit fluidisé	lecho fluidizado
2610	fluidizer	fluidiserare	Wirbelschichter	dispositif de fluidification	fluidificador
2611	flume	ränna	Gerinne, Schwemme, Förderkanal, künstlicher Wasserlauf	canal d'arrivée d'eau	canal
2612	fluorescence	fluorescens	Fluoreszenz	fluorescence	fluorescencia
2613	fluorescent brightener	fluorescerande blekmedel	fluoreszierender Aufheller	éclaircissant fluorescent	abrillantador fluorescente
2614	fluorescent dyes	fluorescerande färger	fluoreszierende Farben	colorants fluorescents	colorantes fluorescentes
2615	fluorescent paper	fluorescerande papper	fluoreszierendes Papier	papier fluorescent	papel fluorescente
2616	fluorescent white	fluorescerande vitt	Leuchtweiss	blanc fluorescent	blanco flúor
2617	fluorine	fluor	Fluor	fluor	flúor
2618	fluorine compound	fluorförening	Fluorverbindungen	composés du fluor	compuestos de flúor
2619	flute	wellpipa	Wellpappenwelle, Flute	cannelure	canaladura
2620	fluted	wellad	gewellt, geriffelt, gerillt, kanneliert	cannelé	ondulado
2621	fluting	vågning	Wellenstoff	cannelure	acanaladura
2622	fluting paper	vågningspapper	Wellenmaterial, Wellenpapier	papier pour cannelure	papel de rizar
2623	fly ash	flygaska	Flugasche	cendres volatiles	ceniza fina
2624	flying paster	flygande klistrare	kontinuierlicher Rollenankleber in der Maschine	système de collage automatique	pegadura rápida
2625	flying splices	flygande skarvar	Klebestellen	collages sur bobine	pegaduras de corrido
2626	fly knife	svävkniv	rotierendes Messer (eines Querschneiders)	couteau rotatif	cuchilla móvil

	ENGLISH	SVENSKA	DEUTSCH	FRANÇAIS	ESPAÑOL
2627	fly roll	svävvals	Losrolle	rouleau de renvoi de la feuille	rodillo flotante
2628	foam	skum	Schaum, schäumen	écume, mousse	espuma
2629	foam breaker	skumsläckare	Schaumbrecher	anti-mousse	rompedor de espuma
2630	foaming	skumning	Schäumen	production de mousse	espumante
2631	foaming agent	skumningsmedel	Schaummittel, Treibmittel	produit provoquant la formation de mousse	agente espumante
2632	foam inhibition	skumdämpning	Schaumverhütung	empêchement de la formation de mousse	impedimento de la espuma
2633	foam inhibitor	skumdämpare	Schaumverhütungsmittel	anti-mousse	inhibidor de espuma
2634	foam killer	skumsläckare	Schaumbekämpfungsmittel	anti-mousse	antiespumante
2635	foam marks	skummarkering	Schaumflecken	taches de mousse	marcas de espuma
2636	foam spots	skumningsfläckar	Schaumflecken	points de mousse	manchas de espuma
2637	fogging	dimbildning	Vernebeln	voile en offset	anublado
2638	foil	foil	Abstreifer	feuille mince de métal	lámina
2639	foil angle	foilvinkel	Abstreiferansatzwinkel	angle de racle d'égouttage	ángulo de lámina
2640	foil baffle	foilbaffel	Foil	déflecteur d'eau	deflector de lámina
2641	foil body	foil-body	Foilkörper (mit mehreren Entwässerungsleisten)	support pour contrecollé	cuerpo de la lámina
2642	foil paper	folie	Folienpapier	papier contrecollé alu	papel metalizado
2643	fold	veck, vika	Falz, Falte, Bruch	pli	pliegue
2644	folded	vikt	gefalzt	plié	plegado
2645	folder gluer	veckad limpåläggare	Faltschachtel-Klebemaschine	plieuse colleuse	encoladora plegadora
2646	folding	vikning	Falten, Zusammenlegen, Kniffen	pliage	plegado
2647	folding board	falskartong	Falzkarton	carton pliant	cartón plegable
2648	folding boxboard	falskartong	Faltschachtelkarton	carton pour boîtes pliantes	cartón duplex y triplex
2649	folding box	pappkartong	Faltschachtel	boîte pliante	caja plegable
2650	folding bristol	bristolkartong	Bristol-Faltschachtelkarton	bristol pliant	bristol plegable
2651	folding carton	pappkartong	Faltschachtel	carton pliant	cartonaje plegable
2652	folding endurance	vikhållfasthet	Dauerbiegefestigkeit, Falzfestigkeit	résistance au pliage	resistencia al plegado
2653	folding endurance tester	vikningsprovare	Dauerbiegeprüfer, Falzgerät	pliographe	plegámetro
2654	folding endurance tests	vikningsprov	Dauerbiegeversuche, Falzversuche	essais de résistance au pliage	pruebas de resistencia al plegado
2655	folding machine	bigningsmaskin	Falzmaschine, Falzer	plieuse	máquina plegadora
2656	folding quality	bigningskvalitet	Falzqualität	aptitude au pliage	calidad del plegado
2657	folding resistance	bigningsmotstånd	Falzwiderstand	résistance au pliage	resistencia al plegado
2658	folding stock	falskartongmassa	Falzmaterial	papier à fibres longues pour support de couche	carpeta
2659	folding strength	bigningshållfasthet	Falzfestigkeit	résistance au pliage	resistencia al plegado
2660	fold tester	bigningsprovare	Falzfestigkeitsprüfer, Doppelfalzer	pliographe	pruebas de pliegue
2661	food board	livsmedelskartong	Lebensmittelkarton	carton pour produits alimentaires	cartón para productos alimenticios
2662	food wrap paper	livsmedelsomslagspapper	Lebensmitteleinwickelpapier	papier d'emballage pour produits alimentaires	papel de embalaje para alimentos
2663	footage	längd	Gesamtlänge (in Fuss)	métrage (en pieds anglais)	longitud en pies
2664	force	kraft	Kraft, Stärke	force	fuerza
2665	foreman	förman	Aufseher, Vorarbeiter, Polier	contremaître	capataz
2666	forest	skog	Wald, Forst	forêt	bosque
2667	forestation	skogshantering	Aufforsten	forestation	arborización
2668	forest canopy	grenverk	Waldkrone	voute des arbres	sobrecielo del bosque
2669	forest fire	skogseld	Waldbrand	incendie de forêt	incendio forestal
2670	forest genetics	skogsgenetik	Forstgenetik	génétique forestière	genética forestal
2671	forest industry	skogsindustri	Waldwirtschaft	industrie forestière	industria forestal
2672	forest litter	skogsförorening	Waldstreu	litière de la forêt, couverture morte	lecho del bosque
2673	forest management	skogsförvaltning	Waldbewirtschaftung, Forstverwaltung	exploitation forestière	administración forestal
2674	forest operation	skogsdrift	Forstwirtschaft	exploitation forestière	operación forestal
2675	forest pathology	skogspatologi	Forstpathologie	pathologie forestière	patología forestal
2676	forest products	skogsprodukter	forstwirtschaftliche Erzeugnisse	produits forestiers	productos forestales
2677	forestry	skogslära	Forstwesen	sylviculture	silvicultura
2678	forklift truck	gaffeltruck	Gabelstapler	chariot élévateur à fourche	carretilla elevadora de horquilla
2679	formability	bildbarhet	Formbarkeit, Formveränderungsvermögen	aptitude au façonnage	formabilidad
2680	formaldehyde	formaldehyd	Formaldehyd	formaldéhyde	formaldehido
2681	formalin	formalin	Formalin	formaline	formol
2682	formamide	formamid	Formamid	formamide	formamida
2683	formates	format	Formate, Formiate	formiates	formiatos
2684	formation	bildning, formation	Bildung, Gestaltung, Struktur	formation, épair	contextura del papel
2685	formation of sheet	arkformning	Blattbildung	constitution de la feuille	formación de la hoja

	ENGLISH	SVENSKA	DEUTSCH	FRANÇAIS	ESPAÑOL
2686	formation tester	arkformningsprovare	Blattbildungsprüfgerät	appareil de mesure de l'épair	probador de contextura
2687	former	formare	Formleiste, Spant, Schablone, Blattbildner	gabarit, calibre	plantilla
2688	formic acid	myrsyra	Ameisensäure	acide formique	ácido fórmico
2689	forming	formning	formen, bilden, formgeben	formation	fabricación
2690	forming board	formbräda	Siebtisch	marbe en tête de machine à papier	mármol de mesa de fabricación
2691	forming fabric	bröstläder	Kunststoffgewebe für die Blattbildung	produit en formation	tela de formación
2692	forming roll	formeringsvals	Einlaufwalze, Egoutteurwalze, Patrizenwalzen	rouleau de formation	rodillo de formación
2693	formula	formel	Formel	formule	fórmula
2694	formulation	formulering	Formulierung	établissement d'une formule	formulación
2695	fortified size	förstärkt lim	spezifische Leimart zur Erzielung erhöhter Wasserfestigkeit in Papier und Karton	collage renforcé	cola reforzada
2696	fortified rosin	förstärkt harts	spezifische Leimart zur Erzielung erhöhter Wasserfestigkeit in Papier und Karton	résine renforcée	resina reforzada
2697	fossil fuel	fossilt bränsle	Erdöl, Steinöl, Petroleum	pétrole brut	combustible fósil
2698	foundation	grund	Fundament, Grundlage	fondation	fundación
2699	fountain	fontän, tråg	Fontäne, Quelle, Reservoir	bassine de mouillage	fuente
2700	fountain coater	slitsbestrykare	Tauchwalzenstreicher	coucheuse à lame à "fontaine"	estucadora de fuente
2701	fountain roll	doppvals	Tauchwalze	rouleau de transmission de sauce	rodillo fuente
2702	fountain solution	smet	Wischwasser (beim Offsetdruck)	solution de mouillage	solución de origen
2703	fourdrinier board	planvirakartong	Langsiebkarton	carton fabriqué sur table plate	mesa plana
2704	fourdrinier machine	planviramaskin	Langsiebmaschine	machine à table plate	máquina de papel de mesa plana
2705	fourdrinier wire	planvira	Langsieb	toile de la machine à table plate	mesa plana
2706	fourdrinier yankee machine	kombinerad pappersmaskin	Selbstabnahme-Langsiebmaschine	machine à table plate avec gros sécheur frictionneur	máquina de papel yankee de mesa plana
2707	fraction	fraktion	Fraktion, Bruchteil, Fragment	fraction	fracción
2708	fractionating	fraktionering	Fraktionieren	fractionnement	fraccionamiento
2709	fractionation	fraktionering	Fraktionierung, Faserfraktionierung	études des pâtes par la méthode du fractionnement	fraccionamiento
2710	fractionator	fraktioneringsapparat	Fraktioniergerät	séparateur	fraccionador
2711	fracture	fraktur, brott	zerbrechen, Bruch, Fraktur	fracture, rupture	rotura
2712	fragmentation	fragmentering	Zersplitterung, Fragmentierung	fragmentation	fragmentación
2713	fragments of fiber	fiberfragment	Faserfragmente	fragments de fibres	trozos de fibra
2714	frame	ram	gestalten, bilden, formen, Stuhlung (am Kalander), Maschinenrahmen	bâti, cadre, châssis	bastidor
2715	frame work	stativ	Stuhlung (am Kalander)	châssis, bâti, cadre	bastidor (en calandra)
2716	Fraxinus (Lat.)	Fraxinus	Esche	frêne	fresno
2717	free-beaten	mager	Röschstoff-Holländer	pâte raffinée maigre	refinado magro
2718	freeness	freeness	Röschheit (ist immer reziprok zum Mahlgrad)	qualité d'une pâte maigre	grado de refino
2719	freeness recorder	freenesskrivare	Mahlgradprüfer, Mahlgradanzeigegerät	enregistreur de degré de raffinage	registrador de grado de refino
2720	freeness tester	freenessprovare	Mahlgradprüfer	essayeur de degré de raffinage	refinómetro
2721	free on board (fob)	free on board	frei an Bord	franco à bord	F.O.B. (franco a bordo)
2722	free pulp (Brit.)	mager massa	röscher Zellstoff	pâte maigre	pasta magra
2723	free sheet	ark av mager mäld	holzfreies Papier, Papier aus röschem Stoff	feuille volante	hoja libre
2724	free stock	mager mäld	röscher Stoff	pâte maigre	pasta magra
2725	frequency	frekvens	Frequenz, Häufigkeit, Schwingungszahl	fréquence	frecuencia
2726	frequency converter	frekvensomvandlare	Frequenzwandler	convertisseur de fréquence	convertidor de frecuencia
2727	fresh water	friskvatten	Frischwasser, Feinwasser, Süsswasser	eau fraîche	agua dulce
2728	friction	friktion	Friktion, Reibung	friction, frottement	fricción
2729	friction barker	friktionsbarkare	Reibungsentrinder	écorceuse à friction	descortezadora de fricción
2730	friction board	friktionsgättad kartong	Lagerkarton, gewalzte Graupappe	carton laminé dur pour poulies	cartón de fricción
2731	friction calender	friktionskalander	Friktionskalander	calandre à friction	calandra de fricción
2732	friction drive	friktionsdrift	Friktionsantrieb, Gleitantrieb	commande par friction	accionamiento por fricción
2733	friction factor	friktionsfaktor	Friktionsfaktor	facteur de frottement	coeficiente de fricción
2734	friction glazed	friktionsglättad	Friktionsglättung	glacé sur calandre à friction	satinado sobre calandra por fricción

	ENGLISH	SVENSKA	DEUTSCH	FRANÇAIS	ESPAÑOL
2735	friction loss	friktionsförlust	Reibungsverlust	perte par frottement	pérdida de carga por fricción
2736	front edge	framkant	Vorderkante	rebord antérieur	borde delantero
2737	front side	framsida, förarsida	Vorderseite	côté conducteur	lado conductor
2738	froth	skum	Schäumen, Schaum, Abschaum	mousse, écume	espuma
2739	froth spots	skumfläckar	Schaumflecken	taches de mousse	marcas de espuma
2740	frozen-food carton	livsmedelskartong	Tiefkühlkarton	carton pour produits alimentaires congelés	envases para alimentos congelados
2741	frozen wood	frusen ved	gefrorenes Holz	bois gelé	madera helada
2742	fruit-wrap tissues	fruktomslagspapper	Obsteinwickelpapier	mousseline à fruits	muselina para embalaje de frutas
2743	fuel	bränsle	Feuerung, Brennmaterial, Treibstoff	combustible	fuel
2744	fuel oil	brännolja	Heizöl, Schweröl	fuel	petróleo
2745	full-chemical pulp	helkemisk massa	Vollzellstoff	pâte entièrement chimique	pasta enteramente química
2746	fuller's earth	fullers jord	Fullererde, Bleicherde	terre à foulon	arcilla de batán
2747	full size	hellim	Planbogen, ungefalztes Papier	grandeur nature	tamaño natural
2748	full-tree logging	helträdfångst	Gesamtbaumschlag	abattage d'arbres entiers	aprovechamiento forestal completo
2749	fumaric acid	fumarsyra	Fumarsäure	acide fumarique	ácido fumárico
2750	fumes	rökgaser	Rauch, Dunst, Dampf	fumées	vapores
2751	functional groups	funktionella grupper	funktionelle Gruppe	groupes fonctionnels	grupos funcionales
2752	fungi	svamp	Schwämme, Pilze	champignons	hongos
2753	fungicides	fungicider	Schleimbekämpfungsmittel	fongicides	fungicidas
2754	fungus proofing	svampsäkring	Pilzprüfung	protection contre les champignons	tratamiento a prueba de hongos
2755	fungus resistance	rötmotstånd	Schleimwiderstandsfähigkeit, pilztötend	résistance aux champignons	resistencia a los hongos
2756	furans	furaner	Furan	furans	furfuranos
2757	furfural	furfurol	Furfurol	furfurol	furfurol
2758	furfurane	furfuran	Furfuran	furfurane	furfurano
2759	furfuryl alcohol	furfurylalkohol	Furfurylalkohol	alcool de furfuryl	alcohol furfuril
2760	furnace	ugn	Ofen, Brennofen, Schmelzofen	four, fourneau	horno
2761	furnish	sats	Stoffeintrag	quantité chargée, composition de fabrication	composición
2762	fuzz	damm	fusseln, stauben, Fussel, feiner Staub	duvet, peluche	pelusa
2763	fuzzability	damningsbarhet	Staubungsneigung	tendance au peluchage	aptitud a la pelusa
2764	fuzziness	damningsbenägenhet	Stauben (von Papier)	peluchage	pelusa

G

	ENGLISH	SVENSKA	DEUTSCH	FRANÇAIS	ESPAÑOL
2765	GE brightness	G.E.-ljushet	General Electric (G.E.) Helligkeitsmessung	blancheur	blancura
2766	gpd (gallons per day)	g.p.d.	Gallonen pro Tag	mesure de capacité = gallons par jour	galones por día
2767	g/m2 (grams per square meter)	g/m2	grams pro quadrat meter (g/m2)	grammes/m2	gramos por metro cuadrado
2768	gable-top carton	gavelspetskartong	Spitzverschluss	carton à fermature pignon	cartonaje con tapas de faldón
2769	gage (gauge)	mätare, mått	messen, eichen, Messgerät, Masstab	jauge, épaisseur	calibrador
2770	galactans	galaktaner	Galaktan	galactans	galactanos
2771	galactomannan	galaktomannan	Galaktomannan	galactomannan	galactómano
2772	galactose	galaktos	Galaktose	galactose	galactosa
2773	gallon	gallon	Gallone	unité de mesure pour liquides	galón
2774	galvanic corrosion	galvanisk korrosion	galvanische Korrosion	corrosion galvanique	corrosión galvánica
2775	galvanize	galvanisera	galvanisieren	galvaniser	galvanizar
2776	galvanizing	galvanisering	Galvanisierung	galvaniser	galvanización
2777	galvanometer	galvanometer	Galvanometer	galvanomètre	galvanómetro
2778	gamma cellulose	gammacellulosa	Gamma-Cellulose	gamma cellulose	gamma celulosa
2779	garland (Brit.)	krans	Girlande	guirlande	guirnalda
2780	gas	gas, bensin	Gas	gaz	gas
2781	gas analysis	gasanalys	Gasanalyse	analyse gazeuse	análisis de gas
2782	gas chromatography	gaskromatografi	Gaschromatographie	chromatographie des gaz	cromatografía de gases
2783	gas discharge	gasurladdning	Gasentladung, Gasausträmung	évacuation des gaz	descarga gaseosa
2784	gaseous	gasformig, gas-	gasförmig, gasartig	gazeux	gaseoso
2785	gas generator	gasgenerator	Gaserzeuger, Gasgenerator	générateur de gaz	generador de gases
2786	gas heater	gasvärmare	Gasheizung	réchauffeur de gaz	calentador de gas

	ENGLISH	SVENSKA	DEUTSCH	FRANÇAIS	ESPAÑOL
2787	gasification	förgasning	Gasbildung, Vergasung	gazéification	gasificación
2788	gasket	packning	Dichtungsring, Dichtung	joint, épissure	junta
2789	gasket board	packningspapp	Dichtungspappe	carton pour joints	cartón para empaquetaduras
2790	gasket paper	packningspapper	Dichtungspapier	papier pour joints	papel para empaquetaduras
2791	gas permeability	gaspermeabilitet	Gasdurchlässigkeit	perméabilité aux gaz	permeabilidad al gas
2792	gas scrubber	gasskrubber	Gaswäscher	laveur de gaz	depuradora de gas
2793	gas scrubbing	gasskrubbning	Gaswäsche	lavage des gaz	depuración de gas
2794	gas turbine engine	gasturbinmotor	Gasturbine	turbine à gaz	turbina a gas
2795	gate	grind, slussventil	Schieber, Schleusentor	porte, vannette	puerta
2796	gate-roll coater	slussvalsbestrykare	Gate-Roll-Streichanlage, Gate-Roll-Coater	coucheuse à rouleau barreur	estucadora con rodillo regulador
2797	gate valve	slussventil	Absperrschieber, Schieberventil	vanne à passage direct	válvula de compuerta
2798	gauge	mått, mätare	messen, eichen, Messgerät, Masstab	jauge, indicateur	indicador, calibrador
2799	gear	växel	antreiben, in Betrieb setzen, ineinandergreifen, Getriebe, Antrieb	engrenage	engranaje
2800	gearbox	växellåda	Getriebekasten, Getriebegehäuse, Wechselgetriebe	boîte d'engrenages	caja de engranajes
2801	gel	gel	Gel	gel (suspension colloïdale)	gel
2802	gelatin	gelatin	Gelatine, Gallerte, Knochenleim	gélatine	gelatina
2803	gelatinized	gelatinerad	geliert, gelatiniert	gélatinisé	gelatinizado
2804	gelatinous layer	gelatinliknande skikt	gelatineartige Schicht	couche gélatineuse	capa gelatinosa
2805	gelation	gelbildning	Gelatinierung, Gelierung	gélification	gelación
2806	general manager	direktör	Generaldirektor	directeur général	director gerente
2807	general superintendent	inspektör	Betriebsleiter, Fabrikationsleiter	directeur de la production	director de producción
2808	generator	generator	Generator, Stromerzeuger, Gleichstromerzeuger	générateur	generador
2809	genetics	genetik	Genetik, Vererbungslehre	génétique	genética
2810	germination	groning	Keimen, Spriessen	germination	germinación
2811	gift-wrap paper	presentomslagspapper	Geschenkeinwickelpapier	papier d'emballage pour cadeaux	papel envoltura para regalos
2812	girdling	bläckning	umgürten	ceinturage, annelation circulaire	cortar un círculo alrededor del tronco
2813	glare	glans	hell leuchten, glänzen, strahlen	reflet, éclat, brillant	brillo
2814	glass fiber paper	glasfiberpapper	Glasfaserpapier	papier de fibres de verre	papel de fibra de vidrio
2815	glass fiber	glasfiber	Glasfaser	fibre de verre	fibra de vidrio
2816	glassine	pergamyn	Glassin, Pergamin	cristal	cristal
2817	glassine paper	pergamynpapper	Pergaminpapier	papier cristal	papel cristal
2818	glaze	glätta	glasieren, glätten, satinieren, Glasur	lustre	lustre
2819	glazed	glättad	satiniert, geglättet	glacé, satiné	satinado
2820	glazed board	glättad kartong	Glanzkarton, Satinierkarton	carton laminé, carton lissé	cartón lustrado
2821	glazed coated book paper	glättbestruket bokpapper	glanzgestrichenes Buchpapier	papier d'édition couché	papel estucado y satinado para libros
2822	glazed paper	glättat papper	satiniertes Papier, Glanzpapier	papier satiné	papel satinado
2823	glazing	glättning	Satinage, Satinieren, Glänzendmachen	calandrage, lissage, glaçage	abrillantamiento
2824	glazing cylinder	glättcylinder	Glättzylinder	cylindre frictionneur	cilindro de satinado
2825	globe digester	kulkokare	Kugelkocher	lessiveur sphérique	lejiadora esférica
2826	gloss	glans	Glanz, Politur, glänzen, polieren, glätten	brillant, lustre	brillo
2827	gloss agent	glansmedel	Glanzmittel, Politurmittel	pouvoir lustrant	agente abrillantador
2828	gloss calender	glättpress	Heissglanzpresse	calandre de satinage	calandra de satinado
2829	gloss calendering	glättpressning	heisskalandrieren	satinage sur calandre	calandrado de satinado
2830	gloss meter	glansmätare	Glanzmesser	glarimètre	satinómetro
2831	gloss paper	glanspapper	Glanzpapier	papier satiné sur calandre	papel brillo
2832	glucomannan	glukomannan	Glukomannan	glucomannan	glucómano
2833	glucose	glukos	Glukose	glucose	glucosa
2834	glucose derivatives	glukosderivat	Glukosederivate	dérivés du glucose	derivados de glucosa
2835	glucose esters	glukosestrar	Glukoseester	esters du glucose	ésteres de glucosa
2836	glucose ethers	glukosetrar	Glukoseäther	éthers du glucose	éteres de glucosa
2837	glucose unit	glukosenhet	Glukoseeinheit	unité de glucose	unidad de glucosa
2838	glucosides	giukosider	Glukoside	glucosides	unión glucósida
2839	glucosidic bond	glukosbindning	Glukosebindung	liaison glucosique	cola
2840	glue	lim	Leim, Klebstoff	colle forte, gélatine	aplicador de cola
2841	glue applicator	limpåläggare	Leimauftrag, Leimauftragsaggregat	encolleuse	probador de adherencia de la cola
2842	glue-bond tester	limfasthetsprovare	Klebeverbindung-Prüfgerät	appareil de mesure de la résistance des joints de collage	
2843	glue-coated paper	limbestruket papper	Papier mit leimhaltigem Strich, Leimstrichpapier	papier couché à la colle	papel estucado encolado
2844	glued joints	limfogar	Anklebestellen	joints de collage	juntas encoladas

	ENGLISH	SVENSKA	DEUTSCH	FRANÇAIS	ESPAÑOL
2845	glue failure	limbrott	Versagen des Leims, Nichthaftung des Leims	défaut de collage	fallo de la cola
2846	glue line	limstreck	Leimauftrag, aufgetragene Leimfläche	ligne de collage	línea de cola
2847	gluer	limmare	Kleber, Klebemaschine	encolleuse	encoladora
2848	glue spreading	limspridning	Leimverteilung	application de la colle	aplicación de la cola
2849	gluing	limning	Leimung, Leimen, Kleben	encoller	encolado
2850	glycerine	glycerin	Glyzerin	glycérine	glicerina
2851	glycerol	glycerol	Glyzerinlösung, Glyzerin	glycérol	glicerol
2852	glycerol stearates	glycerolstearater	Glyzerinstearate	stéarates de glycérol	glicerol estearatos
2853	glycols	glykoler	Glykole	glycols	glicoles
2854	glyoxal	glyoxal	Glyoxal	glyoxal	glioxal
2855	governor	ledare	Regler, Regulator	régulateur	regulador
2856	grab sample	gripprov	Exkavatorprobe	échantillon pour essai, analyse pris au hasard	muestra fortuita
2857	grade	sort, grad	Qualität, Sorte	sorte, degré, qualité	clase de papel
2858	gradient	gradient	Gradient, Neigung	pente	inclinación
2859	grading	gradering	Einstufen, Sortieren, Klassifizieren	classage	clasificación
2860	graft	ympning	aufpfropfen, pfropfen	greffe	injertar
2861	graft copolymers	ymppolymerer	Pfropfcopolymere	copolymères greffés	injertos copolimeros
2862	grain	korn, fiber	Korn, Körnigkeit, Maserung, Struktur, Gefüge	grain (du papier)	granulación
2863	grain direction	fiberriktning	Faserrichtung	sens machine	dirección de la granulación superficial
2864	grained paper	anisotropt papper	genarbtes Papier, gemasertes Papier	papier grainé, papier gaufré	papel granulado
2865	graininess	anisotropi	Körnung	granulation	granulación
2866	grainy	grynig, fransig	körnig	grenu	granoso
2867	grainy edge	fransig kant	Oberflächenstruktur, von der Kante aus in das Papier (die Bahn) hineinverlaufend	bord grainé	borde granoso
2868	grainy paper	kornigt papper	narbiges Papier	papier grainé	papel granulado
2869	gram	gram	Gramm	gramme	gramo
2870	grand fir	grand fir	Küstentanne	sapin élancé de Vancouver	abeto grandis
2871	granite	granit	Granit	granit	granito
2872	granite paper	granitpapper	scheckiges Papier, graumeliertes Papier	papier granité	papel granito
2873	granite roll	granitvals	Granitwalze, Steinwalze	presse en pierre	rodillo de granito
2874	granule	granul, flinga	Körnchen	granule	gránulo
2875	graphic arts	grafisk konst	graphische Kunst	arts graphiques	artes gráficas
2876	grapple	grepp, gripare	festhalten, packen, Greifer, Greifzange	sorte de pince pour saisir les bobines	asir
2877	grapple skidder	lunnare med gripare	Greif-Skidder	débusqueur à pinces	galga aferradora
2878	grate	galler	raspeln, zerreiben, Rost	grille,grillage	rejilla
2879	gravity conveyor	gravitationstransportör	Rollenbahn	système de transport par gravité	transportador de gravedad
2880	gravure coating	gravyrbestrykning	Gravurstreichen	couchage par gravure	estucado para grabado
2881	gravure paper	djuptryckspapper	Tiefdruckpapier	papier hélio	papel de fotograbado
2882	gravure printing	rotogravyr	Tiefdruck	impression hélio	impresión de heliograbado
2883	grease	smörja, fett	schmieren, fetten, ölen, Fett, Schmiere	graisse	grasa
2884	greaseproof	fettäkta	fettdicht	ingraissable	impermeable a las grasas
2885	greaseproof board	fettät kartong	fettdichter Karton	carton ingraissable	cartón simili sulfurizado
2886	greaseproof coating	smärpappersbestrykning	fettdichte Beschichtung	couchage imperméable à la graisse	estucado resistente a las grasas
2887	greaseproofing	fettätning	fettdicht machen, fettdichte Ausrüstung	rendre imperméable à la graisse	impermeabilización a la grasa
2888	greaseproofness	fettäkthet	Fettdichtigkeit	imperméabilité à la graisse	impermeabilidad a la grasa
2889	greaseproof paper	smörpapper	fettdichtes Papier, Butterverpackpapier	simili-sulfurisé ingraissable	papel simili sulfurizado
2890	grease resistance	fettmotstånd	Fettbeständigkeit, Ölbeständigkeit	résistance à la graisse	resistencia a la grasa
2891	grease-resistant board	fettäkta kartong	fettdichter Karton	carton ingraissable	cartón resistente a la grasa
2892	grease-resistant paper	fettsäktert papper	fettdichtes Papier	papier ingraissable	papel resistente a la grasa
2893	grease spots	fettfläckar	Fettflecken	taches de graisse	manchas de grasa
2894	greasing	smörjning	schmieren, ölen, Fettschmierung	graissage	lubricación
2895	greasy	smörjig, fet	fettig, schmierig	pâte engraissée	grasiento
2896	green	grön	grün, grün färben, grün werden	vert	verde
2897	green liquor	grönlut	Grünlauge	solution de salin récupérée, lessive verte	lejía verde
2898	green wood	färskträ	Sommerholz, Frühholz	bois vert	madera verde
2899	greeting card	gratulationskort	Glückwunschkarte	carte de voeux	tarjeta de felicitación
2900	grinder	slipstol	Schleifer, Schleifmaschine	défibreur	desfibradora

	ENGLISH	SVENSKA	DEUTSCH	FRANÇAIS	ESPAÑOL
2901	grinderman	slipare	Schleiferführer	ouvrier défibreur	operador de desfibradora
2902	grinder pit	sliptråg	Schleifertrog	fosse sous le défibreur	tina de desfibradora
2903	grinder pocket	slipficka	Schleifmagazine	défibreur à chambres	prensa de desfibradora
2904	grinder room	sliperi	Schleiferei	salle des défibreurs	sala de desfibradoras
2905	grinder stone	slipsten	Schleifstein	meule de défibreur	muela de desfibradora
2906	grinding	slipning	Schleifen, Mahlen	défibrage	desfibrado
2907	grinding surface	slipyta	Schleifoberfläche	surface de défibrage	superficie de desfibrado
2908	grindstone	slipsten	Schleifstein	meule	muela de desfibradora
2909	grindstone dresser	slipstens	Schleifstein-Schärfvorrichtung, Schärfrolle	mollette pour rhabiller les meules	moleta
2910	gripper	hållare	Greifer, Halter	appareil de préhension	arrastrador
2911	grit	gruskorn	Kies, grobkörniger Sandstein	grain de sable, gravier	arenilla
2912	groove	spår, rilla, räffla	Nut, Nute, Furche, Rille	gorge, rainure	acanaladura
2913	grooved press	räfflad press	Nutenpresse, Rillenwalzenpresse	presse rainurée	prensa ranurada
2914	grooved roll	räfflad vals	gerillte Walze, profilierte Walze	rouleau rainuré	rodillo ranurado
2915	grooving	räffling	Reifung, Rille, Einschleifen	rainer, canneler	acanalado
2916	groundwood	slipmassa	Holzschliff	pâte mécanique	pasta mecánica
2917	groundwood book paper	slipbokpapper	holzschliffhaltiges Buchpapier	papier d'édition avec bois	papel de pasta mecánica para libros
2918	groundwood free	träfri	holzfrei	sans bois	sin pasta mecánica
2919	groundwood from chips	raffinörmekanisk massa	Hackschnitzelschliff	pâte mécanique de copeaux	pasta mecánica de virutas
2920	groundwood mill	träsliperi	Schleiferei, Holzschleiferei	râperie	fábrica de pasta mecánica
2921	groundwood papers	slipmassepapper	holzschliffhaltige Papiere	papiers avec bois	papel de pasta mecánica
2922	groundwood printing papers	sliptryckpapper	holzschliffhaltige Druckpapiere	papiers d'impression avec bois	papel de pasta mecánica para imprimir
2923	groundwood pulp	slipmassa	Holzschliff	pâte mécanique	pasta mecánica
2924	groundwood pulping	slipmassetillverkning	Holzschliffherstellung, Holzaufschluss	fabrication de pâtes mécaniques	elaboración de pasta mecánica
2925	grout	grums	vergiessen, Fugen verstreichen	mortier liquide	mortero líquido
2926	growth	växt, tillväxt	Wachstum, Zuwachs	croissance, accroissement	crecimiento
2927	growth layer	tillväxtlager	Zuwachszone	couche de croissance	lecho de crecimiento
2928	growth rate	tillväxthastighet	Wachstumsrate, Zuwachsrate	taux de croissance	índice de crecimiento
2929	growth ring	årsring	Jahresring, Wachstumsring, Zuwachsring	anneau de croissance	anillo anual
2930	guard	vakt	bewachen, beschützen, Schutzvorrichtung, Wächter, Aufseher	onglet	cartivana
2931	guar gum	guargum	Guar, Verdickungsmittel für Papierverleimung, Flotationszusatzmittel	gomme de Guar	goma de guar
2932	guide	gejder, förare	lenken, steuern, führen, leiten	guide	guía
2933	guide roll	ledvals	Leitwalze, Siebleitwalze, Regulierwalze	rouleau guideur	rodillo guía
2934	guiding of felt	filtföring	Filzführung	guidage du feutre	conducción del fieltro
2935	guiding of paper	banföring	Papierführung	guidage du papier	conducción del papel
2936	guiding of wire	viraföring	Siebführung	guidage de la toile	conducción de la tela metálica
2937	guillotine	giljotin	Planschneider, Formatschneider	massicot	guillotina
2938	guillotine cutter	giljotinskärare	Planschneider	guillotine	guillotina
2939	guillotine knife	giljotinkniv	Planschneidemesser, Maschinenmesser	lame de massicot	cuchilla de guillotina
2940	gum	gum	Gummi, Gummiharz	gomme, caoutchouc	goma
2941	gum arabic	gummi arabicum	Gummiarabikum	gomme arabique	goma arábiga
2942	gummed paper	klisterpapper	gummiertes Papier	papier collant	papel engomado
2943	gummed tape	klistertejp	Klebestreifen	ruban adhésif	cinta engomada
2944	gumming	klistring	Gummierung, Gumbildung	gommage	engomado
2945	gumming machine	klistringsmaskin	Gummiermaschine, Klebemaschine, Rollengummiermaschine	encolleuse	engomadora
2946	gumming paper	klisterpapper	Gummierrohpapier, gummierfähiges Papier	papier support pour gommage	papel para engomar
2947	gum tragacanth	gummidragkant	Tragant, Tragantgummi	gomme adragante	adragante
2948	gun	kanon	Spritzpistole	bronze	pistola
2949	Gurley tester	Gurleyprovare	Gurley-Dichtemesser	appareil Gurley de mesure de la perméabilité à l'air	probador Gurley
2950	gusset	triangelformad plåt	Winkelblech, Anschlussblech, Keil, Seitenfalte einer Tragetasche	soufflet (sac)	fuelle
2951	gymnosperms	gymnosperm	nacktsamige Pflanze	gymnospermes	gimnospermo
2952	gypsum board	gipsplatta	Gipspappe, Gipsbaupappe	carton plâtre	cartón de yeso

PINNACLE PRIME.™

Westvaco's flash-dried hardwood pulp that's got everything you want in a pulp.

If you need a pulp that's extra-clean and strong and bright and uniform,
you need Westvaco Pinnacle Prime 100% hardwood pulp. It's being used throughout
the world to make everything from tissue to coated and uncoated printing and converting
papers. And it's doing it without the problems associated with other flash-dried pulps.

That's because Westvaco has a fully computerized operation. With stringent
temperature and moisture controls. A Kamyr continuous digester for uniformity.
C.E.D.E.D. bleaching for high brightness (88+ G.E.) with minimum fiber degradation.
And tertiary stage cleaning for a high degree of cleanliness.

But you don't have to take our word for it. Find out for yourself. Ask for
our Valley Beater Curves. Get a free 5-pound sample for your lab analysis. Or, let
us assist you with your specific applications.

If you want unbleached softwood pulp, we supply that, too. Westvaco's
high-yield unbleached pulp is especially suited for economical linerboard production.
And, it is available all year round from our warmwater port in Charleston,
South Carolina. The loading dock is adjacent to our mill.

For more information on either Pinnacle Prime bleached hardwood pulp or
unbleached softwood pulp, contact our New York Office or our offices in Brussels,
Belgium, or Sydney, Australia.

New York: Westvaco, Woodpulp and Export Sales, 299 Park Avenue,
New York, New York 10017, Tel: (212) 688-5000, Telex: 234960

Belgium: Westvaco Europe, S.A., 296 D Avenue de Tervueren,
Brussels 15, Belgium, Telex: 23174

Australia: Westvaco Pacific Pty, Ltd., P.O. 273, Milsons Point, N.S.W.,
Australia, 2106, Telex: 21109

Westvāco

69

ENGLISH	SVENSKA	DEUTSCH	FRANÇAIS	ESPAÑOL

H

	ENGLISH	SVENSKA	DEUTSCH	FRANÇAIS	ESPAÑOL
2953	HV-dryer (high-velocity air dryer)	höghastighetstork	Hochleistungstrockner	sécheur à air à grande vitesse	sequeria por aire de alta velocidad
2954	half stuff	halvtyg	Halbzeug, Papierhalbstoff	demi-pâte	semipasta
2955	halides	halider	Halogenid	halogénures	sales haloideas
2956	halogen compounds	halogenföreningar	Halogenverbindungen	composés d'halogène	compuestos halógenos
2957	halogens	halogener	Halogene, Salzbildner	halogènes	halógenos
2958	hammer	hammare	Hammer	marteau	martillo
2959	hammer mill	hammarkvarn	Hammermühle	broyeur à marteaux	molino de martillos
2960	hammer milling	hammarkvarnuppslagning	in der Hammermühle mahlen	broyage	trituración por martillos
2961	handle	handtag, hantera	behandeln, umgehen mit, abwickeln, bedienen, bearbeiten, Griff, Stiel, Henkel, Griffigkeit	poignée, manette	manilla
2962	handmade felt	handgjord filt	handgenähter Filz	feutre fait à la main	fieltro a la forma
2963	handmade finish	manuell finish	geprägte Oberfläche (die dem Papier ein handgeschöpftes Aussehen gibt)	papier fini à la main	acabado a mano
2964	handmade paper	handgjort papper	handgeschöpftes Papier, Büttenpapier	papier à la main	papel a la tina
2965	hand mold	form	Handform	forme à main	forma a mano
2966	handsheet	handgjort pappersark	Handmuster, Probestück (Papier)	feuille faite à la main	hoja de registro
2967	handsheet former	form	Blattbildner für Handmuster	forme pour papiers destinés aux essais de pâtes	plantilla de hojas-registro
2968	hanger	hängare	Aufhängevorrichtung, Hängelager	crochet	soporte
2969	hanging	hängning	Aufhängen, Hängen	papier support tenture	colgamiento
2970	hanging paper	hängtorkad papper	Tapete	papier de tenture	papel pintado
2971	hanging raw stock	handpappersmäld	Tapetenrohpapier	matières premières pour support tenture	soporte para papel pintado
2972	hard beating	rysk malning	stark ausmahlen, trocken mahlen	raffinage poussé	engrasamiento fuerte
2973	hardboard	fiberskiva	Hartpappe, Hartfaserplatte	carton dur	cartón piedra
2974	hard cook	hårt kok	Hartkochung (Kochen bis zu einem geringen Aufschlussgrad oder Ligninabbau)	cuisson poussée	cocción a fondo
2975	hardener	härdare	Härter, Härtemittel	durcisseur	endurecedor
2976	hardening	härdning	Härten, Härtung, Erhärtung	trempe, durcissement	endurecimiento
2977	hardiness	hårdhet	Widerstandsfähigkeit, Unempfindlichkeit, Ausdauer	dureté	robustez
2978	hard lump	hård knut	harter Stoffbatzen	pâton dur	grumo duro
2979	hardness	hårdhet	Härte (z.B. Rockwell-Härte oder Shore-Härte)	dureté	dureza
2980	hardness gauge	hårdhetsprovare	Härtemessung	échelle de dureté	indicador de dureza
2981	hardness tester	hårdhetsprovare	Härteprüfer	duromètre Bekk	durómetro
2982	hardness tests	hårdhetsprov	Härteprüfung	essais de dureté	pruebas de dureza
2983	hard pulp	hård massa	hartgekochter Zellstoff	pâte dure	pasta dura
2984	hard sized	hårdlimmat	vollgeleimt	fortement collé	encolado fuerte
2985	hard-sized paper	hårdlimmat papper	vollgeleimtes Papier	papier fortement collé	papel fuertemente encolado
2986	hard stock	hårdmald massa	Hadernzellstoff (Zellstoff aus Lumpen oder Hadern oder Jute)	pâte de chiffons, pâte de cordages, pâte de jute	pasta dura
2987	hardwood	lövträ	Laubholz	bois de feuillu	madera de frondosas
2988	hardwood pulp	lövträmassa	Laubholzzellstoff	pâte de feuillus	pasta de madera de frondosas
2989	Harper machine	Harper-maskin	Harper-Maschine (Langsiebmaschine mit Selbstabnahme)	machine à papier à table inversée	máquina de papel de mesa invertida
2990	harvester	skördemaskin	Erntemaschine, Mähmaschine	moissonneur	segadora
2991	harvesting	skörd	ernten	moisson	siega
2992	haze	dis	Dunstschleier, Trübung, feiner Nebel	vapeur	niebla
2993	head	huvud, trycknivå	Kopf, Deckel, Haube, Säule, Druckhöhe, Stoffstand, Stoffstauhöhe	tête, hauteur de charge	carga
2994	headbox	inloppslåda	Stoffauflauf	caisse d'arrivée de pâte	caja de entrada
2995	header	huvudledning	Stutzen, Stirnseite, Kantenschutz	collecteur	colector
2996	header pipe	huvudledning	Verteilerrohr, Rohrverbinder	tuyau collecteur	tubo colector
2997	heading paper	täckpapper, lockpapper	–	papier en-tête	papel de avance
2998	head loss	tryckförlust	Verlust an Stoffstandshöhe, Gefälleverlust	perte de charge	pérdida de carga

ENGLISH	SVENSKA	DEUTSCH	FRANÇAIS	ESPAÑOL
2999 heartwood	kärnved	Kernholz	bois de coeur	madera de corazón
3000 heat	värme	heizen, erhitzen, heiss werden, Wärme, Hitze	chaleur	calor
3001 heat accumulator	värmeackumulator	Wärmespeicher	accumulateur de chaleur	acumulador térmico
3002 heat balance	värmebalans	Wärmeausgleich, Wärmebilanz, Wärmehaushalt	bilan calorifique	equilibrio calorífico
3003 heat conductivity	värmeledningsförmåga	Wärmeleitfähigkeit	conductivité calorifique	conductividad térmica
3004 heat comsumption	värmeförbrukning	Wärmeverbrauch	consommation de chaleur	consumo térmico
3005 heat efficiency	värmeverkningsgrad	Wärmeausnutzung	pouvoir calorifique	rendimiento calórico
3006 heat exchanger	värmeväxlare	Wärmeaustauscher	échangeur de chaleur	intercambiador de calor
3007 heat gradient	temperaturgradient	Wärmegefälle	chaleur progressive	pendiente calórica
3008 heating	värmning, upphettning	Heizung, Beheizung	chauffage	calefacción
3009 heating equipment	uppvärmningsanordning	Heizgerät	installation de chauffage	equipo calefactor
3010 heat insulation	värmeisolation	Wärmeisolierung	isolation calorifique	aislamiento térmico
3011 heat loss	värmeförlust	Wärmeverlust	déperdition de chaleur	pérdida de calor
3012 heat of combustion	förbränningsvärme	Verbrennungswärme	chaleur de combustion	calor de combustión
3013 heat recovery	värmeåtervinning	Wärmerückgewinnung	récupération de chaleur	recuperación del calor
3014 heat resistance	värmemotstånd	Hitzebeständigkeit, Wärmefestigkeit, Erhitzungswiderstand	résistance à la chaleur	resistencia al calor
3015 heat seal strength	värmeförseglingshållfasthet	Festigkeit der Heissiegelung	résistance au thermocollage	resistencia del sellado térmico
3016 heat sealing	värmeförsegling	Heissiegelung, Heissverkleben, Heissverschweissen	thermocollable	sellado térmico
3017 heat-sealing machine	värmeförseglingsmaskin	Heissiegelmaschine	machine à thermocoller	selladora térmica
3018 heat-sealing paper	värmeförseglingspapper	Heissiegelpapier	papier thermocollable	papel para sellado térmico
3019 heat-set ink	värmehärdande tryckfärg	heisstrocknende Farbe, für Heisstrocknung geeignete Farbe	encre fixable par la chaleur	tinta fijada por calor
3020 heat setting	värmehärdning	hitzehärtbar, in der Hitze abbindend	réglage de la chaleur	fraguado al fuego
3021 heat transfer	värmeöverföring	Wärmeübertragung, Wärmeübergang	échange thermique	traspaso de calor
3022 heat-transfer coefficient	värmeöverföringskoefficient	Wärmeleitzahl, Wärmeübergangszahl	coefficient d'échange thermique	coeficiente de transferencia térmica
3023 heat transmission	värmeledning	Wärmeabführung, Wärmeübertragung	transmission de chaleur	transmisión de calor
3024 heat treatment	värmebehandling	thermische Behandlung, Wärmevergütung, Wärmebehandlung	traitement thermique	tratamiento térmico
3025 heavy weights	tungvikt	schwergewichtige Papiere	poids lourds	pesos pesados
3026 height	höjd	Höhe, Wellenhöhe (Wellpappe), Höhepunkt, lichte Höhe (Konstr.)	hauteur, flèche	altura
3027 helical	spiralformig, spiral-	schneckenformig, schraubenförmig, schrägverzahnt (mech.)	hélicoïdal	helicoidal
3028 helper drive	hjälpdrift	Hilfsantrieb	commande auxiliaire	mando auxiliar
3029 hemicelluloses	hemicellulosa	Hemicellulose	hémicelluloses	hemicelulosas
3030 hemlock	hemlock	Schierling, Schierlingstanne	sapin du Canada	abeto del Canadá
3031 hemp	hampa	Hanf	chanvre	cáñamo
3032 hemp paper	hamppapper	Hanfpapier, Manilapapier	papier à base de chanvre	papel de manila
3033 heritability	ärftlighet	Vererbbarkeit, Erblichkeit	capabilité d'hériter	heredable
3034 heterocyclic compounds	heterocyklisk förening	heterozyklische Kohlenstoffverbindungen	composés hétérocycliques	compuestos heterocíclicos
3035 heterogeneity	heterogenitet	Ungleichförmigkeit, Heterogenität	hétérogènéité	heterogeneidad
3036 hickies	böjdon	kleine ringförmige Flecken in gedrucktem Material	puces	manguito sujetador
3037 hiding power	opacitet	Deckkraft, Deckvermögen	pouvoir couvrant	capacidad encubridora
3038 high-bulk book paper	högbulkigt bokpapper	voluminöses Buchdruckpapier	papier d'édition très bouffant	papel para libros en fardos grandes
3039 high-bulking	bulkning	stark auftragend	–	enfardado grande
3040 high-bulk paper	högbulkigt papper	voluminöses Papier	papier très bouffant	papel en fardos grandes
3041 high consistency	hög konsistens	hohe Stoffdichte	concentration très élevée	elevada consistencia
3042 high density	hög densitet	hohe Dichte	densité très élevée	alta densidad
3043 high-density bleaching	högkonsistensblekning	Dickstoffbleiche	blanchiment à haute densité	blanqueado de alta densidad
3044 high-density refining	högkonsistensmalning	Mahlen bei hoher Stoffdichte	raffinage à haute densité	refinado de alta densidad
3045 high energy	hög energi	energiereich	haute énergie	gran potencia
3046 high glaze	högglans	scharf satiniert, hoch satiniert	très lustré	muy brillante
3047 high-gloss paper	högglanspapper	Hochglanzpapier	papier très glacé	papel muy abrillantado
3048 high-intensity press	högtryckspress	Presse mit hohem spezifischem Druck	presse à haute intensité	prensa de gran intensidad
3049 high-lead logging	högkabellunning	Holzeinschlag im Kopfhochverfahren	débardage par câble (système "high-lead")	arrastre de troncos por cable aéreo
3050 high machine finish	hög maskinfinish	hohe Maschinenglätte	surcalandré	acabado especial en máquina
3051 high-modulus rayon	högmodulsrayon	hochelastische Viskosefaser	rayonne à haute ténacité	rayón de módulo grande

ENGLISH	SVENSKA	DEUTSCH	FRANÇAIS	ESPAÑOL
3052 high pressure	högtryck	Hochdruck	haute pression	alta presión
3053 high-pressure feeder	högtrycksmatare	Zufuhr unter Hochdruck, Zufuhr unter Pressluft	appareil d'alimentation à haute pression	alimentador de alta presión
3054 high temperature	hög temperatur	warmfest, Hochtemperatur	température élevée	temperatura elevada
3055 high-test liner	high-testliner	Beklebekarton mit besonders hoher Berstfestigkeit	kraftliner	forro a toda prueba
3056 high velocity	hög hastighet	schnellaufend, hochtourig, mit hoher Geschwindigkeit, Hochleistungs-	grande vitesse	alta velocidad
3057 high-velocity drying	höghastighetstorkning	Hochleistungstrocknung	séchage à grande vitesse	secado rápido
3058 high-velocity dryer	höghastighetstork	Hochleistungstrockner, Düsenstrahltrockner, Zwangskonvektionstrockner	cylindre sécheur à grande vitesse	secadora de alta velocidad
3059 high yield	högt utbyte	hohe Ausbeute, hoher Ertrag, ertragreich	haut rendement	elevado rendimiento
3060 high-yield pulp	högutbytesmassa	Hochausbeutezellstoff	pâte à haut rendement	pasta de elevado rendimiento
3061 high-yield pulping	högutbyteskokning	Zellstoffaufschluss mit hoher Ausbeute	fabrication de pâtes à haut rendement	fabricación de pasta de gran rendimiento
3062 hinged-lid container	högbehållare	Behälter mit Scharnierdeckel	récipient avec couvercle à charnière	contenedor con tapas articuladas
3063 hog	korthugga	nach oben krümmen, Jährlingswolle, Rührer, Rührwerk	agitateur	trituradora de madera
3064 hogged fuel	hackved	Brennstoff aus Holzabfällen	déchets de bois écrasé (combustible)	combustible de leña
3065 hog knife	huggkniv	Rührwerksmesser	coupeuse, déchiqueteuse	cuchilla chata
3066 hoist	hissa, höja	hochziehen, hochwinden, heben, Aufzug, Hebewerk, Hebezeug	appareil de levage	torno elevador
3067 holder	hållare	Halter, Halterung, Fassung, Behälter, Inhaber, Besitzer	monture, support	armadura
3068 holding box	sugkammare	Saugkammer an der Presse	–	sifón aspirante de prensa
3069 holding-down roll	nedtryckningsvals	Zugwalze, Andruckrolle	rouleau presseur	rodillo sujetador
3070 holding time	fördröjning	Verweilzeit	temps de prise	tiempo de ocupación
3071 holdout	uthållning	Stand der Farbe, Ausgiebigkeit der Farbe	retenue d'encre (résistance du papier à la pénétration de l'encre)	aguante
3072 hole	hål	lochen, durchlöchern, ein Loch bohren, Loch, Bohrung	trou, orifice	orificio
3073 hole size	hålstorlek	Öffnungsweite, Öffnungsgrösse	dimension d'un orifice	tamaño del orificio
3074 hollander	holländare	Holländer, Stoffmühle	pile défileuse	pila holandesa
3075 hollander beater	holländare	Mahlholländer	pile raffineuse	pila holandesa
3076 holocellulose	holocellulosa	Holocellulose	holocellulose	holocelulosa
3077 holopulping	holocellulosatillverkning	Holoaufschluss (nur Lignin und äusseres Material wird von dem Pflanzengewebe entfernt)	cuisson au chlorite	holodesfibrado
3078 holy roll	hålvals	Lochwalze, Wenzelwalze (Stoffauflauf)	rouleau perforé (caisse d'arrivée)	rodillo perforado
3079 homogeneity	homogenitet	Homogenität, Gleichartigkeit	homogénéité	homogeneidad
3080 homogenization	homogenisering	Homogenisierung, Homogenisation	homogénéisation	homogeneización
3081 homogenizer	homogeniseringsmedel	Homogenisiermaschine	dispositif d'homogénéisation	homogeneizador
3082 honeycomb core	bikakebas	zellenartiger Kern, Wabenkern, wabenförmige Mittellage	mandrin en nid d'abeilles	cono de panal
3083 honeycomb structure	bikakestruktur	zellenartige Struktur	structure en nid d'abeilles	estructura de panal
3084 hood	huv	Haube, Abzug	hotte	campana
3085 hood wrap	huvtäckning	Grad der Haubenabdeckung	–	cubrimiento de la caperuza
3086 hopper	tratt	Trichter, Fülltrichter	trémie	tolva
3087 horizontal porosity	horisontell porositet	Porosität quer über die Bahn gemessen	porosité horizontale	porosidad horizontal
3088 hornbeam	avenbok	Hornbaum, Weissbuche, Hagebuche	charme	haya blanca
3089 horse chestnut	hästkastanj	Rosskastanie	marronnier d'Inde	castaño de indias
3090 horsepower	hästkraft	Pferdestärke (PS)	cheval-vapeur	caballo de fuerza
3091 hose	slang	Schlauch, Wasserschlauch, Luftschlauch	tuyau flexible	manguera
3092 hot alkali extraction	hetalkaliextraktion	Heissalkaliextraktion	extraction de l'alcali à chaud	extracción del álcali en caliente
3093 hot grinding	varmslipning	heisschleifen, heissmahlen	défibrage à chaud	desfibrado en caliente
3094 hot melt	smälta	Heissschmelze	matière thermoplastique	fundición en caliente
3095 hot-melt adhesive	smältbindemedel	Heissschmelzleim	adhésif à base de matière thermoplastique	adhesivo fundido en caliente
3096 hot-melt coating	strängsprutning	Aufschmelzüberzug, Heissschmelzstreichen	couchage par collage à chaud	estucado por fundido en caliente
3097 hot pressing	varmpressning	Heisspressen, Heisssatinieren, Heissglätten	satinage à chaud	prensado en caliente

Wood and Chip Handling Systems
Pulp Mill Systems
Paper Mill Stock Preparation Systems
Paper and Board Machines

Coating Machines and Systems
Converting Machinery
Plastics Processing Systems
Solid State Drive Systems

Black Clawson

	ENGLISH	SVENSKA	DEUTSCH	FRANÇAIS	ESPAÑOL
3098	hot refining	kokarraffinering	Heissmahlung	raffinage à chaud	refinado en caliente
3099	hot roll	varm vals	warmwalzen, satinieren (mittels beheizter Kalanderwalzen)	cylindre chauffé à la vapeur	rodillo caliente
3100	hot stock treatment	varmmassabehandling	Heissschliffbehandlung, Heissmahlung, Heissbehandlung des Stoffs	traitement de la pâte à chaud	tratamiento de la pasta en caliente
3101	hot well	kondens	Kondensatsammler	fosse pour eau condensée	colector de condensamiento
3102	housing	hus	Gehäuse, Unterbringung	carter	cárter
3103	hue	nyans	färben, tönen, Tönung, Farbton, Farbabstufung	nuance, teinte	matiz
3104	humectant	humektant	Befeuchter, Anfeuchter, Benetzungsmittel	humectant	humectante
3105	humidification	befuktning	Anfeuchtung, Befeuchtung	humidification	humidificación
3106	humidifier	befuktningsmedel	Anfeuchter, Befeuchtungsapparat	humecteuse	humectadora
3107	humidity	fuktighet	Feuchte, Feuchtigkeitsgehalt, relative Feuchtigkeit der Luft	humidité	humedad
3108	humidity control	fuktighetskontroll	Feuchtigkeitsregulierung, Feuchtigkeitskontrolle	contrôle de l'humidité	control de humedad
3109	humidity measurement	fuktighetsmätning	Feuchtigkeitsmessung	mesure de l'humidité	medida de la humedad
3110	humidity room	fuktrum	Klimaraum	salle conditionnée	sala húmeda
3111	hybrid	hybrid	Hybride, Bastard, Mischling, Kreuzung	hybride	híbrido
3112	hydrafiner	hydrafiner, konkvarn	Hydrafiner, hochtourig laufendes und kontinuierliches Mahlgerät	hydrafiner (raffineur cônique)	refinador hidráulico
3113	hydrate cellulose	hydratcellulosa	Hydratcellulose	cellulose engraissée	hidrocelulosa
3114	hydrated	hydrerad	wasserhaltig, mit Wasser verbunden	hydraté	hidratado
3115	hydrated stock	hydratmassa	wasserhaltiger Stoff, gequollener Stoff	pâte engraissée	pasta hidratada
3116	hydrates	hydrater	Hydrate	hydrates	hidratos
3117	hydration	hydrering	Hydration, Verbindung mit Wasser, Hydratisierung	hydration, engraissement	hidratación
3118	hydraulic	hydraulisk	hydraulisch	hydraulique	hidráulico
3119	hydraulics	hydraulik	Strömungsverhältnisse, Hydraulik	hydraulique	hidráulica, hidrotecnia
3120	hydraulic barker	hydraulisk barkare	Strahlentrinder, Wasserstrahlentrindung, hydraulische Entrindung	écorceuse hydraulique	descortezadora hidráulica
3121	hydraulic pressure	hydrauliskt tryck	Wasserdruck	pression hydraulique	presión hidráulica
3122	hydraulic system	hydrauliskt system	Hydraulikanlage, Hydroanlage	système hydraulique	sistema hidráulico
3123	hydrazine	hydrazin	Hydrazin, Diamid	hydrazine	hidracina
3124	hydrides	hydrider	Hydride	hydrures	hidruros
3125	hydro barker	hydraulisk barkare	Strahlentrinder, Wasserstrahlentrindung, hydraulische Entrindung	écorceuse hydraulique	hidrodescortezadora
3126	hydrobromic acid	hydrobromsyra	Bromwasserstoffsäure	acide bromhydrique	ácido bromhídrico
3127	hydrocarbons	kolväten	Kohlenwasserstoff	hydrocarbures	hidrocarburos
3128	hydrocellulose	hydrocellulosa	Hydrocellulose	hydrocellulose	hidrocelulosa
3129	hydrochloric acid	klorvätesyra	Chlorwasserstoffsäure	acide chlorhydrique	ácido clorhídrico
3130	hydrocyclone	virvelrenare	Hydrozyklon, Wirbelsichter	épurateur tourbillonnaire	hidrociclón
3131	hydrodynamics	hydrodynamik	Strömungsmechanik, Hydrodynamik	hydrodynamique	hidrodinámica
3132	hydroelectric power	vattenkraft	Wasserkraft	énergie hydroélectrique	fuerza hidroeléctrica
3133	hydroelectric power plant	vattenkraftverk	Wasserkraftwerk	centrale hydroélectrique	planta hidroeléctrica
3134	hydroexpansivity	våtexpansivitet	Feuchtdehnbarkeit	hydrodilatabilité	hidroexpansividad
3135	hydrofluoric acid	fluorvätesyra	Fluorwasserstoffsäure	acide fluorhydrique	ácido fluorhídrico
3136	hydrofoil	strömlinjeformat blad	Tragfläche, Entwässerungsleiste, Rührflügel	racle d'égouttage	superficie de reacción hidráulica
3137	hydrofoil blade	strömlinjeformat blad	Wasserabstreifer	lame de racle d'égouttage	cuchilla hidrodinámica
3138	hydrogenated rosin	hydrogenerad harts	gehärtetes Kollophonium	résine hydrogénée	resina hidrogenada
3139	hydrogenation	hydrogenering	Hydrierung, Wasserstoffanlagerung	hydrogénation	hidrogenación
3140	hydrogen bond	vätebindning	Wasserstoffbindung	liaison hydrogénée	unión de hidrógeno
3141	hydrogen bonding	vätebindning	Wasserstoffverbindung	enchaînement par liaison hydrogénée	unión de hidrógeno
3142	hydrogen compounds	väteföreningar	Wasserstoffverbindungen	composés de l'hydrogène	compuestos de hidrógeno
3143	hydrogen-ion concentration	vätejonkoncentration	Wasserstoffionenkonzentration, Wasserstoffzahl	indice d'acidité ionique	concentración hidrogeniónica
3144	hydrogen peroxide	väteperoxid	Wasserstoffperoxid	eau oxygénée	agua oxigenada
3145	hydrogen sulfide	svavelväte	Schwefelwasserstoff	hydrogène sulfuré	sulfuro de hidrógeno
3146	hydrolysis	hydrolys	Hydrolyse	hydrolyse	hidrólisis
3147	hydrolysis lignins	hydrolyslignin	Ligninhydrolyse	lignines d'hydrolyse	ligninas hidrolíticas
3148	hydrolyzates	hydrolysat	Hydrolysate	hydrolysats	hidrolizados
3149	hydrometer	hydrometer	Hydrometer, Aräometer, Senkwaage	aréomètre	densímetro

	ENGLISH	SVENSKA	DEUTSCH	FRANÇAIS	ESPAÑOL
3150	hydrophilic	hydrofil	wasserbindend, wasseranziehend	hydrophile	hidrófilo
3151	hydrophobic	hydrofob	hydrophob, wassermeidend, wasserabstossend	hydrophobique	hidrófobo
3152	hydrostatic	hydrostatisk	hydrostatisch	hydrostatique	hidrostático
3153	hydrosulfides	hydrosulfider	hydrosulfide, Bisulfide, Ammoniumsulfide	sulphydrates	hidrosulfuros
3154	hydrotropic agent	hydrotropmedel	hydrotropisches Mittel	agent hydrotropique	agente hidrotrópico
3155	hydrotropic lignins	hydrotropiskt lignin	hydrotropische Lignine	lignines hydrotropiques	ligninas hidrotrópicas
3156	hydrotropic pulping	hydrotropisk uppslutning	hydrotropischer Zellstoffaufschluss	cuisson hydrotropique	elaboración de pasta hidrotrópica
3157	hydroxides	hydroxider	Hydroxid	hydrates	hidróxidos
3158	hydroxyalkyl celluloses	hydroxyalkylcellulosa	Hydroxyalkylcellulosen	hydroxyalkyl cellulose	hidroxialquil celulosas
3159	hydroxyalkyl groups	hydroxyalkylgrupper	Hydroxyalkylgruppen	groupes hydroxyalkyl	grupos hidroxialquil
3160	hydroxyethylation	hydroxyetylering	Hydroxyäthylierung	hydroxyéthylation	hidroxietilación
3161	hydroxyethyl cellulose	hydroxyetylcellulosa	Hydroxyäthylcellulose	hydroxyéthyl cellulose	hidroxietil celulosa
3162	hydroxyethyl groups	hydroxyetylgrupper	Hydroxyäthylgruppen	groupes hydroxyéthyl	grupos hidroxietil
3163	hydroxyl	hydroxyl	Hydroxyl	hydroxyle	oxidrilo
3164	hydroxyl bond	hydroxylbinding	Hydroxylbindung	liaison hydroxyle	unión oxidrilo
3165	hydroxyl groups	hydroxylgrupper	Hydroxylgruppen	groupes hydroxyle	grupos de oxidrilo
3166	hydroxyl-ion concentration	hydroxyljonkoncentration	Hydroxylionenkonzentration	concentration hydroxyle-ion	concentración de hidroxilión
3167	hydroxypropyl cellulose	hydroxypropylcellulosa	Hydroxypropylcellulose	hydroxypropyl cellulose	hidroxipropil celulosa
3168	hydroxypropylmethyl cellulose	hydroxypropylmetylcellulosa	Hydroxypropylmethylcellulose	hydroxypropylméthyl cellulose	hidroxipropil-metil celulosa
3169	hygroexpansivity	hygroexpansivitet	Feuchtigkeitsmessung	hygrodilatabilité	higroexpansividad
3170	hygroscopicity	hygroskopicitet	Hygroskopizität, Wasseraufnahmefähigkeit	hygroscopique	higroscopicidad
3171	hypobromites	hypobromiter	Hypobromite	hypobromites	hipobromitos
3172	hypochlorites	hypokloriter	Hypochlorite, unterchlorigsaure Salze	hypochlorites	hipocloritos
3173	hypochlorous acid	underklorsyrlighet	unterchlorige Säure	acide hypochloreux	ácido hipocloroso
3174	hysteresis	hysteres	Hysterese, Hysteresis	hysteresis	histéresis

I

3175	identification	identifiera	Identifizierung, Nachweis, Erkennung	identification	identificación
3176	idle time	mellantid	Leerlaufzeit, Stillstandszeit	temps perdu, arrêt	tiempo muerto
3177	idler roll	mellanvals	Spannrolle, Riemenspannrolle	galet tendeur	rodillo guía
3178	ignition	tändning	Zündung, Entzündung	ignition, inflammabilité	encendido
3179	ilex	järnek	Ilex, Stecheiche	ilex	acebo
3180	imides	imider	Imide	imides	imidas
3181	imines	iminer	Imine	imines	iminas
3182	imitation handmade paper	imiterat handgjort papper	Büttenersatzpapier	imitation de papier à la main	papel imitación barba
3183	imitation leather	imiterat läder	Kunstleder	simili cuir	imitación cuero
3184	imitation leather board	imiterad läderpapp	Halblederpappe, Imitationslederpappe	carton simili cuir	cartón imitación cuero
3185	imitation parchment	imiterat pergament	Pergamentersatz	simili sulfurisé	imitación pergamino
3186	immiscible	ej blandbar	nicht mischbar, unvermischbar	non-miscible	no mezclable
3187	impact mill	kulkvarn, hammarkvarn	Schlagmühle, Stossmühle, Prallzerspannungmaschine	mélangeur à disques	trituradora de martillos
3188	impact strength	stöthållfasthet	Schlagfestigkeit, Schlagzähigkeit	résistance au choc	resistencia al impacto
3189	impact tests	stötprov	Schlagversuche	essais de résistance aux chocs	pruebas de impacto
3190	impedance	impedans	Impedanz, Scheinwiderstand	impédance	impedancia
3191	impeller	omrörare	Laufrad, Ventilatorrad, Flügelrad	propulseur, rotor	propulsor
3192	imperfect sheet	defekt ark	Defektbogen	feuille défectueuse	hoja defectuosa
3193	impermeability	ogenomtränglighet	Undurchlässigkeit, Undurchdringlichkeit	imperméabilitée	impermeabilidad
3194	impervious	ogenomtränglig	dicht, hermetisch, undurchdringlich	imperméable	impermeable
3195	impregnant	impregneringsmedel	imprägnierend, sättigend, porenfüllend	produit d'imprégnation	líquido impregnante
3196	impregnated	impregnerad	imprägniert, durchtränkt, gesättigt	imprégné	impregnado
3197	impregnated papers	impregnerade papper	imprägniertes Papier, getränktes Papier, Lackpapier	papiers imprégnés	papeles impregnados

ENGLISH	SVENSKA	DEUTSCH	FRANÇAIS	ESPAÑOL
3198 impregnating papers	impregneringspapper	imprägnierfähige Papiere	papiers support pour imprégnation	papeles para impregnar
3199 impregnation	impregnering	Imprägnierung, Tränkung	imprégnation	impregnación
3200 impurity	förorening, orenhet	Verunreinigung, Schmutz, Fremdkörper	impureté	impureza
3201 inch-hour	tumtimme	Arbeitsgeschwindigkeit einer oder mehrerer Papiermaschinen in einer bestimmten Zeiteinheit (unter Berücksichtigung der Maschinenbreite aber nicht der Maschinengeschwindigkeit)	pouce par heure	pulgada/hora
3202 inclined screen	lutande vira	Steilsieb	lépurateur incliné	criba inclinada
3203 inclined wire	lutande vira	Steilsieb	toile inclinée	tela inclinada
3204 inclined wire machine	maskin med lutande vira	Steilsiebmaschine	machine à toile inclinée	máquina de tela metálica inclinada
3205 increment	tillskott	Zuwachs, Zunahme	accroissement, croissance, augmentation	incremento
3206 increment core	tillväxthylsa	Bohrspan	mandrin extensible	mandril de incremento
3207 india paper	flygpostpapper	Bibelpapier, Dünndruckpapier	papier bible	papel biblia
3208 indicator	indikator	Anzeiger, Indikator, Zähler, Signallampe, Anzeigevorrichtung	indicateur, appareil de mesure	indicador
3209 indicator paper	indikatorpapper	Indikatorpapier	papier pour la détermination du pH	papel indicador
3210 indigoid dyes	indigoliknande färger	Indigofarbstoffe, indigoide Farbstoffe	colorants à l'indigo	tintes de añilina
3211 indirect heating	indirekt uppvärmning	indirekte Heizung	chauffage indirect	calentamiento indirecto
3212 indirect steaming	indirekt ångbehandling	indirektes Dämpfen	étuvage indirect	tratamiento indirecto en estufa de vapor
3213 induced-draft fan	utsugningsfläkt	Saugzuggebläse	ventilateur à tirage indirect	ventilador aspirador
3214 induction heating	induktionsuppvärmning	Induktionsheizung	chauffage par induction	calentamiento por inducción
3215 industrial engineering	industriell teknologi	Gewerbetechnik, Gewerbeingenieurwesen	ingénierie industrielle	ingeniería industrial
3216 industrial papers	industripapper	Industriepapiere	papiers pour usages industriels	papeles industriales
3217 industrial tissue	industritunnpapper	Tissues	papier mince pour usages industriels	muselina industrial
3218 industrial wipes	industritrasor	Wischtücher	chiffons pour essuyer	soldaduras industriales
3219 infestation	nedlusning	Befall, Plage, Zerstörung (durch Schimmelbildung)	infestation	infestación
3220 infrared	infraröd	Infrarot	infra-rouge	infarrojo
3221 infrared dectector	infraröddetektor	Infrarot-Detektor	détecteur à infra-rouges	detector de infrarrojos
3222 infrared drying	infrarödtorkning	Infrarot-Trocknung	séchage par infra-rouges	secado infarrojo
3223 infrared heater	infrarödvärmare	Infrarot-Heizung, Infrarot-Heizstrecke	réchauffeur à infra-rouges	calentador infarrojo
3224 infrared heating	infrarödvärmare	Infrarot-Heizung, Infrarot-Heizstrecke	chauffage par infra-rouges	calentamiento infarrojo
3225 infrared radiation	infrarödstrålning	infrarote Strahlung	rayonnement infra-rouge	radiación infarroja
3226 ingredient	ingrediens	Bestandteil, Zutat, Einschluss	ingrédient, élément	ingrediente
3227 inhibition	förhinder, inhibering	Inhibition, Hemmung, Hinderung, Verbot	inhibition, empêchement	inhibición
3228 inhibitor	inhibitor	Inhibitor, reaktionshindernde Substanz, Schranke	inhibiteur	inhibidor
3229 initial tearing resistance	inrivningshållfasthet	Einreisswiderstand	résistance initiale à la déchirure	resistencia inicial al rasgado
3230 injection	injektion	Injektion, Einspritzen, Einpumpen	injection	inyección
3231 injection molding	injektionsformning	Spritzgussverfahren	moulage par injection	vaciado por inyección
3232 ink	färg, bläck	mit Tinte schwärzen, Farbe, Druckfarbe, Tinte	encre	tinta
3233 ink absorption	färgabsorption	Farbaufnahme	absorption de l'encre	absorción de la tinta
3234 ink film thickness	färgskikttjocklek	Dicke der Farbschicht	épaisseur du film d'encre	espesor de la película de tinta
3235 ink holdout	inkholdout, färghållning	Stand der Farbe	prise d'encre	toma de tinta
3236 inking system	färgningssystem	Einfärbungsmethode, Einfärbungsverfahren	dispositif d'encrage	sistema de entintado
3237 ink lay	färgpåföring	Druckfarbenauflage	couche d'encre	puesta de tinta
3238 ink receptivity	färgmottaglighet	Farbaufnahmevermögen	imprimabilité	receptividad de la tinta
3239 ink resistance	färgmotstånd	Farbwiderstand	résistance à l'encre	resistencia de la tinta
3240 ink rub resistance	radermotstånd	Farbabriebwiderstand	résistance de l'encre au frottement	resistencia de la tinta al rozamiento
3241 ink show-through	färggenomskinlighet	Durchschlagen der Farbe	apparition de l'encre d'imprimerie à travers le papier	transparencia de la tinta
3242 ink smear tests	smetprov	Tintenwischprüfung	essais à la tache d'encre	pruebas de manchas de tinta
3243 ink transfer	färgöverföring	Farbaufnahme, Farbübertragung	transfert de l'encre	transferencia de tinta
3244 inlet	inlopp	Durchlass, Einlass, Einlassöffnung	arrivée, entrée	toma

	ENGLISH	SVENSKA	DEUTSCH	FRANÇAIS	ESPAÑOL
3245	in-line cutter slitter	in-line rullmaskin	Längs-und Querschneider	coupeuse en bout de machine	cortadora longitudinal en línea
3246	in-line freeness recorder	in-line freenessprovare	in die Stoffzufuhrleitung eingebauter Mahlgradprüfer	appareil de mesure de l'égouttage en continu	registrador del grado de refino en línea
3247	inorganic acids	oorganiska syror	anorganische Säuren	acides inorganiques	ácidos inorgánicos
3248	inorganic compounds	oorganiska föreningar	anorganische Verbindungen	composés inorganiques	compuestos inorgánicos
3249	inorganic fiber	oorganisk fiber	anorganische Faser	fibre inorganique, minérale	fibra inorgánica
3250	inorganic papers	oorganiska papper	anorganische Papiere	papiers inorganiques	papeles inorgánicos
3251	inorganic salts	oorganiska salter	anorganische Salze, Metallsalze	sels inorganiques	sales inorgánicas
3252	inside diameter	inre diameter, innerdiameter	lichte Weite, Innendurchmesser	diamètre intérieur	diámetro interior
3253	inspection	inspektion	Inspektion, Beaufsichtigung, Prüfung, Kontrolle, Untersuchung	inspection, vérification	inspección
3254	installation	installation	Anlage, Einrichtung, Einbau, Installation	installation	instalación
3255	instrument	instrument	Instrument, Gerät, Messgerät	instrument, appareil	instrumento
3256	instrumentation	instrumentering	Ausrüstung mit Messgeräten	instrumentation	instrumentación
3257	insulating board	isolationspapp	Isolierpappe	carton isolant	cartón aislante
3258	insulating material	isoleringsmaterial	Isoliermaterial, Isolierstoff	matériaux isolants	material aislante
3259	insulating sheathing	isoleringshölje	isolierende Verkleidung, Umkleidung, Ummantelung	papier de construction isolant	revestimiento aislante
3260	insulating tissue	isoleringspapper	Isolier-Tissue	papier mince isolant pour câbles	muselina aislante
3261	insulation	isolering	Isolierschutz, Wärmeschutz, Abkleidung, Dämmung	isolation thermique	aislamiento
3262	insulation paper	isoleringspapper	Isolierpapier	papier pour isolation thermique	papel para aislamiento
3263	intake system	intagningssystem	Ansaugsystem	transmission	sistema de admisión
3264	integrated mill	integrerad fabrik	integrierte Fabrik, integrierter Betrieb	usine intégrée	fábrica integrada
3265	integration	integrering	Integrierung	intégration	integración
3266	intensity	intensitet	Intensität, Stärke, Heftigkeit	intensité	intensidad
3267	interchangeability	utbytbarhet	Austauschbarkeit, Auswechselbarkeit	interchangeabilité	intercambiabilidad
3268	interfiber bond	bindning mellan fibrer	Faser-Faser-Bindung, Zwischenfaserbindung	liaison interfibre	unión entre fibras
3269	interleaving	interfoliering	mit Papier durchschiessen, eine Schicht oder Lage einfügen	intercalage d'une feuille entre deux autres	interponer una hoja entre otras dos
3270	interleaving paper	interfolieringspapper	Zwischenlagenpapier, Durchschusspapier	papier pour intercalaires	papel para interponer
3271	intermediate calender	mellanglätt	Glättwerk (zwischen zwei Trockengruppen angeordnet)	calandre intermédiaire	calandra intermedia
3272	intermediate roll	mellanvals	Zwischenwalze, Feuchtglättwerkswalzen	rouleau intermédiaire	rodillo intermedio
3273	intermediate stuff box	mellaninloppslada	Bütte mit Stoffdichteregler	caisse de tête intermédiaire	caja intermedia
3274	intermittent operation	intermittent drift	diskontinuierlicher Betrieb	fonctionnement intermittent	operación intermitente
3275	internal bond strength	inre bindningsstyrka	innere Bindekraft	cohésion interne	resistencia de la adherencia interna
3276	internal bond tester	provare för inre bindning	Prüfgerät zur Messung der inneren Bindekraft	appareil de mesure de la cohésion interne	probador de adherencia interna
3277	internal tearing strength	genomrivhållfasthet	innere Durchreissfestigkeit	résistance au déchirement amorcé	resistencia al rasgado interno
3278	internal sizing	mäldlimning	Masseleimung	collage interne	encolado interno
3279	internal tearing resistance	genomrivmotstånd	innerer Durchreisswiderstand	résistance au déchirement amorcé	resistencia interna al rasgado
3280	internally ribbed yankee dryer	inuti räfflad yankeetorkare	innen gerippter Trockenzylinder	gros sécheur frictionneur nervuré intérieurement	secadora yankee con estriado interno
3281	interstice	mellanrum	kleiner Zwischenraum, Lücke, Spalt	interstice, intervalle	intersticio
3282	inventory	inventering	Inventar, Lagerbestandsverzeichnis	inventaire, état des stocks	inventario
3283	inventory control	inventeringskontroll	Lagerbestandsaufnahme	contrôle des stocks	control del inventario
3284	iodine	jod	Jod	iode	yodo
3285	iodine compounds	jodföreningar	Jodverbindungen	composés de l'iode	compuestos de yodo
3286	ion exchange	jonbyte	Ionenaustausch	échange d'ions	intercambio iónico
3287	ion-exchange paper	jonbytespapper	Ionenaustauschpapier	papier échangeur d'ions	papel de intercambio iónico
3288	ion-exchange resins	jonbyteshartser	Ionenaustauschharze	résines échangeuses d'ions	resinas de intercambio iónico
3289	ion exchanger	jonbytare	Ionenaustauscher	échangeur d'ions	intercambiador iónico
3290	ionic strength	jonstyrka	Ionenstärke	concentration ionique	intensidad iónica
3291	ionization	jonisering	Ionenspaltung, Ionisierung	ionisation	ionización
3292	ionomers	jonomerer	Ionomere	ionomères	ionomeros
3293	ions	joner	Ione	ions	iones
3294	iron	järn	bügeln, Eisen	fer	hierro
3295	iron compounds	järnföreningar	Eisenverbindungen	composés du fer	compuestos de hierro
3296	iron oxides	järnoxider	Eisenoxide	oxydes de fer	óxidos de hierro

ENGLISH	SVENSKA	DEUTSCH	FRANÇAIS	ESPAÑOL
3297 ironwood	järnved	Eisenholz (harte u. feste u. dunkelgefärbte Hölzer aus den Tropen)	bois de fer	madera recia tropical
3298 irradiate	stråla	anstrahlen, bestrahlen, erleuchten	irradier	irradiar
3299 isobutyl alcohol	isobutylalkohol	Isobutylalkohol	alcool isobutyl	alcohol isobutil
3300 isocyanates	isocyanater	Isocyanate	isocyanates	isocianatos
3301 isomers	isomerer	Isomere	isomères	isomeros
3302 isopropyl alcohol	isopropylalkohol	Isopropylalkohol	alcool isopropyl	isopropil alcohol
3303 isotherms	isotermer	Isothermen	isothermes	isotermos
3304 isotopes	isotoper	Isotopen	isotopes	isótopos
3305 isotropy	isotropi	Isotropie	isotropie	isotropía

J

ENGLISH	SVENSKA	DEUTSCH	FRANÇAIS	ESPAÑOL
3306 jacket	mantel	Schutzumschlag, Verkleidung, Mantel, Umhüllung	manchon, chemise	camisa
3307 jack pine	jackpine	Bankskiefer, Zwergkiefer	pin gris	pino del Canadá
3308 jet	stråle	Düsenstrahl, Düse	jet	chorro
3309 jet deckle	spritsdäckel	Formatbegrenzung mittels Düsenstrahlabspritzung	bords barbés	calibrador de chorro
3310 job lot	restparti	Ramsch-Ware	lot refusé à la livraison, lot de papier soldé	partida de papel con rebaja de precio
3311 job lot paper	restpartipapper	Papier zweiter Wahl	papier soldeé	partida de papel con rebaja de precio
3312 jog	knuffa, skaka	rütteln, stossen, geradestossen (Papier), Stoss	taquer (un paquet de feuilles)	igualar
3313 joggle	skaka	rütteln, schütteln, verzahnen, Hervorstehen	remuer	acodar
3314 joint	skarv	Fuge, Naht, Verbindungsstelle	joint	junta
3315 jordan	jordan	Jordan	nom de l'inventeur du raffineur cônique	Jordan
3316 jordaning	jordanmalning	Mahlen in der Jordan-, Kegelstoffmühle	raffinage au raffineur cônique	refinado cónico
3317 jordan plug	jordankubb	Refinerkegel	cône de raffineur cônique	cono de Jordan
3318 jordan refiner	fordankvarn	Jordan-Refiner, Kegelstoffmühle	raffineur cônique	refinador Jordan
3319 journal	axeltapp, tidskrift	Zapfen, Lagerzapfen	tourillon	soporte de cojinete
3320 Juglans (Lat.)	Juglans	Walnuss	Juglans	nuez
3321 jumbo roll	jumbovals	Jumbo-Rolle (besonders grosse Papierrolle)	bobine brute (non rebobinée), bobine mère	bobina bruta
3322 juniper	juniper, en	Wacholderbaum	genèvrier	enebro
3323 Juniperus (Lat.)	Juniperus	Wacholderbaum, Kranawettstrauch	genèvrier	enebro
3324 junk remover	skrottransportör, skrotavskiljare	Entfernung grosser Holzstücke	élévateur à godets	recogedor de desperdicios
3325 jute	jute	Jute	jute	yute
3326 jute bag paper	jutesäckpapper	Jutesackpapier	papier jute pour sacs	papel de yute para bolsas
3327 jute board	jutekartong	Triplexkarton	carton pour boîtes pliantes	cartón de yute
3328 jute fiber	jutefiber	Jutefaser	fibre de jute	fibra de yute
3329 jute liner	juteliner	Karton für die Kaschierung von Versandbehältern (enthält keine Jute)	papier jute pour couverture de caisses	forro de yute
3330 jute linerboard	jutelinerkartong	Karton für die Kaschierung von Versandbehältern (enthält keine Jute)	carton à base de jute	linerboard de yute
3331 jute paper	jutepapper	Papier aus Jutefasern, Tauenpapier	papier jute	papel de yute
3332 jute pulp	jutemassa	Jutezellstoff	pâte de jute	pasta de yute
3333 jute test liner	jutetestliner	Karton für die Kaschierung von Versandbehältern (enthält keine Jute)	couverture bico-kraft	test liner de yute
3334 juvenile wood	ungved	Holz das sich sehr früh (in der Jugend) gebildet hat	bois de printemps	madera joven

ENGLISH	SVENSKA	DEUTSCH	FRANÇAIS	ESPAÑOL

K

3335	Kady mill	Kadykvarn	Kady-Mühle	mélangeur Kady	fábrica Kady
3336	kaolin	kaolin	Kaolin	kaolin	caolin
3337	kappa number	kappatal	Kappazahl	résistance	índice Kappa
3338	Kelvin degree	Kelvingrad	Kelvingrad, Grad Kelvin	degré Kelvin	grado Kelvin
3339	kenaf	kenaf	Kenaf, Hanfeibisch	kénaf	kenaf
3340	keratin	keratin	Keratin, Hornsubstanz	kératine	queratina
3341	kerf	sågskär	Kerbe, Einschnitt, Schnittbreite	trait, entaille	corte
3342	kerosene	fotogen	Kerosin, Leuchtpetroleum	kérosène, pétrole lampant	keroseno
3343	ketenes	ketener	Ketene	cetènes	quetenos
3344	ketones	ketoner	Ketone	cétones	acetonas
3345	kiln	ugn	Ofen, Brennofen, Röstofen, Darre, Trocknungsanlage	four	horno de secar
3346	kinetic energy	kinetisk energi	kinetische Energie, Bewegungsenergie	énergie cinétique, force vive	energía cinética
3347	king roll	kungsvals	Tragwalze, Unterwalze (eines Kalanders)	rouleau inférieur	rodillo inferior de calandra
3348	kiss coater	kisscoater	Auftragsmaschine	coucheuse par séchage (au rouleau)	estucadora de lamedura
3349	kiss coating	kisscoating	Streichen indem die Auftragswalze die Bahn berührt	couchage par séchage	estucado de lamedura
3350	kneader	knådare	Kneter, Knetmaschine	déchiqueteur, broyeur	desmenuzadora
3351	knife	kniv	Messer	couteau, lame	cuchilla
3352	knife barker	knivbarkare	Messerschälmaschine	écorceuse à couteaux	descortezadora de cuchilla
3353	knife coater	bladbestrykare	Messerstreicher, Schaberstreicher	coucheuse à râcles	estucadora de cuchillas
3354	knife coating	bladbestrykning	Messerstreichen, Schaberstreichen	couchage par râcle	estucado de cuchilla
3355	knife edge	knivegg	Schneide, Schneidkante, Messerschneide	bord coupé au couteau	filo de cuchilla
3356	knit fabric	trika	gewirkter Stoff	article tissé	tela de punto
3357	knives	knivar	Holländerbemesserung	couteaux, lames	cuchillas
3358	knockoff shower	avslagningssprits	Abspritzanlage mit Abstelimechanismus	rinceur de choc de cassé	regadera de descontinuo
3359	knot screen	kvistfång	Knotenfänger	trieur de noeuds	criba de nudos
3360	knots	kvistar	Knoten, Ast	noeuds	nudos
3361	knotter	kvistfång	Astfänger, Splitterfänger	trieur de noeuds et buchettes	separador de nudos
3362	knotter pulp	kvistmassa	Aesterstoff	pâte de noeuds	pasta de nudos
3363	knotter screen	kvistfång	Splitterfang	trieur de noeuds	criba de incocidos
3364	knotty pulp	kvistrik massa	knotiger Stoff, knotiger Schliff	pâte bucheteuse	pasta con astillas
3365	kollergang	kollergång	Kollergang	meuleton	muela móvil
3366	kraft	kraftmassa, kraftpapper	Kraftpapier, braunes Packpapier	kraft	kraft
3367	kraft bag paper	kraftsäckpapper	Kraftsackpapier	papier draft pour sacs	papel kraft para sacos
3368	kraft bitumen paper	asfaltkraftpapper	Kraftbitumenpapier	papier draft goudronné	papel kraft embreado
3369	kraft board	kraftkartong	Kraftpappe, Kraftkarton	carton kraft	cartón kraft
3370	kraft corrugating medium	kraftvågpapp	Kraft-Wellpappenmittellage	cannelure kraft	cartón kraft a ondular
3371	kraft liner	kraftliner, krafttäckskiktpapp	Kraftliner	kraft liner	kraft liner
3372	kraft linerboard	kraftlinerkartong	mit Kraftpapier kaschierter Karton	carton kraft liner	kraft linerboard
3373	kraft liquor	sulfatkokvätska	Kraftlauge	lessive kraft	solución kraft
3374	kraft mill	sulfatfabrik	Kraftpapierfabrik	usine kraft	fábrica kraft
3375	kraft paper	kraftpapper	Kraftpapier	papier kraft	papel kraft
3376	kraft pulping	kraftmassakokning	Sulfataufschluss	fabrication de pâtes kraft	elaboración de pasta kraft
3377	kraft pulp	kraftmassa	Sulfatzellstoff	pâte kraft	pasta kraft
3378	kraft test liner	krafttestliner	Kraftdecke, Kraftpapier zum Kaschieren von Karton	couverture forte kraft	test liner kraft
3379	kraft twisting	karamellpapper av kraftmassa	Kraft-Dreheinschlag	papier kraft pour câbles	kraft para hilados
3380	kraft wrapping	kraftpapperemballage	Kraftpackpapier	emballages kraft	embalaje kraft

ENGLISH	SVENSKA	DEUTSCH	FRANÇAIS	ESPAÑOL

L

	ENGLISH	SVENSKA	DEUTSCH	FRANÇAIS	ESPAÑOL
3381	label	etikett	beschriften, bezeichnen, etikettieren, Etikett, Aufklebezettel	étiquette	etiqueta
3382	labeling	etikettpapper	Etikettieren	étiquetage	etiquetado
3383	labeling machine	etiketteringsmaskin	Etikettiermaschine	machine à coller les étiquettes	máquina etiquetadora
3384	label paper	etikettpapper	Etikettenpapier	papier pour étiquettes	papel para etiquetas
3385	labor	arbetskraft	schwer arbeiten, sich abmühen, Arbeitskräfte	travail, effort	mano de obra
3386	laboratory	laboratorium	Laboratorium, Labor	laboratoire	laboratorio
3387	labour (Brit.)	arbetskraft	schwer arbeiten, sich abmühen, Arbeitskräfte	main d'oeuvre	mano de obra
3388	lace paper	girlangpapper	Spitzenpapier, Kuchenspitzenpapier	papier dentelle	encaje·de papel
3389	lacquer	lack	Lack	laque, vernis	laca
3390	lacquered paper	lackerat papper	lackiertes Papier	papier verni	papel barnizado
3391	lacquering	lackbehandling	Lackieren, Lackierung	vernissage	barnizado
3392	lactic acid	mjölksyra	Milchsäure	acide lactique	ácido láctico
3393	lagoon	lagun, damm	Lagune, Haff	fosse de décantation	laguna
3394	lagooning	dammbildning	Wasserlagerung von Holz	décanter	encharcar
3395	laid	utlagd, ribbad	gelegt, gerippt (Papier), mit Wasserlinien versehenes Papier	vergé	verjurado
3396	laid antique	ribbat antikpapper	geripptes Antikpapier	vergé non apprêté	verjurado antiguo
3397	laid dandy roll	ribbad egutör	Spiralripp-Egoutteur	rouleau vergeur	rodillo filigranador
3398	laid finish	ribbfinish	gerippte Oberfläche	fini vergé	acabado filigrana
3399	laid lines	ribbningslinjer	gerippte Wasserzeichen	vergeures	verjurados
3400	laid mold	ribbad vira	gerippte Form	forme vergée	molde verjurado
3401	laid paper	ribbat papper	geripptes Papier	papier vergé	papel verjurado
3402	laid wires	ribbade viror	Verbindungsdraht	fils vergeurs, vergeures	telas verjuradas
3403	laid writing	ribbat skrivpapper	geripptes Schreibpapier	papier écriture vergé	papel filigrana
3404	lake	lackfärg	See, Binnensee	laque	laca
3405	lake pigment	lackfärgspigment	Pigmentfarbe	pigment laqué	pigmento de la laca
3406	lamella	lamell	Lamelle, Blättchen	lamelle	laminilla
3407	lamellar structure	lamellstruktur	lamelliert, blättrig, Lamellenstruktur	structure lamellaire	estructura laminar
3408	laminar flow	laminär strömning	laminare Strömung, Laminarströmung	écoulement laminaire	flujo laminar
3409	laminate	laminat	kaschieren, Schichtstoff	laminer	laminar
3410	laminated	laminerad	kaschiert, beschichtet	stratifié, laminé	laminado
3411	laminated board	laminerad kartong	kaschierter Karton	carton contrecollé	cartón laminado
3412	laminated paper	laminerat papper	kaschiertes Papier, Schichtpapier	papier contrecollé	papel laminado
3413	laminated plastic	laminerad plast	Schichtstoff	complexe plastique	plástico laminado
3414	laminating	laminering	Schichtstoff herstellen, Schichtstoff kaschieren	stratification, contre-collage sous pression	estratificación
3415	laminating paper	lamineringspapper	Kaschierpapier	papier support pour stratifiés	papel a estratificar
3416	lamination	laminering	Beschichtung, Kaschierung, Schichtfolie, Schicht, Lage	stratification	laminación
3417	laminator	laminator	Beschichtungsanlage, Kaschieranlage	machine à doubler par collage	laminador
3418	land fill	landutfyllnad	Erdhinterfüllung	terre de remblai	relleno de tierras
3419	lap	ark	übereinanderlegen, sich überlappen, umwickeln, gefaltete, lagerfertige Lagen Nasszellstoff, versandfertige Lagen Nasszellstoff	recouvrement, balle de pâte humide en feuilles repliées	fardo de pasta húmeda
3420	lap machine	upptagningsmaskin	Bandwickler, Läppmaschine	machine à mettre en balles la pâte humide en feuilles repliées	enfardadora de pasta húmeda
3421	lapped pulp	upptagen massa	entwässerter und in Ballen geschichteter Zellstoff	pâte humide en feuilles repliées	pasta en cartones
3422	lapping	upptagning i ark	Entwässern und Falten von Nasszellstoff zur Herstellung von Ballen	mise en balle de pâte humide en feuilles repliées	plegado en tres
3423	larch	lärkträd	Lärche	mélèze	alerce
3424	Larix	Larix	–	mélèze	alerce
3425	laser	laser	Laser	laser	láser
3426	latency	latens	Latenz, Latenzzeit	obscurité	estado latente
3427	latent image	latent bild	latentes Bild	image latente	imagen latente

ENGLISH	SVENSKA	DEUTSCH	FRANÇAIS	ESPAÑOL
3428 lateral porosity	sidledsporositet	Porosität quer über die Bahn gemessen	porosité latérale	porosidad lateral
3429 late wood	höstved	Spätholz	bois d'automne	madera de otoño
3430 latex	latex	Latex, Gummimilch	latex	látex
3431 latexes	latex	Latex	latex	látexes
3432 latex-treated papers	latexbehandlade papper	mit Latex behandeltes Papier	papiers traités au latex	papeles tratados al látex
3433 layboy	arkbord	Bogenableger	ramasse-feuilles	recoge hojas
3434 layer	skikt, lager, lag	Schicht, Lage	couche	capa
3435 leach	lut	durchsickern, auslaugen, Lauge	lessive	lixiviar
3436 leaching	lutbehandling, luttillverkning	Auslaugen	lixivation	lixiviación
3437 lead	bly	führen, lenken, leiten, durchschiessen, verbleien, Führung, Leitung, Ganghöhe, Blei	plomb, conduite de feuille	plomo
3438 leading edge	framkant	Führungskante	bord d'attaque	borde conductor
3439 leading roll	främre vals	Leitrolle, Leitwalze, erste Walze	rouleau guide rouleau conducteur	rodillo conductor
3440 leaf	blad, ark	Blatt, Bogen, Folie	feuille	hoja
3441 leaf filter	bladfilter	Blattfilter	fibre de chanvre de Manille	filtro de hoja
3442 leaflet	mindre blad, trycksak	Merkblatt, Blättchen	prospectus, notice	hojilla
3443 leakage	läckage	Lecken, Leckwerden, Undichtheit, Durchsickern	fuite	fuga
3444 leakproof container	läcksäker behållare	leckdichter, lecksicherer Behälter, tropfsicherer Behälter	récipient étanche	contenedor a prueba de fugas
3445 leatherboard	läderimitation	Lederpappe	carton simili cuir	cartón cuero
3446 leather or leatherette paper	konstaläderpapper	Kunstlederpapier, Lederimitationspapier	papier simili cuir	papel cuero
3447 ledger paper	bokföringspapper	Geschäftsbücherpapier, oberflächengeleimtes und festes Hadernpapier	papier registre	papel registro
3448 length	längd	Länge, Zeitdauer	longueur	longitud
3449 letterpress ink	boktryckfärg	Farbe für Buchdruck	encre pour typographie	tinta para tipografía
3450 letterpress paper	boktryckpapper	Buchdruckpapier	papier pour typographie	papel para litografía
3451 letterpress printing	boktryckning	Buchdruckverfahren, Hochdruckverfahren	typographie	impresión tipográfica
3452 level	nivå, libell, vattenpass	gleichmachen, planieren, nivellieren, Niveau, Stauhöhe, Höhe, Stand	niveau	nivel
3453 level control	nivåkontroll	Standregelung	contrôle du niveau	control de nivel
3454 level controller	nivåregulator	Stoffstandsregler	contrôleur de niveau	controlador de nivel
3455 level indicator	nivåindakator	Stoffstandsanzeige	indicateur de niveau	indicador de nivel
3456 leveling	avvägning	Egalisieren, Nivellieren	mise de niveau	nivelación
3457 leveling agent	vätska för avvägningsinstrument	Verlaufmittel, Egalisiermittel	produit égalisateur	agente nivelador
3458 leveling screw	nivåskruv	Stellschraube, Nivellierschraube	vis d'ajustement	tornillo nivelador
3459 lever	spak	Hebel	levier	palanca
3460 lever arm	hävarm	Hebelarm	bras de levier	brazo de palanca
3461 lever weighting	hävarmsbelastning	Beschwerung des Hebels	pesage à la bascule	lastrado de palanca
3462 levulinic acid	levulonsyra	Lävulinsäure	acide levulinique	ácido levulínico
3463 Libocedrus (Lat.)	Libocedrus	chilenische Flusszeder	cèdre	Libocedrus (lat.)
3464 lick up	avtagning	Übernahme der Papierbahn durch den Oberfilz ohne Anwendung von Vakuum	prise automatique	toma automática
3465 life	livslängd, liv	Leben, Dauerhaftigkeit, Haltbarkeit, Lebensdauer	durée	duración
3466 lift	hiss, lyft	heben, hochheben, Lift, Aufzug, Hub	ascenseur, monte-charge	elevador
3467 lifting	lyftning	Hochheben, sich Abheben (einer Schicht)	arrachage en surface	alzamiento
3468 lifting equipment	lyftanordning	Hebewerkzeug, Hebevorrichtung	matériel de levage	equipo de elevación
3469 lift truck	lyfttruck	Stapler	chariot élévateur	carretilla elevadora
3470 light	ljus, lätt	zünden, anzünden, entzünden, licht, hell, schwach, Licht	lumière	luz
3471 light fastness	ljusäkthet	Lichtechtheit, Lichtbeständigkeit	solitité à la lumière	estabilidad a la luz
3472 light scattering	ljusspridning	Lichtstreuung	dispersion de la lumière	dispersión de luz
3473 light-sensitive paper	ljuskänsligt papper	lichtempfindliches Papier	papier sensible à la lumière	papel fotosensible
3474 light source	ljuskälla	Lichtquelle	source de lumière	fuente luminosa
3475 lightweight	lättvikt	leichtgewichtig	léger	ligero
3476 lightweight linerboard	lätt linerkartong	leichtgewichtiger Beklebekarton	carton liner léger	linerboard fino
3477 lightweight paper	tunnpapper	leichtgewichtiges Papier	papier mince	papel fino
3478 ligneous	ligninartad, lignin-	ligninhaltig	ligneux	lignoso
3479 lignification	lignifiering	Ligninanreicherung, Verholzung	lignification	lignificación
3480 lignin carbohydrate complexes	lignin-kolhydratkomplex	Lignin-Kohlehydrat-Komplex	lignine hydrate de carbone	complejos lignina-carbohidrato

	ENGLISH	SVENSKA	DEUTSCH	FRANÇAIS	ESPAÑOL
3481	lignin content	ligninhalt	Ligningehalt, Verholzungsgrad	teneur en lignine	contenido de lignina
3482	lignins	ligniner	Lignine	lignines	ligninas
3483	lignocellulose	lignocellulosa	Lignocellulose	lignocellulose	lignocelulosa
3484	lignols	lignoler	–	lignols	lignoles
3485	lignosulfonates	lignosulfonater	Lignosulfonat	lignosulfonates	lignosulfonatos
3486	lignosulfonic acids	lignosulfonsyra	Ligninsulfonsäuren	acides lignosulfoniques	ácidos lignosulfónicos
3487	lime	kalk	Kalk, Linde	chaux	cal
3488	lime kiln	kalkbränningsugn	Kalkofen	four à chaux	hornada de cal
3489	lime milk	kalkmjölk	Kalkmilch	lait de chaux	lechada de cal
3490	lime mud	kalkslam	Kalkschlamm	boue de chaux	lodo de cal
3491	lime slaking	kalksläckning	Kalklöschen	extinction de la chaux	apagado de la cal
3492	lime sludge	kalkslam	Kalkschlamm	boue résiduaire de chaux	fanfo de cal
3493	lime slurry	kalkslam	Kalkbrei, Kalksuspension	savon de chaux	lechada de cal
3494	limestone	kalksten	Kalkstein	terre à chaux, calcaire	piedra caliza
3495	limit switch	gränsbrytare	Endschalter, Begrenzungsschalter	interrupteur	disyuntor automático
3496	linear pressure	linjetryck	Liniendruck	pression linéaire	presión lineal
3497	linear programing	linjärprogrammering	lineare Programmierung	programme linéaire	programación lineal
3498	lined	fodrad	liniert, ausgekleidet, gefüttert, kaschiert	doublé	revestido
3499	lined boards	täckskiktkartong	beklebter und kaschierter Karton	cartons affichés blanchis	cartones encartelados
3500	linen	linne	Leinen, Leinwand	lin	lienzo
3501	linen-faced paper	linnepressat papper	leinenkaschiertes Papier, Papier mit Leinenstruktur	papier entoilé	papel entelado
3502	linen fiber	linnefiber	Leinenfaser, Flachsfaser	fibre de lin	fibra de lienzo
3503	linen finish	linnefinish	Leinenoberfläche, Leinenprägung	gaufrage toile	gofrado tela
3504	linen paper	linnepapper	Leinenpapier	papier de chiffons	papel tela
3505	liner	liner, täckskikt	Deckschicht eines Kartons, Kaschierpapier, Beklebepapier	doublure, liner	forro
3506	linerboard	täckskiktkartong	kaschierter Karton, Beklebekarton, Deckenbahn (bei Wellpappe)	carton double	linerboard
3507	liner paper	täckskiktpapper	Deckenpapier, Beklebepapier	papier à doubler	papel para forrar
3508	lineshaft	huvudaxel	Laufwelle	ligne d'arbres	eje de línea
3509	lining	täckning, fodring	Auskleidung, Futter, Belag	doublure, revêtement, garniture	forro
3510	lining up	uppradning	Ausrichten	alignement	alineación
3511	linoleic acid	linolsyra	Linolsäure	acide linoléique	ácido linoleico
3512	linotype machine	linotypmaskin	Zeilengusssetzmaschine, Zeilensetzmaschine	linotype	linotipia
3513	lint	linter	Staub, Papierstaub, Lint, Lintbaumwolle	charpie, effilochure	hilas
3514	linters	linters	Linters	linters	borra de algodón
3515	linting	linterhantering	Stauben	effilocher	desfibrado de algodón
3516	lip of a headbox	inloppslådeläpp	Stoffauflaufflippe	bord de la caisse à pâte en tête de machine	borde de caja de entrada
3517	liquid	vätska, vätskeformig	flüssig, Flüssigkeit, Flüssigkeitshöhe	liquide	líquido
3518	Liquidambar (Lat.)	Liquidambar	Amberbaum, Satinnussbaum	copalme d'Amérique	copal rojo
3519	liquid permeability	vätskepermeabilitet	Flüssigkeitsdurchlässigkeit	perméabilité aux liquides	permeabilidad a los líquidos
3520	liquidproofing	vätsketätning	Flüssigkeitsdichteprüfung	essai de perméabilité aux liquides	impermeabilización a los líquidos
3521	liquor	lut, kokvätska	Lauge, Flotte	lessive, solution	lejía
3522	liquor evaporating plant	lutindunstningsanläggning	Laugeneindampfungsanlage	installation d'évaporation des lessives	planta de evaporación de lejías
3523	liquor-phase cooking	vätskefaskokning	Kochen in der Flüsigkeitsphase	cuisson en phase liqueur	cocción fase lejía
3524	liquor-to-wood ratio	lut-vedförhållande	Lauge-Holz-Verhältnis	rapport liqueur/bois	proporción lejía-madera
3525	Liriodendron (Lat.)	Liriodendron	Tulpenbaum	tulipier	Liriodendron
3526	litharge	blyoxid	Bleiglätte, Bleioxid	litharge	litargirio
3527	lithium	litium	Lithium	lithium	litio
3528	lithium chloride	litiumklorid	Lithiumchlorid	chlorure de lithium	cloruro de litio
3529	lithium compounds	litiumföreningar	Lithiumverbindungen	composés du lithium	compuestos de litio
3530	litho-coated paper	offsetbestrukna papper	einseitig gestrichenes für den Steindruck ausgerüstetes Papier	papier couché pour lithographie	papel estucado lito
3531	lithograph paper	litografipapper	Lithographiepapier, Steindruckpapier	papier litho	papel para litografía
3532	lithography	litografi	Lithographie, Steindruck	lithographie	litografía
3533	litho paper	litografipapper	Lithographiepapier, Steindruckpapier	papier litho	papel lito
3534	lithopone	litopon	Lithopon, Zinksulfidweiss	lithopone	litopono
3535	litmus paper	lakmuspapper	Lackmuspapier	papier tournesol	papel de girasol
3536	live bottom	skruvbotten	Behälter mit Lochboden	fond mobile	fondo vivo
3537	live-bottom bin	skruvbottenbinge	Behälter mit Lochboden	silo à fond mobile (avec vis transporteuse)	silo de fondo vivo

	ENGLISH	SVENSKA	DEUTSCH	FRANÇAIS	ESPAÑOL
3538	load	last, belastning	beladen, belasten, beschweren, beschicken, Last, Beladung, Belastung, Charge	charge	carga
3539	load cell	belastningscell	Druckmessdose	cellule de pesage	compartimento de carga
3540	load control	lastreglering, last-förlängning	Belastungskontrolle	contrôle de la charge	control de carga
3541	load elongation	belastning	Spannungs-Dehnung	charge-allongement	extensión de la carga
3542	loader	fyllnadsmedel	Verlader, Lader, Ladevorrichtung	chargeur	cargador
3543	loading	belastning	Verladen, Verfrachten, Belasten, Beschicken, Beladen, Einhängen der Papierrollen	chargement	alimentación
3544	loading material	fyllnadsmedel	Füller, Beschwerungsmittel	charge	material de carga
3545	localized watermark	lokaliserat vattenmärke	zentriertes Wasserzeichen, Formatwasserzeichen	filigrane localisé	filigrana localizada
3546	locust	gräshoppa	unechte Akazie, Robinie, Heuschrecke	robinier, faux-acacia	falsa acacia
3547	locust bean gum	Locust Bean Gum	–	gomme de robinier	goma de algarroba
3548	lodgepole pine	lodgepole pine	Murraykiefer	pin de murray	pino alerce
3549	loft	torklada	Dachboden, Speicher	séchoir à l'air	almacén
3550	loft dried	lufttorkad	an der Luft getrocknet	séché à l'air	secado al aire
3551	loft-dried paper	lufttorkat papper	an der Luft getrocknetes Papier	papier séché à l'air	papel secado al aire
3552	loft dryer	lufttorkare	Trockenspeicher	séchoir à l'air	secador aireador
3553	loft drying	lufttorkning	Lufttrocknung	séchage à l'air	secado por aire
3554	log	stock	eintragen, fällen, abholzen, Prügel, Klotz, Block	rondin	madero
3555	log haul-up	lunning	Holzaufzug, Holzklotzschrägaufzug, Stammförderer, Längstransporteur	transporteur-élévateur de rondins	arrastre de maderos
3556	logging	timmerhuggning	Fällen, Holzeinschlag, Holzbringung	tronçonnage du bois en rondins	explotación forestal
3557	logging machine	skogsavverkningsmaskin	Holzerntemaschine	tronçonneuse	máquina taladora
3558	logging residue	rubb	Fällungsrestholz	déchets de tronçonnage	residuos de la tala maderera
3559	log jam	timmerbröt	Holzgletscher	rondins en désordre	atascamiento de maderos
3560	log pond	timmerbassäng	Dämpfkanal, Wasserbecken zur Aufnahme von Schleiferholz	bassin à rondins	estanque para rollos (de madera)
3561	log splitter	timmerklyvning	Stammspaltmaschine	fendeuse à bois	cortadora de madera
3562	long direction	längsriktning	Maschinenrichtung	sens machine	dirección larga
3563	longevity	långt liv	lange Lebensdauer, Langlebigkeit	longévité	longevidad
3564	long-fibered pulp	långfibrig massa	langfaseriger Zellstoff	papier à fibres longues	pasta de fibras largas
3565	long fold	långsveck	Faltung in Längsrichtung	plié en long	pliegue largo
3566	longitudinal fold	långsveck	Längsfalzung, Längsfalz	plié dans le sens longitudinal	pliegue longitudinal
3567	long-leaf pine	långbarrig tall	Gelbkiefer, Sumpfkiefer	pin jaune à feuilles longues	pino amarillo
3568	long stock	långfibrig massa	lange Ware	pâte longue	pasta larga
3569	long ton	long ton	Langtonne	tonne anglaise	tonelada inglesa
3570	look-through	genomsikt	Transparenz, Durchsicht	épair, transparence	transparencia uniforme
3571	loop drying	hängtork	Hängetrocknung, Girlandentrocknung	séchage sur accrocheuse	secado con enganchadora
3572	loose-leaf paper	lösblad	Papier für Loseblatt-Sammlung, Papier für lose Karteiblätter	papier pour feuillets mobiles	papel para hojas reemplazable
3573	loosely wound roll	lösrulle	locker gewickelte Rolle	bobine peu serrée, bobine molle	bobina floja
3574	loss	förlust	Verlust, Schaden, Nachteil	perte	pérdida
3575	low-angle gloss	glans vid strykande infall	Glanz eines gestrichenen Papiers (bei 75grad vertikal gemessen)	brillant à faible angle	abrillantamiento de ángulo reducido
3576	low consistency	låg konsistens	geringe Dichte (Stoffdichte)	faible consistance	baja consistencia
3577	low density	låg densitet	geringe Dichte	faible densité	baja densidad
3578	low finish	låg finish	Grobausrüstung	surface rugueuse	acabado débil
3579	low pressure	lågtryck	Niederdruck, Unterdruck	basse pression	baja presión
3580	low-pressure feeder	lågtrycksmatare	Niederdruckbeschickungsanlage, Niederdruckförderanlage	appareil d'alimentation à basse pression	alimentador de baja presión
3581	low temperature	låg temperatur	Tieftemperatur	basse température	temperatura baja
3582	lowering device	sänkdon	Mechanismus zum Herablassen (Absenken oder Tiefereinstellen)	dispositif de descente	dispositivo de descenso
3583	lubricant	smörjmedel	Schmiermittel, Gleitmittel	lubrifiant	lubricante
3584	lubricating oil	smörjolja	Schmieröl, Maschinenöl	huile de graissage	aceite lubricador
3585	lubricatiòn	smörjning	Schmierung	graissage	engrase
3586	lumber	timmer	Holzbringung, Schnittholz, Bauholz, Nutzholz, Holz fällen und zurichten	bois en grume	madera en troncos
3587	lumber jack	lumberjack, timmerhuggare	Holzfäller, Waldarbeiter	bucheron	leñador
3588	lumen	lumen	Lumen, lichte Weite	lumen	lumen
3589	luminosity	luminositet	Helligkeit	luminosité	brillantez
3590	luminous reflectivity	luminös reflektivitet	Reflexionsvermögen	pouvoir réfléchissant	reflexividad luminosa

	ENGLISH	SVENSKA	DEUTSCH	FRANÇAIS	ESPAÑOL
3591	lump	klump	sich klumpen, zusammenballen Klumpen, Stoffbatzen	mâton, pâton	grumo
3592	lump breaker	klumpupplösare	Gautsch-Gegenwalze	rouleau briseur de mâtons	triturador de grumos
3593	lump breaker roll	guskridvals	Andruckwalze an der Sauggautsche	rouleau écraseur de mâtons	rodillo desgrumador
3594	lump detector	knutdetektor	Detektor für Stoffbatzen	détecteur de mâtons	detector de grumos
3595	luster	lyster	glänzend machen, schimmern, glänzen, Glanz, Lüster, Schein, Schimmer	brillant, éclat	brillo

M

	ENGLISH	SVENSKA	DEUTSCH	FRANÇAIS	ESPAÑOL
3596	MF (machine finished)	maskinfinished, maskinglättad	maschinenglatt, einseitig glatt	apprêté	satinado
3597	MF paper	maskinglättat papper	maschinenglattes Papier, einseitig glattes Papier	papier apprêté	papel satinado
3598	MG (machine glazed)	maskinglättad	maschinenglatt	apprêté	satinado
3599	MG cylinder (Brit.)	glättcylinder	Glättzylinder (für maschinenglattes Papier)	frictionneur (sécheur)	cilindro satinador
3600	MG machine	yankeemaskin, encylindermaskin	Selbstabnahmemaschine	machine frictionneuse	máquina friccionadora
3601	MG paper	MG-papper	maschinenglattes Papier	papier frictionné	papel satinado una cara
3602	macerate	sönderdela	mazerieren, quellen, aufweichen, Schnitzelmaterial (als Fällstoff für Pressmassen)	macérer, amaigrir	macerar
3603	macerated	sönderdelad	aufgeweicht	macéré	macerado
3604	maceration	sönderdelning	Mazeration, Einweichung, Erweichung	macération	maceración
3605	machinability	bearbetbarhet	Bearbeitungsfähigkeit, Verarbeitungsfähig	aptitude au passage sur machine	labrabilidad
3606	machine	maskin	bearbeiten, spanabhebend bearbeiten, Maschine, Anlage	machine	máquina
3607	machine broke	utskott	Maschinenausschuss, Papiermaschinenausschuss	cassés de machine	recorte de máquina
3608	machine calender	maskinglätt	Maschinenkalander	calandre de machine	calandra de máquina
3609	machine chest	maskinkar	Maschinenbütte, Stoffbütte	cuvier de machine	tina de máquina
3610	machine clothing	maskinbeklädnad	alle beweglichen Teile einer Papiermaschine (Langsieb, Filz, Deckelriemen, siebleder, etc.)	habillage de machine	guarniciones
3611	machine coated	maskinbestruken	maschinengestrichen	couché machine	estucado en máquina
3612	machine-coated board	maskinbestruken kartong	maschinengestrichener Karton	carton couché machine	cartón estucado en máquina
3613	machine-coated paper	maskinbestruket papper	maschinengestrichene Papiere	papier couché machine	papel estucado en máquina
3614	machine coating	maskinbestrykning	Streichen in der Maschine	couchage sur machine	estucado en máquina
3615	machine creped	maskinkräppad	maschinengekreppt	crêpé sur machine	rizado en máquina
3616	machine deckle	maskindäckel	nutzbare Arbeitsbreite einer Maschine	largeur de table utilisée	anchura de la tela utilizada
3617	machine design	maskinkonstruktion	Maschinenausführung, Maschinenkonstruktion	plan d'une machine	diseño de máquina
3618	machine direction	maskinriktning	Maschinenrichtung, Längsrichtung	sens machine	dirección de máquina
3619	machine dried	maskintorkad	maschinengetrocknet	séché sur machine	secado en máquina
3620	machine finish	maskinfinish	Maschinenglätte, maschinenglatt	non apprêté, brut de machine	acabado en máquina
3621	machine-finish cover paper	täckpapper för —	Umschlagpapier (das durch ein Trockenglättwerk gelaufen ist)	papier couverture non apprêté	papel de encuadernar satinado
3622	machine finished (MF)	maskinglättad	maschinengeglättet, maschinenglatt	apprêté	satinado
3623	machine-finished paper	maskinglättat papper	maschinenglattes Papier	papier apprêté	papel satinado
3624	machine glazed	maskinglättad	einseitig glatt	frictionné	satinado en máquina
3625	machine-glazed paper	maskinglättat papper	einseitig glattes Papier	papier frictionné	papel satinado en máquina
3626	machine glazing	maskinglättning	Maschinenglätte	frictionné sur machine	satinado una cara
3627	machine hand	maskinsida	Maschinenantriebsseite	aide-conducteur	lado de transmisión de la máquina
3628	machine loading	maskinbelastning	Belastung einer Maschine	chargement d'une machine	carga de la máquina
3629	machine operation	maskindrift	Maschinenbedienung	fonctionnement d'une machine	funcionamiento de la máquina
3630	machine pit	guskgrop	Siebschiff	fosse sous machine	fosa bajo la tela
3631	machine room	maskinrum	Maschinenraum	salle des machines	sala de máquinas
3632	machinery	maskineri	Maschinen, Anlagen	machines, mécanisme	maquinaria
3633	machine shop	maskinverkstad	Maschinenwerkstatt	atelier d'entretien	taller mecánico
3634	machine tender	maskinförare	Maschinenführer	conducteur de machine	conductor de máquina de pap
3635	machine time	maskintid	Maschinenzeit	temps machine	tiempo de máquina

	ENGLISH	SVENSKA	DEUTSCH	FRANÇAIS	ESPAÑOL
3636	machine wire	maskinvira	Maschinensieb	toile de la machine	tela metálica de la máquina de papel
3637	machining	maskinbearbetning	Bearbeiten, spanabhebende Bearbeitung	usinage	labrado
3638	macromolecules	makromolekyler	Makromoleküle	macromolécules	macromoléculas
3639	Madeleine dryer	Madelainetork	Madelaine-Trockner	sécheur type Madeleine	secadora Madeleine
3640	magazine grinder	kannslip	Magazinschleifer	défibreur à magasine	desfibradora de almacén
3641	magazine paper	journalpapper	Zeitschriftenpapier	papier pour magazines	papel para revistas
3642	magazine stock	journalpappersmassa	Stoff für Zeitschriftenpapier, Magazinpapier	papier pour magazines	pasta de periódicos
3643	magnesia	magnesium	Magnesia, Talkerde	magnésie	magnesia
3644	magnesium	magnesium	Magnesium	magnésium	magnesio
3645	magnesium-base liquor	magnesiumbaslut	Magnefit-Lauge	liqueur à base de magnésium	lejía de magnesio
3646	magnesium bisulfite	magnesiumbisulfit	Magnesiumbisulfit	bisulfite de magnésium	bisulfito de magnesio
3647	magnesium carbonate	magnesiumkarbonat	Magnesiumkarbonat	carbonate de magnésie	carbonato magnésico
3648	magnesium chloride	magnesiumklorid	Magnesiumchlorid	chlorure de magnésie	cloruro magnésico
3649	magnesium compounds	magnesiumföreningar	Magnesiumverbindungen	composés de magnésium	compuestos de magnesio
3650	magnesium hydroxide	magnesiumhydroxid	Magnesiumhydroxid	hydrate de magnésium	hidróxido de magnesio
3651	magnesium oxide	magnesiumoxid	Magnesiumoxid, Magnesia	oxyde de magnésium	magnesia
3652	magnesium sulfate	magnesiumsulfat	Magnesiumsulfat, Bittersalz	sulfate de magnésium	sulfato de magnesio
3653	magnesium sulfite	magnesiumsulfit	Magnesiumsulfit	sulfite de magnésium	sulfito de magnesio
3654	magnet	magnet	Magnet	aimant	imán
3655	magnetic equipment	magnetisk utrustning	magnetische Ausrüstung	matériel magnétique	equipo magnético
3656	magnetic field	magnetfält	magnetisches Feld	champ magnétique	campo magnético
3657	magnetic ink	magnetiskt bläck	magnetische Tinte	encre magnétique	tinta magnética
3658	magnetic properties	magnetiska egenskaper	magnetische Eigenschaften	propriétés magnétiques	propiedades magnéticas
3659	magnetic separator	magnetisk separator	Magnetscheider, magnetische Trennvorrichtung	séparateur magnétique	separador magnético
3660	magnetic tape	magnetband	Magnettonband, Tonband, Magnetophonband	bande magnétique	cinta magnética
3661	main press	huvudpress	erste Presse	presse principale	prensa principal
3662	maintenance	underhåll	Wartung	entretien, maintenance	mantenimiento
3663	maintenance shop	underhållsverkstad	Wartungs- und Instandsetzungswerkstatt	atelier d'entretien	taller mecánico
3664	makeready	lappning	Zurichtung	mise en train	puesta a punto
3665	make-up liquor	ersättningslut	Frischlauge	lessive de complément	lejía de complemento
3666	make-up water	tillsatsvatten	Zusatzwasser, Betriebswasser	eau additionnelle	aqua de rellenar
3667	making order	tillverkningsorder	Auftrag nach Kundenspezifikation	commande sur fabrication	orden de fabricación
3668	maleic acid	maleinsyra	Maleinsäure	acide maléique	ácido málico
3669	maleic anhydride	maleinsyreanhydrid	Maleinsäureanhydrid	anhydride maléique	anhidrido málico
3670	malfunction	mankemang	Versagen, Funktionsstörung	mauvais fonctionnement	funcionamiento defectuoso
3671	maltose	maltos	Maltose, Malzzucker	maltose	maltosa
3672	manager	administratör	Betriebsleiter, Direktor, Geschäftsführer	directeur	administrador
3673	mandrel	spindel	Rollstange	mandrin	mandril
3674	manganese	mangan, brunsten	Mangan	manganèse	manganeso
3675	manganese compounds	manganföreningar	Manganverbindungen	composés du manganèse	compuestos de manganeso
3676	manifold	mångfaldigande	mannigfaltig, mehrfach, vielfach, Verteilerstück, Rohrverzweigung	collecteur (de gaz)	colector
3677	manifold paper	kopiepapper	Abzugspapier, Durchschreibpapier, Vervielfältigungspapier	papier pelure pour doubles	papel para copias
3678	manila	manilla	Manila	chanvre de Manille	manila
3679	manila board	manillakartong	Manilakarton	carton brun de pâte mi-chimique	cartón de Manila
3680	manila fiber	manillafiber	Manila, Manilafaser	fibre de chanvre de Manille	fibra de cáñamo de Manila
3681	Manila hemp	manillahampa	Manilahanf	chanvre de Manille	cáñamo de Manila
3682	manila-lined boxboard	manillatäckt lädkartong	Manila-Faltschachtelkarton	carton pour boîtes doublé de papier écru	cartón para cajas forrado de manila
3683	manila paper	manillapapper	Manilapapier	papier de pâte chimique écrue	papel Manila
3684	mannans	mannaner	Mannan (Hemizellulose)	mannans	anhidridos de manosa
3685	mannogalactan	mannogalaktan	Mannogalaktan	mannogalactan	manogalactan
3686	mannose	mannos	Mannose	mannose	manosa
3687	manufacture	tillverkning	Herstellung, Fabrikation, Erzeugnis, Fertigware, Fabrikat	fabrique, manufacture, fabrication	fabricación
3688	manufacturers' joint	fabriks-skarv	Falzkante in Wellpappenbehältern	bande gommée sur toile pour emballage	agrupación de fabricantes
3689	maple	lönn	Ahorn	érable blanc	arce blanco
3690	map paper	kartpapper	Landkartenpapier	papier pour cartes géographiques	papel cartográfico

	ENGLISH	SVENSKA	DEUTSCH	FRANÇAIS	ESPAÑOL
3691	marble paper	marmorerat papper	Marmorpapier	papier marbré	papel para jaspear
3692	marbled paper	marmorerat papper	marmoriertes Papier	papier marbré	papel jaspeado
3693	marbling	marmorering	marmorieren, einen Marmoreffekt schaffen	jaspage	jaspear
3694	mark	märke, markering	markieren, kennzeichnen, auszeichnen (Ware), Markierung, Warenzeichen, Kennzeichnung	marque	marca
3695	marker	markör	Markierung, Markierungsvorrichtung, Anzeiger	marqueur, signet, pavillon	marca de papel
3696	marketing	marknadsföring	Marktwesen, Absatzwesen, Marktversorgung, Absatz, Vertrieb	commercialisation, marketing	marketing
3697	market pulp	avsalumassa	Marktzellstoff	pâte commercialisée	pasta para la venta
3698	market research	marknadsundersökning	Marktforschung	étude de marché	estudio de mercados
3699	marking	markering	Markierung, Kennzeichnung, Signieren	marquage	acción de marcar
3700	marking device	markeringsmedel	Markiervorrichtung	dispositif marqueur	instrumento marcador
3701	marking felt	markerfilt	Markierfilz	feutre marqueur	fieltro marcador
3702	marking roll	markervals	Wasserzeichen-Egoutteur	mollette de marquage	moleta
3703	masking	maskning	Abdeckung, Maskierung, Abdecken bestimmter Flächen beim Aufbringen von Farbe	blocage	obturación
3704	masking agent	maskningsmedel	Abdeckmittel	produit provoquant un blocage	agente de obturación
3705	mass	massa, vikt	ansammeln, anhäufen, Masse, Haufen	masse	masa
3706	Massey coater	Massey-bestrykare	Massey-Streicher	coucheuse type Massey	estucadora Massey
3707	mass spectroscopy	masspektroskopi	Massenspektroskopie	spectroscopie de masse	espectroscopia de la masa
3708	mass transfer	massöverföring	Stoffübergang	transfert de masse	transferencia de la masa
3709	mat	matta, matt	matt, glanzlos, mattiert, Matte, Matrize	matelas (de fibres)	mate
3710	mat board	matt kartong	Karton für Untersetzer, Nut- und Spundbrett	carton pour dessous de verres	cartón mate
3711	match	passa, tändsticka	zusammenpassen, entsprechen, Gegenstück, Zündholz	comparer, échantillonner	fósforo
3712	matching	passning	passend, zusammenpassend	comparaison	comparación
3713	material	material	Material, Stoff, Werkstoff	matériel	material
3714	materials balance	materialbalans	Verhältnis von Stoffen zueinander, Materialienverhältnis	équilibre pondéral	balance de materiales
3715	materials consumption	materialförbrukning	Stoffverbrauchen, Materialverbrauch	consommation de matières premières	consumo de materiales
3716	materials handling	materialhantering	Materialverwaltung, Materialbeförderung	manutention des matières premières	manipulación de materiales
3717	materials testing	materialprovning	Materialprüfung, Stoffprüfung	essais de matières premières	pruebas de materiales
3718	mathematical model	matematisk modell	mathematisches Modell	modèle mathématique	modelo matemático
3719	mat paper	matt papper	mattes Papier, glanzloses Papier, Mattpapier	papier mat	papel mate
3720	matrix paper	matrispapp	Matrizenpapier	papier pour flans	papel para matrices
3721	matte finish	matt finish	matte Oberfläche	fini mat	acabado mate
3722	mature wood	mogen ved	erntereifes Holz	bois mûr	madera madura
3723	maturity	mognad	Reife	maturité	madurez
3724	maximum capacity	maximal kapacitet	maximales Fassungsvermögen, Höchstbelastung (einer Leitung), maximale Aufnahmefähigkeit	capacité de pointe	capacidad máxima
3725	maximum deckle	maximal däckel	maximal nutzbare Siebbreite	largeur de toile utile	anchura útil de la tela
3726	measurement	mätning	Masseinheit, Abmessung, Messung	mesure	medición
3727	measuring instrument	mätinstrument	Messgerät, Messinstrument	instrument de mesure	instrumentos de medida
3728	mechanical classifier	mekanisk separator	mechanischer Sortierer	classeur mécanique pour l'étude des fibres par la méthode du fractionnement	clasificador mecánico
3729	mechanical deckle-edge paper	oskuret trämassepapper	Imitationsbüttenrandpapier (wobei die Papierkanten mechanisch nachbehandelt werden)	papier à bords frangés imitant le papier à la main	papel mecánico de barba
3730	mechanical degradation	sönderdelning	mechanischer Abbau, mechanische Verringerung	dégradation mécanique	degradación mecánica
3731	mechanical drive	mekanisk drift	mechanischer Antrieb	commande mécanique	impulso mecánico
3732	mechanical measurement	mekanisk mätning	mechanische Messung	mesure mécanique	medición mecánica
3733	mechanical properties	mekaniska egenskaper	mechanische Eigenschaften	propriétés mécaniques	propiedades mecánicas
3734	mechanical pulp	mekanisk massa	Holzschliff	pâte mécanique	pasta mecánica
3735	mechanical pulping	mekanisk massatillverkning	mechanischer Aufschluss	fabrication de pâtes mécaniques	elaboración de pasta mecánica
3736	mechanical strain gauge	mekanisk töjningsgivare	mechanischer Dehnungsmesser	jauge mécanique de tension	medidor de formación

	ENGLISH	SVENSKA	DEUTSCH	FRANÇAIS	ESPAÑOL
3737	mechanical straw pulp	mekanisk halmmassa	Strohzellstoff	pâte mécanique de paille	pasta mecánica de paja
3738	mechanical wood pulp	mekanisk trämassa	Holzschliff	pâte de bois mécanique	pasta mecánica
3739	mechanics	mekanik	Mechanik	mécanique	mecánica
3740	mechanization	mekanisering	Mechanisierung	mécanisation	mecanización
3741	medicated paper	medicinskt preparerat papper	antiseptisches Papier	papier antiseptique	papel medicinal
3742	medium finish	mellanfinish	mittlere Oberflächenausrüstung (weder glatt noch rauh)	fini intermédiaire	acabado medio
3743	melamine	melamin	Melamin	mélamine	melamina
3744	melamine-formaldehyde resin	melamin-formaldehydharts	Melaminharze, Melaminformaldehydharze	résine mélamine-formaldéhyde	resina de melamina-formaldehido
3745	melting	smältning	Schmelzen, Schmelzung, Schmelze	fusion	fusión
3746	melting point	smältpunkt	Schmelzpunkt	point de fusion	punto de fusión
3747	membrane	membran	Membran, Häutchen, Diaphragma	membrane	membrana
3748	mensuration	mätning	Abmessung, Vermessung	mesurage	medición
3749	mercaptans	merkaptaner	Merkaptane, Thiolalkohole	mercaptans	mercaptanos
3750	mercapto lignins	merkaptolignin	Merkaptolignine	mercapto lignines	mercapto ligninos
3751	mercerization	mercerisering	Merzerisierung	mercerisation	mercerización
3752	mercury	kvicksilver	Quecksilber	mercure	mercurio
3753	mercury compounds	kvicksilverföreningar	Quecksilberverbindungen	composés du mercure	compuestos de mercurio
3754	merger	fusion	Fusion, Zusammenschluss	fusion, absorption	fusión
3755	meristems	meristems, tillväxtvävnad	Meristeme, Grundgewebe	méristème	meristemos
3756	mesh	maska	ineinandergreifen, Maschenweite (des Siebs), Siebweite, Sieböffnung, Eingriff	maille (toile de machine)	malla
3757	mesomorphous cellulose	mesomorf cellulosa	mesomorphe Cellulose	cellulose mésomorphe	celulosa mesoforma
3758	metacellulose	metacellulosa	Metacellulose	métacellulose	metacelulosa
3759	metal	metall	Metall, Metallegierung	métal	metal
3760	metal compounds	metallföreningar	Metallverbindungen	composés métallifères	compuestos de metal
3761	metal fiber	metallfiber	Metallfaser	fibre métallique	fibra metálica
3762	metal finishing	metallfinish	metallische Ausrüstung	métallisation	acabado metálico
3763	metal foil	metallfolie	Metallfolie, Blattmetall	feuille mince de métal	laminilla metalizada
3764	metalization	metallisering	Metallisierung, Imprägnierung mit Metallsalzen	métallisation	metalización
3765	metallic coating	metallisk bestrykning	Metallbeschichtung, metallischer Strich	couchage métallique (sur caséine ou laque)	estucado metálico
3766	metallic ink	metalliskt bläck	Bronzedruckfarbe	encre métallisée	tinta para metales
3767	metallic paper	metalliskt papper	metallisches Papier, Metallpapier	papier métallisé	papel metálico
3768	metallized paper	metalliserat papper	metallisiertes Papier, mit Metallsalzen imprägniertes Papier	papier métallisé	papel metalizado
3769	meteorology	meteorologi	Meteorologie	météorologie	meteorología
3770	meter	meter, mätare	messen, dosieren, zuteilen, Messinstrument, Zähler, Zählwerk	mètre	contar
3771	metering apparatus	mätapparat	Anzeigegerät, Dosiergerät	appareil mesureur	aparato de medición
3772	metering equipment	mätutrustning	Messvorrichtung, Dosiervorrichtung	matériel de mesure	equipo de medición
3773	metering system	mätsystem	Dosiersystem, Zuteilsystem, Messystem	dispositif mesureur	sistema métrico, sistema de medición
3774	methacrylates	metakrylater	Methacrylate, Methacrylsäureester	méthacrylates	metacrilatos
3775	methacrylic acid	metakrylsyra	Methacrylsäure	acide méthacrylique	ácido metacrílico
3776	methane	metan	Methan	méthane	metano
3777	methanol	metanol	Methanol	méthanol	metanol
3778	methods engineering	metodteknologi	Verfahrenstechnik	technique des méthodes	métodos de dirección
3779	methyl acetate	metylacetat	Methylacetat	acétate de méthyl	metil acetato
3780	methyl acrylate	metylakrylat	Methylacrylat	acrylate de méthyl	metil acrilato
3781	methyl alcohol	metylalkohol	Methylalkohol	alcool méthylique	alcohol metílico
3782	methylation	metylering	Methylierung	méthylation	metilación
3783	methyl cellulose	metylcellulosa	Methylcellulose	méthyl cellulose	metil celulosa
3784	methylene cellulose	metylencellulosa	Methylencellulose	méthylène cellulose	metileno celulosa
3785	methyl ethyl ketone	metyletylketon	Methyl-Äthylketon	méthyl ethyl ketone	metil etil cetona
3786	methyl glucosides	metylglukosider	Methylglukoside	glucosides de méthyl	metil glucósidos
3787	methyl groups	metylgrupper	Methylgruppen	groupes méthyl	grupos de metilo
3788	methyl mercaptan	metylmerkaptan	Methylmerkaptan	méthyl mercaptan	metil mercaptan
3789	methyl methacrylate	metylmetakrylat	Methylmethacrylat	méthacrylate de méthyl	metil metacrilato
3790	methyl orange	metylorange	Methylorange	méthyl orange	anaranjado de metilo
3791	metric system	metriskt system	metrisches System, Dezimalsystem	système métrique	sistema métrico
3792	metric ton	metrisk ton	metrische Tonne	tonne métrique	tonelada métrica
3793	mica	glimmer	Glimmer	mica	mica
3794	mica paper	glimmerpapper	Atlassatinpapier, Glimmerpapier	papier micacé	papel de mica
3795	microanalysis	mikroanalys	Mikroanalyse	micro-analyse	microanálisis
3796	micro-corrugated board	mikrovågpapp	Mikrowellpappe	carton ondulé microcannelé	cartón micro-ondulado
3797	microcrystalline cellulose	mikrokristallin cellulosa	mikrokristalline Cellulose	cellulose microcristalline	celulosa microcristalina

	ENGLISH	SVENSKA	DEUTSCH	FRANÇAIS	ESPAÑOL
3798	microcrystalline wax	mikrokristallint vax	Mikrokristallinwachs	cire microcristalline	cera microcristalina
3799	microfibril	mikrofibrill	Mikrofibrille	microfibrille	microfibrilla
3800	microfiche	mikrofiche	–	micro fiche	microficha
3801	microfilm	mikrofilm	Mikrofilm	micro film	microfilme
3802	micrometer	mikrometer	Mikrometer, Feinmesser, Dickenmesser	micromètre d'épaisseur	micrómetro
3803	microorganism	mikroorganism	Mikroorganismus	microorganisme	microorganismo
3804	microorganism control	mikroorganismkontroll	Kleinlebewesenkontrolle	contrôles des microorganismes	control de microorganismo
3805	microporosity	mikroporositet	Mikroporosität	micro-porosité	microporosidad
3806	microscopy	mikroskopi	Mikroskopie	microscopie	microscopio
3807	microtome	mikrotom	Mikrotom (zur Herstellung von Papierquerschnitten)	microtome	microtomo
3808	microwave spectroscopy	mikrovågspektroskopi	Mikrowellenspektroskopie	spectroscopie aux microondes	espectroscopia de microondas
3809	middle lamellae	mittlamell	Mittellamelle	lamelle moyenne	laminilla media
3810	mid-feather	skiljevägg	Scheidewand, Zwischenwand	cloison centrale (épi d'une pile)	espiga
3811	migration	migration, migrering, vandring	Migration, Wanderung (z.B. des Bindemittels)	migration	migración
3812	mil	mil, tusendels tum	–	millième de pouce	milipulgada
3813	mildew	mögel	modern, Moder	moisissure	moho
3814	milk carton	mjölkförpackning	Milchkarton	carton pour l'emballage du lait	cartón para contenedores de leche
3815	milk of lime	kalkmjölk	Kalkmilch	lait de chaux	lechada de cal
3816	mill	fabrik, kvarn	fräsen, mahlen, walzen, rändeln, Werk, Fabrik	usine, malaxeur	fábrica
3817	mill blanks	bolagsformulär, firmaformulör	Einlagenkarton, beidseitig weisser Karton für Menukarten	carton deux côtés blancs	cartones multiplex
3818	mill boards	handpapp	Gautschpappe, Buchbinderpappe	celloderme à l'enrouleuse	cartones grises
3819	mill brand	handpapp	mit Wasserzeichen versehenes Papier	marque de fabrique	marca de fábrica
3820	mill broke	utskott	Maschinenausschuss, Kollerstoff	cassés de fabrication	recortes de fabricación
3821	mill count	fabriksräkning	Zählung einer Lieferung durch das Werk	comptage à l'usine	cuenta en fábrica
3822	mill cut	fabriksskärning	unbeschnitten	bords barbés	corte de fábrica
3823	mill edge	maskingjord kant	Maschinenrand, rauhe Papierkante	bord non rogné	borde laminado
3824	milled wood lignins	ligninmjöl, MWL	Kantholzlignine	lignines de bois broyé	ligninas de madera trituradas
3825	mill effluent	fabriksavlopp	Fabrikabwasser	effluent d'usine	efluentes de fábrica
3826	mill-finished	slipad	maschinenglatt	apprêté machine	acabado en fábrica
3827	mill manager	fabrikschef	Werksdirektor, Werksleiter	directeur d'usine	director de fábrica
3828	mill roll	kvarnkubb	Maschinenrolle (nicht umgerollt)	bobine brute (non rebobinée)	rollo de máquina
3829	mill wrappers	fabriksomslag	Werkseinschlagpapier, Blaupack (zum Verpacken von Riesen)	macules pour rames	embaladoras en fábrica
3830	millwright	mekaniker	Werksmechaniker	mécanicien d'usine	mecánico
3831	mineral deposit	mineralavsättning	Erzlagerstätte, Mineralienablagerung	dépôt minéral	depósito mineral
3832	mineral fiber	mineralfiber	Mineralfaser	fibre minérale	fibra mineral
3833	mist	dimma	feiner Nebel, feinzerstäubte Flüssigkeit	chiné, granité	neblina
3834	misting	dimbildning	Zerstäuben, Beschlagen	chiner, graniter	empañado
3835	MIT folding tester	MIT dubbelvikngsprovare	MIT-Falzfestigkeitsprüfgerät	appareil "MIT" de mesure de la résistance au pliage	probador MIT de pliegues
3836	Mitscherlich process	Mitscherlichprocess	Mitscherlich-Verfahren	procédé Mitscherlich (bisulfite)	proceso Mitscherlich
3837	mix	blandning	mischen, vermengen, Gemisch	mélanger	mezclar
3838	mixed-base liquor	blandlut	–	lessive composée	lejía con mezcla de bases
3839	mixed papers	blandpapper	unsortiertes Altpapier	papiers mêlés	papeles mezclados
3840	mixed stands	blandskog	Mischwaldbestand	peuplements hétérogènes	soportes mezclados
3841	mixer	blandare	Rührer, Mischmaschine	mélangeur	mezclador
3842	mixing	blandning	Mischen, Mischung	mélange	mezcla
3843	mixing box	blandningslåda	Mischkasten	caisse de mélange	caja de mezcla
3844	mixing chest	blandningskar	Mischbütte	cuvier de mélange	caja de mezcla
3845	mixing pump	blandningspump	Mischpumpe	pompe de mélange	bomba de mezcla
3846	mixture	blanding	Mixtur, Mischung, Gemisch	mélange	mezcla
3847	mode of operation	arbetssätt	Arbeitsweise, Bedienungsweise	mode d'action	sistema operacional
3848	model	modell	Modell, Muster, Vorlage	modèle	moldear
3849	modernization	modernisering	Modernisierung	modernisation	modernización
3850	modified cellulose	modiferad cellulosa	modifizierte Cellulose	cellulose modifiée	celulosa modificada
3851	modified starches	modifierade stärkelser	umgewandelte Stärken	amidons modifiés	almidones modificados

	ENGLISH	SVENSKA	DEUTSCH	FRANÇAIS	ESPAÑOL
3852	modifier	modifieringsmedel	Modifiziermittel, Umwandler, Umsteuergrösse	modificateur	modificador
3853	modifying treatment	modifierande behandling	eine Umwandlung bewirkende Behandlung	traitement modificateur	tratamiento de modificación
3854	modulus	modul	Modul	module	módulo
3855	modulus of elasticity	elasticitetsmodul	Elastizitätsmodul	module d'élasticité	módulo de elasticidad
3856	modulus of rupture	brottmodul	Bruchmodul	module de rupture	módulo de ruptura
3857	moist	fukt	feucht, nass	humide	húmedo
3858	moistener	befuktningsanordning	Anfeuchter, Befeuchter	humecteuse	humectadora
3859	moistening	befuktning	Befeuchten, Anfeuchten	mouillage	humectar
3860	moisture	fukt	Feuchtigkeit	humidité	humedad
3861	moisture content	fukthalt	Feuchtigkeitsgehalt	teneur en eau	contenido de humedad
3862	moisture control	fuktreglering	Feuchtigkeitskontrolle, Feuchtigkeitsregulierung, Feuchtigkeitssteuerung	contrôle de l'humidité	control de humedad
3863	moisture meter	fuktmätare	Feuchtigkeitsmesser, Feuchtemesser	appareil pour mesurer l'humidité	indicador de humedad
3864	moisture pickup	fuktupptagning	Feuchtigkeitsaufnahme	fixateur de l'humidité	toma automática de la humedad
3865	moistureproof	fuktsäker, fukttät	wasserdampfundurchlässig, feuchtigkeitsundurchlässig, mit Feuchtigkeitsschutz	résistant à l'humidité	a prueba de humedad
3866	moistureproofness	fukttäthet	Wasserdampfundurchlässigkeit, Feuchtigkeitsundurchlässigkeit	résistance à l'humidité	resistencia a la humedad
3867	moisture resistance	fuktmotstånd	Beständigkeit gegen Feuchtigkeit	résistance à l'humidité	resistencia a la humedad
3868	moisture setting	fuktinställning	direkt trocknend, in der Feuchtigkeit abbindend	adaptation à l'humidité	regulación de la humedad
3869	moisture tests	fuktprov	Feuchtigkeitsprüfung, Feuchtigkeitsmessung	essais d'humidité	pruebas de humedad
3870	moisture welts	kantband	Feuchtigkeitsschwielen	traces d'humidité	pliegues de humedad
3871	mold	form	Form, Schöpfform, Schimmel	moisissure	forma
3872	moldability	formbarhet	Verpressbarkeit, Verformbarkeit	aptitude à la moisissure	aptitud al moldeado
3873	molded board	rundvirakartong	gezogene Pappe, Ziehpappe	carton moulé	cartón de moldear
3874	molded products	rundviraprodukter	gepresste Sauglinge (aus Faserbrei und Harz), Presstücke, Formteil	produits en pâte moulée	productos moldeados
3875	molded pulp	formgjuten massa	Fasergussmasse	pâte moulée	pasta de moldear
3876	molded pulp package	formgjuten förpackning	Verpackung aus Fasergussmasse	emballage en pâte moulée	paquete de pasta de moldear
3877	molded pulp products	formgjutna massaprodukter	Fasergussprodukte	produits en pâte moulée	productos de pasta moldeada
3878	mold felt	rundvirafilt	Rundsiebfilz	feutre pour forme ronde	fieltro de forma
3879	molding	formning, formgjutning	Formen, Formgebung	moulage	moldear
3880	mold machine	rundviramaskin	Formmaschine	machine à forme ronde	máquina de formas
3881	mold-resistant papers and boards	mögelresistentpapper och kartong	schimmelfestes Papier und Karton	papiers et cartons résistant aux moisissures	papeles y cartones resistentes al moldeo
3882	molecular structure	molekylstruktur	Molekülbau	structure moléculaire	estructura molecular
3883	molecular weight	molekylarvikt	Molekulargewicht, Molgewicht	poids moléculaire	peso molecular
3884	molecules	molekyler	Moleküle	molécules	moléculas
3885	molybdenum	molybden	Molybdän	molybdène	molibdeno
3886	molybdenum compounds	molybdenföreningar	Molybdänverbindungen	composés du molybdène	compuestos de molibdeno
3887	monitor	monitor	überwachen, ständig messen, Monitor, Kontrolleinrichtung, Warngerät, Messeinrichtung	programmateur, appareil de contrôle	monitor
3888	monitoring	monitorera	Kontrolle, Überwachung	programmation, contrôle	monitoraje
3889	monocotyledons	monokotyledoner	einkeimblättrige Pflanzen	monocotylédons	monocotiledones
3890	monofilament	enkeltråd	Einzelfaden, Elementarfaden	monofilament	monofilamento
3891	monolayer	monoskikt	monomolekulare Schicht	couche monomoléculaire	de una capa
3892	monomers	monomerer	Monomere	monomères	monomeros
3893	monotype machine	monotypmaskin	Monotypemaschine	machine monotype	monotipia
3894	montmorillonite	montmorillonit	Montmorillonit	montmorillonite	montmorillonita
3895	mordants	betningsmedel	Ätzmittel, Beizmittel	mordant	mordientes
3896	mortar	bruk	Mörtel	mortier	mortero
3897	motion	rörelse	Bewegung, Gang	mouvement	movimiento
3898	motor	motor	Triebkraft, Motor, Kraftmaschine	moteur	motor
3899	mottle	melerad	scheckig (durch unregelmässige Farbannahme)	marbrure	jaspear
3900	mottled	melerad	gescheckt, marmoriert, meliert, gesprenkelt	marbré	jaspeado
3901	mottled color	melerad färg	fleckig, geschecktfarbig	couleur marbrée	color jaspeado
3902	mottled finish	melerad finish	melierte Oberfläche, scheckige Oberfläche	fini marbré	acabado jaspeado
3903	mottled paper	melerat papper	marmoriertes Papier, meliertes Papier	papier marbré	papel jaspeado
3904	mottling	melering	Scheckigwerden	marbrer	jaspeado
3905	mould (Brit.)	gjutform, viratråg	Form, Schöpfform, Schimmel	moisissure, humus	forma

	ENGLISH	SVENSKA	DEUTSCH	FRANÇAIS	ESPAÑOL
3906	mud	dy	Schmutz, Schlamm, Schlick	boue	lodo
3907	mullen	mullen	Mullen	résistance à l'éclatement d'après Mullen	Mullen
3908	Mullen tester	Mullenprovare	Mullen-Berstfestigkeitsprüfer	éclatomètre Mullen	probador Mullen de reventamiento
3909	multi-fourdrinier machine	flerviramaskin	Mehrfachlangsieb	machine à table plate	máquina múltiple de mesa plana
3910	multicellular suction roll	flercellig sugvals	Vielkammersaugwalze	rouleau aspirant multicellulaire	rodillo aspirante multicelular
3911	multicomponent coating	flerkomponentbestrykning	Mehrkomponentenstreichverfahren	couchage à plusieurs composants	estucado compuesto
3912	multicylinder machine	mångcylindermaskin	Mehrzylindermaschine, Mehrrundsiebmaschine	machine à forme ronde multicylindres	máquina de varios cilindros
3913	multilayer	flerskikt	Mehrlagen	multicouche	multicapa
3914	multilayer board	flerskiktskartong	Mehrlagenkarton	carton multijets	cartón multicapa
3915	multilayer paper board	flerskiktskartong	Mehrlagenkarton	carton composé de plusieurs couches de papier	cartón multicapa
3916	multiple blade unit	flerschaberenhet	Mehrfachabstreifer	–	deflector múltiple
3917	multiple coating	flerskiktsbestrykning	Mehrfachstreichen	couchage multiple	estucado múltiple
3918	multiple-ply former	multiplexformare	Mehrlagenformierungseinrichtung	formeur multijet	plantilla de dos o más capas
3919	multiple-ply sheet	flerlagsark	Mehrlagenbahn	feuille en plusieurs jets	hoja multicapa
3920	multiple use	mångsidig användning	Mehrfachverwendung, Mehrzweckverwendung, Mehrfachbenutzung, Mehrfachabstreifer	utilisations multiples	uso múltiple
3921	multi-ply board	multiplexkartong	Mehrlagenkarton	carton en plusieurs jets	cartón multicapa
3922	multi-ply	flerlags, flerskikt	mehrlagig, mehrschichtig	en plusieurs jets	dos o más capas
3923	multi-ply paper	flerskiktspapper	mehrlagiges Papier, mehrschichtiges Papier	papier en plusieurs jets	papel multicapa
3924	multistage bleaching	flerstegsblekning	Mehrstufenbleiche	blanchiment en plusieurs stades	blanqueado de varios pasos
3925	multistage process	flerstegsprocess	Mehrstufenverfahren	procédé en plusieurs stades	proceso de varios pasos
3926	multistage pump	flerstegspump	Differentialpumpe	pompe multicellulaire	bomba multicelular
3927	multivat board machine	flervirakartongmaskin	Mehrrundsiebmaschine	machine à carton multiforme	máquina de cartón multiforma
3928	multivat machine	flerviramaskin	Mehrrundsiebmaschine	machine multiforme	máquina multiforma
3929	multiwall bag	flerlagssäck	mehrlagige Säcke	sac en plusieurs épaisseurs de papier	saco de varias hojas
3930	multiwall bag kraft paper	flerlagssäckpapper	Mehrlagenbeutelkraftpapier	papier kraft en plusieurs épaisseurs pour sacs G.C.	papel kraft para sacos de varias hojas
3931	multiwall corrugated board	flerlagsvågpapp	Mehrlagenwellpappe	carton ondulé en plusieurs épaisseurs	cartón ondulado de varias hojas
3932	multiwire machine	flerviramaskin	Mehrsiebmaschine, Vielsiebmaschine	machine à plusieurs toiles	máquina con varias telas
3933	municipal waste	tätortsavfall	Stadtmüll	ordures municipales	basura municipal
3934	Murray pine	Murray pine	Murraykiefer	pin de Murray	pino de Murray
3935	mutation	mutation	Umwandlung, Mutation	mutation	mutación

N

3936	NCR (no carbon required)	NCR-papper	NCR-Papier (kein Kohlepapier erforderlich)	auto-copiant	sin necesidad de carbono
3937	NSSC (neutral sulfite semichemical)	neutralsulfitmassa	halbchemisches Neutralsulfitverfahren	N.S.S.C., procédé mi-chimique au sulfite neutre	sulfito neutro semiquímico
3938	naphtha	nafta	Naphtha, Kerosin, Leuchtpetroleum	naphte	nafta
3939	napkin papers	servettpapper	Serviettenpapier	papier pour nappes	papeles para servilletas
3940	napkin tissue	servettpapper	Serviettentissue	papier mince pour nappes	muselina para servilletas
3941	natural browns	naturliga bruna färger	braunes Kraftpackpapier	papiers d'emballage écrus	pardos naturales
3942	natural colored	naturfärgad	naturfarben	coloré naturellement	coloreado natural
3943	natural fiber	naturfiber	Naturfaser	fibre naturelle	fibra natural
3944	natural gas	naturgas	Erdgas	gaz naturel	gas natural
3945	natural reforestation	naturlig skogsåtervväxt	natürliche Wiederaufforstung	reboisement naturel	repoblación forestal natural
3946	natural resource	näturtillgångar	Naturvorkommen, Naturschätze	ressources naturelles	recursos naturales
3947	naval stores	marinens förråd	Bevorratung für die Marine	tous les produits extraits du bois de pin	pertrechos navales
3948	needle bearing	nållager	Nadellager	roulement à aiguilles	rodamiento de agujas
3949	needle valve	nålventil	Nadelventil	robinet à pointeau	llave de punzón
3950	needled felt	nålad filt	genadelter Filz	feutre aiguilleté	fieltro a la aguja
3951	negative crown	negativ bombering	konkave Bombierung	bombé concave	corona negativa

	ENGLISH	SVENSKA	DEUTSCH	FRANÇAIS	ESPAÑOL
3952	neoprene	neopren	Neopren, Polychlorbutadien	néoprène	neopreno
3953	nestable containers	inpassbara behållare	zusammenstellbare Containersätze	emballages qui s'emboîtent les uns dans les autres	contenedores encajables
3954	neutralization	neutralisering	Neutralisation	neutralisation	neutralización
3955	neutral kraft	neutralkraftmassa	säurefreies Kraftpapier	kraft neutre	kraft neutro
3956	neutral size	neutrallim	neutraler Kleber	collage neutre	cola neutra
3957	neutral sulfite liquor	neutralsulfitlut	Neutralsulfitlauge	lessive au sulfite neutre	lejía de sulfito neutro
3958	neutral sulfite semichemical (NSSC) pulp	neutralsulfitmassa	neutraler halbchemischer Sulfitzellstoff	pâte mi-chimique au sulfite neutre (monosulfite)	pasta semiquímica de sulfito neutro
3959	neutral sulfite semichemical pulping	neutralsulfitmassatillverkning	neutraler Sulfitaufschluss	fabrication de pâtes mi-chimiques au monosulfite	elaboración de pasta semiquímica al sulfito neutro
3960	news	tidningspapper, nyheter	Nachrichten, Zeitungsdruckpapier	vieux journaux	papel ordinario de periódicos
3961	newsboard	anslagstavla	Feingraupappe	carton gris de vieux journaux	carton prensa
3962	news bogus paper	imitationstidningspapper	Imitationszeitungsdruck (ausschliesslich aus altem Zeitungsdruck hergestellt)	papier à base de vieux journaux	papel simili para periódicos
3963	news grade	tidningspapperskvalitet	Zeitungsdruckpapierqualität	qualité journal	papel tipo periódico
3964	news-lined board	trämassefodrad kartong	beidseitig mit Zeitungsdruckpapier kaschierte Schrenzpappe	carton recouvert de papier à base de papiers recyclés	cartón forrado de papel periódico
3965	newspaper	tidning	Zeitung	journal	periódico
3966	newsprint	tidningspapper	Zeitungsdruckpapier	papier journal	papel prensa
3967	newsprint sheets	tidningspappersark	Zeitungsdruckpapierbogen	papier journal en feuilles	hojas de papel prensa
3968	news vat-lined chipboard	trämassefodrad gråpapp	beidseitig mit Zeitungsdruckpapier kaschierte Schrenzpappe	carton gris à base de vieux papiers fabriqué sur machine multiforme	cartón gris de recortes forrado en tina
3969	nickel	nickel	Nickel	nickel	níquel
3970	nickel compounds	nickelföreningar	Nickelverbindungen	composés du nickel	compuestos de níquel
3971	ninepoint corrugating material	niopunkt(0009) tums vå gpappersmaterial	Wellpappe von 140-150 g/m2 Flächengewicht	microcannelure	material a ondular del nueve
3972	ninepoint semichemical board	niopunkt(0009) tums halvkemisk kartong	halbchemischer Karton von 140-150 g/cm2	carton de pâte mi-chimiques	cartón semiquímico del nueve
3973	ninepoint strawboard (.009)	niopunkt(0009) tums halmkartong	Strohpappe von 140-150 g/cm2	carton paille	cartón paja del nueve
3974	nip	nyp	Spalt, Berührungsstelle der Walzen (z.B. am Kalander)	ligne de contact entre deux presse, ligne de contact entre deux rouleaux	línea de tangencia
3975	nip guard	nypskydd	Einlaufschutz, Einlaufschutzstange	garde-mains	defensas
3976	nipple	nippel	Nippel, Schmiernippel	raccord	manguito roscado
3977	nip pressure	nyptryck, linjetryck	Liniendruck	pression linéaire	presión lineal
3978	nitrate groups	nitratgrupper	Nitratgruppen	groupes des nitrates	grupos nitrato
3979	nitrates	nitrater	Nitrate	nitrate	nitratos
3980	nitration	nitrering	Nitrieren, Nitrierung	nitration	nitración
3981	nitric acid	salpetersyra	Salpetersäure	acide nitrique	ácido nítrico
3982	nitric acid pulp	salpetersyramassa	Salpetersäurezellstoff	pâte à l'acide nitrique	pasta al ácido nítrico
3983	nitric acid pulping	salpetersyrakokning	Salpetersäureaufschluss	fabrication de pâtes à l'acide nitrique	elaboración de pasta al ácido nítrico
3984	nitrides	nitrider	Nitride	nitrures	nitruros
3985	nitrification	nitrering	Salpeterbildung, Nitrierung	nitrification	nitrificación
3986	nitriles	nitriler	Nitrile	nitriles	nitrilos
3987	nitrites	nitriter	Nitrite, salpetrigsaure Salze	nitrites	nitritos
3988	nitrocellulose	nitrocellulosa	Nitrocellulose, Schiessbaumwolle	nitrocellulose	nitrocelulosa
3989	nitrogen	kväve	Nitrogen, Stickstoff	azote	nitrógeno
3990	nitrogen compounds	kväveföreningar	Stickstoffverbindungen	composés de l'azote	compuestos de nitrógeno
3991	nitrolignins	nitroligniner	Nitrolignin	nitrolignines	nitroligninas
3992	no-carbon-required (NCR) paper	NCR-papper	NCR (kein Kohle-Papier)	papier auto-copiant	papel sin necesidad de carbono
3993	nodule	nodel, flinga	Knollen, Knötchen	nodule	nódulo
3994	noise	buller	Lärm, Geräusch	bruit	ruido
3995	noise control	bullerkontroll	Geräuschkontrolle, Lärmkontrolle	contrôle du bruit	control de ruido
3996	noise meter	bullermätare	Geräuschmesser, Geräuschspannungsmesser	appareil pour mesurer le bruit	indicador de ruido
3997	nominal weight	nominell vikt	Nenngewicht	poids nominal	peso nominal
3998	nonbender	icke böjbar	nicht falzbarer Karton	qui ne se courbe pas	no doblador
3999	noncurling paper	icke-kurlande papper	Papier ohne Rollneigung	papier qui ne roule pas	papel no abarquillable
4000	nondestructive tests	icke-förstörande prov	zerstörungsfreie Materialprüfung	essais non destructifs	pruebas indestructivas
4001	nonlinear programing	nonlinjär programmering	nichtlineare Programmierung	programmation non linéaire	programado no lineal
4002	nonreturnable core	engångshylsa	stationäre Hülsenstange, nicht umkehrbare Hülsenstange	mandrin perdu	cono sin retorno
4003	nonskid	antiglid	rutschfest, profiliert	anti-dérapant	antideslizante
4004	nonstick	icke klibbande	nicht klebend, nicht klebrig	anti-adhérent	antiadherente

	ENGLISH	SVENSKA	DEUTSCH	FRANÇAIS	ESPAÑOL
4005	nontest chip	icke normenlig flis	Schrenzkarton (an den hinsichtlich Festigkeit keine Ansprüche gestellt werden)	intérieur de carton à base de vieux papiers	viruta sin probar
4006	nonwood plant	vedfri växt	Einjahrespflanze	végétal autre que de bois	planta sin pasta mecánica
4007	nonwoven fabric	flor	Faservlies, Vliesstoff	étoffe non tissée	tela no tejida
4008	nonwovens	flor	Nonwovens	non tissés	no tejidos
4009	nozzle	munstycke	Düse, Mundstück, Ausströmöffnung	buse, ajutage	boquilla
4010	Nu number	ny-värde	Besetztzeichen	indice NU	índice Nu
4011	nursery	plantskola	Baumschule, Pflanzenschule	pépinière	semillero
4012	nutrient	näringsmedel	Nährstoff	nourrissant, nutritif	nutritivo
4013	nylon fiber	nylonfiber	Nylonfaser	fibre de nylon	fibra nilón

O

	ENGLISH	SVENSKA	DEUTSCH	FRANÇAIS	ESPAÑOL
4014	o.d. (oven dry)	ugnstorr	absolut trocken	sec absolu	seco absoluto a cien grados
4015	oak tree	ekträd	Eiche	chêne	roble
4016	obsolescence	urmodighet	Veralten, Schwund, Atropie, Überholtsein	tendance à tomber en désuétude	obsolescencia
4017	odor	odör	Geruch, Duft	odeur	olor
4018	odor control	odörbekämpning	Geruchskontrolle	contrôle des odeurs	control del olor
4019	off color	felfärgad	Fehlfarbe, nicht lupenrein	qui n'a pas la teinte voulue	desteñido
4020	off-machine coater	efterbestrykningsmaskin	Separatstreicher	coucheuse hors machine	estucadora fuera de máquina
4021	off-machine coating	efterbestrykning	Streichen ausserhalb der Papiermaschine	couchage hors machine	estucado fuera de máquina
4022	off-machine processing	efterbehandling	separat verarbeiten	fabrication hors machine	tratamiento fuera de máquina
4023	off quality	icke normenlig	Qualitätseinbusse	qui ne répond pas à la qualité voulue	calidad inferior
4024	offset	offset	absetzen, versetzt, Offsetdruck, Ausgleich, Kompensation, Abweichung	offset	offset
4025	offset ink	offsetfärg	Offsetfarben	encre pour impression offset	tinta para offset
4026	offset lithography	offsetlitografi	photographischer Offsetdruck	litho en offset	litografía offset
4027	offset news	offset-tidningspapper, tidnings-offset	Offset-Zeitungsdruckpapier	journal offset	papel periódico offset
4028	offset paper	offsetpapper	Offsetpapier	papier offset	papel para offset
4029	offset printing	offsettryckning	Offsetdruck	impression en offset	offset
4030	offset sheet	offsetark	Offsetbogen, Offsetdruckbogen	feuille de papier offset	hoja offset
4031	off square	sned, vind	nicht ganz quadratisch	mal équerré	desajustado
4032	oil	olja	Öl	huile	aceite
4033	oil absorbency	oljeabsorbens	Ölabsorptionsvermögen	pouvoir absorbant à l'huile	poder absorbente de aceite
4034	oil absorption	oljeabsorbtion	Ölabsorption, Ölaufnahme	absorption de l'huile	absorción del aceite
4035	oil absorption test	oljeabsorbtionsprov	Ölabsorptionstest	essais d'absorption de l'huile	pruebas de absorción de aceite
4036	oiled	oljad	geschmiert, geölt	huilé	lubricado
4037	oiled boards	oljeimpregnerad papp	geölte Deckpappe	cartons huilés	cartones aceitados
4038	oiled paper	oljeimpregnerat papper	Ölpapier, Firnispapier	papiers huilés	papel aceitado
4039	oil filter	oljefilter	Ölfilter, Ölseparator	filtre à huile	filtro de aceite
4040	oil penetration	oljepenetration	Durchdringen von Öl	imperméabilité à l'huile	penetración del aceite
4041	oil penetration tests	oljepenetrationsprov	Öldurchdringtest	essais d'imperméabilité à l'huile	pruebas de penetración del aceite
4042	oilproof paper	oljesäkert papper	öldichtes Papier	papier imperméable à l'huile	papel impermeable al aceite
4043	oil resistance	oljemotstånd	Ölbeständigkeit	résistance à l'huile	resistencia al aceite
4044	oil rub resistance	gnidmotstånd i oljemiljö	Ölabriebwiderstand	résistance au frottement à l'huile	resistencia al frotamiento con aceite
4045	oil spots	oljefläckar	Ölflecken	taches d'huile	manchas de aceite
4046	oil wettability	oljbarhet	Ölbenetzbarkeit	mouillabilité à l'huile	aptitud al engrasamiento
4047	oleates	oleater	Oleate, Oleinate, ölsaures Salz	oléates	oleatos
4048	olefins	olefiner	Olefine	oléfines	olefinas
4049	oleic acid	oljesyra	Ölsäure, Oelinsäure	acide oléique	ácido oléico
4050	oleoresins	oleohartser	Fettharze, Weichharze	oléorésines	oleoresinas
4051	on-machine coating	maskinbestrykning	Streichen in der Maschine	couchage sur machine	estucado sobre máquina
4052	on-machine processing	maskinbearbetning	Verarbeiten in der Maschine, Verarbeiten in einem Arbeitsgang	fabrication sur machine	tratamiento sobre máquina
4053	one-side finish	ensidig finish	einseitige Glätte	fini sur une face	acabado una cara
4054	one-sided coated paper	ensidigt bestruket papper	einseitig gestrichenes Papier	papier couché une face	papel estucado una cara

	ENGLISH	SVENSKA	DEUTSCH	FRANÇAIS	ESPAÑOL
4055	one-time carbon paper	engångskarbonpapper	Einmal-Kohlepapier	papier carbone une fois	papel carbón de una sola vez
4056	onionskin paper	onionskinpapper	Florpost, Zwiebelschaleneffekt papier	pelure surglacée	papel cebolla
4057	opacifier	opacifieringsmedel	Trübungsmittel	opacificateur	que hace opaco
4058	opacimeter	opacimeter	Opazitätsmesser, Trübungsmesser	opacimètre	indicador de opacidad
4059	opacity	opacitet	Opazität, Lichtundurchlässigkeit, Trübung	opacité	opacidad
4060	opaque	opak	opak, gut deckend (Farbe), lichtundurchlässig	opaque	opaco
4061	opaque paper	opakt papper	opakes Papier	papier opaque	papel opaco
4062	opener	öppnare	Öffner, Aufschläger, Zerfaserer	orifice	abridor
4063	open-face calender	öppen kalander	cffener Kalander	calandre à bâtis ouverts	calandra de ancho total
4064	open headbox	öppen inloppslåda	offener Stoffauflauf	caisse à pâte ouverte	caja de entrada abierta
4065	open-top container	öppen behållare	offener Behälter	récipient à dessus ouvrant	contenedor sin tapa
4066	operating	i drift	Arbeiten, Bedienen, Operieren, Betätigen	fonctionnement	funcionamiento
4067	operating cost	driftkostnad	Betriebskosten	coûts de fabrication	gastos de funcionamiento
4068	operating side	driftsida	Bedienungsseite	côté fabrication	lado de mando
4069	operation	drift	Produktionsablauf, Betriebsablauf, Arbeitsweise	opération, fonctionnement, travail manoeuvre, intervention, manipulation	operación
4070	operations research	operationsforskning	Unternehmensforschung	recherche opérationnelle	investigación operacional
4071	operator	operatör, skötare, förare	Betriebsmann, Bedienungspersonal, Maschinist	agent, opérateur	operario
4072	optical bleaching	optisk blekning	optische Bleichmittel, optisches Bleichen	blanchiment optique	blanqueo óptico
4073	optical brightener	optiskt blekmedel	optischer Aufheller	éclaircisseur optique	blanqueador óptico
4074	optical measurement	optisk mätning	optische Messung	mesure optique	medición óptica
4075	optical properties	optiska egenskaper	optische Eigenschaften	propriétés optiques	propiedades ópticas
4076	optical scanner	optisk avsökare	optische Abtastung, optische Tastung	appareil d'exploration optique	explorador optoelectrónico
4077	optical scattering	ljusspridning	Lichtstreuung	dispersion optique	dispersión óptica
4078	optical whitening	optisk blekning	optische Aufhellung	azurage optique	azulamiento óptico
4079	optical whitening agent	optiskt blekmedel	optischer Aufheller	blancophore	agente de azulamiento óptico
4080	optimization	optimering	Optimierung	optimisation	optimización
4081	orange peel	apelsinskal	Orangenschale	pelure d'orange	mondo de naranja
4082	orange peel effect	apelsinskalseffekt	Orangenschaleneffekt	effet pelure d'orange	efecto mondo de naranja
4083	organic acids	organiska syror	organische Säuren	acides organiques	ácidos orgánicos
4084	organic compounds	organiska föreningar	organische Verbindungen	composés organiques	compuestos orgánicos
4085	organosolv lignins	organiskt lösligt lignin	Organosollignine	lignines organo-solubles	ligninas organo-solubles
4086	orientation	orientering	Orientierung, Faserorientierung	orientation	orientación
4087	orifice flow	munstycksflöde	Düsenaustrittsströmung, Düsenaustrittsfliessmenge	diaphragme de débit	flujo de abertura
4088	oscillating	oscillerande, svängande	oszillierend, drehschwingend, hin- und herschwingend	oscillant	oscilante
4089	oscillation	oscillation, svängning	Schwingung, Vibration	oscillation	oscilación
4090	oscillator	oscillator	Oszillator, Schwingungserzeuger	oscillateur	oscilador
4091	oscillograph	oscillogram	Oszillograph, Kurvenschreiber, Schwingungsmesser	oscillographe	oscilógrafo
4092	osmosis	osmos	Osmose	osmose	ósmosis
4093	osmotic pressure	osmotiskt tryck	osmotischer Druck	pression osmotique	presión osmótica
4094	outfall	utfall	Abflussleitung, Vorfluter	chute d'eau	desembocadura
4095	outlet	utlopp	Abfluss, Austritt, Ablass, Auslass, Ausgang	débouché orifice d'évacuation	orificio de salida
4096	output	utgående mängd, produkt	Leistung, Ausstoss, Arbeitsleistung, Produktion	production, débit	rendimiento
4097	oven	ugn	Ofen, Trockenofen	étuve, four	horno
4098	oven dry (o.d.)	ugnstorr	absolut trocken	sec absolu	seco absoluto a cien grados
4099	oven-dry weight	ugnstorr vikt	Darrgewicht (Trocknung bei 105-110 grad C)	poids sec absolu	peso seco absoluto
4100	overcoating	täckning	Überziehen	surcouchage	sobreestucado
4101	overcook	överkok	übermässiges Kochen	surcuisson	sobrecocción
4102	overcooked	överkokad	zu stark gekocht, zu lange gekocht	trop cuite (pâte)	demasiado cocido
4103	overdried	övertorkad	übertrocknet	surséché	resecado
4104	overdry	övertorr	übertrocken	surséché	resecar
4105	overflow	överlopp	überlaufen, überfliessen, überströmen, Überlauf	trop-plein	rebosante
4106	overissue news	överexemplar	nicht verkaufte Exemplare von Druckerzeugnissen	journaux à plat invendus bouillons	periódicos viejos
4107	overlay	överläggning	überlagern, überziehen, zurichten, Zurichtung, Zurichtebogen, Auflegemaske	enduit, overlay	baño de enlucido

ENGLISH	SVENSKA	DEUTSCH	FRANÇAIS	ESPAÑOL
4108 overlay paper	överlagt papper	Overlaypapiere	papier overlay	papel impregnado
4109 overprinting	övertryckning	Überdrucken, Aufdrucken, Eindrucken	surimpression	sobreimpresión
4110 overrun	översvämmad	überfluten, überschwemmen, überschreiten, (Zeilen) hinübernehemen, umbrechen (Druck)	excédent fabriqué sur une commande	excedente de fabricación
4111 oversize	överstorlek	überdimensionieren, Übergrösse, Übermass	ne pouvant être rogné au format prévu	sobretamaño
4112 overweight	övervikt	Übergewicht, Mehrgewicht	qui dépasse le poids commandé	sobrepeso
4113 overweight roll	överviktig rulle	Rolle mit Übergewicht	rouleau plus lourd que celui qui a été commandé	rolfo de sobrepeso
4114 overweight skid	överviktglidning	übergewichtige Palette	pallette qui dépasse le poids de celle qui a été commandée	patinaje por sobrecarga
4115 oxalates	oxalater	Oxalate	oxalates	oxalatos
4116 oxalic acid	oxalsyra	Oxalsäure, Kleesäure	acide oxalique	ácido oxálico
4117 oxidants	oxidationsmedel	Oxidationsmittel, Sauerstoffträger	cxydants	oxidantes
4118 oxidation	oxidation	Oxidierung	oxydation	oxidación
4119 oxidation resistance	oxidationsmotstand	Oxidationsbeständigkeit	résistance à l'oxydation	resistencia a la oxidación
4120 oxides	oxider	Oxide, Oxyde	oxydes	óxidos
4121 oxidizing agent	oxidationsmedel	Oxidationsmittel, Oxydationsmittel	agent oxydant	agente oxidante
4122 oxycellulose	oxicellulosa	Oxicellulose, Oxycellulose	oxycellulose	oxicelulosa
4123 oxygen	syre	Sauerstoff	oxygène	oxígeno
4124 oxygenation	syrsättning	Oxydierung	oxygénation	oxigenación
4125 oxygen compounds	syreföreningar	Sauerstoffverbindungen	composés d'oxygène	compuestos de oxígeno
4126 oxygen demand	syreförbrukning	Sauerstoffbedarf	consommation d'oxygène	demanda de oxígeno
4127 oxystarch	oxiderad stärkelse	oxydativ abgebaute Stärke	amidon oxydé	oxialmidón
4128 ozone	ozon	Ozone	ozone	ozono
4129 ozonization	ozonbehandling, ozonisering	Ozonisierung	ozonisation	ozonización

P

ENGLISH	SVENSKA	DEUTSCH	FRANÇAIS	ESPAÑOL
4130 package	paket, emballage	Paket, Packung, Verpackung	paquet (colis), ensemble	paquete
4131 package design	emballagekonstruktion	Verpackungsform, Verpackungsmuster	conception d'un ensemble	diseño del embalaje
4132 packaging	förpackning	Verpacken	emballage	embalaje
4133 packaging industry	förpackningsindustri	Verpackungsindustrie	industrie de l'emballage	industria del embalaje
4134 packaging machine	förpackningsmaskin, emballagemaskin	Verpackungsmaschine	machine à emballer	máquina empaquetadora
4135 packaging material	förpackningsmaterial	Verpackungsmaterial	matériaux d'emballage	material de embalaje
4136 packaging paper	förpackningspapper	Verpackungspapier, Packpapier	papier d'emballage	papel de embalar
4137 packing	packning	Packen, Verpacken, Dichtung, Dichtpackung, Füllung	empaquetage	empaquetamiento
4138 packing board	förpackningskartong	Packpappe	carton d'emballage	cartón para empaquetar
4139 packing paper	förpackningspapper	Packpapier	papier d'emballage	papel para empaquetar
4140 packing tissue	förpackningspapper	Packseidenpapier	mousseline pour emballage	muselina de embalaje
4141 pad	kudde	auspolstern, füllen, wattieren, Polster, Unterlage	tampon	forrar
4142 pallet	palett	Palette	palette	paleta
4143 palletization	pallning	Palettieren, auf Paletten verpacken	palettisation	paletización
4144 panel	panel	Schalttafel, Täfelung, Tafel, Verkleidung, Holzvertäfelung	panneau	panel
4145 panel board	panelkartong	Schalttafel, Schaltbrett, Kofferpappe, Baupappe	celloderme pour valises, carrosserie	cartón para paneles
4146 paper	papper	Papier	papier	papel
4147 paper bag	papperspåse, papperssäck	Papiertüte, Papierbeutel	sac en papier	bolsa de papel
4148 paper-bag machine	säckmaskin, påsmaskin	Papierbeutelmaschine	machine à sacs en papier	máquina para bolsas de papel
4149 paper-base laminate	papperslaminat	Hartpapier (mit hitzehärtbarem Harz getränkt)	papier support pour stratifiés	laminado con base de papel
4150 paper-base plastics	pappersbasplast	Hartpappe (mit hitzehärtbarem Harz getränkt)	papier support pour plastification	plásticos con base de papel
4151 paper birch	pappersbjörk	Papierbirke	bouleau à papier	abedul para pasta
4152 paperboard	kartong	Kartoń, Pappe	carton gris	cartón
4153 paperboard container	papplåda	Kartonschachtel	récipient en carton gris	contenedor de cartón
4154 paperboard grade	kartongkvalitet	Kartonsorte, Kartonqualität	sorte de carton	clase de cartón
4155 paperboard industry	kartongindustri	Kartonindustrie, Pappenindustrie	cartonnerie	industria cartonera

	ENGLISH	SVENSKA	DEUTSCH	FRANÇAIS	ESPAÑOL
4156	paperboard product	kartongprodukt	Kartonprodukt	article en carton	producto de cartón
4157	paperboard properties	kartongegenskaper	Kartoneigenschaften	propriétés des cartons	propiedades del cartón
4158	paperboard tests	kartongprov	Kartonprüfung	essais des cartons	pruebas del cartón
4159	paper chromatography	papperskromatografi	Papierchromatographie	chromatographie sur papier	cromatografía del papel
4160	paper conversion	pappersförädling	Papierverarbeitung	transformation des papiers	transformación del papel
4161	paper converter	pappersförädlare	Papierverarbeiter	transformateur de papiers	transformador de papel
4162	paper core	pappershylsa	Papierhülse	mandrin en papier	mandril de papel
4163	paper cutter	pappersskärare, kutter	Papierquerschneider	massicot, coupeuse	cortadora de papel
4164	paper grade	papperssort	Papiersorte, Papierqualität	sorte de papier	clase de papel
4165	paper industry	pappersindustri	Papierindustrie	industrie papetière	industria del papel
4166	paper inspector	pappersgranskare	Papierprüfer, Papierkontrolleur	vérificateur de papiers	inspector del papel
4167	paper laminate	papperslaminat	Papierbeschichtung	papier support pour contrecollage, papier support pour stratifiés	laminado de papel
4168	paper machine	pappersmaskin	Papiermaschine	machine à papier	máquina de papel
4169	paper-machine drive	pappersmaskinsdrift	Papiermaschinenantrieb	commande de la machine à papier	mando de la máquina de papel
4170	papermaker ·	papperstillverkare, pappersmakare	Papiermacher	papetier	fabricante de papel
4171	papermakers' alum	pappersmakarens alun	Papiermacheralaun	alun, sulfate d'alumine	alumbre del fabricante de papel
4172	papermakers' felt	pappersfilt	Papiermaschinenfilz	feutre de papeterie	fieltro del fabricante de papel
4173	papermaking	papperstillverkning	Papierproduktion, Papierherstellung	fabrication du papier	fabricación de papel
4174	papermaking equipment	papperstillverkningsutrustning	maschinelle Ausrüstung zur Papierherstellung	matériel pour la fabrication du papier	equipos para fabricar papel
4175	paper manufacturer	papperstillverkare	Papierhersteller	fabricant de papier, papetier	fabricante de papel
4176	paper merchant	pappersförsäljare	Papierhändler	négociant en papiers, distributeur	comerciante en papel
4177	paper mill	pappersbruk	Papierfabrik	papeterie	fábrica de papel
4178	paper napkins	papperservietter	Papierservietten	nappes en papier	servilletas de papel
4179	paper pads	pappersblock, kladdblock	Notizblöcke	séparations en papier (emballage)	almohadillas de papel
4180	paper plates	pappersplåtar	Papierteller	assiettes en papier	platos de papel
4181	paper products	pappersprodukter	Papierprodukte	produits papetiers	productos de papel
4182	paper properties	pappersegenskaper	Papiereigenschaften	propriétés du papier	propiedades del papel
4183	paper scale	pappersvåg	Papierwaage	balance à papier	graduación del papel
4184	paper sheet	papperark	Papierbogen	feuille de papier	hoja de papel
4185	paper size	papperslim	Papierformat, Papierleim	format	tamaño del papel
4186	paper specks	pappersfläckar	Flecken im Papier	boutons (dans le papier)	motas en el papel
4187	paper stock	pappersmäld	Papierstoff, Stoffbrei	pâte à papier	reserva de papel
4188	paper structure	pappersstruktur	Papiergefüge, Papierstruktur	structure, composition d'un papier	estructura del papel
4189	paper substitute	pappersersättning	Papierersatz	produit remplaçant le papier	substituto del papel
4190	paper tests	pappersprov	Papierprüfung	essais de papiers	pruebas del papel
4191	paper textiles	papperstextilier	Papiertextilien	papier à filer	textiles de papel
4192	paper towels	pappershanddukar	Papierhandtücher	serviettes en papier	toallas de papel
4193	paper twine	pappersgarn	Papierkordel	ficelle en papier	cordel de papel
4194	paper web	pappersbana	Papierbahn	feuille de papier en continu	hoja continua de papel
4195	paper yarn	papperssnöre, pappersgarn	Papiergarn	fil en papier	hilo de papel
4196	papyrus	papyrus	Papyrus, Papyrusstaude	papyrus	papiro
4197	paraffin wax	paraffinvax	Paraffinwachs	cire de paraffine	cera de parafina
4198	parallel laminated	parallellaminerad	parallel geschichtet	laminé en parallele	laminado paralelo
4199	parameter	parameter	Parameter	paramètre	parámetro
4200	parchment	pergament	Pergament	parchemin (peau)	pergamino
4201	parchment bond	pergamentpapper	pergamentartiges Schreibpapier	papier écriture simili parcheminé	pergamino bond
4202	parchment finish	pergamentfinish	pergamentähnliche Oberflächenausrüstung	fini parchemin	acabado pergamino
4203	parchmentization	pergamentering	Pergamentierung	parcheminage, sulfurisation	empergaminamiento
4204	parchmentize	pergamentera	pergamentieren	parcheminer	apergaminar
4205	parchmentizing paper	pergamenteringsbaspapper	pergamentierfähiges Papier	papier support pour parcheminage	papel para apergaminar
4206	parchment paper	pergamentpapper	Pergamentpapier	papier sulfurisé	papel pergamino
4207	parchment vellum	pergamentpapper, velinpapper	Pergament mit Velinausrüstung	vélin parcheminé	vitela pergamino
4208	parchment writing	pergamentskrivpapper	Pergamentschreibpapier	papier écriture parcheminé	pergamino para escribir
4209	parenchyma	parenkym	Parenchym	parenchyme	parénquima
4210	parent roll	modervals	Mutterrolle	bobine mère	bobina madre
4211	Parshall flume	parshallränna	künstlicher Wasserlauf nach Parshall	courant d'arrivée d'eau	canal medidor de Parshall

particle

	ENGLISH	SVENSKA	DEUTSCH	FRANÇAIS	ESPAÑOL
4212	particle	partikel	Partikel, Teilchen	particule	partícula
4213	particle board	spånplatta	Spanplatten (aus Stückspänen)	panneau de particules	panel de partículas
4214	particle size	partikelstorlek	Teilchengrösse, Korngrösse, Spangrösse	dimension des particules	dimensión de la partícula
4215	particle-size distribution	partikelstorleksfördelning	Teilchengrössenverteilung	répartition de la taille des particules	distribución de partículas por tamaño
4216	partition chipboard	delningskartong	Zwischenlagenkarton, Zwischenwandkarton, Stegkarton	carton gris à cloisons intérieures	cartón gris de recortes para divisiones
4217	partitioned container	sammansatt behållare	abgeteilter Behälter, Behälter mit Zwischenwänden	récipient à cloisons intérieures	contenedor dividido
4218	partition slotter	slitsfräs	Stegschlitzmaschine	découpeuse de fente des séparateurs	ranuradora divisoria
4219	paste	pasta, klister	Brei, Paste, kleben, kleistern, bekleben	colle	encolar
4220	pasteboard	klisterkartong	kaschierter Karton, beklebter Karton, Klebekarton	carton contrecollé	cartón a encolar
4221	pasted	gummerad, klistrad	geklebt, beklebt, geleimt	collé	encolado
4222	pasted blanks	gummerade formulär	beklebter Karton, gestrichener Karton, ungestrichener Karton	rognures collées	formularios pegados
4223	pasted board	gummerad kartong	kaschierter Karton, beklebter Karton, Klebekarton	carton contrecollé	cartón pegado
4224	pasted chipboard	gummerad gråpapp	beklebte Schrenzpappe	carton gris contrecollé	cartón gris pegado
4225	paster	klistringsmaskin	Anklebevorrichtung, Kleber	encolleuse	pegadura
4226	pasting	gummering, klistring	Kleben, Leimen, Bekleben	contrecollage	contracolado
4227	pasting machine	gummeringsmaskin	Klebemaschine, Anleimmaschine	colleuse	máquina pegadora
4228	patch mark	lappmärke	Markierung einer Flickstelle	marques de plaque	marca del pegamento
4229	patent coated	patentbestruken	Faltschachtelkarton mit Rundsiebdeckschicht	couché à la cuve	estucado patentado
4230	pattern	mönster	Muster, Modell, Vorlage	modèle, patron	modelo
4231	pattern board	mönsterkartong	Musterkarton, Modellpappe	carton pour patrons de découpe	cartón para modelos
4232	pattern paper	mönsterpapper	dessiniertes Papier, Schnittmusterpapier	papier pour patrons	papel para modelos
4233	pattern tissue	mönsterpapper	Schnittmusterseidenpapier	papier mousseline pour patrons	muselina para modelos
4234	pearl filler	pärlfyllnadsmedel	Füllstoff aus wasserfreiem Calciumsulfat	sulfate de calcium précipité (charge)	sulfato de calcio precipitado
4235	pearl starch	pärlstärkelse	Perlstärke, körnige Stärke	amidon perle	almidón perlado
4236	pearl white	pärlvitt	perlweiss, Schwerspat, blanc fixe	carbonate de chaux pur	blanco de perla
4237	pebble mill	kulkvarn	Kugelmühle	raffiner à boulets	molino de piedras
4238	pebbling	kulbildning	Spritznarben, Oberflächenausrüstung	raffiner dans un raffineur à boulets	granulación en molino de piedras
4239	pectin	pektin	Pektin	pectine	pectina
4240	peelability	skalbarhet	Abziehbarkeit, Abstreifbarkeit	aptitude au pelurage	aptitud al descortezado
4241	peeled wood	barkad ved, skalad ved	geschältes Holz, Schälholz	bois écorcé	madera descortezada
4242	peeler	barkare, avskalningsdon	Schälanlage, Handschäler	écorceuse	descortezadora
4243	peeling	skalning	Schälen, Entschalen, Abziehen	pelurage	descortezado
4244	peel test	avskalningsprov	Schäl- und Abhebeprüfung	essais de pelurage	prueba de espito
4245	pellet	tablett	Tablette, Kügelchen, stückiger Kontaktkörper	boulette	bolita
4246	pelleting	tablettbildning	Tablettieren	mise en boulettes	formación de bolitas de papel
4247	pen	penna	Feder, Schreibfeder, Gehege	plume	pluma
4248	penetrant	penetreringsmedel	penetrant, durchdringend	pénétrant	penetrante
4249	penetration	penetrering	Durchdringung, Erweichungstiefe	pénétration	penetración
4250	pentachlorophenol	pentaklorfenol	Pentachlorphenol	pentachlorophénol	pentaclorofenol
4251	pentosans	pentosaner	Pentosan	pentosans	pentosanas
4252	peracetic acid	perättiksyra	Peressigsäure	acide péracétique	ácido peracético
4253	perchlorates	perklorater	Perchlorate, überchlorsaure Salze	perchlorates	percloratos
4254	perchloric acid	överklorsyra	Perchlorsäure, Überchlorsäure	acide perchlorique	ácido perclórico
4255	perforated roll	perforerad vals, hålvals	perforierte Walze, Lochwalze	rouleau perforé, cylindre perforé	rodillo perforado desgotador
4256	perforating	perforering	perforieren, durchbohren, durchlöchern	perforer	perforación
4257	perforating paper	perforeringspapper	gelochtes Papier, perforiertes Papier	papier à bords parallèles après perforation	papel a perforar
4258	perforation	perforering	Lochung, Perforation, Durchbohrung	perforation	perforación
4259	perforator	perforerare	Perforiermaschine	perforeuse, perforateur	perforadora
4260	performance	uppförande, beteende	Vorstellung, Ausführung, Funktionieren, Verrichtung, Leistung	réalisation, exécution	rendimiento

96

	ENGLISH	SVENSKA	DEUTSCH	FRANÇAIS	ESPAÑOL
4261	performance tests	egenskapsprov	Funktionsprüfung, Leistungsprüfung	essais de fonctionnement	pruebas de rendimiento
4262	periderm	periderm	Periderm (Schutzschicht)	périderme	peridermis
4263	periodate lignins	perjodatlignin	Perjodatlignine	lignines au périodate	paryodato ligninas
4264	periodates	perjodater	Perjodate	périodates	paryodatos
4265	periphera¡, peripheric	perifer	peripherisch, Umfangs-	périphérique	periférico
4266	peripheral drive	periferidrift	Peripherantrieb	commande périphérique	mando periférico
4267	permanence	permanens	Dauer-, Dauerhaftigkeit, Permanenz	permanence, stabilité	permanencia
4268	permanent paper	permanent papper	alterungsbeständiges Papier	papier stable	papel duradero
4269	permanganate number	parmanganattal	Permanganatzahl	indice de permanganate	índice de permanganato
4270	permanganates	parmanganater	Permanganate	permanganates	permanganatos
4271	permeability	permeabilitet	Permeabilität, Durchlässigkeit	perméabilité	permeabilidad
4272	peroxides	peroxider	Peroxide	peroxydes	peróxidos
4273	persulfates	persulfater	Persulfate	persulfates	persulfatos
4274	pH	pH	pH	pH, indice d'acidité ionique	pH
4275	pH control	pH-reglering	pH-Messung	contrôle du pH	control de pH
4276	pH value	pH-värde	pH-Wert	indice d'acidité ionique	índice de acidez real
4277	phenol	fenol	Phenol	phénol	fenol
4278	phenol-formaldehyde resin	fenol-formaldehydharts	Phenolformaldehydharz, Phenoplast	résine phénol-formaldéhyde	resina de fenol-formaldehido
4279	phenol groups	fenolgrupper	Phenolgruppen	groupes du phénol	grupos de fenol
4280	phenolic acids	fenolsyror	Phenolsäuren	acides phénoliques	ácidos fenólicos
4281	phenolic resins	fenolhartser	Phenolharze	résines phénoliques	resinas fenólicas
4282	phenol lignins	fenolligniner	Phenollignine	lignines phénoliques	ligninas de fenol
4283	phenolphthalein	fenolftalein	Phenolphthalein	phénolphthaléine	fenolftaleína
4284	phenyl groups	fenylgrupper	Phenylgruppen	groupes phényl	grupos de fenilo
4285	phloem	floem	Phloem, Bastteil (eines Rindenleitstranges)	phloème	floema
4286	phosphates	fosfater	Phosphate	phosphates	fosfatos
4287	phosphorus	fosfor	Phosphor	phosphore	fósforo
4288	phosphorus compounds	fosforföreningar	Phosphorverbindungen	composés du phosphore	compuestos de fósforo
4289	photoelectric cell	fotoelektrisk cell	Photozelle, lichtelektrische Zelle	cellule photoélectrique	célula fotoeléctrica
4290	photoelectric colorimeter	fotoelektrisk kolorimeter	photoelektrischer Farbmesser	colorimètre à cellule photoélectrique	colorímetro fotoeléctrico
4291	photographic paper	fotografiskt papper	photografisches Papier, lichtempfindliches Papier	papier brut pour photo	papel fotográfico
4292	photometer	fotometer	Photometer, Lichtmesser	photomètre	fotómetro
4293	photo-offset	fotooffset	photographischer Offsetdruck	litho offset	foto-offset
4294	photosensitivity	fotokänslighet	Lichtempfindlichkeit	photosensibilité	fotosensibilidad
4295	photostat paper	fotostatpapper	Photokopierpapier	papier pour photocopies	papel para reproducciones
4296	photosynthesis	fotosyntes	Photosynthese	photosynthèse	fotosíntesis
4297	phthalic acid	ftalsyra	Phthalsäure	acide phtalique	ácido ftálico
4298	phthalic anhydride	ftalsyraanhydrid	Phthalsäureanhydrid	anhydride phtalique	anhídrido ftálico
4299	physical properties	fysikaliska egenskaper	physikalische Eigenschaften	propriétés physiques	propiedades físicas
4300	Picea (Lat.)	Picea	Kiefer, Föhre	épicéa	epícea
4301	pick	nappning	rupfen, pflücken, abnehmen	arrachage	arrancamiento
4302	pick, blade scratch	schabernappning	Herausrupfen der Fasern beim Streichen durch Messerkratzer	strie de lame	mordidas de la cuchilla
4303	pick, blister	blåsnappning	Rupfen infolge von Blasenbildung	ampoule	arrancamiento por las ampollas
4304	pick, calender	kalandernappning	Rupfen infolge Hängenbleiben des Papiers am Kalander	arrachage causé par la calandre	arrancamiento en la calandra
4305	pick, coating	bestrykningsnappning	Abheben des Strichs	arrachage du couchage	arrancamiento en el estucado
4306	pick, fiber	fibernappning	Rupfen der Faser, Herausrupfen	arrachage des fibres	arrancamiento en la fibra
4307	picking	nappning, plockning	Rupfen	arrachage	arrancamiento
4308	pick out	plocka ut, välja	herausrupfen	matière d'arrachage	separar
4309	pick resistance	nappningsmotstånd	Rupffestigkeit	résistance à l'arrachage	resistencia al arrancamiento
4310	pick, rupture	nappningsbrott	Rupfen beim Bruch	arrachage, rupture	arrancamiento de ruptura
4311	pick strength	nappningsstyrka	Rupffestigkeit	résistance à l'arrachage	resistencia al arrancamiento
4312	pick tester	nappningsprovare	Rupffestigkeitsprüfgerät	appareil pour essais d'arrachage	probador de arrancamiento
4313	pick tests	nappningsprov	Rupffestigkeitsprüfung	essais d'arrachage de la surface d'un papier	pruebas de arrancamiento
4314	pickup	avtagning	Aufnahme, Abnahme	prise automatique, pick up	toma automática
4315	pickup felt	avtagningsfullt	Abnahmefilz	feutre leveur, feutre preneur	paño tomador
4316	pickup pan	uppsamlingsvanna	Auffangwange	–	artesa tomadora
4317	pickup press	avtagningspress	Pick-up Presse	presse pick up	prensa de toma automática
4318	pickup roll	avtagninsvals	Auftragswalze	rouleau leveur aspirant	rodillo tomador
4319	pigment	pigment	Pigment, Farbkörper, Beschwerungsmittel	pigment	pigmento
4320	pigment formulation	pigmentrecept	Pigmentansatz	–	formulación del pigmento

	ENGLISH	SVENSKA	DEUTSCH	FRANÇAIS	ESPAÑOL
4321	pigmented	pegmenterad	gefärbt, pigmentiert, Pigmentansatz	pigmenté, chiné	pigmentado
4322	pile	stapel	Stapel, Stoss	pile, tas	pila
4323	pilot plant	försöksanläggning	Pilot-Anlage, Technikumsanlage	usine pilote	planta piloto
4324	pin adhesion	nålhäftning	Vernietung von Wellpappenlagen	adhésion de la cannelure à la couverture (essai à l'aiguille)	adherencia de clavija
4325	pine	tall	Kiefer, Föhre, Pinie	pin	pino
4326	pinenes	pinener	Pinen, Sulfatterpentin	pinènes	marchitamiento
4327	pine resin	tallharts	Kiefernharz	résine	resina de pino
4328	pine sulfate pulp	tallsulfatmassa	Fichtensulfatzellstoff	pâte de pin au sulfate	pasta de pino al sulfato
4329	pine sulfite pulp	tallsulfitmassa	Fichtensulfitzellstoff	pâte de pin au sulfite	pasta de pino al bisulfito
4330	pinhole	nålstick	Pore, Loch	trou d'épingle	poros
4331	pinhole free	pinhålfri	porenfrei	sans piqûres	sin poros
4332	pinoresinol	pinoresinol	–	pinorésinol	pinoresinol
4333	Pinus (Lat.)	Pinus	Kiefer, Föhre, Pinie	pin	pino
4334	pipe	rör	Rohr, Rohrleitung	tuyau, conduite	tubo
4335	pipefitter	rörmokare	Installateur, Rohrleger, Rohrschlosser	tuyauterie	tubero
4336	pipe fitting	rörläggning, rörmokeri	Rohrverlegung, Installation	canalisation	ajuste de tubos
4337	piping	rörsystem	Rohrleitungssystem, Leitungsnetz	tuyauterie cordon (défaut du papier)	tubería
4338	pipeline	rörledning	Rohrleitung, Fernleitung	tuyau souterrain	conducción
4339	pit	grop	Kasten (z. B. Siebwasserkasten), Grube	fosse	foso
4340	pitch	beck, kåda, harts	neigen, schrägstellen, aufprallen, Zellpech (im Zellstoff), Zellstoffharz	poix	resina
4341	pitch control	hartskontroll	Zellstoffharzkontrolle, Blattwinkelverstellung	contrôle de la poix, contrôle de la résine	control de la resina
4342	pitch pine	hartstall	Pechkiefer	pin jaune	pino amarillo
4343	pitch spots	hartsfläckar	Harzflecken	taches de résine	marcas de resina
4344	pith	märg	Mark, Holzkern	moelle, parenchyme	médula
4345	pitless grinding	dammlös slipning	trogloses Schleifen	défibrage sans fosse	desfibrado sin foso
4346	Pitot tube	Pitotrör	Pitotrohr, Staurohr	tube de Pitot	tubo de Pitot
4347	pit type grinding	slipning med damm	Trogschleifen	défibrage avec fosse	desfibrado con foso
4348	plain chipboard	slät gråpapp	Schrenzpappe, Graupappe, gedeckter Karton	carton gris compact	cartulina simple de estraza
4349	plain pickup transfer	enkel avtagning	Transfereinrichtung mit glatter Abnahmewalze	–	transferencia con rodillo tomador liso
4350	plain surfaced roll	slätvals	Walze mit glatter Oberfläche	–	rodillo de superficie lisa
4351	plain weave	enkel väv	einfache Webart, Leinwandbindung	toile unie	tela lisa
4352	plane tree	slätt träd	Platane	platane	plátano
4353	plant	planta, fabrik	Pflanze, technische Anlage, Werk, Betriebsanlage	usine, installation	planta
4354	plantation	plantage, plantering	Pflanzung, Plantage, Anbau	plantation	plantación
4355	plant fiber	växtfiber	Pflanzenfaser	fibre végétale (autre que de bois)	fibra vegetal
4356	planting	plantering	Pflanzen, Anpflanzen	plantation	plantación
4357	plant residue	plantrest	pflanzliche Rückstände	résidu végétal	residuos de fabricación
4358	plaster board	gipskartong	Fasergipsplatte	carton plâtre, carton silicaté	cartón-yeso
4359	plastic	plast	Kunststoff	plastique	plástico
4360	plastic coated paper	plastbestruket papper	kunststoffbeschichtetes Papier	papier enduit de matières plastiques	papel estucado plástico
4361	plastic flow	plastisk flytning	plastisches Fliessen	écoulement plastique	deformación plástica
4362	plasticity	plasticitet	Plastizität, Formveränderungsvermögen	plasticité	plasticidad
4363	plasticizer	plasticeringsmedel	Weichmacher, Plastifiziermittel	plastifiant	plastificante
4364	plasticizing	plasticering	weichmachen, plastifizieren	plastifier	plastificación
4365	plastometer	plastometer	Plastometer	plastomètre	plastómetro
4366	plate	platta, plåt, tallrik	plattieren, mit Metalschicht überziehen, Platte, Druckplatte, Scheibe	plaque	plancha
4367	plate finish	plåtfinish	Hochglanzoberflächenausrüstung (das Papier wird dabei unter hohem Druck durch Kupfer- oder Zinkplatten hindurchgeführt)	satiné, laminé à la plaque	acabado a la plancha
4368	plate-finished paper	plåtfinish-papper	Hochglanzpapier	papier satiné, laminé à la plaque	papel acabado a la plancha
4369	plate glazed	plåtglättad	hochsatiniert	laminé à la plaque	laminado entre chapas
4370	plate making	plåttillverkning	Plattenherstellung	taille-douce	fabricación de planchas
4371	plate paper	plåtpapper	Kupferdruckpapier, Stahlstichpapier	papier pour taille-douce	papel a la plancha
4372	plateboard	plåtkartong	Kupfertiefdruckkarton	carton laminé à la plaque	cartón a la plancha
4373	platen	platta	Druckplatte, Drucktiegel, Pressentisch	platine, plateau	platina

Kennen Sie Cellier wirklich?

photos - fontana + thomasset - photos X

CELLIER, der Spezialist für Streichfarben?
Natürlich, und zwar der grösste in der Welt (80 % des Weltmarktes).

Aber wissen Sie auch, dass das Tätigkeitsfeld von CELLIER viel umfangreicher ist?

CELLIER, das heisst auch Anlagen und Verfahren, die sich bewährt haben für die :
– Dispergierung und Lagerung von Pigmenten mit hohem Feststoffgehalt,
– Satinweiss-Herstellung
– Aufbereitung löslicher Bindemittel : Stärke - PVA - Kasein, usw
– Oberflächenbehandlungen durch Spezialbeschichtung :
  Papier für elektrostatische Reproduktion
  Hotmelt-Beschichtung - PVDC - Fungizid - Gummierung
  Silikonpapiere, Antikleber, Selbstklebe- und Trennpapiere
– Färbung auf der Leimpresse
– kontinuierliche Färbung für Dekorpapiere
– Verteilung von Edelpigmenten (TiO2)
– Aufbereitung der Zusatzstoffe zur Masse für Papier und Karton und deren Zudosierung am Stoffauflauf.

TAPETENHERSTELLUNG :
– komplette Produktionseinheiten und neue Methode für die Druckfarbenherstellung
– Messketten : Viskosität - pH - Feststoffgehalt
– Prozesssteuerung für die Folgeregelung und die Verarbeitung des gestrichenen Papiers von der Maschine bis zur Verarbeitungswerkstatt.

Von der einfachsten Einrichtung bis zum kompletten automatisierten Komplex projektiert und realisiert CELLIER auch die "produktefertige" Anlage für Ihren Bedarf.

Cellier Abteilung "Papier"
B.P. 58 - 73102 Aix-les-Bains France
Tél. (79) 35.05.65
Telex : 980053 F INOXEL

Vertretung : Fa. Dr Hansen & Sohn
61 - Darmstadt - Eberstadt,
Im Biengarten 22
Tel. 0 6151 / 55 09 1
Telex : 04 19 455

édipress-grenoble O6BG.

Für den Verkauf und den Kundendienst steht Ihnen in Ihrem Land unser Vertreter zur Verfügung.

	ENGLISH	SVENSKA	DEUTSCH	FRANÇAIS	ESPAÑOL
4374	platen dryer	plattork	Heizplatten (die laufende Papier-/Kartonbahn wird durch Zwei aufeinanderliegende Heizplatten geführt)	séchoir à platines	secadora de platinas
4375	platen press	plattpress	Tiegeldruckpresse, Etagenpresse	presse à plateaux	prensa de platinas
4376	plater	plåtglätt	Hochglanzmaschine (bestehend aus 2 gusseisernen Walzen)	laminoir	bandeja
4377	plater board	plåtglättad kartong	Kupfertiefdruckkarton, Hochglanzkarton	carton laminé à la plaque	cartón para bandejas
4378	plating	plätering	Plattieren, Metallbeschichten	laminage à la plaque	plancheado
4379	plenum chamber	plenisal	Trockenkammer, Speicherraum	chambre plénière	cámara de pleno
4380	pliability	böjlighet	Biegsamkeit, Falzfähigkeit, Geschmeidigkeit	aptitude au pliage	flexibilidad
4381	plies	lag	Falten, Lage, Schicht, Zwischenlage	jets, couches	capas
4382	plug	plugg	Refinerkegel, Hülsenkern, Stöpsel, Stecker, zustopfen, verstopfen	cône de raffineur cônique, bouchon en bois	taco
4383	plugging	pluggning	zustopfen, verstopfen	obstruction par accumulation de matière pompée	orificación
4384	plume	plym	Dampfstrahl	plume	pluma
4385	plunger	kolv, plunger	Kolben, Pumpenkolben	plongeur	pistón inmergible
4386	plus trees	plusträd	Plusbaum	arbres surrégénérés	áboles positivos
4387	ply	lag	falten, ausüben, Lage, Schicht, Strähne	jet, couche	capa
4388	ply adhesion	laghäftning	Haftfestigkeit der Lagen	adhésion des couches	adherencia de las capas
4389	ply bond	laghäftning	Haftfestigkeit, Lagenbindung	adhésion des couches	adherencia de las capas
4390	ply bond strength	laghäftningsstyrka	Spaltfestigkeit	résistance des couches à l'arrachage	resistencia de la adherencia de capas
4391	ply separation	lagseparation	Lagentrennung, Schichttrennung, Lagenspaltung	séparation des couches	separación de las capas
4392	pneumatic equipment	pneumatisk utrustning	pneumatische Ausrüstung	équipement pneumatique	equipos neumáticos
4393	pocket	ficka, fack, kanna	Tasche, Beutel, Schleifermagazin	pouche de défibreur	bolsillo
4394	pocket grinder	kannslip	Magazinschleifer	défibreur à chambres	desfibradora de prensas
4395	pocket ventilation roll	ventilationshålvals	Belüftung mittels Lufttaschen (zwischen Zylinder und Bahn), Blaswalze	rouleau de ventilation des pouches de sécherie	rodillo ventilador del bolsillo
4396	point	punkt, spets	Punkt, Spitze, punktieren, einlegen (Druck)	point	punto
4397	polarity	polaritet	Polarität	polarité	polaridad
4398	polarization	polarisation	Polarisation	polarisation	polarización
4399	pole drying	hängtorkning	Pfahltrocknung	séchage à la perche, séchage à la barre	secado en varas
4400	polish	polera, polering	polieren, Politur, Glätte, Glanz	lustre	brillo
4401	polished drum coating	bestrykning med polervals	eine Art Gusstreichen (erübrigt nachfolgendes Kalandrieren oder Satinieren)	couchage avec séchage sur tambour poli	estucado por tambor lustrado
4402	polishing	polering	Hochglanzgebung, Fertigpolieren	polissage	abrillantado
4403	pollination	polinering	Bestäubung	pollinisation	polinación
4404	pollution	förorening	Verschmutzung, Verunreinigung	pollution	contaminación
4405	pollution control	föroreningskontroll	Umweltschutz	contrôle de la pollution	control de la contaminación
4406	polyacrylamide	polyakrylamid	Polyacrylamid	polyacrylamide	poliacrilamida
4407	polyacrylate	polyakrylat	Polyacrylat	polyacrylate	poliacrilato
4408	polyamide resins	polyamidhartser	Polyamidharze	résines polyamides	resinas de poliamida
4409	polyamides	polyamider	Polyamide	polyamides	poliamidas
4410	polyelectrolytes	polyelektrolyter	Polyelektrolyte	polyélectrolytes	polielectrólitos
4411	polyesters	polyestrar	Polyester	polyesters	poliesteros
4412	polyethylene	polyten	Polyäthylen	polyéthylène	polietileno
4413	polyethylene coated paper	polytenbestruket papper	polyäthylenbeschichtetes Papier	papier couché au polyéthylène	papel estucado al polietileno
4414	polyethylenimine	polyetylimine	Polyäthylenimin	polyéthylénimine	polietilenimina
4415	polyimides	polyimider	Polyimide	polymides	polimidas
4416	polymerization	polymerisation	Polymerisation	polymérisation	polimeración
4417	polymerize	polymerisera	polymerisieren	polymériser	polimerizar
4418	polymers	polymerer	Polymere	polymères	polimeros
4419	polymethacrylate	polymetaakrylat	Polymethacrylat	polyméthacrylate	polimetacrilato
4420	polypropylene	polypropylen	Polypropylen	polypropylène	polipropileno
4421	polystyrene	polystyren	Polystyrol	polystyrène	poliestireno
4422	polysulfide pulp	polysulfidmassa	Polysulfidzellstoff	pâte au polysulfure	pasta al polisulfuro
4423	polysulfide pulping	polysulfidkokning	Polysulfidzellstoffaufschluss	fabrication de pâtes au polysulfure	elaboración de pasta al polisulfuro
4424	polysulfides	polysulfider	Polysulfide	polysulfures	polisulfuros
4425	polyurethanes	polyuretaner	Polyurethane	polyuréthanes	poliuretanos
4426	polyvinyl acetate	polyvinylacetat	Polyvinylacetat	acétate de polyvinyl	polivinil-acetato

	ENGLISH	SVENSKA	DEUTSCH	FRANÇAIS	ESPAÑOL
4427	polyvinyl alcohol	polyvinylalkohol	Polyvinylalkohol	alcool de polyvinyl	alcohol de polivinilo
4428	polyvinyl chloride	polyvinylklorid	Polyvinylchlorid	chlorure de polyvinyl	cloruro de polivinilo
4429	polyvinyl esters	polyvinylestrar	Polyvinylester	esters de polyvinyl	ésteres de polivinilo
4430	polyvinyl ethers	polyvinyletrar	Polyvinyläther	éthers de polyvinyl	éteres de polivinilo
4431	polyvinylidene chloride	polyvinylidenklorid	Polyvinylidenchlorid	chlorure de polyvinylidène	cloruro de polivinilideno
4432	polyvinyls	polyvinyler	Polyvinyle	polyvinyls	polivinilos
4433	pond	damm	Teich, Sumpf, Stau	mare, bassin	tina
4434	pond type trailing blade coater	släpbladsbestrykare med damm	Sumpfschleppräkelstreicher	coucheuse à lame traînante immergée	estucadora con cuchillas de arrastre de bandeja
4435	poorly wound roll	defekt rulle	schlecht gewickelte Rolle	bobine molle, bobine peu serrée	bobina floja
4436	poor man's pickup	fattigmansavtagning	Einfach-Pick-up, Bahnabnahme zwischen Saug-und Umkehrwalze	pick up de pauvre	–
4437	poor man's venta nip	fattigmans ventanip	Schrumpfsiebpresse, Venta-Nip Presse	venta nip du pauvre	–
4438	Pope reel	friktionsrullstol	Tragtrommelwickler, Tragtrommelroller	enrouleuse Pope	enrolladora Pope
4439	poplar	poppel	Pappel	peuplier	álamo
4440	Populus (Lat.)	Populus	Pappel	peuplier	chopo
4441	pore	por	Pore	pore	poro
4442	pore size	porstorlek	Porengrösse	dimension du pore	dimensión del poro
4443	porosimeter	porosimeter	Porosimeter, Luftdurchlässigkeits messer	porosimètre	porosímetro
4444	porosity	porositet	Porosität	porosité	porosidad
4445	porosity tester	porisitetsprovare	Porositätsprüfgerät	porosimètre	probador de porosidad
4446	porous	porös	porig, porös, durchlässig	poreux	poroso
4447	porous material	poröst material	poröses Material, poröse Stoffe	matériaux poreux	material poroso
4448	porous wood	porös ved	poröses Holz	bois poreux	madera porosa
4449	poster board	affichkartong	Plakatkarton	carton pour affiches	cartón para carteles
4450	poster paper	affichpapper	Plakatpapier, Affichenpapier	papier pour affiches	papel para carteles
4451	postrefining	efterraffinering	Nachmahlung	post-raffinage	postrefinado
4452	potassium	kalium	Kalium	potassium	potasio
4453	potassium compounds	kaliumföreningar	Kaliumverbindungen	composés de potassium	componentes de potasio
4454	potassium iodide	kaliumjodid	Jodkalium, Kaliumjodid	iodure de potassium	yoduro de potasio
4455	potassium permanganate	kaliumpermanganat	Kaliumpermanganat	permanganate de potassium	permanganato de potasio
4456	potato starch	potatisstärkelse	Kartoffelstärke, Kartoffelstärkemehl	fécule de pomme de terre	almidón de patata
4457	potentiometer	potentiometer	Potentiometer, Spannungsteiler	potentiomètre	potenciómetro
4458	pouch machine	påsaskin	Beutelherstellmaschine	machine à sachets	máquina de saquitos
4459	pouch pack	planpåse	–	emballage en sachets	embalaje en bolsitas
4460	pounds per point	pund per tusendels tum	Dichtemessung von Papier oder Karton (Flächengewicht: 1/1000 inch Dicke)	pression en livres par point	libras por punto
4461	pour point	flytpunkt	Stockpunkt, Fliesspunkt	température de coulée	punto de colada
4462	powder	pulver	pulverisieren, zu Pulver zerkleinern, Pulver, Puder, Staub	poudre	polvo
4463	powdered	pulvriserad	pulverisiert	pulvérisé	pulverizado
4464	power	kraft, effekt	Kraft, Stärke, Energie, Leistung, Leistungsvermögen, Vollmacht, potenzieren, Strom	énergie, force, puissance	fuerza
4465	power factor	effektfaktor	Leistungsfaktor	facteur de puissance	factor de potencia
4466	power generation	kraftalstring	Krafterzeugung	force motrice	generación de fuerza
4467	power house	kraftstation	Kraftwerk	centrale de force motrice, centrale électrique	central eléctrica
4468	power supply	kraftförsörjning	Energieversorgung	approvisionnement en énergie	suministro de energía
4469	precipitate	fälla, falla	ausfällen, niederschlagen, Niederschlag, Ausfällung	précipité	precipitar
4470	precipitating	fällande	ausfällend, niederschlagend	précipiter	precipitado
4471	precipitation	fällning	Abscheidung, Niederschlagen, Präzipitation	précipitation	precipitación
4472	precipitator	fällningsmedel	Fällungsmittel	précipitateur	precipitante
4473	precision	precision	Präzision, Genauigkeit	précision	precisión
4474	pre-coating	förbestrykning	Vorstreichen	première couche d'un couchage	pre-estucado
4475	pre-dryer	förtork	Vortrockner	pré-sécheur	presecador
4476	prefabrication	prefabrikation	Vorfertigung, Montagebau	préfabrication	prefabricación
4477	preforming	förformning	vorformen	préformation	preformación
4478	prehydrolysis	förhydrolys	Vorhydrolyse	préhydrolyse	prehidrólisis
4479	preparation	preparation, beredning	Vorbereitung, Aufbereitung	préparation	preparación
4480	preservation	preservering	Bewahrung, Erhaltung, Konservierung	conservation, maintien, protection	preservación

	ENGLISH	SVENSKA	DEUTSCH	FRANÇAIS	ESPAÑOL
4481	preservative	preservativ	schützend, erhaltend	stabilisant	protector
4482	press	press	drücken, pressen, Presse, Druckerpresse	presse	prensa
4483	press arrangement	pressarrangemang	Presseneinrichtung	disposition des presses	distribución de prensas
4484	press board	presspan	Presspappe, Pressspan, Stanzpappe	carton comprimé, carton lustré	cartón comprimido
4485	press coater	pressbestrykare	Druckwalzenstreichen	coucheuse sur presse	estucadora de prensa
4486	press coating	pressbestrykning	Pressenstreichen	couchage sur presse	estucado de prensa
4487	press felt	pressfilt	Pressfilz, Nassfilz	feutre coucheur	fieltro de prensa
4488	pressing	pressning	Pressen, Drücken, Stanzen	pression, compression	prensado
4489	press marks	pressmarkeringar	künstliches Wasserzeichen	marques à la molette	estriados
4490	press roll	pressvals	Presswalze, Druckwalze	rouleau de presse coucheuse	rodillo de prensa
4491	pressroom	pressrum, tryckeri	Maschinensaal (Druckerei)	salle des presses	sala de prensas
4492	pressroom superintendent	tryckeriföreståndare	Obermaschinenmeister	chef de fabrication des presses	capataz de la sala de prensas
4493	press section	pressparti	Pressenpartie	département des presses	sección de prensas
4494	press stopper	tryckpresstoppare	Druckmaschinenstopper	butée de presse	tapadero de prensa
4495	pressure	tryck	Druck, Drücken, Pressen, Druckkraft	pression	presión
4496	pressure control	tryckreglering	Druckregulierung	contrôle de la pression	control de presión
4497	pressure drop	tryckfall	Druckabfall	chute de pression	caída de presión
4498	pressure filter	tryckfilter	Druckfilter	filtre sous pression	filtro a presión
4499	pressure gauge	tryckmätare	Druckmesser, Manometer	manomètre	indicador de presión
4500	pressure gradient	tryckgradient	Druckgefälle	foulage	gradiente de presión
4501	pressure headbox	tryckinloppslåda	Druckstoffauflauf	caisse d'arrivée de pâte sous pression	caja de entrada a presión
4502	pressure measurement	tryckmätning	Druckmessung	mesure de la pression	medición de la presión
4503	pressure reduction	tryckminskning	Druckreduzierung	diminution de la pression	reducción de presión
4504	pressure regulator	tryckregulator	Druckregler	régulateur de pression	regulador de presión
4505	pressure roll	anpressvals	Anpresswalze	–	rodillo a presión
4506	pressure-sensitive paper	tryckkänsligt papper	druckempfindliches Papier	papier auto-adhésif	papel sensible a la presión
4507	pressure sensitivity	tryckkänslighet	druckempfindlich	sensibilité à la pression	sensibilidad a la presión
4508	pressurization	tryckpåläggning	Druckfestigkeit, unter Druck setzen	pressurisation	presurización
4509	pressurized equipment	tryckbehandlad utrustning	druckdichte Anlagen, druckdichte Einrichtung	matériel sous pression	equipos presurizados
4510	pressurized refining	tryckraffinering	Mahlen unter Hochdruck	raffinage sous pression	refino bajo presión
4511	pretreatment	förbehandling	Vorbehandlung	pré-traitement	pretratamiento
4512	preventive maintenance	preventivt underhåll	vorbeugende Wartung	entretien préventif	mantenimiento de prevención
4513	primary growth	primärtillväxt	Primärwachstum	croissance primaire	crecimiento primario
4514	primary headbox	primär inloppslåda	Primärstoffauflauf	caisse de tête primaire	caja de entrada primaria
4515	primary press	förstapress	Vorpresse	presse primaire	prensa primaria
4516	primary screening operation	försilning	Vorsortierung, erste Sortierstufe	fonctionnement de l'épuration primaire	operación de cribado primario
4517	primary treatment	förbehandling	Vorbehandlung	traitement primaire	tratamiento primario
4518	primary wall	primärvägg	Primärwand, Primärzellwand	paroi primaire	pared primaria
4519	primer	primer	Grundierlack, Spachtelmasse, Antiquaschrift, Einspritzvorrichtung	amorce	aprestador
4520	printability	tryckbarhet	Bedruckbarkeit	imprimabilité	aptitud a la impresión
4521	printer	tryckare	Drucker	imprimeur	impresor
4522	printing	tryckning	Drucken, Druck	impression, imprimerie	impresión
4523	printing blanket	tryckfilt	Drucktuch, Gummituch	blanchet	mantilla
4524	printing ink	tryckfärg	Druckfarbe	encre d'imprimerie	tinta para imprimir
4525	printing machine	tryckpress	Druckmaschine, Druckpresse, Druckerpresse	machine d'impression	máquina de imprimir
4526	printing paper	tryckpapper	Druckpapier	papier d'impression	papel de imprimir
4527	printing smoothness	tryckglättning	Druckglätte	aptitude à l'impression	igualdad en la impresión
4528	printing tests	tryckbarhetsprov	Bedruckbarkeitsprüfung	essais d'imprimabilité	pruebas de impresión
4529	print-on coating	rotogravyrbestrykning	Auftragsstreichen	couchage par impression	impresión sobre estucado
4530	print roll coater	tryckvalsbestrykare	Druckwalzenstreicher	coucheuse sur le principe d'une presse typo	estucadora de rodillo estampador
4531	print roll coating	tryckvalsbestrykning	Auftragswalzenstreichen	couchage sur le principe d'une presse typo	estucado de rodillo estampador
4532	process	process	verarbeiten, behandeln, Verfahren, Prozess, Verarbeitungsmethode, Verlauf, Ablauf	technique, processus, procédé	procedimiento
4533	process control	processreglering	Prozesskontrolle	contrôle de la fabrication	control de procedimientos
4534	processing	bearbetning, tillverkning	Verarbeiten, Bearbeiten, Veredeln	fabrication	elaboración
4535	process variable	processvariabel	Verfahrensgrösse	variable de traitement	variable de elaboración
4536	process water	processvatten	Betriebswasser	eau de fabrication	agua de elaboración
4537	product	produkt	Produkt, Ergebnis, Erzeugnis	produit	producto

Ross Air Systems Division

Drying and process equipment for the pulp and paper industry

Paper machine air systems:
hoods, pocket ventilation, moisture profiling, caliper control and flutter control

Pulp dryers:
floater dryers, flash dryers and cylinder-dryer air systems

Yankee hoods and air systems

Coater dryers

Heat recovery systems:
systems to reclaim and re-use heat from all major drying and process sources

Mill ventilation and comfort conditioning

All engineered and supplied by Ross Air Systems Division and its licensees around the world. Ross Air Systems Division, Midland-Ross of Canada Limited, 304 St. Patrick Street, LaSalle, Quebec, Canada H8N 2H1, Phone 514/366-5160. Telex 05-268704.

	ENGLISH	SVENSKA	DEUTSCH	FRANÇAIS	ESPAÑOL
4538	product development	produktutveckling	Produktentwicklung	amélioration de la production	desarrollo del producto
4539	production	produktion	Produktion, Herstellung, Bildung	production	producción
4540	production control	produktionskontroll	Produktionskontrolle, Betriebsüberwachung, Fertigungsplanung	contrôle de la production	control de producción
4541	production manager	produktionsledare	Produktionsleiter, Betriebsleiter	directeur de la production	director de producción
4542	production method	produktionsmetod	Produktionsverfahren, Produktionsmethode, Produktionsablauf	méthode de production	método de producción
4543	productivity	produktivitet	Produktivität, Ergiebigkeit	productivité	productividad
4544	profile	profil	Profil	profil	perfil
4545	profiler	profilometer	Profilmessgerät	–	perfilador
4546	profit	profit, förtjänst	Gewinn, Nutzen, Vorteil	profit, bénéfice	beneficio
4547	programed instruction	programmerad instruktion	programmierter Befehl, programmierter Unterricht	macro-instruction	instrucción programada
4548	programing language	programmeringsspråk	Programmiersprache	langage de programmation	idioma de programación
4549	propagation	utbredning, fortplantning	Fortpflanzung	propagation	propagación
4550	propane	propan	Propan	propane	propano
4551	propeller	propeller	Propeller, Schraube	hélice, propulseur	propulsor
4552	property	egenskap, egendom	Eigentum, Besitz, Eigenschaft, Fähigkeit, Vermögen	propriété, qualité	propiedad
4553	propionates	propionater	Propionat	propionates	propionatos
4554	proportioner	proportionerare, fördelare	Regulierkasten	doseur	dosificador
4555	proportioning	proportionering	Dosierung, Zuteilung, Mischung	classement (d'après la granulation)	dosificación
4556	propyl alcohol	propylalkohol	Propylalkohol	alcool propylique	alcohol propílico
4557	protective clothing	skyddsbeklädnad	Schutzkleidung	vêtements de protection	prendas de protección
4558	protective coating	skyddsbeläggning	Schutzschicht, Schutzüberzug	couchage de protection	estucado protector
4559	proteins	proteiner, äggviteämnen	Protein	protéines	proteínas
4560	protolignins	protoligniner	Protolignin	protolignines	protoligninas
4561	protons	protoner	Proton	protons	protones
4562	Prunus (Lat.)	Prunus	Pflaumenbaum	Prunus	ciruelo
4563	pseudoplasticity	pseudoplasticitet	Pseudoplastizität	pseudoplasticité	seudoplasticidad
4564	Pseudotsuga (Lat.)	Pseudotsuga	Pseudotsuga	sapin de Douglas	Pseudotsuga
4565	pucker	skrynkla	sich kräuseln, zusammenziehen	pli	pliegue
4566	pulley	drev	Riemenscheibe, Rolle	poulie	polea
4567	pull roll	dragvals	Zugwalze	rouleau de traction	rodillo de tracción
4568	pulp	massa	Zellstoff, Papierstoff	pâte	pasta
4569	pulpboard	cellulosakartong	Zellstoffpappe, Zellstoffkarton	carton de pâte mécanique et de vieux papiers	cartón de pasta
4570	pulp content	massahalt	Zellstoffgehalt	teneur en pâte	contenido de pasta
4571	pulper	pulper	Stoffauflöser, Zerfaserer, Pulper	broyeur, triturateur	desfibrador
4572	pulp fiber classifier	massafiberfraktionerare	Faser-Fraktioniergerät	appareil pour l'étude des fibres par la méthode du fractionnement	clasificador de la fibra de pasta
4573	pulp industry	massaindustri	Zellstoffindustrie	industrie des pâtes	industria de la pasta
4574	pulping	massatillverkning, kokning	Aufschliessen, Kochen (von Zellstoff), Stoffauflösung	réduire en pâte	elaboración de pasta
4575	pulping liquor	kokvätska, koklut	Kochlauge	lessive pour la fabrication de pâtes	lejía para fabricar pasta
4576	pulp lap	massaark	gefaltene Nasszellstoffbogen	balle de pâte humide	fardo de pasta húmeda
4577	pulp mill	massafabrik	Zellstoffabrik	usine de pâtes	fábrica de pasta
4578	pulp molding	massaformning	Faserguss	emballage en pâte moulée	cartón moldeado
4579	pulp quality	massakvalitet	Zellstoffqualität	qualité d'une pâte	calidad de la pasta
4580	pulp sales	massaförsäljning	Zellstoffabsatz, Zellstoffverkauf, Zellstoffgeschäft	ventes de pâtes	ventas de pasta
4581	pulp screen	massasil	Knotenfänger	épurateur de pâte	depurador de pasta
4582	pulp sheet	massaark	Zellstoffblatt, Zellstoffbogen, Zellstoffprobe	feuille de pâte	hoja de pasta
4583	pulpstone	slipsten	Schleifstein	meule de défibreur	muela desfibradora
4584	pulpwood	massaved	Faserholz	bois de papeterie	madera para pasta
4585	pulsation	pulsering	Pulsieren, Schwingung	pulsation	pulsación
4586	pulverizer	pulvriserare	Zerkleinerer, Feinstmahlanlage, Zerstäuber, Pulverisierer	pulvérisateur	pulverizador
4587	pump	pump	Pumpe	pompe	bomba
4588	pumping	pumpning	Pumpen	pompage	bombeo
4589	punched card	hålkort	Lochkarte	carte perforée	tarjeta perforada
4590	punched tape	hålresma	Lochstreifen, Lochband	bande perforée	cinta perforada
4591	puncture	punktera	Durchbohren, Durchstechen, Durchstoss	perforation	perforación

	ENGLISH	SVENSKA	DEUTSCH	FRANÇAIS	ESPAÑOL
4592	puncture resistance	punkteringsmotstånd	Sticheinreissfestigkeit	résistance à la perforation	resistencia a la perforación
4593	puncture tester	punkteringsprovare	Sticheinreissprüfgerät, Durchstossprüfer, Puncture-Tester	perforamètre	perforámetro
4594	puncture tests	punkteringsprov	Sticheinreissprüfung	essais de perforation	pruebas de perforación
4595	purchase	köpa	Kauf, Hebevorrichtung	achat	comprar
4596	purification	rening	Reinigung, Klärung	purification	depuración
4597	purity	renhet	Reinheit	pureté	pureza
4598	pyrite	pyrit, kis	Pyrit, Schwefelkies	pyrite	pirita
4599	pyrite burning	kisbränning	Pyritrösten	calcination de pyrite	quema de piritas
4600	pyrite burner	kisbrännare	Pyritofen	four à pyrite	quemador de piritas
4601	pyrolysis	pyrolys	Pyrolyse, Hitzebehandlung	pyrolyse	pirólisis
4602	pyroxylin coating	pyroxylinbestrykning	Kollodiumwollbeschichtung	couchage à la pyroxyline (nitrate de cellulose)	estucado a la piroxilina

Q

4603	qualitative analysis	kvalitativ analys	qualitative Analyse	analyse qualitative	análisis cualitativo
4604	quality	kvalitet	Qualität, Eigenschaft, Beschaffenheit	qualité	calidad
4605	quality control	kvalitetskontroll	Qualitätskontrolle	contrôle de la qualité	control de calidad
4606	quantitative analysis	kvantitativ analys	quantitative Analyse	analyse quantitative	análisis cuantitativo
4607	quaternary ammonium compounds	kvartära ammoniumföreningar	Quartärammoniumverbindungen	composés ammoniacaux quaternaires	compuestos cuaternarios del amoniaco
4608	queen roll	drottningvals	zweitunterste Walze im Glättwerk	rouleau supérieur	rodillo superior
4609	Quercus (Lat.)	Quercus	Eiche	chêne	roble
4610	quire	ark	vierundzwanzig Bogen (aus einem 480er Ries - Grobpapier), fünfundzwanzig Bogen (aus einem 500er Ries - Feinpapier)	main de papier	mano de papel

R

4611	racking strength	sträckhållfasthet	Abziehfestigkeit, Abziehwiderstand	résistance à l'allongement	resistencia a la deformación transversal
4612	radiant heating	strålvärme	Strahlungsheizung	chauffage par rayonnement	calefacción radiante
4613	radiation	strålning	Strahlung, Ausstrahlung	radiation, rayonnement	radiación
4614	radical	radikal	radikal, fundamental, grundlegend, Radikal, Wurzel (math.), Rest	radical	radical
4615	radioactive material	radioaktivt material	radioaktiver Stoff	équipement radioactif	material radioactivo
4616	radioactivity	radioaktivitet	Radioaktivität	radioactivité	radioactividad
4617	radius	radie	Radius, Halbmesser	rayon (d'une cercle)	radio
4618	raffinate	raffinat	Raffinat, Raffinationsprodukt	raffiner	capa líquida en sistema de extracción solvente
4619	raft	flotte	flössen, Floss, Fundamentrahmen	radeau	balsa
4620	rafting	flottning	Flössen	flottage des bois	flotación
4621	rag	lump, trasa	Hadern, Lumpen	chiffon	trapo
4622	rag boiler	lumpkokare	Hadernkocher, Lumpenkocher	lessiveur à chiffons	caldera de trapos
4623	rag breaker	lumpuppslagare	Hadernholländer	pile défileuse (à chiffons)	pila filochadora
4624	rag content	lumphalt	Haderngehalt, Hadernanteil	teneur en chiffons	contenido de trapos
4625	rag-content paper	lumphaltigt papper	hadernhaltiges Papier	papier à base de chiffons	papeles con contenido de trapos
4626	rag cutter	lumphuggare	Hadernschneider, Hadernzerreisswolf	coupeuse de chiffons	cortadora de trapos
4627	rag cuttings	halvtyg	Hadernzuschnitte, Hadernabfälle	chiffons coupés	recortes de trapos
4628	rag duster	lumpdammare	Lumpenentstäuber, Hadernstäuber	blutoir à chiffons	cedazo para trapos
4629	rag paper	lumppapper	Hadernpapier	papier de chiffons	papel de trapos
4630	rag pulp	lumpmassa	Hadernzellstoff	pâte de chiffons	pasta de trapos
4631	rag room	lumphanteringsrum	Lumpensaal, Hadernboden	chiffonnerie	trapería
4632	rag sorting	lumpsortering	Lumpensortierung	tri des chiffons	clasificación de los trapos
4633	rag stock	lumpmassa, lumpmäld	Hadernstoff	stock de chiffons	pasta de trapos
4634	rag thrasher	lumpupplösare	Haderndrescher	blutoir à chiffons	trillador de trapos
4635	railroad	järnväg	Eisenbahn	chemin de fer	ferrocarril
4636	railroad car	järnvägsvagn	Eisenbahnwaggon	wagon de chemin de fer	vagón de tren

	ENGLISH	SVENSKA	DEUTSCH	FRANÇAIS	ESPAÑOL
4637	railroad track	järnvägsspår	Eisenbahnspur (auch beim Streichen)	voie de chemin de fer	vía ferroviaria
4638	ramie	ramie	Ramie	ramie	ramio
4639	ramie fiber	ramiefiber	Ramiefaser, Chinagras	fibre de ramie	fibra de ramio
4640	randomly oriented fibers	slumpartat orienterade fibrer	Wirrfaserlage	fibres disposées au hasard	fibras orientadas al azar
4641	randomly oriented web	isotropt ark	Wirrfaservlies	feuille en continu orientée au hasard	orientación de las fibras al azar
4642	raschig rings	raschigringar	Raschigringe	anneaux de Raschig	anillos de Raschig
4643	rate of drying	torkhastighet	Trocknungsgeschwindigkeit, Trocknungsleistung	taux de séchage	grado del secado
4644	ratio	kvot	Verhältnis, Quotient, Mass	rapport, taux	relación
4645	rattle	skramla	Rasseln, Rattern, Klang (Papier), Rascheln	carteux, sonnant	carteo
4646	raw cook	råkok	Kochung bei der die Nichtcellulosesubstanzen nur begrenzt entfernt wurden	cuisson brute	lejiación cruda
4647	raw material	råmaterial	Rohstoff, Rohmaterial, Ausgangsstoff	matières premières	materia prima
4648	raw stock	råmassa	Rohstoff, Rohpapier	matières premières	pasta cruda
4649	raw water	råvatten	Frischwasser	eau brute (ni filtrée ni épurée)	agua cruda
4650	raw weight	bruttovikt	Rohgewicht	poids brut	peso bruto
4651	ray cell	strålcell	Markstrahlzelle, Strahlenzelle	rayon médullaire	célula de radiación
4652	rayon	rayon	Reyon, Kunstseide	rayonne	rayón
4653	rayon fiber	rayonfiber	Reyonfaser	fibre de rayonne	fibra de rayón
4654	rayon pulp	rayonmassa	Reyonzellstoff	pâte pour rayonne	pasta de rayón
4655	rayon reject	rayonutskott	Reyonausschuss	déchets de rayonne	rechazos de rayón
4656	reaction mechanism	reaktionsmekanism	Reaktionsmechanismus	mécanique de la réaction	mecanismo de reacción
4657	reaction site	reaktionsläge	Reaktionsstelle	siège d'une réaction	planta de reacción
4658	reaction time	reaktionstid	Reaktionszeit	durée d'une réaction	tiempo de reacción
4659	reaction wood	reaktionsved	Reaktionsholz, Richtgewebe	bois de réaction	madera de reacción
4660	reactivity	reaktivitet	Reaktionsfähigkeit, Reaktionsvermögen, Reaktivität	réactivité	reactividad
4661	reactor	reaktor	Reaktor	réacteur	reactor
4662	reader	läsare	Leser, Korrektor	lecteur	corrector de pruebas
4663	reagent	reagens	Reagens	réactif	reactivo
4664	ream	ris	Ries (480 Bogen)	rame	resma
4665	ream labels	risetikett	Riesaufkleber	étiquettes pour rames	etiquetas de resmilla
4666	ream markers	rismarkör	Riesmarkierzettel	bandes de papier pour séparer les rames, Pavillons	marcadores de resma
4667	ream sealed	risförseglad	riesverpackt	rame cachetée	sellado en resma
4668	ream weight	risvikt, ytvikt	Riesgewicht	poids d'une rame	peso de la resma
4669	ream wrapped	risomslagen	riesverpackt	emballage sous macule	embalado en resma
4670	ream wrappers	risomslag	Rieseinschlag, Rieseinwickler	macules pour emballage	embalajes de resma
4671	rebuilding	återuppbyggning	Umbau, Wiederaufbau	reconstruction	reconstrucción
4672	receptivity	receptivitet	Aufnahmefähigkeit	réceptivité	receptividad
4673	rechipper	omflishugg	Nachzerspaner, Nachzerkleinerungsmaschine	broyeur à copeaux	retroceadora
4674	reclaimed fiber	återvunnen fiber	rückgewonnene Faser, Sekundärfaser	fibre de récupération	fibra de recuperación
4675	reclamation	återvinning, reklamation	Zurückführung, Wiedergewinnung	récupération	reclamación
4676	record	register, lista, diagram	aufzeichnen, aufschreiben, eintragen, Rekord, Registrierung, Aufschreibung, Bestleistung	donnée, rapport, document, archives	registro
4677	record card	registerkort	Registrierstreifen, Karteikarte	fiche d'enregistrement	ficha de registro
4678	recorder	skrivare	Registriergerät, Schreiber, Aufzeichengerät	enregistreur, archiviste	registrador
4679	recorder chart paper	skrivarpapper	Schreiberdiagrammpapier	papier pour diagrammes d'appareil enregistreur	papel gráfico para registrador
4680	recording	registrering	Aufzeichnen, Aufschreiben, Eintragen, Berichten	enregistrement	registro
4681	recording instrument	skrivande instrument	Schreiber, registrierendes Messgerät	appareil enregistreur	instrumento gráfico
4682	recording paper	skrivarpapper	Registrierpapier	papier pour enregistrements	papel para registro
4683	recovering	återvinning	Rückgewinnen, Rückgewinnung, Regenerieren	récupération	recuperación
4684	recovery	återvinning, återhämtning	Regenerierung, Rückgewinnung	récupération	recuperación
4685	recovery furnace	återvinningsugn	Schmelzofen (Sulfatverfahren)	four de récupération	horno de recuperación
4686	recovery tower	återvinningstorn	Bleichturm	tour de récupération	torre de recuperación
4687	rectifier	likriktare	Gleichrichter	rectificateur, redresseur	rectificador
4688	rectifier roll	vändvals	Lochwalze	rouleau perforé	rodillo perforado
4689	recycling	återcirkulering	in den Kreislauf zurückführen	recyclage	repaso

ENGLISH	SVENSKA	DEUTSCH	FRANÇAIS	ESPAÑOL
4690 red gum	red gum	Amberbaum, Satinnussbaum	copalme d'Amérique	copal
4691 redox potential	redoxpotential	Redox-Potential	potentiel redox	potencial redox
4692 redox reaction	redoxreaktion	Redox-Reaktion, Oxydations-Reduktions-Reaktion	réaction d'oxydo-réduction	reacción redox
4693 reducible sulfur	reducerbart svavel	reduktionsfähiger Schwefel	soufre réductible	sulfuro reducible
4694 reducing agent	reducermedel	Reduktionsmittel	agent réducteur	reductor
4695 reducing valve	reducerventil	Druckminderungsventil, Druckreduzierventil	détendeur	válvula reductora
4696 reduction	reduktion	Reduktion	réduction	reducción
4697 redwood	redwood, rödbok	Rotholz	sequoia toujours vert	secoya
4698 reed	vass	Schilf, Schilfrohr, Halm	roseau	caña
4699 reel	rullmaskin	aufrollen, aufhaspeln, Rolle, Haspel, Spule	rouleau, bobine, enrouleuse en bout de machine	enrolladora
4700 reel band	rullmaskinband	Rollenbanderole	carton paille, carton de vieux papiers pour mandrins	cinta de bobina
4701 reel of board	kartongrulle	Rolle Karton	bobine de carton	bobina de cartón
4702 reel of paper	pappersrulle	Papierrolle	bobine de papier	bobina de papel
4703 reel sample	rullprov	Rollenmusterentnahme	échantillon pris sur la bobine à la machine	muestra tomada sobre bobina
4704 refiner	raffinör	Refiner, Kegelstoff, Jordanmühle	raffineur	refino
4705 refiner disc	raffinörskiva	Refinerscheiben	disque de raffineur	disco de refino
4706 refining	raffinering	Mahlung im Refiner	raffinage	refinado
4707 reflectance	reflektans	Reflexionsstärke, Reflexionsfaktor	réflexion (optique)	reflectancia
4708 reflection	reflektion	Reflexion, Reflektierung, Rückstrahlung	réflexion	reflexión
4709 reflectivity	reflektivitet	Reflexionsvermögen	pouvoir réfléchissant	reflexividad
4710 reflectometer	reflektometer	Reflectometer, Reflexionsmesser	reflectomètre	reflectómetro
4711 reforestation	skogsplantering	Wiederaufforstung	reforestation, reboisement	repoblación forestal
4712 refraction coefficient	refraktion	Brechungsindex, Refraktionskoeffizient	coefficient de réfraction	coeficiente de refracción
4713 refractory	eldfast	feuerfest, feuerbeständig, hitzebeständig, feuerfestes Material	réfractaire	refractario
4714 refuse bag	sopsäck	Müllbeutel	sac à ordures	bolsa de desperdicios
4715 regenerated cellulose	regenererad cellulosa	regenerierte Cellulose, Regeneratcellulose	cellulose régénérée	celulosa regenerada
4716 register	register	Passer	repère	registro
4717 register bond	registerpapper	Geschäftsformularpapier	papier pour registres	bond registro
4718 regulating box	reglerlåda	Regulierkasten, Überlaufkasten	régulateur de densité	caja reguladora
4719 regulator	regulator	Regler	régulateur	regulador
4720 reinforced building paper	armerat byggpapper	zweilagiges Kraftpapier (meist verstärkt mit Asphalt oder Glasfaser oder Jute)	papier de construction renforcé	papel forrado con tela para la construcción
4721 reinforced concrete	armerad betong	Stahlbeton	béton armé	cemento armado
4722 reinforced paper	armerat papper	verstärktes Papier, textilverstärktes Papier	papier entoilé, armé	papel forrado con tela
4723 reinforced plastic	armerad plast	verstärkte Kunststoffe	plastique renforcé	plástico reforzado
4724 reject	rejekt	Auswurf (beim Sortieren), Abfall	déchet	rechazo de depuración
4725 reject refining	utskottsraffinering	Grobstoffbehandlung im Refiner	raffinage des refus	manipulación de los rechazos de depuración en el refino
4726 rejection	rejektering	Ausschussausscheidung, Nicht-Annahme, Zurückweisung	rejet	rechazo
4727 relative humidity	relativ fuktighet	relative Feuchtigkeit, Luftfeuchtigkeit	degré hygrométrique	humedad relativa
4728 relaxation	relaxation	Entspannung, Lockerung, Erholung	relâchement, repos	relajación
4729 release	lätta, lyfta	auslösen, freigeben, entlassen, befreien, Auslösung (z.B. eines Hebels), Freigabe	desserrer, dégager, expulser, libérer (gaz)	soltar
4730 release agent	borttagningsmedel	Trennmittel	anti-adhésif	agente soltador
4731 release coated paper	papper med skyddsskikt	gestrichenes Release-Papier	papier couché anti-adhésif	papel estucado antiadhesivo
4732 release coating	skyddsbeläggning	Release-Streichen	couchage anti-adhésif	estucado antiadhesivo
4733 release papers	skyddspapper	Release-Papiere	papiers anti-adhésifs	papeles antiadhesivos
4734 reliability	tillförlitlighet	Zuverlässigkeit	exactitude, sécurité (de fonctionnement) certitude, fiabilité	seguridad
4735 relief	lättnad, tryckreducering, relief	Entlastung, Relief, Reliefdruck	décharge, échappement, dégazage	desgasificación
4736 relief gas	tryckreduceringsgas	Übertreibgas, Abgas	gaz d'échappement	gases de escape
4737 relief paper	reliefpapper	Reliefpapier	papier de sécurité	papel relieve
4738 relief valve	reducerventil, säkerhetsventil	Druckbegrenzungsventil	vanne de décharge	válvula de desahogo
4739 relieving	trycknedsättning	entlasten (z.B. Kalanderwalzen)	dégazage	desgasificación
4740 remote control	fjärrkontroll	Fernsteuerung, Fernkontrolle	contrôle (ou commande) à distance	control remoto

ENGLISH	SVENSKA	DEUTSCH	FRANÇAIS	ESPAÑOL
4741 removable fourdrinier part	utdragbart viraparti	ausfahrbare Langsiebpartie	table de fabrication amovible	mesa plana amovible
4742 repellent	repelleringsmedel	abweisend, wasserabstossend, wasserabstossender Stoff	répulsif	repelente
4743 reproducibility	reproducerbarhet	Reproduzierbarkeit	aptitude à la reproduction	que puede reproducirse
4744 reproduction paper	reproduceringspapper	Vervielfältigungspapier, Kopierpapier	papier pour reproductions	papel para reproducir
4745 reprography	reprografi	Reprographie	reprographie	reprografía
4746 repulping	återupplösning av massa	Wiederanfeuchten und Kochen von Zellstoff	remise en pâte	hacer pasta de nuevo
4747 research	forskning	Forschung	recherche	investigación
4748 reservoir	reservoar	Reservoir, Behälter, Speicher, Becken	réservoir	reserva
4749 residual lignin content	restligninhalt	Restligningehalt	teneur en lignine résiduelle	contenido residual de lignina
4750 residue	rest, återstod	Rückstand, Rest	résidu	residuo
4751 resilience	resiliens	Rückprall, Rückfederung, Elastizität	rebondissement, élasticité	resiliencia
4752 resiliency	resiliens	Rückprall, Rückfederung, Elastizität	résilience, rebondissement	resiliencia
4753 resins	hartser	Harz, Kunstharz	résines	resinas
4754 resin acids	hartssyror	Harzsäure	acides résiniques	ácidos de resina
4755 resinates	resinater	Resinat	résinates	resinatos
4756 resin ducts	hartskanaler	Harzgang, Harzkanal	canaux résinifères	canales resiníferos
4757 resinous wood	hartshaltig ved	Kienholz, harzreiches Holz	bois de résineux	madera resinosa
4758 resistance	motstånd, resistans	Widerstand, Widerstandsfähigkeit, Beständigkeit, Festigkeit	résistance	resistencia
4759 resistance to penetration by a liquid	vätskepenetrationsmotstånd	Durchdringungswiderstand gegen Flüssigkeiten	résistance à la pénétration par un liquide	impermeabilidad
4760 resistance to wear	slitagemotstånd	Verschleissfestigkeit	résistance à l'usure	resistencia al desgaste
4761 resistivity	resistivitet	spezifischer Widerstand	résistance, résistibilité	resistividad
4762 resistivity, electrical	elektrisk resistivitet	elektrikscher Widerstand	résistance électrique	resistividad eléctrica
4763 resonance	resonans	Resonanz	résonnance	resonancia
4764 resorcinol	resorcinol	Dioxybenzol, Resorzin	resorcinol	resorcina
4765 response time	svarstid	Einstellzeit, Reaktionszeit	temps de réponse	tiempo de repuesta
4766 restraint	återhållning	Zurückhaltung, Beschränkung, Einschränkung	contrainte, restriction	fijación
4767 retardant	fördröjningsmedel	Verzögerungsmittel	relentissant	retardador
4768 retardation	retardation	Verzögerung, Verlangsamung	retardement	retardación
4769 retention	retention	Retention, Verhaltung, Zurückhalten	rétention, conservation	retención
4770 retention aid	retentionsmedel	Retentionsmittel	adjuvant de rétention	ayuda retentiva
4771 reticulate	nätformig	netzartig	réticulé	reticular
4772 retrogradation	retrogradering	Rückläufigkeit	rétrogradation, recul	retrogradación
4773 return on investment	räntabilitet	Kapitalrendite	rentabilité de capital, rendement de l'actif	reinversión
4774 reuse	ateranvändning	wiederverwenden	réutilisation, réemploi	reutilizar
4775 reverse drive	omvänd drift	Wendeantrieb	commande inversée	mando de retroceso
4776 reverse osmosis	omvänd osmos	reversible Osmose	osmose inverse	ósmosis inversa
4777 reverse press	vändpress	Wendepresse	presse montante	prensa montante
4778 reverse press felt	vändpressfilt	Wendepressenfilz	feutre montant	fieltro montante
4779 reverse roll coater	vändvalsbestrykare	Gegenlaufwalzenstreicher, Umkehrwalzenbeschichter	coucheuse à rouleaux marchant en sens inverse	estucadora con rodillos invertidos
4780 reverse roll coating	vändvalsbestrykning	Umkehrwalzenbeschichtung	couchage sur coucheuse à rouleaux marchant en sens inverse	estucado por rodillos invertidos
4781 reversion	omvändning	Umkehrung, Umsteuerung	retour, inversion	inversión
4782 revolution	varv	Umdrehung, Tour, Umlauf	révolution, tour	vuelta
4783 revolutions per minute	varv per minut	Umdrehungen pro Minute	tours par minute	revoluciones por minuto
4784 revolving knife	roterkniv	rotierendes Messer, Kreismesser	couteau rotatif	cuchilla móvil
4785 rewet	återväta	wiederanfeuchten	réhumidifier	rehumectar
4786 rewind	omrulla	rollen, umrollen	rebobiner	rebobinar
4787 rewinder	omrullningsmaskin	Rollmaschine	rebobineuse	rebobinadora
4788 rewinding	omrullning	Aufrollen, Umrollen	rebobinage	rebobinado
4789 rheological properties	reologiska egenskaper	rheologische Eigenschaften	propriétés rhéologiques	propiedades reológicas
4790 rheology	reologi	Rheologie	rhéologie	reología
4791 rice paper	rispapper	Reisstrohpapier	papier de Chine, papier de riz	papel jaspeado
4792 rice starch	risstärkelse	Reisstärke	amidon de riz	almidón de arroz
4793 rice straw	rishalm	Reisstroh	paille de riz	paja de arroz
4794 rich white water	fiberhaltigt bakvatten	stoffreicher Siebwasser	eau collée	agua colada enriquecida
4795 rider roll	ridvals	leichte Presswalze, Egoutteur	rouleau presseur	rodillo prensor
4796 ridge	rygg, ås, veck	Siebfalte	arête	filo
4797 riding roll	ridvals	leichte Presswalze, Egoutteur	rouleau presseur sur bobine	rodillo prensor
4798 riffler	sandfång	Sandfang	sablier	arenero

	ENGLISH	SVENSKA	DEUTSCH	FRANÇAIS	ESPAÑOL
4799	riffling	rening från sand	Reinigen (im Sandfang)	épuration par sablier	decantación por arenero
4800	rigidity	styvhet	Steifigkeit	rigidité	rigidez
4801	rim	kant	einfassen, Rand, Randzone, Randfassung, Felge	jante	llanta
4802	ring	ring	läuten, schallen, umkreisen, umringen, Ring, Reif, Jahresring (Holz)	anneau, bague	anillo
4803	ring barker	ringbarkare	ringförmiges Schälen der Rinde	écorceuse à anneau	descortezadora de anillos
4804	ring crush tests	ringkrossprov	Ringstauchversuch	essais d'écrasement en anneau	pruebas de trituración por anillos
4805	ring porous wood	ringporig ved	ringporiges Holz	bois à pores en anneaux	madera de porosidad decreciente hacia el extremo de los anillos
4806	ring stiffness	ringstyvhet	Papierkanteneindrückwiderstand, Kanteneindrückbeständigkeit	rigidité en anneau	rigidez del anillo
4807	ripple finish	krusfinish	geriffelte Oberfläche	état de surface grossier obtenu par laminage	acabado en rizo
4808	rippling	krusning	Riffelung, Wellen, Kräuseln	onduler, rider	ondulado
4809	river	flod	Fluss, Strom	rivière, fleuve	río
4810	river driving	flodfart	Holztransport auf Flüssen, Fluss-Schwimmtransport	flottage des bois	flotación
4811	roadside processing	fältdrift	Verabeitung des Holzes am Wegerand	façonnage en bord de route	manipulación junto al camino
4812	Rockwell hardness	rockwellhårdhet	Rockwell-Härte	dureté en degrés Rockwell	dureza Rockwell
4813	rod	stav, stång	Stab, Stange	tige, barre, baguette	varilla
4814	rod coater	valsrakelbestrykare	Stabstreicher	coucheuse à barre	estucado de rodillos
4815	rod mill	stångkvarn	Rohrmühle, Stabmühle	broyeur à barres	molino de rodillos
4816	roll	vals	Rolle, Walze	rouleau, bobine, cylindre	rodillo
4817	roll clamps	rullklämmor	Rollenklammern	colliers de rouleau	grapas de bobina
4818	roll coater	valsbestrykare	Walzenstreicher	coucheuse à rouleaux	estucadora de rodillos
4819	roll coating	valsbestrykning	Walzenstreichen	couchage par rouleaux	estucado de rodillos
4820	roll crown	valsbombering	Walzenbombierung	rouleau bombé	bombeado del cilindro
4821	roll doctor	rullskaber	Walzenabstreifer	rouleau égaliseur	deflector de rodillo
4822	rolled edge	valsad kant	gerollte Papierkante	bord roulé	borde laminado
4823	roll end	valsände, valsgavel	Stirnblatt (als Rollenschutz)	fin de bobine	extremidad del rodillo
4824	roller bearing	rullager	Walzenlager, Rollenlager	roulement à rouleaux	cojinete de rulemán
4825	roll kiss coater	valsbestrykare	Walzenauftragsmaschine	coucheuse par léchage	estucadora de rodillo lamedor
4826	roll lowerator	valspålägggare	Walzen-Absenkvorrichtung	mécanisme pour retirer un rouleau	abatidor de rodillo
4827	roll-out fourdrinier	utdragbart viraparti	ausfahrbare Langsiebpartie	table plate sortante	máquina de mesa plana para desenrollar
4828	roll paper	valspapper	Rollenpapier	rouleau de papier	papel en rollo
4829	roll protector	valsskydd	Rollenschutz	dispositif pour protéger les bobines expédiées	protector del rodillo
4830	roll set	rullsätt	Rollneigung des Papiers	jeu (ensemble de rouleaux)	juego de rodillos
4831	roll stand	valsstativ	Rollengerüst	dévidoir support de bobines	juego de laminadores
4832	roofing felt	takpapp	Dachpappe	carton feutre pour toitures	fieltro para tejados
4833	roofing paper	takpapp	Dachpappe	papier pour toitures	papel embreado para tejados
4834	root	rot	Wurzel	racine	raíz
4835	rope	rep	Seil, Aufführseil	câble, cordage	cuerda
4836	rope carrier	repbärare	Seilführung	câble pour passer la feuille (dans la sècherie)	cuerda (para pasar la hoja en la sequería)
4837	rope feed control	repmatningsreglering	Seilaufführung	contrôle par câble	guía del cable
4838	rope marks	repmarkering	Seilmarkierung	marques de câble	marcas de la cuerda
4839	rope paper	reppapper	Tauenpapier	papier de chanvre de Manille	papel para cordel
4840	roping	repbestyckning	Tauwerk	production de cordons sur une bobine	ahilamiento
4841	rosin	harts	Colophonium, harzförmiger Rückstand aus der Rohterpentindestillation	résine (colophane)	resina
4842	rosin size	hartslim	Harzleim	collage à la résine	cola de resina
4843	rosin sized	hartslimmad	harzgeleimt	collé à la résine	encolado a la resina
4844	rosin sizing	hartslimning	Harzleimung	collage à la résine	encolado a la resina
4845	rosin specks	hartsfläckar	Harzflecken, Harzstippen	taches de résine	motas de resina
4846	rosin spots	hartsfläckar	Harzflecken, Harzstippen	taches de résine	marcas de resina
4847	rotameter	rotameter	Kurvenmesser	rotomètre (indicateur de débit)	rotámetro
4848	rotary cutter	roterande kniv	Rotationsquerschneider	coupeuse rotative	cortadora rotativa
4849	rotary digester	roterande kokare	Drehkocher	lessiveur rotatif	lejiadora
4850	rotary filter	roterande filter	Drehfilter	filtre rotatif	filtro
4851	rotary pump	roterande pump	Rotationspumpe, Umlaufpumpe	pompe rotative	bomba rotatoria

ENGLISH	SVENSKA	DEUTSCH	FRANÇAIS	ESPAÑOL
4852 rotary screen	roterande sil	Rundsortierer	épurateur rotatif	criba rotativa
4853 rotary sheeting	skärning med kutter	rotierendes Querschneiden	coupe en feuilles rotative	puesta en hojas en rotativa
4854 rotary slitter	kutter	Rotationslängsschneider	coupeuse rotative	hendedora
4855 rotary viscometer	roterande viskosimeter	rotierender Konsistenzmesser, Rotationsviskosimeter	viscosimètre rotatif	viscosímetro rotativo
4856 rotation	rotation	Rotation, Drehung, Umdrehung, Kreislauf, Umlauf	rotation, roulement	rotación
4857 rotogravure paper	rotogravyrpapper	Tiefdruckpapier	papier hélio	papel para huecograbado
4858 rotogravure printing	rotogravyr	Tiefdruck	impression en héliogravure	impresión en huecograbado
4859 rotor	rotor	Rotor, Läufer	cône, rotor	rotor
4860 rotor blade	rotorblad	Laufradschaufel, Rotorblatt	lame de rotor	cuchilla del rotor
4861 roughness	råhet	Rauhigkeit, Unebenheit	rugosité	aspereza
4862 roundness	rundhet	Rundheit, Rundung	rondeur	redondez
4863 routine control procedure	rutinkontroll procedur	routinemässiges Kontrollverfahren, routinemässige Kontrolle	opération de contrôle de routine	control de rutina
4864 rubber	gummi	Gummi, Kautschuk	caoutchouc, tablier	goma
4865 rubber-covered roll	gummiklädd vals	gummiüberzogene Walze	rouleau garne de caoutchouc	rodillo revestido de caucho
4866 rubber latex	gummilatex	Gummimilch	latex	goma látex
4867 rubber roll	gummivals	Gummiwalze, Gummirolle	rouleau en caoutchouc	rodillo de caucho
4868 rubber spots	gummifläckar	Gummiflecken	taches de caoutchouc	manchas de la goma
4869 rubbing	gnidning	reiben, Abrieb, Reibung	frottement, friction	frotación
4870 runnability	körbarhet	Lauffähigkeit, Laufeigenschaft	comportement	comportamiento
4871 runoff	utkörd, bortkörd, avkörd	ablaufen, abfliessen, Oberflächenabfluss	sous-produit rejeté, écoulement, entraînement	imprimir
4872 run-out type fourdrinier	utdragbart viraparti	ausfahrbare Langsiebpartie	–	máquina run-out de mesa plana
4873 rupture modulus	brottmodul	Bruchmodul	modèle de rupture	módulo de ruptura
4874 rust	rost	rosten, rostig werden, Rost	rouille	herrumbre
4875 rust inhibitor	rosthinder	Rostschutzmittel	anti-oxydant	anticorrosivo
4876 rust spots	rostfläckar	Rostflecken	taches de rouille	manchas de herrumbre
4877 rutile	rutil	Rutil, Titandioxid	brillant	rutilo
4878 rye straw	råghalm	Roggenstroh	paille de seigle	paja de centeno

S

4879 sack paper	säckpapper	Sackpapier	papier à sacs	papel para sacos
4880 safety	säkerhet	Sicherheit, Sicherung	sécurité	seguridad
4881 safety cock	säkerhetskok	Sicherheitshahn	robinet de sécurité	grifo de seguridad
4882 safety equipment	säkerhetsutrustning	Sicherheitsanlagen, Sicherheitsausrüstung	matériel sécurité	equipos de seguridad
4883 safety paper	säkerhetspapper	Sicherheitspapier	papier de sûreté, infalsifiable	papel infalsificable
4884 sag	säcka, hänga ned	durchhängen, sich durchbiegen, Absackung, Durchbiegung	fléchir	pandeo
4885 sale	försäljning	Verkauf	vente	venta
4886 salinity	salthalt, salinitet	Salzgehalt, Salzhaltigkeit, Salzigkeit	salinité	salinidad
4887 Salix (Lat.)	Salix	Weide	saule, osier	sauce
4888 salt cake	natriumsulfat	Natriumsulfat, Glaubersalz	sulfate de calcium brut	sulfato de sodio 'sin refinar
4889 salt water	saltvatten	Salzwasser, Sole	eau salée	agua salobre
4890 salvage	återstod, tillvaratagande	Bergung, Rettung, Nutzung, Aufarbeitung, wiederverwerten	sauvetage	recuperación
4891 sample	prov	Muster, Probe	échantillon	muestra
4892 sample preparation	provberedning	Probenvorbereitung	échantillonnage	preparación de muestras
4893 sampler	provtagare	Zerkleinerungsmaschine zur Probeentnahme, Mustermacher	modèle, patron	sacamuestras
4894 sampling	provtagning	Probeentnahme	échantillonnage	preparación de muestrario
4895 sand filter	sandfilter	Sandfilter	filtre à sable	filtro de arena
4896 sand trap	sandfång	Sandfang	sablier (de pile)	arenero
4897 sanitary napkins	sanitetsbindor	Binden	serviettes hygiéniques	servilletas higiénicas
4898 sanitary paper	sanitetspapper	Hygienepapier	papier hygiénique	papel higiénico
4899 sanitary tissue	sanitetspapper	Hygiene-Tissue, Toilettentissue	papier hygiénique	muselina higiénica
4900 sap	sav	entsaften, Saft, Pflanzensaft	sève	savia
4901 sapling	ungträd	Schössling, junger Baum	jeune arbre	renuevo
4902 saponification	förtvälning	Verseifung	saponification	saponificación
4903 saponification number	förtvålningstal	Verseifungszahl	indice de saponification	índice de saponificación
4904 sap peeling	savskalning	am wachsenden Stamm entrinden	écorçage de jeunes arbres	espitado de la savia

	ENGLISH	SVENSKA	DEUTSCH	FRANÇAIS	ESPAÑOL
4905	sapwood	splintved	Splintholz	aubier	albura
4906	sassafrass	sassafrass	Sassafras	sassafrass	sasafrás
4907	satin finish	satängfinish, atlasfinish	Satinoberfläche	lissé, uni	acabado satinado
4908	satin white	satängvitt	Satinweiss	blanc satin	blanco satino
4909	saturated	mättad	gesättigt, getränkt	imprégné, saturé	saturado
4910	saturating felts	impregnerfilt	imprägnierbarer Filz, Imprägnierfilz	feutres à impregner	fieltros de impregnación
4911	saturating papers	impregneringsbaspapper	Rohpapiere für Imprägnierungen, imprägnierbare Papiere, Imprägnierpapiere	papiers support pour imprégnation	papeles de impregnación
4912	saturating properties	mättningsegenskaper	Imprägniereigenschaften	propriétés d'imprégnation	propiedades de saturación
4913	saturation	mättnad	Sättigung, Durchtränkung, Imprägnierung	imprégnation, saturation	impregnación
4914	saveall	fiberåtervinnare	Stoffänger, Faserstoffrückgewinnungsanlage	ramasse-pâte	recogepasta
4915	save-all box	bakvattenlåda	Abwasserkasten	ramasse-pâte	caja de aguas residuales
4916	saveall tray	fiberåtervinningsskepp	Stoffängerauffangwanne	dalle à eaux blanches, bacholle	bandeja
4917	saw	såg	sägen, zersägen, absägen, Säge	scie	sierra
4918	sawdust	sågspan	Sägespäne	sciure	serrín
4919	sawing	sågning	Sägen	sciage	aserradura
4920	sawmill	sågverk	Sägewerk, Sägemühle	scierie	aserradero
4921	sawmill residue	sågverksavfall	Sägewerksabfall	déchets de scierie	residuos de aserradero
4922	scale	skala, våg, inkrust	ablösen, abschaben, abblättern, wiegen, Skala, Masstab, Gradeinteilung, Ablagerung	graduation, échelle, écaille	escala
4923	scale, dryer	torkcylinderinkrust	Trockengradeinteilung	écaille de sécheur	sedimento en la secadora
4924	scale paper (Brit.)	skalpapper	Millimeterpapier, Profilpapier	papier millimétré	papel cuadriculado
4925	scanning	avsökning, svep	Abtasten der Papierbahn, Bildabtastung	équipement de guidage électronique du papier	exploración
4926	scarification	rivning	Einritzen der Baumrinde, Einkerben der Baumrinde	scarification	escarificación
4927	scattering	spridning	Zerstreuen, Streuung, Streueffekt	éparpiller	desparramiento
4928	scheduling	tidsplanering	Planen	programmation	programación
4929	Schopper Riegler freeness	malgrad	Schopper-Riegler-Mahlgrad	indice d'égouttage Schopper	grado de refino de Schopper-Riegler
4930	scoop	skopa	schöpfen, Schöpfeimer	écope, nacelle	canjilón
4931	score	skåra, tjog, konto	ritzen, einschneiden, zählen, die Punktzahl, Grenzlinie, Markierungslinie	pli rainé	entalladura
4932	score break	rekord	Abriss einer Rollenbahn	cassure au pli rainé	ruptura por entalladuras
4933	score cut	skåra	Rollenschnitt	coupe par molettes sur cylindre en acier trempé	corte por moleta
4934	score cutter	huggjärn	Rollenschneidmaschine	coupeuse à molettes	cortadora de moleta
4935	scored	registrerat	eingeritzt, gefurcht	rainé	rayado
4936	scorer	markör	Ritzmaschine	marqueur	marcador
4937	scoring	märkning	Ritzen, Rillen	action de rainer	entalladura
4938	scotch pine	scotch pine	Föhre	pin d'Ecosse, pin sylvestre	pino silvestre
4939	scouring	bykning	scheuern, reinigen, entfetten, entbasten	dégraissage d'une feutre	desengrase
4940	scrap	skrot	verschrotten, Beschnitt, Papierschnitzel, Rest, Ausschuss	morceau, fragment	desechos
4941	scraper board (Brit.)	skräpkartong	Schabpapier, Schabkarton	papier "procédé"	papel para dibujo al raspado
4942	scratch, blade	schaberrepa	Kratzer (vom Messer oder der Rakel)	strie de lame	estría de cuchilla
4943	scratch board	skisskartong	Schabpapier, Schabkarton	papier "procédé"	papel para dibujo al raspado
4944	scratches	repor	Kratzer	stries, égratignures	rayas
4945	screen	sil	sieben, sortieren, Knotenfänger (am Stoffauflauf), Raster, Sortierer, Schirm, Abschirmung	tamis, épurateur, assortisseur, classeur	depurador
4946	screen angle	silvinkel	Siebwinkel (nur an der Streichmaschine)	orientation de trame	ángulo de la criba
4947	screening	silning	Sortieren, Sichten	épuration, classage	cribado
4948	screenings	silrest	Sortierstoff, Siebrückstand	refus d'épuration, déchets	rechazo de depuración
4949	screen room	sileri	Sortiersaal	salle d'épuration	sala de depuración
4950	screw	skruv	anschrauben, anziehen, Schraube, Schnecke	vis	tornillo
4951	screw conveyor	skruvtransportör, skruvmatare	Schneckenförderer, Förderschnecke	transporteur à vis sans fin	transportador de tornillo
4952	screw press	skruvpress	Spindelpresse	presse à vis	prensa de tornillo
4953	scrim	häftgas	Polsterfutter	grille de renforcement (pour non tissé)	tela de forro

	ENGLISH	SVENSKA	DEUTSCH	FRANÇAIS	ESPAÑOL
4954	scrubber	skrubber	Gaswäscher, Berieselungsturm, Wäscher	laveur de gaz	lavador de gas
4955	scrubbing tower	tvättorn	Waschturm	tour de lavage	torre de lavado
4956	scuffing	hasning	Abreiben	frottement, friction	fricción
4957	scuff resistance	hasningsmotstånd	Nassabriebfestigkeit, Abnutzungsfestigkeit	résistance au frottement	resistencia à la fricción
4958	scum	skum	abschäumen	mousse, écume	espuma
4959	seal	tätning, försegling	verdichten, absperren, plombieren, versiegeln, Dichtung, Absperrung, Verschluss	cacheter, sceller	sellar
4960	sealability	tätbarhet	Versiegelbarkeit, Verschweissbarkeit	aptitude à l'adhésion	aptitud al sellado
4961	sealant	tätningsmedel	Dichtungsmittel	adhésif	sellador
4962	sealing	tätning	verdichten, versiegeln, verschweissen, Sealing (Kraftpapiersorte)	cachet (pour sceller ou clore)	sellado
4963	sealing paper	tätningspapper	Siegelpapier, siegelfähiges Papier	papier de cellulose à la soude	papel sellador
4964	sealing machine	tätningsmaskin	Beutelschliessmaschine, Verschlussmaschine	machine à sceller	máquina selladora
4965	seal pit	tätningsvattenlåda	Saugerwasserkasten	fosse couverte	caja aspirante
4966	seal tank	tätningsvattentank	Fallwasserkasten	réservoir couvert	tina de aspiración
4967	seam	söm	Naht, Saum	couture, jonction, raccordement	costura
4968	seaming	sömnad	Besäumen, Säumen	coudre, raccorder, souder	cosido
4969	seasonal variation	säsongsvariationer	jahreszeitlich bedingte Schwankungen	variation saisonnière	variación estacional
4970	secondary forest	sekundaskog	Sekundärwald	forêt secondaire	bosque secundario
4971	secondary growth	sekundaväxt	Nachwuchs (Wald)	croissance secondaire	árboles renacidos
4972	secondary headbox	andra inloppslåda	Sekundärstoffauflauf	seconde caisse d'arrivée de pâte	caja de entrada secundaria
4973	secondary stock	andramassa	Sekundärstoff	fibres de récupération	pasta secundaria
4974	secondary treatment	vidarebehandling	Nachbehandlung	traitement secondaire	tratamiento secundario
4975	secondary wall	sekundärvägg	Sekundärzellwand	paroi secondaire	pared secundaria
4976	second press	andrapress	zweite Presse	seconde presse	segunda prensa
4977	seconds	sekunder	Produkt zweiter Wahl, zweite Qualität	deuxième choix	segunda calidad
4978	sectional drive	sektionsdrift	Gruppenantrieb	commande sectionnelle	mando seccional
4979	sediment	sediment	Sediment, Bodensatz, Niederschlag	sédiment	sedimento
4980	sedimentation	sedimentation	Absetzen, Niederschlagen	sédimentation	sedimentación
4981	seed	frö	säen, Samen, Saat	graine, semence	grano, semilla
4982	seed bed	frödling	Saatbeet	pépinière	lecho de simiente
4983	seeding	frösådd	Besamen	germe	sembrar
4984	seedling	ungplanta	Sämling	semis	planta criada de semilla
4985	seepage	vattenvandring	Durchsickern, Tropfen, Lecken	égouttage, suintement	chorreo
4986	selection	selektion, urval	Wahl, Auswahl, Selektion	sélection, choix	selección
4987	selectivity	selektivitet	Selektivität, Trennschärfe	sélectivité	selectividad
4988	semibending chip	semibendingflis	Faltschachtelkarton aus Schrenzpappe	carton de vieux papiers pour boîtes plaintes ordinaires	viruta de semidoblamiento
4989	semibleached	halvblekt	halbgebleicht	mi-blanchi	semiblanqueado
4990	semibleached pulp	halvblekt massa	halbgebleichter Zellstoff	pâtes mi-blanchies	pasta semiblanqueada
4991	semichemical board	halvkemisk kartong	Pappe aus Halbzellstoff	carton de pâtes mi-chimiques	cartón semiquímico
4992	semichemical corrugating medium	halvkemisk vågpapp	Wellpappenwelle aus Halbzellstoff	cannelure mi-chimique	cartón a ondular semiquímico
4993	semichemical pulp	halvkemisk massa	halbchemischer Zellstoff	pâte mi-chimique	pasta semiquímica
4994	semichemical pulping	halvkemisk kokning	halbchemischer Aufschluss	fabrication de pâtes mi-chimiques	elaboración de pasta semiquímica
4995	semiconductor	halvledare	Halbleiter	semi-conducteur	semiconductor
4996	semipermeable membrane	semipermeabelt membran	halbdurchlässige Membran	membrane semi-imperméable	membrana semipermeable
4997	sensitivity	känslighet	Sensitivität, Sensibilität, Empfindlichkeit	sensibilité	sensibilidad
4998	sensitization	sensibilisering	Sensibilisierung, Sensitivierung	sensibilisation	sensibilización
4999	sensitized paper	sensibiliserat papper	lichtempfindliches Papier, Photopapier	papier photo sensibilé	papel sensibilizado
5000	sensitizer	sensibiliseringsmedel	Sensibilisator	sensibilisateur	sensibilizador
5001	sensitizing paper	sensibiliseringspapper	Lichtpauspapier	papier support photo	papel para sensibilzar
5002	sensor	sensor, känselkropp, givare, sond	Messwertgeber, Messfühler, Tastelement	détecteur, sondeur	sensor
5003	separation	separation	Trennung, Abscheidung, Abspaltung	séparation	separación

	ENGLISH	SVENSKA	DEUTSCH	FRANÇAIS	ESPAÑOL
5004	separator	separator	Trennvorrichtung, Abscheider	séparateur, décanteur	separador
5005	sequence	följd	Folge, Reihenfolge, Aufeinanderfolge	suite, série, succession	serie
5006	sequencing	ordna i följd	Folgesteuerung, Ordnen	sérier	arreglo en series
5007	sequoia	Sequoia	Mammutbaum	sequoia	secoya
5008	serrated	sågkantad	ausgezackt, gesägt	serré	dentado
5009	service	service, tjänst	Dienstleistung, Dienst, Kundendienst, Bedienung	service, utilité, usage	servicio
5010	service life	livslängd	Lebensdauer, Haltbarkeit	durée d'utilisation	duración de servicio
5011	servomechanism	servomekanism	Servomechanismus	servomécanisme	servomecanismo
5012	set	sätt	einstellen, regulieren, richten, einfassen, Satz, Sortiment, Garnitur, Gruppe, Drucksatz	jeu, série, assortiment	juego
5013	set off	överföra	auslösen (eines Alarmsystems), Ausgleich, Kontrast, Gegensatz, Abschmutzen	éliminer, déclecher	maculado
5014	setting	sättning	Einstellen, Abbinden, Festwerden	montage	regulación
5015	setting time	sättningstid	Abbindezeit, Härtezeit	temps de montage	tiempo de ajuste
5016	settling	sedimentering	Absetzen, Sedimentieren	dépôt, décantation	decantación
5017	settling pond	sedimenteringsbassäng	Klärsumpf, Absetzteich, Absetzbecken	bassin de décantation	arenero
5018	settling tank	sedimenteringstank	Absetztank, Klärbecken	bassin de décantation	tanque de decantación
5019	setup boxboard	–	Aufstellschachtel, Festkartonage	carton pour boîtes montées	cartón para montar
5020	setup boxes	–	Aufstellschachtel, Festkartonage	boîtes montées en carton	cajas de cartón para montar
5021	sewage	avlopp	Abwasser	épandage	aguas usadas
5022	sewer	avloppsbrunn	Abwasserkanal, Abzugskanal, Strassenkanal, Gully	égout	conducto de desagüe
5023	sewer loss	avloppsförlust	Abwasserverlust	perte à l'égout	pérdida del vertedero
5024	shade	skugga	nuancieren, schattieren, Farbabstufung, Farbtönung, Schattierung	teinte	tinte
5025	shadow marks	skuggmarkering	Schattenmarkierung	ombres	marcas de sombra
5026	shaft	axel	Schaft, Welle, Achse	arbre (transmission)	eje
5027	shake	skak, skaka	schütteln, rütteln, Siebschüttelung, Vibration	branlement, secousse	traqueo
5028	shaker	skak	Schüttelapparatur, Schüttelvorrichtung	agitateur à secousses, trembleur	sacudidor
5029	shaker knot screen	kvistfång med skak	Schüttelsieb, Schüttelsortierung	trieur de noeuds à secousses	criba vibrante
5030	shake unit	skakenhet	Schüttelzeit	système de branlement	unidad sacudidora
5031	shaking	skakande, skakning	Erschütterung, Schütteln	agitation, secousse, branlement	sacudimiento
5032	sharpening	skärpning	Schärfung, Schliff (Messer), Schärfen	affutage, piquage (d'une meule)	afilado
5033	sharpness	skärpa	Schärfe	intensité	agudeza
5034	shavings	spån	Holzschnitzel, Papierschnitzel, Holzspäne, Bohrenabfall, Hobelnabfall, Feilenabfall	rognures, chutes	recortes
5035	shearing	skjuvning	Scheren, Abscheren, Schub	cisaillement	cortadura
5036	shearing strength	skjuvhållfasthet	Scherfestigkeit, Scherkraft	résistance au cisaillement	resistencia al corte
5037	shear rate	skjuvhastighet	Schubzahl	vitesse de rupture	velocidad de corte
5038	shear strength	skjuvhållfasthet	Scherfestigkeit, Scherkraft	résistance à la rupture	fuerza cortante
5039	shear stress	skjuvspänning	Scherkraft, Scherbeanspruchung, Torsionsspannung	tension de rupture	esfuerzo de corte
5040	shear tests	skjuvprov	Scherprobe	essais de rupture	pruebas de corte
5041	sheathing paper	överdragspapper	Umkleidungspapier, Ummantelungspapier	papier de construction	papel para revestir
5042	sheet	ark	Bogen, Blatt	feuille	hoja
5043	sheet cutter	arkskärare	Bogenschneider, Querschneider	coupeuse en feuilles	conductor de guillotina
5044	sheet fed	arkmatad	mit Bogenzufuhr	presse feuille à feuille	alimentado de hoja
5045	sheet feeding	arkmatning	Bogenzuführung, Bogenanlegen	alimentation en feuilles	alimentación de hojas
5046	sheet former	arkform	Blattbildner	machine de formation de feuille	plantilla de hojas
5047	sheet forming	arkformning	Blattbildung, Blattformierung	formation de la feuille	formación de hojas
5048	sheet guides	arkledare	Bahnführung, Blattführung	guide feuille	guía de la hoja
5049	sheeting	arkformning	Aufeinanderlegen, Verschalung, Folienmaterial, Aufschneiden in Bogen, Schneiden von Bogen aus Bahn oder Rolle	mise en feuille	puesta en hojas
5050	sheeting equipment	arkformningsutrustning	Bogenstapler, Bogenschneider	matériel de coupe	equipo para la puesta en hojas
5051	sheet mold	arkform	Handschöpfrahmen	forme pour la formation de feuille	molde de hoja
5052	sheet of paper	pappersark	Papierbogen, Papierblatt	feuille de papier	hoja de papel

ENGLISH	SVENSKA	DEUTSCH	FRANÇAIS	ESPAÑOL
5053 sheet roll	pappersvals	Rolle einer kontinuierlichen Papierbahn	rouleaux porteurs de la feuille	rodillo de hojas
5054 sheet size	arkformat	Bogenformat	format de la feuille	tamaño de la hoja
5055 Sheffield tester	Sheffieldprovare	Sheffield-Prüfgerät	appareil Sheffield de mesure de la perméabilité à l'air	probador Sheffield
5056 shelf life	lagringslivslängd	Lagerbeständigkeit	durée de stockage	vida del entrepaño
5057 shell	skal, mussla	Schale, Hülle, Gehäuse	coquille, enveloppe, virole de sécheur	cubierta
5058 shield	sköld	Schild, Schutz, Abschirmung	écran, revêtement	pantalla de protección
5059 shift	skift	Schichtarbeit, Wechsel, Verlagerung, Verschiebung	équipe, faction	turno
5060 shiner	mynt	glänzende Fehlerstelle im fertigen Papier	particule de charge transparente après calandrage, point lustré	abrillantador
5061 shipping container	transportemballage	Versandbehälter	caisse pour transport	contenedor de transporte
5062 shive	spet	Splitter, Holzsplitter	bûchette	astilla
5063 shock	stöt	Schock, Erschütterung	choc	sacudimiento
5064 shock drying	chocktorkning	Schocktrocknung	séchage excessif de la feuille	secado excéntrico
5065 shock resistance	stötmotstånd	Stossfestigkeit, Schlagbiegefestigkeit	résistance aux chocs	resistencia a los golpes
5066 shoe board	skokartong	Schuhpappe	carton cuir pour chaussures	cartón cuero para calzado
5067 Shore durometer	shore-durometer	Shore-Härtemesser	duromètre Shore	durómetro Shore
5068 Shore hardness	shore-hårdhet	Shore-Härte	dureté shore	dureza Shore
5069 shortage	brist	Mangel, Knappheit	manque, insuffisance, pénurie	escasez
5070 shortening of fibers	fiberförkortning	Faserverkürzung, Faserschneiden	raccourcissement des fibres	acortado de las fibras
5071 short-fibered	kortfibrig	kurzfaserig	à fibres courtes	de fibras cortas
5072 shortleaf pine	kortfibrig tall	Glattkiefer, Gelbkiefer	pin à feuilles courtes	pino de hojas cortas
5073 short stock	kort mäld	Zellstoff aus Kurzfasern, Kurzholzware (weniger als 1.80 m)	pâte courte	madero corto
5074 short ton	kort ton	neunhundertsieben comma einhundertfünfundachtzig kg (907,185 kg)	tonne américaine	tonelada americana
5075 short wood harvester	kortvedavverkare	Kurzholzerntemaschine	moissonneur à rondins	cortadora de árboles bajos
5076 short wood logging	kortvedavverkning	Zersägen eines Baumes in kurze Längen	abattage du bois en petites dimensions	cortado de árboles bajos
5077 shower	sprits	bespritzen, abbrausen, Spritzrohr, Schaueranlage	rinceur	regadera
5078 show-through	genomlysning	Durchscheinen	aspect du papier vu par transparence	apertura de inspección
5079 shredder	rivare	Zerreissmühle	déchiqueteur	desfibradora
5080 shredding	rivning	zerreissen, zerkleinern, zerschnitzeln	déchiquetage	desfibrado
5081 shrink	krympa	schrumpfen, einlaufen, einlaufen lassen, Schrumpf	contracter (se) rétrécir	encoger
5082 shrinkage	krympning	Schwund, Schrumpfen	retrait, contraction	encogido
5083 shutdown	avstängning	Stillsetzen, Stillstand, Stillegung, Arbeitseinstellung	arrêt (machine), fermeture (usine)	parar (una máquina)
5084 shutoff valve	avstängningsventil	Absperrventil	vanne d'arrêt	válvula de cierre
5085 shute wire	inslagsträd	Schussdrähte am Papiersieb	trame d'une toile métallique	tela metálica con trama de seda
5086 side roll	kantrulle	Nebenrolle	bobineau	rodillo lateral
5087 siderun	kantrulle	Nebenbahn, Randabschnitte	à côté de bobine	marcha lateral
5088 sieve	sikt	sieben, Sieb	tamis, crible	cribar, tamiz
5089 sieve analysis	siktanalys	Siebanalyse	analyse granulométrique par tamisage	análisis de tamiz
5090 silicates	silikater	Silikate	silicates	silicatos
5091 silicon	kisel	Silikon, Silizium	silicium	silicio
5092 silicon compounds	kiselföreningar	Siliziumverbindungen	composés de silicium	compuestos de silicio
5093 silicon rectifier	kisellikriktare	Silizium-Flächengleichrichter	redreseur au silicium	rectificador de silicio
5094 silicone resins	silikonharts	Silikonharz	résines de silicium	resinas de silicio
5095 silk screen printing	silkscreen-tryckning	Seidensiebdruck	impression à l'écran de soie	impresión sobre pantalla de seda
5096 silt	silt	versanden, sedimentieren	vase	sedimentos
5097 silver compounds	silverföreningar	Silberverbindungen	composés de l'argent	compuestos de plata
5098 silver fir	silvergran	Silbertanne, Coloradotanne	sapin argenté	abeto plateado
5099 silviculture	skogsvård, silvikultur	Waldbau	sylviculture	silvicultura
5100 single coated	enkelbestruken	einfach gestrichen, einmal gestrichen	couché une face	estucado una cara
5101 single-coated paper	enkelbestruket papper	einfach gestrichenes Papier, einmal gestrichenes Papier	papier couché une face	papel estucado una cara
5102 single-cylinder machine	encylindermaskin	Einzylindermaschine	machine monocylindrique	máquina de cilindro único
5103 single-faced corrugated board	ensidigt täckt vågpapp	einseitig kaschierte Wellpappe	carton ondulé simple face	cartón ondulado una cara

	ENGLISH	SVENSKA	DEUTSCH	FRANÇAIS	ESPAÑOL
5104	single-faced roll	rulle av ensidig wellpapp	einseitig kaschierte Wellpappe	carton ondulé à base de microcannelure	rodillo para una cara
5105	single facer	maskin för ensidig wellpapp	einseitige Wellpappe	machine à onduler simple face	de una cara
5106	single fiber	enkelfiber	Einzelfaser	monofibre	fibra sencilla
5107	single kraft-lined chipboard	gråpapp med ett kraftmassatäckskikt	einseitig kaschierte Schrenzpappe	carton gris recouvert d'une feuille de papier kraft	cartón gris forrado una cara kraft
5108	single lined	ensidigt täckt	einseitig kaschiert	doublé sur une seule face	blanqueado una cara
5109	single-lined board	ensidigt täckt kartong	einseitig kaschierter Karton	carton duplex	cartón blanqueado una cara
5110	single manila-lined news	tidningspapper med ett manillatäckskikt	einseitig mit Zeitungsdruck kaschiertes Manila	carton de vieux papiers recouvert d'une feuille de papier d'emballage	papel periódico forrado manila una cara
5111	single ply	enlag	einlagig	un jet	una capa
5112	single-ply board	enlagskartong	einlagiger Karton	carton un jet	cartón una capa
5113	single-sheet cutter	enkelarkskutter	Querschneider für Wasserzeichenpapiere	coupeuse simplex	cortadora de una hoja
5114	single white vat-lined chipboard	ensidigt täckt gråpapp	weisser einseitig kaschierter Schrenzkarton	carton gris recouvert d'une feuille de papier à la main	cartón gris blanqueado en tina
5115	siphon	sifon	Siphon, Saugheber	siphon	sifón
5116	siphon type condensate removal	sifonkondensatavskiljning	Kondensatabsauger, Kondensatabführung	système d'élimination des condensats du type siphon	extracción de condensados por sifón
5117	sisal	sisal	Sisal	sisal	sisal
5118	sisal hemp	sisalhampa	Sisalhanf	chanvre de sisal	cáñamo sisal
5119	site preparation	lokalberedning	Vorbereitung des Geländes, Vorbereitung der Baustelle	préparation du site	preparación del emplazamiento
5120	Sitka spruce	Sitka spruce	Sitkafichte	épicéa de Sitka	epícea Sitka
5121	size	lim	leimen, kalibrieren, zurichten, Grösse, Mass, Format	colle (collage du papier), format	cola
5122	sized and supercalendered	limmad och kalandrerad	geleimt und superkalandriert	collé et surcalandré	encolado y satinado
5123	sized paper	limmat papper	geleimtes Papier	papier collé	papel encolado
5124	size emulsion	limemulsion	Harzemulsion, Harzmilch, Leimmilch	lait de colle	lechada de cola
5125	size press	limpress	Leimpresse	presse encolleuse, size press	prensa encoladora
5126	size roll	limpressvals	Leimauftragswalze	cylindre de format	cilindro formato
5127	size specks	limfläckar	Leimflecken	taches de résine	motas de cola
5128	sizing	limning	Oberflächenleimung	collage	encolado
5129	sizing tester	limningsprovare	Leimungsprüfer, Leimungsgradprüfer	collagimètre	colagímetro
5130	sizing tests	limningsprov	Leimungsgradprüfung	essais de collage	pruebas de encolado
5131	skating	glidning	Streifenbildung im Stoff (auf dem Sieb) hervogerufen durch Fehler im Stoffauflauf	formation de traînée	patinaje
5132	skid	glidning	Ablagebock, Ladegestell, Markierungen durch überfliessendes Material	coin, patin, benne	arrastrar (troncos)
5133	skidder	glidsko	Skidder	débusqueur, débardeur	arrastrador (de troncos)
5134	skidding	glidning	Rücken von Holz (vom Fällungsort zum Holzlagerplatz)	débusquage, débardage	arrastre (de troncos)
5135	skid of paper	pappersstapel	Papierstapel	–	pila de papel
5136	skimmer	skumborttagare	Schaumüberlauf (beim Voith-Stoffauflauf), Entschäumer	écumoire	espumadera
5137	skimming	skummande	Abschäumen, Abschöpfen	écumage	despumación
5138	skip scoring	rejektmängd	Fehlstellenanzeige	rainurage	marcaje de salto
5139	skips	rejekt	ausgelassene Stellen (beim Streichen)	saut	cartón para modelos, papel para modelos
5140	skyline logging	kalhuggning	–	débardage par câble (système skyline)	tala de árboles por cable
5141	slab	platta	Platte, Fliese, Tafel	bûchette, feuille de pâte humide	costero
5142	slack edge	lös kant	schlaffe, lose Ränder, flatternde Ränder	bord détendu	borde aflojado
5143	slack sized	svaglimmat	schwach geleimt	peu collé	encolado ligero
5144	slaked lime	släckt kalk	Kalkschlamm	chaux éteinte	cal apagada
5145	slaker	släckningsmedel	Lㅗscher	dissolveur à chaux	apagador
5146	slaking	släckning	Löschen, Kalk löschen	extinction de la chaux	deleznamiento
5147	slash	sprätta	Schlagabraum, Reisig	coupure	trocha
5148	slasher	sprättare	Brennholzsägemaschine	tronçonneuse à scies multiples	cortador
5149	sleeve	manchett, hylsa	Dichtmanschette	manchon	manguito
5150	sleeve bearing	hylslager	Gleitlager	palier à manchon	cojinete de manguito
5151	sleeve press	hylspress	Schrumpfsiebpresse	presse manchonnée	prensa de manguito
5152	slewing	svängbar	schwenkbar, drehbar	pivoter	barrido

ENGLISH	SVENSKA	DEUTSCH	FRANÇAIS	ESPAÑOL
5153	slewing roller	styrhjul	Papierwerfwalze	rouleau pivotant	rodillo de barrido
5154	slice	slits, öppning, linjal	Auslauf, Staulatte	règle (table de fabrication)	regla
5155	slice lip	inloppssläpp	Auslaufflippen	lèvre de la caisse de tête	labio de la caja de alimentación
5156	slice marks	linjalmarkering	Markierungen durch den Stau	marques de la lèvre	marcas de regla
5157	slice position	läppinställning	Stellung der Spindel am Stoffauflaug	angle de coupe	posición de la regla
5158	slice velocity	utströmningshastighet	Austrittsgeschwindigkeit	vitesse de coupe	velocidad de la regla
5159	slick finish	slät finish	glatte Oberfläche, glitschige Oberfläche	fini lisse et uni	acabado liso
5160	slime	slem	Schleim, Schlamm	vase, boue, dépôt	poso
5161	slime hole	slemhål	durch Schlamm hervorgerufene Löcher, durch Schleimflecke hervorgerufene Löcher	trou du à la présence de boues	orificio para posos
5162	slime spots	slemfläckar	Schleimflecken, Schlammflecken	taches de boue	manchas de posos
5163	slimicides	slembekämpningsmedel	Schleimbekämpfungsmittel	limoneux, vaseux	mucilagocidas
5164	slip	glida, glidning, slip	sich verschieben, rutschen, Schlupf, Zettel	sauce (de couchage)	baño de estucado
5165	slippage	glidning, slirning	Gleiten, Schlupf	ripage, glissement	deslizamiento
5166	slipped roll	slirad vals	entkeilte Walze	bobine qui a glissé sur son mandrin	rodillo desplazado
5167	slitter	slitsverk	Rollenschneider, Längsschneider	coupeuse en long, refendeuse	hendedora
5168	slitter dust	slitsningsdamm	Schneidstaub	poussière aux couteaux circulaires	polvo de la hendedora
5169	slitter edge	knivegg	Schneidkante	effilochure	borde de la hendedora
5170	slitter operator	slitsverksförare	Schneidmaschinenbedienungsmann	coupeur (ouvrier)	cortador
5171	slitter-rewinder	rullmaskin	Schneid- und Aufwickelmaschine	bobineuse-refendeuse	bobinadora cortadora
5172	slitting	längsskärning, slitsning	Schlitzen, Längsschneiden	trancher	henedura
5173	slitting machine	slitsverk	Längsschneider	bobineuse trancheuse	máquina hendedora
5174	sliver	spån	Splitter	bûchette, incuit	incocido
5175	slope	sluttning	Gefälle, Schräge, Abhang	pente	pendiente
5176	slot	hål, slits	Schlitz, Einschnitt, Kerbe	fente, rainure	ranura
5177	slotter	slitsningsmaskin	Schlitzmaschine	slotter, mortaiseuse	ranurador
5178	slotting	slitsning	Schlitzen, Kerben	formation d'entaille, rainurage	ranurado
5179	slowness	dränagemotstånd	Schmierigkeit (des Stoffs)	qualité d'une pâte grasse	retraso
5180	slow stock	smörjig mäld	schmieriger Stoff	pâte grasse	pasta engrasada
5181	sludge	slam	Schlamm	boue, dépôt	lodo
5182	sludge disposal	slamborttagning	Schlammbeseitigung	evacuation des boues	disposición del lodo
5183	sluice	sluss	Schleuse	vannette à pâte	canal
5184	slurry	slamma	zu einer Suspension verflüssigen, Suspension, pumpfähige Masse	pâte liquide	pasta lechada
5185	slurrying	slamning	zu einer Suspension verflüssigen	pâte liquide en fabrication	mezclado pastoso
5186	slush	slam	im Pulper auflösen	boue liquide, pâte liquide	desfibrar
5187	slusher	slammare	Auflöser	pile défileuse	pila desfibradora
5188	slushing	slamning	Auflösen	défilage, désintégration	desfibrado
5189	slushing equipment	slamningsutrustning	Auflöseapparatur, Auflöseanlage	triturateur	triturador
5190	slush pulp	otorkad massa	pumpfähiger Zellstoff	pâte liquide	pasta engrasada
5191	small-dimensioned wood	klenved	schwaches Rundholz, Schwachholz	bois de petite dimension	maderos pequeños
5192	smelt	smälta	schmelzen, Stint, Schmelze	salin	sales fundidas
5193	smelt furnace	smältugn	Schmelzofen	four de fusion	horno de fundición
5194	smelt tank	smälttank	Schmelztank	réservoir à salin	tanque de fundición
5195	smoke	rök	rauchen, Rauch, Qualm	fumée	humo
5196	smooth-finish board	glättad kartong	Karton mit glatter Oberfläche	carton satiné	cartón calandrado
5197	smoothing	glättning	Glätten, Schlichten	satinage	alisadura
5198	smoothing press	glättpress	Glättpresse	presse offset	prensa offset (sobre máquina de papel)
5199	smoothing roll	glättvals	Verstreichwalze, Glättwalze	rouleau de lisse	prensa offset
5200	smoothing roll coater	glättvalsbestrykning	Glättwalzen-Streicher	coucheuse à rouleaux	estucadora de rodillo alisador
5201	smoothness	glättning, glätthet	Glätte	lissé, état de surface d'un papier lisse et uni	alisado
5202	smoothness tester	glättningsprovare	Glätteprüfer	appareil pour déterminer l'état de surface d'un papier	probador del alisado
5203	snailing	krypning	durch Luftblasen verursachte Streifen im Papier	cloques provoquées par le rouleau égoutteur	acaracolado
5204	snap	snäppa	Klang, Rascheln (von Papier)	craquelure	mordedura
5205	soaking	blötläggning	Einweichen, Tränken	trempage	empapado
5206	soap	tvål	Seife	savon	jabón
5207	soap skimmer	skumsläckare	Seifenentschäumer	écumoire à savon	espumadora del jabón
5208	soap wrap	tvålomslag	Seifeneinwickelpapier	emballages pour savons	envoltura para jabón

	ENGLISH	SVENSKA	DEUTSCH	FRANÇAIS	ESPAÑOL
5209	soda	soda	Soda, Natriumkarbonat	soude, carbonate de sodium	sosa
5210	soda ash	sodaaska	technisches Natriumkarbonat	carbonate de soude	carbonato de sosa
5211	soda liquor	sodalut, natronlut	Natronlauge	lessive	lejía sódica
5212	soda process	sodaprocess	Natronaufschlussverfahren	procédé à la soude	proceso sódico
5213	soda pulp	sodamassa	Natronzellstoff	pâte à la soude	pasta sódica
5214	soda pulping	sodakokning	Natronaufschluss	fabrication de pâtes à la soude	elaboración de pasta sódica
5215	sodium	natrium	Natrium	sodium	sodio
5216	sodium alginate	natriumalginat	Natriumalginat	alginate de sodium	alginato sódico
5217	sodium aluminate	natriumaluminat	Natriumaluminat	aluminate de sodium	aluminato sódico
5218	sodium-base liquor	natriumkoklut	Lauge auf Natriumbase	liqueur à base de sodium	lejía con base de sodio
5219	sodium bicarbonate	natriumbikarbonat	Natriumbikarbonat	bicarbonate de soude	bicarbonato de soda
5220	sodium bisulfite	natriumbisulfit	Natriumbisulfit	bisulfite de sodium	bisulfito de sodio
5221	sodium borate	natriumborat	Natriumborat, Borax	borate de sodium	borato de sodio
5222	sodium borohydride	natriumborohydrid	Natriumborhydrid	borohydrure de sodium	hidruro de boro sódico
5223	sodium carbonate	natriumkarbonat	Natriumkarbonat, Soda	carbonate de soude	carbonato de sosa
5224	sodium chlorate	natriumklorat	Natriumchlorat	chlorate de sodium	clorato de sodio
5225	sodium chloride	natriumklorid	Natriumchlorid, Kochsalz	chlorure de sodium	cloruro de sodio
5226	sodium chlorite	natriumklorit	Natriumchlorit	chlorite de sodium	clorito de sodio
5227	sodium compounds	natriumföreningar	Natriumverbindungen	composés du sodium	compuestos de sodio
5228	sodium hydrosulfide	natriumhydrosulfid	Natriumsulfhydrid, Natriumsulfhydrat	hydrosulfure de sodium	sulfhidrato sódico
5229	sodium hydrosulfite	natriumhydrosulfit	Natriumhydrosulfit, Fixiernatron	bisulfite de sodium	hidrosulfito de sodio
5230	sodium hydroxide	natriumhydroxid	Natriumhydroxyd, Ätznatron	hydrate de sodium	sosa cáustica
5231	sodium hypochlorite	natriumhypoklorit	Natriumhypochlorit	hypochlorite de soude	hipoclorito de sodio
5232	sodium peroxide	natriumperoxid	Natriumperoxid	peroxyde de sodium	peróxido de sodio
5233	sodium silicate	natriumsilikat	Natriumsilikat	silicate de soude	silicato de sodio
5234	sodium stearate	natriumstearat	Natriumstearat	stéarate de soude	estearato de sodio
5235	sodium sulfate	natriumsulfat	Natriumsulfat, Schwefelsaures Natrium	sulfate de soude	sulfato de sodio
5236	sodium sulfide	natriumsulfid	Natriumsulfid	sulfure de sodium	sulfuro de sodio
5237	sodium sulfite	natriumsulfit	Natriumsulfit	sulfite de sodium	sulfito de sodio
5238	sodium thiocyanate	natriumtiocyanat	Natriumthiocyanat, Natriumsulfozyanid	thiocyanate de sodium	tiocianato de sodio
5239	sodium thiosulfate	natriumtiosulfat	Natriumthiosulfat	thiosulfate de sodium	tiosulfato de sodio
5240	sodium xylenesulfonate	natriumselensulfonat	Natriumxylolsulfonat	xylènesulfonate de sodium	xilenosulfonato de sodio
5241	soft	mjuk	weich, geschmeidig	mou, tendre, doux	blando
5242	softener	mjukgörare	Weichmacher	adoucisseur	ablandador
5243	softening	mjukgöring	weichmachen	adoucissement, ramollissement	ablandamiento
5244	softening point	mjukpunkt	Erweichungspunkt	point de ramollissement	punto de reblandecimiento
5245	soften water	mjukgöta vatten	Wasser enthärten	eau douce	suavizar el agua
5246	soft fold	löst veck	umgeklappt, nicht scharf gefaltet	pliage à la main de plusieurs feuilles à la fois	pliegue suave
5247	soft lump	lös knut	weicher Stoffbatzen	pâton mou	grumo blando
5248	soft maple	mjuklönn	rotor Ahorn	érable rouge	arce rojo
5249	softness	mjukhet	Weichheit	moelleux, douceur	blandura
5250	soft paper	mjukpapper	saugfähiges Papier, weiches Papier	papier mince et doux	papel suave
5251	soft pulp	mjuk massa	lappiger Stoff (Zellstoff), weicher Stoff (Zellstoff)	pâte molle	pasta blanda
5252	soft roll	lösrulle	schlecht gewickelte Rolle	bobine peu serrée	rodillo blando
5253	soft sized	svagt limmad	schwach geleimt	peu collé	encolado ligero
5254	softwood	barrträd, barrved	Nadelholz	bois de conifères, bois de résineux	madera de coníferas
5255	softwood pulp	barrträdsmassa, barrvedsmassa	Nadelholzzellstoff	pâte de conifères, pâte de résineux	pasta de madera de coníferas
5256	soil	jord	beschmutzen, verunreinigen, Boden, Grund, Erde	sol, terrain	terreno
5257	sol	sol	kolloide Lösung, Sol	sol	sol
5258	soldering	lödning	Löten	soudure	soldadura
5259	sole plate	grundplatta	Fundamentplatte, Grundplatte	plaque d'assise, plaque de fondation	placa de soporte
5260	solid	solid, fast	Feststoff, dicht, fest, stark, solide	matière solide (restant dans une solution)	sólido
5261	solid board	hårdboard	Vollpappe	carton compact	cartón consistente
5262	solidification	stelning	Erstarrung, Verfestigung, Festwerden	solidification	solidificación
5263	solid phase	fast fas	feste Phase, Festsubstanz	phase solide	fase fija
5264	solid pulp board	homogenpapp	Vollzellstoffkarton	carton compact à base de pâtes	cartón de pasta real

CAMERON®

FOR QUALITY SLITTING AND ROLL WINDING

WALDRON FOR QUALITY COATING

CAMERON®
Two-Drum Winders for producing top quality rolls of paper and paperboard in widths to 400'' and speeds to 7000 F.P.M.
CAMERON®
MULTIWIND® and Duplex shaft and shaftless winders for producing top quality rolls of light weight coated and uncoated papers.

WALDRON Coaters for producing top quality coating on paper and paperboard in width to 350'' and speeds to 4000 F.P.M. These systems include "TURRETAIR" MICROJET®, levelon, reverse roll, gravure and blade coaters.

CAMERON/WALDRON
Research broadly based on 70 years experience gained in the manufacture of thousands of

Two-Drum surface Winders, Duplex combination center-surface winders and coaters works with you to meet new demands, to help you attain your most ambitious objective. For complete information write or call:
CAMERON MACHINE SA, a subsidiary of Midland-Ross Corp., Rue de Douvrain, 17 7410 GHLIN-lez-MONS, Belgium. Telex 57183, Phone 065.31.11.71

	ENGLISH	SVENSKA	DEUTSCH	FRANÇAIS	ESPAÑOL
5265	solid state equipment	halvledarutrustning	fester Aggregatzustand	équipement transistorisé	equipos de primera calidad
5266	solid waste	fast avfall	Feststoffabfall	résidus solides	desperdicios sólidos
5267	solids content	torrhalt	Feststoffanteil, Feststoffgehalt	teneur en matières solides	contenido de sólidos
5268	solids flow	flöde av torrt material	Fliesseigenschaft der Feststoffteilchen	écoulement de matières solides	flujo de sólidos
5269	solubility	löslighet	Löslichkeit	solubilité	solubilidad
5270	solubilization	solubilisering	Löslichmachung	solubilisation	solubilización
5271	soluble-base liquor	kokvätska med löslig bas	–	liqueur à base soluble	lejía con base soluble
5272	solution	lösning	Lösung	solution	solución
5273	solvation	lösning, solvering	Solvation, Solvatisierung	solvation	solvatación
5274	solvent	lösningsmedel	Lösungsmittel	solvant, disolvant	disolvente
5275	solvent sizing	limning med hartslösning	Lösungsmittelleimung	solution de résine dissoute dans un solvant appliqué sur un papier non collé, le solvant étant récupéré par évaporation	encolado disolutivo
5276	soot	sot	verrussen, Russ	suie	hollín
5277	sorbent	sorbent	saugfähig, aufsaugend	sorbent	absorbente
5278	sorption	sorption	Adsorption	sorption	absorción
5279	sort	sortera, sort	sortieren, sichten, klassifizieren, Sorte, Art	trier, classer	clasificar
5280	sorter	sorterare	Sortierer, Ordner	trieur, assortisseur	clasificador
5281	sorters' bench	sorterarbänk	Papierprüftisch	banc de trieuse de chiffons	banco de clasificadora de trapos
5282	sorting	sortering	Sortieren, Klassifizieren	triage, classage	escogido
5283	sound wave	ljudvåg	Schallwellen	onde sonore (lumineuse)	onda sonora
5284	sour	sur	sauer, sauer werden, säuern, ablaugen	acide, aigre, laver à l'eau acidulée	ácido
5285	souring	surgörning	Aussäuerung	passer la toile à l'acide	acidulación
5286	sourness	surhet	Sauerkeit, Säure	acidité	acidez
5287	southern pine	southern pine, sydstatstall	Gelbkiefer, Sumpfkiefer	pin du sud	pino austral
5288	soy proteins	sojaproteiner	Sojaproteine	protéines de soja	proteínas de soya
5289	spacer	mellanstycke	Leiste, Steg, Holzkeil, Distanzhalter	entre-lames d'un cylindre de pile, entre-lames d'un rotor	separador de cuchillas
5290	spacing	mellanrum	Zwischenraum, Abstand	écartement	separación
5291	spare felt	reservfilt	Schonfilz, Ersatzfilz	feutre de rechange	filtro de recambio
5292	sparger	dusch	Zerstäuber, Sprenggerät	tambour à lattes	rociador
5293	special food board	speciellivsmedelkartong	Speziallebensmittelkarton	carton spécial pour produits alimentaries	cartón para alimentos especiales
5294	specialties	specialitet	Spezialität, Besonderheit	spécialités	especialidades
5295	specialty board	specialkartong	Spezialkarton	carton pour usages spéciaux	cartón para usos especiales
5296	specialty corrugation	specialvågpapp	Spezialwellpappe	ondulation pour usages spéciaux	ondulado para uso especial
5297	specialty paper	specialpapper	Spezialpapiere	papier pour usages spéciaux	papel especial
5298	specialty pulp	specialmassa	Spezialzellstoff, Sonderzellstoff	pâte pour usages spéciaux	pasta para usos especiales
5299	specification	specifikation	Spezifizierung, genaue Angabe	spécification, description	especificación
5300	specific gravity	specifik vikt, densitet	spezifiches Gewicht, Eigengewicht, Stoffgewicht	poids spécifique	peso específico
5301	specific heat	specifik värme, värmeenergitet	spezifische Wärme	chaleur spécifique	calor específico
5302	specific surface	specifik yta	spezifische Oberfläche	surface spécifique	superficie específica
5303	specific volume	specifik volym, volymitet	spezifisches Volumen	volume spécifique	volumen específico
5304	speck	fläck	Stippe, Fleck	tache, parcelle, poivre	grano
5305	speckle	fläck	sprenkeln, Fleck, Tupfen	moucheture	mota
5306	specks, binder	bindemedelsfläckar	Stippen (im Bindemittel)	grains d'adhésif	motas del ligante
5307	spectral reflectance	spektral reflektans	spektrale Reflexion	pouvoir réfléchissant spectral	espectrorreflectancia
5308	spectral reflectivity	spektral reflektivitet	spektrales Reflexionsvermögen	réflexion spectrale	espectrorreflexividad
5309	spectrophotometer	spektrofotometer	Spektralphotometer	spectrophotomètre	espectrofotómetro
5310	spectroscope	spektoskop	Spektroskop	spectroscope	espectroscopio
5311	spectroscopy	spektroskopi	Spektroskopie	spectroscopie	espectroscopia
5312	specular gloss	speglande glans	Speckglanz	poli spéculaire	pulido especular
5313	specular reflection	speglande reflektion	Spiegelreflexion, Rückspiegelung	réflexion spéculaire	reflexión especular
5314	speed indicator	hastighetsindikator	Geschwindigkeitsmesser, Umdrehungsanzeiger	tachymètre	indicador de velocidad
5315	speed reducer	hastighetsminskare, broms	Reduktionsgetriebe	réducteur de vitesse	reductor de velocidad
5316	speed up	accelerera	beschleunigen	accélérer	acelerar
5317	spent liquor	avlut	Ablauge	lessive résiduaire (noire)	lejía agotada
5318	spent sulfite liquor	sulfitavlut	Sulfitablauge	lessive résiduaire de bisulfite	lejía al sulfito agotada

ENGLISH	SVENSKA	DEUTSCH	FRANÇAIS	ESPAÑOL
5319	spinning	spinnpapper	Spinnen, Umwickeln	filature	filatura
5320	spinning tissue	spinnpapper	Spinngewebe	papier mince à filer	muselina para hilar
5321	spiral chipper	spiralflishugg	spiralförmiger Holzspaner	coupeuse en spirale	tronzadora helicoidal
5322	spiral laid	lagd i spiral	spiralverlegt	vergeur en spirale	puesta en espiral
5323	spiral tube	spiralrör	Rohrschlange	tube en hélice	tubo helicoidal
5324	spiral-wound dandy	spiralegutör	spiralförmiger Egoutteur	rouleau vergeur en spirale	rodillo desgotador bobinado en espiral
5325	splice	skarv, skarva	Klebestelle	collure, épissure, ajouture	pegadura
5326	splice tag	skarvmarkering	Anklebefahne	pavillon pour indiquer l'emplacement d'une collure dans une bobine	etiqueta de pegar
5327	splicing	skarvning	Verklebung eines Rollenendes während der laufenden Produktion	collage au bobinage	empalme
5328	splicing tissue	skarvpapper	verstärktes Tissue	papier mince pour collage au bobinage	muselina aisladora
5329	spline	krysskil	Schweissdraht, Feder für Keilnut	cannelure d'entraînement	lengüeta postiza
5330	split collar	delad krage	gespaltene Gewindemuffe	bague fendue	anillo partido
5331	splitter	delare	Spaltmaschine, Plattenspalter	fendeur	partidor
5332	splitting	delning	Spaltung, Zerlegung, Auflösen (Holzgefüge)	fendre	hendedura
5333	spongy	svampig	schwammig, porös	spongieux	esponjoso
5334	spool	spole	Spule	bobine (pour fil)	devanadera
5335	spot, backing roll	motvalsfläck	Flecken, Schmutz	tache de rouleau de soutien	mancha del rodillo de sostén
5336	spot bonding	punktvis fästning	Punktbindung, Punktverleimung	tache de liant	unión por puntos
5337	spot, coating color	bestrykningsfläck	Flecken (durch Streichfarbe)	tache de colorant de couchage	manchas en el color de estucado
5338	spotting	nedfläckning	Fleckigwerden, Fleckenbildung	tacheter	maculación
5339	spouting	sprutande, sprutning	Ausfliessen, Hervorsprudeln, Spritzen	jaillissement	borbor
5340	spouting velocity	spruthastighet	Ausflussgeschwindigkeit	vitesse de jaillissement	velocidad de borbor
5341	spray	spritsa, spraya	zerstäuben, versprühen, sprühen, zerstäubte Flüssigkeit	pulvériser, atomiser	pulverizar
5342	spray coater	spritsbestrykare	Sprühauftragsmaschine, Spritzauftragsmaschine	coucheuse par pulvérisation	estucadora por aspersión
5343	spray coating	spritsbestrykning	Sprühauftrag, Spritzauftrag	couchage par pulvérisation	estucado de aspersión
5344	spray damper	spraybefuktare	Feuchtapparat, Anfeuchter	mouilleuse à pulvarisation	humectadora por pulverización
5345	spray drying	spraytorkning	Sprühtrocknung, Zerstäubungstrocknung	séchage par pulvérisation	secado de pulverización
5346	spray dyeing	sprutmålning	Färbung durch Zerstäubung der Farbe	coloration au pulvérisateur	coloración por pulverización
5347	sprayer	sprayare	Zerstäuber	pulvérisateur	pulverizador
5348	spraying	sprayning	Zerstäuben	pulvérisation	pulverización
5349	spreader	spridare	Ausstreichwalze, Streckwalze, Verteiler	répartisseur de pâte	dispositivo de repartición de pasta
5350	spreader bar	spridarbom	Ausstreichleiste	barre déplisseuse	barra de desplisar
5351	spreader roll	spridarvals	Breitstreckwalze	rouleau déplisseur	rodillo desplisador
5352	spreading	spridning	Ausbreitung, Verbreitung, Verteilung	épandage	repartición
5353	springback	elastisk återgång	Zurückfedern	être rejeté en arrière	retroceder
5354	spring roll	fjädervals	Federwalze (am Glättwerk)	rouleau monté sur ressorts	rodillo de resorte
5355	spring wood	vårved	Frühholz, weitlumiges Holz	bois de printemps	madera de primavera
5356	sprinkler	sprinkler	Sprinkler, Berieselungsanlage, Sprühanlage, Feuerlöschbrause	gicleur	regadera
5357	sprinkling	sprinkling	Berieseln, Besprühen	aspersion	aspersión
5358	sprout	grodd	spriessen, keimen	germer, bourgeonner	tallo
5359	spruce	gran	Fichte	épicéa	epícea
5360	spun bonding	spunnen bindning	spun-bonded	résilience, rebondissement	unión con hilado
5361	squared	rätskuren	kariert	carré, équerré	encuadrado
5362	squareness	grad av isotropi	Längs-/Querverhältnis einer Bahn	d'équerre	relación largo/ancho de una pista
5363	square rule paper	rutat papper	kariertes Papier	papier quadrillé	papel cuadriculado
5364	square sheet	styrkeisotropt papper, kvadratiskt papper	Papier oder Karton mit gleichen Zug- und Fortreissfestigkeitswerten in Maschinen- und Querrichtung	feuille dont la résistance est la même dans les deux sens	hoja cuadrada
5365	squeeze roll	urpressningsvals	Vorgautschwalze	rouleau essoreur	rodillo escurridor
5366	squeeze roll coater	valsbestrykare	Streichmaschine mit Abquetschwalze	coucheuse à rouleau essoreur	estucadora con rodillo escurridor
5367	stability	stabilitet	Stabilität, Beständigkeit, Resistenz	stabilité	estabilidad
5368	stability of pulp	stabilitet hos massa	Zellstoffbeständigkeit (gegen verschiedenste Einflüsse)	stabilité d'une pâte	estabilidad de la pasta
5369	stabilization	stabilisering	Stabilisierung	stabilisation	estabilización

	ENGLISH	SVENSKA	DEUTSCH	FRANÇAIS	ESPAÑOL
5370	stabilized	stabiliserad	stabilisiert, Papier dessen Feuchtigkeitsgehalt im Gleichgewicht steht mit der Luftfeuchte	stabilisé, conditionné	estabilizado
5371	stabilizer	stabilisator	Stabilisator	stabilisateur, stabilisant	estabilizador
5372	stack	stapel	Stapel, Stoss, Haufen, Glättwerk	pile	almiar
5373	stacker	staplare	Stapler, Stapelvorrichtung	gerbeur, empileur de bois	apilador
5374	stacking	stapling	Stapeln, Schichten	gerbage, empilage	apilamiento
5375	stain	fläcka	beflecken, fleckig werden, Farbflecken, Schmutzflecken	tache, coloration	mancha
5376	stained paper	fläckigt papper	farbbeflecktes Papier	papier peint	papel con manchas
5377	staining	fläckning	Beflecken, Anfärben	coloration	tintado
5378	stainless steel	rostfritt stal	rostfreier Stahl	acier inoxydable	acero inoxidable
5379	stain resistance	rostmotstånd	Korrosionswiderstand, Rostwiderstand	résistance aux taches	resistencia al manchado
5380	stamping	stämpling	Stampfen, Pressen, Prägen	timbrage, impression	estampado
5381	stamp paper	frimärkspapper	Briefmarkenpapier	papier pour timbres-poste	papel para timbres
5382	standard	standard	normal, Standard, Norm	standard, normal, type étalonné, normalisé	norma
5383	standard newsprint	standard tidningspapper	Normalzeitungsdruck	papier journal standard	papel prensa standard
5384	staple fiber	stapelfiber	Stapelfaser	fibre (de coton ou de textiles artificiels)	fibras corrientes
5385	stapling	häftning	Heften	agraffage	clasificación de fibras
5386	starch	stärkelse	Stärke	amidon, fécule	almidón
5387	starch conversion	stärkelsekonvertering	Stärkeumwandlung	conversion de l'amidon	conversión del almidón
5388	starch cooker	stärkelsekokare	Stärkekocher	cuiseur d'amidon	cocedora de almidón
5389	starch derivatives	stärkelsederivat	Stärkederivat	dérivés de l'amidon	derivados del almidón
5390	starch xanthate	stärkelsexantat	Stärkexanthat, Stärkexanthogenat	xanthate d'amidon	xantato de almidón
5391	starch xanthide	stärkelsexantid	Stärkexanthid	xanthure d'amidon	xanturo de almidón
5392	starred roll	stjärnmönstrad vals	fehlerhafte (sternförmig) auslaufende Rollenwicklung	bobine détériorée aux extrémités	bobina estropeada en los bordes
5393	startup	starta	Anlauf, Inbetriebnahme, Anfahren	mise en marche, démarrage	puesta en marcha
5394	static electricity	statisk elektricitet	statische Elektrizität	electricité statique	electricidad estática
5395	static inhibitor	inhibitor för statisk elektricitet	statischer Inhibitor, Schranke	anti-statique	inhibidor estático
5396	static pressure	statisk tryck	statischer Druck	pression statique	presión estática
5397	static tests	statiska prov	Statiktests, statische Versuche, Belastungsproben	essais statiques	pruebas estáticas
5398	statistical analysis	statistisk analys	statistische Analyse	analyse statistique	análisis estadísticos
5399	statistics	statistik	Statistiken	statistiques	estadística
5400	stator	stator	Stator, Ständer	stator	estátor
5401	steady state	stationärt tillstånd	Dauerzustand, stabiler Zustand, Beharrungszustand	état de régime	estado permanente
5402	steam	ånga	mit Dampf behandeln, dämpfen, Dampf, Wasserdampf, Dunst	vapeur	vapor
5403	steam coils	ångspiraler	Dampferhitzer	serpentins de chauffage	calentador de vapor
5404	steam electric power	ångkraft	Elektrodampfkraft	énergie électrique thermique	fuerza eléctrica generada por vapor
5405	steam finish	ångfinish	Dampfbehandlung (eines hochsatinierten Papiers)	humectage à la vapeur	acabado al vapor
5406	steam heating	ångvärmning	Dampfheizung	chauffage à la vapeur	calentamiento por vapor
5407	steaming	ångning	Dämpfung, Verdampfung	étuvage	tratamiento en estufa de vapor
5408	steam plant	pannhus	Dampfkraftwerk	installation thermique	planta generadora de vapor
5409	steam roll	ångvals	dampfbeheizte Kalanderwalze	rouleau chauffé à la vapeur	rodillo a vapor
5410	steam stripping	ångavdrivning	Dampfabzug	entraînement à la vapeur	separación de hojas por vapor
5411	steam trap	ångfälla	Kondenstopf	purgeur à vapeur	purgador
5412	steam turbine	ångturbin	Dampfturbine	turbine à vapeur	turbina de vapor
5413	stearates	stearater	Stearate	stéarates	estearatos
5414	stearic acid	stearinsyra	Stearinsäure	acide stéarique	ácido esteárico
5415	steel	stål	Stahl	acier	acero
5416	steeping	mercerisering	Eintauchen, Umwandlung von Cellulose in Alkalicellulose	macération	maceración
5417	stele	stele	Stele	tige (veget.)	estela
5418	stem	stam	Stiel, Stengel, Halm, Stamm	tige	tallo
5419	stencil	stencil	Schablone, Matrize	stencil, patron, pochoir	estarcido
5420	stencil board	stencilkartong	Schablonenkarton	carton pour pochoir	cartón para stencil
5421	stencil paper	stencilpapper	Matrizenpapier	papier stencil	papel stencil
5422	stepped belt cone	trappremskiva	Stufenriemenkegel	cône à étages	polea escalonada
5423	stereoscope	stereoskop	Stereoskop	stéréoscope	estereoscopio
5424	stick	klibba	kleben, haften	bâton, branche (arbre)	palo
5425	stiffness	styvhet	Steifigkeit, Biegefestigkeit	rigidité	rigidez
5426	stiffness tester	styvhetsprovare	Steifigkeitsprüfer	flexiomètre	probador de rigidez

	ENGLISH	SVENSKA	DEUTSCH	FRANÇAIS	ESPAÑOL
5427	stiffness tests	styvhetsprov	Steifigkeitsprüfung	essais de rigidité	pruebas de rigidez
5428	stippling	punktering	Punktierung	pointillé	graneo
5429	stirring	omrörning	Umrühren, Rühren	remuer, agiter, agitation	agitado
5430	stock	mäld	aufbewahren, speichern, lagern, Stoff, Bestand	matières premières, pâte liquide, stock, capital, action (finances)	pasta (tratada)
5431	stock chest	mäldkar	Stoffbütte	cuvier à pâte	tina de pasta
5432	stock cutting	lagerreducering	Lagerabbau, Laggerreduzierung	abattage des arbres	corte de pasta
5433	stock flow	mäldflöde	Stofffluss	arrivée de pâte sur machine	flujo de la pasta
5434	stock inlet	mäldinlopp	Stoffauflauf	arrivée de pâte sur machine	entrada de pasta
5435	stock preparation	mäldberedning	Stoffaufbereitung	traitement de la pâte	preparación de la pasta
5436	stock pump	mäldpump, massapump	Stoffpumpe	pompe à pâte	bomba para pasta
5437	stock regulator	mäldregulator	Stoffregulierer, Stoffregler	régulateur de pâte	regulador de pasta
5438	stock sizes	mäldlim	Lagerbestände in Papier- und Kartonstandardabmessungen	formats en stock	tamaños corrientes
5439	stoker	eldare	Heizer	chargeur, chauffeur	cargador
5440	stonite	stonit	Stonit	stonite	estonita
5441	storage	lagring	Lagerung, Lagern	stockage, conservation, accumulation	almacenaje
5442	storage tank	lagringstank	Vorratsbehälter	bassin de réserve	tanque almacenador
5443	straightness	rätlinjighet	Geradlinigkeit	raideur	derechura
5444	strain	sila, töja, töjning	anspannen, beanspruchen, verbiegen, Spannung, Beanspruchung	effort, tension	colar
5445	strainer	sil	Sieb, Filter	épurateur, crépine	depurador
5446	strain gauge	töjningsgivare	Spannungsprüfer	jauge de contrainte	medidor de deformación
5447	strap	rem	Metallband	courroie, bande	correa
5448	strapping	remning	Verpackung mit Stahlband, mit Stahlband versehen	cerclage	zunchamiento
5449	strapping of rolls	remning av rullar	Papierrollen mit Stahlband versehen	cerclage de bobines avec du feuillard	zunchamiento de los rollos
5450	stratification	skiktning	Schichtung, Blattschichtung	stratification	estratificación
5451	stratified paper	skiktat papper	geschichtetes Papier	papier pour stratifiés	papel estratificado
5452	straw	halm	Stroh	paille	paja
5453	strawboard	halmkartong	Strohpappe	carton paille	cartón paja
5454	straw pulp	halmmassa	Strohzellstoff	pâte de paille	pasta de paja
5455	streak	streck, rand	Streifen, Schliere	traînée	reguero
5456	streak, coating	bestrykningsrand	Streifenfehler, Fehler beim Streichen	traînée de couchage	huellas del estucado
5457	streaking	randning	Streifenbildung	strier	raspado
5458	stream flow	laminär strömning	laminare Strömung	force du courant	caudal
5459	strength	styrka, hållfasthet	Festigkeit, Stärke, Kraft	force, résistance	fuerza
5460	strength tests	styrkeprov	Festigkeitsprüfung	essais de résistance	pruebas de resistencia
5461	stress	spänning	Druckspannung, Belastung	effort, pression	tensión
5462	stress analysis	spänningsanalys	Druckanalyse	analyse de la tension de retrait	análisis de la tensión
5463	stressing	påfrestning	Beanspruchung, Betonung	application de contrainte	someter a tensión
5464	stress-strain properties	spännings-töjningsegenskaper	Spannungs-Dehnungs-Eigenschaften	propriétés d'extension	propiedades tenso - deformadoras
5465	stretch	sträckning	einen Bogen auf den Zylinder aufziehen, Dehnung, Bruchdehnung	allongement, extension	alargamiento
5466	stretchable paper	töjbart papper	dehnbares Papier	papier extensible	papel extensible
5467	stretcher	sträckare	Streckvorrichtung, Spanner	tendeur	tensor
5468	stretching	sträckning	Dehnen, Strecken, sich Erstrecken	tendre, allonger	estiraje
5469	striated	strierad	gestreift, streifig	strié	estriado
5470	strike-in	färginslag	Wegschlagen (einer Druckfarbe)	rentrer	juntarse
5471	strike-through	färggenomslag	Durchschlagen (einer Druckfarbe)	traverser, percer	calar
5472	strip	remsa, avdriva	schmaler Streifen einer Bahn	ruban, bande, bandelettre	cinta
5473	strip coater	ridåbestrykare	Streifenlackiermaschine, Streifenstreichmaschine	coucheuse par bandes	desestucadora
5474	strip coating	ridåbestrykning	Streifenlackierung, Streifenstreichen	couchage par bandes	desestucado
5475	stripping	avdrivning	Abziehen, Ablösen	épuration, purification	desmonte
5476	strong acid	stark syra	starke Säure	lessive forte (bisulfite)	ácido concentrado
5477	structural engineering	byggnadsteknik	Bauwesen, Bautechnik	technique des structures	ingeniería de estructuras
5478	structure	byggnad, struktur	Struktur, Gefüge, Bau	structure, édifice	estructura
5479	stub roll	valsände	kurze Restrollen	à côté de bobine	rodillo de tope
5480	stuff	mäld	Stoff, Material, Papiermasse	pâte	pasta (tratada)
5481	stuff box	inloppslåda	Stoffüberlaufkasten	presse-étoupe	tina de máquina
5482	stuff chest	mäldkar	Stoffbütte	cuvier à pâte	tina de máquina
5483	stuff consistency	mäldkonsistens	Stoffdichte	densité de la pâte	consistencia de la pasta

	ENGLISH	SVENSKA	DEUTSCH	FRANÇAIS	ESPAÑOL
5484	stuffing box	inloppslåda	Stopfbüchse	presse-étoupe	prensaestopas
5485	stuffing gland	tätningsring	Stopfbüchsenbrille	chapeau de presse-étoupe	tapa de prensaestopas
5486	stuff sizing	mäldlimning	Stoffleimung	collage en pâte, collage dans la masse	encolado en pasta
5487	stump	stubbe, stump	roden, Baumstumpf	souche (d'arbre)	cepa
5488	stumpage	stubbar	Holzeinschlagsabgabe	arrachage de souches	madera en pie
5489	styrene	styren	Styrol	styrolène	estireno
5490	subjective gloss	subjektiv glans	einseitiger Glanz	poli subjectif	brillo subjetivo
5491	sublimation	sublimering	Sublimierung	sublimation	sublimación
5492	submersion	nedföring i vätska, blötläggning	Eintauchen, Untertauchen	immersion	sumersión
5493	substance	substand, ytvikt	Substanz, Flächengewicht, Masse	grammage	substancia
5494	substitute	substitut	Ersatzstoff	substituant, succédané, produit de remplacement	substituto
5495	substrate	substrat	Substrat, Schichtträger, Trägerpapier, Streichrohpapier	couche inférieure	substrato
5496	sucrose	sukros	Sucrose, Saccharose	saccharose	sucrosa
5497	suction	sug	Saugung, Saugen, Ansaugen	aspiration	aspiración
5498	suction box	suglåda	Siebsaugkasten	caisse aspirante	caja aspirante
5499	suction box covers	suglådelock	Siebsaugerbeläger	châssis de caisse aspirante	cubiertas de la caja aspirante
5500	suction-box marks	suglådemarkering	Siebsaugermarkierung, Sauglöchermarkierung	marques de la caisse aspirante	marcas del sifón
5501	suction-box slot	suglådeslits	Saugerschlitz	rainure de la caisse aspirante	ranuras del sifón
5502	suction couch roll	sugguskvals	Sauggautschwalze	cylindre aspirant de la toile	cilindro aspirante
5503	suction dusting	dammsugning	Staubabsaugung an den Rollenenden oder Seiten eines Papierstapels	aspiration de poussières	desempolvado por succión
5504	suction feed	sugmatning	Vakuumzufuhr, Zufuhr mittels Vakuum	alimentation par aspiration	alimentado por aspiración
5505	suction flatbox	plansuglåda	Flachsauger	caisse plate aspirante	sifón aspirador
5506	suction former	sugarkformare	Vakuumformierung	forme sous vide	plantilla aspiradora
5507	suction pickup roll	sugavtagningsvals	Vakuumabnahme	rouleau leveur aspirant	rodillo tomador de succión
5508	suction pipes	sugrör	Rohrsauger	tuyaux aspirants	tubos de succión
5509	suction press	sugpress	Saugpresse	presse coucheuse aspirante	prensa aspirante
5510	suction press roll	sugpressvals	Saugpresswalze	cylindre de presse aspirante	rodillo de prensa aspirante
5511	suction primary press	sugförpress	Saugvorpresse	première presse coucheuse aspirante	rodillo tomador aspirante
5512	suction pump	sugpump	Saugpumpe, Vakuumpumpe	pompe aspirante	bomba aspirante
5513	suction roll	sugvals	Saugwalze	rouleau aspirant	rodillo aspirante
5514	suction-roll mark	sugvalsmarkering	Saugwalzenmarkierung, Lochwalzenmarkierung	marque du rouleau aspirant	marca del rodillo aspirante
5515	suction straight-through press	sugrakpress	Saugliegepresse	–	prensa aspirante
5516	sugarcane	sockerrör	Zuckerrohr	canne à sucre	caña de azúcar
5517	sugarcane fiber	sockerrörsfiber	Bagassefaser	fibre de bagasse	fibra de caña de azúcar
5518	sugar maple	sockerlönn	Zuckerahorn	érable rouge	arce de azúcar
5519	sulfamates	sulfamater	Sulfamate, sulfamidsaure Salze	sulfamates	sulfamatos
5520	sulfamic acid	sulfaminsyra	Sulfamidsäure	acide sulfamique	ácido sulfámico
5521	sulfate board	sulfatkartong	Sulfatpappe, Kraftzellstoffpappe	carton au sulfate(kraft)	cartón sulfato
5522	sulfate liner	sulfattäckskikt	Sulfatdeckschicht (eines Kartons)	carton liner au sulfate (kraft)	forro sulfático
5523	sulfate liquor	sulfatlut	Sulfatlauge, Sulfatablauge	lessive au sulfate (kraft)	lejía sulfática
5524	sulfate paper	sulfatpapper	Sulfatpapier, Natronkraftpapier	papier au sulfate (kraft)	papel sulfato
5525	sulfate process	sulfatprocess	Sulfatverfahren, Sulfataufschluss	procédé au sulfate	procedimiento de sulfato
5526	sulfate pulp	sulfatmassa	Sulfatzellstoff, Kraftzellstoff	pâte au sulfate (kraft)	pasta al sulfato
5527	sulfate turpentine	sulfatterpentin	Sulfatterpentin	essence de térébenthine provenant de liqueur kraft	trementina sulfática
5528	sulfates	sulfater	Sulfate	sulfates	sulfatos
5529	sulfidation	sulfidering	Sulfidierung	sulfuration	sulfuración
5530	sulfides	sulfider	Sulfide	sulfures	sulfuros
5531	sulfidity	sulfiditet	Sulfidität (%-Verhältnis von Na-sulfid zu den wirksamen Alkalien)	aptitude à la conversion en sulfure	contenido en sulfitos
5532	sulfite liquor	sulfitlut	Sulfitlauge	lessive de bisulfite	lejía al sulfito
5533	sulfite paper	sulfitpapper	Sulfitpapier	papier sulfite	papel sulfito
5534	sulfite pulp	sulfitmassa	Sulfitzellstoff	pâte au bisulfite	pasta al bisulfito
5535	sulfite pulping	sulfitkokning	Sulfitaufschluss	fabrication de pâtes au bisulfite	elaboración de pasta al bisulfito
5536	sulfites	sulfiter	Sulfit	sulfites	sulfitos
5537	sulfo groups	sulfogrupper	Sulfogruppen	groupes sulfo	sulfogrupos
5538	sulfonates	sulfonater	Sulfonate	sulfonates	sulfonatos
5539	sulfonyl groups	sulfonylgrupper	Sulfonylgruppen	groupes sulfonyl	grupos sulfonil

	ENGLISH	SVENSKA	DEUTSCH	FRANÇAIS	ESPAÑOL
5540	sulfoxides	sulfoxider	Sulfoxide	sulfoxydes	óxidos de azufre
5541	sulfur	svavel	Schwefel	soufre	azufre
5542	sulfur burning	svavelbränning	Schwefelziehen	combustion de soufre	combustión de azufre
5543	sulfur compounds	svavelföreningar	Schwefelverbindungen	composés du soufre	compuestos de azufre
5544	sulfur dioxide	svaveldioxid	Schwefeldioxide	anhydride sulfureux	anhídrido sulfuroso
5545	sulfuric acid	svavelsyra	Schwefelsäure	acide sulfurique	ácido sulfúrico
5546	sulfurous acid	svavelsyrlighet	schwefelige Säure	acide sulfureux	ácido sulfúrico
5547	sulphur (Brit.)	svavel	Schwefel	soufre	azufre
5548	summer wood	höstved	Sommerholz, Spätholz, englumiges Holz	bois d'été	madera de verano
5549	sump	sump	Sumpf	puisard	fondo de cárter
5550	supercalender	kalander	Superkalander	supercalandre	supercalandra
5551	supercalendered finish	kalanderfinish	Hochglanzausrüstung, Hochglanzoberfläche	surcalandré	satinación
5552	supercalendered paper	kalandrerat papper	hochglänzendes Papier, superkalandriertes Papier	papier surcalandré, papier surglacé	papel satinado
5553	supercalendering	kalandrering	Superkalandrieren, Hochsatinieren	surglaçage	satinación
5554	super finish	kalanderfinish	Hochglanz	surcalandré, surglacé	papel glaseado
5555	super glazed	kalanderglättat	hoch satiniert, stark satiniert	surglacé	papel glaseado
5556	super-glazed finish	kalanderfinish	hochsatinierte Oberflächenausrüstung	fini surglacé	acabado glaseado
5557	superheated	överhettad	überheizt, überhitzt	surchauffé	recalentado
5558	superheating	överhettning	Überhitzung, Dampfüberhitzung	surchauffe	recalentamiento
5559	superintendent	inspektör	Verwalter, Inspektor	chef de fabrication	jefe de fabricación
5560	supernatant	moder-	Überstand, das Überstehende einer Probe	surnageant	sobrenadante
5561	super news	supernews	Zeitungsdruckpapier mit glatter Oberfläche	journal amélioré	papel especial de periódicos
5562	superstandard news	superstandard news	maschinengeglättetes Zeitungsdruckpapier (für Beilagen)	magazine amélioré	papel prensa especial
5563	supervision	överinseende	Aufsicht, Überwachung, Kontrolle	supervision, surveillance, inspection	administración
5564	supplier	leverantör	Lieferant, Lieferfirma, Zulieferer	fournisseur	abastecedor
5565	supply	leverera	versorgen, speisen, zuführen, liefern	fourniture, approvisionnement	suministro
5566	supply room	förråd	Lager	salle des fournitures	sala de suministros
5567	support	stödja	Stütze, Unterstützung, Träger, Unterlage	soutien, appui. support	soporte
5568	surface	yta	Oberfläche, Fläche	surface	superficie
5569	surface-active agent	ytaktivt medel	oberflächenaktives Mittel, oberflächenaktivierendes Mittel	produit agissant en surface	agente activo de superficie
5570	surface application	anbringning på ytan	Oberflächenauftrag	enduction	impregnación
5571	surface bonding strength	ythållfasthet	Oberflächenfestigkeit, Rupffestigkeit	résistance à l'arrachage en surface	resistencia al arrancamiento
5572	surface coated	bestruken	oberflächengestrichen, oberflächenbeschichet, oberflächenimprägniert	couché une ou deux faces	estucado en superficie
5573	surface coloring	ytfärgning	Oberflächenfärbung	coloration en surface	coloración en superficie
5574	surface dyed	ytfärgad	oberflächengefärbt	coloré en surface	coloreado en superficie
5575	surface finish	ytfinish	Oberflächenausrüstung, Oberflächenstruktur	état de surface	estado de superficie
5576	surface finishing	ytbehandling	Oberflächenveredlung	apprêt de papier	acabado en superficie
5577	surface hardening	ythärdning	Oberflächenhärtung	durcissement en surface	endurecimiento en superficie
5578	surface lifting	ytskiktslyftning	Oberflächenabhebung	arrachage en surface	levantamiento superficial
5579	surface peeling	ytavskalning	Abblättern, Abspalten der Oberfläche	pelurage en surface	descortezado superficial
5580	surface properties	ytegenskaper	Oberflächeneigenschaften	propriétés de surface	propiedades de la superficie
5581	surface resistance	ytmotstånd	Oberflächenwiderstand	résistance en surface	resistencia superficial
5582	surface sized	ytlimmad	oberflächengeleimt	gélatiné	encolado en superficie
5583	surface-sized paper	ytlimmat papper	oberflächengeleimtes Papier	papier gélatiné	papel encolado en superficie
5584	surface-size press	limpress	Oberflächenleimpresse	presse encolleuse	prensa encoladora de superficie
5585	surface sizing	ytlimning	Oberflächenleimung	collage en surface	encolado en superficie
5586	surface strength	ythållfasthet	Oberflächenfestigkeit	résistance superficielle	resistencia superficial
5587	surface tension	ytspänning	Oberflächenspannung	tension superficielle	tensión superficial
5588	surface tinted	ytfärgad	oberflächengefärbt, oberflächengetönt	teinté en surface	coloreado en superficie
5589	surface treatment	ytbehandling	Oberflächenbehandlung	traitement en surface	tratameinto de superficie
5590	surface water	ytvatten	Oberflächenwasser	eau de surface	agua superficial
5591	surface wettability	ytvätbarhet	Oberflächenenbenetzbarkeit	mouillabilité en surface	aptitud de la superficie a la humectación
5592	surface wettability tests	ytvätbarhetsprov	Oberflächenbenetzbarkeitsprüfung	essais de mouillabilité en surface	pruebas de humectación superficial

	ENGLISH	SVENSKA	DEUTSCH	FRANÇAIS	ESPAÑOL
5593	surfactant	ytaktivt medel	Oberflächenwirksames Mittel, Oberflächenbehandlungsmittel	agent tensio-actif	pulidora
5594	surge tank	utjämningsbehållare	Zwischenbehälter	réservoir de compensation	tanque igualador
5595	suspension of pulp	massasuspension	Aufschlämmung von Zellstoff	suspension de pâte	suspensión de la pasta
5596	sustained yield	upprätthållet utbyte	Nachhaltigkeit (Forstwirtschaft)	rendement constant	rendimiento estable
5597	sweat cylinder	kylcylinder	Kühlzylinder	cylindre refroidisseur	cilindro enfriador
5598	sweat dryer	foktigcylinder	Nachfeuchtzylinder, Kühlzylinder	cylindre refroidisseur	secadora enfriadora
5599	sweat roll	kylvals	Kühlwalze	rouleau refroidisseur	cilindro enfriador
5600	sweetener	filtrerfiber	Stoff für die Filterschicht des Stoffängers	adoucisseur	suavizador
5601	sweet gum	sweet gum	Amberbaum	copalme d'Amérique	copal
5602	swelling	svällning	quellen, Quellung	gonflement	hinchamiento
5603	swelling agent	svällningsmedel	Quellmittel	produit gonflant	agente esponjador
5604	swimming roll	svävvals	Schwimmwalze	rouleau plongeur	rodillo natatorio
5605	switch	brytare, omkopplare	Schalter, einschalten, umschalten	interrupteur	conmutador
5606	sycamore	sykamor	Sykomore, Platane	sycomore	sicamoro
5607	symmetry	symmetri	Symmetrie	symétrie	simetría
5608	synchronism	synkronism	Synchronismus, Gleichlauf, Gleichzeitigkeit	synchronisme	sincronismo
5609	synchronous	synkron	synchron, gleichlaufend	synchrone	sincrónico
5610	synchronous motor	synkronmotor	Synchronmotor	moteur synchrone	motor sincrónico
5611	synergism	synergism	Synergismus, synergetischer Effekt	synergisme	sinergismo
5612	synthesis	syntes	Synthese, Aufbau	synthèse	síntesis
5613	synthetic fiber	syntetfiber	synthetische Faser, Chemiefaser, Kunstfasern	fibre synthétique	fibra sintética
5614	synthetic polymers	syntetpolymerr	synthetische Polymere	polymères synthétiques	polimeros sintéticos
5615	synthetic resin	syntetharts	Kunstharz	résine synthétique	resina sintética
5616	synthetic rubber	syntetgummi	synthetischer Gummi	caoutchouc synthétique	goma sintética
5617	syringyl groups	syringylgrupper	Syringylgruppen	groupes syringyl	grupos celindros
5618	system	system	System, Methode, Verfahren	système	sistema
5619	systems engineering	systemteknologi	Verfahrenstechnik	technique des ordinateurs	ingeniería de sistemas

T

5620	table roll	registervals	Registerwalzen	pontuseau	rodillo desgotador
5621	table-roll baffle	registerskärm	Registerwalzen-Staublech, Siebspritzblech	pontuseau	deflector del rodillo desgotador
5622	table-roll deflector	registerskärm	Registerwalzenabstreifer	déflecteur de pontuseau	deflector del rodillo desgotador
5623	table-roll rail	registerskena	Registerlineal	rail de pontuseau	riel del rodillo desgotador
5624	tables	tabeller	Schablonen, Listen, Auflagetisch	tableaux	tablas
5625	tabulating card stock	registerkortmassa	Lochkartenmaterial	papier pour cartes statistiques	almacén de fichas de tabulación
5626	tabulation	tabulering	Tabellarisierung	mise en tableaux, classification	tabulación
5627	tachometer	tachometer	Tachometer, Geschwindigkeitsmesser, Tourenzähler	tachymètre	tacómetro
5628	tack	klibbighet, nubb	Klebrigkeit	attacher	tachuela
5629	tackle	garnityr	Bemesserung (Holländer), Mahlkörper	lamage d'une pile	guarnición
5630	tacky	klibbig	klebrig, strenge Farbe	collant	pegajoso
5631	tag	etikett, lapp	Anhängsel, mit einem Etikett oder Anhängezettel versehen	étiquette, fiche	etiqueta
5632	tag board	etikettkartong	Anhängerkarton	carton pour étiquettes	cartulina para etiquetas
5633	tail gluer	spetsklistrare	Bahnankleber	encolleuse	encolador de cola
5634	tailing	utskott	Spuckstoff, Sauerkraut	déchet, buchette, rejet	rechazo de depuración
5635	tail water	viravatten	Siebwasser (z.B. in der Rundsiebpartie)	eau résiduaire	agua de descarga
5636	talc	talk	Talkum	talc	talco
5637	tall oil	tallolja	Tallöl	tall oil, huile de pin	tall oil
5638	tall oil soap	talloljetvål	Tallölseife	savon à l'huile de pin	jabón de tall oil
5639	tamarack	tamarack	Lyalls-Lärche	mélèze d'Amérique	alerce
5640	tamarix	tamarix	Tamariske	tamaris	tamarisco
5641	tank	tank	Tank, Behälter, Reservoir, Gefäss	réservior, citerne	tanque
5642	tantalum compounds	tantalföreningar	Tantalverbindungen	composés du tantale	compuestos de tántalo
5643	tape	remsa, tejp	Band, Streifen, Klebstreifen	bande, ruban	cinta
5644	tape paper	tejppapper	Bandpapier	papier en bandes	papel para cintas

	ENGLISH	SVENSKA	DEUTSCH	FRANÇAIS	ESPAÑOL
5645	tapered roller bearing	koniskt rullager	Kegelrollenlager	palier à rouleaux côniques	cojinete de rodillos ahusados
5646	tapering	konisk, avsmalnande	sich verjüngend, Konizität, Verjüngung	cônique	adelgazado
5647	taping	tejpning, avsmalnande	Umwicklung mit Band	fermer avec une bande gommée	encintado
5648	tapioca starch	tapiokastärkelse	Tapiokastärke	amidon de tapioca	almidón de tapioca
5649	tapping	tappning, knackning	Anzapfen, Harzgewinnung	taraudage	enrosque de hembra
5650	tar	tjära	Teer, teeren	goudron, brai	alquitrán
5651	tare	tara	Tara, Verpackungsgewicht, Eigengewicht	tare	tara
5652	tar paper	tjärpapper	Teerpapier	papier goudronné	papel alquitranado
5653	Taxus (Lat.)	Taxus	Eibe, Eibenbaum	if	árbol
5654	tea-bag paper	tepåspapper	Teebeutelpapier	papier pour sachets de thé	papel para bositas de té
5655	tear	riva	Fortreissfestigkeit	déchirure	rasgar
5656	tear factor	rivfaktor	Zerreissfaktor	facteur de déchirure	factor de desgarre
5657	tearing	rivning	Reissen, Zerreissen	déchirure	ragado
5658	tearing resistance	rivmotstånd	Einreissfestigkeit, Zerreissfestigkeit	résistance à la déchirure	resistencia al rasgado
5659	tearing strength	rivhållfasthet	Durchreissfestigkeit, Weiterreissfestigkeit	résistance à la déchirure	resistencia al rasgado
5660	tear out	riva ut	ausreissen, herausreissen	arracher	arrancar
5661	tear ratio	rivfaktor	Verhältnis von Längs-zu Querrichtung in Zerreissfestigkeit	taux de déchirure	índice del rasgado
5662	tear strength	rivhållfasthet	Fortreissfestigkeit	résistance à la déchirure	resistencia al rasgado
5663	tear tester	rivprovare	Durchreissfestigkeitsprüfer, Fortreissfestigkeitsprüfer	appareil pour mesurer la résistance à la déchirure	dechirómetro
5664	tear tests	rivprov	Fortreissfestigkeitsprüfung, Durchreissfestigkeitsprüfung	essais de résistance	pruebas de rasgado
5665	technique	teknik	Technik, Verfahren, Methode	technique	técnica
5666	telescoped	teleskopformad	ineinandergeschoben	pouvant emboîter (s'), encastrer (s')	enchufado
5667	telescoped roll	teleskopformad vals	in sich verschobene Papierrolle	bobine foirée	bobina corrida
5668	telescoping container	utdragbar behållare	Teleskopbehälter, Teleskopfutteral	récipient s'emboîtant dans un autre	contenedor enchufador
5669	television	television	Fernsehen	télévision	televisión
5670	temperature	temperatur	Temperatur	température	temperatura
5671	temperature control	temperaturreglering	Temperaturregler, Temperaturregelung	contrôle de la température	control de temperatura
5672	tender	förare, sköare	Maschinenführer	appel d'offres	ablandar
5673	tending aisle	maskinsalsgolv	führerseitiger Gang	côté conducteur	pasillo conductor
5674	tending side	förarsida	Bedienungsseite	côté conducteur	lado conductor
5675	tensile	sträck-, drag-	dehnbar, streckbar, Zug-	tension, traction	de tensión
5676	tensile energy absorption	sträckningsenergi	Reissarbeit	absorption de l'énergie de rupture par traction	absorción de la energía tensora
5677	tensile properties	draghållfasthetsegenskaper	Zugeigenschaften, Dehnungseigenschaften	propriétés de traction	propiedades tensoras
5678	tensile strength	draghållfasthet	Reissfestigkeit, Bruchfestigkeit	résistance à la rupture par traction	resistencia a la tracción
5679	tensile strength tester	draghållfasthetsprovare	Zugfestigkeitsprüfer	dynamomètre	dinamómetro de tracción
5680	tensile stress	dragspänning	Zugspannung, Zugbeanspruchung	effort de tension	esfuerzo de tensión
5681	tensile stretch	sträckning	Zerreissdehnung, Bruchdehnung	allongement	estiramiento tensor
5682	tensile tester	dragprovare	Zerreissprüfgerät	dynamomètre	probador de tracción
5683	tensile tests	dragprov	Reissprobe	essais de tension	ensayos a la tracción
5684	tensiometer	tensiometer, spänningsmätare	Spannungsmesser, Dehnungsmesser	tensiomètre	tensiómetro
5685	tension	spänning	Spannung, Zug	tension, effort	tensión
5686	tension control	spänningsreglering	Zugregelung	contrôle de la tension	control de tensión
5687	tension roller	spännvals	Spannwalze, Federwalze	rouleau tendeur	rodillo tensor
5688	tension wood	spänningsved	Zugholz, Weissholz	bois de tension	madera rámea
5689	terpenes	terpener	Terpen	terpènes	terpenos
5690	terpinenes	terpinener	Terpinen	terpinènes	terpinenos
5691	tertiary lamella	tertiärlamell	Tertiärlamelle	lamelle tertiaire	laminilla terciaria
5692	tertiary treatment	tertiärbehandling	Tertiärbehandlung	traitement tertiaire	tratamiento terciario
5693	tertiary wall	tertiärvägg	Tertiärzellwand	paroi tertiaire	pared terciaria
5694	test bench	provbänk	Prüfstand, Prüftisch	banc d'essai	banco de pruebas
5695	test board	testboard	Prüfschrank, Behälterkarton mit besonders hoher Berstfestigkeit	carton simili kraft	cartón de prueba
5696	testing	provning	Prüfung, Untersuchung, Erprobung	essai, épreuve, examen	prueba
5697	test linerboard	testlinerkartong	Test-Deckenbahn (Wellpappe)	couverture forte	testlinerboard
5698	test method	provningsmetod	Prüfmethode, Messverfahren	méthode d'essai	método de prueba
5699	test papers	provpapper	Papier für Prüfzwecke	papiers réactifs	papeles reactivos
5700	test sheet	provark	Probebogen, Prüfbogen	feuille d'essai	hoja de prueba

	ENGLISH	SVENSKA	DEUTSCH	FRANÇAIS	ESPAÑOL
5701	textile fiber	textilfiber	Textilfaser, Spinnfaser	fibre textile	fibra textil
5702	texture	textur	Struktur, Aufbau, Gefüge	texture, formation	textura
5703	thermal conductivity	värmekonduktivitet	Wärmeleitfähigkeit	conductivité thermique	conductividad térmica
5704	thermal expansion	termisk utvidgning	Wärmeausdehnung	dilatation thermique	expansión térmica
5705	thermal insulation	värmeisolering	Wärmeisolierung	isolation thermique calorifugeage	aislamiento térmico
5706	thermal pollution	termisk miljöförstöring	Verschmutzung durch Hitze, Verschmutzung durch Wärme	pollution thermique	contaminación térmica
5707	thermal properties	termiska engenskaper	Verhalten in der Wärme	propriétés thermiques	propiedades térmicas
5708	thermal sensitivity	värmekänslighet	Wärmeempfindlichkeit, Hitzeempfindlichkeit	sensibilité thermique	sensibilidad térmica
5709	thermal stability	värmestabilitet	Wärmebestandigkeit, Hitzebeständigkeit	stabilité thermique	estabilidad térmica
5710	thermal stress	värmespänning	Temperaturspannung, Wärmespannung	tension thermique	resistencia térmica
5711	thermocolor pencil	värmeskrivare	Wärmeanzeiger, Thermocolorstift	–	lápiz de color térmico
5712	thermocouple	termoelement	Thermoelement, Thermopaar	thermocouple	pila termoeléctrica
5713	thermodynamics	termodynamik	thermodynamische Wärmelehre	thermodynamique	termodinámica
5714	thermographic paper	värmeskrivarpapper	Schreibthermometerpapier	papier pour impression en relief	papel termográfico
5715	thermomechanical pulping	termomekanisk massatillverkning	thermomechanischer Aufschluss, Thermorefiner-Aufschluss	fabrication de pâtes thermomécaniques	elaboración termomecánica de pasta
5716	thermometer	termometer	Thermometer	thermomètre	termómetro
5717	thermoplastics	termoplast	thermoplastischer Stoff, Thermoplasten	thermoplastiques	termoplásticos
5718	thermosetting	termosättning	in der Wärme abbindend, in der Hitze abbindend	thermodurcissement	fraguado térmico
5719	thickener	förtjockare	Eindicker, Eindickzylinder, Verdickungsmittel	épaississeur	espesador
5720	thickening	förtjockning	Eindicken	épaississement	espesamiento
5721	thickening agent	förtjockningsmedel	Dickmittel	agent épaississeur	agente espesador
5722	thickness	tjocklek	Dicke, Stärke	épaisseur	espesor
5723	thickness gauge	tjockleksmätare	Dickenmesser	micromètre d'épaisseur	micrómetro
5724	thin film	tunnfilm	Feinfolie	film mince	película fina
5725	thinning	gallring	Durchforstung, Ausforstung, Verdünnung	dilution (de la pâte), éclaircie (forêt)	aclarado
5726	thin paper	tunnpapper	Dünnpost, Dünndruckpapier	papier mince	papel delgado
5727	thiocarbamates	tiokarbamater	Thiocarbamate	thiocarbamates	tiocarbamatos
5728	thiocarbonates	tiokarbonater	Thiocarbonate	thiocarbonates	tiocarbonatos
5729	thiolignins	tioligniner	Thiolignine	thiolignines	tioligninas
5730	thiosulfates	tiosulfater		thiosulfates	tiosulfatos
5731	third hand	tredje hand	dritter Geselle	aide-conducteur	tercera mano
5732	third press	tredjepress	dritte Presse	troisième presse	prensa auxiliar
5733	thixotropy	tixotropi	Thixotropie	thixotropie	tixotropía
5734	threading	trädning, spetsdragning	Einführen der Papierbahn	filetage, engagement du papier dans la machine	embarcamiento
5735	three ply	trelags-	dreilagig, dreischichtig	multijet (trois couches)	de tres capas
5736	three-pocket grinder	trekannslip	Dreipressenschleifer	défibreur à trois chambres (presses)	desfibradora de tres cámaras
5737	throttle valve	klaffventil	Drosselventil, Drosselklappe	vanne d'admission dans une machine à vapeur	válvula de estrangulación
5738	through drying	genomtorkning	Umlufttrockner	séchage à fond	secado directo
5739	Thuja (Lat.)	Thuja	Lebensbaum, Atlaszypresse, Thuja	thuya	tuya

	ENGLISH	SVENSKA	DEUTSCH	FRANÇAIS	ESPAÑOL
5740	ticket board	biljettkartong	Fahrkartenkarton	carton pour billets de chemin de fer	cartulina para billetaje
5741	tight roll	hård rulle	straff gewickelte Rolle	bobine serrée	bobina dura
5742	tile	kakelplatta	auskacheln, mit Ziegel bedecken, Kachel	tuile, carreau	baldosa
5743	Tilia (Lat.)	Tilia	Linde	tilleul	tilo
5744	tilting digester	stjälpbar kokare	Sturzkocher	lessiveur incliné	lejiadora basculante
5745	timber	timmer	Bauholz, Nutzholz	bois (de charpente), bois en grume	madera en troncos
5746	time	tid	Zeit, Zeitdauer, Frist	temps, durée	tiempo
5747	timing device	tidsinställningsanordning	Zeitmesser	échéancier	aparato de cronometraje
5748	tin	tenn, burk	Dose, Konservendose, Zinn	étain	estaño
5749	tin compounds	tennföreningar	Zinnverbindungen	composés d'étain	compuestos de estaño
5750	tint	nyans	tönen, leicht färben, schattieren	teinte	tinte
5751	tinted	färgjusterad	getönt	teinté	coloreado
5752	tinted white	nyanserat vitt	Mischweiss, Weisspigment	blanc nuancé	teñido blanco
5753	tinting	nyansering	Aufhellen, Tönen	coloration	coloración
5754	tire cord	däckskord	Reifencord	rayonne pour pneus	cuerda para cubiertas
5755	tissue machine	tunnpappersmaskin	Tissuepapiermaschine	machine à papier mousseline	máquina de papel de seda
5756	tissue paper	tunnpapper	Tissuepapier, Seidenpapier	papier mince, mousseline, pelure	papel muselina
5757	titanium	titan	Titan	titane	titanio
5758	titanium compounds	titanföreningar	Titanverbindungen	composés de titane	compuestos de titanio
5759	titanium dioxide	titandioxid	Titandioxid	bioxyde de titane	dióxido de titanio
5760	titratable acid	titrebar syra	titrierbare Säure	acide titrable	ácido titulable
5761	titratable alkali	titrerbart alkali	titrierbares Alkali	alcali titrable	álcali titulable
5762	titration	titrering	Titrierung, Titration	titrage	análisis volumétrico
5763	toilet paper	toalettpapper	Toilettenpapier	papier hygiénique	papel higiénico
5764	toilet roll	toalettrulle	Toilettenpapierrolle	rouleau de papier hygiénique	rollo de papel higiénico
5765	toilet roll cutter and perforator	toalettrullskärare och perforator	Toilettenpapierschneide-und Perforiermaschine	machine à couper et à perforer le papier hygiénique	cortadora perforadora para papel higiénico
5766	toilet tissue	toalettpapper	Toiletten-Tissue	papier hygiénique	papel higiénico de muselina
5767	tolerance	tolerans	Toleranz, zulässige Abweichung, Fehlergrenze	tolérance	tolerancia
5768	toluene	toluen	Toluol	toluène	tolueno
5769	ton	ton	Tonne	tonne	tonelada
5770	toner	nyanseringsmedel	Toner, Tönungsmittel	colorant pour aviver	colorante (para avivar un matiz)
5771	tool	verktyg	Werkzeug, Gerät, maschinell bearbeiten, mit Werkzeugen arbeiten	outil	herramienta
5772	tooth	tand	Zahn, Zinke, Zacken	dent (d'engrenage), grain du papier	diente
5773	top	topp	Oberseite, Oberfläche, Decke, Deckel, oberes Ende, Kopfende, Belag	sommet, haut, côté feutre de la feuille	cumbre
5774	top couch roll	ridvals	obere Gautschwalze	rouleau supérieur de presse humide	rodillo superior de prensa húmeda
5775	top felt press	överfiltpress	obere Filzpresse	feutre preneur supérieur	prensa de fieltro superior
5776	topliner	övre liner	oberste Dickschicht eines Kartons, Kartondecke	couche supérieure de la doublure recouvrant un carton, couche supérieure de la doublure recouvrant une boîte	capa superior
5777	top press roll	överpressvals	obere Presswalze	rouleau de presse supérieure	rodillo de prensa superior
5778	top roll	övervals	Oberwalze	rouleau supérieur	rodillo superior
5779	top side	översida	Oberseite, Filzseite des Papiers	côté supérieur, côté feutre	cara fieltro (de un papel)
5780	top sizing	ytlimning	Oberflächenleimung	collage en surface	encolado superior
5781	top slice	övre läpp	Oberlippe des Einlaufspalts	–	labio superior de la caja de alimentación
5782	torn deckle	riven kant	Imitationsbüttenrand	bord imitant celui d'un papier à la main	marco rasgado
5783	torn sheet	rivet ark	beschädigtes Riespapier	feuille défectueuse	hoja desgarrada
5784	torque	vridmoment	Drehmoment	couple (mouvement de rotation)	par motor
5785	torsion	torsion	Drehung, Windung	torsion	torsión
5786	total alkali	total alkali	Gesamtalkaligehalt	alcali total	álcali total
5787	total transmittance	totaltransmittans	Totaldurchgang, Gesamtübertragung	transmission totale (de la lumière à travers le papier)	transmitencia total
5788	toughness	seghet	Zähigkeit, Zähflüssigkeit	flexibilité	tenacidad

	ENGLISH	SVENSKA	DEUTSCH	FRANÇAIS	ESPAÑOL
5789	tour	vakt	Schicht	tour, faction	turno
5790	tour boss	vaktchef	Schichtführer	contremaître de faction	capataz de servicio
5791	towel	handduk	Handtuch	serviette	servilleta
5792	toweling	handdukspapper	Handtuchstoff	essuie-mains	tejido esponja
5793	toweling paper	handdukspapper	Handtuchpapier, Handtuchkrepp	papier pour essuie-mains	papel para toallas
5794	tower	torn	Turm, Säureturm	tour	torre
5795	toxicity	toxicitet, giftighet	Toxizität, Giftigkeit	toxicité	toxicidad
5796	trace elements	spårelement	Spurenelemente	éléments traceurs	elementos de calco
5797	tracheids	trakeider	Tracheide	trachéides	traqueidas
5798	tracing paper	kalkerpapper	Pauspapier	papier calque	papel calco
5799	tragacanth gum	gummidragkant	Tragantgummi	gomme adragante	adragante
5800	trailing blade coater	slåpbladsbestrykare	Schlepprakelstreicher	coucheuse à lame traînante	estucadora de cuchillas de arrastre
5801	trailing blade coating	slåpbladsbestrykning	Schlepprakelauftrag	couchage à lame traînante	estucado por cuchillas de arrastre
5802	training equipment	övningsutrustning	Ausbildungsgeräte	matériel d'apprentissage	equipos didácticos
5803	transducer	givare, transduktor	Umwandler, Energieumwandler	transducteur (de pression)	transductor
5804	transfer	överföra	Übergang, Übertragung, Umdruck, Abzug	transfert, report, décalque	reporte litográfico
5805	transfer paper	kalkerpapper	Umdruckpapier	papier à reports, papier autographique	papel autográfico
5806	transfer press	överföringspress	Übergabepresse	presse à décalquer, presse à report	prensa de reporte
5807	transformer	transformator	Umspanner, Umformer	transformateur	transformador
5808	transistor	transistor	Transistor	transistor	transistor
5809	transition point	transitionspunkt	Übergangspunkt, Reflexionswelle (einer Leitung)	point de transition, point de passage	punto de transición
5810	translucence	genomskinlighet	Lichtdurchlässigkeit	transparence	translucidez
5811	translucency	genomskinlighet	Durchsichtigkeit, Durchscheinen	translucidité	translucidez
5812	translucent coating	genomskinlig bestrykning	lichtdurchlässige Beschichtung, optisch dünner Strich	couchage translucide	estucado translúcido
5813	transmission line	transmissionsledning	Hochspannungsleitung, Überlandleitung, Übertragungsleitung	ligne de transmission	línea de transmisión
5814	transmittance	transmittans	Durchlassgrad	transmission (de la lumière à travers un papier)	transmitencia
5815	transparence	transparens	Transparenz	transparence	transparencia
5816	transparency	transparens	Durchsichtigkeit, Transparenz	transparence	transparencia
5817	transparency ratio	transparenskvot	Grad der Transparenz	coefficient de transparence	relación de transparencia
5818	transparent	transparent, genomsynlig	transparent	transparent	transparente
5819	transparent cellulose	transparent cellulosa	Klarsichtfolie	cellulose transparente	celulosa transparente
5820	transparentizing	att göra genomsynlig	Durchsichtigmachen	rendre transparent	hacer transparente
5821	transpiration	transpiration	Austreten (von Gasen), Transpiration, Schwitzen	transpiration	transpiración
5822	transplantation	transplantation	Transplantation, Verpflanzung, Umpflanzung	transplantation	trasplante
5823	transport	transport	transportieren, befördern, übertragen (Druck), Transport, Beförderung, Verfrachtung	transport	transporte
5824	transversal flow press	tvärsflödespress	Transversalflusspresse	presse à écoulement transversal	prensa de flujo transversal
5825	transverse fiber	tvärsfiber	Querfaser	fibre transversale	fibra transversal
5826	transverse porosity	tvärsporositet	querverlaufende Porosität	porosité transversale	porosidad transversal
5827	trap	fälla	Abscheider, Klappe, eingeschlossen, festgehalten	trappe, purgeur (vapeur)	purgador
5828	trash discharge	avfallstömning	Schmutzschleuse	décharge d'impuretés	descarga de residuos
5829	tray	viraskepp	Schale, Trog	cuvette, dalle pour recueillir les eaux sous toile	bandeja
5830	tree	träd	Baum	arbre	árbol
5831	tree cone	kotte	Koniferenzapfen, Nadelbaumzapfen, Tannenzapfen	cône (arbre)	piña de árbol
5832	tree farm	trädplantering	Baumschule	pépinière	vivero de árboles
5833	tree-length log	stock av trädlängd	ganzer Baumstamm	rondin en longueur d'arbre	tronco entero
5834	tree shear	trädsax	–	abattage	corte de árboles
5835	trial run·	provkörning	Probelauf	essai préliminaire	funcionamiento de prueba
5836	trickling filter	droppfilter	Tropfilter	filtre à ruissellement	filtro de escurrimiento
5837	trickling condenser	droppkondensor	Sickerkühlung	condensateur à ruissellement	condensador del escurrimiento
5838	trim	renskära	die Maschine voll ausnutzen	largeur rognée d'une machine à papier	recortar
5839	trimmed	renskuren	beschnitten	rogné	recortado
5840	trimmed paper	renskuret papper	Papier mit Randbeschnitt, zugeschnittenes Papier	papier rogné	papel recortado

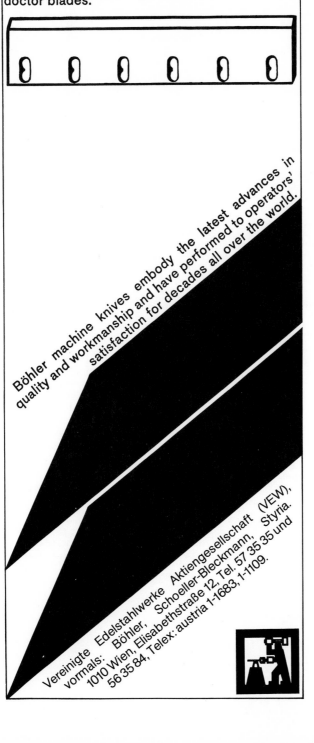

	ENGLISH	SVENSKA	DEUTSCH	FRANÇAIS	ESPAÑOL
5841	trimmed size	renskuren storlek	Endformat, Schnittmass, Schneidmass	format fini	formato acabado
5842	trimmed splice	renskuren skarv	zugeschittene Klebestelle	collure rognée, ajouture rognée	empalme recortado
5843	trimmed width	renskuren bredd	beschnittene Arbeitsbreite (der Papierbahn)	largeur rognée	ancho del recorte
5844	trimmer	renskärare	Beschneidmaschine	massicot	guillotina
5845	trimmer press	justerpress	Planschneider	massicot	prensa recortadora
5846	trimming	renskärning	Planschneiden, Besäumen, Beschneiden der Kanten	rognure	recortadrua
5847	trimming knife	renskärningskniv	Planschneidemesser	lame de massicot	cuchilla recortadora
5848	trimming machine	justermaskin	Planschneider, Formatschneider	massicot	máquina recortadora
5849	trim shower	gusksprits	Randstreifenabspritzrohr	bordeur	chorro de orillo
5850	triple fourdrinier machine	trippelviramaskin	Dreifach-Langsiebmaschine	machine à table plate à trois toiles	máquina de papel triple de mesa plana
5851	triplex	triplex	Triplex, dreifach, dreilagig, dreischichtig	triplex	tríplex
5852	triplex board	trelagskartong	Triplex-Karton	carton triples	cartón tríplex
5853	trisodium phosphate	trinatriumfosfat	Trinatriumphosphat	phosphate de trisodium	fosfato trisódico
5854	tropical plant	tropisk växt	tropische Pflanze	plante tropicale	planta tropical
5855	truck	truck, lastbil	Lastkraftwagen, Laster, offener Güterwagen	camion, chariot, wagonnet	carro
5856	truncate	stympa	kürzen, abstutzen, abstumpfen	tronquer	troncar
5857	trunnion	tapp	Lagerzapfen	tourillon	muñón
5858	Tsuga (Lat.)	Tsuga	Tsuga	pruche du Canada	Tsuga
5859	tub coloring	färgning i pappersmaskin	in der Bütte färben, Büttenfärbung	coloration au plongé	coloración en tina
5860	tub liners	trågfodring	Büttenauskleidung	revêtement d'une cuve	liners de tina
5861	tub sized	ytlimmad	in der Bütte geleimt	collé à la cuve	encolado en tina
5862	tub-sized paper	ytlimmatt papper	oberflächengeleimtes Papier	papier collé à la cuve	papel encolado en tina
5863	tub-size press	limpress	Büttenleimpresse	presse encolleuse au plongé	prensa encoladora de tina
5864	tub sizing	ytlimning	Büttenleimung, Oberflächenleimung	collage en cuve	encolado en tina
5865	tube	rör, tub, hylsa	Röhre, Rohr, Hülse, Schlauch	tube, lampe	tubo
5866	tube board	tubkartong	Hülsenpappe, Wickelpappe	carton pour tubes	cartón tubo
5867	tube stock	tubmassa	Hülsenmaterial	carton pour tubes	pasta de mandriles
5868	tulip tree	tulpanträd	Tulpenbaum	tulipier	tulipero de Virginia
5869	tung oil	kinesisk träolja	Tungöl	huile d'abrasin, huile de tung	aceite de palo
5870	tungsten compounds	volframföreningar	Wolframverbindungen	composés du tunstène	compuestos de tungsteno
5871	tunnel dryer	tunneltork	Kanaltrockner	séchoir à tunnel	secadora de túnel
5872	turbid	turbid	trübe	trouble, vaseux	turbio
5873	turbidimeter	turbidimeter	Trübungsmesser	opacimètre	turbidímetro
5874	turbidity	turbiditet, grumlighet	Trübung	trouble, opacité	turbieza
5875	turbine	turbin	Turbine	turbine	turbina
5876	turbogenerator	turbogenerator	Turbogenerator	turbogénérateur	turbogenerador
5877	turbulence	turbulens	Turbulenz	turbulence, trouble	turbulencia
5878	turbulent flow	turbulent flöde	Wirbelströmung	écoulement turbulent	flujo turbulento
5879	turned edge	vikt kant	umgelegte Kante, umgeknickte Kante	bord retourné	borde torneado
5880	turnkey plant	startklar fabrik	schlüsselfertige Anlage	usine clés en main	planta llave en mano
5881	turn over	omsättning	umwenden, Umsatz	retourner, inverser, feuilleter (un livre)	dar la vuelta
5882	turpentine	terpentin	Terpentine	térébenthine	aguarrás
5883	turpentine test	terpentinprov	Terpentin-Test (auf Fettdichtigkeit)	essais à l'essence de térébenthine	pruebas del aguarrás
5884	twig	kvist	Zweig	petite branche, brindille	ramita
5885	twill weave	fyrskaftad väv	Köperbindung	étoffe croisée	tela sargada
5886	twin wire machine	tvillingviramaskin	Twin-wire Maschine, Zweisiebmaschine, Doppelsiebmaschine	machine à tables multiples	máquina de papel de telas gemelas
5887	twin wire paper	tvillingvirapapper	Duplex-Papier	papier fabriqué sur machine à tables multiples	papel dos telas
5888	twisting	vridning	Drehen, Winden, Drall	câblage, torsion	ensortijado
5889	twisting paper	karamellpapper	Bonbondreheinschlag	papier paraffiné pour confiserie, papier à filer	papel para hilados
5890	twisting tissue	karamellpapper	Dreheinschlagseidenpapier	papier à paraffiner (ou à filer)	muselina para parafinar
5891	two ply	tvålags-	zweilagig, zweischichtig	duplex	duplex
5892	two-ply board	tvålagskartong	zweilagiger Karton, Duplex-Karton	carton duplex	cartón duplex
5893	two-side coater	tvåsidesbestrykare	Streichanlage für beidseitiges Streichen	coucheuse deux faces	estucadora de ambos lados

ENGLISH	SVENSKA	DEUTSCH	FRANÇAIS	ESPAÑOL
5894 two-side coated paper	tvåsidigt bestruket papper	beidseitig gestrichenes Papier	papier couché deux faces	papel estucado dos caras
5895 two-side coating	tvåsidig bestrykning	zwei- oder beidseitiges Streichen	couchage sur deux faces	estucado dos caras
5896 two-sided	tvåsidig	zweiseitig	deux faces	dos caras
5897 two-sidedness	tvåsidighet	Zweiseitigkeit	envers du papier (coloration)	revés (de un papel)
5898 tyloses	tyloser	Tylose, Methylcellulose	tyloses	tilosis
5899 tympan paper	däckelpapper	Pressdeckelpapier	papier pour mise en train (impr.)	papel tímpano
5900 type setting	typsättning	Schriftsetzen	composition	composición

U

ENGLISH	SVENSKA	DEUTSCH	FRANÇAIS	ESPAÑOL
5901 Uhle box	Uhle-låda	Saugkasten	caisse aspirante Uhle	caja Uhie
5902 Ulmus (Lat.)	Ulmus	Ulme	orme	olmo
5903 ultraviolet radiation	ultraviolett strålning	ultraviolette Strahlung	rayonnement ultraviolet	radiación ultravioleta
5904 unbleached	oblekt	ungebleicht	écru	sin blanquear
5905 unbleached paper	oblekt papper	ungebleichtes Papier	papier écru, papier bulle	papel no blanqueado
5906 unbleached pulp	oblekt massa	ungebleichter Zellstoff	pâte écrue	pasta cruda
5907 uncalendered	oglättad	nicht kalandriert	brut, sans apprêt	no calandrado
5908 uncoated	obestruken	ungestrichen	non couché	no estucado
5909 uncoated book paper	obestruket boktryckpapper	ungestrichenes Buchpapier, Naturbuchpapier	papier d'édition non couché	papel sin estucar para libros
5910 uncoated paper	obestruket papper	Naturpapier	papier non couché	papel sin estucar
5911 uncoated printing paper	obestruket tryckpapper	ungestrichenes Druckpapier	papier d'impression non couché	papel sin estucar para imprimir
5912 uncoated weight	obestruken vikt	Gewicht ohne Strichauftrag	poids du support de couche	peso sin estucado
5913 undercooked	underkokad	schlecht aufgeschlossen, ungenügend aufgeschlossen	pâte incuite	cocido incompleto
5914 underliner	underliner	Schonschicht (eines Kartons)	sous-couche	rayador
5915 under-run	otillräckligt körd	Untergewicht, unterfahren, nicht erfüllte Produktionsleistung	quantité fabriquée en moins sur une commande	correr por debajo
5916 unfinished	oglättad	unsatiniert, unfertig, nicht ausgerüstet	brut, sans apprêt	sin satinar
5917 unglazed	oglättad	unsatiniert	apprêté	sin abrillantar
5918 uniform formation	likformig arkbildning	gleichmässige Formierung, homogene Blattbildung	épair régulier	formación uniforme
5919 uniformity	likformighet	Gleichmässigkeit	uniformité, régularité	uniformidad
5920 unit	enhet	Einheit, Masseinheit	unité	unidad
5921 unitized handling	enhetlig hantering	rationalisiertes Arbeiten	manutention de charges unitaires	manipulación unificada
5922 unloaded	obelastad, ofylld	unbelastet, entlastet, entladen, abgeladen, unbeschwert, füllstoffrei	déchargé	descargado
5923 unloader	avlastare	Entlader	déchargeur	descargador
5924 unloading	avlastning	Abladen, Ausladen, Entladen	déchargement	descargar
5925 unsized	olimmad	ungeleimt	non collé	sin cola
5926 unsized paper	olimmat papper	ungeleimtes Papier	papier non collé	papel sin cola
5927 untrimmed paper	icke renskuret papper	unbeschnittenes Papier	papier non rogné	papel sin desbarbar
5928 unwind	avrulla	abrollen	débobiner	desenrollar
5929 unwinder	avrullningsmaskin	Abroller, Abwickler	débobineuse	desenrolladora
5930 unwinding	avrullning	Abrollen, Abwickeln	débobinage, déroulement	desenrollado
5931 unwinding stand	avrullningsstativ	Abrollung, Abrollvorrichtung	dévidoir	devanadera
5932 upgrading	uppgradering	aufbessern	améliorer, faire passer dans une qualité supérieure	mejora de la calidad
5933 uphill wire	uppåtgående vira	Steilsieb	câble de montée	tela ascendente
5934 urea	urea	Harnstoff	urée	urea
5935 urea-formaldehyde resins	ureaformaldehydharts	Harnstoffformaldehyd-Harz, Karbamidharz	résines urée-formaldéhyde	resinas de urea-formaldehido
5936 urea resins	ureahartser	Harnstoffharze	résines urée	resinas de urea
5937 urethanes	uretaner	Urethan	uréthanes	uretanos
5938 utilities	arbetskraft	Betriebskraft	manoeuvres	servicio público
5939 utilization	utnyttjande	Verwertung, Nutzbarmachung	utilisation	utilización

ENGLISH	SVENSKA	DEUTSCH	FRANÇAIS	ESPAÑOL

V

5940	vacuum	vakuum	Vakuum, luftleerer Raum	vide	vacío
5941	vacuum box	suglåda	Vakuum-Flachsaugkasten, Vakuumflachsauger	caisse aspirante	caja aspirante
5942	vacuum deposition	avsättning under vakuum	Vakuumumlader, Niederschlag durch Vakuumabsaugung	nettoyage par le vide	depósitos del vacío
5943	vacuum filter	sugfilter	Vakuumfilter	filtre à vide	filtro al vacío
5944	vacuum filtration	sugfiltrering	Vakuumfiltrierung	filtration sous vide	filtración al vacío
5945	vacuum forming	sugformning	Vakuumverformung, Blattbildung unter Vakuumabsaugung	formation sous vide	fabricación al vacío
5946	vacuum metalizing	vakuummetallisering	Hochvakuum-Metallisierung	métallisation sous vide	metalización al vacío
5947	vacuum pickup	sugavtagning	Vakuumabnahme	levage à vide	toma automática de vacío
5948	vacuum pump	vakuumpump	Absaugpumpe	pompe à vide	bomba de vacío
5949	vacuum roll	sugvals	Absaugwalze	rouleau aspirant	rodillo aspirante
5950	vacuum transfer	sugöverföring	Vakuumbahnüberführung	transvaser sous vide	transferencia de vacío
5951	valence	valens	Wertigkeit	valence	valencia
5952	valve	ventil	Ventil, Schieber	valve, vanne, soupape, robinet	válvula
5953	valve bag	ventilsäck	Ventilsack	sac G.C. à valve	bolsa de papel a válvula
5954	vane	blad	Rotorflügel, Rotorschaufel	ailette, pale	paleta
5955	vanillin	vanillin	Vanillin	vanilline	vanillina
5956	vanillin derivatives	vanillinderivat	Vanillinderivate	dérivés de la vanilline	derivados de vanillina
5957	vapor	ånga	Dampf, Dunst, Schwaden	vapeau, buée	vapor
5958	vapor barrier	ångspärr	Dampfschranke, Flüssigkeitsschranke	anti-buée	barrera de vapor
5959	vaporization	förångning	Verdampfung	vaporisation, nébulisation	vaporización
5960	vapor permeability	ångpermeabilitet	Dampfdurchlässigkeit	perméabilité à la vapeur	permeabilidad al vapor
5961	vapor phase	ångfas	Dampfphase	en phase vapeur	fase de vapor
5962	vapor phase pulping	ångfaskokning	Dampfphasenaufschluss	fabrication de pâtes en phase vapeur	elaboración de pasta con fase de vapor
5963	vapor pressure	ångtryck	Dampfdruck	pression de la vapeur	presión del vapor
5964	vaporproof	ångtät	dampfdicht	étanche à la vapeur	estanco al vapor
5965	vaporproof paper	ångtätt papper	dampfdichtes Papier	papier étanche à la vapeur	papel estanco al vapor
5966	vapour (Brit.)	ånga	Dampf, Dunst, Schwaden	vapeur, buée	vapor
5967	variability	variabilitet	Veränderlichkeit	variation	variabilidad
5968	variable	variabel	variabel, veränderlich, Variable, unbeständig	variable	variable
5969	variable-speed drive	drift med variabel hastighet	stufenloser Antrieb	commande à vitesses variables	accionamiento de velocidad variable
5970	variance	varians	Abweichung	variation, modification	variación
5971	variety	varietet, variant	Vielfalt, Verschiedenheit, Auswahl, Sortiment	variété, diversité	variedad
5972	varnish	fernissa	lackieren, firnissen, Lack, Firnis, Lasurlack	vernis	barniz
5973	varnishability	fernissbarhet	Lackierbarkeit	aptitude au vernissage	aptitud al barnizado
5974	varnishing	fernissning	Lackieren, Aufpolieren, Firnissen	vernissage	barnizado
5975	vat	tråg	Bütte, Rundsiebtrog, Rundsiebbütte	cuve	tina
5976	vat dyes	trågfärger	Küpenfarbstoffe, Pigmentfarbstoffe	colorants de cuve	colorantes a la tina
5977	vat lined	med guskat ytskikt	gegautscht	carton doublé d'une feuille faite sur forme ronde	blanqueado a la tina
5978	vat-lined board	papp med guskat ytskikt	gegautschter Karton	carton doublé d'une feuille faite sur forme ronde	cartón encartelado a la tina
5979	vat machine	rundviramaskin	Rundsiebzylindermaschine	machine à forme ronde	tina
5980	vat papers	rundvirpapper	Büttenpapiere, Rundsiebpapiere	papiers à la main	papeles a la tina
5981	vegetable parchment paper	vegetabiliskt pergamentpapper	Echtpergament-Papiere	papier sulfurisé véritable	pergamino vegetal
5982	vehicle	lösningsmedel	flüssiger Anteil der Druckfarbe, Fahrzeug, Farbträger (Druck)	véhicule, voiture	vehículo
5983	vellum	velängpapper	Velin, feines Pergament	vélin	vitela
5984	vellum board	velängkartong	Velinkarton	carton vélin	cartón vitela
5985	vellum finish	velängfinish	Velinausrüstung, Velinoberfläche	fini vélin	acabado vitela
5986	vellum paper	velängpapper	Velinpapier	papier vélin	papel vitela
5987	velocity	hastighet	Geschwindigkeit	vitesse	velocidad
5988	velocity control	hastighetsreglering	Geschwindigkeitsregelung, Geschwindigkeitskontrolle	contrôle de la vitesse	control de velocidad

	ENGLISH	SVENSKA	DEUTSCH	FRANÇAIS	ESPAÑOL
5989	velocity measurement	hastighetsmätning	Geschwindigkeitsmessung	mesure de la vitesse	medición de la velocidad
5990	velour paper	velourpapper	Velourspapier	papier velours	papel terciopelo
5991	velvet finish	sammetsfinish	mattsatiniert	fini velours	acabado terciopelo
5992	veneer	faner	Furnier, Furnierblatt, Furnierholz	feuille de placage, contreplaqué	hoja de madera
5993	veneer paper	fanerpapper	Furnierpapier, Möbelfolienpapier	papier imitant le bois	papel de revestimiento
5994	veneer waste	fanerutskott	Furnierabfall	déchets de contreplaqué	desechos del enchapado
5995	vent	ventil, sprund	durchlüften, entlüften, ventilieren, Loch, Öffnung, Luftloch	ouverture, passage, trou	respiradero
5996	ventilation	venitlering, ventilation	Belüftung, Entlüftung, Lüftung	ventilation	ventilación
5997	ventilator	ventilator	Luftgebläse, Lüfter, Ventilator	ventilateur, aérateur	ventilador
5998	venting	ventilering	Belüften, Entlüften	purge (d'air ...)	descarga
5999	Venturi tube	venturirör	Venturi-Düse, Saugrohr, Saugdüse	tube venturi	tubo de Venturi
6000	vertical digester	vertikal kokare	stehender Zellstoffkocher, Vertikall-Zellstoffkocher	lessiveur vertical	lejiadora vertical
6001	vertical flow press	vertikalflödespress	Vertikalpresse	presse à écoulement vertical	prensa de flujo vertical
6002	vertical forming machine	vertikalformningsmaskin	Vertikalformierungseinrichtung	machine à formation verticale	mesa de formación vertical
6003	vessel	kärl	Gefäss	vase, récipient	recipiente
6004	vibrating screen	vibrerande sil	Vibrationssortierer, Schüttelsortierer	épurateur à vibrations	depurador vibrante
6005	vibration	vibration	Vibration, Oszillation, Schwingung	vibration	vibración
6006	vibrator	vibrator	Schüttelmaschine, Schwingungserzeuger, Oszillator	vibreur	vibrador
6007	Vickery felt conditioner	vickeryfilttvätt	Vickery-Filzinstandhalter	conditionneur de feutre type vickery	lavador de fieltro Vickery
6008	vinyl acetate	vinylacetat	Vinylacetat	acétate de vinyl	acetato de vinilo
6009	vinyl cellulose	vinylcellulosa	Vinylcellulose	vinyl cellulose	vinil celulosa
6010	vinyl chloride	vinylklorid	Vinylchlorid	chlorure de vinyl	cloruro de vinilo
6011	vinyl compounds	vinylföreningar	Vinylverbindungen	composés du vinyl	compuestos de vinilo
6012	vinylidene chloride	vinylidenklorid	Vinylidenchlorid	chlorure de vinyldiène	cloruro de viniliden
6013	virgin stands (virgin timber)	ungträd	Jungholz	peuplements vierges	madera vírgen
6014	virgin stock	färskmassa	unbearbeiteter Stoff	pâte vierge, pâte neuve	pasta virgen
6015	viscoelasticity	viskoelasticitet	Viskoelastizität	viscoélasticité	viscoelasticidad
6016	viscometer	viskometer	Viskometer, Konsistenzmesser	viscosimètre	viscómetro
6017	viscometry	viskometri	Viskosimetrie	viscosimétrie	medición de la viscosidad
6018	viscose	viskos	Viskose	viscose	viscosa
6019	viscose fiber	viskosfiber	Viskosefaser	fibre de viscose	fibra viscosa
6020	viscose pulp	viskomassa	Viskosezellstoff	pâte viscose	pasta viscosa
6021	viscose rayon	viskosrayon	Viskosefaser	rayonne viscose	seda artificial viscosa
6022	viscosimeter	viskosimeter	Viskometer, Konsistenzmesser	viscosimètre	viscosímetro
6023	viscosity	viskositet	Viskosität, Zähflüssigkeit	viscosité	viscosidad
6024	viscosity control agent	viskositetsmedel	Viskositäts-Reguliermittel	agent de contrôle de la viscosité	agente de control de la viscosidad
6025	viscosity index	viskositetsindex	Viskositäts-Index	taux de viscosité	índice de viscosidad
6026	viscous flow	viskös flytning	zähflüssig Fliessen, zäher Fluss	courant visqueux	flujo viscoso
6027	vitrification	vitrifiering, förglasning	Sinterung, Verglasung	vitrification	vitrificación
6028	void	hålrum	leer, Hohlraum, Lücke, Leere, Fehlstelle, Pore (im Papier), unausgefüllt	vide	hueco
6029	void fraction	hålvolym	Verhältnis zwischen Poren/Luftblasenvolumen zum Gesamtvolumen des Papiers	fraction de vide	fracción de huecos
6030	void ratio	hålrumskvot	Porenziffer	rapport de vide	relación de vacío
6031	void volume	hålvolym	Porenziffer, Porenvolumen	volume de vide	volumen de vacío
6032	volatility	flyktighet	Flüchtigkeit, Verflüchtigung	volatilité	volatilidad
6033	voltmeter	voltmeter	Voltmeter, Spannungsmesser	voltmètre	voltímetro
6034	volume	volym	Volumen, Fassungsvermügen	volume	volumen
6035	vortex	virvel	Wirbel, Strudel	tourbillon	vórtice
6036	vulcanized fiber	vulkanfiber	Vulkanfiber	fibre vulcanisée	fibra vulcanizada
6037	vulcanized fiber stock	vulkanfibermassa	Vulkanfiberstoff	fibre vulcanisée	pasta de fibra vulcanizada
6038	vulcanizing	vulkanisering	Vulkanisieren	vulcaniser	vulcanizado.
6039	vulcanizing paper	vulkaniseringsbaspapper	vulkanisierfähiges Papier	papier à vulcaniser	papel para vulcanizar

ENGLISH	SVENSKA	DEUTSCH	FRANÇAIS	ESPAÑOL

W

	ENGLISH	SVENSKA	DEUTSCH	FRANÇAIS	ESPAÑOL
6040	wad	vadd	Papierknäuel, Batzen, Wattebausch	botte (de paille)	guata
6041	wadding	vaddpapper	Wattierung, Stosschutz, Schutzpolster	ouate de cellulose	guata
6042	wadding stock	vaddmassa	Watteeinlagestoff, Packwatte	ouate de cellulose	pasta para guata
6043	wall board	panel	Baupappe, Isolierpappe, Dämmpappe	carton de construction	hojas de fibra prensada
6044	wall effect	väggeffekt	Wandeffekt	effet de paroi	efecto de pared
6045	wallpaper	tapet	Tapeten	papier tenture	papeles pintados
6046	warp	varp	Kette (Gewebe), Kettfaden	chaîne (d'une toile métallique)	urdimbre
6047	warping	varpning, skevning	Verwerfen	gauchissement, gondolage	doblado
6048	warp threads	varptrådar	Kettfäden	fils de chaîne	hilos de urdimbre
6049	warp wire	varptråd	Kettdraht, Kettendrähte am Papiersieb	fils de chaîne	urdimbre
6050	washboard	tvättbräde	Setzbord	planche à laver	rodapié
6051	washboard marks	tvättbrädesmarkering	Verwerfen (von Papier), Welligkeit	marques de planche à laver	marcas de rodapié
6052	washer	tvätt, packning	Halbzeugholländer, Waschanlage, Berieselungsturm	pile laveuse	pila lavadora
6053	washing	tvättning	Waschen	lavage	lavado
6054	washing drum	tvättrumma	Waschtrommel	tambour laveur	tambor lavador
6055	wash up	diska	abwaschen	laver	lavado
6056	waste	avfall	verschwenden, verlorengehen, unbrauchbar, Abfall, Ausschuss	déchet, rebut	desperdicio
6057	waste disposal	avfallshantering	Abfallbeseitigung	évacuation des déchets	eliminación de desperdicios
6058	waste liquor	avlut	Ablauge	lessive résiduaire	lejía residual
6059	wastepaper	makulatur	Altpapier	vieux papiers	papelote
6060	wastepaper dealer	makulaturhandlare	Altpapierhändler	négociant en vieux papiers	negociante en papeles viejos
6061	wastepaper stock	makulaturmassa	Altpapierstoff	vieux papiers	pasta de recortes
6062	waste water	avloppsvatten	Abwasser	eau résiduaire	aguas residuales
6063	water	vatten	Wasser, wässern	eau	agua
6064	water absorbent	vattenabsorberande	wasseraufnahmefähig	qui absorbe l'eau	hidrófilo
6065	water absorption	vattenabsorbtion	Wasseraufnahme	absorption de l'eau	absorción del agua
6066	water analysis	vattenanalys	Wasseruntersuchung, Wasseranalyse	analyse de l'eau	análisis del agua
6067	water bleeding	vattenläckning	Ablassen, Ausschwitzen, Auslaufen von Wasser	fuite d'eau	exudación acuosa
6068	water box	vattenlåda	Wasserkasten	caisse à eau	caja de agua
6069	water box doctor	vattenschaber	Wasserabstreifer	docteur à eau	doctor de agua
6070	water box equipped stack	stapel med vattenschaber	Feuchtglättwerk	calandre humide	calandra húmeda
6071	water consumption	vattenförbrukning	Wasserverbrauch	consommation d'eau	consumo de agua
6072	water-cooled spring roll	vattenkyld fjädervals	wassergekühlte Federwalze (am Glättwerk)	rouleau de tension refroidi à l'eau	rodillo de resorte refrigerado por agua
6073	water creped	vattenkräppad	nassgekreppt	crêpage à l'état humide	rizado húmedo
6074	water finish	vattenfinish	Feuchtglätte	calandrage humide	calandrado por doctor de agua
6075	water finished	vattenfinished	feuchtgeglättet, feuchtsatiniert	calandré avec docteur à eau	calandrado húmedo
6076	water-finished paper	våtglättat papper	feuchtgeglättetes Papier	papier calandré par docteur à eau	papel calandrado húmedo
6077	water finishing	våtglättning	Feuchtglätten	calandrage humide	calandrado húmedo
6078	water flow	vattenströmning	Wasserführung	cours d'eau	flujo de agua
6079	water jet	vattenstråle	Wasserstrahl	jet d'eau	chorro de agua
6080	water leaf	provark från mälderi	ungeleimtes Papier	papier à la main sec, papier noncollé	papel absorbente
6081	water leg	sugben	Wassersäule	colonne barométrique	columna barométrica
6082	water-lined paper	vattenrandat papper	mit Wasserlinien versehenes Papier	papier avec lignes d'eau	papel encartelado al agua
6083	water management	vattenhantering	Wasseramt	gestion de l'eau	administración hidráulica
6084	watermark	vattenmärke	Wasserzeichen	filigrane	filigrana
6085	watermark dandy	vattenmärkningsegutör	Wasserzeichen-Egoutteur	rouleau filigraneur	rodillo filigranador
6086	watermark laid dandy	vattenmärkningsegutör	liegender Wasserzeichen-Egoutteur	rouleau égoutteur vergé	rodillo filigranador
6087	watermark roll	vattenmärkningsvals	Wasserzeichenprägewalze	rouleau filigraneur	rodillo desgotador
6088	watermark wove dandy	vattenmärkningsegutör	Wasserzeichenwebegoutteur	rouleau filigraneur vergeur	rodillo filigranador de vitela
6089	watermarked paper	papper med vattenmärke	mit Wasserzeichen versehenes Papier, Wasserzeichenpapier	papier filigrané	papel filigranado

Secondary waste water treatment system including pure oxygen activated sludge and sludge disposal for a boxboard mill in Florida.

Dravo guarantees your wastewater treatment plant will work. Definitely.

A Dravo designed and constructed wastewater treatment plant is guaranteed to work. Everything's in writing. Your plant will take a specified influent and yield an effluent of agreed-upon purity. There's no hedging.

If necessary, Dravo will modify or rebuild your treatment plant, as required, to meet the agreement. At no cost to you. And no risk to you. That's why your treatment plant will work. Definitely.

We know that most people file away performance guarantees unless a problem develops. And we believe that everyone is entitled to enjoy that kind of confidence. That's why Dravo receives monthly process performance reports throughout the warranty protection

period. We try to avoid potential problems before they develop. It's all part of our warranty.

We think our guarantee is the best around. If you can find one that's better, send us a copy. In the meantime, why not take a look at ours. Write: Louis Eckert, Industrial Sales Manager, Water and Waste Treatment Division, Dravo Corporation, One Oliver Plaza, Pittsburgh, Pa. 15222.

A company of uncommon enterprise

	ENGLISH	SVENSKA	DEUTSCH	FRANÇAIS	ESPAÑOL
6090	watermarking dandy	vattenmärkningsegutör	Wasserzeichen-Egoutteur	rouleau filigraneur	rodillo filigranador
6091	water permeability	vattenpermeabilitet	Wasserdurchlässigkeit	perméabilité à l'eau	permeabilidad
6092	water pollution	vattenförorening	Wasserverschmutzung, Wasserverseuchung	pollution de l'eau	contaminación del agua
6093	waterproof	vattentät	wasserdicht	imperméable à l'eau	impermeable
6094	waterproof board	vattentät kartong	wasserdichter Karton	carton imperméable à l'eau	cartón impermeable
6095	waterproofing	vattentätning	Wasserdichtmachen, Imprägnieren	imperméabilisation	impermeabilización
6096	waterproofness	vattentäthet	Wasserdichtigkeit	imperméabilité, étachéité à l'eau	impermeabilidad
6097	waterproof paper	vattentätt papper	wasserdichtes Papier	papier imperméable à l'eau	papel impermeable
6098	water quality	vattenkvalitet	Wassergüte, Wasserqualität	qualité de l'eau	calidad del agua
6099	water reclamation	vattenåtervinning	Wasserverwertung	récupération de l'eau	restauración del agua
6100	water removal	vattenborttagning, avvattning	Wasserbeseitigung, Wasserentfernung	évacuation de l'eau, vidange de l'eau	evacuación del agua
6101	water repellence	vattenavstötning	wasserabweisender Effekt	hydrophobie	impermeabilidad
6102	water resistance	vattenmotstånd	Wasserfestigkeit, Wasserbeständigkeit, Flüssigkeitswiderstand	résistance à l'eau	resistencia al agua
6103	water resource	vattentäkt	Wasserquelle	ressources en eau	recursos hidráulicos
6104	water retention	vattenretention	Zurückhalten des Wassers, Wasserretention	conservation de l'eau	retención de agua
6105	water reuse	vattenåteranvändning	Wiederverwendung des Wassers	recyclage de l'eau	reutilización del agua
6106	water seal	vattenförsegling	Dichtungsabschluss, Wasserabdichtung, wasserdichter Abschluss	joint hydraulique	cierre hidráulico
6107	watershed	vattendelare	Wasserscheide, Wassereinzugsgebiet	ligne de partage des eaux	colector de agua
6108	water softening	mjukgöring av vatten	Wasserenthärtung	épuration de l'eau	suavizador del agua
6109	water soluble	vattenlöslig	wasserlöslich	soluble dans l'eau	soluble en agua
6110	water spots	vattenfläckar	Wasserflecken	tache d'eau	manchas de agua
6111	water streaks	vattenränder	Wasserstreifen	traînées d'eau sur la feuille humide	corrientes de agua
6112	water supply	vattenförsörjning	Wasserversorgung	approvisionnements en eau	suministro de agua
6113	water treatment	vattenbehandling	Wasseraufbereitung	traitement de l'eau	tratamiento del agua
6114	water tube boiler	vattenrörkokare	Siederohrkessel	chaudière à eau multitubulaire	tubo hervidor
6115	water turbine	vattenturbin	Wasserturbine	turbine hydraulique	turbina hidráulica
6116	water vapor	vattenånga	Wasserdampf	vapeur d'eau	vapor de agua
6117	water vapor permeability	ångpermeabilitet	Wasserdampfdurchlässigkeit	perméabilité à la vapeur d'eau	permeabilidad al vapor de agua
6118	waterwheel	vattenhjul	Wasserrad, Wasserturbine	roue hydraulique	rueda hidráulica
6119	waviness	vågighet	Welligkeit	aptitude à l'ondulation	ondulación
6120	wavy edge	vågig kant	wellige Kante	bord gondolé	borde ondeado
6121	wax	vax	Wachs	cire, paraffine	cera
6122	wax coating	vaxbestrykning	Wachsbeschichtung	par affinage	estucado cera
6123	waxed board	vaxad kartong	gewachster Karton, Wachskarton	carton paraffiné	cartón encerado
6124	waxed glassine	vaxad pergamyn	Wachspergamin	papier cristal paraffiné	papel cristal encerado
6125	waxed kraft	vaxat kraftpapper	Wachskraftpapier	papier kraft paraffiné	kraft encerado
6126	waxed paper	vaxat papper	gewachstes Papier, Wachspapier	papier paraffiné	papel encerado
6127	waxed tissue	vaxat tunnpapper	gewachstes Tissue, Wachstissue	papier mince paraffiné	muselina parafinada
6128	wax emulsion	vaxemulsion	Wachsemulsion	émulsion aqueuse de paraffine	emulsión de cera
6129	waxing	vaxning	Wachsen, mit Wachs überziehen	paraffinage	parafinado
6130	waxing machine	vaxningsmaskin	Wachskaschiermaschine, Wachsauftragsmaschine	paraffineuse	máquina de parafinar
6131	waxing paper	vaxbaspapper	Wachsbeschichtungspapier, Wachsimprägnierpapier	papier support pour paraffinage	papel para encerar
6132	waxing tissue	vaxbaspapper	Wachsimprägniertissue, Rohtissue für die Wachsbeschichtung	papier mince pour paraffinage	muselina para encerar
6133	wax size	vaxlim	Wachsleim	collage à la paraffine	cola a la cera
6134	wax-sized paper	vaxlimmat papper	wachsgeleimtes Papier	papier collé à la paraffine	papel encolado a la cera
6135	wax spot	vaxfläck	transparente Flecken im Papier	tache de paraffine	manchas de cera
6136	wax test	vaxprov	Wachstest	essai de paraffine	pruebas de cera
6137	weak acid	svag syra	schwache Säure	lessive faible (bisulfite)	ácido débil
6138	wear	slitage	Abnutzung, Verschleiss	usure	desgaste
6139	wear tester	slitageprovare	Verschleissprüfgerät, Abnutzungsprüfgerät	appareil d'essais à l'usure	probador de desgaste
6140	wear tests	slitageprov	Verschleissprüfung	essais d'usure	pruebas de desgaste
6141	weatherproof	väderbeständig	wetterbeständig, witterungsbeständig, wetterfest	résistant aux intempéries	a prueba de intemperie
6142	weatherproofing	väderbeständigande	wetterfest ausrüsten	rendre résistant aux intempéries	hacer a prueba de intemperie

	ENGLISH	SVENSKA	DEUTSCH	FRANÇAIS	ESPAÑOL
6143	weather resistance	väderbeständighet	Wetterbeständigkeit	résistance aux intempéries	resistencia a las intemperies
6144	weave	väva, väv	Webart, Gewebe, Webstruktur	tisser, tresser	tejer
6145	weaving	vävning	Weben, Wirken	tissage	tejido
6146	web	bana	Bahn (Papierbahn)	feuille de papier en continu	hoja continua
6147	web break	banbrott	Bahnabriss	rupture de la feuille sur machine	rotura
6148	web calendered	maskinglättad	rollensatiniert, in Rollen geglättet	calandré en continu	calandrado de hoja continua
6149	web embossing	maskinpräglad	kontinuierliche Rollenprägung	gaufrage en continu	estampado de hoja continua
6150	web feeding	banmatning	Bahneinführung	alimentation de la bande	alimentación del rollo de hoja continua
6151	web glazing	maskinglättning	kontinuierliche Bahn-/Rollensatinage	satinage en continu	abrillantado de la hoja continua
6152	web guides	banstyrningsanordningar	Bahnführung, Bahnsteuerungsgerät	guide-feuille	guías de la hoja continua
6153	web news	tidningspapper i rulle	Zeitungsdruckpapierrollen (aus einer kontinuierlichen Bahn)	papier journal en continu	rollo de papel para periódico
6154	web offset paper	rulloffsetpapper	Rollenoffsetpapier	papier offset en continu	papel offset en rollo
6155	web offset printing	rulloffsettryckning	Rollenoffsetdruck	papier d'impression offset en continu	impresión offset sobre rollo
6156	web paper	rullpapper	Papierbahn, Papierrollen	papier en continu	papel en hoja continua
6157	web sizing	rullimning	Oberflächenleimung	satinage humide	encolado en rollo
6158	web spreader	arkspridare, banspridare	Streckwalze, Breitstreckwalze	rouleau déplisseur	desplisador de la hoja continua
6159	web tension	banspänning	Bahnspannung	tension de la feuille en continu	tensión de la hoja continua
6160	web threader	banpåförare	Bahneinführmechanismus	système d'engagement de la feuille	embarcadora de la hoja continua
6161	web transfer	banöverföring	Bahnüberführung, Bahnübergabe	transfert de la feuille	cambiadora de rollos de hoja continua
6162	web turner	banvändare	Bahnwendevorrichtung	changement de direction de feuille	volteadora de rollos de hoja continua
6163	weed tree	sly	Wildwuchsbäume	rebut (arbre)	maleza
6164	weft	väft, inslag	Schuss (Gewebe)	trame	trama
6165	weft wire	inslagstråd	Schussdrähte	fil de trame d'une toile métallique	tela de trama
6166	weft yarn	inslagstråd	Schussfaden, Schussgarn, Einschlagfaden	fil de trame	hilo de trama
6167	weighing	vägning	Wiegen, Wägen	pesage, pesée	pesada
6168	weighing hopper	vågtratt	Wiegetrichter	trémie de pesage	tolva de pesadas
6169	weighing scales conveyor	vågskålstransportör	Bandwaage	convoyeur à pesage	transportador de pesos
6170	weight	vikt, tyngd	Gewicht, Schwere	poids	peso
6171	weight indicator	viktvisare	Gewichtsanzeige	balance, bascule	indicador de peso
6172	weight tolerance	vikttolerans	Gewichtstoleranz	tolérance de poids	tolerancia de peso
6173	weir	skibord	Wehr	déversoir, barrage	vertedero
6174	welded joint	svetsfog	Schweissnaht	raccord soudé	unión soldada
6175	welding	svetsning	Schweissen	soudure	soldadura
6176	well	brunn	Brunnen, Schacht, Bohrloch	puits	pozo
6177	well-closed formation	väl sluten formering	gut geschlossene Blattbildung, gut geschlossene Oberfläche	papier bien garni	formación compacta
6178	well sized	vällimmad	gut geleimt	papier bien collé	bien encolado
6179	welt	fåll	Einfassung, Kante	bord	reborde
6180	wet	våt	nass, feucht	humide	húmedo
6181	wet board machine	våtkartongmaskin	Handpappenmaschine	enrouleuse à carton	enrolladora
6182	wet broke	våtutskott	Nassausschuss	cassés humides	recortes húmedos
6183	wet draw	våtsträckning	Nasszug (Spannungszustand der Bahn in der Nasspartie)	tirage dans la partie humide	tiro húmedo
6184	wet end	våtända	Nasspartie	partie humide	sección húmeda
6185	wet-end additive	tillsats för våtända	in der Nasspartie zugegebene Papierhilfsstoffe	adjuvant introduit dans la masse	aditivo de la sección húmeda
6186	wet-end finish	finish i våtändan	Oberflächenausrüstung in der Nasspartie	fini vélin en fin de machine	acabado en la sección húmeda
6187	wet felt	våtfilt	Nassfilz	feutre coucheur	fieltro húmedo
6188	wet finish	våtfinish	Nassappretur	fini vélin sur machine	acabado húmedo
6189	wet lapped	våtark	Nassfaserbündelung	plié en trois à l'état humide	plegado en tres húmedo
6190	wet lay system	våtsystem	nass gelegt, auf dem nassen Weg	procédé humide	sistema de estucado húmedo
6191	wet machine	upptagningsmaskin	Siebtischentwässerungsmaschine	presse-pâte	prensa-pastas
6192	wet machine board	våtupptagen kartong	Handpappe, Wickelpappe	carton à l'enrouleuse	enrolladora de cartón
6193	wet Mullen	våtmullen	Nassberstfestigkeit nach Mullen	résistance à l'éclatement à l'état humide	húmedo Mullen
6194	wetness	smörjighet	Feuchtigkeit, Nässe, Schmierigkeit des Stoffs	degré d'engraissement	grado de engrase
6195	wet press	våtpress	Nasspresse	presse coucheuse	prensa húmeda
6196	wet pressing	våtpressning	Nasspressen	pressage humide	prensado húmedo

	ENGLISH	SVENSKA	DEUTSCH	FRANÇAIS	ESPAÑOL
6197	wet pulp	våtmassa	Nasszellstoff, Nasschnitzel	pâte grasse	pasta engrasada
6198	wet rub	våtgnidning	Nassabrieb	frottement humide	frotación húmeda
6199	wet rub resistance	våtgnidningshållfasthet	Nassabriebfestigkeit	résistance au frottement humide	resistencia a la frotación húmeda
6200	wet streaks	våtränder	nasse Streifen	traînées humides	manchas de humedad
6201	wet strength	våthållfasthet, våtstyrka	Nassfestigkeit	résistance à l'état humide	resistencia en estado húmedo
6202	wet-strength agent	våtförstärkningsmedel	Nassverfestigungsmittel	produit rendant résistant à l'état humide	agente de resistencia en estado húmedo
6203	wet-strength broke	våtstarkt utskott	nassfester Ausschuss	cassés de papier résistant à l'état humide	recortes resistentes en estado húmedo
6204	wet-strength paper	våtstarkt papper	nassfestes Papier	papier résistant à l'état humide	papel resistente en estado húmedo
6205	wet-strength resins	våtförstärkningshartser	nassfeste Harze	résines résistant à l'état humide	resinas resistentes en estado húmedo
6206	wet-strength retention	våtstyrkeretention	Nassfestigkeits-Rückhaltevermögen	indice de résistance à l'état humide	índice de resistencia en estado húmedo
6207	wettability	vätbarhet	Vernetzbarkeit, Benetzbarkeit	mouillabilité	aptitud a la humectación
6208	wet tensile strength	våtdraghållfasthet	Nasszugfestigkeit	résistance à la traction à l'état humide	resistencia a la tracción en estado húmedo
6209	wetting	vätning	Benetzung	mouillant	remojo
6210	wetting agent	vätningsmedel	Netzmittel	produit mouillant	agente humectador
6211	wet-waxed paper	våtvaxat papper	nassgewachstes Papier	papier paraffiné en surface	papel encerado húmedo
6212	wet web	våt bana	feuchte Bahn, nasse Bahn	feuille continue humide	hoja continua húmeda
6213	wet weight	våtvikt	Nassgewicht	poids humide	peso húmedo
6214	wet wood	vått trä	nasses Holz, Nassholz	bois humide	madera húmeda
6215	wet wrinkle	våtrand	Nassfalte	pli humide	pliegue falso húmedo
6216	wheat starch	vetestärkelse	Weizenstärke	amidon	almidón de trigo
6217	wheat straw	vetehalm	Weizenstroh	paille de blé	paja de trigo
6218	white birch	vitbjörk	weisse Birke, Weissbirke	bouleau blanc	abedul
6219	white fir	vittall	Weisstanne	sapin concolore	abeto blanco
6220	white liquor mud	bakvattenslam	Weisslaugenschlamm	boue de lessive neuve	lodo de lejía fresca
6221	white liquor	bakvatten	Kochlauge	lessive neuve	lejía fresca
6222	whiteness	vithet	Weisse, Weissgehalt, Weissgrad	blancheur	blancura
6223	white paper	vitt papper	weisses Papier, grafische Papiere	papier blanc	papel blanco
6224	white spruce	vitgran	Weissfichte	épicéa du Canada	epícea del Canadá
6225	white water	bakvatten	Siebwasser, Siebabwasser	eau collée, eau blanche	agua colada
6226	whiting	vitgörning	Schlämmkreide	blanchir	blanco de yeso
6227	wicking	tamponering	Korbflechten	pénétration capillaire, ascension capillaire	empaquetadura de algodón
6228	width	vidd, bredd	Breite	largeur	ancho
6229	wild	vild	wild, wolkig	irrégulier, pauvre	desigual
6230	wild formation	vild arkformning	wolkige Blattbildung	épair irrégulier	transparencia desigual
6231	wild look	vilt utseende	schlechte Durchsicht, starke Wolkigkeit	épair nuageux	transparencia desigual
6232	wildness	vildhet	Wolkigkeit (Pap.)	inculte	páramo
6233	willow	pil	Weide	saule, blutoir	sauce
6234	wind	rulla	wickeln, umwickeln, Windung, Wicklung	enrouler, bobiner	bobinar
6235	winder	rullmaskin	Rollenschneider	bobineuse	bobinadora
6236	winder welts	rullningsfåll	Schwielen (durch den Rollenscheider verursacht)	plis au bobinage, cordons au bobinage	rebordes de bobinadora
6237	winder wrinkles	rullningsveck	Falten (durch den Rollenschneider verursacht)	plis au bobinage, cordons au bobinage	pliegues falsos de bobinadora
6238	winding	upprullning	Aufhaspeln, Wickeln, Aufrollen	enroulage	bobinado
6239	winding drum	bärvals	Tambour	tambour de bobinage	tambor de enrollar
6240	winding shaft	rullspindel	Welle	broche d'enroulage	barra de bobinadora
6241	winding tightness	rullhårdhet	Aufrollfestigkeit	tension d'enroulage	apretado en el bobinado
6242	wire	vira	Sieb, Draht	fil (électrique ou métallique), toile métallique	tela metálica
6243	wire cloth	viraväv	Siebtuch	toile métallique	tela metálica
6244	wire end	våtända	Siebende, Siebpartie	table de fabrication	mesa de fabricación
6245	wire guides	virariktare	Sieblaufregler	guide-toile	guías-tela
6246	wire hole	virahål	Siebloch, Siebmaschenweite	trou de toile	orificio en la tela
6247	wire life	viralivslängd	Sieblaufzeit	durée d'une toile	duración de la tela
6248	wire lines	viraränder	Siebmarkierung, Sieblinien, Siebleitlinie	lignes de toile	cables de acero
6249	wire loading	virabelastning	Siebbelastung	mise en place d'une toile	cargamento de la tela
6250	wire mark	viramarkering	Siebmarkierung	marques de la toile	marcas de la tela
6251	wire mold	rundvira	Papierform, Drahtform	moisissure de la toile	forma de tela
6252	wire pit	viragrav, viragrop	Siebschiff, Siebwasserkasten	fosse sous toile	fosa bajo la tela
6253	wire roll	viravals	Siebleitwalze, Siebspannwalze	rouleau guide-toile	rodillo de tela

	ENGLISH	SVENSKA	DEUTSCH	FRANÇAIS	ESPAÑOL
6254	wire section	viraparti	Siebpartie	section toile	sección de telas
6255	wire side	virasida	Siebseite	côté toile, envers	cara inferior
6256	wire spot	virafläck	Siebflecken	tache de toile	mancha de tela
6257	wire stretcher	virasträckare	Siebspannwalze	tendeur de toile	tensor de tela
6258	wire turning roll	viravändvals	Siebwendewalze, Siebumkehrwalze	rouleau de retour de toile, rouleau de commande de toile	rodillo de retorno de tela
6259	wire-wound rod coater	viravalsbestrykning	spiralförmig mit Draht umwickelter Rotationsstabstreicher	coucheuse à barre filetée	estucadora de rodillos armados
6260	wood	trä	Holz, Wald, Forst, Waldgebiet	bois	madera
6261	wood chips	träflis	Hackschnitzel, Chips	copeaux de bois	virutas de madera
6262	wood containing paper	trähaltigt papper	holzhaltiges Papier	papier avec bois	papel con pasta mecánica
6263	wooden core	trähylsa	Hülse aus Holz, hölzerne Hülse	mandrin en bois	mandril de madera
6264	wooden plug	träplugg	Hülsenkern, Refinerkegel	bouchon en bois	taco
6265	wood fiber	träfiber	Holzfaser	fibre de bois	fibra de madera
6266	wood free	träfri	holzfrei (h'f)	sans bois	sin pasta mecánica
6267	wood-free paper	träfritt papper	holzfreies Papier	papier sans bois	papel sin pasta mecánica
6268	wood grain	ådrighet	Holzmaserung	grain du bois	grano de madera
6269	wood grinder	träslipare	Holzschleifer	défibreur	desfibradora
6270	wood handling	vedhantering	Bearbeitung von Holz, Holztransport, Arbeiten mit Holz	manutention des bois	manipulación de la madera
6271	wood measurement	tumning	Holzabmessung	arpentage, cubage	medición de la madera
6272	wood particle	vedpartikel	Holzpartikel, Holzteilchen	particules de bois	partícula de madera
6273	wood pile	vedtrave	Holzhaufen, Holzstapel	pile de bois	pila de leña
6274	wood preparation	vedpreparation	Holzaufbereitung, Holzzubereitung, Holzvorbereitung	préparation du bois	preparación de la madera
6275	wood products	vedprodukter, träprodukter	Holzprodukte	produits forestiers	productos madereros
6276	wood pulp	trämassa	Holzschliff	pâte de bois	pasta de madera
6277	wood-pulp board	trämassekartong	Holzschliffkarton, Holzschliffpappe	carton bois	cartón de pasta de madera
6278	wood resin	vedharts	Holzharz	résine	resina de la madera
6279	wood room	vedhanteringsrum	Holzreinigungsanlage	salle de préparation du bois	taller de maderas
6280	wood structure	vedstruktur	Holzstruktur	structure du bois	estructura de madera
6281	wood waste	träavfall, vedavfall	Holzabfall	déchets de bois	desperdicios de la madera
6282	wood wool	träull	Holzwolle	laine de bois	virutas finas de madera
6283	wood yard	vedgård	Holzlagerplatz	parc à bois	maderería
6284	woof	väft, inslag	Einschuss (Textilien), Einschlag	trame de toile métallique	trama
6285	wool	ull, ylle	Wolle	laine	lana
6286	woolen	yllene, ull-	aus Wolle	laineux	lanoso
6287	woolen paper	papper av ylle	Kalanderwalzenpapier, Wollpapier	papier laineux pour calandres	papel lanoso
6288	wool felt	yllefilt	Wollfilz	feutre de laine	fieltro de lana
6289	worm drive	skruvdrift	Schneckengetriebe	commande par vis sans fin	transmisión por tornillo sin fin
6290	wound roll	lindad vals	gewickelte Rolle	bobine terminée	bobina terminada
6291	wove	velängpapper	Velinpapier	vélin	vitela
6292	wove dandy	velängegutör	Velin-Egoutteur	rouleau vergeur	desgotador de vitela
6293	wove mold	velängform	Velinpapier-Schöpfform	forme vélin	matriz vitela
6294	woven fabric	vävt tyg	Webstoff, Wirkstoff	tissu tissé	tejido
6295	wove paper	velängpapper	Velinpapier, Papier mit wolkiger Durchsicht	papier vélin	papel vitela
6296	wrapper	emballage	Papierumschlag, Schutzhülle	enveloppe, couverture, emballeur	envoltura
6297	wrapping	emballage	Verpackung, Packpapier, Einwickelpapier	emballage	embalaje
6298	wrapping machine	emballagmaskin	Verpackungsmaschine	machine à emballer	máquina embaladora
6299	wrapping paper	omslagspapper	Verpackungspapier, Packpapier, Einwickelpapier	papier d'emballage	papel para envolver
6300	wringer roll	filtpressvals	Filzwaschwalze	rouleau essoreur	batán
6301	wrinkle	rynka, veck	Falte	faux-pli	falso pliegue
6302	writing paper	skrivpapper	Schreibpapier	papier écriture	papel de escritura

ENGLISH	SVENSKA	DEUTSCH	FRANÇAIS	ESPAÑOL

X

	ENGLISH	SVENSKA	DEUTSCH	FRANÇAIS	ESPAÑOL
6303	xanthate groups	xantatgrupper	Xanthatgruppen	groupes xanthate	grupos de xantato
6304	xylem	xylem	Xylan, Holzgummi	xylem	tejido leñoso
6305	xylenes	xylener	Xylole	xylènes	xilenos
6306	xylenesulfonates	xylensulfonater	Xylolsulfonate	sulfonate de xylène	sulfonatos de xileno
6307	xylose	xylos	Xylose	xylose	xilosa
6308	x-ray analysis	röntgenanalys	röntgenografischer Nachweis, Röntgenstrahlenanalyse	analyse aux rayons X	pruebas por rayos X
6309	x rays	röntgenstrålar	Röntgenstrahlen	rayons X	rayos X
6310	x-spectrometer	rötgenspektrometer	Röntgenstrahlenmesser	spectromètre à rayons X	espectrógrafo de rayos X

Y

	ENGLISH	SVENSKA	DEUTSCH	FRANÇAIS	ESPAÑOL
6311	yankee dryer	yankeetork	Selbstabnahmetrockenzylinder	gros sécheur frictionneur	secador friccionador
6312	yankee machine	yankeemaskin	Selbstabnahmemaschine	machine à papier avec gros cylindre frictionneur	máquina friccionadora
6313	yardage	storlek i yard	Länge (in yards)	cubage	yardaje
6314	yardman	vedgårdsarbetare	Hofarbeiter	manoeuvre	obrero de patio
6315	yarn	garn	Garn, Faden	fil textile	hilo textil
6316	yeast	jäst	Hefe	levure de lessive	levadura
6317	yellowing	gulning	Vergilben, gelb werden	jaunissement (du papier)	amarilleamiento
6318	yellow poplar	gulpoppel	Tulpenbaum	peuplier jaune	álamo amarillo
6319	yield	utbyte	Ausbeute, Ertrag	rendement, rentabilité	rendimiento
6320	yield point	sträckgräns, flytgräns	Streckgrenze, Dehnungsgrenze, Fliessgrenze	seuil de plasticité	punto de deformación
6321	yield tables	utbytestabeller	Ertrags-Aufstellung, Ertrags-Liste, Ausbeuteeintragung	tableaux de rendement	índice de rendimiento

Z

	ENGLISH	SVENSKA	DEUTSCH	FRANÇAIS	ESPAÑOL
6322	Z-direction strength	Z-riktningsstyrka	Zugkraft pro Flächeneinheit um eine Papier- oder Kartonprobe zu zerreissen (Kraft wirkt senkrecht auf Probe)	résistance dans le sens Z	resistencia de dirección Z
6323	zeolites	zeoliter	Zeolite	zéolites	zeolitas
6324	zero span tensile strength	nollinspänningsstyrka	Nullpunktsreissfestigkeit	résistance à la traction à mâchoires jointes	resistencia a la tensión de espacio nulo
6325	zeta potential	zetapotential	Zetapotential	potentiel électrocinétique	potencial zeta
6326	zinc	zink	Zink	zinc	cinc
6327	zinc compounds	zinkföreningar	Zinkverbindungen	composés du zinc	compuestos de cinc
6328	zinc hydrosulfite	zinhydrosulfit	Zinkhydrosulfit	hydrosulfite de zinc	hidrosulfito de cinc
6329	zinc oxide	zinkoxid	Zinkoxid, Zinkasche	oxyde de zinc	óxido de cinc
6330	zinc stearate	zinkstearat	Zinkstearat	stéarate de zinc	estearato de cinc
6331	zinc sulfate	zinksulfat	Zinksulfat	sulfate de zinc	sulfato de cinc
6332	zinc sulfide	zinksulfid	Zinksulfid	sulfure de zinc	sulfuro de cinc
6333	zirconium compounds	zirkonföreningar	Zirkonverbindungen	composés du zirconium	compuestos de circonio

english

svenska

deutsch

français

español

SVENSKA

A

abaca, 103
abaca-massa, 104
Abies, 105
abietinsyra, 106
absolut fuktighet, 113
absolut torr, 498, 824
absolut torr vikt, 825
absorbator, 118
absorbera, 114
absorberande, 116, 125
absorberande papper, 117
absorptionsförmåga, 115, 126, 127
absorbtionskoefficient, 121
absorptionskudde, 119
absorbtionsmätning, 122
absorbtionsprovare, 123
absorbtionstorn, 124
absorption, 120
absorbtionskoefficient, 1426
accelerera, 5316
accelereringsmedel, 131
accepterad flis, 135
accepterad mäld, 136
accepterat material, 137
accepterbarhet, 132
accessibilitet, 138
Acer, 144
acetaldehyd, 145
acetaler, 146
acetater, 150
acetatfiber, 147
acetatmassa, 148
acetatsilke, 149
aceton, 153
acetonitril, 154
acetylering, 155
acetylgrupper, 156
ackumulator, 142
ackumulering, 141
adderingsmaskinpapper, 200
additionspolymer, 201
adhesion, 204
adiabatiska förhållanden, 209
administratör, 3672
adresslappapper, 203
adsorbat, 215
adsorbent, 216
adsorbera, 214
adsorption, 217
adsorptionsfömåga, 218
adsorptivitet, 219
aerob process, 224
affichkartong, 4449
affichpapper, 686, 4450
affinitet, 225
affärsformulär, 981
agalit, 227
agar, 228
agatpapper, 229
agglomerat, 233
agglomeration, 234
aggregat, 235
airfoil, 267
akasia, 128
akasiagum, 129
akrylamid, 187
akrylater, 188
akrylfiber, 191
akrylföreningar, 190
akrylnitril, 192
akrylsyra, 189
aktiverat slam, 194
aktiverat slam-process, 195
aktivt alkali, 196
aktivt klor, 197
aktivt kol, 193, 1228
aktivt svavel, 198
aktuell vikt, 199

akustikplatta, 183
akustiska egenskaper, 185
al, 293
albuminer, 286
albumpapper, 287
aldehyder, 292
aldehydgrupper, 291
alfacellulosa, 322
alfamassa, 325
alfaprotein, 324
alfastrålningsmätare, 323
alger, 294
algicider, 295
alginater, 296
alifatiska föreningar, 298
alkali, 299
alkalibehandling, 307
alkalicellulosa, 300
alkaliextraktion, 301
alkalifast, 314
alkalifast papper, 315
alkalifast tvålförpackningkartong, 316
alkalihärdig cellulosa, 304
alkaliligniner, 302
alkalilöslig, 305
alkaliresistans, 303
alkalisk limning, 312
alkalisk massa, 310
alkalisk process, 309
alkaliskt fyllmedel, 308
alkalisk uppslutning, 311
alkalitet, 313, 602
alkoholer, 290
alkoholgrupper, 288
alkoholligniner, 289
alkylarylsulfonater, 317
alkylering, 318
alkylgrupper, 319
alm, 2246
Alnus, 321
aluminater, 328
aluminium, 329, 330
aluminiumfolie, 332
aluminiumföreningar, 331
aluminiumpapper, 333
aluminiumstearat, 334
aluminiumsulfat, 335
alun, 327
alunfläckar, 336
amalys, 357
amfoter, 356
amider, 337
aminer, 339
aminoetylcellulosa, 341
aminogrupper, 342
aminopropylcellulosa, 343
aminosyror, 340
aminpolymerer, 338
ammoniak, 345
ammoniumbisulfat, 347
ammoniumföreningar, 348
ammoniumhydroxid, 349
ammoniumlut, 346
ammoniumstearat, 350
ammoniumsulfat, 351
ammunitionspapper, 352
amorf, 353
amorf cellulosa, 354
amorft material, 355
ampermeter, 344
amylos, 358
anaerobisk process, 359
analog, 360
analogdator, 361
analog-digitalomvandlare, 363
analogsystem, 362
analys, 364
analytiskt filterpapper, 366
anatas, 367
anblick, 417
anbringare, 419
anbringning på ytan, 5570

andra inloppslåda, 4972
andramassa, 4973
andrapress, 4976
anemometer, 369
angiospermer, 370
anhydrider, 371
anhydrit, 372
anilin, 373
anilinfärgämnen, 374
anilintryckning, 375
anisotropi, 2865
anisotropt papper, 2864
anjonbytare, 379
anjonbyte, 378
anjoner, 382
anjonisk, 380
anjoniska föreningar, 381
ankarbult, 368
anod, 386
anpressvals, 4505
ansiktsservett, 2374
anslagstavla, 3961
antaglighet, 132
antiflockningsmedel, 1873
antifärger, 390
antiglid, 4003
antikbokpapper, 398
antikbristol, 399
antikfinish, 401
antikfinpapper, 403
antikglättat papper, 402
antikpapper, 404
antioxideringsmedel, 397
antisyra, 387
apelsinskal, 4081
apelsinskalseffekt, 4082
A-pipa, 102
apparent viskositet, 416
applikator, 419
arabinos, 428
arbetskraft, 3385, 3387, 5938
arbetssätt, 3847
ark, 3419, 3440, 4610, 5042
ark av mager mäld, 2723
arkbord, 3433
arkform, 5046, 5051
arkformat, 5054
arkformning, 2685, 5047, 5049
arkformningsprovare, 2686
arkformningsutrustning, 5050
arkledare, 5048
arkmatad, 5044
arkmatning, 5045
arkskärare, 5043
arkspridare, 6158
armerad betong, 4721
armerad plast, 4723
armerat byggpapper, 4720
armerat papper, 4722
armeringspapper, 430
aromatiska föreningar, 432
aromatiska grupper, 433
aromatiska syror, 431
arrowroot-stärkelse, 434
arsenikpapper, 435
artikulationspapper, 437
asbest, 446
asbestartad, 445
asbestcementpapper, 448
asbestfiber, 450
asbestfilt, 449
asbestkartong, 447
asbestpapper, 451
asbesttakpapp, 452
asbesttätningspapper, 453
asexuell reproduktion, 455
asfalt, 462
asfaltemulsion, 463
asfaltfilt, 464
asfaltisoleringspapper, 468
asfaltkartong, 469, 470
asfaltkraftpapper, 3368

asfaltlaminerat papper, 465
asfaltpapper, 466, 471
asfalttakpapp, 467
ask, 460
aska, 456
askfri, 458
askfritt filterpapper, 459
askhalt, 457
askorbinsyra, 454
asp, 461
assimilering, 473
asymmetri, 475
ateranvändning, 4774
atlasfinish, 4907
atmosfär, 476
atmosfäriskt tryck, 477
atombindning, 478
attapulgit, 480
att göra genomsynlig, 5820
autoklav, 482
automation, 487
automatisk kontroll, 484
automatisk ledvals, 485
automatiskt inställbar läpp, 489
automatisk viraledvals, 486
autotypipapper, 491
avbarkningsmaskin, 579
avdriva, 5472
avdrivning, 5475
avdunsta, 2331
avdunstning, 2332
avenbok, 3088
avfall, 6056
avfallshantering, 6057
avfallstömning, 5828
avfärgning, 1857
avgas, 2341
avgasare, 1877
avgasning, 1878
avgiftning, 1929
avhartsning, 1912
avjonisering, 1892
avkörd, 4871
avlastare, 5923
avlastning, 5924
avlopp, 2193, 5021
avloppsbehandling, 2194
avloppsbrunn, 5022
avloppsförlust, 5023
avloppsvatten, 6062
avlufta, 1834
avluftare, 1836
avluftning, 1835
avlut, 5317, 6058
avrulla, 5928
avrullning, 5930
avrullningsmaskin, 5929
avrullningsstativ, 5931
avsalumassa, 3697
avskalningsdon, 4242
avskalningsprov, 4244
avskrivare, 1916
avslagningssprits, 3358
avslitningshållfasthet, 891
avslitningslängd, 890
avslitningsmotstånd, 892
avsmalnande, 5646, 5647
avstängning, 5083
avstängningsventil, 5084
avsvärta, 1889
avsvärtad massa, 1890
avsvärtning, 1891
avsätta, 1915
avsättning, 1914
avsättning under vakuum, 5942
avsökning, 4925
avtagning, 3464, 4314
avtagningsfullt, 4315
avtagningspress, 4317
avtagningsvals, 4318
avtunning, 2401
avvattna, 1931, 2085

Svenska

avvattning, 1849, 1886, 1932, 2086, 6100
avvaxning, 1933
avvägning, 3456
axel, 495, 5026
axeltapp, 3319
axialflödeslump, 494
axiell, 493
azpföreningar, 496

B

babbitmetall, 502
badrumspapper, 611
baffel, 521
baffelbräda, 522
baffelplatta, 523
bagasse, 527
bagassfiber, 530
bagassmassa, 532
bagasspapper, 531
bagassskärmaskin, 528
bak-, 503
bakgrund, 505
bakgrundspapp, 506
bakgrundspapper, 507, 510
baksida, 515
baktericider, 520
bakterie, 518
bakterieräkning, 519
bakvatten, 6221, 6225
bakvattenlåda, 4915
bakvattenslam, 6220
bal, 545
balad, 546
balans, 543
bald cypress, 544
ballongtimring, 555
balningsmaskin, 548
balpapper, 550
balpress, 551
balsamgran, 557
balsampoppel, 558
baltillverkning, 549
baltråd, 552
balupplösare, 547
bambu, 559
bambumassa, 561
bambupapper, 560
bana, 6146
banbrott, 6147
bandmassa, 564
bandning, 562
bandningsmaskin, 563
bandtransportör, 1563
banföring, 2935
bankpappersmassa, 567
banmatning, 6150
banpåförare, 6160
banspridare, 6158
banspänning, 6159
banstyrningsanordningar, 6152
banvändare, 6162
banöverföring, 6161
barium, 571
bariumkarbonat, 572
bariumsulfat, 573
bark, 574, 1621
barka, 1837
barkad ved, 4241
barkare, 4242
barkfläckar, 581
barkning, 577, 1839
barkningsmaskin, 576, 1838
barkpress, 580
barktorkare, 575
barktrumma, 578
barometerben, 582
barometertryck, 583
barriär, 585
barrträ, 1510
barrträd, 1509, 5254
barrträdsmassa, 5255
barrved, 5254
barrvedsmassa, 5255

baryt, 590
baryter, 593
barytkartong, 591
barytpapper, 592
bas, 595
basaltgarnityr, 594
basmassa, 598
basmäld, 794
baspapper, 596
bastfiber, 606
bastpapper, 607
bastträ, 605
batistfinish, 1067
batteripanel, 612
batteripapper, 613
Bauer-McNett fraktioneringsapparat, 614
bearbetbarhet, 3605
bearbetning, 4534
beck, 4340
befuktning, 3105, 3859
befuktningsanordning, 3858
befuktningsmedel, 3106
behandling i autoklav, 483
behandling satsvis, 609
behållarbotten, 1525
behållare, 689, 1523
beklädnad, 1364
belastning, 3538, 3541, 3543
belastningscell, 3539
Ben Day-sil, 657
bensin, 669, 2780
bensoater, 670
bensoesyra, 671
bensylcellulosa, 672
bensylgrupper, 673
bentonit, 668
beredning, 4479
beryllium, 674
bestruken, 1384, 5572
bestruken bristolkartong, 1391
bestruken kartong, 1387, 1390
bestruken pergamyn, 1392
bestruket boktryckpapper, 1389
bestruket etikettpapper, 1397
bestruket finpapper, 1388
bestruket journalpapper, 1393
bestruket offsetpapper, 1394
bestruket tryckpapper, 1396
bestrukna ark, 1386
bestrukna papper, 1395
bestrykare, 1398
bestrykbarhet, 1383
bestrykning, 1382, 1399
bestrykning med polervals, 4401
bestrykningsband, 1401
bestrykningsblandning, 1411
bestrykningsfel, 1407
bestrykningsfläck, 5337
bestrykningsklump, 1409
bestrykningskärl, 1412
bestrykningslera, 1404, 1408
bestrykningsmaskin, 1410
bestrykningsmiss, 1415
bestrykningsnappning, 1413, 4305
bestrykningsrand, 5456
bestrykningsrecept, 1406
bestrykningsslipp, 1416
bestrykningssmet, 1405
bestrykningstillsats, 1400
bestrykningsvikt, 1417
beständig färg, 2388
besättning, 1693
betacellulosa, 675
betastrålar, 678
betastrålningsmätare, 676, 677
beteende, 4260
betningsmedel, 3895
Betula, 679
bevillning, 423
B-flute, 501
bibanor, 1238
bibelpapper, 682
bifenyl, 706
bigningshållfasthet, 2659
bigningskvalitet, 2656
bigningsmaskin, 2655
bigningsmotstånd, 2657
bigningsprovare, 2660
bikakebas, 3082

bikakestruktur, 3083
bikarbonater, 684
bikromattal, 1425
bildbarhet, 2679
bildning, 2684
biljettkartong, 5740
bilklädsel, 1125, 1126
bilklädselkartong, 488
billast, 1127
bilplats, .1128
bindemedel, 691, 820
bindemedelsfläckar, 5306
bindemedelsvandring, 207
bindning, 694, 817, 819
bindning mellan fibrer, 3268
bindningsremsa, 696
bindningsstyrka, 695, 821, 823
bindningsyta, 818
binge, 689
binärt system, 690
biocider, 698
biodegraderbar, 699
biokemisk syreförbrukning (BS), 697
biologiska prov, 704
biologisk avfallsbehandling, 702
biologisk behandling, 705
biologisk kontroll, 701
biologisk syreförbrukning, 703
biosönderfall, 700
biprodukt, 1000
bisulfater, 709
bisulfiter, 713
bisulfitlut, 710
bisulfitmassa, 711
bisulfituppslutning, 712
biträdande förman, 474
bitumen, 714
bitumenemulsion, 716
bitumenpapp, 715
björk, 707
björkmassa, 708
blad, 723, 3440, 5954
bladbestrykare, 724, 3353
bladbestrykning, 725, 3354
bladfilter, 3441
bladsnitt, 1765
blanc fix, 726
blandare, 757, 3841
blandlut, 3838
blandning, 213, 756, 3837, 3842, 3846
blandningskar, 3844
blandningslåda, 3843
blandningspump, 2387, 3845
blandpapper, 3839
blandskog, 3840
blankpolering, 965, 967
bleka, 733
blekare, 739
blekbar, 735
blekbarhet, 734
blekbar massa, 736
blekeri, 740, 747
blekeriavlopp, 742
blekmedel, 895
blekning, 733, 741
blekningsbehov, 748
blekningsefterfrågan, 737
blekningsgrad, 1881
blekningsresistenta papper, 749
blekningsskala, 750
blekpulver, 744
blekt massa, 738
blektorn, 745
blekvätska, 743, 746
blindskriftpapper, 875
blisterbestrykning, 759
blisterförpackning, 762
blisterskiktseparation, 763
blisterutskott, 760
blixtrande, 2546
blockningshindrande medel, 389
blodtätt papper, 767
blush coating, 787
blushing, 788
bly, 3437
blyoxid, 3526
blå, 783
blåglasmetod, 784

blåglasprov, 784
blåkopiepapper, 785, 786
blåsa, 758, 771
blåsa bort, 777
blåsare, 772
blåsbildning, 761
blåsledning, 774
blåsledningsrening, 775
blåsnappning, 4303
blåsning, 773
blåsningslut, 776
blåsningsprover, 781
blåsningstank, 780
blåsningsvärde, 782
blåsvals, 779
blått ribbat skrivpapper, 497
bläck, 3232
bläckning, 2812
blöda, 751
blödning, 752
blödningsmotstånd, 753, 768
blötläggning, 5205, 5492
bocköga, 933
bok, 649, 826
bokbaspapper, 827, 829
bokbindaravfall, 693
bokbindarkartong, 692
bokbindarpapper, 830
bokfoderpapper, 828
bokföringsmaskinpapper, 834
bokföringspapper, 3447
bokkartong, 831
bokmassa, 650
bokomslag, 833, 840
bokpapper, 837, 839
bokpapperspån, 838
bokpärmpapper, 832
boktryckfärg, 3449
boktryckning, 3451
boktryckpapper, 3450
bolagsformulär, 3817
bom, 568, 841
bomberad vals, 1065, 1713
bombering, 1712, 1715
bomberingsinlägg, 1714
bomull, 1624
bomullscellulosa, 1625
bomullshaltigt papper, 1626
bomullslinter, 1627
bomullsmassa, 1628
bor, 848
borater, 842
borax, 843
borföreningar, 849
borhydrider, 847
borstbestrykare, 916
borstbestrykning, 917
borste, 915
borstemaljpapper, 919
borstfinish, 920
borstfinishbestrykning, 921
borstfuktare, 918
borstglättning, 922
borstmarkering, 925
borstning, 923
borstpolering, 926
borstpoleringsmaskin, 927
borstvals, 928
borsyra, 845
bortkörd, 4871
borttagningsmedel, 4730
brand, 2521
branddetektor, 2523
brandhindrande, 2541
brandhämmande, 2530
brandmotstånd, 2529
brandskydd, 2524, 2528
brandsäker, 2539
brandsäkert papper, 2540
brandtegel, 2522
bredblad, 903
bredd, 6228
brick-kartong, 1381
brinellhårdhet, 898
Brinellprovare, 899
brist, 5069
bristning, 1667, 2383
bristol, 901
bristolkartong, 900, 2650

brokadpapper, 904
brom, 910
bromföreningar, 911
broms, 5315
bromsar, 876
bromsning, 877
brons, 912
bronsering, 913
brott, 2383, 2711
brottförlängning, 2248
brottmodul, 3856, 4873
bruk, 3896
brunn, 6176
brunsten, 3674
bruten kant, 907
brutet ris, 908
bruttovikt, 4650
brytare, 5605
bränd, 968
brännare, 964
brännbarhet, 2542
brännolja, 2744
bränsle, 2743
brödomslag, 542, 880, 881
brödpåspapper, 879
bröstläder, 424, 2691
bröstvals, 894
bröstvalslåda, 893
BS (biokemisk syreförbrukning), 499
BTU, 500
bubbelbestrykning, 930
bubbla, 929
bubbling, 931
buckligt papper, 1756
buffer, 937
buffert, 936
bulk, 945
bulkbokpapper, 951
bulkhantering, 947
bulkig, 957
bulkindex, 949
bulkkartong, 950
bulklagring, 956
bulkmoduler, 955
bulkning, 3039
bulkpapper, 952
bulktjocklek, 954
bulktryck, 953
buller, 3994
bullerkontroll, 3995
bullermätare, 3996
bundet vatten, 857
bunt, 959
buntning, 960
burk, 5748
butadien, 982
butaner, 983
butylacetat, 994
butylgrupper, 995
butylgummi, 996
butyrater, 997
butyrylgrupper, 999
byggnad, 5478
byggnadspapper, 940
byggnadsteknik, 5477
bykning, 4939
byrett, 962
bägare, 617, 1747
bägarformare, 1740
bägarkartong, 1739
bägarpapper, 1741
bärvals, 1133, 6239
bäverpapp, 645
böja, 1870
böjare, 658
böjdon, 3036
böjd vals, 858
böjhållfasthet, 663, 665, 2566
böjkvalitet, 664
böjlighet, 4380
böjmotstånd, 2565
böjning, 659, 2563
böjningsmaskin, 662
böjprov, 666
böjutmattningsprovare, 661

C

Candian Standard freeness, 1069
Carpinus, 1131
carragen-pektin, 1132
cash flow, 1146
Catalpa, 1155
ceder, 1173
Cedrus, 1174
cell, 1177
cellofan, 1178
cellpapp, 1180
cellstruktur, 1179, 1181
cellulosa, 1182
cellulosaacetat, 1183
cellulosabutyrat, 1184
cellulosacellvägg, 1199
cellulosaderivat, 1188
cellulosaestrar, 1189
cellulosaetrar, 1190
cellulosafiber, 1191
cellulosafilm, 1192
cellulosahalt, 1185
cellulosakartong, 4569
cellulosalösningsmedel, 1195
cellulosanedbrytning, 1187
cellulosanitrat, 1193
cellulosaplast, 1194
cellulosapolymerer, 1186
cellulosastruktur, 1196
cellulosavadd, 1197
cellulosaxantat, 1198
cellväggtjocklek, 1200
celsiusskala, 1201
cementsäckpapper, 1202
centerskikt, 1203
centerskiktmassa, 1204
centervalsrullning, 1205
centipoise, 1206
centrifug, 1214
centrifugal, 1207
centrifugalblåsare, 1208
centrifugalfiltrering, 1210
centrifugalfraktionerare, 1209
centrifugalkompressor, 1212
centrifugalsil, 1211
centrifugering, 1213
centripetal, 1215
C-flute, 1004
Chamaecyparis, 1224
championbestrykare, 1225
checklista, 1235
checkpapper, 1236, 1264
chefskemist, 1269
chemigroundwood, 1262
cheviot, 1268
china clay, 1275
chipboard, 1278
chocktorkning, 5064
chuck, 1325, 1435
cigarrettpapper, 1326, 1327
cirkelkniv, 2021
cirkulationsholländare, 1784
cirkulär skärarkniv, 1330
cirkulärt diagram, 1329
clippersöm, 1353
Cobb-prov, 1418
cockle, 1420
cockle-finish, 1423
cockle-papper, 1422
cockle-utskott, 1421
cockling, 1424
concertina fold, 1489
Concoraprov, 1490
conifer, 1509
containerkartong, 1524
containerliner, 1527
containerprovning, 1528
cortex, 1621
cottonwood, 1629
coulometri, 1639
crackershell board, 1661
crescentformare, 1691
critical-path metoder, 1699

cuam-rayon, 1734
cunit, 1738
Cupressus, 1744
curling, 1754
curl-prov, 1755
cyanamider, 1777
cyanater, 1778
cyanider, 1779
cyanoetylcellulosa, 1781
cyanoetylering, 1780
cyanuronföreningar, 1782
cyclodextriner, 1787
cykel, 1783
cyklahexan, 1788
cykliska föreningar, 1785
cyklisk belastning, 1786
cyklon, 1789
cyklonseparator, 1790
cylinder, 1791, 2110
cylinderflishugg, 2112
cylinderpress, 1799
cylinderrullning, 2116
cylindertork, 2113
cylindertorkad, 1794
cylindertvätt, 2115
cymener, 1801
cypress, 1802

D

daggpunkt, 1930
dagsljus, 1830
damm, 1804, 2164, 2598, 2762,
   3393, 4433
dammbildning, 3394
dammfilter, 2167
dammformig massa, 2599
dammlös slipning, 4345
dammsugning, 5503
dammuppsamlare, 2165
damning, 2168
damningsbarhet, 2763
damningsbenägenhet, 2764
damningskontroll, 2166
damningstendens, 2600
data, 1821
databearbetning, 1823
datainsamlare, 1822
datalagring, 1827
dataregistrerare, 1824
dataregistrering, 1825
datatabeller, 1828
dataåtervinning, 1826
dataöverföring, 1829
dator, 1483
datorpapper, 1485
datorprogram, 1484
debreringsprocess, 1868
decalcomaniapapper, 1840
deceleration, 1844
dedwood, 1833
defekt, 1861
defekt ark, 3192
defekt rulle, 4435
defibrator, 1867
defibrera, 1862, 1864, 2463
defibrerad massa, 1865
defibrering, 1863, 1866
deflektera, 1870
deflektion, 1871
deflektionshindrande, 392
deflektor, 1872
deflockulering, 1874
deformation, 1876
deformerad rulle, 1721
degradering, 1879
dehydrator, 1887
dehydrera, 1885
dehydrering, 1886, 1888
dekorationspapper, 1859
dekulator, 1860
delad krage, 5330
delad press, 2046
delaminera, 1893
delaminering, 1894

delare, 2047, 5331
delning, 5332
delningskartong, 4216
dendrologi, 1899
denier, 1900
Dennisons vax-prov, 1901
densitet, 1905, 5300
densitetsmätare, 1907
densitetsmätning, 1906
densitometer, 1903, 1908
densitometri, 1904
Denvercell, 1909
deodorering, 1910
deoxidering, 1911
depolarisation, 1913
desensibilisering, 1920
desickant, 1921
desickator, 1922
desorption, 1925
destillat, 2038
destillation, 2039
destillerat vatten, 2040
detektering, 1927
dextriner, 1935
dextrintillsats, 1934
dextros, 1936
diafragma, 1944
diagnos, 1937
diagram, 1231, 1938, 4676
diagrampapper, 1232
dialdehydcellulosa, 1940
dialdehydstärkelse, 1941
dialys, 1942
diameter, 1943
diatomacejordart, 1945
diatomacekisel, 1946
diatomer, 1947
dicyanidamid, 1953
dielektricitetskonstant, 1960
dielektrikumpapper, 1962
dielektrisk, 1959
dielektriska egenskaper, 1964
dielektrisk förlust, 1961
dielektrisk hållfasthet, 1965
dielektrisk uppvärmning, 1963
dietyleter, 1966
differentialdrift, 1967
differentialväxel, 1968
diffraktion, 1969
diffusion, 1974
diffusör, 1971
diffusörtvätt, 1972
diffusörtvättning, 1973
digitaldator, 1983
digitalreglering, 1984
digitalsystem, 1985
diisocyanater, 1986
dikloretan, 1950
diklormetan, 1951
dikotyledoner, 1952
dilatans, 1987
dilatant, 1988
dimbildning, 2637, 3834
dimension, 1992
dimensionsstabilisering, 1994
dimensionsstabilitet, 1993
dimer, 1995
dimetyldisulfid, 1996
dimetylformamid, 1997
dimetylsulfat, 1998
dimetylsulfid, 1999
dimetylsulfon, 2000
dimetylsulfoxid, 2001
dimhindranade medel, 396
dimma, 3833
diod, 2002
dipol, 2005
direkta färgämnen, 2007
direktupphettning, 2008
direkt ångtillförsel, 2010
direktör, 2806
dis, 2992
disintegrator, 2025
diska, 6055
dispergering, 2028
dispergeringsmedel, 2027
dispersion, 2029
dissolvering, 2035
dissolveringmassa, 2036

dissolveringtank, 2037
distorsion, 2041
distribution, 2042
disulfider, 2044
diversifiering, 2045
djup, 1917
djuptryckspapper, 2881
djurlim, 376
djurlom, 377
dolomit, 2060
doppbestrykare, 2003
doppbestrykning, 2004
doppvals, 2701
dosering, 2062
Douglasgran, 2083
drag, 2096
drag-, 5675
draghållfasthet, 5678
draghållfasthetsegenskaper, 5677
draghållfasthetsprovare, 5679
dragprov, 5683
dragprovare, 5682
dragspelsbälg, 140
dragspänning, 5680
dragvals, 4567
draperbarhet, 2094
drev, 4566
drift, 2103, 2104, 4069
driftkostnad, 4067
drift med variabel hastighet, 5969
driftsida, 4068
drivande, 2106
driven vals, 2105
drivsida, 2107
droppfilter, 5836
droppkondensor, 5837
drottningvals, 4608
dränage, 2086, 2088
dränagefaktor, 2087
dränagelåda, 2091
dränagemotstånd, 2089, 5179
dränera, 2085
dränering, 2093
dräneringsmassa, 2092
dräneringstid, 2090
dualpress, 2147
dubbelbestruken, 2065
dubbelbestrykare, 2066
dubbelbindning, 2063
dubbel delad press, 2070
dubbeldäckel, 2068
dubbelglättad, 2064
dubbelmönstrare, 2076
dubbelmönstrat papper, 2075
dubbelsidig bestrykning, 1002
dubbelskivraffinör, 2069
dubbeltäckt, 2081
dubbeltäckt gråpapp, 2078
dubbeltäckt slipmassekartong, 2079
dubbelväggig låda, 2082
dumpning, 2151
duplex, 2153
duplexbestrykare, 2156
duplexbristolkartong, 2155
duplexfilt, 2157
duplexfinish, 2158
duplexkartong, 2154
duplexpapper, 2159
duplex rullmaskin, 2160
dupliceringspapper, 2161
durometer, 2163
dusch, 5292
dy, 3906
dynamiska prover, 2174
dynamitpatronpapper, 732
däckel, 1850
däckelbord, 1851
däckelbyte, 1226
däckelfläckat, 1855
däckelkant, 1852
däckelkantad kartong, 1853
däckelkantat papper, 1854
däckelpapper, 5899
däckelrem, 1856
däckskord, 5754
däcksomslag, 490
dämpning, 1763, 1810
dämpningssträckning, 1812
dämpvals, 1811

dämpveck, 1813
dödande filt, 1831
dödning, 2430

E

ecotyp, 2181
effekt, 4464
effektfaktor, 4465
effektivitet, 2191
effektivt alkali, 2189
efterbehandling, 4022
efterbestrykning, 4021
efterbestrykningsmaskin, 4020
eftergulning, 897, 1453
efterraffinering, 4451
eftertork, 226
egendom, 4552
egenskap, 4552
egenskapsprov, 4261
egotör, 1816, 1820
egotörbestrykare, 1817
egotörmarkering, 1818
egotörnappning, 1819
egoutör med sluten lagring, 1360
ej blandbar, 3186
ejektor, 2198
ekologi, 2178
ekonomiser, 2180
ekonomisk analys, 2179
ekträd, 4015
ekvation, 2298
ekvivalent vikt, 2301
elasticitet, 2200
elasticitetsmodul, 3855
elastisk hållfasthet, 2201
elastisk kalandervals, 2199
elastisk återgång, 5353
elastomer, 2202
eld, 2521
eldare, 5439
eldfast, 4713
eldfast kräpp, 2525
eldfast papper, 2527
eldsäkring, 2526
elektriska egenskaper, 2208
elektrisk anslutning, 2211
elektrisk apparat, 2226
elektrisk drift, 2216
elektrisk impedans, 2218
elektrisk isolering, 2219
elektrisk konduktivitet, 2204
elektrisk kraft, 2221
elektrisk kraftdistribution, 2222
elektrisk kraftledning, 2223
elektrisk krets, 2209
elektrisk ledare, 2210
elektrisk motor, 2220
elektrisk omvandlare, 2213
elektrisk reglering, 2212
elektrisk resistivitet, 4762
elektrisk ström, 2214
elektrisk säkring, 2217
elektriskt motstånd, 2225
elektriskt relä, 2224
elektrisk urladdning, 2215
elektrod, 2227
elektrofores, 2237
elektrofotografiskt papper, 2238
elektrolys, 2228
elektrolyt, 2229
elektrolytisk cell, 2230
elektromagnetiskt fält, 2231
elektron, 2236
elektronik, 2235
elektronisk reglering, 2232
elektroniskt instrument, 2234
elektronisk utrustning, 2233
elektropapper, 2207
elektrostatik, 2244
elektrostatisk bestrykning, 2240
elektrostatisk kopiering, 2241
elektrostatisk laddning, 2239
elektrostatiskt fällningsmedel, 2243
elektrostatisk utfällning, 2242

elektroteknik, 2205
elementarfibrill, 2245
elisoleringspapper, 2206
eluera, 2249
eluering, 2250
elutriat, 2251
elutriering, 2252
emalj, 2266
emaljerad, 2267
emaljerat papper, 2268
emballage, 4130, 6296, 6297
emballagekonstruktion, 4131
emballagemaskin, 4134, 6298
emission, 2260
emissionsspektroskopi, 2261
emulgator, 2263
emulgera, 2264
emulgering, 2262
emulsion, 2265
en, 3322
encylindermaskin, 3600, 5102
endosperm, 2274
endotermisk, 2275
endotermisk reaktion, 2276
energi, 2277
energibalans, 2278
energiförbrukning, 2279
energiomvandling, 2280
Engelmanngran, 2281
engelsk finish, 2285
engångshylsa, 4002
engångskarbonpapper, 4055
enhet, 5920
enhetlig hantering, 5921
enkelarkskutter, 5113
enkel avtagning, 4349
enkelbestruken, 5100
enkelbestruket papper, 5101
enkelfiber, 5106
enkeltråd, 3890
enkel väv, 4351
enlag, 5111
enlagskartong, 5112
ensidig bestrykning, 1001
ensidig finish, 4053
ensidigt bestruket papper, 4054
ensidigt täckt, 5108
ensidigt täckt gråpapp, 5114
ensidigt täckt kartong, 5109
ensidigt täckt vågpapp, 5103
entalpi, 2286
entomologi, 2287
entropi, 2289
enzymer, 2294
epibromohydrin, 2295
epiklorohydrin, 2296
E-pipa, 2175
epoxihartser, 2297
erosion, 2304
ersättningslut, 3665
espartmassa, 2307
esparto, 2305
espartopapper, 2306
estergrupper, 2308
estrar, 2310
etage, 1830
etan, 2311
etanol, 2312
etanolaminer, 2313
etergrupper, 2314
etikett, 3381, 5631
etiketteringsmaskin, 3383
etikettkartong, 5632
etikettpapper, 3382, 3384
etrar, 2316
etylacetat, 2317
etylakrylat, 2318
etylamin, 2319
etylcellulosa, 2321
etylendiamin, 2323
etylengrupper, 2322
etylenimin, 2324
etylering, 2320
etylgrupper, 2325
etylmerkaptan, 2326
eukalyptus, 2327
eutrofering, 2328
evakuera, 2329

excelsiorpapper, 2336
exoterm reaktion, 2343
expanderlåda, 2344
expanderpapper, 2346
expanderspindel, 2345
expandervals, 2347
expansion, 2348
experimentpappersmaskin, 2349
exploderade fibrer, 2351
exportmassa, 2352
extraktion, 2357
extraktionskammare, 2358
extraktionsmedel, 2361
extraktionsplatta, 2359
extraktiv, 2360
extruder, 2362
extrudering, 2363

F

fabrik, 3816, 4353
fabriksavlopp, 3825
fabrikschef, 3827
fabriksomslag, 3829
fabriksräkning, 3821
fabriks-skarv, 3688
fabriksskärning, 3822
fack, 4393
Fagus, 2382
fakturapapper, 687, 688
falla, 4469
falsbar gråpapp, 660
falsk, 795
falskartong, 2647, 2648
falskartongmassa, 2658
faner, 5992
fanerpapper, 5993
fanerutskott, 5994
fast, 5260
fast avfall, 5266
fast fas, 5263
fast kostnad, 2536
fat, 584, 1068
fattigmansavtagning, 4436
fattigmans ventanip, 4437
fel, 2557
feldetektor, 2558
felfärgad, 4019
fenol, 4277
fenol-formaldehydharts, 4278
fenolftalein, 4283
fenolgrupper, 4279
fenolhartser, 4281
fenolligniner, 4282
fenolsyror, 4280
fenylgrupper, 4284
fernissa, 5972
fernissbarhet, 5973
fernissning, 5974
fet, 2895
fett, 2883
fettfläckar, 2893
fettmotstånd, 2890
fettsyror, 2395
fettsäktert papper, 2892
fettäkta, 2884
fettäkta kartong, 2891
fettäkthet, 2888
fettät kartong, 2885
fettätning, 2887
fiber, 2445, 2473, 2862
fiberanalys, 2446
fiberbindning, 2448
fibercementprodukter, 2451
fiberdiameter, 2457
fiberdimensioner, 2458
fiberfibriller, 2479
fiberfiller, 2480
fiberflätning, 2460
fiberfragment, 2713
fiberfraktion, 2461
fiberfraktionering, 2452
fiberfraktioneringsapparat, 2453
fiberförkortning, 5070
fiberhalt, 2455

fiberhaltigt bakvatten, 4794
fiberknippe, 2449
fiberknutar, 2464
fiberlängd, 2465
fiberlängdsfördelning, 2466
fibermatta, 2467
fibermättnadspunkt, 2470
fibernappning, 4306
fiberriktning, 2863
fibersammansättning, 2454
fiberskiva, 939, 2447, 2973
fibersnittning, 2456
fiberstruktur, 2471
fibersträng, 2472
fiberåtervinnare, 4914
fiberåtervinning, 2468, 2469
fiberåtervinningsskepp, 4916
fibriller, 2474, 2478
fibrillera, 2475
fibrillering, 924, 2476, 2477
ficka, 4393
film, 2498
filmbildare, 2499
filt, 2418, 2430, 2431
filter, 2500
filterbädd, 2503
filterkaka, 2504
filtermassa, 2506
filtermedel, 2502
filterpapper, 2507
filtfinish, 2423
filtföring, 2934
filthår, 2424
filtkonditionerare, 2419
filtkonditionering, 2420
filtmarkering, 728, 2426, 2427
filtning, 2425
filtpapper, 729, 2428
filtpressvals, 6300
filtrat, 2508
filtrerbarhet, 2501
filtrerfiber, 5600
filtrering, 2505, 2509
filtriktningsmarkering, 2421
filtsida, 2432
filtsträckare, 2433
filtsträckning, 2434
filtsträckvals, 2435
filttorkare, 2422
filttvätt, 2436
filtvirapress, 2372
finfördelning, 479
finhet, 2510
finish, 2516
finish i våtändan, 6186
finishprocess, 2517
finishsal, 2519
finishutskott, 2518
finmaterial, 2512
finmaterialborttagning, 2515
finpapper, 566, 822, 2511
finsil, 2513
finsilning, 2514
firmaformulör, 3817
fiskögon, 2534
fjädervals, 5354
fjäderviktpapper, 2402
fjärrkontroll, 4740
flagning, 2537
flamtorkning, 2538
flercellig sugvals, 3910
flerkomponentbestrykning, 3911
flerlags, 3922
flerlagsark, 3919
flerlagssäck, 3929
flerlagssäckpapper, 3930
flerlagsvågpapp, 3931
flerschaberenhet, 3916
flerskikt, 3913, 3922
flerskiktsbestrykning, 3917
flerskiktskartong, 3914, 3915
flerskiktspapper, 3923
flerstegsblekning, 3924
flerstegsprocess, 3925
flerstegspump, 3926
flervirakartongmaskin, 3927
flerviramaskin, 3909, 3928, 3932
flexibilitet, 2562
flexografi, 2564

flinga, 2874, 3993
flingtorkad massa, 2544
flingtorkning, 2545
flintpapper, 2567
flis, 1276
flisbinge, 1277
flisdimensioner, 1285
flisgrop, 778
flishugg, 1291
flishuggkniv, 1292
flishuggning, 1294
flishög, 1293
flisklassificering, 1281
fliskross, 1284
flislastare, 1279
flismatare, 1286
flismätare, 1288
flispackning, 1290
flisschakt, 1280
flissilo, 1296
flisslip, 1287
flissåll, 1282, 1295
flistransportör, 1283
floating knife-bestrykare, 2569
flock, 2575, 2579
flocka, 2571
flockig arkbildning, 1369
flockning, 2573, 2577
flockningsmedel, 2570, 2572, 2574
flockpapper, 2578
flod, 4809
flodfart, 4810
floem, 4285
flor, 4007, 4008
flotering, 2583
floteringsavsvärtning, 2585
floteringsmedel, 2584
floteringsåtervinnare, 2586
flotte, 4619
flottning, 4620
fluid, 2601
fluidik, 2606
fluidiserad bädd, 2609
fluidiserare, 2610
fluidisering, 2608
fluiditet, 2607
fluor, 2617
fluorescens, 2612
fluorescerande blekmedel, 2613
fluorescerande färger, 2614
fluorescerande papper, 2615
fluorescerande vitt, 2616
fluorförening, 2618
fluorvätesyra, 3135
flygande klistrare, 2624
flygande skarvar, 2625
flygaska, 2623
flygpostpapper, 3207
flyktighet, 6032
flytförmåga, 961
flytgräns, 6320
flytpunkt, 4461
fläck, 755, 5304, 5305
fläcka, 5375
fläckbildning i glätt, 1027
fläckigt papper, 5376
fläckning, 5377
fläkt, 2384
fläktblad, 2385
fläktning, 2386
fläns, 2543
flödat nyp-bestrykare, 2581
flöde, 2588, 2595
flöde av torrt material, 5268
flödesdiagram, 2591
flödesmätare, 2594
flödesmätning, 2593
flödesreglering, 2592
flödesspridare, 2596
flödning, 2582
flödningsmedel, 2589
F.O.B., 2369
fodrad, 3498
fodring, 3509
foil, 2638
foilbaffel, 2640
foil-body, 2641
foilvinkel, 2639
foktigcylinder, 5598

folie, 2642
fontän, 2699
fordankvarn, 3318
form, 2965, 2967, 3871
formaldehyd, 2680
formalin, 2681
formamid, 2682
formare, 2687
format, 2683
formation, 2684
formbarhet, 1507, 3872
formbräda, 2690
formel, 2693
formeringsvals, 2692
formgjuten förpackning, 3876
formgjuten massa, 3875
formgjutna massaprodukter, 3877
formgjutning, 3879
formning, 2689, 3879
formulering, 2694
forskning, 4747
fortplantning, 4549
fosfater, 4286
fosfor, 4287
fosforföreningar, 4288
fossilt bränsle, 2697
fotoelektrisk cell, 4289
fotoelektrisk kolorimeter, 4290
fotogen, 3342
fotografiskt papper, 4291
fotokänslighet, 4294
fotometer, 4292
fotooffset, 4293
fotostatpapper, 4295
fotosyntes, 4296
fragmentering, 2712
fraktion, 2707
fraktionering, 2708, 2709
fraktioneringsapparat, 2710
fraktur, 2711
framkant, 2736, 3438
frammatningskontroll, 2408
framsida, 2737
fransig, 2866
fransig kant, 2867
Fraxinus, 2716
freeness, 2718
freenesskrivare, 2719
freenessprovare, 2720
free on board, 2721
frekvens, 2725
frekvensomvandlare, 2726
friktion, 2728
friktionsbarkare, 2729
friktionsbehandling, 408
friktionsdrift, 2732
friktionsfaktor, 2733
friktionsförlust, 2735
friktionsglättad, 2734
friktionsgättad kartong, 2730
friktionskalander, 2731
friktionsrullstol, 4438
frimärkspapper, 5381
friskvatten, 2727
fruktomslagspapper, 2742
frusen ved, 2741
främmande material, 1539
främre vals, 3439
fräsa, 1955
frö, 4981
fröodling, 4982
frösådd, 4983
ftalsyra, 4297
ftalsyraanhydrid, 4298
fukt, 3857, 3860
fuktare, 1807
fukthalt, 3861
fuktig, 1806
fuktighet, 3107
fuktighetskontroll, 3108
fuktighetsmätning, 3109
fuktinställning, 3868
fuktlåda, 1057
fuktmotstånd, 3867
fuktmätare, 3863
fuktning, 1808
fuktningsmaskin, 1809
fuktprov, 3869
fuktreglering, 3862

fuktrum, 3110
fuktsäker, 3865
fukttät, 3865
fukttäthet, 3866
fuktupptagning, 3864
fullers jord, 2746
fumarsyra, 2749
fungicider, 2753
funktionella grupper, 2751
fura, 2520
furaner, 2756
furfuran, 2758
furfurol, 2757
furfurylalkohol, 2759
fusion, 3754
fyllande, 2497
fylld, 2485
fylld bristolkartong, 2487
fylld filt, 2488
fylld kartong, 2486, 2491
fylld slipmassekartong, 2489
fyllmedel, 2493
fyllmedelsretention, 2496
fyllmedeltillsats vid malning, 632
fyllnadsmedel, 3542, 3544
fyllning, 2484
fyllt papper, 2490
fyrskaftad väv, 5885
fysikaliska egenskaper, 4299
fåll, 6179
fållande rems, 754
fälla, 4469, 5827
fällande, 4470
fällarmede, 2416
fällning, 2414, 2417, 4471
fällningsmedel, 4472
fältdrift, 4811
färg, 1439, 1456, 2170, 3232
färgabsorption, 3233
färgad, 1440, 2171
färgade papper, 1441
färgbad, 1449
färgfläckar, 1455
färggenomskinlighet, 3241
färggenomslag, 5471
färghållning, 3235
färginslag, 5470
färgjusterad, 5751
färgjustering, 1451, 1452
färgklump, 1450
färgmaterial, 1447
färgmotstånd, 3239
färgmottaglighet, 3238
färgning, 1446, 2172
färgning i pappersmaskin, 5859
färgningssystem, 3236
färgpigment, 1448
färgpåföring, 3237
färgskikttjocklek, 3234
färgspecifikation, 1454
färgäkthet, 1442, 2389
färgämne, 2170, 2173
färgöverföring, 3243
färskmassa, 6014
färskträ, 2898
följd, 5005
förare, 2932, 4071, 5672
förarsida, 2737, 5674
förbehandling, 4511, 4517
förbestrykning, 4474
förbränning, 1466
förbränningsgas, 2597
förbränningskammare, 1467
förbränningsprodukter, 1468
förbränningsvärme, 3012
fördel, 667
fördelare, 4554
fördelningsvals, 2043
fördelning, 2042
fördröjning, 3070
fördröjningsmedel, 4767
förestring, 2309
företring, 2315
förfalskningssäkert papper, 394
förfogande, 2031
förformning, 4477
förgasning, 2787
förglasning, 6027
förhinder, 3227

förhydrolys, 4478
förkolning, 1100
förlust, 3574
förlängare, 2353
förlängning, 2247
förman, 2665
förorena, 1529
förorening, 1530, 3200, 4404
föroreningskontroll, 4405
förpackning, 4132
förpackningsindustri, 4133
förpackningskartong, 4138
förpackningsmaskin, 4134
förpackningsmaterial, 4135
förpackningspapper, 4136, 4139, 4140
förråd, 5566
försegling, 4959
försilning, 4516
första huvudpress, 2532
första press, 2533
förstapress, 4515
första torkcylindern, 2531
förstärkt harts, 2696
förstärkt lim, 2695
försvagning, 481
försäljning, 4885
försämras, 1841
försättspapper, 835
försök, 2350
försöksanläggning, 4323
förtjockare, 5719
förtjockning, 5720
förtjockningsmedel, 5721
förtjänst, 4546
förtork, 4475
förtryckt botten, 1720
förtunnare, 1989
förtvålningstal, 4903
förtvälning, 4902
förångning, 5959
förädlat papper, 1558
förädlingsfabrik, 1559
förädlingsindustri, 1556
förädlingsprocess, 1560

G

g/m2, 2767
gaffeltruck, 2678
galaktaner, 2770
galaktomannan, 2771
galaktos, 2772
galler, 2878
gallon, 2773
gallring, 5725
galvanisera, 2775
galvanisering, 2776
galvanisk korrosion, 2774
galvanometer, 2777
gammacellulosa, 2778
garn, 2482, 6315
garnityr, 5629
gas, 2780
gas-, 2784
gasanalys, 2781
gasformig, 2784
gasgenerator, 2785
gaskromatografi, 2782
gaspermeabilitet, 2791
gasskrubber, 2792
gasskrubbning, 2793
gasturbinmotor, 2794
gasurladdning, 2783
gasvärmare, 2786
gavelspetskartong, 2768
gejder, 2932
gel, 2801
gelatin, 2802
gelatinerad, 2803
gelatinliknande skikt, 2804
gelbildning, 2805
G.E.-ljushet, 2765
generator, 2808
genetik, 2809
genomlysning, 5078

genomrivhållfasthet, 3277
genomrivmotstånd, 3279
genomsikt, 3570
genomskinlig bestrykning, 5812
genomskinlighet, 5810, 5811
genomsynlig, 5818
genomtorkning, 5738
giftighet, 5795
giljotin, 2937
giljotinkniv, 2939
giljotinskärare, 2938
gipskartong, 4358
gipsplatta, 2952
girlangpapper, 3388
givare, 5002, 5803
gjutbestruket papper, 1149
gjutbestrykare, 1150
gjutbestrykning, 1151
gjutform, 3905
gjutjärn, 1153
gjutning, 1152
glans, 2195, 2813, 2826
glansmedel, 2827
glansmätare, 2830
glanspapper, 2831
glans vid strykande infall, 3575
glasfiber, 2815
glasfiberpapper, 2814
glida, 5164
glidning, 5131, 5132, 5134, 5164, 5165
glidningsfri bestrykning, 407
glidsko, 5133
glimmer, 3793
glimmerpapper, 3794
glukomannan, 2832
glukos, 2833
glukosbindning, 2839
glukosderivat, 2834
glukosenhet, 2837
glukosestrar, 2835
glukosetrar, 2836
glukosider, 2838
glycerin, 2850
glycerol, 2851
glycerolstearater, 2852
glykoler, 2853
glyoxal, 2854
glätt, 1025
glätta, 2818
glättad, 1035, 2819
glättad kartong, 2820, 5196
glättat papper, 1036, 2822
glättbestruket bokpapper, 2821
glättcylinder, 2824, 3599
glättfinish, 1032
glättflagor, 1046
glättfläckning, 1054
glättfärgad, 1034
glätthet, 5201
glättlimning, 1050
glättlåda, 1028
glättning, 2823, 5197, 5201
glättningsfläckar, 1051
glättningsfärgad, 1030
glättningskrossad, 1031
glättningsprovare, 5202
glättningssvärtad, 1026
glättningsutskott, 1029
glättpapper, 1049
glättparti, 1048
glättpress, 2828, 5198
glättpressning, 2829
glättrems, 1033
glättsmulor, 1053
glättstapel, 1052
glättvals, 5199
glättvalsbelägg, 1047
glättvalsbestrykning, 5200
glättvalspapper, 2495
glättvalsspärrning, 765
glättveck, 1055
gnidmotstånd i oljemiljö, 4044
gnidning, 4869
godkännandeprov, 134
g.p.d., 2766
grad, 2857
grad av isotropi, 5362
gradering, 2859

gradient, 2858
grafisk konst, 2875
gram, 2869
gran, 2520, 5359
grand fir, 2870
granit, 2871
granitpapper, 2872
granitvals, 2873
granul, 2874
gratulationskort, 2899
gravitationstransportör, 2879
gravyrbestrykning, 2880
grenverk, 2668
grepp, 2876
grind, 2795
gripare, 2876
gripprov, 2856
grodd, 5358
groning, 2810
grop, 4339
grov, 1377
grov finish, 963
grov sil, 1380
grovsil, 958
grovt papper, 1379
grumlighet, 5874
grums, 2100, 2925
grund, 2698
grundformat, 603
grundfärgämnen, 601
grundplatta, 597, 5259
gruskorn, 2911
grynig, 2866
gråpapp med ett kraftmassatäckskikt, 5107
gränsbrytare, 3495
gräshoppa, 3546
grön, 2896
grönlut, 2897
guargum, 2931
gulning, 6317
gulpoppel, 6318
gum, 2940
gummerad, 4221
gummerade formulär, 4222
gummerad gråpapp, 4224
gummerad kartong, 4223
gummering, 4226
gummeringsmaskin, 4227
gummi, 4864
gummi arabicum, 427, 2941
gummidragkant, 2947, 5799
gummifläckar, 4868
gummiklädd vals, 4865
gummilatex, 4866
gummivals, 4867
Gurleyprovare, 2949
gusk, 1630
guskfilt, 1631, 1633
guskgrop, 1635, 3630
gusk-knekt, 1767
guskknekt, 1638
guskmarkering, 1634
guskning, 1632
guskpress, 1636
guskridvals, 3593
gusksprits, 5849
guskvals, 851, 1637
gymnosperm, 2951
gödning, 2438
gödningsmedel, 2439

H

hackved, 3064
halider, 2955
halm, 5452
halmkartong, 5453
halmmassa, 5454
halogener, 2957
halogenföreningar, 2956
halvblekt, 4989
halvblekt massa, 4990
halvkemisk kartong, 4991
halvkemisk kokning, 4994

halvkemisk massa, 4993
halvkemisk vågpapp, 4992
halvledare, 4995
halvledarutrustning, 5265
halvmatt antikpapper, 400
halvtyg, 2954, 4627
halvtygsholländare, 1918
hammare, 2958
hammarkvarn, 2959, 3187
hammarkvarnuppslagning, 2960
hampa, 3031
hamppapper, 3032
handduk, 5791
handdukspapper, 5792, 5793
handgjord filt, 2962
handgjort papper, 2964
handgjort pappersark, 2966
handpapp, 3818, 3819
handpappersmäld, 2971
handtag, 2961
hantera, 2961
Harper-maskin, 2989
harts, 4340, 4841
hartser, 4753
hartsfläckar, 4343, 4845, 4846
hartshaltig ved, 4757
hartskanaler, 4756
hartskontroll, 4341
hartslim, 4842
hartslimmad, 4843
hartslimning, 4844
hartssyror, 4754
hartstall, 4342
hasning, 4956
hasningsmotstånd, 4957
hastighet, 5987
hastighetsindikator, 5314
hastighetsminskare, 5315
hastighetsmätning, 5989
hastighetsreglering, 5988
helkemisk massa, 2745
hellim, 2747
helträdfångst, 2748
hemicellulosa, 3029
hemlock, 3030
hetalkalieextraktion, 3092
heterocyklisk förening, 3034
heterogenitet, 3035
high-testliner, 3055
hiss, 3466
hissa, 3066
hjälpdrift, 3028
holländare, 3074, 3075
holländraförare, 633
holocellulosa, 3076
holocellulosatillverkning, 3077
homogenisering, 3080
homogeniseringsmedel, 3081
homogenitet, 3079
homogenpapp, 5264
horisontell porositet, 3087
huggjärn, 4934
huggkniv, 3065
humektant, 3104
hundhår, 2059
hus, 3102
huv, 3084
huvtäckning, 3085
huvud, 2993
huvudaxel, 3508
huvudledning, 2995, 2996
huvudpress, 3661
hybrid, 3111
hydrafiner, 3112
hydratcellulosa, 3113
hydrater, 3116
hydratmassa, 3115
hydraulik, 3119
hydraulisk, 3118
hydraulisk barkare, 3120, 3125
hydrauliskt system, 3122
hydrauliskt tryck, 3121
hydrazin, 3123
hydrerad, 3114
hydrering, 3117
hydrider, 3124
hydrobromsyra, 3126
hydrocellulosa, 3128
hydrodynamik, 2602, 3131

hydrofil, 3150
hydrofob, 3151
hydrogenerad harts, 3138
hydrogenering, 3139
hydrolys, 3146
hydrolysat, 3148
hydrolyslignin, 3147
hydromekanik, 2604
hydrometer, 3149
hydrostatisk, 3152
hydrosulfider, 3153
hydrotropiskt lignin, 3155
hydrotropisk uppslutning, 3156
hydrotropmedel, 3154
hydroxider, 3157
hydroxyalkylcellulosa, 3158
hydroxyalkylgrupper, 3159
hydroxyetylcellulosa, 3161
hydroxyetylering, 3160
hydroxyetylgrupper, 3162
hydroxyl, 3163
hydroxylbinding, 3164
hydroxylgrupper, 3165
hydroxyljonkoncentration, 3166
hydroxypropylcellulosa, 3167
hydroxypropylmetylcellulosa, 3168
hygroexpansivitet, 3169
hygroskopicitet, 3170
hylsa, 1585, 5149, 5865
hylskartong, 1586
hylslager, 5150
hylsmaskin, 1592
hylsmassa, 1590
hylspapper, 1587
hylsplugg, 1588
hylspress, 5151
hylsspindel, 1589
hylsutskott, 1591
hypobromiter, 3171
hypokloriter, 3172
hysteres, 3174
hål, 3072, 5176
hålkort, 4589
hållare, 2910, 3067
hållfasthet, 5459
hålresma, 4590
hålrum, 6028
hålrumskvot, 6030
hålstorlek, 3073
hålvals, 525, 3078, 4255
hålvolym, 6029, 6031
hårdboard, 5261
hårdhet, 2977, 2979
hårdhetsprov, 2982
hårdhetsprovare, 2980, 2981
hård knut, 2978
hårdlimmat, 2984
hårdlimmat papper, 2985
hårdmald massa, 2986
hård massa, 2983
hård rulle, 5741
hårfint snitt, 1766
hårt kok, 2974
häftgas, 4953
häftning, 5385
hänga ned, 4884
hängare, 2968
hängning, 2969
hängning i slingor, 2444
hängtork, 2442, 3571
hängtorkad, 2441
hängtorkad papper, 2970
hängtorkning, 2443, 4399
härdare, 2975
härdig massa, 2390
härdning, 2976
hästkastanj, 3089
hästkraft, 3090
hävarm, 3460
hävarmsbelastning, 3461
högbehållare, 3062
högbulkigt bokpapper, 3038
högbulkigt papper, 3040
hög densitet, 3042
hög energi, 3045
högglans, 3046
högglanspapper, 3047
hög hastighet, 3056
höghastighetstork, 2953, 3058

höghastighetstorkning, 3057
högkabellunning, 3049
hög konsistens, 3041
högkonsistensblekning, 3043
högkonsistensmalning, 3044
hög maskinfinish, 3050
högmodulsrayon, 3051
hög temperatur, 3054
högtryck, 3052
högtrycksmatare, 3053
högtryckspress, 3048
högt utbyte, 3059
högutbyteskokning, 3061
högutbytesmassa, 3060
höja, 3066
höjd, 3026
höstved, 3429, 5548

I

icke böjbar, 3998
icke-förstörande prov, 4000
icke klibbande, 4004
icke-kurlande papper, 3999
icke normenlig, 4023
icke normenlig flis, 4005
icke renskuret papper, 5927
identifiera, 3175
i drift, 4066
imider, 3180
iminer, 3181
imitationsbristol, 798
imitationsduplex, 801
imitationsetikett, 810
imitationsfodring, 803
imitationsförpackningspapper, 805, 811
imitationskartong, 797
imitationsklisterpapper, 807
imitationskopiepapper, 809
imitationskraftpapper, 802
imitationsläskpapper, 808
imitationsmanilla, 804
imitationspapper, 806
imitationsritpapper, 800
imitationstidningspapper, 3962
imitationsvell, 799
imiterad fodring, 796
imiterad läderpapp, 3184
imiterat handgjort papper, 3182
imiterat läder, 3183
imiterat pergament, 3185
impedans, 3190
impregnerad, 3196
impregnerade papper, 3197
impregnerfilt, 4910
impregnering, 3199
impregneringsbaspapper, 4911
impregneringsmedel, 3195
impregneringspapper, 3198
inbyggd spänning, 942
inbyggd töjning, 941
inbäddning, 2253
indigoliknande färger, 3210
indikator, 3208
indikatorpapper, 3209
indirekt uppvärmning, 3211
indirekt ångbehandling, 3212
indragning, 2288
induktionsuppvärmning, 3214
indunstare, 2333
indunstning, 2332
industriell teknologi, 3215
industripapper, 3216
industritrasor, 3218
industritunnpapper, 3217
inflöde, 421
inflödessystem, 422
infraröd, 3220
infraröddetektor, 3221
infrarödstrålning, 3225
infrarödtorkning, 3222
infrarödvärmare, 3223, 3224
ingenjör, 2282
ingrediens, 3226
inhibering, 3227

inhibitor, 3228
inhibitor för statisk elektricitet, 5395
injektion, 3230
injektionsformning, 3231
inkapsling, 2269
inkholdout, 3235
inklädd kokare, 1333
inklädning, 1334
inklädningsmetall, 1335
inkrust, 4922
inkrustborttagning, 1919
inkrustering, 2270
in-line freenessprovare, 3246
in-line rullmaskin, 3245
inlopp, 3244
inloppslåda, 2590, 2994, 5481, 5484
inloppslådeläpp, 3516
inloppsläpp, 5155
innerdiameter, 3252
inpassbara behållare, 3953
inre bindningsstyrka, 3275
inre diameter, 3252
inriktning, 297
inrivhållfasthet, 2188
inrivningshållfasthet, 3229
inslag, 6164, 6284
inslagstråd, 5085, 6165, 6166
inspektion, 3253
inspektör, 2807, 5559
installation, 3254
instrument, 3255
instrumentering, 3256
intagningssystem, 3263
integrerad fabrik, 3264
integrering, 3265
intensitet, 3266
interfoliering, 3269
interfolieringspapper, 3270
intermittent drift, 3274
inuti räfflad yankeetorkare, 3280
inventering, 3282
inventeringskontroll, 3283
isobutylalkohol, 3299
isocyanater, 3300
isolationspapp, 3257
isolering, 3261
isoleringshölje, 3259
isoleringsmaterial, 3258
isoleringspapper, 3260, 3262
isomerer, 3301
isopropylalkohol, 3302
isotermer, 3303
isotoper, 3304
isotropi, 3305
isotropt ark, 4641

J

jackpine, 3307
jod, 3284
jodföreningar, 3285
jonbytare, 3289
jonbyte, 3286
jonbyteshartser, 3288
jonbytespapper, 3287
joner, 3293
jonisering, 3291
jonomerer, 3292
jonstyrka, 3290
jord, 5256
jordan, 3315
jordankubb, 3317
jordanmalning, 3316
journalpapper, 3641
journalpappersmassa, 3642
Juglans, 3320
jumbovals, 3321
juniper, 3322
Juniperus, 3323
justeringsfärg, 211
justermaskin, 5848
justerpress, 5845
justerskruv, 212
jute, 3325
jutefiber, 3328

jutekartong, 3327
juteliner, 3329
jutelinerkartong, 3330
jutemassa, 3332
jutepapper, 3331
jutesäckpapper, 3326
jutetestliner, 3333
jämnåriga plantor, 2334
jämvikt, 2299
jämviktspunkt, 888
järn, 3294
järnek, 3179
järnföreningar, 3295
järnoxider, 3296
järnved, 3297
järnväg, 4635
järnvägsspår, 4637
järnvägsvagn, 4636
jäst, 6316

K

kabel, 1005
kabelloggning, 1006
kabelpapper, 1007
kadmium, 1008
kadmiumföreningar, 1009
Kadykvarn, 3335
kakelplatta, 5742
kalander, 5550
kalanderfinish, 5551, 5554, 5556
kalanderförare, 1039, 1042
kalanderglättat, 5555
kalanderlimmat papper, 1049
kalanderlimning, 1050
kalandermarkerad, 1040
kalandermarkering, 1041
kalandernappning, 1043, 4304
kalandervals, 1044
kalanderveck, 1670
kalandrerat papper, 5552
kalandrering, 1037, 1038, 5553
kalcinering, 1010
kalcium, 1011
kalciumbisulfit, 1014
kalciumföreningar, 1017
kalciumhydroxid, 1018
kalciumhypoklorit, 1019
kalciumkarbonat, 1015
kalciumklorid, 1016
kalciumlignosulfonat, 1020
kalciumoxid, 1021
kalciumsulfat, 1022
kalciumsulfit, 1013, 1023
kalhuggning, 5140
kalibrera, 1058
kalibrering, 1059
kalium, 4452
kaliumföreningar, 4453
kaliumjodid, 4454
kaliumpermanganat, 4455
kalk, 3487
kalkaktigt utseende, 1223
kalkbränningsugn, 3488
kalkerpapper, 2095, 5798, 5805
kalkmjölk, 3489, 3815
kalkning, 1222
kalkslam, 3490, 3492, 3493
kalksläckning, 3491
kalksten, 3494
kall kaustisk soda, 1429
kallslipning, 1431
kallsodamassa, 1433
kallsodaprocess, 1432
kallsodauppslutning, 1434
kalorimeter, 1062
kalorimetri, 1063
kam, 1064
kambium, 1066
kanister, 1071
kanisterpåse, 1704
kanna, 4393
kannslip, 3640, 4394
kanon, 2948
kant, 2183, 4801

kantad por, 844
kantband, 3870
kantdeformation, 2185
kantilever, 1073
kantileverkonstruktion, 1074
kantileverviraparti, 1075
kantkrossprov, 2184
kantrulle, 5086, 5087
kantskydd, 2187
kantskärare, 2186
kantsprits, 874
kaolin, 3336
kapacitans, 1076, 1080
kapillär, 1082
kapilläritet, 1081
kapital, 1083
kappatal, 3337
kaprolaktan, 1086
kapsel, 1087
kapsylkartong, 850
kapsylpapper, 1084, 1085
kar, 1266
karakteristisk, 1227
karamellpapper, 5889, 5890
karamellpapper av kraftmassa, 3379
karbamater, 1089
karbamider, 1090
karbider, 1091
karbonater, 1094
karbonering, 1095
karbonfläckar, 1104
karbonisering, 1097, 1100
karbonpapper, 1103
karbonråpapper, 1102
karbontetraklorid, 1105
karbonyler, 1107
karbonylgrupper, 1106
karbonylsulfid, 1108
karborundumpapper, 1109
karboxyalkylcellulosa, 1111
karboxy-alkylering, 1110
karboxyalkylgrupper, 1112
karboxyetylcellulosa, 1114
karboxyetylering, 1113
karboxyetylgrupper, 1115
karboxylering, 1116
karboxylgrupper, 1117
karboxylsyror, 1118
karboxymetylcellulosa, 1120
karboxymetylering, 1119
karboxymetylgrupper, 1121
karduspapper, 1139
karnubavax, 1129
kartong, 789, 1123, 4152
kartongbruk, 793
kartonggegenskaper, 4157
kartongfinish, 1124
kartongfoder, 790, 791
kartongindustri, 4155
kartongkvalitet, 4154
kartongmaskin, 792
kartongprodukt, 4156
kartongprov, 4158
kartongrulle, 4701
kartpapper, 3690
kasein, 1142
kassaregisterpapper, 1147
kassera, 1735
kassun, 1427
kastanjeträd, 1267
katalogpapper, 1154, 2011
katalysator, 1156
katjon, 1159
katjonbytare, 1161
katjonbyte, 1160
katjonföreningar, 1163
katjonisk, 1162
katjonstärkelse, 1164
katod, 1157
katodstrålerör, 1158
kausticering, 1170
kausticitera, 1168
kausticiterare, 1169
kausticitet, 1167
kaustik, 1166
kaustisk soda, 1171
kavitering, 1172
kedjedisintegrator, 1220
kedjetransportör, 1221

kelater, 1239
kelatering, 1241
kelatmedel, 1240
kelit, 1176
Kelvingrad, 3338
kemifilterpapper, 1248
kemikalie/vedkvot, 1259
kemikalieförbrukning, 1244
kemikalier, 1257
kemiska egenskaper, 1251
kemiska prov, 1258
kemisk barkning, 1242, 1245
kemisk behandling, 1260
kemisk bindning, 1243
kemisk förlust, 1249
kemisk massa, 1252, 1261
kemisk-mekanisk massa, 1263
kemisk motståndskraft, 1256
kemisk nedbrytning, 1246
kemisk reaktion, 1254
kemisk syreförbrukning, 1003, 1250
kemisk uppslutning, 1253
kemisk återvinning, 1255
kenaf, 3339
keramik, 1218
keramisk bestrykning, 1216
keramisk fiber, 1217
keramisk suglådebeläggning, 1219
keratin, 3340
ketener, 3343
ketoner, 3344
kilkant, 2397
kilkantdäckel, 2399
kilkantpapp, 2398
kilkantpapper, 2400
kimrök, 1096
kinesisk träolja, 5869
kinetisk energi, 3346
kis, 4598
kisbrännare, 4600
kisbränning, 4599
kisel, 5091
kiselföreningar, 5092
kisellikriktare, 5093
kisscoater, 3348
kisscoating, 3349
kladdblock, 4179
klaffventil, 5737
klara, 1341
klargörare, 1340
klarläggande, 1339
klassare, 1343
klassning, 1342
klenved, 5191
klibba, 5424
klibbig, 5630
klibbighet, 5628
klister, 4219
klisterkartong, 4220
klisterpapper, 205, 2942, 2946
klistertejp, 208, 2943
klistrad, 4221
klistring, 2944, 4226
klistringsmaskin, 2945, 4225
klon, 1355
klor, 1303
klorater, 1297
klorbehov, 1311
klorcell, 1304
klorerade ligniner, 1299
klorerad stärkelse, 1300
klorering, 1301
kloreringsprocess, 1310
kloreringssteg, 1302
klorföreningar, 1305
klorgas, 1307
klorider, 1298
kloroxid, 1306
kloriter, 1312
klorligniner, 1314
klormonoxid, 1308
kloros, 1315
klortal, 1309
klorvätesyra, 3129
klorättiksyra, 1313
klotfinish, 1363
klump, 1370, 3591
klumpupplösare, 3592
klämma, 873, 1337

klämmningmärken, 1338
knackning, 5649
knappfläckar, 993
knippare, 2415
knippe, 959
kniv, 3351
knivar, 3357
knivbarkare, 3352
knivegg, 3355, 5169
knivrepad, 1768
knopp, 935
knoväxel, 680
knuffa, 3312
knutdetektor, 3594
knutlösare, 1869
knådare, 3350
knäckning, 934
koagulera, 1373
koagulerande, 1374
koagulerat material, 1373
koagulering, 1376
koaguleringsmedel, 1375
koagulermedel, 1372
kohesion, 1428
kok, 1565
koka, 1975
kokare, 812, 1566, 1976
kokarförare, 1980
kokarinmurning, 1979
kokarmatarvatten, 813
kokarraffinering, 3098
kokarsats, 1977
kokeri, 814, 1978
koklut, 4575
kokning, 1567, 1981, 1982, 4574
kokningscykel, 1569
kokpunkt, 815
koksyra, 1568
koktemperatur, 1570
kokvätska, 1572, 3521, 4575
kokvätska med löslig bas, 5271
kokvätska på kalciumbas, 1012
kokvätskesammansättning, 1571
kol, 1093
koldisulfid, 1098
kolfiber, 1099
kolhydrater, 1092
kollapsad kärna, 1719
kollergång, 3365
kolloidal, 1436
kolloider, 1438
kolloidkvarn, 1437
kolorimeter, 1443
kolorimetri, 1445
kolorimetrisk renhet, 1444
kolpapper, 1230
kolritpapper, 1229
kolv, 4385
kolvals, 1101
kolväten, 3127
kombinationsfilt, 1460
kombinationsglätt, 1464
kombinationskartong, 1458, 1459
kombinationspress, 1462
kombinerad kartong, 1461
kombinerad pappersmaskin, 2706
kombinerad svaveldioxid, 1463
kombustibelt papper, 1465
kommunikationspapper, 1471
kompatibilitet, 1474
kompressibilitet, 1477
kompression, 1478
kompressionshållfasthet, 1479
kompressionsprov, 1480
kompressor, 1482
komprimerad luft, 1476
kon, 1505
koncentration, 1487
koncentrator, 1488
kondens, 1491
kondensat, 1491
kondensation, 1492
kondensationsharts, 1494
kondensationspolymerisation, 1493
kondensator, 1079, 1495
kondensatorpapper, 1077, 1078, 1496, 1497
kondensor, 1495
konditionerad, 1498

konditionerat papper, 1499
konditionering, 1501
konditioneringsapparat, 1500
kondrev, 1506
konduktivt papper, 1503
konisk, 5646
konisk raffinör, 1508
koniskt rullager, 5645
konistensskrivare, 1515
konkavisering, 2023
konkavslipad vals, 1486
konkvarn, 3112
konservering, 1511
konsistens, 1512
konsistensreglering, 1513
konsistensregulator, 1514, 1516
konstaläderpapper, 3446
konstant hastighet, 1517
konstgjort pergament, 440
konstläder, 438
konstläderpapper, 439
konstruktion, 1923
konstruktionsmaterial, 1518
konstruktionspapper, 1519
konsttryckaffichpapper, 443
konsttryckpapper, 441, 1385
konsttryckpergament, 442
konsttrycksomslag, 436
konsumtion, 1520
kontaktfri mätning, 1522
kontaktvinkel, 1521
kontinuerlig, 1531
kontinuerliga formar, 1533
kontinuerlig freenessprovare, 1534
kontinuerlig kokare, 1532
kontinuerlig kokning, 1536
kontinuerlig process, 1535
konto, 4931
kontrastkvot, 1540
kontroll, 1234, 1541
kontrollerad atmosfär, 1544
kontrollpanel, 1543
kontrollprovtagning, 133
kontrollsystem, 1547
kontrollventil, 1237
konvektion, 1548
konvektionstork, 1549
konvektionsvärmning, 1550
konvertbestrykning, 1552
konverter, 1554
konverterad stärkelse, 1553
konvertering, 1555
konverteringsmaskin, 1557
kopiepapper, 1582, 1583, 3677
kopieringsmaskin, 1578
kopolymerer, 1579
koppar, 1580
kopparammoniumhydroxid, 1742
kopparammoniumviskositet, 1743
kopparetylendiaminhydroxid, 1745
kopparetylendiaminviskositet, 1746
koppartal, 1581
koppling, 1371, 1651
kopplingsmedel, 1652
kord, 1584
korn, 2862
kornigt papper, 2868
korona, 1594
koronaurladdning, 1595
korrodera, 1596
korrosion, 1597
korrosionsbenägenhet, 1605
korrosionshinder, 1600
korrosioninhibitor, 1598
korrosionsmekanism, 1599
korrosionsmotstånd, 1602
korrosionsprodukt, 1601
korrosionsprov, 1603
korrosiv, 1604
korrugerad, 1606
korrugering, 1614, 1618
korrugeringsmaskin, 1619
korrugeringsvals, 1620
korskräppad, 1705
kort, 1122
kortfibrig, 5071
kortfibrig tall, 5072
korthugga, 3063
kort mäld, 5073

kort ton, 5074
kortvedavverkare, 5075
kortvedavverkning, 5076
kostnad, 1622
kostnadsstrukturering, 1623
kotte, 1505, 5831
kovalent bindning, 1653
krackelera, 1663
kraft, 2664, 4464
kraftalstring, 4466
kraftförsörjning, 4468
kraftkartong, 3369
kraftliner, 3371
kraftlinerkartong, 3372
kraftmassa, 3366, 3377
kraftmassaokning, 3376
kraftpapper, 3366, 3375
kraftpappersemballage, 3380
kraftstation, 4467
kraftsäckpapper, 3367
krafttestliner, 3378
krafttäckskiktpapp, 3371
kraftvågpapp, 3370
kran, 1419, 1664
krans, 2779
krater, 1665
kraterbildning, 1666
kresoler, 1692
kretsbrytare, 1328
krimp, 1695
krimpning, 1696
kristall, 1728
kristallincellulosa, 1729
kristallinitet, 1731
kristallint område, 1730
kristallisering, 1732
kristallstruktur, 1733
kritisk hastighet, 1702
kritisk temperatur, 1701
kritiskt flöde, 1698
kritiskt tryck, 1700
kritvit, 1832
krom, 1320
kromater, 1316
kromaticitet, 1317
kromatografipapper, 1319
kromatpapper, 1318
kromföreningar, 1321
kromkartong, 1323
kromoxid, 1322
krompapper, 1324
krossad, 1718
krossare, 1722
krossmotstånd, 1726
krossning, 914, 1717, 1723
krossningsstyrka, 1724
krossprov, 1725
krusfinish, 4807
krusning, 4808
kryll, 1694
krympa, 5081
krympning, 5082
kryogenik, 1727
krypning, 1678, 5203
krypningshindrande medel, 391
krypprovning, 1679
krysskil, 5329
kräppad, 1681
kräppad vadd, 1683
kräppapper, 1685
kräppat papper, 1682
kräppat tunnpapper, 1690
kräppmaskin, 1689
kräppning, 1680, 1687
kräppningsfinish, 1684
kräppningskvot, 1686
kräppschaber, 1688
krökning, 1751
krökt vals, 1762
kudde, 4141
kulbildning, 4238
kulkokare, 2825
kulkvarn, 554, 3187, 4237
kullager, 553
kultivering, 1737
kulventil, 556
kungsvals, 3347
kurlator, 1753
kurlering, 1752

kurva, 1761
kutter, 4163, 4854
kuvert, 2292
kuvertmaskin, 2290
kuvertpapper, 2291
kvadratiskt papper, 5364
kvalitativ analys, 4603
kvalitet, 4604
kvalitetskontroll, 4605
kvantitativ analys, 4606
kvarn, 624, 3816
kvarnkniv, 626
kvarnkubb, 635, 3828
kvartära ammoniumföreningar, 4607
kvicksilver, 3752
kvicksilverföreningar, 3753
kvist, 5884
kvistar, 3360
kvistfång, 3359, 3361, 3363
kvistfång med skak, 5029
kvistmassa, 3362
kvistning, 1896
kvistrik massa, 3364
kvot, 4644
kväve, 3989
kväveföreningar, 3990
kylare, 1573
kylcylinder, 1575, 5597
kyld vals, 1272
kylning, 1574
kylsystem, 1576
kylvals, 1271, 5599
kylvatten, 1577
kåda, 4340
känsel, 2413
känselkropp, 5002
känsla, 2413
känslighet, 4997
kärl, 6003
kärnved, 2999
köpa, 4595
körbarhet, 4870
körsbär, 1265
köttomslag, 984, 985, 986

L

laboratorium, 3386
lack, 3389
lackbehandling, 3391
lackerat papper, 3390
lackfärg, 3404
lackfärgspigment, 3405
lag, 3434, 4381, 4387
lagd i spiral, 5322
lager, 618, 3434
lagerhus, 619
lagerreducering, 5432
laghäftning, 4388, 4389
laghäftningsstyrka, 4390
lagring, 5441
lagringslivslängd, 5056
lagringstank, 5442
lagseparation, 4391
lagun, 3393
lakmuspapper, 3535
lamell, 3406
lamellstruktur, 3407
laminat, 3409
laminator, 3417
laminerad, 3410
laminerad kartong, 3411
laminerad plast, 3413
laminerat papper, 3412
laminering, 3414, 3416
lamineringspapper, 3415
laminär strömning, 3408, 5458
landutfyllnad, 3418
lapp, 5631
lappmärke, 4228
lappning, 3664
Larix, 3424
laser, 3425
last, 3538
lastbil, 5855

last-förlängning, 3540
lastreglering, 3540
latens, 3426
latent bild, 3427
latex, 3430, 3431
latexbehandlade papper, 3432
ledare, 2855
ledning, 1502
ledningsförmåga, 1504
ledvals, 2933
legering, 320
lera, 1344
lerbestruken, 1345
lerbestruken kartong, 1347
lerbestrukna ark, 1346
lerfyllmedel, 2494
lerfyllt papper, 1348
lerklump, 1349
leverans, 1898
leverantör, 5564
leverera, 5565
levulonsyra, 3462
libell, 3452
Libocedrus, 3463
lignifiering, 3479
lignin-, 3478
ligninartad, 3478
ligniner, 3482
ligninhalt, 3481
lignin-kolhydratkomplex, 3480
ligninmjöl, 3824
lignocellulosa, 3483
lignoler, 3484
lignosulfonater, 3485
lignosulfonsyra, 3486
likformig arkbildning, 5918
likformighet, 5919
likriktare, 4687
likström, 1803, 2006
lim, 2840, 5121
limbestruket papper, 2843
limbrott, 2845
limemulsion, 5124
limfasthetsprovare, 2842
limfläckar, 5127
limfogar, 2844
limmad och kalandrerad, 5122
limmare, 2847
limmat papper, 5123
limning, 2849, 5128
limning med hartslösning, 5275
limningsprov, 5130
limningsprovare, 5129
limpress, 5125, 5584, 5863
impressvals, 5126
limpåläggare, 2841
limspridning, 2848
limstreck, 2846
lin, 2559
lindad vals, 6290
liner, 3505
linjal, 524, 5154
linjalmarkering, 5156
linjetryck, 3496, 3977
linjärprogrammering, 3497
linne, 3500
linnefiber, 3502
linnefinish, 3503
linnekartong, 2560
linnepapper, 3504
linnepressat papper, 2371, 3501
linolsyra, 3511
linotypmaskin, 3512
linter, 3513
linterhantering, 3515
linters, 3514
Liquidambar, 3518
Liriodendron, 3525
lista, 4676
litium, 3527
litiumföreningar, 3529
litiumklorid, 3528
litografi, 3532
litografipapper, 3531, 3533
litopon, 3534
liv, 3465
livslängd, 3465, 5010
livsmedelskartong, 2661, 2740
livsmedelsomslagspapper, 2662

ljudisolering, 184
ljudvåg, 5283
ljus, 3470
ljushet, 896
ljuskälla, 3474
ljuskänsligt papper, 3473
ljusspridning, 3472, 4077
ljusäkthet, 3471
lockpapper, 2997
Locust Bean Gum, 3547
lodgepole pine, 3548
lokalberedning, 5119
lokaliserat vattenmärke, 3545
long ton, 3569
ludd, 2598
luft, 242
lufta, 220
luftblad, 244
luftblåsare, 245, 269
luftborste, 244
luftborstebestrykare, 247
luftborstebestrykning, 248
luftbubblor, 249
luftburen, 246
luftdäckel, 256
luftfilter, 263
luftfiltreringspapper, 264
luftflöde, 266
luftförorening, 282
luftgap, 250
lufthalt, 252
luftinblåsning, 270
luftindragning, 262
luftintag, 271
luftkammare, 251
luftklocka, 243
luftkniv, 273
luftknivbestrykare, 274
luftknivbestrykning, 275
luftknivsmärke, 276
luftkuddetork, 2568
luftkuddetorkare, 265
luftkvalitet, 283
luftkyld, 253
luftkylning, 254
luftlagd, 277
luftmunstycke, 280
luftning, 221, 472
luftningsgarvning, 222
luftningsmedel, 223
luftpermeabilitet, 281
luftpostpapper, 279
luftridå, 255
luftrulle, 284
luftrum, 251
luftstråle, 272
lufttorkad, 257, 3550
lufttorkare, 3552
lufttorkat papper, 258, 3551
lufttorkning, 261, 3553
lufttorr, 101, 259
luftvärmare, 268
lumberjack, 3587
lumen, 3588
luminositet, 3589
luminös reflektivitet, 3590
lump, 4621
lumpdammare, 4628
lumphalt, 4624
lumphaltigt papper, 4625
lumphanteringsrum, 4631
lumphuggare, 4626
lumpkokare, 4622
lumpmassa, 4630, 4633
lumpmäld, 4633
lumppapper, 4629
lumpsortering, 4632
lumptakpapp, 2429
lumpupplösare, 4634
lumpuppslagare, 4623
lunnare med gripare, 2877
lunning, 3555
lut, 1171, 3435, 3521
lutande vira, 3202, 3203
lutbehandling, 3436
lutindunstningsanläggning, 3522
luttillverkning, 3436
luttorkare, 260
lut-vedförhållande, 3524

lyft, 3466
lyfta, 4729
lyftanordning, 3468
lyftning, 3467
lyfttruck, 3469
lyster, 3595
låda, 862, 1140, 1148
låddelar, 872
lådemaljpapper, 869
lådetiketter, 1136, 1143
lådfoderpapper, 1137, 1144
lådfodring, 870
lådförpackat skrivpapper, 868
lådförseglingspapper, 1138
lådkartong, 863
lådkomprimering, 866
lådmaskin, 871, 1135
lådmängd, 1145
lådomslagspapper, 867
låg densitet, 3577
låg finish, 3578
låg konsistens, 3576
låg temperatur, 3581
lågtryck, 3579
lågtrycksmatare, 3580
långbarrig tall, 3567
långfibrig massa, 3564, 3568
långsveck, 3565, 3566
långt liv, 3563
läcka a.o., 751
läckage, 3443
läckning a.o., 752
läckningsmotstånd, 753
läcksäker behållare, 3444
läderimitation, 3445
längd, 2663, 3448
längsriktning, 3562
längsskärning, 5172
läppinställning, 5157
lärkträd, 3423
läsare, 4662
läskkartong, 769
läskpapper, 683, 770
lätt, 3470
lätta, 4729
lätt linerkartong, 3476
lättlöslighet, 2177
lättnad, 4735
lättvikt, 3475
lödning, 5258
lönn, 3689
lösa, 2032
lösblad, 3572
lös kant, 5142
lös knut, 5247
löslighet, 5269
lösning, 5272, 5273
lösningsmedel, 5274, 5982
lösrulle, 3573, 5252
löst syre, 2033
löst veck, 5246
lövträ, 1847, 2987
lövträd, 1846
lövträmassa, 2988
löv- årligen avfallande, 1845

M

Madelainetork, 3639
mager, 2717
mager massa, 2722
mager mäld, 2724
magnesium, 3643, 3644
magnesiumbaslut, 3645
magnesiumbisulfit, 3646
magnesiumföreningar, 3649
magnesiumhydroxid, 3650
magnesiumkarbonat, 3647
magnesiumklorid, 3648
magnesiumoxid, 3651
magnesiumsulfat, 3652
magnesiumsulfit, 3653
magnet, 3654
magnetband, 3660
magnetfält, 3656

magnetiska egenskaper, 3658
magnetisk separator, 3659
magnetiskt bläck, 3657
magnetisk utrustning, 3655
majsstärkelse, 1593
makromolekyler, 3638
makulatur, 6059
makulaturhandlare, 6060
makulaturmassa, 6061
mala, 620
malbarhet, 621
mald, 622
maleinsyra, 3668
maleinsyreanhydrid, 3669
malgrad, 1880, 4929
malning, 641
malningsgarnityr, 639
malningshastighet, 643
malningskar, 627
malningsprov, 640
malningsprovare, 623
malningstid, 644
malningstillstånd, 642
malningstryck, 647
maltos, 3671
manchett, 5149
mangan, 3674
manganföreningar, 3675
manilla, 3678
manillafiber, 3680
manillahampa, 3681
manillakartong, 3679
manillapapper, 3683
manillatäckt lädkartong, 3682
mankemang, 3670
mannaner, 3684
mannogalaktan, 3685
mannos, 3686
mantel, 3306
manuell finish, 2963
marinens förråd, 3947
markerfilt, 3701
markering, 3694, 3699
markeringsmedel, 3700
markervals, 3702
marknadsföring, 3696
marknadsundersökning, 3698
markör, 3695, 4936
marmorerat papper, 3691, 3692
marmorering, 3693
maska, 3756
maskin, 3606
maskinbearbetning, 3637, 4052
maskinbeklädnad, 3610
maskinbelastning, 3628
maskinbestruken, 3611
maskinbestruken kartong, 3612
maskinbestruket papper, 3613
maskinbestrykning, 3614, 4051
maskindrift, 3629
maskindäckel, 3616
maskineri, 3632
maskinfinish, 3620
maskinfinished, 3596
maskinförare, 3634
maskin för ensidig wellpapp, 5105
maskingjord kant, 3823
maskinglätt, 3608
maskinglättad, 3596, 3598, 3622,
3624, 6148
maskinglättat papper, 3597, 3623,
3625
maskinglättning, 3626, 6151
maskinkar, 3609
maskinkonstruktion, 3617
maskinkräppad, 3615
maskin med lutande vira, 3204
maskinpräglad, 6149
maskinriktning, 3618
maskinrum, 3631
maskinsalsgolv, 5673
maskinsida, 3627
maskintid, 3635
maskintorkad, 3619
maskinverkstad, 3633
maskinvira, 3636
maskning, 3703
maskningsmedel, 3704
massa, 3705, 4568

massaark, 4576, 4582
massafabrik, 4577
massafiberfraktionerare, 4572
massaformning, 4578
massa från kall kaustisk soda, 1430
massa för malning, 631
massaförsäljning, 4580
massahalt, 4570
massaindustri, 4573
massakvalitet, 4579
massapump, 5436
massasil, 4581
massasuspension, 5595
massatillverkning, 4574
massaved, 4584
Massey-bestrykare, 3706
masspektroskopi, 3707
massöverföring, 3708
matare, 2407
matematisk modell, 3718
material, 3713
materialbalans, 3714
materialförbrukning, 3715
materialhantering, 3716
materialprovning, 3717
matning, 2403, 2411
matningstransportör, 2406
matningstratt, 2410
matningsvatten, 2412
matrispapp, 2580, 3720
matt, 3709
matta, 3709
mattbestruket papper, 2149
matt finish, 2150, 3721
matt kartong, 3710
matt papper, 3719
mattväv, 1130
maximal däckel, 3725
maximal kapacitet, 3724
medel, 232, 2375
medel mot statisk elektricitet, 409
med guskat ytskikt, 5977
medicinskt preparerat papper, 3741
mekanik, 3739
mekaniker, 3830
mekanisering, 3740
mekaniska egenskaper, 3733
mekanisk drift, 3731
mekanisk halmmassa, 3737
mekanisk massa, 3734
mekanisk massatillverkning, 3735
mekanisk mätning, 3732
mekanisk separator, 3728
mekanisk trämassa, 3738
mekanisk töjningsgivare, 3736
melamin, 3743
melamin-formaldehydharts, 3744
melerad, 3899, 3900
melerad finish, 3902
melerad färg, 3901
melerat papper, 3903
melering, 3904
mellanfinish, 3742
mellanglätt, 3271
mellaninloppslada, 3273
mellanrum, 3281, 5290
mellanstycke, 5289
mellantid, 3176
mellanvals, 3177, 3272
membran, 1944, 3747
mercerisering, 3751, 5416
meristems, 3755
merkaptaner, 3749
merkaptolignin, 3750
mesomorf cellulosa, 3757
metacellulosa, 3758
metakrylater, 3774
metakrylsyra, 3775
metall, 3759
metallfiber, 3761
metallfinish, 3762
metallfolie, 3763
metallföreningar, 3760
metalliserat papper, 3768
metallisering, 3764
metallisk bestrykning, 3765
metalliskt bläck, 3766
metalliskt papper, 3767
metan, 3776

metanol, 3777
meteorologi, 3769
meter, 3770
metodteknologi, 3778
metrisk ton, 3792
metriskt system, 3791
metylacetat, 3779
metylakrylat, 3780
metylalkohol, 3781
metylcellulosa, 3783
metylencellulosa, 3784
metylering, 3782
metyletylketon, 3785
metylglukosider, 3786
metylgrupper, 3787
metylmerkaptan, 3788
metylmetakrylat, 3789
metylorange, 3790
MG-papper, 3601
migration, 3811
migrering, 3811
mikroanalys, 3795
mikrofibrill, 3799
mikrofiche, 3800
mikrofilm, 3801
mikrokristallin cellulosa, 3797
mikrokristallint vax, 3798
mikrometer, 3802
mikroorganism, 3803
mikroorganismkontroll, 3804
mikroporositet, 3805
mikroskopi, 3806
mikrotom, 3807
mikrovågpapp, 3796
mikrovågspektroskopi, 3808
mil, 3812
miljö, 2293
mindre blad, 3442
mineralavsättning, 3831
mineralfiber, 3832
missfärgning, 2018
missfärgningskartong, 411
missfärgningskyddande, 410
missfärgningskyddande papper, 412
missfärgningskyddande tunnpapper, 413
MIT dubbelvikngsprovare, 3835
Mitscherlichprocess, 3836
mittlamell, 3809
mjuk, 5241
mjukgörare, 5242
mjukgöring, 5243
mjukgöring av vatten, 6108
mjukgöta vatten, 5245
mjukhet, 5249
mjuklönn, 5248
mjuk massa, 5251
mjukpapper, 5250
mjukpunkt, 5244
mjöl, 2587
mjölkförpackning, 3814
mjölksyra, 3392
modell, 3848
moder-, 5560
modernisering, 3849
modervals, 4210
modiferad cellulosa, 3850
modifierade stärkelser, 3851
modifierande behandling, 3853
modifieringsmedel, 3852
modul, 3854
mogen ved, 3722
mogna, 1748
mognad, 1749, 3723
mognadsmedel, 1750
molekylarvikt, 3883
molekyler, 3884
molekylstruktur, 3882
molnig, 1368
molnighet, 1367
molnighetseffekt, 1366
molybden, 3885
molybdenföreningar, 3886
monitor, 3887
monitorera, 3888
monokotyledoner, 3889
monomerer, 3892
monoskikt, 3891
monotypmaskin, 3893
montmorillonit, 3894

motark, 1648
motor, 3898
motskär, 634, 648
motskärsbom, 646
motström, 1643, 1644, 1646
motströms, 1538
motströmsbestrykare, 1537
motströmsprocess, 1645
motstånd, 4758
motstånd mot alkalifärgning, 306
mottryck, 514
motvals, 1647
motvalsfläck, 5335
mullen, 3907
Mullenprovare, 3908
multiplexformare, 3918
multiplexkartong, 3921
munstycke, 4009
munstycksflöde, 4087
Murray pine, 3934
mussla, 5057
mutation, 3935
MWL, 3824
mynt, 5060
myrsyra, 2688
mångcylindermaskin, 3912
mångfaldigande, 3676
mångsidig användning, 3920
mått, 2769, 2798
mäld, 2462, 5430, 5480
mäldberedning, 5435
mälderi, 636
mäldflöde, 5433
mäldfyllning, 630
mäldfärgad, 628
mäldfärgning, 629
mäldinlopp, 5434
mäldkar, 5431, 5482
mäldkonsistens, 5483
mäldlim, 5438
mäldlimnad, 637
mäldlimning, 638, 2284, 3278, 5486
mäldpump, 5436
mäldregulator, 5437
mäldtillsatsmedel, 625
märg, 4344
märgupplösare för bagass, 529
märke, 3694
märkning, 4937
mässing, 878
mätapparat, 3771
mätare, 2769, 2798, 3770
mätinstrument, 3727
mätning, 3726, 3748
mätsystem, 3773
mättad, 4909
mättnad, 4913
mättning, 2431
mättningsegenskaper, 4912
mätutrustning, 3772
mögel, 3813
mögelresistentpapper och kartong, 3881
mönster, 4230
mönsterkartong, 4231
mönsterpapper, 4232, 4233

N

nafta, 3938
nappning, 4301, 4307
nappningsbrott, 4310
nappningsmotstånd, 4309
nappningsprov, 4313
nappningsprovare, 4312
nappningsstyrka, 4311
natrium, 5215
natriumalginat, 5216
natriumaluminat, 5217
natriumbikarbonat, 5219
natriumbisulfit, 5220
natriumborat, 5221
natriumborohydrid, 5222
natriumföreningar, 5227
natriumhydrosulfid, 5228
natriumhydrosulfit, 5229

natriumhydroxid, 5230
natriumhypoklorit, 5231
natriumkarbonat, 5223
natriumklorat, 5224
natriumklorid, 5225
natriumklorit, 5226
natriumkoklut, 5218
natriumperoxid, 5232
natriumselensulfonat, 5240
natriumsilikat, 5233
natriumstearat, 5234
natriumsulfat, 4888, 5235
natriumsulfid, 5236
natriumsulfit, 5237
natriumtiocyanat, 5238
natriumtiosulfat, 5239
natronlut, 5211
naturfiber, 3943
naturfärgad, 3942
naturgas, 3944
naturliga bruna färger, 3941
naturlig skogsåtervärt, 3945
naturtillgångar, 3946
NCR-papper, 3936, 3992
nedbrytning, 1858, 1879
neddragning, 2097
neddragningsprov, 2098
nedfall, 2109
nedfläckning, 5338
nedföring i vätska, 5492
nedlusning, 3219
nedsotning, 1703
nedtryckningsvals, 3069
negativ bombering, 3951
neopren, 3952
neutralisering, 3954
neutralkraftmassa, 3955
neutrallim, 3956
neutralsulfitlut, 3957
neutralsulfitmassa, 3937, 3958
neutralsulfitmassatillverkning, 3959
nickel, 3969
nickelföreningar, 3970
niopunkt(0009) tums halmkartong, 3973
niopunkt(0009) tums halvkemisk kartong,
3972
niopunkt(0009) tums
vågpappersmaterial, 3971
nippel, 3976
nitrater, 3979
nitratgrupper, 3978
nitrering, 3980, 3985
nitrider, 3984
nitriler, 3986
nitriter, 3987
nitrocellulosa, 3988
nitroligniner, 3991
nivå, 3452
nivåindakator, 3455
nivåkontroll, 3453
nivåregulator, 3454
nivåskruv, 3458
nodel, 3993
noggrannhet, 143
nollinspänningsstyrka, 6324
nominell vikt, 3997
nonlinjär programmering, 4001
nubb, 5628
nyans, 3103, 5750
nyanserat vitt, 5752
nyansering, 5753
nyanseringsmedel, 5770
nyheter, 3960
nylonfiber, 4013
nyp, 3974
nypskydd, 3975
nyptryck, 3977
ny-värde, 4010
nålad filt, 3950
nålhäftning, 4324
nållager, 3948
nålstick, 4330
nålventil, 3949
näringsmedel, 4012
nätformig, 4771
nöta, 107
nötning, 108
nötningsförmåga, 110
nötningsmotstånd, 109

O

obaland, 681
obelastad, 5922
obestruken, 5908
obestruken vikt, 5912
obestruket boktryckpapper, 5909
obestruket papper, 5910
obestruket tryckpapper, 5911
oblekt, 5904
oblekt massa, 5906
oblekt papper, 5905
odör, 4017
odörbekämpning, 4018
offererad hastighet, 1924
offset, 4024
offsetark, 4030
offsetbestrukna papper, 3530
offsetfärg, 4025
offsetlitografi, 4026
offsetpapper, 4028
offset-tidningspapper, 4027
offsettryckning, 4029
ofylld, 5922
ogenomtränglig, 3194
ogenomtränglighet, 3193
oglättad, 5907, 5916, 5917
oleater, 4047
olefiner, 4048
oleohartser, 4050
olimmad, 5925
olimmat papper, 5926
olja, 4032
oljad, 4036
oljbarhet, 4046
oljeabsorbens, 4033
oljeabsorbtion, 4034
oljeabsorbtionsprov, 4035
oljefilter, 4039
oljefläckar, 4045
oljeimpregnerad papp, 4037
oljeimpregnerat papper, 4038
oljemotstånd, 4043
oljepenetration, 4040
oljepentrationsprov, 4041
oljesyra, 4049
oljesäkert papper, 4042
omfalsning, 616
omflishugg, 4673
omkopplare, 5605
omrulla, 4786
omrullning, 4788
omrullningsmaskin, 4787
omröra, 239
omrörare, 241, 2338, 3191
omrörning, 240, 5429
omslagspapper, 6299
omsättning, 5881
omvandling, 1551
omvänd drift, 4775
omvändning, 4781
omvänd osmos, 4776
onionskinpapper, 4056
oorganiska föreningar, 3248
oorganiska papper, 3250
oorganiska salter, 3251
oorganiska syror, 3247
oorganisk fiber, 3249
opacifieringsmedel, 4057
opacimeter, 4058
opacitet, 3037, 4059
opak, 4060
opakt papper, 4061
operationsforskning, 4070
operatör, 4071
optimering, 4080
optiska egenskaper, 4075
optisk avsökare, 4076
optisk blekning, 4072, 4078
optisk mätning, 4074
optiskt blekmedel, 4073, 4079
ordna i följd, 5006
orenhet, 3200
organiska föreningar, 4084

organiska syror, 4083
organiskt lösligt lignin, 4085
orientering, 4086
orienteringsgrad, 1882
oscillation, 4089
oscillator, 4090
oscillerande, 4088
oscillogram, 4091
oskuret trämassepapper, 3729
osmos, 4092
osmotiskt tryck, 4093
otillräckligt körd, 5915
otorkad massa, 5190
otryckt tidningspapper, 730
oxalater, 4115
oxalsyra, 4116
oxicellulosa, 4122
oxidation, 4118
oxidationsmedel, 4117, 4121
oxidationsmotstand, 4119
oxider, 4120
oxiderad stärkelse, 4127
ozalidkopiering, 1948
ozalidpapper, 1949
ozon, 4128
ozonbehandling, 4129
ozonisering, 4129

P

packare, 1473
packning, 1472, 1902, 2788, 4137,
6052
packningspapp, 2789
packningspapper, 2790
paket, 4130
palett, 4142
pallning, 4143
panel, 4144, 6043
panelkartong, 4145
pannhus, 5408
papp, 789
pappbehållare, 2459
pappburk, 2450
papper, 4146
papperark, 4184
papper av kemisk fiber, 1247
papper av ylle, 6287
papper med skyddsskikt, 4731
papper med vattenmärke, 6089
pappersark, 5052
pappersbana, 4194
pappersbasplast, 4150
pappersbjörk, 4151
pappersblock, 4179
pappersbruk, 4177
pappersegenskaper, 4182
pappersersättning, 4189
papperserviietter, 4178
pappersfilt, 4172
pappersfläckar, 4186
pappersförsäljare, 4176
pappersförädlare, 4161
pappersförädling, 4160
pappersgarn, 4193, 4195
pappersglättvals, 2492
pappersgranskare, 4166
pappershanddukar, 4192
pappershylsa, 4162
pappersindustri, 4165
papperskromatografi, 4159
papperslaminat, 4149, 4167
pappersslim, 4185
pappersmakare, 4170
pappersmakarens alun, 4171
pappersmaskin, 4168
pappersmaskindrift, 4169
pappersmäld, 4187
papperspåtar, 4180
pappersprodukter, 4181
pappersprov, 4190
papperspåse, 4147
pappersrulle, 4702
pappersskärare, 4163
pappersnöre, 4195

papperssort, 4164
pappersstapel, 5135
pappersstruktur, 4188
papperssäck, 4147
papperstextilier, 4191
papperstillverkare, 4170, 4175
papperstillverkning, 4173
papperstillverkningsutrustning, 4174
pappersvals, 1045, 5053
pappersvåg, 4183
pappkartong, 2649, 2651
papplåda, 864, 1134, 4153
papp med guskat ytskikt, 5978
papyrus, 4196
par, 1650
paraffinvax, 4197
parallellaminerad, 4198
parameter, 4199
parenkym, 4209
parmanganater, 4270
parmanganattal, 4269
parshallränna, 4211
partikel, 4212
partikelfälla, 887
partikelstorlek, 4214
partikelstorleksfördelning, 4215
passa, 3711
passare, 517
passivitet, 2061
passning, 2535, 3712
pasta, 4219
patentbestruken, 4229
pegmenterad, 4321
pektin, 4239
pelarhållfasthet, 1457
penetrering, 4249
penetreringsmedel, 4248
penna, 4247
pentaklorfenol, 4250
pentosaner, 4251
perforerad vals, 4255
perforerare, 4259
perforering, 4256, 4258
perforeringspapper, 4257
pergament, 4200
pergamentera, 4204
pergamentering, 4203
pergamenteringsbaspapper, 4205
pergamentfinish, 4202
pergamentpapper, 4201, 4206, 4207
pergamentskrivpapper, 4208
pergamyn, 2816
pergamynpapper, 2817
periderm, 4262
perifer, 4265
periferidrift, 4266
periferihastighet, 1332
perifert register, 1331
perjodater, 4264
perjodatlignin, 4263
perklorater, 4253
permanens, 4267
permanent papper, 4268
permeabilitet, 4271
peroxider, 4272
persulfater, 4273
perättiksyra, 4252
pH, 4274
pH-reglering, 4275
pH-värde, 4276
Picea, 4300
pigment, 4319
pigmentrecept, 4320
pil, 6233
pinener, 4326
pinhålfri, 4331
pinoresinol, 4332
Pinus, 4333
Pitotrör, 4346
plan finish, 2552
planhet, 2553
plankrossningsmotstånd, 2550
plankrossningsprov, 2551
planpress, 2548
planpåse, 4459
plansil, 2556
planskikt, 2377
planskiktpapper, 2376
plansuglåda, 5505

planta, 4353
plantage, 4354
plantering, 4354, 4356
plant papper, 2554
plantrest, 4357
plantskola, 4011
planvira, 2705
planvirakartong, 2703
planviramaskin, 2704
plast, 4359
plastbestruket papper, 4360
plasticering, 4364
plasticeringsmedel, 4363
plasticitet, 4362
plastisk flytning, 4361
plastometer, 4365
platta, 4366, 4373, 5141
plattork, 4374
plattpress, 4375
plenisal, 4379
plocka ut, 4308
plockning, 4307
plugg, 4382
pluggning, 4383
plunger, 4385
plusträd, 4386
plym, 4384
plåt, 4366
plåtfinish, 4367
plåtfinish-papper, 4368
plåtglätt, 4376
plåtglättad, 4369
plåtglättad kartong, 4377
plåtkartong, 4372
plåtpapper, 4371
plåttillverkning, 4370
plätering, 4378
pneumatisk spänningsanordning, 278
pneumatisk utrustning, 4392
polarisation, 4398
polaritet, 4397
polera, 4400
polerare, 966
polering, 4400, 4402
polinering, 4403
polyakrylamid, 4406
polyakrylat, 4407
polyamider, 4409
polyamidhartser, 4408
polyelektrolyter, 4410
polyestrar, 4411
polyetylimine, 4414
polyimider, 4415
polymerer, 4418
polymerisation, 4416
polymerisationsgrad, 1883
polymerisera, 4417
polymetaakrylat, 4419
polypropylen, 4420
polystyren, 4421
polysulfider, 4424
polysulfidkokning, 4423
polysulfidmassa, 4422
polyten, 4412
polytenbestruket papper, 4413
polyuretaner, 4425
polyvinylacetat, 4426
polyvinylalkohol, 4427
polyvinyler, 4432
polyvinylestrar, 4429
polyvinyletrar, 4430
polyvinylidenklorid, 4431
polyvinylklorid, 4428
poppel, 4439
Populus, 4440
por, 4441
porisitetsprovare, 4445
porosimeter, 4443
porositet, 4444
porslinskartong, 1274
porslinslera, 1275
porstorlek, 4442
porös, 4446
poröst material, 4447
poröst trä, 1970
porös ved, 4448
potatisstärkelse, 4456
potentiometer, 4457
precision, 4473

prefabrikation, 4476
preparation, 4479
presentomslagspapper, 2811
preservativ, 4481
preservering, 4480
press, 4482
pressarrangemang, 4483
pressbestrykare, 4485
pressbestrykning, 4486
pressfilt, 4487
pressmarkeringar, 4489
pressning, 4488
presspan, 4484
pressparti, 4493
pressrum, 4491
pressvals, 4490
preventivt underhåll, 4512
prickräkning, 2013
primer, 4519
primär inloppslåda, 4514
primärtillväxt, 4513
primärvägg, 4518
process, 4532
processreglering, 4533
processvariabel, 4535
processvatten, 4536
produkt, 4096, 4537
produktion, 4539
produktionskontroll, 4540
produktionsledare, 4541
produktionsmetod, 4542
produktivitet, 4543
produktutveckling, 4538
profil, 4544
profilometer, 4545
profit, 4546
programmerad instruktion, 4547
programmeringsspråk, 4548
propan, 4550
propeller, 4551
propionater, 4553
proportionerare, 4554
proportionering, 4555
propylalkohol, 4556
proteiner, 4559
protoligniner, 4560
protoner, 4561
prov, 4891
provare för inre bindning, 3276
provark, 5700
provark från mälderi, 6080
provberedning, 4892
provbänk, 5694
prov genom accelererad åldring, 130
provkörning, 5835
provning, 5696
provningsmetod, 5698
provpapper, 5699
provtagare, 4893
provtagning, 4894
Prunus, 4562
pråm, 570
präglad, 2254
präglare, 2256
präglat papper, 2255
prägling, 1958, 2257
präglingskalander, 2258
präglingspapper och -kartong, 2259
pseudoplasticitet, 4563
Pseudotsuga, 4564
pulper, 2025, 4571
pulperknivar, 884
pulsering, 4585
pulver, 4462
pulvriserad, 4463
pulvriserare, 4586
pump, 4587
pumpning, 4588
pund per tusendels tum, 4460
punkt, 4396
punktera, 4591
punktering, 5428
punkteringsmotstånd, 4592
punkteringsprov, 4594
punkteringsprovare, 4593
punktvis fästning, 5336
pyrit, 4598
pyrolys, 4601
pyroxylinbestrykning, 4602

påfrestning, 5463
påläggningsvals, 420
påsaskin, 4458
påse, 526
påsmaskin, 4148
pärlbestrykare, 615
pärlfyllnadsmedel, 4234
pärlstärkelse, 4235
pärlvitt, 4236
pösig, 536
pösighet, 533

Q

Quercus, 4609

R

raderbarhet, 2302
raderbart pergament, 2303
radermotstånd, 3240
radie, 4617
radikal, 4614
radioaktivitet, 4616
radioaktivt material, 4615
raffinat, 4618
raffinering, 4706
raffinör, 4704
raffinörmekanisk massa, 2919
raffinörskiva, 4705
ram, 2714
ramie, 4638
ramiefiber, 4639
rand, 5455
randning, 5457
raschigringar, 4642
rayon, 4652
rayonfiber, 4653
rayonmassa, 4654
rayonutskott, 4655
reagens, 4663
reaktionsläge, 4657
reaktionsmekanism, 4656
reaktionstid, 4658
reaktionsved, 4659
reaktivitet, 4660
reaktor, 4661
receptivitet, 4672
red gum, 4690
redoxpotential, 4691
redoxreaktion, 4692
reducerbart svavel, 4693
reducermedel, 4694
reducerventil, 4695, 4738
reduktion, 4696
redwood, 4697
reflektans, 4707
reflektion, 4708
reflektivitet, 4709
reflektometer, 4710
refraktion, 4712
regenererad cellulosa, 4715
register, 4676, 4716
registerkort, 4677
registerkortmassa, 5625
registerpapper, 4717
registerskena, 5623
registerskärm, 5621, 5622
registervals, 5620
registrerat, 4935
registrering, 4680
reglerbar hastighet, 210
reglerdon, 1545
reglering, 1541
reglerlucka, 1543
reglerlåda, 4718
reglerutrustning, 1542
reglervals, 1814, 1815
regulator, 4719
rejekt, 4724, 5139
rejektering, 4726

rejektmängd, 5138
reklamation, 4675
rekord, 4932
relativ fuktighet, 4727
relaxation, 4728
relief, 4735
reliefpapper, 4737
rem, 5447
remband, 653
rembytare, 656
remdrift, 655
remning, 5448
remning av rullar, 5449
remsa, 5472, 5643
remtransportör, 654
renare, 1350
renhet, 4597
rening, 1351, 4596
rening från sand, 4799
renskuren, 5839
renskuren bredd, 5843
renskuren skarv, 5842
renskuren storlek, 5841
renskuret papper, 5840
renskära, 5838
renskärare, 5844
renskärning, 1352, 5846
renskärningskniv, 5847
reologi, 4790
reologiska egenskaper, 4789
rep, 4835
repbestyckning, 4840
repbärare, 4836
repelleringsmedel, 4742
repmarkering, 4838
repmatningsreglering, 4837
repor, 4944
reppapper, 4839
reproducerbarhet, 4743
reproduceringspapper, 4744
reprografi, 4745
reservfilt, 5291
reservoar, 4748
resiliens, 4751, 4752
resinater, 4755
resistans, 4758
resistivitet, 4761
resonans, 4763
resorcinol, 4764
rest, 4750
restligninhalt, 4749
restparti, 3310
restpartipapper, 3311
retardation, 4768
retention, 4769
retentionsmedel, 4770
retrogradering, 4772
ribbad, 3395
ribbad egutör, 3397
ribbade viror, 3402
ribbad vira, 3400
ribbat antikpapper, 3396
ribbat papper, 3401
ribbat skrivpapper, 3403
ribbfinish, 3398
ribbningslinjer, 3399
ridvals, 4795, 4797, 5774
ridåbestrykare, 1759, 5473
ridåbestrykning, 1760, 5474
riktning, 2009
rilla, 2912
rimlighet, 2396
ring, 4802
ringar, 385
ringbarkare, 4803
ringformig, 384
ringkrossprov, 4804
ringporig ved, 4805
ringstyvhet, 4806
ris, 4664
risetikett, 4665
risförseglad, 4667
rishalm, 4793
rismarkör, 4666
risomslag, 4670
risomslagen, 4669
rispapper, 4791
risstärkelse, 4792
risvikt, 4668

ritpapper, 2099
riva, 5655
rivare, 5079
riva ut, 5660
riven kant, 5782
rivet ark, 5783
rivfaktor, 5656, 5661
rivhållfasthet, 5659, 5662
rivmotstånd, 5658
rivning, 4926, 5080, 5657
rivprov, 5664
rivprovare, 5663
rockwellhårdhet, 4812
rost, 4874
rostfläckar, 4876
rostfritt stal, 5378
rosthinder, 4875
rostmotstånd, 5379
rostskyddande, 405
rostskyddande papper, 406
rot, 4834
rotameter, 4847
rotation, 4856
roterande filter, 4850
roterande kniv, 4848
roterande kokare, 4849
roterande pump, 4851
roterande sil, 4852
roterande viskosimeter, 4855
roterkniv, 4784
rotogravyr, 2882, 4858
rotogravyrbestrykning, 4529
rotogravyrpapper, 4857
rotor, 4859
rotorblad, 4860
rubb, 3558
rulla, 6234
rullager, 4824
rulle av ensidig wellpapp, 5104
rullhårdhet, 6241
rullimning, 6157
rullklämmor, 4817
rullmaskin, 4699, 5171, 6235
rullmaskinband, 4700
rullningsfåll, 6236
rullningsveck, 6237
rulloffsetpapper, 6154
rulloffsettryckning, 6155
rullpapper, 6156
rullprov, 4703
rullschaber, 4821
rullspindel, 1589, 6240
rullstol med dubbel bärcylinder, 2071, 2072, 2073
rullsätt, 4830
rundhet, 4862
rundvira, 6251
rundvirabristol, 1793
rundvirafilt, 3878
rundviraformare, 1797
rundvirakartong, 1792, 3873
rundvirakraftliner, 1795
rundviramaskin, 1796, 1800, 3880, 5979
rundvirapapper, 1798
rundviraparti utan tråg, 2146
rundviraprodukter, 3874
rundvirpapper, 5980
rutat papper, 5363
rutil, 4877
rutinkontroll procedur, 4863
rutten ved, 1843
rygg, 4796
rynka, 6301
rysk malning, 2972
rå, 1377
råalun, 600
rådande denistet, 1758
råghalm, 4878
råhet, 4861
råkok, 4646
råmassa, 4648
råmaterial, 4647
råolja, 1716
råpapper för bestrykning, 1402, 1403, 1414
råvatten, 4649
rå yta, 1378
räffla, 2912

räfflad press, 2913
räfflad vals, 2914
räffling, 2915
räknare, 1641
räknarpanel, 1642
räknemaskin, 1024
räkning, 1640, 1649
ränna, 2611
räntabilitet, 4773
rätlinjighet, 5443
råtskuren, 5361
rödbok, 4697
rök, 5195
rökgaser, 2750
röntgenanalys, 6308
röntgenstrålar, 6309
rör, 4334, 5865
rörelse, 3897
rörledning, 4338
rörläggning, 4336
rörmokare, 4335
rörmokeri, 4336
rörsystem, 4337
rötgenspektrometer, 6310
rötmotstånd, 2755

S

salinitet, 4886
Salix, 4887
salpetersyra, 3981
salpetersyrakokning, 3983
salpetersyramassa, 3982
salthalt, 4886
saltvatten, 4889
sammansatt behållare, 4217
sammansatt burk, 1475
sammetsfinish, 5991
sampolymerer, 1579
sandfilter, 4895
sandfång, 4798, 4896
sanitetsbindor, 4897
sanitetspapper, 4898, 4899
sassafrass, 4906
sats, 2761
satspulper, 610
satsvis process, 608
satängfinish, 4907
satängvitt, 4908
sav, 4900
savskalning, 4904
saxverk med cirkelkniv, 2019
schaber, 2048
schaberblad, 2049
schaberdamm, 2051
schaberhållare, 2052, 2053
schabermarkering, 2055
schabernappning, 4302
schaberrepa, 4942
schaberutskott, 2050
schabervals, 2057
schaberveck, 2056
schabring, 2058
scotch pine, 4938
sedelpapper, 565, 1757
sediment, 4979
sedimentation, 4980
sedimentering, 5016
sedimenteringsbassäng, 5017
sedimenteringstank, 5018
seghet, 5788
sektionsdrift, 4978
sekundaskog, 4970
sekundaväxt, 4971
sekunder, 4977
sekundärvägg, 4975
selektion, 4986
selektivitet, 4987
semibendingflis, 4988
semipermeabelt membran, 4996
sensibiliserat papper, 4999
sensibilisering, 4998
sensibiliseringsmedel, 5000
sensibiliseringspapper, 5001
sensor, 5002

separation, 5003
separator, 5004
Sequoia, 5007
servettpapper, 3939, 3940
service, 5009
servomekanism, 5011
Sheffieldprovare, 5055
shore-durometer, 5067
shore-hårdhet, 5068
sidledsporositet, 3428
sifon, 5115
sifonkondensatavskiljning, 5116
sikt, 5088
siktanalys, 5089
sil, 4945, 5445
sila, 5444
sileri, 4949
silikater, 5090
silikonharts, 5094
silkscreen-tryckning, 5095
silning, 4947
silrest, 4948
silt, 5096
silverföreningar, 5097
silvergran, 5098
silvikultur, 5099
silvinkel, 4946
sisal, 5117
sisalhampa, 5118
Sitka spruce, 5120
sköare, 5672
skada, 1805, 1861
skak, 5027, 5028
skaka, 3312, 3313, 5027
skakande, 5031
skakenhet, 5030
skakning, 5031
skal, 5057
skala, 4922
skalad ved, 4241
skalbarhet, 4240
skalning, 4243
skalpapper, 4924
skarv, 3314, 5325
skarva, 5325
skarvdon, 992
skarvmarkering, 5326
skarvning, 5327
skarvpapper, 5328
skenbar bulk, 415
skenbar densitet, 414
skevning, 6047
skibord, 6173
skift, 5059
skikt, 3434
skiktat papper, 5451
skiktning, 5450
skiljevägg, 3810
skisskartong, 4943
skiva, 2014, 2026
skivfilter, 2020
skivraffinör, 2022
skjuvhastighet, 5037
skjuvhållfasthet, 5036, 5038
skjuvning, 5035
skjuvprov, 5040
skjuvspänning, 5039
skjuvströmning, 2605
skog, 2666
skogsavverkningsmaskin, 3557
skogsdrift, 2674
skogseld, 2669
skogsförorening, 2672
skogsförvaltning, 2673
skogsgenetik, 2670
skogshantering, 2667
skogsindustri, 2671
skogslära, 2677
skogspatologi, 2675
skogsplantering, 4711
skogsprodukter, 2676
skogsvård, 5099
skokartong, 5066
skoning, 2437
skopa, 4900
skoptransportör, 932
skorsten, 1273
skott, 948
skramla, 4645

skrivande instrument, 4681
skrivare, 4678
skrivarpapper, 4679, 4682
skrivpapper, 6302
skrot, 4940
skrotavskiljare, 3324
skrottransportör, 3324
skrubber, 4954
skruv, 4950
skruvbotten, 3536
skruvbottenbinge, 3537
skruvdrift, 6289
skruvmatare, 4951
skruvpress, 4952
skruvtransportör, 4951
skrymdensitet, 946
skrynkla, 4565
skräpkartong, 4941
skugga, 5024
skuggmarkering, 5025
skum, 2628, 2738, 4958
skumbekämpningsmedel, 1875
skumbildning, 2190
skumborttagare, 5136
skumdämpare, 2633
skumdämpning, 2632
skumfläckar, 2739
skummande, 5137
skummarkering, 2635
skumning, 2630
skumningsfläckar, 2636
skumningsmedel, 2631
skumsläckande medel, 395
skumsläckare, 2629, 2634, 5207
skyddsbeklädnad, 4557
skyddsbeläggning, 4558, 4732
skyddspapper, 4733
skåra, 4931, 4933
skärmaskin, 1770
skärmaskinförare, 1774
skärmaskinsdamm, 1772
skärmaskin-sorterare, 1773
skärmaskinsutskott, 1771
skärning, 1775
skärning med kutter, 4853
skärpa, 969, 5033
skärpning, 970, 5032
skärptal, 971
skärverktyg, 1776
sköld, 5058
skörd, 2991
skördemaskin, 2990
skötare, 4071
slam, 5181, 5186
slamborttagning, 5182
slamma, 5184
slammare, 5187
slamning, 5185, 5188
slamningsutrustning, 5189
slang, 3091
slem, 5160
slembekämpningsmedel, 5163
slemfläckar, 5162
slemhål, 5161
slinga, 2440
slip, 5164
slipad, 3826
slipare, 2901
slipbokpapper, 2917
sliperi, 2904
slipficka, 2903
slipmassa, 2916, 2923
slipmassepapper, 2921
slipmassetillverkning, 2924
slipning, 2906
slipning med damm, 4347
slippapper, 111
slippulver, 112
slipsten, 2905, 2908, 4583
slipstens, 2909
slipstol, 2900
sliptryckpapper, 2922
sliptråg, 2902
slipyta, 2907
slirad vals, 5166
slirning, 5165
slitage, 6138
slitagemotstånd, 4760
slitageprov, 6140

slitageprovare, 6139
slits, 5154, 5176
slitsbestrykare, 2700
slitsfräs, 4218
slitsning, 5172, 5178
slitsningsdamm, 5168
slitsningsmaskin, 5177
slitsverk, 5167, 5173
slitsverksförare, 5170
slumpartat orientarade fibrer, 4640
sluss, 5183
slussvalsbestrykare, 2796
slussventil, 2795, 2797
sluten arkbildning, 1362
sluten inloppslåda, 1361
sluten krets, 1356
sluten kåpa, 1357
slutet system, 1359
sluttning, 5175
sly, 6163
slåpbladsbestrykare, 5800
slåpbladsbestrykning, 5801
släckning, 5146
släckningsmedel, 5145
släckt kalk, 5144
släpbladsbestrykare med damm, 4434
slät finish, 5159
slät gråpapp, 4348
slätt träd, 4352
slätvals, 2555, 4350
smet, 2702
smetprov, 3242
smuts, 2012
smälta, 3094, 5192
smälta bort, 1887
smältbindemedel, 3095
smältning, 3745
smältpunkt, 3746
smälttank, 5194
smältugn, 5193
smärpappersbestrykning, 2886
smörja, 2883
smörjig, 2895
smörjighet, 6194
smörjig mäld, 5180
smörjmedel, 5583
smörjning, 2894, 3585
smörjolja, 3584
smöromslag, 991
smörpapper, 2889
smörsyra, 998
sned, 4031
snedbelastning, 681
snitt, 1764
snittstorlek, 1769
snäppa, 5204
sockerlönn, 5518
sockerrör, 5516
sockerrörsfiber, 5517
soda, 5209
sodaaska, 5210
sodakokning, 5214
sodalut, 5211
sodamassa, 5213
sodaprocess, 5212
sojaproteiner, 5288
sol, 5257
solid, 5260
solubilisering, 5270
solvering, 5273
sond, 5002
sopsäck, 4714
sorbent, 5277
sorption, 5278
sort, 2857, 5279
sortera, 5279
sorterarbänk, 5281
sorterare, 5280
sortering, 5282
sot, 5276
southern pine, 5287
spak, 3459
specialitet, 5294
specialkartong, 5295
specialmassa, 5298
specialpapper, 5297
specialvågpapp, 5296
speciellivsmedelkartong, 5293
specifikation, 5299

specifik vikt, 5300
specifik volym, 5303
specifik värme, 5301
specifik yta, 5302
speglande glans, 5312
speglande reflektion, 5313
spektoskop, 5310
spektral reflektans, 5307
spektral reflektivitet, 5308
spektrofotometer, 5309
spektroskopi, 5311
spet, 5062
spets, 4396
spetsdragning, 5734
spetsklistrare, 5633
spindel, 3673
spinnpapper, 5319, 5320
spiral-, 3027
spiralegutör, 5324
spiralflishugg, 5321
spiralformig, 3027
spiralrör, 5323
spjällventil, 990
splintved, 4905
spole, 5334
spraya, 5341
sprayare, 5347
spraybefuktare, 5344
sprayning, 5348
spraytorkning, 5345
spricka, 1659
sprick-kant, 1660
sprickning, 1662
spridarbom, 5350
spridare, 5349
spridarvals, 5351
spridning, 4927, 5352
sprinkler, 5356
sprinkling, 5357
sprits, 5077
spritsa, 5341
spritsbestrykare, 5342
spritsbestrykning, 5343
spritsdäckel, 3309
sprund, 5995
sprutande, 5339
spruthastighet, 5340
sprutmunstycke, 2366
sprutmålning, 5346
sprutning, 5339
sprängfaktor, 973
spränghållfasthet, 975, 978
sprängindex, 977
sprängning, 974
sprängtryck, 972
sprängtrycksprov, 980
sprängtrycksprovare, 976, 979
sprätta, 5147
sprättare, 5148
sprödhet, 902
spunnen bindning, 5360
spån, 5034, 5174
spånplatta, 4213
spår, 2912
spårelement, 5796
späda, 1990
spädning, 1991
spänning, 5461, 5685
spänningsanalys, 5462
spänningsmätare, 5684
spänningsreglering, 5686
spännings-töjningsegenskaper, 5464
spänningsved, 5688
spännvals, 5687
spärr, 585
spärrmaterial, 586
spärrning, 589, 764
spärrningsmotstånd, 766
spärrskikt, 588
stabilisator, 5371
stabiliserad, 5370
stabilisering, 5369
stabiliseringsvals, 393
stabilitet, 5367
stabilitet hos massa, 5368
stagnationstryck, 285
stam, 5418
standard, 5382
standard tidningspapper, 5383

stapel, 4322, 5372
stapelfiber, 5384
stapel med vattenschaber, 6070
staplad dubbelpress, 886
staplare, 5373
stapling, 5374
stark syra, 5476
starta, 5393
startklar fabrik, 5880
stationärt tillstånd, 5401
statiska prov, 5397
statisk elektricitet, 5394
statisk tryck, 5396
statistik, 5399
statistisk analys, 5398
stativ, 2715
stator, 5400
stav, 4813
stavbestrykare, 569, 2054
stearater, 5413
stearinsyra, 5414
stele, 5417
stelning, 5262
stencil, 5419
stencilkartong, 5420
stencilpapper, 5421
stereoskop, 5423
stillestånd, 2084
stjälpbar kokare, 5744
stjärnmönstrad vals, 5392
stock, 3554
stock av trädlängd, 5833
stonit, 5440
stoppare för matningsregulator, 2409
storlek i yard, 6313
streck, 5455
strierad, 5469
struktur, 5478
stråla, 3298
strålcell, 4651
stråle, 3308
strålning, 4613
strålvärme, 4612
sträck-, 5675
sträckare, 5467
sträckgräns, 6320
sträckhållfasthet, 4611
sträckning, 5465, 5468, 5681
sträckningsenergi, 5676
strängsprutning, 3096
strängsprutningsbad, 2367
strängsprutningsbestrykare, 2364
strängsprutningsbestrykning, 2365
strömlinjeformat blad, 3136, 3137
stubbar, 5488
stubbe, 5487
stump, 5487
stumskarv, 988
stuvningsgods, 2152
stympa, 5856
styren, 5489
styrhjul, 5153
styrka, 5459
styrkeisotropt papper, 5364
styrkeprov, 5460
styrning, 1541
styrning med återkoppling, 1358
styvhet, 4800, 5425
styvhetsprov, 5427
styvhetsprovare, 5426
stål, 5415
stålklädd platta, 1336
stång, 4813
stångkvarn, 4815
stämpel, 1954
stämpelfräs, 1956
stämpelfräsning, 1957
stämpling, 5380
stärkelse, 5386
stärkelsederivat, 5389
stärkelsekokare, 5388
stärkelsekonvertering, 5387
stärkelsexantat, 5390
stärkelsexantid, 5391
stödja, 5521
stöd-liner, 512, 513
stödvals, 508
stödvalsmarkering, 509
stödvira, 511

stöt, 5063
stöthållfasthet, 3188
stötmotstånd, 5065
stötprov, 3189
subjektiv glans, 5490
sublimering, 5491
substand, 5493
substitut, 5494
substitutionsgrad, 1884
substrat, 5495
sug, 5497
sugarkformare, 5506
sugavtagning, 5947
sugavtagningsvals, 5507
sugben, 2108, 6081
sugfilter, 5943
sugfiltrering, 5944
sugformning, 5945
sugförpress, 5511
sugguskvals, 5502
sugkammare, 3068
suglåda, 2549, 5498, 5941
suglådelock, 5499
suglådemarkering, 5500
suglådeslits, 5501
sugmatning, 5504
sugpress, 5509
sugpressvals, 5510
sugpump, 5512
sugrakpress, 5515
sugrör, 5508
sugvals, 5513, 5949
sugvalsmarkering, 5514
sugöverföring, 5950
sukros, 5496
sulfamater, 5519
sulfaminsyra, 5520
sulfater, 5528
sulfatfabrik, 3374
sulfatkartong, 5521
sulfatkokvätska, 3373
sulfatlut, 5523
sulfatmassa, 5526
sulfatpapper, 5524
sulfatprocess, 5525
sulfatterpentin, 5527
sulfattäckskikt, 5522
sulfider, 5530
sulfidering, 5529
sulfiditet, 5531
sulfitavlut, 5318
sulfiter, 5536
sulfitkokning, 5535
sulfitlut, 5532
sulfitmassa, 5534
sulfitpapper, 5533
sulfogrupper, 5537
sulfonater, 5538
sulfonylgrupper, 5539
sulfoxider, 5540
sump, 5549
supernews, 5561
superstandard news, 5562
sur, 5284
sura färgämnen, 159
sura halider, 164
sura klorider, 158
surgörning, 166, 5285
surhet, 167, 5286
sur sulfit, 180
sur (syra-), 165
surt fägêmne, 160
surt lim, 178
svaglimmat, 5143
svag syra, 6137
svagt limmad, 5253
svamp, 2752
svampig, 5333
svampsäkring, 2754
svarstid, 4765
svartalbumpapper, 717
svart aska, 718
svartgran, 722
svartlut, 720
svartning, 719
svartpoppel, 721
svavel, 5541, 5547
svavelbränning, 5542
svaveldioxid, 5544

svavelföreningar, 5543
svavelsyra, 5545
svavelsyrlighet, 5546
svavelväte, 3145
svep, 4925
svetsfog, 6174
svetsning, 6175
svällning, 5602
svällningsmedel, 5603
svängande, 4088
svängbar, 5152
svängning, 4089
svävkniv, 2626
svävvals, 2627, 5604
sweet gum, 5601
sydstatstall, 5287
sykamor, 5606
symmetri, 5607
synergism, 5611
synkron, 5609
synkronism, 5608
synkronmotor, 5610
syntes, 5612
syntetfiber, 5613
syntetgummi, 5616
syntetharts, 5615
syntetpolymerr, 5614
syra, 157
syrabehandling, 181
syrafabrik, 169
syrafast, 170
syrafast fodring, 172
syrafast lim, 179
syrafast murning, 172
syrafast tegel, 171
syrafri, 161
syrafritt papper, 162
syragrupper, 163
syrahärdig, 176
syrahärdigt papper, 173, 177
syramotstånd, 174
syramotståndsprov, 175
syraskyddande manillapapper, 388
syratillverkning, 168
syratvätt, 182
syre, 4123
syreförbrukning, 4126
syreföreningar, 4125
syringylgrupper, 5617
syrsättning, 4124
system, 5618
systemteknologi, 5619
såg, 4917
sågkantad, 5008
sågning, 4919
sågskär, 3341
sågspan, 4918
sågverk, 4920
sågverksavfall, 4921
säck, 526
säcka, 4884
säck-liner, 537
säckmaskin, 535, 538, 539, 4148
säckpapper, 540, 4879
säckslutningsmaskin, 541
säcktillverkning, 534
säkerhet, 4880
säkerhetskok, 4881
säkerhetspapper, 4883
säkerhetsutrustning, 4882
säkerhetsventil, 4738
sämskskinnimitation, 938
sänkdon, 3582
säsongsvariationer, 4969
sätt, 5012
sättning, 5014
sättningstid, 5015
söm, 4967
sömnad, 4968
sönderdela, 882, 3602
sönderdelad, 3603
sönderdelning, 3604, 3730

T

tabeller, 5624
tablett, 4245
tablettbildning, 4246
tabulering, 5626
tachometer, 5627
takpapp, 1175, 4832, 4833
talk, 5636
tall, 2520, 4325
tallharts, 4327
tallolja, 5637
talloljetvål, 5638
tallrik, 4366
tallsulfatmassa, 4328
tallsulfitmassa, 4329
tamarack, 5639
tamarix, 5640
tamponering, 6227
tand, 5772
tank, 859, 5641
tank för spontan avdunstning, 2561
tantalföreningar, 5642
tapet, 6045
tapiokastärkelse, 5648
tapp, 5857
tappning, 5649
tara, 5651
Taxus, 5653
tejp, 5643
tejpning, 5647
tejppapper, 5644
teknik, 5665
teknologi, 2283
teleskopformad, 5666
teleskopformad vals, 5667
television, 5669
temperatur, 5670
temperaturgradient, 3007
temperaturreglering, 5671
tenn, 5748
tennföreningar, 5749
tensiometer, 5684
tepåspapper, 5654
termiska engenskaper, 5707
termisk miljöförstöring, 5706
termisk utvidgning, 5704
termodynamik, 5713
termoelement, 5712
termomekanisk massatillverkning, 5715
termometer, 5716
termoplast, 5717
termosättning, 5718
terpener, 5689
terpentin, 5882
terpentinprov, 5883
terpinener, 5690
tertiärbehandling, 5692
tertiärlamell, 5691
tertiärvägg, 5693
testboard, 5695
testlinerkartong, 5697
textilfiber, 5701
textur, 5702
Thuja, 5739
tid, 5746
tidning, 3965
tidnings-offset, 4027
tidningspapper, 3960, 3966
tidningspapper i rulle, 6153
tidningspapper med ett manillatäckskikt, 5110
tidningspappersark, 3967
tidningspapperskvalitet, 3963
tidsinställningsanordning, 5747
tidskrift, 3319
tidsplanering, 4928
Tilia, 5743
tillbehör, 139
tillförlitlighet, 4734
tillgänglikt klor, 492
tillsatser, 202
tillsats för våtända, 6185
tillsatsmedel, 206

tillsatsvatten, 3666
tillskott, 3205
tillvaratagande, 4890
tillverkning, 3687, 4534
tillverkningsorder, 3667
tillväxt, 2926
tillväxthastighet, 2928
tillväxthylsa, 3206
tillväxtlager, 2927
tillväxtvävnad, 3755
tillämpning, 418
timmer, 3586, 5745
timmerbassäng, 3560
timmerbröt, 3559
timmerhuggare, 3587
timmerhuggning, 3556
timmerklyvning, 3561
tiokarbamater, 5727
tiokarbonater, 5728
tioligniner, 5729
tiosulfater, 5730
titan, 5757
titandioxid, 5759
titanföreningar, 5758
titrebar syra, 5760
titrerbart alkali, 5761
titrering, 5762
tixotropi, 5733
tjocklek, 1060, 5722
tjockleksmätare, 1061, 5723
tjockända, 987, 989
tjog, 4931
tjänst, 5009
tjära, 5650
tjärpapper, 5652
toalettpapper, 5763, 5766
toalettrulle, 5764
toalettrullskärare och perforator, 5765
tolerans, 5767
toluen, 5768
tomrum, 731
ton, 5769
topp, 5773
torkare, 2122
torkcylinder, 2122, 2138
torkcylinderinkrust, 4923
torkeffektivitet, 2192
torkficka, 2128
torkfilt, 2123, 2131
torkfiltmarkering, 2124
torkfinish, 2132
torkhastighet, 4643
torkhuv, 2125
torkkapacitet, 2136
torkkäpp, 2130
torklada, 3549
torkmedel, 2135
torknappningsmarkering, 2127
torkning, 2134
torkparti, 2126, 2129
torksprickor, 2137
torkugn, 2140
torkvind, 2139
torn, 5794
torr, 2117
torrbestrykning, 2119
torrformningsprocess, 2133
torrgnuggmotstånd, 2142
torrhalt, 5267
torrhet, 2141
torrhållfasthet, 2144
torrhållfasthetsmedel, 2145
torrkräppning, 2120
torrutskott, 2118
torrvaxat papper, 2143
torrände, 2121
torsion, 5785
total alkali, 5786
totaltransmittans, 5787
toxicitet, 5795
trakeider, 5797
transduktor, 5803
transformator, 5807
transistor, 5808
transitionspunkt, 5809
transmissionsledning, 5813
transmittans, 5814
transparens, 5815, 5816
transparenskvot, 5817

transparent, 5818
transparent cellulosa, 5819
transpiration, 5821
transplantation, 5822
transport, 1561, 5823
transportemballage, 5061
transportskruv, 1564
transportör, 1562
trappremskiva, 5422
trasa, 4621
tratt, 3086
tredje hand, 5731
tredjepress, 5732
trekannslip, 5736
trelags-, 5735
trelagskartong, 5852
triangelformad plåt, 2950
trika, 3356
trinatriumfosfat, 5853
triplex, 5851
trippelviramaskin, 5850
tropisk växt, 5854
truck, 5855
trumbarkare, 2111
trumfilter, 2114
trumma, 2110, 2148
tryck, 4495
tryckare, 4521
tryckbarhet, 4520
tryckbarhetsprov, 4528
tryckbehandlad utrustning, 4509
tryckeri, 4491
tryckeriföreståndare, 4492
tryckfall, 4497
tryckfilt, 4523
tryckfilter, 4498
tryckfärg, 4524
tryckförlust, 2998
tryckglättning, 4527
tryckgradient, 4500
tryckinloppslåda, 4501
tryckkänslighet, 4507
tryckkänsligt papper, 4506
tryckminskning, 4503
tryckmätare, 4499
tryckmätning, 4502
trycknedsättning, 4739
tryckning, 4522
trycknivå, 2993
tryckpapper, 4526
tryckpress, 4525
tryckpresstoppare, 4494
tryckpåläggning, 4508
tryckraffinering, 4510
tryckreducering, 4735
tryckreduceringsgas, 4736
tryckreglering, 4496
tryckregulator, 4504
trycksak, 3442
tryckvalsbestrykare, 4530
tryckvalsbestrykning, 4531
tryckved, 1481
tråd, 2481
tråg, 2699, 5975
trågfodring, 5860
trågfärger, 5976
trä, 6260
träavfall, 6281
träd, 5830
trädning, 5734
trädplantering, 5832
trädsax, 5834
trädstam, 816
träfiber, 6265
träflis, 6261
träfri, 2918, 6266
träfritt papper, 6267
trähaltigt papper, 6262
trähylsa, 6263
trämassa, 6276
trämassefodrad gråpapp, 3968
trämassefodrad kartong, 3964
trämassekartong, 6277
träplugg, 6264
träprodukter, 6275
träslipare, 6269
träsliperi, 2920
träull, 6282
Tsuga, 5858

tub, 5865
tubkartong, 5866
tubmassa, 5867
tulpanträd, 5868
tumning, 6271
tumtimme, 3201
tungvikt, 3025
tunna, 1068
tunneltork, 5871
tunnfilm, 5724
tunnpapper, 1470, 3477, 5726, 5756
tunnpappersmaskin, 5755
turbid, 5872
turbidimeter, 5873
turbiditet, 5874
turbin, 5875
turbogenerator, 5876
turbulens, 5877
turbulent flöde, 5878
tusendels tum, 3812
tvêrs fibrerna, 186
tvillingviramaskin, 5886
tvillingvirapapper, 5887
tvåfiltpress, 2077
tvåkompnentfiber, 685
tvål, 5206
tvålags-, 5891
tvålagskartong, 5892
tvålomslag, 5208
tvåsidesbestrykare, 5893
tvåsidig, 5896
tvåsidig bestrykning, 5895
tvåsidighet, 5897
tvåsidig limning, 2080
tvåsidigt bestruket papper, 5894
tvåvåningstork, 2067
tvåwellpapp, 2074
tvärbindning, 1708
tvärbrygga, 1165
tvärnitt, 1711
tvärprofil, 1709
tvärsektion, 1711
tvärsfiber, 5825
tvärsflöde, 1706
tvärsflödespress, 5824
tvärsporositet, 5826
tvärsriktning, 1707
tvärs-återhämtning, 1710
tvätt, 6052
tvättbräde, 6050
tvättbrädesmarkering, 6051
tvättmedel, 1928
tvättning, 6053
tvättorn, 4955
tvättrumma, 6054
tyg, 2370
tyloser, 5898
tyngd, 6170
typsättning, 5900
täckning, 1654, 1655, 1656, 3509, 4100
täckningsförmåga, 1657
täckpapper, 1658, 2997
täckpapper för —, 3621
täckskikt, 3477
täckskiktkartong, 3499, 3506
täckskiktpapper, 3507
tändning, 3178
tändsticka, 3711
tändsticksaskpapper, 1469
tändsticksplånkartong, 836
täppa, 1354
tätbarhet, 4960
tätning, 4959, 4962
tätningsmaskin, 4964
tätningsmedel, 4961
tätningspapper, 587, 4963
tätningsring, 5485
tätningsvattenläda, 4965
tätningsvattentank, 4966
tätortsavfall, 3933
töja, 5444
töjbarhet, 2354
töjbart papper, 2355, 5466
töjning, 5444
töjningsgivare, 5446
töjningsmätare, 2356
tömma, 2015, 2339
tömningsfläkt, 2340

tömningstryck, 2016
tömningsventil, 2017

U

ugn, 2760, 3345, 4097
ugnstorr, 4014, 4098
ugnstorr vikt, 4099
Ûhle-låda, 5901
ull, 6285
ull-, 6286
Ulmus, 5902
ultraviolett strålning, 5903
umbärlig, 2030
underfilt, 852
underfiltpress, 853
underhåll, 3662
underhållsverkstad, 3663
underklorsyrlighet, 3173
underkokad, 5913
underliner, 5914
underpress, 854
under pressvals, 855
understöd, 516
undervals, 856
ungplanta, 4984
ungträd, 4901, 6013
ungved, 3334
uppehållstid, 2169
uppförande, 4260
uppgradering, 5932
upphettning, 3008
upplösa, 2032
upplösare, 2034
upplösarknivar, 884
upplösning, 889, 2035
uppradning, 3510
upprullning, 6238
upprätthållet utbyte, 5596
uppsamlingsvanna, 4316
uppslagning, 2024
uppslagningsholländare, 885
uppslagningsmaskin, 883
uppslutning, 1895
upptagen massa, 3421
upptagning i ark, 3422
upptagningsmaskin, 1848, 3420, 6191
uppvärmningsanordning, 3009
uppåtgående vira, 5933
urblekning, 2379
urblekningsmätare, 2378
urblekningsprov, 2381
urborrning, 846
urea, 5934
ureaformaldehydharts, 5935
ureahartser, 5936
uretaner, 5937
urladda, 2015
urmodighet, 4016
urpressningsvals, 5365
urval, 4986
utblekningsmotstånd, 2380
utbredning, 4549
utbuktning, 944
utbuktningsmotstånd, 943
utbytbarhet, 3267
utbyte, 6319
utbytestabeller, 6321
utdragbar behållare, 5668
utdragbart viraparti, 4741, 4827, 4872
utfall, 4094
utgående mängd, 4096
uthållning, 3071
utjämningsbehållare, 5594
utkörd, 4871
utlagd, 3395
utlopp, 4095
utmattning, 2391
utmattningsbrott, 2392
utmattningsmotstånd, 2393
utmattningsprov, 2394
utnyttjande, 5939
utrustning, 2300
utskott, 905, 1736, 3607, 3820, 5634
utskottsknippe, 906

utskottsmassa, 1088
utskottspulper, 909
utskottsraffinering, 4725
utströmningshastighet, 5158
utsugningsfläkt, 3213
utsvettning, 2368
uttorkad spänning, 2102
uttorkad töjning, 2101
utveckling, 2335
utvärdering, 2330

V

vadd, 6040
vaddbaspapper, 599
vaddmassa, 6042
vaddpapper, 6041
vakt, 2930, 5789
vaktchef, 5790
vakuum, 5940
vakuummetallisering, 5946
vakuumpump, 5948
valens, 5951
vals, 4816
valsad kant, 4822
valsbestrykare, 4818, 4825, 5366
valsbestrykning, 4819
valsbombering, 4820
valsgavel, 4823
valspapper, 4828
valspåläggare, 4826
valsrakelbestrykare, 4814
valsskydd, 4829
valssmatter, 1233
valsstativ, 4831
valsände, 4823, 5479
valvbildning, 429
vandring, 3811
vanillin, 5955
vanillinderivat, 5956
varaktighet, 2162
variabel, 5968
variabilitet, 5967
varians, 5970
variansanalys, 365
variant, 5971
varietet, 5971
varmmassabehandling, 3100
varmpressning, 3097
varmslipning, 3093
varm vals, 3099
varp, 6046
varpning, 6047
varptråd, 6049
varptrådar, 6048
varv, 4782
varv per minut, 4783
vass, 4698
vatten, 6063
vatten-, 426
vattenabsorberande, 6064
vattenabsorbtion, 6065
vattenanalys, 6066
vattenavstötning, 6101
vattenbehandling, 6113
vattenborttagning, 6100
vattendelare, 6107
vattenfinish, 6074
vattenfinished, 6075
vattenfläckar, 6110
vattenförbrukning, 6071
vattenförorening, 6092
vattenförsegling, 6106
vattenförsörjning, 6112
vattenhantering, 6083
vattenhjul, 6118
vattenkraft, 3132
vattenkraftverk, 3133
vattenkräppad, 6073
vattenkvalitet, 6098
vattenkyld fjädervals, 6072
vattenlåda, 6068
vattenläckning, 6067
vattenlöslig, 6109
vattenmotstånd, 6102

vattenmärke, 6084
vattenmärkningsegutör, 6085, 6086, 6088, 6090
vattenmärkningsvals, 6087
vattenpass, 3452
vattenpermeabilitet, 6091
vattenrandat papper, 6082
vattenretention, 6104
vattenränder, 6111
vattenrörkokare, 6114
vattenschaber, 6069
vattenstråle, 6079
vattenströmning, 6078
vattenturbin, 6115
vattentäkt, 6103
vattentät, 6093
vattentäthet, 6096
vattentät kartong, 6094
vattentätning, 6095
vattentätt papper, 6097
vattenvandring, 4985
vattenånga, 6116
vattenåteranvändning, 6105
vattenåtervinning, 6099
vax, 6121
vaxad kartong, 6123
vaxad pergamyn, 6124
vaxat kraftpapper, 6125
vaxat papper, 6126
vaxat tunnpapper, 6127
vaxbaspapper, 6131, 6132
vaxbestrykning, 6122
vaxemulsion, 6128
vaxfläck, 6135
vaxlim, 6133
vaxlimmat papper, 6134
vaxning, 6129
vaxningsmaskin, 6130
vaxprov, 6136
veck, 1669, 2643, 4796, 6301
veckad, 1697
veckad limpåläggare, 2645
veckare, 1672
veckhållfasthet, 1677
veckning, 1675
veckningsmotståndsprovare, 1676
veckretention, 1674
vecksäkring, 1671
veckåterhämtning, 1673
vedavfall, 6281
vedfri växt, 4006
vedgård, 6283
vedgårdsarbetare, 6314
vedhantering, 6270
vedhanteringsrum, 6279
vedharts, 6278
vedpartikel, 6272
vedpreparation, 6274
vedprodukter, 6275
vedstruktur, 6280
vedtrave, 6273
vegetabiliskt pergament för konsttryck, 444
vegetabiliskt pergamentpapper, 5981
velinpapper, 4207
velourpapper, 5990
veluriserat papper, 2576
velängegutör, 6292
velängfinish, 5985
velängform, 6293
velängglättning, 1056
velängkartong, 5984
velängpapper, 5983, 5986, 6291, 6295
venitlering, 5996
ventil, 5952, 5995
ventilation, 5996
ventilationshålvals, 4395
ventilator, 5997
ventilering, 5998
ventilsäck, 5953
venturirör, 5999
verktyg, 5771
vertikalflödespress, 6001
vertikalformningsmaskin, 6002
vertikal kokare, 6000
vetehalm, 6217
vetestärkelse, 6216
vibration, 6005

vibrator, 6006
vibrerande sil, 6004
vickeryfilttvätt, 6007
vidarebehandling, 4974
vidd, 6228
vika, 2643
vikbarhet, 1668
vikhållfasthet, 2652
vikmapp, 2483
vikning, 2646
vikningsprov, 2654
vikningsprovare, 2653
vikt, 2644, 3705, 6170
vikt kant, 5879
vikttolerans, 6172
viktvisare, 6171
vild, 6229
vild arkformning, 6230
vildhet, 6232
vilt utseende, 6231
vind, 4031
vinylacetat, 6008
vinylcellulosa, 6009
vinylföreningar, 6011
vinylidenklorid, 6012
vinylklorid, 6010
vira, 6242
virabelastning, 6249
virabord, 425
virafläck, 6256
viraföring, 2936
viragrav, 6252
viragrop, 6252
virahål, 6246
viralivslängd, 6247
viramarkering, 6250
viraparti, 6254
virariktare, 6245
viraränder, 6248
virasida, 6255
viraskepp, 5829
virasträckare, 6257
viratrag, 3905
viravals, 6253
viravalsbestrykning, 6259
viravatten, 5635
viravändvals, 6258
viraväv, 6243
virvel, 6035
virvelrenare, 3130
virvelström, 2182
visare, 1939
viskoelasticitet, 6015
viskomassa, 6020
viskometer, 6016
viskometri, 6017
viskos, 6018
viskosfiber, 6019
viskosimeter, 6022
viskositet, 6023
viskositetsindex, 6025
viskositetsmedel, 6024
viskosrayon, 6021
viskös flytning, 6026
vitbjörk, 6218
vitgran, 6224
vitgörning, 6226
vithet, 6222
vitrifiering, 6027
vittall, 6219
vitt papper, 6223
volframföreningar, 5870
voltmeter, 6033
voluminös, 957
volym, 6034
volymitet, 5303
vridmoment, 5784
vridning, 5888
vulkanfiber, 6036
vulkanfibermassa, 6037
vulkanisering, 6038
vulkaniseringsbaspapper, 6039
vulstning, 616
våg, 4922
vågighet, 6119
vågig kant, 6120
vågning, 2621
vågningspapper, 2622
vågskålstransportör, 6169

vågtratt, 6168
vårved, 2176, 5355
våt, 6180
våtark, 6189
våt bana, 6212
våtdraghållfasthet, 6208
våtexpansivitet, 3134
våtfilt, 6187
våtfinish, 6188
våtförstärkningshartser, 6205
våtförstärkningsmedel, 6202
våtglättat papper, 6076
våtglättning, 860, 6077
våtgnidning, 6198
våtgnidningshållfasthet, 6199
våthållfasthet, 6201
våtkartongmaskin, 6181
våtmassa, 6197
våtmullen, 6193
våtpress, 6195
våtpressning, 6196
våtrand, 6215
våtränder, 6200
våtstarkt papper, 6204
våtstarkt utskott, 6203
våtsträckning, 6183
våtstyrka, 6201
våtstyrkeretention, 6206
våtsystem, 6190
vått trä, 6214
våtupptagen kartong, 6192
våtutskott, 6182
våtvaxat papper, 6211
våtvikt, 6213
våtända, 6184, 6244
väderbeständig, 6141
väderbeständigande, 6142
väderbeständighet, 6143
väft, 6164, 6284
väggeffekt, 6044
vägning, 6167
välja, 4308
vällimmad, 6178
väl sluten formering, 6177
vändpress, 4777
vändpressfilt, 4778
vändvals, 4688
vändvalsbestrykare, 4779
vändvalsbestrykning, 4780
värme, 3000
värmeackumulator, 3001
värmebalans, 3002
värmebehandling, 3024
värmeenergitet, 5301
värmeförbrukning, 3004
värmeförlust, 3011
värmeförsegling, 3016
värmeförseglingshållfasthet, 3015
värmeförseglingsmaskin, 3017
värmeförseglingspapper, 3018
värmehärdande tryckfärg, 3019
värmehärdning, 3020
värmeisolation, 3010
värmeisolering, 5705
värmekonduktivitet, 5703
värmekänslighet, 5708
värmeledning, 3023
värmeledningsförmåga, 3003
värmemotstånd, 3014
värmeskrivare, 5711
värmeskrivarpapper, 5714
värmespänning, 5710
värmestabilitet, 5709
värmeverkningsgrad, 3005
värmeväxlare, 3006
värmeåtervinning, 3013
värmeöverföring, 3021
värmeöverföringskoefficient, 3022
värmning, 3008
vätbarhet, 6207
vätebindning, 3140, 3141
väteföreningar, 3142
vätejonkoncentration, 3143
väteperoxid, 3144
vätning, 6209
vätningsmedel, 6210
vätska, 3517
vätska för avvägningsinstrument, 3457
vätskefaskokning, 3523

vätskeflöde, 2603
vätskeformig, 3517
vätskeförpackningskartong, 1070
vätskepenetrationsmotstånd, 4759
vätskepermeabilitet, 3519
vätsketätning, 3520
väv, 6144
väva, 6144
vävning, 6145
vävt tyg, 6294
växel, 2799
växellåda, 2800
växelström, 100, 326
växt, 2926
växtfiber, 4355

W

wellad, 2620
wellförslutning, 1608
wellkartong, 1615
wellmaskin, 1616
wellpapp, 1607
wellpappemballage, 1613
wellpapper, 1617
wellpapp för lådor, 1610
wellpapplåda, 1609
wellpapprulle, 1611
wellpipa, 2619
wellskikt, 1612

X

xantatgrupper, 6303
xylem, 6304
xylener, 6305
xylensulfonater, 6306
xylos, 6307

Y

yankeemaskin, 3600, 6312
yankeetork, 6311
ylle, 6285
yllefilt, 6288
yllene, 6286
ympning, 2860
ymppolymerer, 2861
yta, 2373, 5568
ytaktivt medel, 5569, 5593
ytavskalning, 5579
ytbehandling, 5576, 5589
ytegenskaper, 5580
ytfinish, 5575
ytfärgad, 5574, 5588
ytfärgning, 5573
ythållfasthet, 5571, 5586
ythärdning, 1141, 5577
ytlimmad, 5582, 5861
ytlimmat papper, 5583
ytlimmatt papper, 5862
ytlimning, 5585, 5780, 5864
ytmotstånd, 5581
ytskiktslyftning, 5578
ytspänning, 5587
ytvatten, 5590
ytvikt, 604, 4668, 5493
ytvätbarhet, 5591
ytvätbarhetsprov, 5592

Z

zeoliter, 6323
zetapotential, 6325
zinhydrosulfit, 6328
zink, 6326
zinkföreningar, 6327
zinkoxid, 6329
zinkstearat, 6330
zinksulfat, 6331
zinksulfid, 6332
zirkonföreningar, 6333
Z-riktningsstyrka, 6322

Å

ådrighet, 6268
ålder, 230
åldras, 1841
åldring, 231, 236
åldringsmotstånd, 237, 1842
åldringsprov, 238

ånga, 5402, 5957, 5966
ångavdrivning, 5410
ångfas, 59 1
ångfaskokning, 5962
ångfinish, 5405
ångfälla, 5411
ångkraft, 5404
ångkåpa, 1365, 2342
ångning, 5407
ångpermeabilitet, 5960, 6117
ångspiraler, 5403
ångspärr, 5958
ångtryck, 5963
ångturbin, 5412
ångtät, 5964
ångtätt papper, 5965
ångvals, 5409
ångvärmning, 5406
årsring, 383, 2929
ås, 4796
återcirkulering, 4689
återflöde, 504
återhållning, 4766
återhämtning, 4684
återkoppling, 2404
återkopplingskontroll, 2405
återstod, 4750, 4890
återuppbyggning, 4671
återupplösning av massa, 4746
återvinning, 4675, 4683, 4684
återvinningstorn, 4686

återvinningsugn, 4685
återvunnen fiber, 4674
återväta, 4785

Ä

äggkartonger, 2196
äggskalsfinish, 2197
äggviteämnen, 4559
äldre, 2203
ändband, 2271
ändlös vira, 2272
ändlös vävd filt, 2273
ärftlighet, 3033
ättiksyra, 151
ättiksyreanhydrid, 152

Ö

ölfilterpapper, 651
ölglaspapp, 652
öppen behållare, 4065
öppen inloppslåda, 4064
öppen kalander, 4063
öppnare, 4062
öppning, 5154
överdragspapper, 5041
överexemplar, 4106

överfiltpress, 5775
överflöd av bestrykning, 2337
överföra, 5013, 5804
överföringspress, 5806
övergång till containerhantering, 1526
överhettad, 5557
överhettare, 1926
överhettning, 5558
överingenjör, 1270
överinseende, 5563
överklorsyra, 4254
överkok, 4101
överkokad, 4102
överlagt papper, 4108
överlopp, 4105
överläggning, 4107
överpressvals, 5777
översida, 5779
överstorlek, 4111
översvämmad, 4110
övertorkad, 4103
övertorr, 4104
övertryckning, 4109
övertändningspunkt, 2547
övervals, 5778
övervikt, 4112
överviktglidning, 4114
överviktig rulle, 4113
övningsutrustning, 5802
övre liner, 5776
övre läpp,

english

svenska

deutsch

français

español

DEUTSCH

A

Abaka, 103
Abakazellstoff, 104
Abbau, 1841, 1858, 1879
Abbinden, 5014
Abbindezeit, 5015
abblasen, 777
Abblasen (der Papierbahn vom Filz), 773
Abblasetank, 780
abblättern, 4922
Abblättern, 2537, 5579
abbrausen, 5077
abbürsten, 915
Abdämpfen, 1810
Abdecken bestimmter Flächen beim
    Aufbringen von Farbe, 3703
Abdeckmittel, 3704
Abdeckung, 1656, 3703
Abdrängung, 2103
Abfall, 4724, 6056
Abfallbeseitigung, 6057
abfallen, 2109
Abfallen der Stoffbahn vom Filz, 2109
Abfärben, 1703
Abfedern, 1763
abfliessen, 4871
abfliessen lassen, 2085
Abfluss, 2085, 2193, 4095
Abflussleitung, 4094
Abgabe, 1898
Abgas, 2341, 2597, 4736
Abgasvorwärmer, 2180
abgeladen, 5922
abgepackte Packpapierformate, 1648
abgeschrägte Kante, 2399
abgestossene Rollenränder, 907
abgeteilter Behälter, 4217
abgraten, 969
Abguss, 1152
Abhang, 5175
abhauen, 1764
Abheben des Strichs, 4305
abholzen, 3554
Abietinsäuren, 106
Abkleidung, 3261
Abkreiden, 1222
abladen, 2015
Abladen, 2151, 5924
Ablagebock, 5132
ablagern, 230, 1914
Ablagerung, 1915, 4922
Ablage (von Papierbogen), 1898
Ablass, 4095
ablassen, 2015
Ablassen, 752, 2015, 6067
Ablauf, 4532
ablaufen, 2085, 4871
Ablauge, 720, 776, 5317, 6058
ablaugen, 5284
Ableeren, 2151
Ableerkammer (beim Pulper), 2358
Ablenkblech, 1872
ablenken, 1870
Ablenkplatte, 523
ablösen, 4922
Ablösen, 5475
Abmessung, 1992, 3726, 3748
Abnahme, 4314
Abnahmefilz, 852, 4315
Abnahmeprüfung, 134
abnehmen, 4301
Abnehmen der Geschwindigkeit, 1844
Abnutzung, 108, 6138
Abnutzungsfestigkeit, 4957
Abnutzungsprüfgerät, 6139
abreiben, 107
Abreiben, 4956
Abrieb, 108, 4869
Abriebeffekt, 110

Abriebfestigkeit, 109
Abriebgrad, 110
Abriss einer Rollenbahn, 4932
abrollen, 5928
Abrollen, 5930
Abroller, 5929
Abrollständer, 516
Abrollung, 5931
Abrollvorrichtung, 5931
Absackung, 4884
absägen, 4917
Absatz, 1914, 3696
Absatzwesen, 3696
absaugen, 2339
Absauger, 2340
Absaugpumpe, 5948
Absaughaube, 5948
Absaugwalze, 5949
abschaben, 107, 4922
Abschaum, 2738
abschäumen, 4958
Abschäumen, 5137
Abscheider, 5004, 5827
Abscheidung, 1914, 1915, 4471, 5003
Abscheren, 5035
Abscheuerung, 108
Abschirmung, 4945, 5058
abschlämmen, 2251
abschleifen, 107
Abschmutzen, 5013
Abschöpfen, 5137
Abschwächung, 481
Absetzbecken, 5017
Absetzbütte, 2091
absetzen, 4024
Absetzen, 1914, 4980, 5016
Absetzkasten, 2091
Absetztank, 5018
Absetzteich, 5017
absolute Feuchtigkeit, 113
absolutes Trockengewicht, 825
absolut trocken, 498, 824, 4014, 4098
absorbieren, 114
absorbierfähiges Papier, 683
Absorption, 120
Absorptionsgerät, 118
Absorptionskoeffizient, 121, 1426
Absorptionskohle, 193
Absorptionsmessung, 122
Absorptionsmittel, 118
Absorptionsturm, 124
Absorptionsvermögen, 115, 126
Abspalten der Oberfläche, 5579
Abspaltung, 5003
Abspeervorrichtung, 1419
absperren, 4959
Absperrschieber, 2797
Absperrung, 4959
Absperrventil, 1237, 5084
Absplittern, 1294
Abspritzanlage mit Abstellmechanismus,
    3358
Abstand, 5290
Abstauben, 2168
absteifen, 873
Abstreichmesser, 2048
Abstreifbarkeit, 4240
Abstreifen überschüssiger Streichfarbe für
    Testzwecke, 2097
Abstreifer, 1872, 2048, 2638
Abstreiferansatzwinkel, 2639
Abstreifwalze, 2057
abstumpfen, 5856
abstutzen, 5856
Abtasten der Papierbahn, 4925
Abtragung, 2304
abtreiben, 2103
abwaschen, 6055
Abwasser, 2193, 5021, 6062
Abwasseraufbereitung, 2194
Abwasser-Belüfter, 223
Abwasserkanal, 5022
Abwasserkasten, 4915

Abwasserklärung, 1339, 2194
Abwasserreinigung, 2194
Abwasserverlust, 5023
abweichen, 1870
Abweichung, 4024, 5970
abweisend, 4742
abwickeln, 2961
Abwickeln, 5930
Abwickler, 5929
Abziehbarkeit, 4240
Abziehbilderpapier, 1840
Abziehen, 4243, 5475
Abziehfestigkeit, 4611
Abziehwiderstand, 4611
Abzug, 3084, 5804
Abzughaube, 2342
Abzugskammer, 2358
Abzugskanal, 5022
Abzugspapier, 2161, 3677
Acetal, 146
Acetaldehyd, 145
Acetate, 150
Acetatfaser, 147
Acetatseide, 149
Acetatzellstoff, 148
Acetonitril, 154
Acetylgruppen, 156
Acetylierung, 155
Achatimitationspapier, 229
Achse, 495, 5026
achsial, 493
Acidität, 167
acqua destillat, 2040
Acrylamid, 187
Acrylat, 188
Acrylfaser, 191
Acrylnitril, 192
Acrylsäure, 189
Acrylsäureester, 188
Acrylverbindungen, 190
Additionsmaschinenpapier, 200
Additionspolymer, 201
Adhäsion, 204
Adhäsivpapier, 205
adiabatische Zustände, 209
Adressbuchpapier, 2011
Adressenaufkleberpapier, 203
Adsorbat, 215
Adsorbens (der aufnehmende Stoff), 216
adsorbieren, 214
adsorbierter Stoff, 215
adsorbierte Substanz, 215
Adsorption, 217, 5278
Adsorptionsfähigkeit, 218, 219
Adsorptionsvermögen, 218
aerobes Verfahren, 224
Aesterstoff, 3362
Affichenpapier, 4450
Affinität, 225
Agalit, 227
Agar, 228
Agens, 232
Agglomerat, 234
agglomerieren, 233
Aggregat, 235
Ahorn, 144, 3689
Akazie, 128
Akaziengummi, 129
Akazin, 129
aktiviertes Kohlepapier, 193
Akku, 142
Akkumulatorenpapier, 613
aktiver Schwefel, 198
aktives Alkali, 196
aktives Chlor, 492
Aktives Chlorgas, 197
Aktivkohle, 193
Akustikplatte, 183
akustische Eigenschaften, 185
Akzeptierbarkeit, 132
Alaun, 327
Alaunflecken, 336

Albumin, 286
Albumpapier, 287
Aldehyde, 292
Aldehydgruppen, 291
Algen, 294
Algenbekämpfungsmittel, 295
Alginat, 296
aliphatische Verbindungen, 298
Alkaleszenz, 313
Alkali, 299
Alkalibeständigkeit, 303
Alkalicellulose, 300
alkalifest, 314
alkalifeste Cellulose, 304
alkalifester Seifenschachtelkarton, 316
Alkaligewinnung, 301
Alkalilignine, 302
alkalilöslich, 305
Alkalinität, 313
alkalische Behandlung, 307
alkalische Leimung, 312
alkalischer Füllstoff, 308
alkalischer Prozess, 309
alkalischer Zellstoffaufschluss, 311
alkalisches Verfahren, 309
Alkalität, 313
Alkalizellstoff, 310
Alklarylsulfonat, 317
Alkohole, 290
Alkoholgruppen, 288
Alkylgruppen, 319
Alkylierung, 318
alle beweglichen Teile einer
    Papiermaschine (Langsieb, Filz-
    Deckelriemen-Siebleder, etc.) 3610
Alphacellulose, 322
Alphamessgerät, 323
Alphaprotein, 324
Alphazellstoff, 325
als Wahl aussortiertes Papier, 1736
Alter, 230
altern, 230
Altern, 231, 236
Alterung, 231, 236
alterungsbeständiges Papier, 4268
Alterungsbeständigkeit, 237, 1842
Alterungsprüfung, 238
Altpapier, 6059
Altpapierhändler, 6060
Altpapierstoff, 6061
Alufolie, 332
Aluminat, 328
Aluminium, 329, 330
Aluminiumpapier, 333
Aluminiumstearat, 334
Aluminiumsulfat, 335
Aluminiumverbindungen, 331
Amberbaum, 3518, 4690, 5601
amerikanischer Schwarzlindenholz, 605
Amide, 337
Aminoäthylcellulose, 341
Aminogruppen, 342
Aminopropylcellulose, 343
Aminosäuren, 339
Aminpolymere, 338
Ammon, 345
Ammoniak, 345
Ammonium-Bisulfit, 347
Ammonium-Bisulfitlauge, 346
Ammoniumhydroxid, 349
Ammoniumstearat, 350
Ammoniumsulfat, 351
Ammoniumsulfide, 3153
Ammoniumverbindungen, 348
amorph, 353
amorphe Cellulose, 354
amorphes Material, 355
amorphes Silika (Füllstoff), 1946
Amperemeter, 344
amphoter, 356
am wachsenden Stamm entrinden, 4904

Deutsch

Amylase, 357
Amylose, 358
anaerobes Verfahren, 359
analog, 360
Analog-Digital-Umsetzer, 363
Analogrechner, 361
Analogsystem, 362
Analyse, 364
Analysen-Filterpapier, 366
Anatas (Titandioxid), 367
Anbau, 1737, 4354
anblasen, 2384
an den Presschabern bei Abriss
   anfallender Ausschuss, 2050
an den Rändern befleckt, 1855
an den Rändern gefärbt, 1855
an der Luft getrocknet, 3550
an der Luft getrocknetes Papier, 3551
an der Luft trocknen, 261
Andruckrolle, 3069
Andruckwalze an der Sauggautsche,
   3593
Anfachung, 2386
Anfahren, 5393
Anfärben, 5377
Anfeuchten, 1808, 3859
Anfeuchter, 1807, 3104, 3106, 3858,
   5344
Anfeuchtung, 3105
anfühlen, 2413
Angaben, 1821
angebrochener Ballen, 906
angebrochener Pack, 906
angebrochenes Ries, 908
angetriebe Walze, 2105
Angleichung, 473
Anhängerkarton, 5632
Anhängsel, 5631
anhäufen, 235, 3705
Anhäufung, 141, 234, 235
Anhydride, 371
Anhydrite, 372
Anilin, 373
Anilindruck, 375
Anilinfarbstoffe, 374
Anionen, 382
anionenaktiv, 380
Anionenaustausch, 378
anionisch, 380
anionische Verbindungen, 381
Ankerbolzen, 368
Ankerschraube, 368
Anklebefahne, 5326
Anklebestellen, 2844
Anklebevorrichtung, 4225
Anlage, 2300, 2375, 3254,
   3606
Anlagen, 3632
Anlauf, 5393
Anlaufen, 788
Anlaufschutz, 410
Anlaufschutzkarton, 411
Anleimmaschine, 4227
Anode, 386
anorganische Faser, 3249
anorganische Papiere, 3250
anorganische Salze, 3251
anorganische Säuren, 3247
anorganische Verbindungen, 3248
Anpassung, 473
Anpassungsfähigkeit, 1507
Anpassungsvermögen, 1507
Anpflanzen, 4356
Anpresswalze, 4505
Anreicherung, 1487
Anreicherungsapparat, 1488
ansammeln, 3705
Ansammlung, 141
Ansäuern, 166
Ansäuerung, 166
ansaugen, 214
Ansaugen, 5497
Ansaugsystem, 3263
Ansaugung, 472
Anschlag, 2409
Anschlussblech, 2950
anschrauben, 4950
anspannen, 5444
anstossen, 987

anstrahlen, 3298
Antiblockmittel (das ein unerwünschtes
   Kleben oder Zusammenbacken
   verhindert), 389
Antichlor, 390
Anticrawlmittel, 391
Antikausrüstung (rauhe Oberfläche), 401
Antikbristol, 399
Antikbuchpapier, 398
Antikpapier, 404
Antikpapier mit rauher Oberfläche
   (Eierschaleneffekt), 400
Antilleneiche, 1155
Antioxidationsmittel, 397
Antiquaschrift, 4519
antiseptisches Papier, 3741
Antistatikmittel, 409
antreiben, 2104, 2799
Antreiben, 2106
Antrieb, 2104, 2799
Antriebsseite, 2107
Antriebsseite (der Papiermaschine), 515
Antriebswalze, 2105
Anwendung, 418
Anzapfen, 5649
Anzeigegerät, 3771
Anzeiger, 3208, 3695
Anzeigevorrichtung, 3208
anziehen, 4950
anzünden, 2521, 3470
Apparatur, 2535
Appretur, 1363
Äquivalentgewicht, 2301
Arabinose, 428
Aräometer, 3149
Arbeiten, 4066
Arbeiten mit Holz, 6270
Arbeitseinstellung, 5083
Arbeitsgang, 1783
Arbeitsgeschwindigkeit einer oder
   mehrerer Papiermaschinen in einer
   bestimmten Zeiteinheit (unter
   Berücksichtigung der Maschinenbreite
   aber nicht der Maschinengeschwindig-
   keit), 3201
Arbeitskräfte, 3385, 3387
Arbeitsleistung, 4096
Arbeitsplan, 2591
Arbeitsweise, 3847, 4069
aromatische Gruppen, 433
aromatische Säuren, 431
aromatische Verbindungen, 432
arsenhaltiges Papier, 435
arsensaures Papier, 435
Art, 5279
Asbest, 446
asbestartig Magnesiumsilikat, 445
Asbest-Dachpappe, 452
Asbestfaser, 450
Asbestfilz, 449
Asbestpapier, 451
Asbestpappe, 447
Asbestzementpappe, 448
Asche, 456
aschefrei, 458
aschefreies Filterpapier, 459
Aschegehalt, 457
aschehaltiges Papier, 2490
aschehaltige Zellstoffpappe, 2491
Ascorbinsäure, 454
Asphalt, 462
Asphalt-Dachabdeckung, 467
Asphaltemulsion, 463
Asphaltfilz, 464
asphalt-imprägnierter Asbestfilz (zum
   Abdichten), 453
Asphalt-Isolierpappe, 468
Asphaltpapier, 465, 466
Asphaltpappe, 469
Asphalt-Rohpapier, 471
Asphaltrohpappe, 470
Assimilation, 473
Ast, 3360
Astfänger, 3361
Asymmetrie, 475
Äthan, 2311
Äthanol, 2312
Äthanolamin, 2313
Äther, 2316

Ätherbildung, 2315
Äthergruppen, 2314
Äthylacetat, 2317
Äthylacrylat, 2318
Äthylamin, 2319
Äthylcellulose, 2321
Äthylendiamin, 2323
Äthylengruppen, 2322
Äthylenimin, 2324
Äthylgruppen, 2325
Äthylierung, 2320
Äthylmercaptan, 2326
Atlassatinpapier, 3794
Atlaszypresse, 5739
Atmosphäre, 476
Atmosphärendruck, 477
Atombindung, 478
Atropie, 4016
AT-Zellulose, 2321
ätzend, 1166, 1604
Ätzmittel, 1166, 1604, 3895
Ätznatron, 1171, 5230
Aufarbeitung, 4890
Aufbau, 5612, 5702
Aufbereitung, 667, 4479
aufbessern, 5932
aufbewahren, 5430
Aufblitzen, 2546
auf dem nassen Weg, 6190
auf der Rückseite verstärken, 503
auf die Faser aufziehen (von Farbstoffen
   und Chemikalien), 2339
Aufdrucken, 4109
Aufeinanderfolge, 5005
Aufeinanderlegen, 5049
Auffangwange, 4316
Auffindung, 1927
Aufforsten, 2667
Aufführseil, 4835
auffüllen, 2484
Auffüllung, 2497
aufgelöster Sauerstoff, 2033
aufgeschlagener Zellstoff, 1865
aufgetragene Leimfläche, 2846
aufgewandte Sorgfalt, 143
aufgeweicht, 3603
Aufhängen, 2969
Aufhängevorrichtung, 2968
aufhaspeln, 4699
Aufhaspeln, 6238
Aufhellen, 5753
Aufheller, 895
aufhören, 2516
Aufklebekarton, 506
Aufklebezettel, 3381
Auflager, 618
Auflagetisch, 5624
Auflaufrahmen, 1850
Auflegemaske, 4107
auflockern, 220
Auflodern, 2546
Auflöseanlage, 5189
Auflöseapparatur, 5189
Auflösebehälter, 2034, 2037
Auflöseholländer, 885
auflösen, 2032
Auflösen, 2035, 5188
Auflösen (Holzgefüge), 5332
Auflöser, 2025, 2037, 5187
Auflösung, 1858, 2024
Aufnahme, 4314
aufnahmefähig, 125
Aufnahmefähigkeit, 127, 1080,
   4672
aufnehmen, 114, 214
auf Paletten verpacken, 4143
aufpfropfen, 2860
Aufpolieren, 5974
Aufprägung, 2257
Aufprall, 270
aufprallen, 4340
aufrollen, 4699
Aufrollen, 4788, 6238
Aufrollfestigkeit, 6241
aufsaugen, 114
aufsaugend, 5277
aufschichten, 1584
Aufschläger, 4062
Aufschlaggerät, 2025

Aufschlämmung von Zellstoff, 5595
aufschliessen, 1975
Aufschliessen, 1981, 4574
aufschliessen (Holz), 1565
Aufschluss, 1567, 1982
Aufschmelzüberzug, 3096
Aufschneiden in Bogen, 5049
aufschreiben, 826, 4676
Aufschreiben, 4680
Aufschreibung, 4676
Aufschwung, 841
Aufseher, 2665, 2930
Aufsicht, 5563
aufspalten (z. B. bei Karton in einzelne
   Schichten), 1893
Aufspaltung in Schichten oder Lagen,
   1894
Aufstellschachtel, 5019, 5020
Auftrag, 418, 1382
auftragend, 957
Auftrag nach Kundenspezifikation, 3667
Auftragsfähigkeit, 1383
Auftragsmaschine, 3348
Auftragsstreichen, 4529
Auftragsvorrichtung, 419
Auftragswalze, 420, 4318
Auftragswalzenstreichen, 4531
Auftreffen der Luft, 270
Aufwallen, 2190
Aufwand, 1520
aufweichen, 3602
Aufzeichengerät, 4678
aufzeichnen, 4676
Aufzeichnen, 4680
Aufzug, 3066, 3466
Ausbauchen, 944
Ausbauchung, 944
Ausbeulen, 944
Ausbeulfestigkeit, 943
Ausbeute, 6319
Ausbeuteeintragung, 6321
Ausbildungsgeräte, 5802
Ausblasesdhieber (Kocher), 782
Ausbreitung, 2029, 2348, 5352
Ausdauer, 2977
Ausdehnung, 1992, 2247, 2348
Ausdehnvermögen, 1987
Ausdrückvorrichtung, 2198
ausfahrbare Langsiebpartie, 4741,
   4827, 4872
Ausfall, 2383
ausfällen, 4469
ausfällend, 4470
Ausfällung, 4469
Ausfliessen, 2015, 5339
ausflocken, 2571
Ausflockung, 1374, 1376, 1874, 2573
Ausflockungsmittel, 1873, 2570
Ausfluchtung, 297
Ausflussgeschwindigkeit, 5340
Ausforstung, 5725
Ausführbarkeit, 2396
Ausführung, 1923, 4260
Ausgang, 4095
Ausgangsstoff, 4647
ausgekleidet, 3498
ausgelassene Stellen (beim Streichen),
   5139
ausgezackt, 5008
Ausgiebigkeit der Farbe, 3071
Ausgleich, 2299, 4024, 5013
Ausgleichgetriebe, 1968
aushärten, 230, 1748
Aushärtung, 1749
Aushärtmittel, 1750
Ausheber, 2198
auskacheln, 5742
Auskippen, 2151
Auskleiden, 1334
Auskleidung, 1148, 3509
Auskleidung des Digesters, 1979
Auskragung, 1073
Ausladen, 5924
Auslass, 1898, 2015, 4095
Auslassen, 2151
auslassen (Dampf), 777
Auslassventil, 2017
Auslauf, 5154
Auslaufen der Druckfarbe, 751

Auslaufen eines Tintenstrichs auf ungeleimtem Papier, 2401
Auslaufen von Wasser, 6067
Auslaufkopf, 2016
Auslaufflippen, 5155
auslaugen, 3435
Auslaugen, 3436
auslaugend, 2360
Auslaugung, 2357
auslesen, 1735
auslösen, 4729
auslösen (eines Alarmsystems), 5013
Auslösung (z.B. eines Hebels), 4729
auspolstern, 4141
Auspuffgas, 2341
auspumpen, 2339
ausreissen, 5660
Ausrichten, 297, 3510
Ausrichtung, 297
Ausrückhebel, 656
Ausrückvorrichtung, 2198
Ausrüstung, 2300, 2517
Ausrüstung mit Messgeräten, 3256
Ausrüstungssaal, 2519
Aussäuerung, 5285
Ausscheidung, 2368
Ausschuss, 905, 4940, 6056
Ausschussausscheidung, 4726
Ausschuss bei der Ausrüstung, 2518
Ausschuss-Pulper, 909
Ausschwitzen, 1222, 6067
Ausschwitzung, 2368
Aussehen, 417
aus Sekundärstoff unbedrucktes Zeitungsdruckpapier zur Herstellung von Faltschachtelkarton, 730
Aussenluftdruck, 583
aussortieren, 1735
ausstanzen, 1955
Ausstattung, 2535
Ausstoss, 4096
ausstossen, 777
Ausstossen, 2260
Ausstrahlung, 2260, 4613
Ausstreichen, 923
Ausstreichleiste, 568, 5350
Ausstreichwalze, 2043, 5349
Ausströmen, 2260
ausströmen lassen, 2015
Ausströmöffnung, 4009
Austauschbarkeit, 3267
Austreten (von Gasen), 5821
Austritt, 4095
Austrittsgeschwindigkeit, 5158
ausüben, 4387
Auswahl, 4986, 5971
Auswaschung, 2252
Auswechselbarkeit, 3267
Auswerfer, 2198
Auswertung, 2330
aus Wolle, 6286
Auswurf (beim Sortieren), 4724
auszeichnen (Ware), 3694
ausziehbare Rollstange, 2345
ausziehend, 2360
aus 2 Elementen bestehendes System, 690
Autoklav, 482
Automation, 487
automatische Regelung, 2405
automatische Regulierwalze, 485
Automatisierung, 487
Autoreifeneinschlag, 490
Autoreifenverpackung, 490
Autotypiepapier (zur Vervielfältigung), 491
A-Welle (der Wellpappe), 102
axial, 493
axiale Festigkeit (bei Rohren), 1724
Axialpumpe, 494
Azetitril, 154
Azeton, 153
Azetylcellulose, 1183
Azoverbindungen, 496

B

Babbittmetall, 502
Bagasse-Entmarkungsmaschine, 529
Bagassefaser, 530, 5517
Bagassefaserreinigung, 1912
Bagassepapier, 531
Bagasseschneider, 528
Bagassezellstoff, 532
Bagasse (Zuckerrohrrückstände), 527
Bahnabnahme zwischen Saug-und Umkehrwalze, 4436
Bahnabriss, 6147
Bahnankleber, 5633
Bahnanklebevorrichtung, 992
Bahneinführmechanismus, 6160
Bahneinführung, 6150
Bahnführung, 5048, 6152
Bahn (Papierbahn), 6146
Bahnriss, 972
Bahnriss in der Falzlinie, 760
Bahnspannung, 6159
Bahnsteuergerät, 6152
Bahnüberführung, 6161
Bahnübergabe, 6161
Bahnwendevorrichtung, 6162
Bakterien, 518
bakterientötende Mittel, 520
Bakterienzahl (in einem bestimmten Papier oder Karton), 519
Bakterizide, 520
Balance, 543
Ballen, 545
Ballenauflöser, 547
Balleneinpackpapier, 550
Ballenpacker, 548
Ballenpresse, 548, 551
Ballenzerreisser, 547
ballig drehen, 1712
Ballung, 234
Balsampappel, 558
Balsamtanne, 557
Bambus, 559
Bambuspapier, 560
Bambuszellstoff, 561
Band, 653, 817, 5643
Banderolenpapier, 564
Banderolenstoff, 564
Banderoliermaschine, 563
Bandpapier, 5644
Bandtransport, 654
Bandwaage, 6169
Bandwickler, 3420
Banknotenpapier, 565, 567, 822, 1757
Bankpostpapier, 566, 822
Bankskiefer, 3307
Barium, 571
Bariumkarbonat, 572
Bariumoxid, 590
Bariumsulfat, 573, 593, 726
Barriere, 585
Baryt, 590
Barytkarton, 591
Barytpapier, 592
Barytpappe, 591
Basalteinsatz (im Refiner), 594
Basaltzugwinde, 594
Base (chem.), 595
basieren auf, 595
Basis, 595
basischer Farbstoff, 601
Basität, 602
Bassin, 859
Bast, 605
Bastard, 3111
Bastfaser, 606
Bastpapier, 607
Bastteil (eines Rindenleitstranges), 4285
Batzen, 6040
Bau, 5478
Bauart, 1923
bauchig (aufgrund zu loser Wicklung), 536
Bauer-McNett Fraktioniergerät, 614

Bauer-McNett-Sortierer, 614
Bauholz, 3586, 5745
Baum, 5830
Baumaterial, 1518
Baumkunde, 1899
Baumschule, 4011, 5832
Baumstamm, 816
Baumstumpf, 5487
Baumwollcellulose, 1625, 1628
Baumwolle, 1624
baumwollfaserhaltiges Papier, 1626
Baumwollinters, 1627
Baupappe, 645, 939, 4145, 6043
Bautechnik, 5477
Bauwesen, 5477
beanspruchen, 5444
Beanspruchung, 5444, 5463
bearbeiten, 2961, 3606
Bearbeiten, 3637, 4534
Bearbeitungsfähigkeit, 3605
Bearbeitung von Holz, 6270
Beaufsichtigung, 3253
Becher, 1747
Becherformmaschine, 1740
Becherglas, 617
Becherkarton, 1739
Becherwerk, 932
Becken, 4748
bedecken, 1654
bedecktsamige Pflanzen, 370
bedienen, 2961
Bedienen, 4066
Bedienung, 5009
Bedienungspersonal, 4071
Bedienungspult, 1546
Bedienungsseite, 4068, 5674
Bedienungsweise, 3847
bedruckbarer doppelseitiger Karton (dessen Mittellage aus altem Zeitungsdruck besteht), 2079
Bedruckbarkeit, 4520
Bedruckbarkeitsprüfung, 4528
bedrucktes Papier entfärben, 1889
beeinflussen, 681
beenden, 2516
befeuchten, 1806
Befeuchten, 1808, 3859
Befeuchter, 1807, 3104, 3858
Befeuchtung, 3105
Befeuchtungsapparat, 3106
beflecken, 5375
Beflecken, 5377
befördern, 5823
Beförderung, 5823
befreien, 4729
Befunde, 1821
Begrenzungsschalter, 3495
Behälter, 689, 1140, 1523, 3067, 4748, 5641
Behälter aus Karton, 864
Behälter aus Pappe-Papier-Film-oder Folie), 1071
Behälterboden, 1525
Behälterkarton mit besonders hoher Berstfestigkeit, 5695
Behälter mit Lochboden, 3536, 3537
Behälter mit Scharnierdeckel, 3062
Behälter mit Zwischenwänden, 4217
behandeln, 2961, 4532
Behandlung mit Säure, 181
Beharrungszustand, 5401
Beheizung, 3008
behöfter Tüpfel, 844
beidseitig gestrichen, 1002
beidseitig gestrichenes Papier, 5894
beidseitig mit Zeitungsdruckpapier kaschierte Schrenzpappe, 3964, 3968
beidseitig weisser Karton für Menukarten, 3817
Beimischung, 213
beim Streichen ausgelassene Stelle im Papier, 1415
beissend, 1166
Beize, 1166
Beizmittel, 3895
Beklebekarton, 506, 3506
Beklebekarton aus Sekundärfaserstoff,

807
Beklebekarton mit besonders hoher Berstfestigkeit, 3055
bekleben, 4219
Bekleben, 4226
Beklebepapier, 2376, 3505, 3507
Beklebepapier aus Sekundärfaserstoff, 807
beklebt, 4221
beklebter Karton, 4220, 4222, 4223
beklebter und kaschierter Karton, 3499
beklebte Schrenzpappe, 4224
Bekleidung, 1364
beladen, 3538
Beladen, 3543
Beladung, 3538
Belag, 2498, 3509, 5773
belasten, 3538
Belasten, 3543
Belastung, 1758, 3538, 5461
Belastung einer Maschine, 3628
Belastungskontrolle, 3540
Belastungsproben, 5397
belaubt, 1845
Belebtschlamm, 194
Belebtschlammverfahren, 195
Belegschaft, 1693
Belegung, 1655
Belüften, 5998
Belüftung, 221, 5996
Belüftung mittels Lufttaschen (zwischen Zylinder und Bahn), 4395
Belüftungstank, 222
Bemessung (Holländer), 2497, 5629
Ben Day Sortierer, 657
Benetzbarkeit, 6207
Benetzung, 6209
Benetzungsmittel, 3104
Benzoate, 670
Benzoesäure, 671
Benzol, 669
Benzylcellulose, 672
Benzylgruppen, 673
Berechner, 1024
berechnete Geschwindigkeit, 1924
Berechnung, 2330
Bergung, 4890
Berichten, 4680
Berieseln, 5357
Berieselungsanlage, 5356
Berieselungsturm, 4954, 6052
Berstdruckprüfer, 976, 979
Berstdruckprüfung, 980
bersten, 972
Bersten, 974
Bersten durch Übertrocknen, 2137
Berstfaktor, 973
Berstfestigkeit, 975, 978
Berstindex, 977
beruhen auf, 595
berührungslose Messung, 1522
Berührungsstelle der Walzen (z.B. am Kalander), 3974
Berührungswinkel, 1521
Beryllium, 674
Besamen, 4983
Besäumen, 4968, 5846
Besäummaschine, 2186
beschädigen, 1805
beschädigte Hülse, 1719
beschädigtes Riespapier, 5783
Beschaffenheit, 4604
beschichtet, 3410
Beschichtung, 3416
Beschichtungsanlage, 3417
beschicken, 3538
Beschicken, 3543
Beschickung, 2403
Beschickung des Digesters, 1977
Beschickungsanlage, 2407
Beschickungstrichter, 2410
Beschlagen, 3834
beschleunigen, 5316
Beschleuniger, 1156
Beschleunigungsmittel, 131
beschmutzen, 5256
Beschneiden der Kanten, 5846
Beschneidmaschine, 5844
Beschnitt, 4940

beschnitten, 5839
beschnittene Arbeitsbreite (der Papierbahn), 5843
Beschränkung, 4766
beschriften, 3381
beschützen, 2930
beschweren, 3538
beschwert, 2485
Beschwerung des Hebels, 3461
Beschwerungsmittel, 3544, 4319
Beseitigung, 2031
Besetztzeichen, 4010
Besitz, 4552
Besitzer, 3067
Besonderheit, 1227, 5294
Bespannung, 1364
bespritzen, 5077
Besprühen, 5357
Bestand, 5430
Beständigkeit, 4758, 5367
Beständigkeit gegen Feuchtigkeit, 3867
Bestandteil, 3226
Bestäubung, 4403
Bestimmung, 364
Bestleistung, 4676
bestrahlen, 3298
Betacellulose, 675
Betastrahlen, 678
Betastrahlenmessgerät, 677
Betätigen, 4066
Betätigung, 1541
Betonit, 668
Betonung, 5463
Betriebsablauf, 4069
Betriebsanlage, 4353
Betriebsassistent, 474
Betriebskosten, 4067
Betriebskraft, 5938
Betriebsleiter, 2807, 3672, 4541
Betriebsmann, 4071
Betriebsüberwachung, 4540
Betriebswasser, 3666, 4536
Betriebswasservorwärmer, 2180
Beugen, 2563
Beugung, 1969
Beule, 1420
Beuligwerden, 1424
Beutel, 526, 4393
Beutelherstellmaschine, 538, 539, 4458
beutelig, 536
Beuteligkeit, 533
Beutelmaschine, 541
Beutelpapier, 540
Beutelschliessmaschine, 4964
Bevorratung für die Marine, 3947
bewachen, 2930
Bewahrung, 4480
bewegen, 239
Bewegung, 240, 3897
Bewegungsenergie, 3346
Bewertungsliste, 1235
bezeichnen, 3381
Bezug, 1148, 1656
Bibelpapier, 682, 3207
Bichromatzahl, 1425
Biegeeigenschaft, 664
Biegefestigkeit, 663, 665, 2566, 5425
Biegemaschine, 662
biegen, 1761
Biegen, 659, 2563
Biegequalität, 664
Biegeschutz, 392
Biegesteifigkeit, 2565
Biegeversuche, 666
Biegsamkeit, 2562, 4380
biegungsfähiger Karton, 658
Bierdeckelkarton, 652
Bierdeckelpappe, 1381
Bierfiltrierpapier, 651
Bierfilz, 1381
Bikarbonate, 684
Bildabtastung, 4925
bilden, 2689, 2714
Bildung, 2684, 4539
Bildung kleiner Eindruckstellen im gestrichen Papier durch das Aufbrechen von Luftblasen im Strich, 1666

binäres System, 690
Bindefestigkeit, 823
Bindekraft, 695, 821, 1428
Bindemittel, 691, 820
Bindemittelwanderung, 207
binden, 817
Binden, 694, 819, 4897
Binderdraht, 552
Bindevermögen, 695
Bindung, 694, 817
Binnensee, 3404
biochemischer Sauerstoffbedarf, 499, 697
biologisch abbaubar, 699
biologische Abwasserreinigung, 702
biologische Aufbereitung, 705
biologische Kontrolle, 701
biologische Prüfung, 704
biologischer Abbau, 700
biologischer Sauerstoffbedarf, 703
Biozide (toxische Stoffe zur mikro-biologischen Kontrolle) 698
Biphenyle, 706
Birke, 679, 707
Birkenzellstoff, 708
Bisulfate, 709
Bisulfide, 3153
Bisulfit, 180
Bisulfitaufschluss, 712
Bisulfitaufschlussverfahren, 712
Bisulfite, 713
Bisulfitlauge, 710
Bisulfitzellstoff, 711
Bittersalz, 3652
Bitumen, 714
Bitumenemulsion, 716
Bitumenpappe, 715
bitumige Emulsion, 716
blanc fixe, 4236
Blanc fixe, 726
Bläschen, 929
Bläschenstreichen, 930
Bläschenstreicher, 759
Blase, 758
blasen, 771
Blasen bilden, 929
Blasenbildung, 1172
Blasenpackung, 761
Blasenwalze (zur Beseitigung von Luftblasen im Papier), 284
blasig werden, 758
Blaswalze, 779, 4395
Blatt, 3440, 5042
Blattbildner, 2687, 5046
Blattbildner für Handmuster, 2967
Blattbildung, 2685, 5047
Blattbildungsprüfgerät, 2686
Blattbildung unter Vakuumabsaugung, 5945
Blättchen, 3406, 3442
Blattfilter, 3441
Blattformierung, 5047
Blattführung, 5048
Blattmetall, 3763
blättrig, 3407
Blattschichtung, 5450
Blattwinkelverstellung, 4341
blau, 783
bläuen, 783
blau färben, 783
Blauglasprüfung, 784
bläuliches Schreibpapier, 497
Blaupack (zum Verpacken von Riesen), 3829
Blaupauspapier, 785
blau werden, 783
Blei, 3437
Bleichanlage, 740, 747
bleichbar, 735
Bleiche, 733, 741
bleichen, 733
Bleichen, 741
Bleicher, 739
Bleicherei, 2746
Bleicherei, 740
Bleichereiabwasser, 742
bleichfähig, 735
bleichfähiger Zellstoff, 736
bleichfähiges Papier, 749

Bleichfähigkeit, 734
Bleichgrad, 1881
Bleichlösung, 743, 746
Bleichmittelbedarf, 737
Bleichmittelbedarf (zur Erreichung eines bestimmten Weissgehalts im Zellstoff), 748
Bleichpulver, 744
Bleichturm, 745, 4686
Bleiglätte, 3526
Bleioxid, 3526
Blende, 524, 1944
Blindenschriftpapier, 875
Blisterpackung, 762
Block, 3554
Blockfestigkeit, 766
Blockierung der Kalanderwalze, 765
blutundurchlässiges Papier, 767
Blutundurchlässigkeit, 768
Boden, 5256
Bodenbelagunterlage, 1831
Bodendrähte, 511
Bodensatz, 1915, 2100, 4979
Bogen, 3440, 5042
Bogenableger, 3433
Bogenanlegen, 5045
Bogenformat, 5054
Bogenschneider, 5043, 5050
Bogenstapler, 5050
Bogenzuführung, 5045
Bohren, 846
Bohrenabfall, 5034
Bohrloch, 6176
Bohrspan, 3206
Bohrung, 846, 3072
bombieren, 1712
bombierte Walze, 858, 1065, 1713
Bombierung, 1715
Bonbondreheinschlag, 5889
Bor, 848
Borate, 842
Borax, 843, 5221
Bördel, 2543
bördeln, 2543
Bördeln, 616
Borhydrid, 847
Borke, 574
Borsäure, 845
Borstenmarkierung, 925
börteln, 1695
Börteln, 1696
Borverbindungen, 849
Brand, 2521
Brandschutz, 2528
brandsicheres Papier, 2527
braunes Kraftpackpapier, 3941
braunes Packpapier, 3366
Braun'sche Röhre, 1158
Brechen, 889, 1662
brechen (Hanf), 876
Brecherrippen, 884
Brechungsindex, 4712
Brei, 4219
Breite, 6228
Breitstreckwalze, 2347, 5351, 6158
bremsen, 876
Bremsen, 876, 877
Bremsung, 877
Bremsvorrichtung, 876
Bremswirkung, 1844
Brennen, 1010
Brenner, 964
Brennholzsägemaschine, 5148
Brennkammer, 1467
Brennmaterial, 2743
Brennofen, 2760, 3345
Brennschacht, 1467
Brennstoff aus Holzabfällen, 3064
Brett, 789
Briefmarkenpapier, 5381
Briefumschläge, 2292
Briefumschlagmaschine, 2290
Briefumschlagpapier, 2291
Brinell-Härte, 898
Brinell-Härteprüfer, 899
Bristol-Faltschachtelkarton, 2650
Bristolimitation, 798
Bristolkarton, 798, 900
Bristolkarton (dessen Mittellage aus einer

anderen Faserzusammensetzung besteht als die Decklagen), 2487
Brokatpapier, 904
Brom, 910
Bromverbindungen, 911
Bromwasserstoffsäure, 3126
Bronze, 912
Bronzedruckfarbe, 3766
bronzieren, 912
Bronzieren, 913
Brotbeutelpapier, 879
Broteinschlagpapier, 880, 881
Bruch, 882, 2643, 2711
Bruchdehnung, 2247, 2248, 5465, 5681
Bruchfestigkeit, 891, 5678
brüchiges Papier, 968
Brüchigkeit, 902
Bruchmodul, 3856, 4873
Bruchteil, 2707
Bruchwiderstand, 892
Brunnen, 6176
Brustwalze, 894
BTU (0, 252 kcal), 500
Buch, 826
Buchbeschnitt, 754, 838
Buchbinderbeschnitt, 693
Buchbinderchrenz, 693
Buchbinderpappe, 692, 831, 3818
Buchdruckpapier, 837, 3450
Buchdruckverfahren, 3451
Buche, 649
Bucheinbindepapier, 840
Buchenzellstoff, 650
Buchhülle, 833
Buchpapier, 837
Buchrücken, 503
Buchrückenbeklebepapier, 828
Buchrückenpapier, 513, 827
Buchse, 2437
Büchsenkarton, 1070
Büchsenkartonstoff, 1072
bügeln, 3294
Bund, 959
Bündel, 545, 959
Bündelung, 960
Bunker, 689
Bunsenbrenner, 964
Buntpapiere, 1441
Bürette, 962
Bürste, 915
bürsten, 915
Bürsten, 923
Bürstenfeuchter, 918
Bürstenglättung, 921, 922, 926
Bürstenmarkierung, 925
Bürstenstreichen, 917
Bürstenstreicher, 916
Bürstenstrich, 917
Bürstenwalze, 928
Butadien, 982
Butan, 983
Bütte, 1266, 5975
Bütte mit Stoffdichteregler, 3273
Büttenauskleidung, 5860
Büttenersatzpapier, 3182
Büttenfärbung, 5859
Büttenleimpresse, 5863
Büttenleimung, 5864
Büttenpapier, 2964
Büttenpapiere, 5980
büttenrandiger Karton, 2398
büttenrandiges Papier, 2400
Büttenrandkarton, 1853
Büttenrandpapier, 1854
Buttereinwickelpapier, 991
Buttersäure, 998
Butterverpackpapier, 2889
Butylacetat, 994
Butylgruppen, 995
Butylkautschuk, 996
Butyrat, 997
Butyrylgruppen, 999
B-Welle (der Wellpappe), 501

C

Calcium, 1011
Calciumbisulfit, 1014
Calciumcarbonat, 1015
Calciumchlorid, 1016
Calciumhydroxid, 1018
Calciumhypochlorit, 1019
Calciumlauge, 1012
Calciumlignosulfonat, 1020
Calciumoxid, 1021
Calciumsulfat, 1022
Calciumsulfit, 1013, 1023
Calciumverbindungen, 1017
Cambium, 1066
Carpinus, 1131
cash flow, 1146
Celit, 1176
Cellophan, 1178
Cellulose, 1182
Celluloseabbau, 1187
Celluloseacetat, 1183
Cellulosebutyrat, 1184
Cellulose-Copolymere, 1186
Cellulosederivate, 1188
Celluloseester, 1189
Cellulosefaser, 1191
Cellulosegehalt, 1185
Cellulosenitrat, 1193
Cellulosestruktur, 1196
Celluloseviskosität in
    Kupferoxidammoniak, 1743
Cellulosexanthogenat, 1198
Celluloseäther, 1190
Celsiusskala, 1201
Centipoise, 1206
Chamaecyparis, 1224
Champion-Streicher, 1225
charakteristisch, 1227
Charge, 3538
Chargenbetrieb, 608
Chargenpulper, 610
Chefchemiker, 1269
Chelatbildung, 1241
Chelate, 1239
Cheliermittel, 1240
Chelierung, 1241
Chemiefaser, 5613
Chemiefaserpapier, 1247
Chemiezellstoff, 2036
Chemikalien, 1257
Chemikalienbeständigkeit, 1256
Chemikalienrückgewinnung, 1255
chemische Aktivkohle, 1228
chemische Basizität, 602
chemische Behandlung, 1260
chemische Beständigkeit, 1256
chemische Bindung, 1243
chemische Eigenschaften, 1251
chemische Entrindung, 1245
chemische Holzentrindung, 1242
chemische Prüfung, 1258
chemischer Abbau, 1246
chemischer Bedarf, 1244
chemische Reaktion, 1254
chemischer Holzschliff, 1262
chemischer Sauerstoffbedarf,
chemischer Sauerstoffbedarf (COD),
    1003, 1250
chemischer Verbrauch, 1244
chemischer Verlust, 1249
chemischer Zellstoff, 1252
chemischer Zellstoffaufschluss, 1253
chemisch reiner Filterpapier, 1248
chilenische Flusszeder, 3463
China Clay, 1275
Chinagras, 4639
China-Pappe (meist gefärbt z.B. für
    Theaterbillets), 1274
Chipförderanlage, 1283
Chipklassiergerät, 1282
Chipklassierung, 1281
Chips, 6261
Chiptransport, 1286

Chipzugabe, 1286
Chlor, 1303
Chlorat, 1297
Chlorbedarf, 1311
Chlordioxid, 1306
Chloressigsäure, 1313
Chlorgas, 1307
Chlorid, 1298
chloriertes Lignin, 1299
chlorierte Stärke, 1300
Chlorierung, 1301
Chlorierungsstufe, 1302
Chlorierverfahren, 1310
Chlorite, 1312
Chlorlauge, 746
Chlorlignine, 1314
Chlormonoxid, 1308
Chlorose, 1315
Chlorverbindungen, 1305
Chlorwasserstoffsäure, 3129
Chlorzahl, 1309
Chlorzelle, 1304
Chrom, 1320
Chromat, 1316
Chromatizität, 1317
chromatographisches Papier, 1319
Chromgrün, 1322
Chromokarton, 1323
Chromopapier, 1324
Chromoxid, 1322
Chromverbindungen, 1321
Clay, 1344
clay-gestrichen, 1345
Clippernaht, 1353
Cobb-Prüfung des Leimungsgrads, 1418
Cobb-Test, 1418
Colophonium, 4841
Coloradotanne, 5098
Computer, 1483
Computerpapier, 1485
Computerprogramm, 1484
Concora-Test, 1490
Concora-Versuch, 1490
Container, 1523
Containerprüfung, 1528
Container-Reederei, 1527
Copolymere, 1579
Cortex, 1621
CSF (Mahlgrad), 1069
Cuproxam, 1742
Curlator (Kräuselmaschine), 1753
C-Welle (der Wellpappe), 1004
Cyanamide, 1777
Cyanate, 1778
Cyanide, 1779
Cyanoäthylcellulose, 1781
Cyanoäthylierung, 1780
Cyclohexan, 1788
Cymol, 1801

D

Dachboden, 3549
Dachpappe, 2429, 2430, 4832, 4833
Damm, 1804
Dämmpappe, 1175, 1831, 6043
Dämmung, 3261
Dampf, 2750, 5402, 5957, 5966
Dampfabzug, 5410
Dampfbehandlung (eines hochsatinierten
    Papiers), 5405
dampfbeheizte Kalanderwalze, 5409
dampfdicht, 5964
dampfdichtes Papier, 5965
Dampfdruck, 5963
Dampfdurchlässigkeit, 5960
dämpfen, 936, 1806, 5402
Dämpfen, 1763, 1808
Dampfentspanner, 1926
Dämpfer, 1807
Dampferhitzer, 5403
Dampfheizung, 5406
Dämpfkanal, 3560
Dampfkraftwerk, 5408
Dampfphase, 5961

Dampfphasenaufschluss, 5962
Dampfschranke, 5958
Dampfstrahl, 4384
Dampfturbine, 5412
Dampfüberhitzung, 5558
Dämpfung, 5407
Darre, 3345
Darrgewicht (Trocknung bei 105-110
    grad C), 4099
das Paar, 1650
Das Rupfen der klebenden
    Fasern einer Papierbahn
    an den Kalanderwalzen, 1043
das Überstehende einer Probe, 5560
Datenarchiv, 1826
Datenaufzeichengerät, 1824
Datenaufzeichner, 1824
Datenaufzeichnung, 1825
Datenerfassung, 1822
Datenlisten, 1828
Datenregistrierung, 1825
Datenspeicherung, 1827
Datensuchprogramm, 1826
Datentabellen, 1828
Datenübertragung, 1829
Datenverarbeitung, 1823
Dauer-, 4267
Dauerbiegefestigkeit, 2652
Dauerbiegeprüfer, 2653
Dauerbiegeprüfgerät, 661
Dauerbiegeversuche, 2654
Dauerbruch, 2392
Dauerhaftigkeit, 1842, 3465,
    4267
Dauerschwingversuch, 2394
Dauerversuch, 2174
Dauerzustand, 5401
Decke, 1654, 5773
Deckel, 1654, 1656, 2993, 5773
Deckelriemen, 1856
Deckenbahn (bei Wellpappe), 3506
Deckenpapier, 3507
Deckkraft, 1657, 3037
Deckschicht eines Kartons, 3505
Deckvermögen, 3037
Defekt, 1861, 2557
Defektbogen, 3192
Defibrator, 1867
Defibratormethode, 1868
Defibratorverfahren, 1868
defibrieren, 1864, 2463
Defibrierung, 1863, 1866
Deformierung, 1876
Degradation, 1879
dehnbar, 5675
dehnbares Papier, 5466
Dehnbarkeit, 2200, 2354
Dehnen, 5468
Dehnung, 2247, 5465
Dehnungseigenschaften, 5677
Dehnungsfähigkeit, 2354
Dehnungsgrenze, 6320
Dehnungsmesser, 2356, 5684
Dehnungsverhältnis in Zellstoffwatte,
    1686
dehydratisieren, 1885
Dehydratisierung, 1886
Dehydrierung, 1888
de-inken, 1889
De-inking, 1891
de-inkter Stoff, 839
de-inkter Zellstoff, 1890
dekantieren, 2251
dekapieren, 1919
Dekorationspapier, 1859
Dekulator, 1860
Dendrologie, 1899
Denier (Garnzahl), 1900
Dennison Wachsprüfung, 1901
Densitometer, 1903
Densitometrie, 1904
Densometer, 1908
Denver-Zelle (Bauart einer Flotationszelle
    für den De-inking Prozess), 1909
Depolarisation, 1913
der Luft aussetzen, 242
Desensibilisierung, 1920
Desorption, 1925
dessiniertes Papier, 4232

Destillat, 2038
Destillation, 2039
Destillieren, 2039
destilliertes Wasser, 2040
Detektor für Stoffbatzen, 3594
Detergens, 1928
Dextrin, 1935
Dextrinleim, 1934
Dextrose, 1936
Dezimalsystem, 3791
Diagnose, 1937
diagonal, 681
Diagramm, 1938
Diagrammpapier, 1232
Dialdehydcellulose, 1940
Dialdehydstärke, 1941
Dialyse, 1942
Diamid, 3123
Diaphragma, 1944, 3747
Diäthyläther, 1966
Diatom, 1947
Diatomeenerde, 1945
Diazokopierung, 1948
Diazopapiere, 1949
Dichloräthan, 1950
Dichlormethan, 1951
dicht, 3194, 5260
Dichte, 1905
Dichtemesser, 1907
Dichtemessung, 1906
Dichtemessung von Papier oder Karton
    (Flächengewicht: 1/1000 inch Dicke),
    4460
Dichtmanschette, 5149
Dichtpackung, 4137
Dichtung, 2788, 4137, 4959
Dichtungsabschluss, 6106
Dichtungsmittel, 4961
Dichtungspapier, 588, 2790
Dichtungspappe, 2789
Dichtungsring, 2788
Dickdruckpapier, 952
Dicke, 1905, 5722
Dicke der Farbschicht, 3234
Dicke eines Blattes oder eines
    Papierstapels von bestimmter
    Bogenzahl, 954
dicke Griffigkeit, 945
Dickenmesser, 1061, 3802, 5723
Dickenmessgerät, 1061
Dickenmessgerät (für
    Flächengewichtsmessung), 676
dickes Einschlagpapier, 805
dickes Endstück (Baum), 989
dickes Papier zum Schutz der
    Papierrollenenden während des
    Transports, 2271
Dickmittel, 5721
Dickstoffbleiche, 3043
Dicyandiamid, 1953
dielektrisch, 1959
dielektrische Eigenschaften, 1964
dielektrische Erwärmung, 1963
dielektrische Festigkeit, 1965
dielektrischer Verlust, 1961
dielektrisches Papier, 1962
Dielektrizitätskonstante, 1960
die Maschine voll ausnutzen, 5838
Dienst, 5009
Dienstleistung, 5009
die Punktzahl, 4931
die unter spezifische Bedingungen
    gemessene Dicke einer Bahn, 1060
Differentialantrieb, 1967
Differentialgetriebe, 1968
Differentialpumpe, 3926
Diffraktion, 1969
Diffuseur, 778, 1971
Diffuseurwaschanlage, 1972
Diffusion, 1974
digerieren, 1981
Digerieren, 1981
Digestercharge, 1977
Digitalcomputer, 1983
Digitalrechner, 1983
Digitalsystem, 1985
Diisocyanat, 1986
dikotylen, 1952
dilatant, 1988

Dimension, 1992
Dimensionsstabilität, 1993
dimensionsstabil machen, 1994
Dimer, 1995
Dimethyldisulfid, 1996
Dimethylformamid, 1997
Dimethylsulfat, 1998
Dimethylsulfid, 1999
Dimethylsulfon, 2000
Dimethylsulfoxid, 2001
Diode, 2002
Dioxybenzol, 4764
Dipol, 2005
Direcktor, 3672
Direktdämpfung, 2010
direktes Dämpfen, 2010
direktgeheizt, 2008
direkt trocknend, 3868
diskontinuierliche Arbeitsweise, 609
diskontinuierlicher Betrieb, 609, 3274
dispergierend, 2028
Dispergiermittel, 2027
Dispersion, 2029
Distanzhalter, 5289
Disulfid, 2044
Diversifizierung, 2045
Dokumentenpergament, 442
Dolomit, 2060
Doppelbindung, 2063
Doppelfalzer, 2660
Doppelfilzpresse, 2070
Doppelkaschieranlage, 2076
Doppelleimung, 2080
Doppelpresse, 2147
Doppelscheibenrefiner, 2069
doppelseitige Wellpappe, 2074
doppelseitig gestrichen, 2065
doppelseitig in der Maschine kaschiert, 2081
doppelseitig kaschierte Maschinenschrenzpappe, 2078
Doppelsiebmaschine, 5886
Doppelstreichanlage, 2066
Doppeltragtrommelroller, 2071
Doppeltragtrommelumroller, 2072
Doppeltragwalze, 2073
doppelt satiniert, 2064
doppelwandige Kartonschachtel, 2082
Dose, 5748
Dosenkarton, 1070
Dosenkartonstoff, 1072
dosieren, 3770
Dosiergerät, 3771
Dosiersystem, 3773
Dosierung, 2062, 4555
Dosiervorrichtung, 3772
Dosierwalze, 2057
Doublierkalander, 1464
Douglastanne, 2083
Draht, 2481, 6242
Drahtform, 6251
Drainage, 2086, 2093
Drainagefaktor, 2087
Drall, 5888
Drapierfähigkeit, 2094
drehbar, 5152
Dreheinschlagseidenpapier, 5890
Drehen, 5888
Drehfilter, 4850
Drehkocher, 4849
Drehmoment, 5784
drehschwingend, 4088
Drehung, 4856, 5785
dreifach, 5851
Dreifach-Langsiebmaschine, 5850
dreilagig, 5735, 5851
Dreipressenschleifer, 5736
dreischichtig, 5735, 5851
Drift, 2103
dritte Presse, 5732
dritter Geselle, 5731
Drosselklappe, 5737
Drosselventil, 990, 5737
Drosselklappe, 990
Druck, 1478, 4495, 4522
Druckabfall, 4497
Druckanalyse, 5462
Druckbegrenzungsventil, 4738
druckdichte Anlagen, 4509

druckdichte Einrichtung, 4509
Druckdose, 1087
druckempfindlich, 4507
druckempfindliches Papier, 4506
drücken, 4482
Drücken, 4488, 4495
Drucken, 4522
Drucker, 4521
Druckerpresse, 4482, 4525
Druckerschwärze, 1096
Druckfalzfestigkeit, 1677
Druckfarbe, 3232, 4524
Druckfarbenauflage, 3237
Druckfehler (das Farbpigment wird von der Papieroberfläche weggewischt), 1222
Druckfestigkeit, 1478, 1479, 1724, 1726, 4508
Druckfilter, 4498
Druckform, 2580
Druckgefälle, 4500
Druckglätte, 4527
Druckhöhe, 2016, 2993
Druckholz, 1481
Druckkraft, 4495
Druckluft, 1476
Druckmahlung nach der Ausblasstation, 775
Druckmaschine, 4525
Druckmaschinenstopper, 4494
Druckmessdose, 3539
Druckmesser, 4499
Druckmessung, 4502
Druckminderungsventil, 4695
Druckpapier, 4526
Druckplatte, 4366, 4373
Druckpresse, 4525
Druckprobe, 1480, 1725
Druckreduzierung, 4503
Druckreduzierventil, 4695
Druckregler, 4504
Druckregulierung, 4496
Drucksatz, 5012
Druckspannung, 5461
Druckstoffauflauf, 4501
Drucktiegel, 4373
Drucktuch, 4523
Druckversuch, 1725
Druckwalze, 4490
Druckwalzenstreichen, 4485
Druckwalzenstreicher, 4530
Druckwiderstand eines Kartons, 866
Duft, 4017
Düngemittel, 2439
Dünger, 2439
Düngung, 2438
Dünndruckpapier, 3207, 5726
Dünnflüssigkeit, 2607
Dünnpost, 5726
Dunst, 1806, 2750, 5402, 5957, 5966
Dunstschleier, 2992
Duplex, 2153
Duplex-Bristol, 2155
Duplexfilz, 2157
Duplex-Karton, 2154, 5892
Duplexoberflächenausrüstung (beide Oberflächen verschieden ausgerüstet), 2158
Duplex-Papier, 2159, 5887
Duplex-Papier (auf der Vorderseite eine andere Farbe als auf der Rückseite), 2075
Duplexstreicher, 2156
Duplex-Super, 2160
Durchbiegen, 659
Durchbiegung, 1871, 4884
durchbiegungsfreie Walze, 393
Durchblasen, 931
durchbohren, 4256
Durchbohren, 4591
Durchbohrung, 4256
durchdringend, 4248
Durchdringen von Öl, 4040
Durchdringung, 4248
Durchdringungswiderstand gegen Flüssigkeiten, 4759
durch Eigenverbrauch gebundene Zellstoffmenge, 1088
Durchfahrbetrieb, 1535

Durchfluss, 2588, 2595
Durchforstung, 5725
Durchführbarkeit, 2396
durchgehend, 4884
durchhängen, 4884
Durchhängen der Bahn, 533
Durchlass, 3244
Durchlassgrad, 5814
durchlässig, 4446
Durchlässigkeit, 4271
durchlöchern, 3072, 4256
durch Luft befördert, 246
durch Luftblasen verursachte Streifen im Papier, 5203
durchlüften, 5995
durch Luft verlegt (Fasern), 277
Durchmesser, 1943
Durchperlen, 931
Durchreissfestigkeit, 5659
Durchreissfestigkeitsprüfer, 5664
Durchreissfestigkeitsprüfung, 5663, 5664
durchrühren, 239
Durchsackgeschwindigkeit, 1702
Durchsatz, 2595
Durchsatzgeschwindigkeit, 2595
Durchsatzregelung, 2592
Durchscheinen, 5078, 5811
durchschiessen, 3437
Durchschlagen der Farbe, 752, 3241
Durchschlagen (einer Druckfarbe), 5471
Durchschlagfestigkeit, 753
Durchschlagpapier, 1582
Durchschlagsfestigkeit, 1965
durch Schlamm hervorgerufene Löcher, 5161
durch Schleimflecke hervorgerufene Löcher, 5161
Durchschnitt, 1943
Durchschreibpapier, 3677
Durchschusspapier, 3270
Durchsicht, 3570
Durchsichtigkeit, 5811, 5816
Durchsichtigmachen, 5820
durchsickern, 3435
Durchsickern, 3443, 4985
Durchstechen, 4591
Durchstoss, 4591
Durchstossprüfer, 4593
durchtränkt, 3196
Durchtränkung, 4913
Dürrholz, 1833
Düse, 3308, 4009
Düsenaustrittsfliessmenge, 4087
Düsenaustrittsströmung, 4087
Düsenstrahl, 3308
Düsenstrahltrockner, 3058
dynamische Prüfung, 2174
dynamische Untersuchung, 2394

E

Echtdokumentenpergament, 444
echtes Alaun (im Vergleich zum Papiermacheralaun), 600
Echtheit, 2389
Echtpergament-Papiere, 5981
effektiver Alkaligehalt, 2189
Egalisieren, 3456
Egalisiermittel, 3457
Egoutteur, 1816, 4795, 4797
Egoutteur-Markierung, 1818
Egoutteur-Streicher, 1817
Egoutteur-Walze, 1820
Egoutteurwalze, 2692
Egoutteur-Wasserzeichen, 1818
Eibe, 5653
Eibenbaum, 5653
Eiche, 4015, 4609
eichen, 1058, 2769, 2798
Eichung, 1059
Eierkarton, 2196
Eierschachtel, 2196
Eierschalenmattierung, 2197
Eigengewicht, 5300, 5651

Eigenschaft, 1227, 4552, 4604
Eigentum, 4552
Eignung, 132
Eignungstest, 134
ein aus verschiedenem Material zusammengesetzter Behälter, 1475
Einband, 694
Einbau, 2535, 3254
Einbauhöhe, 1830
Einbettung, 2253
eindampfen, 2331
Eindampfer, 1488
Eindampfung, 2332
Eindicken, 1849, 5720
Eindicker, 1488, 1848, 5719
Eindickzylinder, 1848, 5719
Eindrücke im Papier infolge defekter Kalanderwalzen oder darauf haftender Fremdstoffe, 1040
Eindrucken, 4109
eine Art Gusstreichen (erübrigt nachfolgendes Kalandrieren oder Satinieren), 4401
eine in Wasser aufgelöste Sauerstoffmenge, 2033
einen Bogen auf den Zylinder aufziehen, 5465
Einenger, 2333
eine Schicht oder Lage einfügen, 3269
eine Umwandlung bewirkende Behandlung, 3853
einfache Webart, 4351
einfach gestrichen, 5100
einfach gestrichenes Papier, 5101
Einfach-Pick-up, 4436
Einfärbungsmethode, 3236
Einfärbungsverfahren, 3236
einfassen, 4801, 5012
Einfassung, 6179
Einführen der Papierbahn, 5734
Einführen der Chips in den Kocher, 1290
Einfülltrichter, 2410
eingearbeitete Druckbelastung, 942
eingedrückte Papierrolle, 1721
eingerissene Rollenränder, 907
eingeritzt, 4935
eingeschlossen, 5827
eingeknetete Druckbeanspruchung, 2102
eingetrocknete Spannung, 2101
Eingriff, 3756
Einhängen der Papierrollen, 3543
Einheit, 5920
Einjahrespflanze, 4006
Einkapselung, 2269
einkeimblättrige Pflanzen, 3889
Einkerben der Baumrinde, 4926
Einlage, 1203
Einlagenkarton, 3817
Einlagenpapier, 2495
einlagig, 5111
einlagiger Karton, 5112
Einlass, 3244
Einlassöffnung, 3244
einlaufen, 5081
einlaufen lassen, 5081
Einlaufschutz, 3975
Einlaufschutzstange, 3975
Einlauftrichter, 2410
Einlaufwalze, 2692
einlegen (Druck), 4396
ein Loch bohren, 3072
einmal gestrichen, 5100
einmal gestrichenes Papier, 5101
Einmal-Kohlepapier, 4055
ein- oder beidseitig gestrichener Karton, 1386
ein- oder beidseitig pigmentgestrichener Karton, 1346
ein- oder doppelseitige Wellpappe zum Auskleiden von Fässern oder Körben, 1608
Einordnung, 1342
Einpumpen, 3230
Einreissfestigkeit, 5658
Einreisswiderstand, 3229
Einrichtung, 2300, 2375, 3254
Einritzen der Baumrinde, 4926

einsacken, 526
Einsacken, 534
Einsackmaschine, 535
Einsatz, 418
Einsatzhärten, 1141
einschalten, 5605
Einschlag, 6284
Einschlagfaden, 6166
Einschleifen, 2915
Einschluss, 3226
einschneiden, 4931
Einschnitt, 3341, 5176
Einschränkung, 4766
Einschuss (Textilien), 6284
einseitg gestrichen, 1001
einseitige Glätte, 4053
einseitiger Glanz, 5490
einseitige Wellpappe, 5105
einseitige Wirkung, 681
einseitig gefärbter Duplex-Karton, 801
einseitig gestrichenes für den Steindruck
    ausgerüstetes Papier, 3530
einseitig gestrichenes Papier, 4054
einseitig gestrichenes und hochsatiniertes
    Kartonagenpapier (für Kaschierzwecke),
    869
einseitig glatt, 3596, 3624
einseitig glattes Papier, 3597, 3625
einseitig hochsatiniertes Antikpapier,
    402
einseitig kaschiert, 5108
einseitig kaschierter Karton, 5109
einseitig kaschierte Schrenzpappe, 5107
einseitig kaschierte Wellpappe, 5103,
    5104
einseitig mit Zeitungsdruck kaschiertes
    Manila, 5110
einseitig oder beidseitig gestrichenes und
    vor dem Satinieren gebürstetes Papier,
    919
Einspritzen, 3230
Einspritzvorrichtung, 4519
einstellbare Geschwindigkeit, 210
einstellen, 5012
Einstellen, 5014
Einstellen der Farbe, 211
Einstellspindel, 212
Einstellzeit, 4765
Einstufen, 2859
Einstufung, 1342
Eintauchen, 5416, 5492
eintragen, 826, 3554, 4676
Eintragen, 4680
Eintragsvorrichtung, 2407
Einwegprodukt, 2030
Einweichen, 5205
Einweichung, 3604
Einwickelpapier, 6297, 6299
Einzelfaden, 3890
Einzelfaser, 5106
Einziehen von Luft (z. B. in den Stoff),
    2288
Einzug, 2288
ein zum Streichen geeigneter Clay, 1408
Einzylindermaschine, 5102
Eisen, 3294
Eisenbahn, 4635
Eisenbahnspur (auch beim Streichen),
    4637
Eisenbahnwaggon, 4636
Eisenholz (harte u. feste u. dunkelgefärbte
    Hölzer aus den Tropen), 3297
Eisenoxide, 3296
Eisenverbindungen, 3295
Eiweissstoffe, 286
elastische Kalanderwalze, 2199
Elastizität, 2200, 4751, 4752
Elastizitätsmodul, 955, 3855
Elastomere, 2202
elektrikscher Widerstand, 4762
elektrische Eigenschaften, 2208
elektrische Entladung, 2215
elektrische Isolierung, 2219
elektrische Leitfähigkeit, 2204
elektrischer Antrieb, 2216
elektrischer Relais, 2224
elektrischer Gasreiniger, 2243
elektrischer Leiter, 2210
elektrischer Motor, 2220

elektrischer Strom, 2214
elektrischer Widerstand, 2225
elektrische Sicherung, 2217
Elektrizitätsleiter, 2210
Elektroantrieb, 2221
Elektrodampfkraft, 5404
Elektrode, 419, 2227
Elektrofilterung, 2242
Elektrofilter zur Staubabscheidung, 2243
Elektrolyse, 2228
Elektrolysenbad, 2230
Elektrolyt, 2229
elektromagnetisches Feld, 2231
Elektromotor, 2220
Elektron, 2236
Elektronengerät, 2233
Elektronenphysik, 2235
Elektronensteuerung, 2232
Elektronenstrahlröhre, 1158
Elektronentechnik, 2235
Elektronik, 2235
elektronische Ausrüstung, 2234
elektronische Regelung, 2232
Elektrophorese, 2237
Elektrostatik, 2244
elektrostatische Aufladung, 2239
elektrostatische Belastbarkeit, 1076
elektrostatische Beschichtung, 2240
elektrostatische Kapazität, 1076
elektrostatisches Kopierverfahren, 2241
Elektrotechnik, 2205
Elementarfaden, 3890
Elementarfibrille, 2245
Elution, 2250
elutrieren, 2251
Emaille, 2266
emaillieren, 2266
emailliert, 2267
Emission, 2260
Emissionsspektroskopie, 2261
empfinden, 2413
Empfindlichkeit, 4997
Emulgator, 2263
emulgieren, 2264
Emulgierung, 2262
Emulgierungsmittel, 2263
Emulsion, 2265
Endformat, 5841
endloses Garn, 2482
Endlosfaden, 2482
Endlosformular, 1533
Endlosformulardruck (EFD), 1533
Endlosformularsatz, 1533
Endosperm, 2274
endotherm, 2275
endotherme Reaktion, 2276
Endschalter, 3495
Endverarbeitung, 2517
Energie, 2277, 4464
Energiebilanz, 2278
energiereich, 3045
Energieübertragung, 2280
Energieumwandler, 5803
Energieverbrauch, 2279
Energieversorgung, 4468
Engelmann-Fichte, 2281
englumiges Holz, 5548
en gros Lagerung, 956
entästen, 1896
entbasten, 4939
Entdeckung, 1927
Entfärben, 1891
entfärbter Stoff, 839
Entfärbung, 1857, 2018
Entfernung grosser Holzstücke, 3324
Entfernung von kurzen
    Faserbestandteilen, 2515
Entfernung von pulverigem Material,
    2515
entfetten, 4939
Entflammbarkeit, 2542
Entflammungspunkt, 2547
Entgasen, 1878
Entgaser, 1877
Entgasung, 1100
Entgasungsvorrichtung, 1877
Entgiften, 1878
Entgiftung, 1929
Enthalpie, 2286

Entionisierung, 1892
entkeilte Walze, 5166
entladen, 2015, 5922
Entladen, 5924
Entlader, 5923
entlassen, 4729
entlasten (z.B. Kalanderwalzen), 4739
entlastet, 5922
Entlastung, 4735
entleeren, 2085, 2329
Entlignifizierung, 1895
entlüften, 1834, 5995
Entlüften, 5998
entlüften, 1836, 2340
Entlüfter (zwischen Sortierer und
    Einlaufkasten), 1860
Entlüftung, 1835, 5996
Entmarkung der Bagasse, 1912
Entomologie, 2287
entrinden, 574, 1837
Entrinden, 577, 1839
Entrindungstrommel, 578, 1838, 2111
Entropie, 2289
entsaften, 4900
Entschalen, 4243
Entschäumer, 1875, 5136
entschlacken, 1919
entschwärzen, 1889
Entschwärzung, 1891
Entspannung, 4728
Entspannungsgefässe, 2561
entsprechen, 3711
Entstauben, 2168
entstippen, 1864
Entstipper, 1869
Entwachsen, 1933
entwässern, 1885, 2085
Entwässern, 1849, 2093
entwässern (besonders durch Pressen),
    1931
Entwässern und Falten von Nasszellstoff
    zur Herstellung von Ballen, 3422
entwässerter und in Ballen geschichteter
    Zellstoff, 3421
entwässertes Aluminiumsilikatmineral,
    480
Entwässerung, 1886, 1932, 2085
Entwässerungsfaktor, 2087
Entwässerungsleiste, 2088, 3136
Entwässerungswiderstand, 2089
Entwässerungszeit, 2090
entweichen, 2339
Entwicklung, 2335
Entwurf, 1923
Entzündbarkeit, 2542
entzünden, 2521, 3470
entzundern, 1919
Entzündung, 3178
Enzym, 2247
Epibromhydrin, 2295
Epichlorhydrin, 2296
Epoxydharze, 2297
Epoxyharz, 2297
erbauen, 2282
Erblichkeit, 3033
Erde, 5256
Erdgas, 3944
Erdhinterfüllung, 3418
Erdöl, 2697
erfassen, 1371
Ergebnis, 4537
Ergiebigkeit, 4543
ergreifen, 1371
erhaltend, 4481
Erhaltung, 1511, 4480
Erhärtung, 2976
erhitzen, 3000
Erhitzungswiderstand, 3014
Erholung, 4728
Erkennung, 3175
Erle, 293, 321
erleuchten, 3298
Ermüdung, 2391
Ermüdungsbeständigkeit, 2393
Ermüdungsbruch, 2392
Erntemaschine, 2990
ernten, 2991
erntereifes Holz, 3722
Erosion, 2304

Erprobung, 5696
Erreger, 2338
Erregermaschine, 2338
Erregung, 240
errichten, 2282
Ersatzfilz, 5291
Ersatzstoff, 5494
Erscheinen, 417
Erschütterung, 5031, 5063
Erstarrung, 5262
erste Hauptpresse, 2532
erste Presse, 2533, 3661
erster Trockenzylinder, 2531
erste Sortierstufe, 4516
erste Walze, 3439
Ertrag, 6319
ertragreich, 3059
Ertrags-Aufstellung, 6321
Erweichung, 3604
Erweichungspunkt, 5244
Erweichungstiefe, 4249
Erweiterung, 2348
Erzeugnis, 3687, 4537
Erzlagerstätte, 3831
Esche, 456, 2716
Eschenbaum, 460
Esparto, 2305
Espartopapier, 2306
Espartozellstoff, 2307
Espe, 461
Essigäther, 2317
Essigsäure, 151
Essigsäureanhydrid, 152
Essigsäureäthylester, 2317
Ester, 2310
Estergruppen, 2308
Etagenpresse, 4375
Etikett, 3381
Etikettenpapier, 3384
etikettieren, 3381
etikettieren, 3382
Etikettiermaschine, 3383
Etui, 1140
Eukalyptus, 2327
evakuieren, 2329
Evolution, 2335
E-Welle, 2175
Exkavatorprobe, 2856
exotherme Reaktion, 2343
Experimentierung, 2350
Exportzellstoff, 2352
Exsikkator, 1922
Extraktion, 2250, 2357
Extraktionsapparat, 2361
extraktiv, 2360
Extraktor, 2361
Extruder, 2362
Extrudieren, 2363
Extrusion, 2363
Extrusionsbeschichtung, 2365
Extrusionstreicher, 2364

F

Fabric-Presse, 2372
Fabrik, 3816
Fabrikabwasser, 3825
Fabrikat, 3687
Fabrikation, 3687
Fabrikationsfehler, 2557
Fabrikationsleiter, 2807
Faden, 2481, 6315
Fadeometer, 2378
Fagus, 2382
Fähigkeit, 4552
fahren, 2104
Fahren, 2106
Fahrkartenkarton, 5740
Fahrzeug, 5982
fällen, 2414, 3554
Fällen, 2417, 3556
Fallprobe, 2174
Fallrohr, 2108
Fällungsmittel, 4472

Fällungsrestholz, 3558
Fallwasserkasten, 4966
Fallwasserrohr, 582
falsch, 795
falsche Haare (Textilien), 2059
Fälschungsschutzpapier, 394
Falte, 2643, 6301
falten, 1669, 4387
Falten, 1421, 2646, 4381
Faltenbildung, 1669, 1675
Falten (durch den Rollenschneider
    verursacht), 6237
faltiges Papier, 1422
Faltschachtel, 2649, 2651
Faltschachtelkarton, 2648
Faltschachtelkarton aus Schrenzpappe,
    660, 4988
Faltschachtelkarton mit
    Rundsiebdeckschicht, 4229
Faltschachtel-Klebemaschine, 2645
Faltung in Längsrichtung, 3565
Falz, 2643
Falzbeständigkeit (Verharren eines Papiers
    im gefalzten Zustand), 1674
Falzer, 2655
falzfähiger Karton, 658
Falzfähigkeit, 1668, 4380
Falzfestigkeit, 2652, 2659
Falzfestigkeitsprüfer, 2660
Falzgerät, 2653
Falzkante in Wellpappenbehältern, 3688
Falzkarton, 2647
Falzmaschine, 2655
Falzmaterial, 2658
Falzqualität, 2656
Falzversuche, 2654
Falzwiderstand, 2657
Farbabriebwiderstand, 3240
Farbabstimmung, 1452
Farbabstufung, 3103, 5024
Farbanstrich, 1446
Farbart, 1317
Farbaufnahme, 3233, 3243
Farbaufnahmevermögen, 3238
farbbeflecktes Papier, 5376
Farbbeständigkeit, 1442, 2380
Farbbestimmung, 1454
Farbe, 1439, 1456, 2170, 3232
Farbechtheit, 1442, 2380
Farbe für Buchdruck, 3449
färben, 2170, 3103
Färben, 2172
färbend, 1446
Farbfestigkeitsprüfung, 2381
Farbflecken, 1455, 5375
farbig, 1440
farbige Papiere, 1441
Farbklumpen, 1450
Farbkörper, 4319
Farbmesser, 1443
Farbmessung, 1445
Farbnäpfchen (Tiefdruck), 1177
Farbpigment, 1448
Farbstoff, 1447, 2170
Farbstoffe, 2173
Farbsumpf, 1449
Farbton, 3103
Farbtönung, 5024
Farbträger (Druck), 5982
Farbübertragung, 3243
Farbumkehrung, 1453
Färbung, 1446
Färbung durch Zerstäubung der Farbe,
    5346
Farbwiderstand, 3239
Faser, 2445, 2473
Faserabmessungen, 2458
Faseranalyse, 2446
Faserbindung, 2448
Faserbündel, 2449
Faserdurchmesser, 2457
Fasererholung, 2468
Faser-Faser-Bindung, 3268
Faserfilz, 2467
Faserfragmente, 2713
Faserfraktion, 2461
Faser-Fraktioniergerät, 2453, 4572
Faserfraktionierung, 2452, 2709
Fasergehalt, 2455

Fasergipsplatte, 4358
Faserguss, 4578
Fasergussmasse, 3875
Fasergussprodukte, 3877
Faserholz, 4584
faserig, 2479
faseriger Füllstoff, 2480
faseriges Magnesiumsilikat, 445
faseriges Mg-Silikat, 227
faserige Zusatzstoffe für Zement, 2451
Faserkalk, 227
Faserknoten, 2464
Faserlänge, 2465
Faserlängenverteilung, 2466
Fasermatte, 2467
Faseroberflächenbindung, 818
Faserorientierung, 4086
Faserrichtung, 2863
Faserrückgewinnung, 2469
Fasersättigungspunkt, 2470
Faserschneiden, 5070
Faserschnitt, 2456
Faserstoffeintrag, 2462
Faserstoffplatte, 2447
Faserstoffrückgewinnungsanlage, 4914
Faserstruktur, 2471
Faserverflechtung, 2460
Faserverkürzung, 5070
Faservlies, 4007
Faserwerg, 2472
Faserzusammensetzung, 2454
Fass, 584
Fass aus Fasermaterial, 2459
Fassung, 3067
Fassungsvermögen, 6034
fast reines Magnesiumsilikat, 445
faules Holz, 1843
Feder, 4247
Federflocke, 2598
Feder für Keilnut, 5329
Federkraft, 2200
federleichtes Papier, 2402
Federwalze, 5687
Federwalze (am Glättwerk), 5354
Fehler, 755, 1861
Fehleranzeige, 2558
Fehler beim Streichen, 5456
Fehlergrenze, 5767
fehlerhafte (sternförmig) auslaufende
    Rollenwicklung, 5392
Fehlfarbe, 4019
Fehlstelle, 1415, 1861, 6028
Fehlstelle beim Streichen, 1416
Fehlstellenanzeige, 5138
Fehlstellensucher, 2558
Feilenabfall, 5034
feiner Nebel, 2992, 3833
feiner Staub, 2598, 2762
feines Pergament, 5983
Feinfolie, 1812
Feingehalt, 2510
feingeripptes Dokumentenpapier, 403
Feingraupappe, 3961
Feinheit, 2510
Feinmesser, 3802
Feinpapiere, 2511
Feinsortierer, 2513
Feinsortierung, 2514
feinster Faseranteil während der
    Refinermahlung, 1694
Feinstmahlanlage, 4586
Feinstoff, 2512
Feinstoffsortierer, 2513
Feinwasser, 2727
Feinwelle, 501
feinzerstäubte Flüssigkeit, 3833
Feinzerstäubung, 479
Felge, 4801
Fernkontrolle, 4740
Fernleitung, 4338
Fernsehen, 5669
Fernsteuerung, 4740
Fertigpolieren, 4402
fertigstellen, 2516
Fertigstellung, 4540
Fertigware, 3687
fest, 5260
festbinden, 1584
feste Phase, 5263

fester Aggregatzustand, 5265
festgehalten, 5827
festhalten, 2876
Festigkeit, 4758, 5459
Festigkeit der Heissiegelung, 3015
Festigkeitsprüfung, 5460
Festkartonage, 5019, 5020
Festkosten, 2536
Festoffabfall, 5266
Festoffgehalt, 5267
Feststoff, 5260
Feststoffanteil, 5267
Feststubstanz, 5267
Festwerden, 1374, 1376, 5014, 5262
Fett, 2883
Fettbeständigkeit, 2890
fettdicht, 2884
fettdichte Ausrüstung, 2887
fettdichte Beschichtung, 2886
fettdichter Karton, 2885, 2891
fettdichtes Papier, 2889, 2892
fettdichtes Papier zur
    Kartoninnenauskleidung, 1137
Fettdichtigkeit, 2888
fettdicht machen, 2887
fetten, 2883
Fettflecken, 2893
Fettharze, 4050
fettig, 2895
Fettsäuren, 2395
Fettschmierung, 2894
feucht, 3857, 6180
Feuchtapparat, 5344
Feuchtdehnbarkeit, 3134
Feuchtdehnung, 1812
Feuchte, 1806, 3107
feuchte Bahn, 6212
Feuchtemesser, 3863
feuchtgeglättet, 6075
feuchtgeglättetes Papier, 1032, 6076
Feuchtglätte, 1811, 6074
Feuchtglätten, 6077
Feuchtglättwerk, 886, 6070
Feuchtglättwerkswalzen, 3272
Feuchtigkeit, 3860, 6194
Feuchtigkeitsaufnahme, 3864
Feuchtigkeitsgehalt, 3107, 3861
Feuchtigkeitskontrolle, 3108, 3862
Feuchtigkeitsmesser, 3863
Feuchtigkeitsmessung, 3109, 3169,
    3869
Feuchtigkeitsprüfung, 3869
Feuchtigkeitsregulierung, 3108, 3862
Feuchtigkeitsschwielen, 3870
Feuchtigkeitssteuerung, 3862
Feuchtigkeitsstreifen, 1813
feuchtigkeitsundurchlässig, 3865
Feuchtigkeitsundurchlässigkeit, 3866
Feuchtmaschine, 1809
Feuchtpresse, 1811
feuchtsatiniert, 6075
Feuer, 2521
Feueranzeiger, 2523
feuerbeständig, 4713
feuerfest, 4713
feuerfest ausrüsten, 2526
feuerfester Ziegel, 2522
feuerfestes Krepp-Papier, 2525
feuerfestes Material, 4713
Feuerfestigkeit, 2529
feuerfest machen, 2526, 2539
feuerhemmend, 2530
Feuerlöschbrause, 5356
Feuermelder, 2523
Feuerung, 2743
Feuerverhütung, 2524
Fiber, 2445
Fiberkanne, 2450
Fibrille, 2478
Fibrillen, 2474
Fibrillierung, 924
Fichte, 5359
Fichtenholz, 557
Fichtensulfatzellstoff, 4328
Fichtensulfitzellstoff, 4329
Film, 2498
Filmbildner, 2499
Filter, 2500, 5445
Filterbett, 2503

Filterkuchen, 2504
Filtermasse, 2506
Filtrat, 2508
Filtrations-Hilfsmittel, 2502
Filtrierbarkeit, 2501
Filtrieren, 2505
Filtriermasse, 2506
Filtrierpapier, 2507
filtrierte Flüssigkeit, 2508
Filzierung, 2509
Filz, 2418
Filzfehler, 2427
Filzführung, 2934
Filzhaare, 2424
Filzinstandhalter, 2419
Filzinstandhaltung, 2420
Filzkonditionierung, 2420
Filzmarkierung, 2427
Filzmarkierung (in den Filz hineingewebtes
    Muster), 2426
Filzpapier, 2428
Filzprägung (Papieroberflächenausrüstung
    in der Nasspresse), 2423
Filzreiniger, 2419
Filzrichtungsmarkierung, 2421
Filzschlauch, 2273
Filzseite des Papiers, 5779
Filzspanner, 2433
Filzspannung, 2434
Filzspannwalze, 2435
Filztrockenzylinder, 2422
Filztrockner, 2422
Filztuchmarkierung, 728
Filzwäsche, 2436
Filzwaschwalze, 6300
Firnis, 5972
Firnispapier, 4038
firnissen, 5972
Firnissen, 5974
Fischaugen (Materialfehler), 2534
Fixiernatron, 5229
Fixkosten, 2536
Flachdruckpresse, 2548
Fläche, 2373, 5568
Flächengewicht, 604, 5493
Flachliegen, 2553
flachliegendes Papier, 2554
Flachs, 2559
Flachsauger, 2549, 5505
Flachsortierer, 2556
Flachsfaser, 3502
Flachspappe, 2560
Flachstauchfestigkeit, 2550
Flachstauchtest, 2551
Flamme, 2521
flammhemmend, 2541
Flammpunkt, 2547
flammsicher, 2539
flammsicheres Papier, 2540
Flammstrahltrocknung, 2538
Flansch, 2543
flanschen, 2543
Flaschenkapsel, 1085
Flaschenverschlusspapier, 1084
flatternde Ränder, 5142
Flaum, 2598
Fleck, 5304, 5305
Flecken, 5335
Fleckenbildung, 5338
Flecken durch das Bleichmittel
    hervorgerufen, 750
Flecken (durch Streichfarbe), 5337
Flecken im Papier, 4186
fleckig, 3901
fleckiges Papier, 1318
fleckig werden, 5375
Fleckigwerden, 5338
Fleischeinwickelpapier, 767, 986
Flexibilität, 2562
Flexodruck, 375
Flexographie, 2564
Flieder, 2203
Fliese, 5141
Fliessbett, 2609
Fliessbild, 2591
Fliessschema, 2591
Fliessdiagramm, 2591
Fliesseigenschaft der Feststoffteilchen,
    5268

Fliessen, 2588
Fliessfähigkeit, 2607
Fliessgrenze, 6320
Fliessmittel, 2589
Fliesspunkt, 4461
Flintpapier, 2567
Flöckchen, 2579
Flocken, 2537
Flockenstreichen, 2576
Flockenzellstoff, 2599
flockig werden, 2598
Flokulator, 2574
Flockungsbecken, 2574
Flockungsmittel, 2572
Florpost, 4056
Floss, 4619
flössen, 4619
Flössen, 4620
Flotation, 2583
Flotations-De-Inken, 2585
Flotationsmittel, 2584
Flotationsstoffänger, 2586
Flotationszusatzmittel, 2931
Flotte, 3521
Flüchtigkeit, 6032
Fluchtlinie, 297
Flugasche, 2623
Flügelrad, 3191
Flügelradpumpe, 2387
Fluidik, 2606
Fluor, 2617
Fluoreszenz, 2612
fluoreszierende Farben, 2614
fluoreszierender Aufheller, 2613
fluoreszierendes Papier, 2615
Fluorverbindungen, 2618
Fluorwasserstoffsäure, 3135
Fluss, 4809
flüssig, 2601, 3517
Flüssigbett, 2609
flüssiger Anteil der Druckfarbe, 5982
Flüssigkeit, 2601, 3517
Flüssigkeitsdichteprüfung, 3520
Flüssigkeitsdurchlässigkeit, 3519
Flüssigkeitsgrad, 2607
Flüssigkeitshöhe, 3517
Flüssigkeitssäule, 582
Flüssigkeitsscherkraft, 2605
Flüssigkeitsschranke, 5958
Flüssigkeitswiderstand, 6102
Fluss-Schwimmtransport, 4810
Flute, 2619
Föhre, 4300, 4325, 4333, 4938
Foil, 2088, 2640
Foilkörper (mit mehreren
    Entwässerungsleisten), 2641
Folge, 5005
Folgesteuerung, 5006
Folie, 2498, 3440
Folienmaterial, 5049
Folienpapier, 2642
Fond, 1083
Fontäne, 2699
Förderbahn, 1563
Förderband, 654, 1562
Förderdruck, 2016
Förderer, 1562
Fördergerät, 2406
Förderkanal, 2611
Förderkette, 1221
fördern, 1561
Förderschnecke, 1564, 4951
Förderung, 1898
Form, 3871, 3905
Formaldehyd, 2680
Formalin, 2681
Formamid, 2682
Format, 5121
Formatbegrenzung mittels
    Düsenstrahlabspritzung, 3309
Formatbegrenzungsleiste, 1852
Formate, 2683
Formatneueinstellung, 1226
Formatschneiden, 1769
Formatschneider, 1774, 2937, 5848
Formatwasserzeichen, 3545
Formbarkeit, 2679
Formbeständigkeit, 1993
formbeständig machen, 1994

Formel, 2693
formen, 2689, 2714
Formen, 3879
formgeben, 2689
Formgebung, 3879
Formiate, 2683
Formleiste, 2687
Formmaschine, 3880
Formteil, 3874
Formulierung, 2694
Formveränderung, 1876
Formveränderungsvermögen, 2679,
    4362
Forschung, 4747
Forst, 2666, 6260
Forstgenetik, 2670
Forstpathologie, 2675
Forstverwaltung, 2673
Forstwesen, 2677
Forstwirtschaft, 2674
forstwirtschaftliche Erzeugnisse, 2676
Fortleitung, 1502
Fortpflanzung, 1548, 4549
Fortreissfestigkeit, 5655, 5662
Fortreissfestigkeitsprüfer, 5663, 5664
Fragment, 2707
Fragmentierung, 2712
Fraktion, 2707
Fraktionieren, 2708
Fraktioniergerät, 1343, 2710
Fraktionierung, 2709
Fraktur, 2711
fräsen, 3816
frei an Bord, 2369, 2721
Freigabe, 4729
freigeben, 4729
Freiharzleim, 178
freitragend, 1073
freitragende Ausführung, 1074
Freiwicklung, 1205
Fremdkörper, 3200
Fremdstoffe, 1539
Frequenz, 2725
Frequenzwandler, 2726
Friktion, 2728
Friktionsantrieb, 2732
Friktionsfaktor, 2733
Friktionsglättung, 860, 2734
Friktionskalander, 2731
Frischlauge, 3665
Frischwasser, 2727, 4649
Frist, 5746
Fruchtbarmachung, 2438
Frühholz, 2176, 2898, 5355
Fuge, 3314
Fugen verstreichen, 2925
fühlen, 2413
führen, 2932, 3437
Führerseite, 2107
führerseitiger Gang, 5673
Führung, 3437
Führungskante, 3438
füllen, 2484, 4141
Füller, 3544
Fullererde, 2746
Füllmasse, 2497
Füllstoff, 2493
Füllstoff aus wasserfreiem Calciumsulfat,
    4234
Füllstoffclay, 2494
füllstofffrei, 5922
Füllstoffretention, 2496
Füllstoffzugabe im Holländer, 632
Fülltrichter, 3086
Füllung, 1204, 4137
Fumarsäure, 2749
Fundament, 2698
fundamental, 4614
Fundamentanker, 368
Fundamentplatte, 597, 5259
Fundamentrahmen, 4619
fünfhundert Count = fünfhundert Blatt,
    1640
fünfundzwanzig Bogen (aus einem 500er
    Ries - Feinpapier), 4610
funktionelle Gruppe, 2751
Funktionieren, 4260
Funktionsprüfung, 4261

Funktionsstörung, 3670
Furan, 2756
Furche, 2912
Furfuran, 2758
Furfurol, 2757
Furfurylalkohol, 2759
für Heisstrocknung geeignete Farbe,
    3019
Furnier, 5992
Furnierabfall, 5994
Furnierblatt, 5992
Furnierholz, 5992
Furnierpapier, 5993
Fusion, 3754
Fussel, 2762
fusseln, 2762
Futter, 3509

G

Gabelstapler, 2678
Galaktan, 2770
Galaktomannan, 2771
Galaktose, 2772
Gallerte, 2802
Gallone, 2773
Gallonen pro Tag, 2766
galvanische Korrosion, 2774
galvanische Verbindung, 2211
galvanisieren, 2775
Galvanisierung, 2776
Galvanometer, 2777
Gamma-Cellulose, 2778
Gang, 2148, 3897
Ganghöhe, 3437
ganzer Baumstamm, 5833
Garn, 6315
Garnitur, 5012
Gas, 2780
Gasanalyse, 2781
gasartig, 2784
Gasausströmung, 2783
Gasbildung, 2787
Gasblase, 929
Gaschromatographie, 2782
Gasdurchlässigkeit, 2791
Gasentladung, 2783
Gaserzeuger, 2785
gasförmig, 2601, 2784
Gasgenerator, 2785
Gasheizung, 2786
Gasturbine, 2794
Gaswäsche, 2792
Gaswäscher, 2792, 4954
Gate-Roll-Coater, 2796
Gate-Roll-Streichanlage, 2796
Gautschbruchbütte, 1635
Gautsche, 1630
gautschen, 1630
Gautschen, 1632
Gautschfilz, 1631
Gautsch-Gegenwalze, 3592
Gautschknecht, 1638, 1767
Gautschpappe, 3818
Gautschpresse, 1636
Gautschwalze, 1637
Gautschwalzenbezug, 1633
Gautschwalzenmarkierung, 1634
Gebläse, 772, 2384
Gebläseflügel, 2385
Gebläseventil, 782
gebleichter Zellstoff, 738
gebrannter Kalk, 1021
gebündelt, 546
gebundenes Schwefeldioxid, 1463
gebundenes Wasser, 857
gebürstet, 920
gedeckter Karton, 4348
gedeckte Wellpappe, 1180
Gefälle, 5175
Gefälleverlust, 2998
gefaltene Nasszellstoffbogen, 4576
gefaltet, 3419
gefalzt, 2644
gefärbt, 1440, 2171, 4321

gefärbte Papiere, 1441
Gefäss, 5641; 6003
gefrorenes Holz, 2741
Gefüge, 2862, 5478, 5702
Gefühl, 2413
gefüllt, 2485
gefurcht, 4935
gefüttert, 3498
gegautscht, 5977
gegautschter Bristolkarton, 1793
gegautschter Karton, 5978
Gegendruck, 514
Gegendruckwalze, 508
Gegenfluss, 1538
Gegengewicht, 543
Gegenlauf, 1643
gegenläufig, 1643
gegenläufige Fliessrichtung, 1644
Gegenlauf-Streichanlage, 1537
Gegenlaufwalzenstreicher, 4779
Gegenrolle, 1647
Gegensatz, 5013
Gegenstrom, 1538, 1643, 1646
Gegenströmung, 1646
Gegenstromverfahren, 1645
Gegenstück, 3711
Gegenwalze, 1647
geglättet, 2819
Gehänge, 2440
gehärtetes Kollophonium, 3138
Gehäuse, 1140, 1148, 3102, 5057
Gehege, 4247
geklebt, 4221
gekräuselt, 1697
gekräuseltes Papier, 1756
gekreppt, 1681
gekreppte Oberfläche, 1684
gekrepptes Papier, 1682
gekrümmte Walze, 858, 1762
Gel, 2801
Gelatine, 2802
gelatineartige Schicht, 2804
gelatiniert, 2803
Gelatinierung, 2805
Gelbblättrigkeit, 1315
Gelbkiefer, 3567, 5072, 5287
gelb werden, 6317
gelegt, 3395
geleimt, 4221
geleimtes Fleischeinwickelpapier, 985
geleimtes Papier, 5123
geleimtes Zeichenpapier aus
    Sekundärfaserstoff, 800
geleimt und superkalandriert, 5122
geliert, 2803
Gelierung, 2805
gelochte Abzugsplatte (beim Pulper),
    2359
gelochtes Papier, 4257
gemahlen, 622
gemasertes Papier, 2864
Gemisch, 3837, 3846
genadelter Filz, 3950
genarbtes Papier, 2864
genaue Angabe, 5299
Genauigkeit, 143, 4473
Generaldirektor, 2806
General Electric (G.E.) Helligkeitsmessung,
    2765
Generator, 2808
Genetik, 2809
geölt, 4036
geölte Deckpappe, 4037
geprägt, 2254
geprägte Oberfläche (die dem Papier ein
    handgeschöpftes Aussehen gibt),
    2963
geprägtes Papier, 2255
gepresste Sauglinge (aus Faserbrei und
    Harz), 3874
gequollener Stoff, 3115
geradestossen (Papier), 3312
Geradlinigkeit, 5443
Gerät, 3255, 5771
Geräusch, 3994
Geräuschanalyse, 3995
Geräuschkontrolle, 3995
Geräuschmesser, 3996
Geräuschspannungsmesser, 3996
geriffelt, 1606, 2620

geriffelte Oberfläche, 4807
Geriffelte Platte (im Holländer), 648
geriffelte Walze, 1272, 1611
gerillt, 2620
gerillte Walze, 2914
geringe Dichte, 3577
geringe Dichte (Stoffdichte), 3576
Gerinne, 2611
Gerinnen, 1376
Gerinnung, 1374
Gerinnungsmittel, 1375
gerippt, 1606
gerippte Antik-Feinpost, 403
gerippte Form, 3400
gerippte Oberfläche, 3398
geripptes Antikpapier, 3396
geripptes Papier, 3401
geripptes Schreibpapier, 497, 3403
gerippte Wasserzeichen, 3399
gerippt (Papier), 3395
gerollte Papierkante, 4822
Geruch, 4017
Geruchsbeseitigung, 1910
Geruchskontrolle, 4018
gesägt, 5008
Gesamtalkaligehalt, 5786
Gesamtbaumschlag, 2748
Gesamtlänge (in Fuss), 2663
Gesamtübertragung, 5787
gesättigt, 3196, 4909
Geschäftsbücherpapier, 834, 3447
Geschäftsformular, 981
Geschäftsformularpapier, 4717
Geschäftsführer, 3672
geschältes Holz, 4241
gescheckt, 3900
gescheckfarbig, 3901
Geschenkeinwickelpapier, 2811
geschichtetes Papier, 5451
Geschiebe, 647
geschlossene Blattbildung, 1362
geschlossener Regelkreis, 1358
geschlossener Stoffauflauf, 1361
geschlossener Stromkreis, 1356
geschlossenes System, 1359
geschlossene Trockenhaube
   (Trockenpartie), 1357
geschmeidig, 5241
Geschmeidigkeit, 4380
geschmiert, 4036
Geschwindigkeit, 5987
Geschwindigkeitskontrolle, 5988
Geschwindigkeitsmesser, 5314, 5627
Geschwindigkeitsmessung, 5989
Geschwindigkeitsregelung, 5988
Gesichtstücher, 2374
gespaltene Gewindemuffe, 5330
gesprenkelt, 3900
gestalten, 2714
Gestaltung, 2684
gestreift, 5469
gestrichen, 1384
gestrichen Anhängeetiketten, 1397
gestrichen Bristol, 1391
gestrichene Druckpapiere, 1396
gestrichene Papiere, 1395
gestrichener Karton, 1387, 4222
gestrichener Kartonagenkarton, 1390
gestrichenes Banknotenpapier, 1388
gestrichenes Bankpostpapier, 1388
gestrichenes Buchdruckpapier, 1389
gestrichenes Illustrationsdruckpapier,
   1393
gestrichenes Kunstdruckpapier, 1385
gestrichenes Offsetdruckpapier, 1394
gestrichenes Pergamin, 1392
gestrichenes Release-Papier, 4731
geteilte Presse, 2046
getönt, 5751
getränkt, 4909
getränktes Papier, 3197
getrennte Fasern (durch Behandlung mit
   Dampf), 2351
Getriebe, 2104, 2799
Getriebegehäuse, 2800
Getriebekasten, 2800
gewachster Karton, 6123
gewachstes Papier, 6126
gewachstes Papier zur

Kartoninnenauskleidung, 1137
gewachstes Tissue, 6127
gewalzte Graupappe, 2730
Gewebe, 2370, 6144
Gewebebandpresse, 2372
gewellt, 1606, 2620
Gewerbeingenieurwesen, 3215
Gewerbetechnik, 3215
Gewicht, 6170
Gewicht ohne Strichauftrag, 5912
Gewichtsanzeige, 6171
Gewichtstoleranz, 6172
gewickelte Rolle, 6290
Gewinn, 4546
gewirkter Stoff, 3356
gezogene Pappe, 3873
Giessbeschichtung, 1760
Giessbeschichtungsanlage, 1759
Giessen, 1152
Giessform, 2580
Giessling, 1152
Giftigkeit, 5795
Gilben, 2018
Gips, 1022
Gipsbaupappe, 2952
Gipspappe, 2952
Girlande, 2440, 2779
Girlandentrocknung, 2443, 3571
Glanz, 2195, 2826, 3595, 4400
Glanzbürstmaschine, 927
Glanz eines gestrichenen Papiers (bei
   75grad vertikal gemessen), 3575
glänzen, 2813, 2826, 3595
glänzende Fehlerstelle im fertigen Papier,
   5060
glänzend machen, 3595
Glänzendmachen, 2823
glanzgestrichenes Buchpapier, 2821
Glanzkarton, 2820
glanzlos, 3709
glanzlose Glätte, 2552
glanzloses Papier, 3719
Glanzmesser, 2830
Glanzmittel, 2827
Glanzpapier, 2822, 2831
Glasfaser, 2815
Glasfaserpapier, 2814
glasieren, 2818
glasige Flecken, 2534
Glassin, 2816
Glasur, 2818
Glätte, 4400, 5201
glätten, 2818, 2826
Glätten, 967, 5197
glatte Oberfläche, 5159
Glätteprüfer, 5202
glatte Walze, 2555
Glattkiefer, 5072
Glättpresse, 5198
Glättwalze, 5199
Glättwalzen-Streicher, 5200
Glättwerk, 1025, 5372
Glättwerk (zwischen zwei Trockengruppen
   angeordnet), 3271
Glättzylinder, 2824
Glättzylinder (für maschinenglattes
   Papier), 3599
Glaubersalz, 4888
gleichaltriger Baumbestand, 2334
gleichaltriger Waldbestand, 2334
Gleichartigkeit, 3079
Gleichgewicht, 543, 2299
Gleichlauf, 5608
gleichlaufend, 5609
gleichmachen, 3452
Gleichmässigkeit, 5919
gleichmässige Formierung, 5918
Gleichrichter, 2002, 4687
Gleichstrom, 1803, 2006
Gleichstromerzeuger, 2808
Gleichung, 2298
Gleichzeitigkeit, 5608
Gleitantrieb, 2732
Gleiten, 5165
Gleitfähigkeit, 766
Gleitlager, 5150
Gleitmittel, 3583
Gleitschutzbeschichtung, 407
Gleitschutzpräparation, 408

Glimmentladung, 1594
Glimmer, 3793
Glimmerpapier, 3794
glitschige Oberfläche, 5159
Glückwunschkarte, 2899
Glückwunschkartenkarton, 436
Glukomannan, 2832
Glukose, 2833
Glukoseäther, 2836
Glukosebindung, 2839
Glukosederivate, 2834
Glukoseeinheit, 2837
Glukoseester, 2835
Glukoside, 2838
Glykole, 2853
Glyoxal, 2854
Glyzerin, 2850, 2851
Glyzerinlösung, 2851
Glyzerinstearate, 2852
Grad der Haubenabdeckung, 3085
Grad der Transparenz, 5817
Grad des Aufschlusses, 1570
Gradeinteilung, 4922
Gradient, 2858
Grad Kelvin, 3338
grafische Papiere, 6223
Gramm, 2869
grams pro quadrat meter (g/m2), 2767
Granit, 2871
Granitimitation, 1268
Granitwalze, 2873
graphische Darstellung, 1231, 1938
graphische Kunst, 2875
Grat, 969
graumeliertes Papier, 2872
Graupappe, 1278, 4348
Grauverdrücken am Kalander, 719,
   1027
grauverdrückt (Satinierfehler), 1026
Gravurstreichen, 2880
Greifer, 2876, 2910
Greif-Skidder, 2877
Greifzange, 2876
Grenzfläche, 2603
Grenzgeschwindigkeit, 1702
Grenzlinie, 4931
Griff, 2413, 2961
Griffigkeit, 2413, 2961
grob, 1377
Grobausrüstung, 3578
grobe Riffelung, 2911
grobkörniger Sandstein, 2911
Grobsortierer, 1380
Grobstoffbehandlung im Refiner, 4725
Grobzerkleinerungsmaschine, 1722
Grösse, 1992, 5121
grosser Anfangsbuchstabe, 1083
Grube, 4339
grün, 2896
Grund, 5256
Grundbelastung, 647
Grundfläche, 595
Grundgewebe, 3755
Grundierlack, 4519
Grundlage, 2698
grundlegend, 4614
Grundplatte, 597, 5259
Grundwerk, 648
grün färben, 2896
Grünlauge, 2897
grün werden, 2896
Gruppe, 5012
Gruppenantrieb, 4978
Guar, 2931
Gully, 5022
Gumbildung, 2944
Gummi, 2940, 4864
Gummi arabicum, 427
Gummiarabikum, 2941
gummiartige Stoffe, 2202
gummierfähiges Papier, 2946
Gummiermaschine, 2945
Gummierrohpapier, 2946
gummiertes Papier, 2942
Gummierung, 2944
Gummiflecken, 4868
Gummiharz, 2940
Gummimilch, 3430, 4866
Gummirolle, 4867

Gummituch, 4523
Gummituchmarkierung, 728
gummiüberzogene Walze, 4865
Gummiwalze, 4867
Gurley-Dichtemesser, 2949
Gurt, 653
Gürtel, 653
Gusseisen, 1153
gusseisern, 1153
gussgestrichenes Papier, 1149
Gussstreichen, 1151
Gussstreicher, 1150
Gut-Chips, 135
gut deckend (Farbe), 4060
gut geleimt, 6178
gut geschlossene Blattbildung, 6177
gut geschlossene Oberfläche, 6177
Gutstoff, 136, 137

H

haarfein, 1082
haarförmig, 1082
Haarriss, 1766
Hacker, 1291
Hackmaschine, 1291
Hackmesser, 1292
Hackschnitzel, 1276, 6261
Hackschnitzeldosierer, 1288
Hackschnitzel-Gutstoff, 135
Hackschnitzelmühle, 1284
Hackschnitzelschliff, 2919
Hackschnitzelsilo, 1277, 1296
Hackschnitzelstapel, 1293
Hackschnitzeltransport, 1286
Hackschnitzelzuteiler, 1279
Hackspan, 1276
Hackzerkleinerungsmaschine, 1284
Hadern, 4621
Hadernabfälle, 4627
Hadernanteil, 4624
Hadernboden, 4631
Haderndrescher, 4634
Haderngehalt, 4624
hadernhaltiges Papier, 4625
Hadernholländer, 4623
Hadernkocher, 4622
Hadernpapier, 4629
Hadernschneider, 4626
Hadernstäuber, 4628
Hadernstoff, 4633
Hadernzellstoff, 4630
Hadernzellstoff (Zellstoff aus Lumpen
   oder Hadern oder Jute), 2986
Hadernzerreisswolf, 4626
Hadernzuschnitte, 4627
Hafensperre, 841
Haff, 3393
haften, 5424
Haftfähigkeit, 204
Haftfestigkeit, 204, 821, 823, 4389
Haftfestigkeit der Lagen, 4388
Haftpapier, 205
Hagebuche, 3088
halbchemischer Aufschluss, 4994
halbchemischer Karton von 140-150
   g/cm^2, 3972
halbchemischer Zellstoff, 4993
halbchemisches Neutralsulfitverfahren,
   3937
halbdurchlässige Membran, 4996
halbgebleicht, 4989
halbgebleichter Zellstoff, 4990
halbglänzend, 2285
Halblederpappe, 3184
Halbleiter, 4995
Halbmesser, 4617
Halbzeug, 2954
Halbzeugholländer, 883, 6052
Halm, 4698, 5418
Halogene, 2957
Halogenid, 2957
Halogenverbindungen, 2956
Hals (masch.), 1435
Haltbarkeit, 1842, 2162, 3465, 5010

Halter, 2910, 3067
Halterung, 3067
Haltevorrichtung, 1435
Hammer, 2958
Hammermühle, 2959
Handelsformat, 603
handelsübliches Tissue, 1470
Handform, 2965
handgenähter Filz, 2962
handgeschöpftes Papier, 2964
Handmuster, 2966
Handpappe, 6192
Handpappenmaschine, 6181
Handschäler, 4242
Handschöpfrahmen, 5051
Handtuch, 5791
Handtuchkrepp, 5793
Handtuchpapier, 5793
Handtuchstoff, 5792
Hanf, 3031
Hanfeibisch, 3339
Hanfpapier, 3032
Hängelager, 2968
Hängen, 2969
Hängetrockner, 2442
Hängetrocknung, 2443, 3571
Harnstoff, 1090, 5934
Harnstoffharze, -5936
Harnstoffformaldehyd-Harz, 5935
Harper-Maschine (Langsiebmaschine mit
    Selbstabnahme), 2989
Härtemesser, 2163
Härtemessung, 2980
Härtemittel, 2975
Härten, 2976
Härteprüfer, 2163, 2981
Härteprüfung, 2982
Härter, 2975
harter Stoffbatzen, 2978
Härte (z.B. Rockwell-Härte oder Shore-
    Härte), 2979
Härtezeit, 5015
Hartfaserplatte, 645, 2447, 2973
hartgekochter Zellstoff, 2983
Hartgusswalze, 1272
Hartkochung (Kochen bis zu einem
    geringen Aufschlussgrad oder
    Ligninabbau), 2974
Hartpapier (mit hitzehärtbarem Harz
    getränkt), 4149
Hartpappe, 2973
Hartpappe (mit hitzehärtbarem Harz
    getränkt), 4150
Härtung, 2976
Harz, 4753
Harzemulsion, 5124
Harzflecken, 4343, 4845, 4846
harzförmiger Rückstand aus der
    Rohterpentindestillation, 4841
Harzgang, 4756
harzgeleimt, 4843
Harzgewinnung, 5649
Harzkanal, 4756
Harzleim, 4842
Harzleimung, 4844
Harzmilch, 5124
harzreiches Holz, 4757
Harzsäure, 4754
Harzstippen, 4845, 4846
Haspel, 4699
Haube, 2993, 3084
hauen, 2414
Haufen, 2575, 3705, 5372
Häufigkeit, 2725
haupt-, 1083
Häutchen, 3747
Hebel, 3459
Hebelarm, 3460
heben, 3066, 3466
Hebevorrichtung, 3468, 4595
Hebewerk, 3066
Hebewerkzeug, 3468
Hebezeug, 3066
Hefe, 6316
Heften, 5385
Heftigkeit, 3266
Heissalkaliextraktion, 3092
Heissatinieren, 3097
Heissbehandlung des Stoffs, 3100

heissschleifen, 3093
Heissschliffbehandlung, 3100
Heissschmelze, 3094
Heissschmelzleim, 3095
Heissschmelzstreichen, 3096
heissgesiegeltes Papier für
    Lebensmittelkarton, 1138
Heissglanzpresse, 2828
Heissglätten, 3097
Heisssiegelmaschine, 3017
Heisssiegelpapier, 3018
Heisssiegelung, 3016
heisskalandrieren, 2829
heissmahlen, 3093
Heissmahlung, 3098, 3100
Heisspressen, 3097
Heissstauchen, 2004
heisstrocknende Farbe, 3019
Heissverkleben, 3016
Heissverschweissen, 3016
heiss werden, 3000
heizen, 3000
Heizer, 5439
Heizgerät, 3009
Heizöl, 2744
Heizplatten (die laufende Papier-
    /Kartonbahn wird durch zwei
    aufeinanderliegende Heizplatten
    geführt), 4374
Heizung, 3008
hell, 3470
Helligkeit, 896, 3589
hell leuchten, 2813
Hemicellulose, 3029
hemmen, 876
Hemmung, 3227
Henkel, 2961
herausreissen, 5660
herausrupfen, 4308
Herausrupfen, 4306
Herausrupfen der Fasern beim Streichen
    durch Messerkratzer, 4302
hermetisch, 3194
Herstellung, 3687, 4539
Hervorsprudeln, 5339
Hervorstehen, 3313
Heterogenität, 3035
heterozyklische Kohlenstoffverbindungen,
    3034
Heuschrecke, 3546
Hilfsantrieb, 3028
Hindernis, 1354
Hinderung, 3227
hin-und herschwingend, 4088
Hitze, 3000
Hitzebehandlung, 4601
hitzebeständig, 4713
Hitzebeständigkeit, 3014, 5709
Hitzeempfindlichkeit, 5708
hitzehärtbar, 3020
Hobelnabfall, 5034
Hobelspan, 1294
Hochausbeutezellstoff, 3060
Hochdruck, 3052
Hochdruckverfahren, 3451
hochelastische Viskosefaser, 3051
Hochglanz, 5554
Hochglanzausrüstung, 5551
hochglänzendes Papier, 5552
Hochglanzgebung, 4402
Hochglanzkarton, 4377
Hochglanzmaschine (bestehend aus 2
    gusseisernen Walzen), 4376
Hochglanzoberfläche, 5551
Hochglanzoberflächenausrüstung (das
    Papier wird dabei unter hohem Druck
    durch Kupfer- oder Zinkplatten
    hindurchgeführt), 4367
Hochglanzpapier, 2268, 3047, 4368
Hochglanzpolierkalander, 966
Hochglanzsatinage, 965
Hochglanzsatinierkalander, 966
hochheben, 3466
Hochheben, 3467
Hochkonjunktur, 841
Hochleistungs-, 3056
Hochleistungshaube, 250
Hochleistungstrockner, 2953, 3058
Hochleistungstrocknung, 3057

Hochsatinieren, 5553
hoch satiniert, 3046, 5555
hochsatiniert, 4369
hochsatinierte Oberflächenausrüstung,
    5556
Hochspannungsleitung, 5813
Höchstbelastung (einer Leitung), 3724
Hochtemperatur, 3054
hochtourig, 3056
hochtourig laufendes und kontinuierliches
    Mahlgerät, 3112
Hochvakuum-Metallisierung, 5946
hochvoluminöses Papier, 951
hochwertiges Schreibpapier, 566
hochwinden, 3066
Hochwölben der Ränder eines
    Papierstapels, 2023
hochziehen, 3066
Hofarbeiter, 6314
Höhe, 3026, 3452
hohe Ausbeute, 3059
hohe Dichte, 3042
hohe Maschinenglätte, 3050
Höhepunkt, 3026
hoher Ertrag, 3059
hohe Stoffdichte, 3041
Hohlraum, 6028
Hohlraumbildung, 1172
Hohlschliffwalze, 1486
Holländer, 624, 3074
Holländerbemessung, 630, 639, 3357
Holländerbeschickung, 632
Holländerbütte, 627
Holländereinsatz, 630
Holländergrundwerk, 634
Holländermahler, 633
Holländermahlung, 626
Holländermüller, 633
Holländersaal, 631
Holländertrog, 627
Holländerwalze, 635
Holländerzusatz, 625
Holoaufschluss (nur Lignin und äusseres
    Material wird von dem Pflanzengewebe
    entfernt), 3077
Holocellulose, 3076
Holunder, 2203
Holz, 6260
Holzabfall, 6281
Holzabmessung, 6271
Holzaufbereitung, 6274
Holzaufschluss, 2924
Holzaufzug, 3555
Holzbringung, 3556, 3586
Holz das sich sehr früh (in der Jugend)
    gebildet hat, 3334
Holzeinbringung mittels Fesselballon,
    555
Holzeinschlag, 3556
Holzeinschlag im Kopfhochverfahren,
    3049
Holzeinschlagsabgabe, 5488
hölzerne Hülse, 6263
Holzerntemaschine, 3557
Holz fällen und zurichten, 3586
Holzfäller, 3415, 3587
Holzfäller-Skidder, 2416
Holzfaser, 6265
Holzfaserstoff, 289
holzfrei, 2918
holzfreies Papier, 2723, 6267
holzfrei (h'f), 6266
Holzgletscher, 3559
Holzgummi, 6304
holzhaltiges Papier, 6262
Holzharz, 6278
Holzhauer, 2415
Holzhaufen, 6273
Holzkeil, 5289
Holzkern, 4344
Holzklotzschrägaufzug, 3555
Holzkohle, 1228
Holzlagerplatz, 6283
Holzmaserung, 6268
Holzpartikel, 6272
Holzprodukte, 6275
Holzreinigungsanlage, 6279
Holzschlag, 2417
Holzschleifer, 6269

Holzschleiferei, 2920
Holzschliff, 2916, 2923, 3734, 3738,
    6276
holzschliffhaltige Druckpapiere, 2922
holzschliffhaltige Papiere, 2921
holzschliffhaltiges Buchpapier, 2917
Holzschliffherstellung, 2924
Holzschliffkarton, 6277
Holzschliffpappe, 6277
Holzschnitzel, 5034
Holzspäne, 5034
Holzsplitter, 5062
Holzstapel, 6273
Holzstruktur, 6280
Holzteilchen, 6272
Holztransport, 6270
Holztransport auf Flüssen, 4810
Holztransport mittels Fesselballon, 555
Holztransport mittels Kabelseil, 1006
Holzvertäfelung, 4144
Holzvorbereitung, 6274
Holzwolle, 6282
Holzzellstoff, 1261
Holzzubereitung, 6274
homogene Blattbildung, 5918
Homogenisation, 3080
Homogenisiermaschine, 3081
Homogenisierung, 3080
Homogenität, 3079
Hornbaum, 3088
Hornsubstanz, 3340
Hub, 3466
Hülle, 1140, 5057
Hüllen, 2292
Hülse, 1585, 5865
Hülse aus Holz, 6263
Hülsenkern, 1585, 4382, 6264
Hülsenmaterial, 5867
Hülsenpapier, 1587
Hülsenpappe, 1590, 5866
Hülsenpfropfen, 1588
Hülsenpund, 1588
Hülsenstange, 1589
Hülsenwickelmaschine, 1592
Hülsenwickler, 1592
Hybride, 3111
Hydrafiner, 3112
Hydracellulose, 3113
Hydrate, 3116
Hydration, 3117
Hydratisierung, 3117
Hydraulik, 3119
Hydraulikanlage, 3122
hydraulisch, 3118
hydraulische Entrindung, 3120, 3125
Hydrazin, 3123
Hydride, 3124
Hydrierung, 3139
Hydroanlage, 3122
Hydrocellulose, 3128
Hydrodynamik, 3131
Hydrolysate, 3148
Hydrolyse, 3146
Hydrometer, 3149
hydrophob, 3151
hydrostatisch, 3152
hydrosulfide, 3153
hydrotropische Lignine, . 3155
hydrotropischer Zellstoffaufschluss,
    3156
hydrotropisches Mittel, 3154
Hydroxid, 3157
Hydroxyalkylcellulosen, 3158
Hydroxyalkylgruppen, 3159
Hydroxyäthylcellulose, 3161
Hydroxyäthylgruppen, 3162
Hydroxyäthylierung, 3160
Hydroxyl, 3163
Hydroxylbindung, 3164
Hydroxylgruppen, 3165
Hydroxylionenkonzentration, 3166
Hydroxypropylcellulose, 3167
Hydroxypropylmethylcellulose, 3168
Hydrozyklon, 3130
Hygienepapier, 4898
Hygiene-Tissue, 611, 4899
Hygroskopizität, 3170
Hypobromite, 3171
Hypochlorite, 3172

Hysterese, 3174
Hysteresis, 3174

I

Identifizierung, 3175
Ilex, 3179
im Autoklav dämpfen, 483
im Autoklav kochen, 483
im Hängetrockner getrocknet (z. B. Streichpapier), 2441
Imide, 3180
Imine, 3181
Imitationsanhänger, 810
Imitationsbüttenrand, 5782
Imitationsbüttenrandpapier (wobei die Papierkanten mechanisch nachbehandelt werden), 3729
Imitations-Etikettenpapier, 810
Imitationslederpappe, 3184
Imitationszeitungsdruck (ausschliesslich aus altem Zeitungsdruck hergestellt), 3962
Imitation (Zweitfertigung einer bestimmten Qualität wobei diese nicht 100%ig dem Original entsprechen muss), 1469
imitierte Rückseitenverstärkung, 796
im Kalander ausgerüstet, 1037
im Kalander gefärbt, 1034
im Kalander oberflächeniniert, 1037
im Kalander verdrückt, 1031
Impedanz, 2218, 3190
imprägnierbare Papiere, 4911
imprägnierbarer Filz, 4910
Imprägniereigenschaften, 4912
Imprägnieren, 6095
imprägnierend, 3195
imprägnierfähige Papiere, 3198
Imprägnierfilz, 4910
Imprägnierpapiere, 4911
imprägniert, 3196
imprägnierter Filz, 2431
imprägnierte Papiere, 3197
Imprägnierung, 3199, 4913
Imprägnierung mit Metallsalzen, 3764
im Pulper auflösen, 5186
in Ballen verpacken, 549
in Ballen verpackt, 546
Inbetriebnahme, 5393
in Betrieb setzen, 2799
in Beutel füllen, 526
in Beutel verpacken, 534
in das Papier eingearbeitete Spannung, 941
in den Kreislauf zurückführen, 4689
in der Bütte färben, 5859
in der Bütte geleimt, 5861
in der Feuchtigkeit abbindend, 3868
in der Hammermühle mahlen, 2960
in der Hitze abbindend, 3020, 5718
in der Masse geleimt, 637
in der Nasspartie zugegebene Papierhilfsstoffe, 6185
in der Wärme abbindend, 5718
in die Stoffzufuhrleitung eingebauter Mahlgradprüfer, 3246
Indigofarbstoffe, 3210
indigoide Farbstoffe, 3210
Indikator, 3208
Indikatorpapier, 3209
indirekte Heizung, 3211
indirektes Dämpfen, 3212
Induktionsheizung, 3214
Induktionsstrom, 2182
Industriepapiere, 3216
ineinandergeschoben, 5666
ineinandergreifen, 2799, 3756
in einen eutropischen Zustand versetzen, 2328
in Einzelfasern zerlegen, 1864
Infrarot, 3220
Infrarot-Detektor, 3221
infrarote Strahlung, 3225
Infrarot-Heizstrecke, 3223, 3224

Infrarot-Heizung, 3223, 3224
Infrarot-Trocknung, 3222
Ingenieur, 2282
Inhaber, 3067
in Haufen, 2577
Inhibition, 3227
Inhibitor, 3228
Injektion, 3230
in Karton verpacktes Schreibpapier, 868
in Laugenform überführen, 1168
Innendurchmesser, 3252
innen gerippter Trockenzylinder, 3280
innere Bindekraft, 3275
innere Durchreissfestigkeit, 3277
innerer Durchreisswiderstand, 3279
innerste Schicht mehrlagiger Tüten oder Säcke, 537
in Rollen geglättet, 6148
in Schachtel verpacken, 862
in Scharen, 2577
Insektenkunde, 2287
in sich verschobene Papierrolle, 5667
Inspektion, 3253
Inspektor, 5559
Installateur, 4335
Installation, 2535, 3254, 4336
Instrument, 3255
Instrumentenbrett, 1546
integrierte Fabrik, 3264
integrierter Betrieb, 3264
Integrierung, 3265
Intensität, 3266
Inventar, 3282
Ione, 3293
Ionenaustausch, 3286
Ionenaustauscher, 3289
Ionenaustauschharze, 3288
Ionenaustauschpapier, 3287
Ionenspaltung, 3291
Ionenstärke, 3290
Ionisierung, 3291
Ionomere, 3292
Isobutylalkohol, 3299
Isocyanate, 3300
isolierende Verkleidung, 3259
Isoliermaterial, 3258
Isolierpapier, 430, 587, 2206, 3262
Isolierpappe, 3257, 6043
Isolierschutz, 3261
Isolierstoff, 2219, 3258
Isolier-Tissue, 3260
Isomere, 3301
Isopropylalkohol, 3302
Isothermen, 3303
Isotopen, 3304
Isotropie, 3305

J

Jahresring, 383, 2929
Jahresring (Holz), 4802
jahreszeitlich bedingte Schwankungen, 4969
Jährlingswolle, 3063
jede Art von steifer Pappe mit glatter und bedruckbarer Oberfläche, 731
Jod, 3284
Jodkalium, 4454
Jodverbindungen, 3285
Jordan, 3315
Jordanmühle, 4704
Jordan-Refiner, 3318
Jumbo-Rolle (besonders grosse Papierrolle), 3321
junger Baum, 4901
Jungholz, 6013
Jute, 3325
Jutefaser, 3328
Jutesackpapier, 3326
Jutezellstoff, 3332

K

Kabel, 1005
Kabelisolierpapier, 388
Kabelpapier, 1007
Kachel, 5742
Kadmium, 1008
Kadmium-Verbindungen, 1009
Kady-Mühle, 3335
Kahlschlag, 1352
Kahn, 570
Kalander, 1025
Kalanderausschuss, 1029
Kalanderfalten, 1033, 1670
Kalanderflecken, 1051
Kalanderführer, 1039, 1042
Kalandergefärbt, 1030
kalandergeleimtes Papier, 1049
Kalanderkasten, 1028
Kalanderleimung, 1050
Kalandermarkierung infolge defekter oder unsauberer Kalanderwalzen, 1041
Kalandern, 967
Kalanderpartie, 1048
Kalanderschäden, 1051
Kalanderspäne, 1047
Kalanderstaub, 1047
Kalanderwalze, 859, 1044
Kalanderwalze (aus Papier), 2492
Kalanderwalzenpapier, 861, 1045, 6287
Kalanderwasserkasten, 1057
kalandrieren, 860
Kalandrieren, 1038
kalandriert, 1035
kalandriertes Papier, 1036
kalibrieren, 1058, 5121
Kalibrierung, 1059
Kalium, 4452
Kaliumjodid, 4454
Kaliumpermanganat, 4455
Kaliumverbindungen, 4453
Kalk, 3487
Kalkbrei, 3493
Kalk löschen, 5146
Kalklöschen, 3491
Kalkmilch, 3489, 3815
Kalkofen, 3488
Kalkschlamm, 3490, 3492, 5144
Kalkstein, 3494
Kalksuspension, 3493
Kalorimeter, 1062
Kalorimetrie, 1063
Kaltalkali, 1429
kaltalkalibehandelter Zellstoff, 1430
Kälteerzeugung, 1727
Kaltnatronverfahren, 1432
Kaltnatronzellstoff, 1433
Kaltnatronzellstoffaufschluss, 1434
Kaltschleifen, 1431
Kalzinierung, 1010
Kalzium, 1011
Kalziumhypochlorit, 744
Kamin, 1273
kanadische Pappel, 1629
Kanal, 2148
Kanaltrockner, 5871
Kanister, 1071
kanneliert, 2620
Kante, 2183, 6179
Kantendruckfestigkeitstest, 2184
Kanteneindrückbeständigkeit, 4806
Kanteneinreissfestigkeit, 2188
Kantenschutz, 2187, 2995
Kantholzlignine, 3824
Kaolin, 1275, 3336
Kapazität, 1080
kapillar, 1082
Kapillarität, 1081
Kapillarwirkung, 1081
Kapital, 1083
Kapitalrendite, 4773
Kappazahl, 3337
Kaprolactam, 1086

Kapsel, 1087
Karbamat, 1089
Karbamid, 1090
Karbamidharz, 5935
Karbid, 1091
Karbonat, 1094
Karbonsäuren, 1118
Karbonschwärze, 1096
Karbonyle, 1107
Karbonylgruppen, 1106
Karbonylsulfid, 1108
Karborundpapier, 1109
Karboxalkylierung, 1110
Karboxyalkylcellulose, 1111
Karboxyalkylgruppen, 1112
Karboxyäthylcellulose, 1114
Karboxyäthylgruppen, 1115
Karboxyäthylierung, 1113
Karboxylgruppe, 163
Karboxylgruppen, 1117
Karboxylierung, 1116
Karboxymethylcellulose, 1120
Karboxymethylgruppen, 1121
Karboxymethylierung, 1119
Karde, 1122
karden, 1122
kariert, 5361
kariertes Papier, 5363
Karnaubawachs, 1129
Karosseriepappe, 488
Karte, 1122, 1231
Karteikarte, 4677
Kartoffelstärke, 4456
Kartoffelstärkemehl, 4456
Karton, 789, 1123, 1134, 4152
Kartonagenkarton, 863
Kartonaufkleber, 1136
Kartonausrüstung, 1124
Karton aus Sekundärfaserstoff, 797
Kartondecke, 5776
Kartoneigenschaften, 4157
Kartonfabrik, 793
Karton für die Kaschierung von Versandbehältern (enthält keine Jute), 3329, 3330, 3333
Karton für Flaschenkapseln, 850
Karton für Süsswarenverpackung, 1661
Karton für Untersetzer, 3710
Kartonherstellmaschine, 1135
Kartonindustrie, 4155
Kartonkaschierung (hauptsächlich aus Schrenzpappe), 1124
Kartonmaschine, 792
Karton mit glatter Oberfläche, 5196
Karton mit Schrenzlage, 1459
Kartonoberfläche, 1124
Kartonprodukt, 4156
Kartonprüfung, 4158
Kartonqualität, 4154
Kartonrückschicht, 512
Kartonschachtel, 4153
Kartonsorte, 4154
Kartonveredlung, 1124
Karton (0.15 mm und stärker), 901
Kartuschenpapier, 352, 1139
Kaschieranlage, 791, 3417
kaschieren, 3409
Kaschierlage, 791
Kaschierpappen, 3415, 3505
Kaschierpapier aus Sekundärstoff, 803
kaschiert, 3410, 3498
kaschierter Karton, 3411, 3506, 4220, 4223
kaschiertes Papier, 3412
kaschierte Wellpappe, 2074
Kaschierung, 3416
Kasein, 1142
Kastanienbaum, 1267
Kasten, 862, 1134, 1266
Kasten (z. B. Siebwasserkasten), 4339
Katalogpapier, 1154
Katalysator, 1156
Kathode, 1157
Kathodenstrahlröhre, 1158
Kation, 1159
Kationenaustausch, 1160
Kationenaustauscher, 1161
kationisch, 1162
kationische Stärke, 1164

kationische Verbindungen, 1163
Kauf, 4595
kaustifizieren, 1168
Kaustifizieren, 1170
kaustisch, 1166
Kaustizität, 1167
Kautschuk, 4864
Kavitation, 1172
Kegel, 1505
Kegelradantrieb, 680
Kegelradgetriebe, 680
Kegelrefiner, 1508
Kegelrollenlager, 5645
Kegelstoff, 4704
Kegelstoffmühle, 3316, 3318
Keil, 2950
keimen, 5358
Keimen, 2810
Kelvingrad, 3338
Kenaf, 3339
kennzeichnen, 3694
Kennzeichnung, 3694, 3699
Keramik, 1218
Keramikfaser, 1217
keramische Beschichtung, 1216
keramische Siebsaugerbeläge, 1219
Keratin, 3340
Kerbe, 3341, 5176
Kerben, 5178
Kernholz, 2999
Kerosin, 3342, 3938
Kessel, 812
Kesselhaus, 814
Kesselspeisewasser, 813
Ketene, 3343
Ketone, 3344
Kettdraht, 6049
Kette (Gewebe), 6046
Kettendrähte am Papiersieb, 6049
Kettenentrindungsmaschine, 1220
Kettenförderer, 1221
Kettfäden, 6048
Kettfaden, 6046
Kiefer, 4300, 4325, 4333
Kiefernharz, 4327
Kienholz, 4757
Kies, 2911
Kieselalgen, 1947
Kieselgur, 1945
kinetische Energie, 3346
Kirschbaumholz, 1265
Kiste, 862, 1140
Kistenausschlagpapier, 1144
Klafter (Holz), 1584
klaftern, 1584
Klammer, 873, 1337
Klammermarkierung, 1338
Klang, 5204
Klang (Papier), 4645
Klappe, 5827
Klärapparat, 1340
Klärbecken, 5018
klären, 1341
klar machen, 1341
Klärmittel, 1340
Klarsichtfolie, 1192, 5819
Klärsumpf, 5017
Klärung, 1339, 4596
Klassiergerät, 1343
klassifizieren, 5279
Klassifizieren, 2859, 5282
Klassifizierung, 1342
Klebeband, 208
Klebefestigkeit, 821
Klebekarton, 4220, 4223
Klebemaschine, 2847, 2945, 4227
kleben, 4219, 5424
Kleben, 2849, 4226
Kleber, 2847, 4225
Klebestelle, 5325
Klebestellen, 2625
Klebestreifen, 208, 696, 2943, 5643
Klebeverbindung-Prüfgerät, 2842
Klebkraft, 695
klebrig, 5630
Klebrigkeit, 5628
Klebstoff, 2840
Klebstoffe, 206
Kleesäure, 4116

kleine ringförmige Flecken in gedrucktem
   Material, 3036
kleiner Zwischenraum, 3281
Kleinlebewesenkontrolle, 3804
kleistern, 4219
Klemme, 1337
Klimaraum, 3110
klimatisch ausgerüstetes Papier, 1499
Klimatisieren, 1501
klimatisiert, 1498
Klinge, 723
Klon, 1355
Klotz, 3554
Klumpen, 1370, 3591
klumpig werden, 1354
Knappheit, 5069
knarren, 1663
Kneter, 3350
Knetmaschine, 3350
Knick, 882
knicken, 1659
Knickfestigkeit (phys.), 1457
Knickwalze, 1762
kniffen, 1669
Kniffen, 2646
knistern, 1663
Knitterfestigkeitsprüfung, 1676
knitterfest machen, 1671
Knochenleim, 2802
Knollen, 3993
Knopfflecken (von Hadern herrührend),
   993
Knospe, 935
Knospenruhe, 2061
Knötchen, 3993
Knoten, 3360
Knotenfänger, 3359, 4581
Knotenfänger (am Stoffauflauf), 4945
knotiger Schliff, 3364
knotiger Stoff, 3364
Koagulation, 2573
Koagulationsmittel, 1372
koagulieren, 1373
Koagulierungsmittel, 1372, 1375
Kochdauer, 1569
kochen, 1565, 1975
Kochen, 1567, 1981
Kochen in der Flüssigkeitsphase, 3523
Kochen (von Zellstoff), 4574
Kocher, 812, 1566, 1980
Kocherfüllapparatur, 1279, 1289
Kochergrube, 778, 1972
Kochgebäude, 1978
Kochlauge, 1572, 4575, 6221
Kochlaugenzusammensetzung, 1571
Kochprozess, 1569
Kochraum, 1978
Kochsalz, 5225
Kochsäure, 1568
Kochung, 1982
Kochung bei der die Nichtcellulosesub-
   stanzen nur begrenzt entfernt wurden,
   4646
Kofferpappe, 4145
Kohäsion, 1428
Kohle, 1093
Kohlebeschichtung für Kohlepapier,
   1097
Kohleflecken, 1104
Kohlenhydrat, 1092
Kohlensäuresättigung, 1095
kohlensaures Barium, 572
kohlensaures Salz, 1094
Kohlenstoff, 1093
Kohlenstoffaser, 1099
Kohlenstoff beschichtete Walze, 1101
Kohlenstoffdisulfid, 1098
Kohlenwasserstoff, 3127
Kohlepapier, 1093, 1103
Kohlerohpapier, 1102
Kohlezeichenpapier, 1229, 1230
Kolben, 4385
Kollergang, 3365
Kollerstoff, 905, 3820
Kollodiumwollbeschichtung, 4602
Kolloid, 1438
kolloidal, 1436
kolloide Lösung, 5257
Kolloidmühle, 1437

Kolorimeter, 1443
Kolorimetrie, 1445
kolorimetrische Reinheit, 1444
Kombikarton, 1461
Kombinationsfilz, 1460
Kombipresse, 1462
kombustibles Papier, 1465
Kommunikationspapier (z.B. für
   Fernschreibgeräte sowie
   Lochstreifenpapier etc.), 1471
Kompensation, 4024
Kompressibilität, 1477
Kompressionsmodul, 955
Kompressionswiderstand eines Kartons,
   866
Kompressor, 1482
Komprimierbarkeit, 1477
Kondensat, 1491
Kondensatabführung, 5116
Kondensatabsauger, 5116
Kondensationsharz, 1494
Kondensationspolymerisation, 1493
Kondensator, 1079, 1495
Kondensatorpapier, 1077, 1496, 2206,
   2207
Kondensatorseidenpapier, 1497
Kondensatortissue, 1078, 1497
Kondensatsammler, 3101
Kondensierung, 1492
Kondensseidenpapier, 1078
Kondenstopf, 5411
Konditionierapparat, 1500
Konditionieren, 1501
Konditioniergerät, 1500
konditioniert, 1498
konditioniertes Papier, 1499
Konifere, 1509
Koniferenzapfen, 5831
Konizität, 5646
konkave Bombierung, 3951
konkav geschliffene Walze, 1486
Konservendose, 5748
Konservierung, 1511, 4480
Konsistenz, 1512
Konsistenzmesser, 6016, 6022
Konsolenstützbalken, 874
konstante Geschwindigkeit, 1517
konstruieren, 2282
Konstruktionsart, 1923
Konstruktionsmaterial, 1518
Konstruktionspapier, 940, 1519
Konsum, 1520
kontinuierlich, 1531
kontinuierlich arbeitender Mahlgradprüfer,
   1534
kontinuierliche Bahn-/Rollensatinage,
   6151
kontinuierlicher Kocher, 1532
kontinuierliche Rollenprägung, 6149
kontinuierlicher Rollenankleber in der
   Maschine, 2624
kontinuierliches Verfahren, 1535
kontinuierliche Zellstoffkochung, 1536
Kontrast, 5013
Kontrastverhältnis, 1540
Kontrolle, 1234, 1541, 3253, 3888,
   5563
Kontrolleinrichtung, 3887
Kontroller, 2212
Kontrolleur, 1545
Kontrollinstrument, 1545
Konus, 1505
Konvektion, 1548
Konvektionstrockner, 1549
Konverter, 1554, 2213
Konzentration, 1487
Köperbindung, 5885
Kopf, 2993
Kopfende, 5773
Kopiergerät, 1578
Kopierpapier, 1582, 2161, 4744
Kopierseidenpapier, 1583
koppeln, 1650
Korbflechten, 6227
Kordel, 1584
Korn, 2862
Körnchen, 2874
Korngrösse, 4214
körnig, 2866

körnige Stärke, 4235
Körnigkeit, 2862
Körnung, 2865
Korona, 1595
Koronaentladung, 1594
Körperlänge (einer Walze), 2373
Korrektor, 4662
korrodieren, 1596
korrodierend, 1604
Korrosion, 1597
Korrosions-, 1604
Korrosionsanfälligkeit, 1605
Korrosionsbeständigkeit, 1602
Korrosionsmechanismus, 1599
Korrosionsprodukte, 1601
Korrosionsprüfung, 1603
Korrosionsschutz, 1600
Korrosionsschutzmittel, 1598
Korrosionsverhütung, 1600
Korrosionsversuch, 1603
Korrosionswiderstand, 5379
Korrosivität, 1605
Kosmetik-Tissue, 611
kosten, 1622
Kosten, 1622
Kostenplanung, 1623
kovalente Bindung, 1653
Krachen, 1662
Kraft, 2277, 2664, 4464, 5459
Kraft-Beklebekarton, 1795
Kraftbitumenpapier, 3368
Kraftdecke, 3378
Kraft-Dreheinschlag, 3379
Krafterzeugung, 4466
Kraftimitation, 802
Kraftkarton, 3369
Kraftlauge, 3373
Kraftliner, 3371
Kraftmaschine, 3898
Kraftpackpapier, 3380
Kraftpapier, 3366, 3375
Kraftpapierfabrik, 3374
Kraftpapier-Schutzdecke, 1125
Kraftpapier zum Kaschieren von Karton,
   3378
Kraftpappe, 3369
Kraftsackpapier, 3367
Kraftspeicher, 142
Kraftstrom, 2221
Kraftübertragung, 2223
Kraft-Wellpappenmittellage, 3370
Kraftwerk, 4467
Kraftzellstoff, 5526
Kraftzellstoffpappe, 5521
Kran, 1664
Kranawettstrauch, 3323
Krater, 1665
Krater (im Oberflächenanstrich), 1665
Kratzer, 4944
Kratzer (vom Messer oder der Rakel),
   4942
Kräuseleffekt, 1423
Kräuseligwerden, 1424
kräuseln, 1695
Kräuseln, 1696, 1752, 4808
kreidiges Aussehen, 1223
Kreisblatt (für Kreisblattschreiber), 1329
Kreislauf, 1783, 4856
Kreismesser, 2021, 4784
Kreismesser (z. B. am Längsschneider),
   2019
Kreisringe, 385
Krempel, 1122
krempeln, 1122
Kreppausrüstung, 1684
kreppen, 1680
Kreppen, 1687
Kreppmaschine, 1689
Krepp-Papier, 1685
Kreppschaber, 1688
Kreppseidenpapier, 1690
Kreppverhältnis, 1686
Kresol, 1692
Kreuzbodenbeutel, 1704
Kreuzung, 3111
kriechen, 1678
Kristall, 1728
Kristallbildung, 1732
kristalline Cellulose, 1729

kristalliner Bereich, 1730
Kristallinität, 1731
Kristallisation, 1732
Kristallstruktur, 1733
kritische Geschwindigkeit, 1702
kritischer Druck, 1700
kritische Strömungsgeschwindigkeit, 1698
kritische Temperatur, 1701
krümmen, 1761
Krümmen, 659
krümmend, 429
Krümmung, 1761
Kuchenspitzenpapier, 3388
Kügelchen, 4245
Kugelkocher, 2825
Kugellager, 553
Kugelmühle, 554, 4237
Kugelventil, 556
Kühlapparat, 1573
Kühldüsen (am Glättwerk), 245
Kühlen, 1574
Kühler, 1573
Kühlsystem, 1576
Kühlung, 1574
Kühlwalze, 1271, 5599
Kühlwasser, 1577
Kühlzylinder, 1575, 5597, 5598
Kultur, 1737
Kundendienst, 5009
Kunstdruckkarton, 436
Kunstdruckpapier, 441, 2268
Kunstfasern, 5613
Kunstharz, 4753, 5615
Kunstleder, 438, 3183
Kunstlederpapier, 439, 3446
künstlicher Wasserlauf, 2611
künstlicher Wasserlauf nach Parshall, 4211
künstliches Wasserzeichen, 4489
Kunstseide, 4652
Kunststoff, 4359
kunststoffbeschichtetes Papier, 4360
Kunststoffe auf Cellulosebasis, 1194
Kunststoffgewebe für die Blattbildung, 2691
Küpenfarbstoffe, 5976
Kupfer, 1580
Kupferäthylendiaminhydroxid, 1745
Kupferäthylendiamin-Viskosität, 1746
Kupferdruckpapier, 4371
Kupfertiefdruckkarton, 4372, 4377
Kupferzahl, 1581
kuppeln, 1371, 1650
Kupplung, 1371, 1651
Kupplungsstück, 1651
Kuprafaser, 1734
Kupraseide, 1734
Kurve, 1761
Kurvenmesser, 4847
Kurvenschreiber, 4091
kürzen, 5856
kurze Restrollen, 5479
kurzfaserig, 5071
Kurzholzerntemaschine, 5075
Kurzholzware (weniger als 1.80 m), 5073
Küstentanne, 2870
küstliches Pigment, 1449

L

Labor, 3386
Laboratorium, 3386
Lack, 2266, 3389, 5972
Lackierbarkeit, 5973
lackieren, 5972
Lackieren, 3391, 5974
lackiertes Papier, 3390
Lackiertrommel, 1412
Lackierung, 3391
Lackmuspapier, 3535
Lackpapier, 3197
Ladebaum, 841

Ladegestell, 5132
Lader, 3542
Ladevorrichtung, 3542
Lage, 3416, 3434, 4381, 4387
Lagenbindung, 4389
Lagenspaltung, 4391
Lagentrennung, 4391
Lager, 618, 5566
Lagerabbau, 5432
Lagerbestände in Papier- und Kartonstandardabmessungen, 5438
Lagerbeständigkeit, 5056
Lagerbestandsaufnahme, 3283
Lagerbestandsverzeichnis, 3282
lagerfertige Lagen Nasszellstoff, 3419
Lagergehäuse, 619
Lagerkarton, 2730
Lagermetall, 502
lagern, 5430
Lagern, 5441
Lagerung, 5441
Lagerzapfen, 3319, 5857
Laggerreduzierung, 5432
Lagune, 3393
Lamelle, 2014, 2026, 3406
Lamellenlochwalze, 525
Lamellenstruktur, 3407
lamelliert, 3407
laminare Strömung, 3408, 5458
Laminarströmung, 3408
Landkartenpapier, 1232, 3690
Länge, 3448
Länge (in yards), 6313
lange Lebensdauer, 3563
lange Ware, 3568
Langfalzmaschine, 662
langfaseriger Zellstoff, 3564
Langlebigkeit, 3563
Längs-/Querverhältnis einer Bahn, 5362
Längsfalz, 3566
Längsfalzung, 3566
Langsieb, 2272, 2705
Langsiebkarton, 2703
Langsiebmaschine, 2704
Langsiebmaschine mit Ausfahrvorrichtung, 1075
Längsrichtung, 3618
Längsschneiden, 5172
Längsschneider, 5167, 5173
Längstransporteur, 3555
Längs-und Querschneider, 3245
Langtonne, 3569
lappiger Stoff (Zellstoff), 5251
Läppmaschine, 3420
Lärche, 3423
Lärm, 3994
Lärmkontrolle, 3995
Laser, 3425
Last, 3538
Laster, 5855
Lastkahn, 570
Lastkraftwagen, 5855
Lasurlack, 2932, 3437
latentes Bild, 3427
Latenz, 3426
Latenzzeit, 3426
Latex, 3430, 3431
Laubbaum, 903
Laubbaum (der jährlich sein Laub abwirft), 1846
Laubholz, 1847, 2987
Laubholzzellstoff, 2988
laubtragend, 1845
Laufeigenschaft, 4870
Läufer, 4859
Lauffähigkeit, 4870
Laufgang, 1165
Laufpfeil im Filz, 2421
Laufrad, 3191
Laufradschaufel, 4860
Laufrichtung, 2009
Laufrolle, 1133
Laufsteg, 1165
Laufwelle, 3508
Lauge, 3435, 3521
Lauge auf Natriumbase, 5218
Lauge-Holz-Verhältnis, 3524
laugenbeständige Cellulose, 304
laugenbeständiges Papier, 315

Laugeneindampfungsanlage, 3522
läuten, 4802
Lävulinsäure, 3462
Leben, 3465
Lebensbaum, 5739
Lebensdauer, 3465, 5010
Lebensmitteleinwickelpapier, 2662
Lebensmittelkarton, 2661
leckdichter, 3444
Lecken, 3443, 4985
lecksicherer Behälter, 3444
Leckwerden, 3443
Lederimitationspapier, 3446
Lederpappe, 3445
leer, 6028
Leere, 6028
Leerlaufzeit, 3176
Legierung, 320
Lehm, 1344
leichtbrennbares Papier, 1465
leichte Presswalze, 4795, 4797
leicht färben, 5750
leichtgewichtig, 3475
leichtgewichtiger Beklebekarton, 3476
leichtgewichtiges Broteinwickelpapier, 542
leichtgewichtiges Deckpapier, 2376
leichtgewichtiges Kekspapier, 542
leichtgewichtiges Papier, 3477
Leim, 2840
Leimauftrag, 2841, 2846
Leimauftragsaggregat, 2841
Leimauftragswalze, 5126
leimen, 2842
Leimen, 2849, 4226
Leimflecken, 5127
Leimmilch, 5124
Leimpresse, 5125
Leim-Streckmittel, 2353
Leimstrichpapier, 2843
Leimung, 2849
Leimungsgradprüfer, 5129
Leimungsgradprüfung, 5130
Leimungsprüfer, 5129
Leimverteilung, 2848
Lein, 2559
Leinen, 3500
Leinenfaser, 3502
leinengeprägte Oberfläche, 1067
leinenkaschiertes Papier, 3501
Leinenoberfläche, 3503
Leinenpapier, 2371, 3504
Leinenprägung, 3503
Leinwand, 3500
Leinwandbindung, 4351
Leiste, 5289
Leistung, 4096, 4260, 4464
Leistungsbilanz, 2222
Leistungsfähigkeit, 1080, 2191
Leistungsfaktor, 4465
Leistungsprüfung, 4261
Leistungsvermögen, 4464
Leitblech, 521
leiten, 2932, 3437
leitfähiges Papier, 1503
Leitfähigkeit, 1504
Leitkanal, 2148
Leitrolle, 3439
Leitung, 2148, 3437
Leitungsfähigkeit, 1502
Leitungsnetz, 4337
Leitvermögen, 1504
Leitwalze, 2933, 3439
lenken, 2932, 3437
Leser, 4662
Leuchtpetroleum, 3342, 3938
Leuchtweiss, 2616
licht, 3470
Licht, 3470
lichtbeständige Farbstoffe, 159
Lichtbeständigkeit, 3471
lichtdurchlässige Beschichtung, 5812
Lichtdurchlässigkeit, 5810
lichtechte Farbe, 2388
lichter Farbstoff, 2388
lichtechtes Reklamepapier, 686
Lichtechtheit, 3471
Lichtechtheitsprüfer, 2378
lichte Höhe (Konstr.), 3026
lichtelektrische Zelle, 4289

lichtempfindliches Papier, 3473, 4291, 4999
Lichtempfindlichkeit, 4294
lichte Weite, 3252, 3588
Lichtmesser, 4292
Lichtpauspapier, 785, 5001
Lichtpauspapiere, 1949
Lichtquelle, 3474
Lichtstreuung, 3472, 4077
lichtundurchlässig, 4060
Lichtundurchlässigkeit, 4059
Lieferant, 5564
Lieferfirma, 5564
liefern, 5565
Lieferung, 1898
liegender Wasserzeichen-Egoutteur, 6086
Lift, 3466
Ligninanreicherung, 3479
Lignine, 3482
Ligningehalt, 3481
ligninhaltig, 3478
Ligninhydrolyse, 3147
Lignin-Kohlenhydrat-Komplex, 3480
Ligninsulfonsäuren, 3486
Lignocellulose, 3483
Lignosulfonat, 3486
Linde, 605, 3487, 5743
lineare Programmierung, 3497
Liniendruck, 3496, 3977
liniert, 3498
Linolsäure, 3511
Lint, 3513
Lintbaumwolle, 3513
Linters, 3514
Listen, 5624
Lithium, 3527
Lithiumchlorid, 3528
Lithiumverbindungen, 3529
Lithographie, 3532
Lithographiepapier, 3531, 3533
Lithopon, 3534
Loch, 3072, 4330, 5995
Lochband, 4590
Lochblende, 1944
lochen, 3072
Lochkarte, 4589
Lochkartenmaterial, 5625
Lochstreifen, 4590
Lochung, 4258
Lochwalze, 3078, 4255, 4688
Lochwalzenmarkierung, 5514
locker gewickelte Rolle, 3573
Lockerung, 4728
Löschen, 5146
Löscher, 5145
Löschpapier, 683, 770
Löschpapier (schwergewichtiges), 769
lose Aufbewahrung, 956
lösen, 2032
Lösen, 2035
Löslichkeit, 5269
Löslichmachung, 5270
Losrolle, 1814, 2627
Lösung, 5272
Lösungsmittel, 5274
Lösungsmittelleimung, 5275
Löten, 5258
Lücke, 882, 3281, 6028
Luft, 242
Luftabscheider, 1835
Luftansaugstutzen, 271
luftbetätigte Spannungskontolle, 278
Luftbewegung, 266
Luftblase, 243, 929
Luftblasen, 249
Luftbürste, 244
Luftbürstenstreicher, 247
Luftbürstenstrichauftrag, 248
Luftdruck, 477, 583
Luftdurchlässigkeit, 281
Luftdurchlässigkeitsmesser, 4443
Luftdurchlässigkeitsprüfgerät, 1908
Luftdüse, 272, 280
Lufteinlass, 271
Lufteinzug, 262
lüften, 220, 242
Lüfter, 772, 2384, 5997

Lufterhitzer, 268
Luftfeuchtigkeit, 4727
Luftfilter, 263
Luftfilterpapier, 264
Luftgebläse, 5997
Luftgehalt, 252
luftgekühlt, 253
luftgetrocknet, 257
luftgetrocknetes Papier, 258
Lufthahn, 1419
Luftkammer, 251
Luftkühlung, 254
luftleerer Raum, 5940
Luftloch, 5995
Luftmesser, 244, 273, 275
Luftpinsel (zum Verteilen der
    überschüssigen Streichfarbe), 269
Luftpostpapier, 279
Luftrakel, 244, 273
Luftrakelstreicher, 274
Luftschaberstreichen, 275
Luftschlauch, 3091
Luftschleier, 255
Luftstrahl, 272
Luftstrahl zur Begrenzung der Papierbahn
    auf der Siebpartie, 256
Luftstrom, 266
lufttrocken, 101, 259
Lufttrockner, 260
Lufttrocknung, 3553
Lüftung, 221, 5996
Luftverschmutzung, 282
Luftverunreinigung, 282
Luftwalze, 284
Lumen, 3588
Lumpen, 4621
Lumpenentstäuber, 4628
Lumpenkocher, 4622
Lumpensaal, 4631
Lumpensortierung, 4632
Lüster, 3595
Lyalls-Lärche, 5639

M

Madelaine-Trockner, 3639
Magazinpapier, 3642
Magazinschleifer, 3640, 4394
Magazintrockner, 2128
Magnefit-Lauge, 3645
Magnesia, 3643, 3651
Magnesium, 3644
Magnesiumbisulfit, 3646
Magnesiumchlorid, 3648
Magnesiumhydroxid, 3650
Magnesiumkarbonat, 3647
Magnesiumoxid, 3651
Magnesiumsulfat, 3652
Magnesiumsulfit, 3653
Magnesiumverbindungen, 3649
Magnet, 3654
magnetische Ausrüstung, 3655
magnetische Eigenschaften, 3658
magnetisches Feld, 3656
magnetische Tinte, 3657
magnetische Trennvorrichtung, 3659
Magnetophonband, 3660
Magnetscheider, 3659
Magnettonband, 3660
Mahlbarkeit, 621
mahlen, 620, 1862, 3816
Mahlen, 641, 2906
Mahlen bei hoher Stoffdichte, 3044
Mahlen in der Jordan-, 3316
Mahlen unter Hochdruck, 4510
Mahlfähigkeit, 621
Mahlfläche, 648
Mahlgeschwindigkeit, 643
Mahlgrad, 642
Mahlgradanzeigegerät, 2719
Mahlgradprüfer, 623, 2719, 2720
Mahlgradprüfung, 640
Mahlholländer, 3075
Mahlkörper, 5629
Mahlung, 641

Mahlung im Refiner, 4706
Mahlungsgrad, 1880
Mahlzeit, 644
Mähmaschine, 2990
Maisstärke, 1593
Makel, 755
Makromoleküle, 3638
Maleinsäure, 3668
Maleinsäureanhydrid, 3669
Maltose, 3671
Malzzucker, 3671
Mammutbaum, 5007
Mangan, 3674
Manganverbindungen, 3675
Mangel, 1861, 5069
Manila, 3678, 3680
Manila-Faltschachtelkarton, 3682
manilafarbenes Fleischeinwickelpapier,
    984
Manilafaser, 3680
Manilahanf, 103, 3681
Manilaimitation, 804
Manilakarton, 3679
Manilapapier, 3032, 3683
Mannan (Hemizellulose), 3684
mannigfaltig, 3676
Mannogalaktan, 3685
Mannose, 3686
Mannschaft, 1693
Manometer, 4499
Mantel, 3306
Marantastärke, 434
Mark, 4344
markieren, 3694
Markierfilz, 3701
Markierung, 3694, 3695, 3699
Markierung der Kartonoberfläche zur
    Vereinfachung der Falzung, 1768
Markierung durch Luftmesser verursacht,
    276
Markierung einer Flickstelle, 4228
Markierung durch den Stau, 5156
Markierungen durch überfliessendes
    Material, 5132
Markierungslinie, 4931
Markierungsvorrichtung, 3695
Markierung von Papier oder Pappe durch
    an den Walzen haftende
    Pigmentteilchen, 1046
Markiervorrichtung, 3700
Markstrahlzelle, 4651
Marktforschung, 3698
Marktversorgung, 3696
Marktwesen, 3696
Marktzellstoff, 3697
marmorieren, 3693
marmoriert, 3900
marmorierte Schachtelverkleidung, 1365
marmoriertes Papier, 3692, 3903
Marmorpapier, 3691
Maschenweite (des Siebs), 3756
Maschine, 3606
maschinell bearbeiten, 5771
maschinelle Ausrüstung zur
    Papierherstellung, 4174
Maschinen, 3632
Maschinenantriebsseite, 3627
Maschinenausführung, 3617
Maschinenausschuss, 3607, 3820
Maschinenbau, 2283
Maschinenbedienung, 3629
Maschinenbütte, 3609
Maschinenführer, 3634, 5672
maschinengeglättet, 3622
maschinengeglättetes
    Zeitungsdruckpapier (für Beilagen),
    5562
maschinengekreppt, 3615
maschinengestrichen, 3611
maschinengestrichene Papiere, 3613
maschinengestrichener Karton, 3612
maschinengetrocknet, 1794, 3619
maschinenglatt, 2132, 3596, 3598,
    3620, 3622, 3826
Maschinenglätte, 3620, 3626
maschinenglattes Papier, 3597, 3601,
    3623
Maschinenkalander, 3608
Maschinenkonstruktion, 3617

Maschinenmesser, 2939
Maschinenöl, 3584
Maschinenrahmen, 2714
Maschinenrand, 3823
Maschinenraum, 3631
Maschinenrichtung, 3562, 3618
Maschinenrolle (nicht umgerollt), 3828
Maschinensaal (Druckerei), 4491
Maschinensieb, 3636
Maschinenwerkstatt, 3633
Maschinenzeit, 3635
Maschinist, 4071
Maserung, 2862
Maskierung, 3703
Mass, 4644, 5121
Masse, 3705, 5493
Massefärbung, 629
massegefärbt, 628
massegeleimt, 637
Masseinheit, 3726, 5920
Masseinheit für Faserholz, 1738
Masseleimung, 638, 2284, 3278
Massenabwicklung, 947
Massenspektroskopie, 3707
Massey-Streicher, 3706
Massstab, 2769, 2798, 4922
Material, 3713, 5480
Materialbeförderung, 3716
Materialienverhältnis, 3714
Materialprüfung, 3717
Materialverbrauch, 3715
Materialverwaltung, 3716
mathematisches Modell, 3718
Matrize, 1954, 3709, 5419
Matrizenpapier, 507, 2580, 3720,
    5421
matt, 3709
Matte, 3709
matte Oberfläche, 2150, 3721
mattes Papier, 3719
mattgestrichenes Papier, 2149
Mattglanz, 2150
mattiert, 3709
Mattpapier, 3719
mattsatiniert, 2285, 5991
maximale Aufnahmefähigkeit, 3724
maximales Fassungsvermögen, 3724
maximal nutzbare Siebbreite, 3725
Mazeration, 3604
mazerieren, 3602
Mechanik, 3739
mechanische Eigenschaften, 3733
mechanische Messung, 3732
mechanischer Abbau, 3730
mechanischer Antrieb, 3731
mechanischer Aufschluss, 3735
mechanisches Dehnungsmesser, 3736
mechanischer Sortierer, 3728
mechanische Verringerung, 3730
Mechanisierung, 3740
Mechanismus zum Herablassen (Absenken
    oder Tiefereinstellen), 3582
Mehlstoff (des Holzschliffs), 2587
mehrfach, 3676
Mehrfachabstreifer, 3916, 3920
Mehrfachbenutzung, 3920
Mehrfachlangsieb, 3909
Mehrfachstreichen, 3917
Mehrfachverwendung, 3920
Mehrgewicht, 4112
Mehrkomponentenstreichverfahren, 3911
Mehrlagen, 3913
Mehrlagenbahn, 3919
Mehrlagenbeutelkraftpapier, 3930
Mehrlagenformierungseinrichtung, 3918
Mehrlagenkarton, 1458, 2486, 3914,
    3915, 3921
Mehrlagenwellpappe, 3931
mehrlagig, 3922
mehrlagige Säcke, 3929
mehrlagiges Papier, 3923
Mehrrundsiebmaschine, 3912, 3927,
    3928
mehrschichtig, 3922
mehrschichtiges Papier, 3923
Mehrsiebmaschine, 3932
Mehrstufenbleiche, 3924
Mehrstufenverfahren, 3925
Mehrzweckverwendung, 3920

Mehrzylindermaschine, 3912
Melamin, 3743
Melaminformaldehydharze, 3744
Melaminsäure, 3744
meliert, 3900
melierte Oberfläche, 3902
melierte Oberfläche (hervorgerufen durch
    starke Druckausübung in der
    Nasspartie), 1720
melierte Schachtelkaschierung, 1365
meliertes Papier, 3903
Membran, 1944, 3747
Menge, 2575
Meristeme, 3755
Merkaptane, 3749
Merkaptolignine, 3750
Merkblatt, 3442
Merkmal, 1227
Merzerisierung, 3751
mesomorphe Cellulose, 3757
Messeinrichtung, 3887
messen, 2769, 2798, 3770
Messer, 723, 3351
Messerauftragsmaschine, 2054
Messerblock, 648
Messerschälmaschine, 3352
Messerschneide, 3355
Messerschnitt, 1765
Messerstreichen, 3354
Messerstreicher, 724, 3353
Messfühler, 5002
Messgerät, 2769, 2798, 3255, 3727
Messing, 878
Messinstrument, 3727, 3770
Mess und Versuchswerte, 1821
Messung, 3726
Messverfahren, 5698
Messvorrichtung, 3772
Messwertgeber, 5002
Messsystem, 3773
Metacellulose, 3758
Metall, 3759
Metallband, 5447
Metallbeschichten, 4378
Metallbeschichtung, 3765
Metallegierung, 3759
Metallfaser, 3761
Metallfolie, 3763
metallische Ausrüstung, 3762
metallischer Strich, 3765
metallisches Papier, 3767
metallisiertes Papier, 3768
Metallisierung, 3764
Metallpapier, 3767
Metallsalze, 3251
Metallverbindungen, 3760
Meteorologie, 3769
Methacrylate, 3774
Methacrylsäure, 3775
Methacrylsäureester, 3774
Methan, 3776
Methanol, 3777
Methode, 5618, 5665
Methylacetat, 3779
Methylacrylat, 3780
Methylalkohol, 3781
Methyl-Äthylketon, 3785
Methylcellulose, 3783, 5898
Methylencellulose, 3784
Methylglukoside, 3786
Methylgruppen, 3787
Methylierung, 3782
Methylmerkaptan, 3788
Methylmethacrylat, 3789
Methylorange, 3790
metrisches System, 3791
metrische Tonne, 3792
Migration, 3811
Mikroanalyse, 3795
Mikrofibrille, 3799
Mikrofilm, 3801
mikrokristalline Cellulose, 3797
Mikrokristallinwachs, 3798
Mikrometer, 3802
Mikroorganismus, 3803
Mikroporosität, 3805
Mikroskopie, 3806
Mikrotom (zur Herstellung von
    Papierquerschnitten), 3807

Mikrowellenspektroskopie, 3808
Mikrowellpappe, 3796
Milchkarton, 3814
Milchsäure, 3392
Millimeterpapier, 4924
Mindestfrachtmenge an Papier, 1128
Mineralfaser, 3832
Mineralienablagerung, 3831
Mischbütte, 3844
mischen, 756, 3837
Mischen, 3842
Mischer, 757
Mischgerät, 757
Mischkasten, 3843
Mischling, 3111
Mischmaschine, 3841
Mischpumpe, 2387, 3845
Mischung, 3842, 3846, 4555
Mischwaldbestand, 3840
Mischweiss, 5752
mit Asphalt beschichtetes Papier, 465
mit Asphalt imprägnierte Dachpappe, 464
mit Asphalt zu imprägnierende Pappe, 470
mit Banderole versehen, 562
mit Bogenzufuhr, 5044
mit Büttenrand versehenes aber in der
   Maschine hergestelltes Papier, 1798
mit Dampf behandeln, 5402
mit dem Messer aufstreichen, 2058
mit einem Etikett oder Anhängezettel
   versehen, 5631
MIT-Falzfestigkeitsprüfgerät, 3835
mit Feuchtigkeitsschutz, 3865
mit Füllstoff gearbeitete Graupappe,
   2489
mit Füllstoffpigment angereichertes
   Papier, 1348
mit hoher Geschwindigkeit, 3056
mit Isolierlack beschichtetes Papier für
   Lebensmittelkarton, 1138
mit Kohlendioxid saturieren, 1094
mit Kraftpapier kaschierter Karton, 3372
mit Lack überzogen, 2267
mit Latex behandeltes Papier, 3432
mit Metallsalzen imprägniertes Papier,
   3768
mit Metallschicht überziehen, 4366
mit Papier durchschiessen, 3269
Mitscherlich-Verfahren, 3836
mit Schmelz überzogen, 2267
mit Stahlband versehen, 5448
Mittel, 232
Mittellage, 1204
Mittellamelle, 3809
Mittelschicht einer Pappe, 1203
Mittelschichtmaterial, 1590
mittels Luft befördert, 246
Mittelwellenrohpapier, 1617
Mittel zur Haltbarmachung, 1750
Mittel zur Verhinderung der Wanderung
   (z.B. von Lösungsmitteln), 391
Mittel zur Verhinderung elektrostatischer
   Aufladung, 409
mit Tinte schwärzen, 3232
mittlere Oberflächenausrüstung (weder
   glatt noch rauh), 3742
mittlere Riffelung, 1004
mit Wachs überziehen, 6129
mit Wasserlinien versehenes Papier,
   3395, 6082
mit Wasser verbunden, 3114
mit Wasserzeichen versehenes Papier,
   3819, 6089
mit Wellpappe ausgeschlagene
   Behälter, 1609
mit Wellpappe ausgeschlagene Container,
   1609
mit Werkzeugen arbeiten, 5771
mit Ziegel bedecken, 5742
Mixtur, 3846
Möbelfolienpapier, 5993
Modell, 3848, 4230
Modellpappe, 4231
Moder, 3813
modern, 3813
Modernisierung, 3849
Modifiziermittel, 3852
modifizierte Cellulose, 3850
Modul, 3854

Molekulargewicht, 3883
Molekülbau, 3882
Moleküle, 3884
Molgewicht, 3883
Molybdän, 3885
Molybdänverbindungen, 3886
Monitor, 3887
Monomere, 3892
monomolekulare Schicht, 3891
Monotypemaschine, 3893
Montage, 2535
Montagebau, 4476
Montmorillonit, 3894
Mörtel, 3896
Motor, 3898
Muffe, 2437
Müllbeutel, 4714
Mullen, 3907
Mullen-Berstfestigkeitsprüfer, 3908
Mundstück, 4009
Murraykiefer, 3548, 3934
Muster, 3848, 4230, 4891
Musterkarton, 4231
Mustermacher, 4893
Mutation, 3935
Mutterrolle, 4210

N

nach Abwicklung der Rolle an der Hülse
   haftendes Papier, 1591
nachbehandeln, 1748
Nachbehandeln, 1749
Nachbehandlung, 4974
Nachdunklung von gebleichten Stoffen,
   897
Nachfeuchtzylinder, 5598
nachgemacht, 795
Nachhaltigkeit (Forstwirtschaft), 5596
nachlassen, 1954
Nachmahlung, 4451
nach oben krümmen, 3063
Nachprüfen, 1234
Nachrichten, 3960
Nachsortierer, 2513
Nachsortierung, 2514
Nachteil, 3574
Nachtrockner, 226
Nachweis, 1927, 3175
Nachwuchs (Wald), 4971
Nachzerkleinerungsmaschine, 4673
Nachzerspaner, 4673
nacktsamige Pflanze, 2951
Nadelbaumzapfen, 5831
Nadelholz, 1510, 5254
Nadelholzzellstoff, 5255
Nadellager, 3948
Nadelventil, 3949
Nährgewebe, 2274
Nährstoff, 4012
Naht, 969, 3314, 4967
Nahtzahl, 971
Napf, 859
Naphtha, 3938
narbiges Papier, 2868
nass, 3857, 6180
Nassabrieb, 6198
Nassabriebfestigkeit, 4957, 6199
Nassappretur, 6188
Nassausschuss, 6182
Nassberstfestigkeit nach Mullen, 6193
Nasschnitzel, 6197
Nässe, 6194
nasse Bahn, 6212
nasses Holz, 6214
nasse Streifen, 6200
Nassfalte, 6215
Nassfaserbündelung, 6189
nassfeste Harze, 6205
nassfester Ausschuss, 6203
nassfestes Papier, 6204
Nassfestigkeit, 6201
Nassfestigkeits-Rückhaltevermögen,
   6206
Nassfilz, 4487, 6187

nassgekreppt, 6073
nass gelegt, 6190
nassgewachstes Papier, 6211
Nassgewicht, 6213
Nassholz, 6214
Nasspartie, 6184
Nasspresse, 2532, 6195
Nasspresse mit doppelter Bespannung,
   2077
Nasspressen, 6196
Nassverfestigungsmittel, 6202
Nasszellstoff, 6197
Nasszugfestigkeit, 6208
Nasszug (Spannungszustand der Bahn in
   der Nasspartie), 6183
Natrium, 5215
Natriumalginat, 5216
Natriumaluminat, 5217
Natriumbikarbonat, 5219
Natriumbisulfit, 5220
Natriumborat, 5221
Natriumborhydrid, 5222
Natriumchlorat, 5224
Natriumchlorid, 5225
Natriumchlorit, 5226
Natriumhydrosulfid, 5228
Natriumhydrosulfit, 5229
Natriumhydroxid, 5230
Natriumhydroxyd, 5230
Natriumhypochlorit, 5231
Natriumkarbonat, 5209, 5223
Natriumperoxid, 5232
Natriumsilikat, 5233
Natriumstearat, 5234
Natriumsulfat, 4888, 5235
Natriumsulfhydrat, 5228
Natriumsulfid, 5236
Natriumsulfit, 5237
Natriumsulfozyanid, 5238
Natriumthiocyanat, 5238
Natriumthiosulfat, 5239
Natriumverbindungen, 5227
Natriumxylolsulfonat, 5240
Natronaufschluss, 5214
Natronaufschlussverfahren, 5212
Natronkraftpapier, 5524
Natronlauge, 5211
Natronzellstoff, 5213
Naturbuchpapier, 5909
naturfarben, 3942
Naturfaser, 3943
natürliche Wiederaufforstung,
   3945
Naturpapier, 5910
Naturschätze, 3946
Naturvorkommen, 3946
NCR (kein Kohle-Papier), 3992
NCR-Papier (kein Kohlepapier
   erforderlich), 3936
Nebelschutzmittel, 396
Nebenbahn, 5087
Nebenbahnen, 1238
Nebenprodukt, 1000
Nebenrolle, 5086
neigen, 4340
Neigung, 2858
Nenngewicht, 3997
Neopren, 3952
netzartig, 4771
Netzmittel, 6210
Netzplan, 1699
Netzplantechnik, 1699
907,185 kg 5074
neutraler halbchemischer Sulfitzellstoff,
   3958
neutraler Kleber, 3956
neutraler Sulfitaufschluss, 3959
neutrales Weiss, 1832
Neutralisation, 3954
neutralisierend, 387
Neutralsulfitlauge, 3957
Nicht-Annahme, 4726
nicht ausgerüstet, 5916
nicht brennbar, 2539
nicht erfüllte Produktionsleistung,
   5915
nicht falzbarer Karton, 3998
nicht ganz quadratisch, 4031
Nichthaftung des Leims, 2845

nicht kalandriert, 5907
nicht klebend, 4004
nicht klebrig, 4004
nicht kristallin, 353
nichtleitend, 1959
nichtlineare Programmierung, 4001
nicht lupenrein, 4019
nicht mischbar, 3186
nichtrostend, 405
nicht scharf gefaltet, 5246
nicht umkehrbare Hülsenstange, 4002
nicht verkaufte Exemplare von
   Druckerzeugnissen, 4106
Nickel, 3969
Nickelverbindungen, 3970
Niederdruck, 3579
Niederdruckbeschickungsanlage, 3580
Niederdruckförderanlage, 3580
Niederschlag, 1492, 4469, 4979
Niederschlag durch Vakuumabsaugung,
   5942
niederschlagen, 4469
Niederschlagen, 4471, 4980
niederschlagend, 4470
Nippel, 3976
Nitrate, 3979
Nitratgruppen, 3978
Nitride, 3984
Nitrieren, 3980
Nitrierung, 3980, 3985
Nitrile, 3986
Nitrite, 3987
Nitrocellulose, 1193, 3988
Nitrogen, 3989
Nitrolignin, 3991
Niveau, 3452
nivellieren, 3452
Nivellieren, 3456
Nivellierschraube, 3458
Nocke, 1064
Nockenscheibe, 1064
Nonwovens, 4008
Norm, 5382
normal, 5382
Normalformat, 603
Normalzeitungsdruck, 5383
notieren, 826
Notizblöcke, 4179
nuancieren, 5024
Nullpunktsreissfestigkeit, 6324
Nut, 2912
Nute, 2912
Nutenpresse, 2913
Nut- und Spundbrett, 3710
nutzbare Arbeitsbreite einer Maschine,
   3616
nutzbares Chlor, 492
Nutzbarmachung, 5939
Nutzeffekt, 2191
Nutzen, 4546
Nutzholz, 3586, 5745
Nutzung, 4890
Nutzungswert, 2191
Nylonfaser, 4013

O

obere Filzpresse, 5775
obere Gautschwalze, 5774
obere Presswalze, 5777
oberes Ende, 5773
Oberfläche, 2373, 5568, 5773
Oberflächenabfluss, 4871
Oberflächenabhebung, 5578
oberflächenaktives Mittel, 5569
oberflächenaktivierendes Mittel, 5569
Oberflächenauftrag, 5570
Oberflächenauftragstest, 2098
Oberflächenausrüstung, 2516, 4238,
   5575
Oberflächenausrüstung in der Nasspartie,
   6186
Oberflächenbehandlung, 5589
Oberflächenbehandlungsmittel, 5593
Oberflächenbenetzbarkeit, 5591

Oberflächenbenetzbarkeitsprüfung, 5592
Oberflächenbeschaffenheit, 2516
oberflächenbeschichtet, 5572
Oberflächeneigenschaften, 5580
Oberflächenfärbung, 5573
Oberflächenfestigkeit, 5571, 5586
oberflächengefärbt, 1030, 5574, 5588
oberflächengeleimt, 5582
oberflächengeleimtes Papier, 5583, 5862
oberflächengeleimtes und festes Hadernpapier, 3447
oberflächengestrichen, 5572
oberflächengetönt, 5588
Oberflächenhärtung, 5577
oberflächenimprägniert, 5572
Oberflächenleimpresse, 5584
Oberflächenleimung, 5128, 5585, 5780, 5864, 6157
Oberflächenspannung, 5587
Oberflächenstruktur, 2867, 5575
Oberflächenveredlung, 5576
Oberflächenverfärbung, 752
Oberflächenwasser, 5590
Oberflächenwiderstand, 5581
Oberflächenwirksames Mittel, 5593
Oberflächenzustand, 2516
Oberingenieur, 1270
Oberlippe des Einlaufspalts, 5781
Obermaschinenmeister, 4492
Oberseite, 2432, 5773, 5779
oberste Dickschicht eines Kartons, 5776
Oberwalze, 5778
Obsteinwickelpapier, 2742
Oelinsäure, 4049
Ofen, 2760, 3345, 4097
offener Behälter, 4065
offener Güterwagen, 5855
offener Kalander, 4063
offener Stoffauflauf, 4064
öffentliches Kraftwerk, 2226
Öffner, 4062
Öffnung, 5995
Öffnungsgrösse, 3073
Öffnungsweite, 3073
Offsetbogen, 4030
Offsetdruck, 4024, 4029
Offsetdruckbogen, 4030
Offsetfarben, 4025
Offsetpapier, 4028
Offset-Zeitungsdruckpapier, 4027
Ökologie, 2178
Ökotyp, 2181
okulieren, 935
Öl, 4032
Ölabriebwiderstand, 4044
Ölabsorption, 4034
Ölabsorptionstest, 4035
Ölabsorptionsvermögen, 4033
Ölaufnahme, 4034
Ölbenetzbarkeit, 4046
Ölbeständigkeit, 2890, 4043
öldichtes Papier, 4042
Öldurchdringtest, 4041
Oleate, 4047
Olefine, 4048
Oleinate, 4047
ölen, 2883, 2894
Ölfilter, 4039
Ölflecken, 4045
Ölpapier, 4038
Ölsäure, 4049
ölsaures Salz, 4047
Ölseparator, 4039
opak, 4060
opakes Papier, 4061
Opazität, 4059
Opazitätsmesser, 4058
Operieren, 4066
Optimierung, 4080
optisch dünner Strich, 5812
optische Abtastung, 4076
optische Aufhellung, 4078
optische Bleichmittel, 4072
optische Eigenschaften, 4075
optische Messung, 4074
optischer Aufheller, 4073, 4079
optisches Bleichen, 4072
optische Tastung, 4076

Orangenschale, 4081
Orangenschaleneffekt, 4082
Ordnen, 5006
Ordner, 2483, 5280
Organisationsprogramm, 1547
organische Säuren, 4083
organische Verbindungen, 4084
Organosollignine, 4085
Orientierung, 4086
Orientierungsgrad, 1882
Osmose, 4092
osmotischer Druck, 4093
Oszillation, 6005
Oszillator, 4090, 6006
oszillierend, 4088
Oszillograph, 4091
Overlaypapiere, 4108
Oxalate, 4115
Oxalsäure, 4116
Oxicellulose, 4122
Oxidationsbeständigkeit, 4119
Oxidationsmittel, 4117, 4121
Oxide, 4120
Oxidierung, 4118
Oxycellulose, 4122
Oxydationsmittel, 4121
Oxydations-Reduktions-Reaktion, 4692
oxydativ abgebaute Stärke, 4127
Oxyde, 4120
Oxydierung, 4124
Ozone, 4128
Ozonisierung, 4129

P

Packdraht, 552
packen, 2876
Packen, 4137
Packpapier, 4136, 4139, 6297, 6299
Packpapier aus Sekundärfaserstoff, 811
Packpappe, 4138
Packpresse, 548
Packseidenpapier, 4140
Packung, 4130
Packwatte, 6042
Paket, 4130
Paketaufkleber, 1136
Palette, 4142
Palettieren, 4143
Panne, 2383
Papier, 4146
Papier aus aufbereitetem Altpapier, 809
Papier aus Jutefasern, 3331
Papier aus röschem Stoff, 2723
Papier aus Sortierstoff, 809
Papierbahn, 4194, 6156
Papierbeschichtung, 4167
Papierbeutel, 4147
Papierbeutelmaschine, 4148
Papierbirke, 4151
Papierblatt, 5052
Papierbogen, 4184, 5052
Papierchromatographie, 4159
Papier dessen Feuchtigkeitsgehalt im Gleichgewicht steht mit der Luftfeuchte, 5370
Papiereigenschaften, 4182
Papiere mit rauher Oberfläche, 1379
Papierersatz, 4189
Papierfabrik, 4177
Papierform, 6251
Papierformat, 2384, 4185
Papierführung, 2935
Papier für Isolationszwecke, 2207
Papier für Loseblatt-Sammlung, 3572
Papier für lose Karteiblätter, 3572
Papier für Prüfzwecke, 5699
Papiergarn, 4195
Papiergefüge, 4188
Papierhalbstoff, 2954
Papierhändler, 4176
Papierhandtücher, 4192
Papierhersteller, 4175
Papierherstellung, 4173
Papierhilfsstoffe, 202

Papierhülse, 4162
Papierindustrie, 4165
Papierkanteneindrückwiderstand, 4806
Papierknäuel, 6040
Papierkontrolleur, 4166
Papierkordel, 4193
Papierleim, 4185
Papierleitwalze, 1815
Papiermacher, 4170
Papiermacheralaun, 4171
Papiermaschine, 4168
Papiermaschinenantrieb, 4169
Papiermaschinenausschuss, 3607
Papiermaschinenfilz, 4172
Papiermaschinengehilfe, 517
Papiermasse, 5480
Papier mit hoher Dehnung, 2355
Papier mit leimhaltigem Strich, 2843
Papier mit Leinenstruktur, 3501
Papier mit Randbeschnitt, 5840
Papier mit wolkiger Durchsicht, 6295
Papier mit zwei Büttenrändern, 2068
Papier oder Karton mit gleichen Zug- und Fortreissfestigkeitswerten in Maschinen- und Querrichtung, 5364
Papier ohne Rollneigung, 3999
Papierprodukte, 4181
Papierproduktion, 4173
Papierprüfer, 4166
Papierprüftisch, 5281
Papierprüfung, 4190
Papierqualität, 4164
Papierquerschneider, 4163
Papierrolle, 4702
Papierrollen, 6156
Papierrollen mit Stahlband versehen, 5449
Papierschnitzel, 4940, 5034
Papierservietten, 4178
Papiersorte, 4164
Papierstapel, 5135
Papierstaub, 3513
Papierstoff, 4187, 4568
Papierstruktur, 4188
Papierteller, 4180
Papiertextilien, 4191
Papiertüte, 4147
Papierumschlag, 6296
Papierverarbeiter, 4161
Papierverarbeitung, 4160
Papierwaage, 4183
Papierwerfwalze, 5153
Papierwolle, 2336
Papier zweiter Wahl, 1735, 3311
Pappe, 789, 1123, 4152
Pappe aus Halbzellstoff, 4991
Pappel, 4439, 4440
Pappenfabrik, 793
Pappenindustrie, 4155
Pappe oder Papier zum Abtrennen von Füllgut (z. B. in Keks- oder Pralinenschachteln), 2047
Papyrus, 4196
Papyrusstaude, 4196
Paraffinieren, 1933
Paraffinwachs, 4197
parallel geschichtet, 4198
Parameter, 4199
Parenchym, 4209
partielle Herstellung, 609
Partikel, 4212
passend, 3712
Passer, 4716
Paste, 4219
Patrizenwalzen, 2692
Patronenhülsenpapier, 352
Pauspapier, 5798
Pauspapier (für Zeichnungen), 786
Pechkiefer, 4342
Pektin, 4239
penetrant, 4248
Pentachlorphenol, 4250
Pentosan, 4251
Perchlorate, 4253
Perchlorsäure, 4254
Peressigsäure, 4252
Perforation, 4258
perforieren, 4256
Perforiermaschine, 4259

perforiertes Papier, 4257
perforierte Walze, 4255
Pergament, 4200
pergamentähnliche Oberflächen – ausrüstung, 4202
pergamentartiges Schreibpapier, 4201
Pergamentersatz, 440, 3185
pergamentieren, 4204
pergamentierfähiges Papier, 4205
Pergamentierung, 4203
Pergament mit Velinausrüstung, 4207
Pergamentpapier, 4206
Pergamentschreibpapier, 4208
Pergamin, 2816
Pergaminpapier, 2817
Periderm (Schutzschicht), 4262
Peripherantrieb, 4266
peripherisch, 4265
Perjodate, 4264
Perjodatlignine, 4263
perlen, 929
Perlstärke, 4235
perlweiss, 4236
Permanenz, 4267
Permanganate, 4270
Permanganatzahl, 4269
Permeabilität, 4271
Peroxide, 4272
Personal, 1693
Persulfate, 4273
Petroleum, 2697
Pfahltrocknung, 4399
Pfeilwurzmehl, 434
Pferdestärke (PS), 3090
Pflanze, 4353
Pflanzen, 4356
Pflanzenfaser, 4355
Pflanzensaft, 4900
Pflanzenschule, 4011
Pflanzenschutzpapier, 1084, 1085
pflanzliche Rückstände, 4357
Pflanzung, 4354
Pflaumenbaum, 4562
pflücken, 4301
Pfropfcopolymere, 2861
pfropfen, 2860
pH, 4274
Phenol, 4277
Phenolformaldehydharz, 4278
Phenolgruppen, 4279
Phenolharze, 4281
Phenollignine, 4282
Phenolphthalein, 4283
Phenolsäuren, 4280
Phenoplast, 4278
Phenylgruppen, 4284
Phloem, 4285
pH-Messung, 4275
Phosphate, 4286
Phosphor, 4287
Phosphorverbindungen, 4288
photoelektrischer Farbmesser, 4290
photografisches Papier, 4291
photographischer Offsetdruck, 4026, 4293
Photokopierpapier, 4295
Photometer, 4292
Photopapier, 4999
Photosynthese, 4296
Photozelle, 4289
Phthalsäure, 4297
Phthalsäureanhydrid, 4298
pH-Wert, 4276
physikalische Eigenschaften, 4299
Pick-up Presse, 4317
Pigment, 4319
Pigmentansatz, 4320, 4321
Pigmentfarbe, 4303
Pigmentfarbstoffe, 5976
pigmentgestrichen, 1345
pigmentgestrichener Faltschachtelkarton, 1347
pigmentiert, 4321
Pigmentklumpen, 1349
Pigmentstück, 1349
Pilot-Anlage, 4323
Pilze, 2752
Pilzprüfung, 2754
pilztötend, 2755

Pinen, 4326
Pinie, 4325, 4333
Pitotrohr, 4346
Plage, 3219
Plakatkarton, 4449
Plakatkarton (häufig ein- oder beidseitig verstärkt), 443
Plakatpapier, 686, 4450
Planbogen, 2747
Planen, 4928
planieren, 3452
Planlage (des Papiers), 2553
Planscheibe, 2026
Planschneidemesser, 2939, 5847
Planschneiden, 5846
Planschneider, 2937, 2938, 5845, 5848
Plantage, 4354
plastifizieren, 4364
Plastifiziermittel, 4363
plastisches Fliessen, 4361
Plastizität, 4362
Plastometer, 4365
Platane, 4352, 5606
Platte, 2014, 4366, 5141
Plattenherstellung, 4370
Plattenspalter, 5331
plattgedrückte Papierrolle, 2555
plattieren, 4366
Plattieren, 1334, 4378
plattierter Kocher, 1333
plattiertes Metall, 1335
platzen, 972, 1659
Platzen, 974, 2137
plombieren, 4959
Plusbaum, 4386
pneumatische Ausrüstung, 4392
Polarisation, 4398
Polarität, 4397
Polier, 2665
polieren, 2826, 4400
Polieren, 967
Politur, 2826, 4400
Politurmittel, 2827
Polster, 4141
Polsterfutter, 4953
Polyacrylamid, 4406
Polyacrylat, 4407
Polyamide, 4409
Polyamidharze, 4408
Polyäthylen, 4412
polyäthylenbeschichtetes Papier, 4413
Polyäthylenimin, 4414
Polychlorbutadien, 3952
Polyelektrolyte, 4410
Polyester, 4411
Polyimide, 4415
Polymere, 4418
Polymerisation, 4416
Polymerisationsgrad, 1883
polymerisieren, 4417
Polymethacrylat, 4419
Polypropylen, 4420
Polystyrol, 4421
Polysulfide, 4424
Polysulfidzellstoff, 4422
Polysulfidzellstoffaufschluss, 4423
Polyurethane, 4425
Polyvinylacetat, 4426
Polyvinylalkohol, 4427
Polyvinyläther, 4430
Polyvinylchlorid, 4428
Polyvinyle, 4432
Polyvinylester, 4429
Polyvinylidenchlorid, 4431
Pore, 4330, 4441
Pore (im Papier), 6028
porenfrei, 4331
porenfüllend, 3195
Porengrösse, 4442
Porenvolumen, 6031
Porenziffer, 6030, 6031
porig, 4446
porös, 4446, 5333
poröses Holz, 4448
poröses Material, 4447
poröse Stoffe, 4447
Porosimeter, 4443
Porosität, 4444

Porosität quer über die Bahn gemessen, 3087, 3428
Porositätsprüfgerät, 4445
Potentiometer, 4457
potenzieren, 4464
Prägekalander, 2256, 2258
Prägen, 1958, 2257, 5380
Prägen von Papier und Karton, 2259
Prägepapier, 2255
Prägestempel, 1954
Prallblech, 1872
Prallzerspannungsmaschine, 3187
Präzipitation, 4471
Präzision, 143, 4473
Preis, 1622
Pressdeckelpapier, 5899
Presse, 4482
Presse mit hohem spezifischem Druck, 3048
Presse mit zwei Walzenpaaren, 2046
pressen, 4482
Pressen, 4488, 4495, 5380
Pressenausschuss, 2050
Presseneinrichtung, 4483
Pressenpartie, 4493
Pressenstreichen, 4486
Pressentisch, 4373
Pressfilz, 4487
Pressform, 1954
Pressluft, 1476
Presspan, 4484
Presspappe, 4484
Pressstücke, 3874
Presswalze, 4490
primäre Rinde, 1621
Primärstoffauflauf, 4514
Primärwachstum, 4513
Primärwand, 4518
Primärzellwand, 4518
Probe, 4891
Probebogen, 5700
Probeentnahme, 4894
Probelauf, 5835
Probenvorbereitung, 4892
Probestück (Papier), 2966
Produkt, 4537
Produktentwicklung, 4538
Produktion, 4096, 4539
Produktionsablauf, 4536, 4542
Produktionskontrolle, 4540
Produktionsleiter, 4541
Produktionsmethode, 4542
Produktionsverfahren, 4542
Produktivität, 4543
Produkt zweiter Wahl, 4977
Profil, 4544
profiliert, 4003
profilierte Walze, 2914
Profilmessgerät, 4545
Profilpapier, 4924
Programmiersprache, 4548
programmierter Befehl, 4547
programmierter Unterricht, 4547
Propan, 4550
Propeller, 4551
Propionat, 4553
Propylalkohol, 4556
Protein, 4559
Protolignin, 4560
Proton, 4561
Prozess, 4532
Prozesskontrolle, 4533
Prüfbogen, 5700
Prüfgerät zur Messung der inneren Bindekraft, 3276
Prüfmethode, 5698
Prüfschrank, 5695
Prüfstand, 5694
Prüftisch, 5694
Prüfung, 1234, 3253, 5696
Prügel, 3554
Pseudoplastizität, 4563
Pseudotsuga, 4564
Puder, 4462
Puffer, 936
Pufferlösung, 937
puffern, 936
Pulper, 4571
Pulsieren, 4585

Pulver, 4462
pulverisieren, 4462
Pulverisierer, 4586
pulverisiert, 4463
Pumpe, 4587
Pumpen, 4588
Pumpenkolben, 4385
pumpfähige Masse, 5184
pumpfähiger Zellstoff, 5190
Puncture-Tester, 4593
Punkt, 4396
Punktbindung, 5336
punktieren, 4396
Punktierung, 5428
Punktverleimung, 5336
Pyrit, 4598
Pyritofen, 4600
Pyritrösten, 4599
Pyrolyse, 4601

Q

Qualität, 2857, 4604
qualitative Analyse, 4603
Qualitätseinbusse, 4023
Qualitätskontrolle, 4605
Qualm, 5195
quantitative Analyse, 4606
Quartärammoniumverbindungen, 4607
Quecksilber, 3752
Quecksilberverbindungen, 3753
Quelle, 2699
quellen, 3602, 5602
Quellen des Gummituchs beim Offsetdruck, 2257
Quellmittel, 5603
Quellung, 5602
Querfaser, 5825
Querfestigkeit, 663, 665
quergekreppt, 1705
Querprofil, 1709, 1711
Querrichtung, 1707
Querschneider, 1770, 5043
Querschneider für Wasserzeichenpapiere, 5113
Querschnitt, 1711
Querstreifigkeit im Papier, 589
Querströmung, 5602
Querstromverteiler, 2596
querverlaufende Porosität, 5826
quer zur Faser, 186
quer zur Laufrichtung, 1707
Quotient, 4644

R

Radierbarkeit, 2302
radierfestes Echtpergament, 2303
Radierfestigkeit, 2302
radikal, 4614
Radikal, 4614
radioaktiver Stoff, 4615
Radioaktivität, 4616
Radius, 4617
Raffinat, 4618
Raffinationsprodukt, 4618
Rakel, 2048
Rakelauftragsmaschine, 2054
Rakelhalterung, 2053
rakeln, 2058
Rakelstreichen, 725
Rakelstreicher, 724
Ramie, 4638
Ramiefaser, 4639
Ramsch-Ware, 3310
Rand, 2183, 4801
Randabschnitte, 5087
Randbegrenzung, 1850
rändeln, 3816
Randfassung, 4801
Randschneidmaschine, 2186

Randsleiste, 1850
Randstreifenabspritzrohr, 5849
Randwelligkeit, 2023, 2185
Randwinkel, 1521
Randzone, 4801
Rascheln, 4645
Rascheln (von Papier), 5204
Raschigringe, 4642
raspeln, 2878
Rasseln, 4645
rationalisiertes Arbeiten, 5921
Raster, 4945
Rattern, 4645
Rauch, 2750, 5195
rauchen, 5195
Rauchgas, 2597
rauh, 1377
rauhe Kante, 2399
rauhe Kante (Metall), 969
rauhe Oberfläche, 1378
rauhe Papierkante, 3823
Rauhigkeit, 4861
rauhrandiger Karton, 1851
räumen, 2329
Reagens, 4663
Reaktionsfähigkeit, 4660
reaktionshindernde Substanz, 3228
Reaktionsholz, 4659
Reaktionsmechanismus, 4656
Reaktionsmittel, 232
Reaktionsstelle, 4657
Reaktionsvermögen, 4660
Reaktionszeit, 4658, 4765
Reaktivität, 4660
Reaktor, 4661
Rechenmaschinenpapier, 687
Rechner, 1024, 1483
Redox-Potential, 4691
Redox-Reaktion, 4692
Reduktion, 4696
reduktionsfähiger Schwefel, 4693
Reduktionsgetriebe, 5315
Reduktionsmittel, 4694
Refiner, 4704
Refinerkegel, 3317, 4382, 6264
Refinerscheiben, 4705
Refinerschliff, 1263
Reflectometer, 4710
Reflektierung, 4708
Reflexion, 4708
Reflexionsfaktor, 4707
Reflexionsmesser, 4710
Reflexionsstärke, 4707
Reflexionsvermögen, 3590, 4709
Reflexionswelle (einer Leitung), 5809
Refraktionskoeffizient, 4712
Regelgerät, 1542, 2212
Regelschieber, 1543
Regelsystem, 1547
Regelung, 1541
Regeneratcellulose, 4715
Regenerieren, 4683
regenerierte Cellulose, 4715
Regenerierung, 4684
Registerlineal, 5623
Registerpapier, 834
Registerwalzen, 5620
Registerwalzenabstreifer, 5622
Registerwalzen-Staublech, 5621
registrierendes Messgerät, 4681
Registriergerät, 4678
Registrierkassenpapier, 1147
Registrierpapier, 4682
Registrierstreifen, 4677
Registrierung, 4676
Regler, 2855, 4719
Regulator, 2855
regulierbare Geschwindigkeit, 210
regulieren, 5012
Regulierkasten, 4554, 4718
Regulierwalze, 2933
reiben, 4869
Reibung, 2728, 4869
Reibungsentrinder, 2729
Reibungsverlust, 2735
Reif, 4802
Reife, 3723
reifen, 230
Reifen, 231, 236

Reifencord, 5754
Reifung, 2915
Reihenfolge, 5005
Reinheit, 4597
Reinheitsgehalt der Luft, 283
reinigen, 1341, 4939
Reinigen, 1351
Reinigen (im Sandfang), 4799
Reinigung, 4596
Reinigung durch Säure, 182
Reinigungsmittel, 1928
Reisig, 5147
Reissarbeit, 5676
reissen, 882, 1659
Reissen, 1662, 5657
Reissfestigkeit, 891, 5678
Reisslänge, 890
Reissprobe, 5683
Reisstärke, 4792
Reisstroh, 4793
Reisstrohpapier, 4791
Reisswiderstand, 892
Rekord, 4676
relative Feuchtigkeit, 4727
relative Feuchtigkeit der Luft, 3107
Release-Papiere, 4733
Release-Streichen, 4732
Relief, 4735
Reliefdruck, 4735
Reliefpapier, 4737
Rentabilitätsschwelle, 888
Reproduzierbarkeit, 4743
Reprographie, 4745
Reservoir, 2699, 4748, 5641
Resinat, 4755
resistent gegen alkalische Verfärbung, 306
Resistenz, 5367
Resonanz, 4763
Resorzin, 4764
Rest, 4614, 4750, 4940
Restligningehalt, 4749
Retention, 4769
Retentionsmittel, 4770
Rettung, 4890
reversible Osmose, 4776
Reyon, 4652
Reyonausschuss, 4655
Reyonfaser, 4653
Reyonzellstoff, 4654
Rheologie, 4790
rheologische Eigenschaften, 4789
richten, 5012
Richtgewebe, 4659
Richtlinie, 495
Richtung, 2009
Riefelung, 1618
Riemen, 653
Riemenantrieb, 655
Riemenausrücker, 656
Riemenscheibe, 4566
Riemenspannrolle, 3177
Riesaufkleber, 4665
Rieseinschlag, 4670
Rieseinwickler, 4670
Riesgewicht, 4668
Riesmarkierzettel, 4666
riesverpackt, 4667, 4669
Ries (480 Bogen), 4664
Riffeln, 1618
Riffelung, 4808
Riffelwalze (einer Wellpappenmaschine), 1620
Rille, 2912, 2915
Rillen, 1618, 4937
Rillenwalzenpresse, 2913
Rinde, 574
Rindenflecken (im Papier), 581
Rindenpresse, 580
Rindenschälmaschine, 576, 579
Rindentrockner, 575
Ring, 4802
Ringe, 385
ringförmig, 384
ringförmiges Schälen der Rinde, 4803
ringporiges Holz, 4805
Ringstauchversuch, 4804
Ringverbindungen, 1785
Riss, 882, 1659

rissige Kante einer Bahn, 1660
rissig werden, 1667
Riss im Papier, 2557
ritzen, 4931
Ritzen, 4937
Ritzmaschine, 4936
Robinie, 3546
Rockwell-Härte, 4812
roden, 5487
Roggenstroh, 4878
Rohgewicht, 4650
Rohmaterial, 4647
Rohöl, 1716
Rohpapier, 596, 4648
Rohpapiere für Imprägnierungen, 4911
Rohr, 4334, 5865
Röhre, 2148, 5865
Rohrleger, 4335
Rohrleitung, 4334, 4338
Rohrleitungssystem, 4337
Rohrmühle, 4815
Rohrsauger, 5508
Rohrschlange, 5323
Rohrschleuder, 1350
Rohrschlosser, 4335
Rohrverbinder, 2996
Rohrverlegung, 4336
Rohrverzweigung, 3676
Rohstoff, 598, 794, 4647, 4648
Rohtissue für die Wachsbeschichtung, 6132
Rolle, 4566, 4699, 4816
Rolle einer kontinuierlichen Papierbahn, 5053
Rolle Karton, 4701
Rolle mit Übergewicht, 4113
rollen, 1751, 4786
Rollen, 1754
Rollenbahn, 2879
Rollenbanderole, 4700
Rollengerüst, 4831
Rollengestell, 516
Rollengummiermaschine, 2945
Rollenklammern, 4817
Rollenkonus, 1325
Rollenlager, 4824
Rollenmusterentnahme, 4703
Rollenoffsetdruck, 6155
Rollenoffsetpapier, 6154
Rollenpapier, 4828
Rollenrest, 989
rollensatiniert, 6148
Rollenschneider, 5167, 6235
Rollenschneidmaschine, 4934
Rollenschnitt, 4933
Rollenschutz, 4829
Rollmaschine, 4787
Rollneigung, 1751
Rollneigung des Papiers, 4830
Rollneigungsprüfung, 1755
Rollstange, 3673
röntgenografischer Nachweis, 6308
Röntgenstrahlen, 6309
Röntgenstrahlenanalyse, 6308
Röntgenstrahlenmesser, 6310
röscher Stoff, 2724
röscher Zellstoff, 2722
Röschheit (ist immer reziprok zum Mahlgrad), 2718
Röschstoff-Holländer, 2717
Rosskastanie, 933, 3089
Rost, 2878, 4874
rosten, 4874
Rösten, 1010
Rostfestigkeit, 1602
Rostflecken, 4876
rostfrei, 405
rostfreier Stahl, 5378
rostig werden, 4874
Röstofen, 3345
Rostschutzkarton, 411
Rostschutzmittel, 4875
Rostschutzpapier, 406
Rostwiderstand, 5379
Rotation, 4856
Rotationslängsschneider, 4854
Rotationspumpe, 4851
Rotationsquerschneider, 4848
Rotationsviskosimeter, 4855

Rotguss, 912
Rotholz, 1481, 4697
rotierender Konsistenzmesser, 4855
rotierendes Messer, 4784
rotierendes Messer (eines Querschneiders), 2626
rotierendes Querschneiden, 4853
Rotor, 4859
rotor Ahorn, 5248
Rotorblatt, 4860
Rotorflügel, 5954
Rotormühle, 5954
Rotorschaufel, 5954
routinemässige Kontrolle, 4863
routinemässiges Kontrollverfahren, 4863
Rückbildung von Falten (Erholzeit), 1673
Rücken, 503
Rücken von Holz (vom Fällungsort zum Holzlagerplatz), 5134
Rückfederung, 4751, 4752
Rückfluss, 504
Rückgewinnen, 4683
Rückgewinnung, 4683, 4684
rückgewonnene Faser, 4674
Rückkopplung, 2404
Rücklauf, 504
Rückläufigkeit, 4772
Rückprall, 4751, 4752
Rückschicht, 505
Rückschichtenpapier geringer Festigkeit (aus Sekundärfaserstoff), 806
Rückschlagventil, 1237
Rückseite, 503
Rückspiegelung, 5313
Rückstand, 4750
Rückstrahlung, 4708
Rückwand, 503
rühren, 239, 620
Rühren, 240, 5429
Rührer, 241, 3063, 3841
Rührflügel, 3136
Rührwerk, 241, 3063
Rührwerksmesser, 3065
rundgewebter Filz, 2273
Rundheit, 4862
Rundschneidemaschine, 1330
Rundsiebbütte, 5975
Rundsiebfilz, 3878
Rundsiebkarton, 1792
Rundsiebkarton für Streichholzbriefe, 836
Rundsiebmaschine, 1796, 1800
Rundsiebpapiere, 5980
Rundsiebpresse, 1799
Rundsiebzylinder, 1797
Rundsiebtrog, 5975
Rundsiebzylindermaschine, 5979
Rundsortierer, 4852
Rundung, 4862
Runzelbildung (Farbe), 2444
rupfen, 4301
Rupfen, 4301
Rupfen beim Bruch, 4310
Rupfen der Faser, 4306
Rupfen des Egoutteurs, 1819
Rupfen infolge Hängenbleiben des Papiers am Kalander, 4304
Rupfen infolge von Blasenbildung, 4303
Rupffestigkeit, 4309, 4311, 5571
Rupffestigkeitsprüfgerät, 4312
Rupffestigkeitsprüfung, 4313
Rupfstelle in der Bahn (entstanden am Trockenzylinder infolge zu hoher Bahnfeuchtigkeit), 2127
Russ, 1096, 5276
Rutil, 4877
Rutsche für die Hackschnitzel, 1280
rutschen, 5164
rutschfest, 4003
rutschfeste Ausrüstung, 408
rütteln, 3312, 3313, 5027

S

Saat, 4981
Saatbeet, 4982

Saccharose, 5496
Sack, 526
Sackherstellmaschine, 538, 539
Sackleinensatinage, 963
Sackpapier, 540, 4879
Sackschliessmaschine, 541
Sackwand, 534
säen, 4981
Saft, 4900
Säge, 4917
Sägemühle, 4920
sägen, 4917
Sägen, 4919
Sägespäne, 4918
Sägewerk, 4920
Sägewerksabfall, 4921
Salmiakgeist, 346
Salpeterbildung, 3985
Salpetersäure, 3981
Salpetersäureaufschluss, 3983
Salpetersäurezellstoff, 3982
salpetrigsaure Salze, 3987
Salzbildner, 2957
Salzgehalt, 4886
Salzhaltigkeit, 4886
Salzigkeit, 4886
Salzwasser, 4889
Samen, 4981
Samenruhe, 2061
Sämling, 4984
Sammelbehälter, 142
Samttapete, 2578
Sandfang, 4798, 4896
Sandfilter, 4895
santiniertes Papier, 1036
säurewidrig, 387
Sassafras, 4906
Satinage, 2823
Satinagestaub, 1053
satinieren, 860, 2818
Satinieren, 1038, 2823
satinieren (mittels beheizter Kalanderwalzen), 3099
Satinierfalten, 1033
Satinierfärbung, 1054
Satinierflecken, 1051
Satinierkarton, 2820
Satinierstreifen, 1055
satiniert, 2819
satiniertes Papier, 2822
Satinnussbaum, 3518, 4690
Satinoberfläche, 4907
Satinweiss, 4908
sättigend, 3195
Sättigung, 4913
Satz, 5012
Säubern, 1351
sauer, 157, 5284
Sauerkeit, 5286
Sauerkraut, 5634
säuern, 5284
Sauerstoff, 4123
Sauerstoffbedarf, 4126
Sauerstoffentzug, 1911
Sauerstoffionenaustauscher, 379
Sauerstoffträger, 4117
Sauerstoffverbindungen, 4125
sauer werden, 5284
Saugdüse, 5999
Saugen, 5497
Saugerschlitz, 5501
Saugerwasserkasten, 4965
saugfähig, 116, 125, 5277
saugfähiges Papier, 117, 5250
Saugfähigkeit, 115, 127
Sauggautschwalze, 5502
Saugheber, 5115
Saughöhenmessgerät, 123
Saugkammer an der Presse, 3068
Saugkasten, 5901
Saugliegepresse, 5515
Sauglöchermarkierung, 5500
Saugpolster, 119
Saugpresse, 5509
Saugpresswalze, 5510
Saugpumpe, 5512
Saugrohr, 5999
Saugung, 5497

Saugventilator, 2340
Saugvorpresse, 5511
Saugwalze, 5513
Saugwalzenmarkierung, 5514
Saugzuggebläse, 3213
Säule, 2993
Saum, 4967
Säumen, 4968
Säure, 157, 5286
säurebeständig, 170, 176
säurebeständiger Backstein, 171
säurebeständiger Leim, 179
säurebeständiger Mauerstein, 171
säurebeständiger Ziegelstein, 171
Säurebeständigkeit, 174
Säurebeständigkeitsprüfung, 175
Säurechloride, 158
saure Farben, 159
saure Farbstoffe, 159, 160
säurefest, 170, 176
säurefeste Auskleidung, 172
säurefestes Papier, 173, 177
säurefrei, 161
säurefreies Kraftpapier, 3955
säurefreies Papier, 162
Säuregrad, 167
saure Gruppen, 163
Säurehalogenid, 164
säurehaltig, 165
Säurehaltigkeit, 167
Säureherstellung, 168
Säuren entgegenwirkend, 387
saures Sulfit, 180
Säureturm, 5794
Säurewäsche, 182
Schaber, 723, 2048, 2049
Schaberhalterung, 2053
Schabermarkierung, 2055
Schaberstoff, 2051
Schaberstreichen, 3354
Schaberstreicher, 3353
Schabkarton, 4941, 4943
Schablone, 2687, 5419
Schablonen, 5624
Schablonenkarton, 5420
Schabpapier, 4941, 4943
Schacht, 6176
Schachtel, 862, 1140
Schachtelabdeckpapier, 867
Schachtelaufkleber, 1136
Schachtelbeschnitt, 865
Schachteletiketten, 1143
Schachtelfutter, 870
Schachtelmaschine, 871
Schachtelpappe, 863, 1524
Schachteltrennwand (Steg), 872
Schaden, 1805, 3574
Schaft, 5026
Schälanlage, 4242
Schale, 859, 5057, 5829
schälen, 574
Schälen, 577, 1839, 4243
Schalen, 1747
Schälholz, 4241
Schalldämpfungsplatte, 183
schallen, 4802
Schallisolierung, 184
Schallwellen, 5283
Schälmaschine, 576
Schaltbrett, 4145
Schalter, 1641, 5605
Schalttafel, 1546, 4144, 4145
Schäl- und Abhebeprüfung, 4244
Schamottestein, 2522
Schar, 2575
Schärfe, 5033
scharfe Kante, 2397
Schärfen, 5032
Schärfen (des Schleifsteins), 970
Schärfrolle, 2909
scharf satiniert, 3046
Schärfung, 5032
Schattenmarkierung, 5025
schattieren, 5024, 5750
Schattierung, 5024
Schätzung, 2330
Schaueranlage, 5077
Schaum, 2628, 2738
Schaumbekämpfungsmittel, 2634

Schaumbrecher, 2629
schäumen, 2628
Schäumen, 2190, 2630, 2738
Schaum entgegenwirkend, 395
Schaumflecken, 2635, 2636, 2739
Schaumlatte, 523
Schaummittel, 2631
Schaumüberlauf (beim Voith-Stoffauflauf), 5136
Schaumverhütung, 2632
Schaumverhütungsmittel, 395, 2633
scheckig (durch unregelmässige Farbannahme), 3899
scheckige Oberfläche, 3902
scheckiges Papier, 1318, 2872
Scheckigwerden, 3904
Scheckpapier, 688, 1236, 1264
Scheibe, 2014, 2026, 4366
Scheibenfilter, 2020, 2114
Scheibenrefiner, 2022
Scheidewand, 3810
Schein, 3595
scheinbare Dichte, 414
Scheinwiderstand, 2218, 3190
Schema, 1938
Scherbeanspruchung, 5039
Scheren, 5035
Scherfestigkeit, 5036, 5038
Scherkraft, 5036, 5038, 5039
Scherprobe, 5040
Scheuerfestigkeit, 109
scheuern, 4939
Schicht, 2498, 3416, 3434, 4381, 4387, 5789
Schichtablösung, 1894
Schichtarbeit, 5059
Schichten, 5374
Schichtfolie, 3416
Schichtführer, 5790
Schichtpapier, 3412
Schichtstoff, 3409, 3413
Schichtstoff herstellen, 3414
Schichtstoff kaschieren, 3414
Schichtträger, 5495
Schichttrennung, 4391
Schichttrennung im Karton infolge von Blasen, 763
Schichtung, 5450
Schieber, 2795, 5952
Schieberventil, 2797
schief, 681
Schierling, 3030
Schierlingstanne, 3030
Schiessbaumwolle, 3988
Schild, 5058
Schilf, 4698
Schilfrohr, 4698
Schimmel, 3871, 3905
schimmelfestes Papier und Karton, 3881
Schimmer, 3595
schimmern, 3595
Schirm, 4945
schlaffe, 5142
Schlag, 771
Schlagabraum, 5147
Schlagbiegefestigkeit, 5065
schlagen, 620
Schlagen, 641
schlagen (von Holz), 2414
Schlagen von Rollen, 1233
Schlagen von Walzen, 1233
Schlagfestigkeit, 3188
Schlagmühle, 3187
Schlagversuche, 3189
Schlagzähigkeit, 3188
Schlamm, 3906, 5160, 5181
Schlammbeseitigung, 5182
schlämmen, 2251
Schlämmen, 2252
Schlammflecken, 5162
Schlämmkreide, 6226
Schlauch, 3091, 5865
schlecht aufgeschlossen, 5913
schlechte Durchsicht, 6231
schlecht gewickelte Rolle, 4435, 5252
Schleifen, 2906
Schleifer, 2900
Schleiferei, 2904, 2920
Schleiferführer, 2901

Schleifermagazin, 4393
Schleifertrog, 2902
Schleifmagazine, 2903
Schleifmaschine, 2900
Schleifmittel, 112
Schleifoberfläche, 2907
Schleifpapiere, 111
Schleifstein, 2905, 2908, 4583
Schleifstein-Schärfvorrichtung, 2909
Schleim, 5160
Schleimbekämpfungsmittel, 2753, 5163
Schleimflecken, 2534, 5162
Schleimwiderstandsfähigkeit, 2755
Schlepprakelauftrag, 5801
Schlepprakelstreicher, 5800
Schleudergebläse, 1208
Schleudermühle, 1722
schleudern, 1214
Schleudersortierer, 1211
Schleuderung, 1213
Schleuse, 5183
Schleusentor, 2795
Schlichten, 5197
Schlick, 3906
Schliere, 5455
Schliff aus Chips, 1287
Schliff aus Hackschnitzel, 1287
Schliff (Messer), 5032
Schlitz, 5176
Schlitzen, 5172, 5178
Schlitzmaschine, 5177
Schlupf, 5164, 5165
schlüsselfertige Anlage, 5880
schmaler Streifen einer Bahn, 5472
Schmelze, 3745, 5192
schmelzen, 1897, 5192
Schmelzen, 3745
Schmelzglasur, 2266
Schmelzlöser, 1169
Schmelzofen, 2760, 5193
Schmelzofen (Sulfatverfahren), 4685
Schmelzpunkt, 3746
Schmelztank, 5194
Schmelzung, 3745
Schmiere, 2883
schmieren, 2883, 2894
schmierig, 2895
schmieriger Stoff, 5180
Schmierigkeit (des Stoffs), 5179
Schmierigkeit des Stoffs, 6194
Schmiermittel, 3583
Schmiernippel, 3976
Schmieröl, 3584
Schmierung, 3585
Schmirgelpapier, 1109, 2567
Schmirgelpapiere, 111
Schmirgelpapier (für Lederindustrie), 938
Schmutz, 2012, 3200, 3906, 5335
Schmutzflecken, 5375
Schmutzfleckenzahl, 2013
Schmutzschleuse, 5828
Schnalle (im Papier), 1420
Schnallen, 2109
Schnallenbildung, 934, 1424
Schnalle (Papierfehler), 243
Schnecke, 4950
Schneckenförderer, 4951
schneckenformig, 3027
Schneckengetriebe, 6289
Schneide, 3355
schneiden, 1764
Schneiden, 1775
Schneiden von Bogen aus Bahn oder Rolle, 5049
Schneidkante, 3355, 5169
Schneidmaschinenbedienungsmann, 5170
Schneidmass, 5841
Schneidstaub, 5168
Schneid- und Aufwickelmaschine, 5171
Schneidwerkzeug, 1776
Schnellalterungstest, 130
schnellaufend, 3056
schnell entwässerter Zellstoff, 2390
schnelles Auflösevermögen, 2177
schnellgetrockneter Zellstoff, 2544
Schnellhefter, 2483
Schnelligkeit, 2389
Schnellkocher, 482

Schnelltrocknung, 2545
Schnittabfall, 1771
Schnittbreite, 3341
Schnittholz, 3586
Schnittmass, 5841
Schnittmusterpapier, 4232
Schnittmusterseidenpapier, 4233
Schnittstaub, 1772
Schnitzel, 1276
Schnitzelmaterial (als Fällstoff für Pressmassen), 3602
Schnur, 1584
Schock, 5063
Schocktrocknung, 5064
Schonfilz, 5291
Schonschicht (eines Kartons), 5914
Schöpfeimer, 4930
schöpfen, 4930
Schöpfform, 3871, 3905
Schopper-Riegler-Mahlgrad, 4929
Schössling, 4901
Schott, 948
schräg, 681
Schräge, 5175
schrägstellen, 4340
schrägzahnt (mech.), 3027
Schranke, 585, 3228, 5395
Schraube, 4551, 4950
schraubenförmig, 3027
Schreiber, 4678, 4681
Schreiberdiagrammpapier, 4679
Schreibfeder, 4247
Schreibpapier, 6302
Schreibthermometerpapier, 5714
Schrenzkarton (an den hinsichtlich Festigkeit keine Ansprüche gestellt werden), 4005
Schrenzpappe, 1278, 4348
Schriftsetzen, 5900
Schrotpatronenpapier, 599
Schrumpf, 5081
schrumpfen, 5081
Schrumpfen, 5082
Schrumpfsiebpresse, 4437, 5151
Schub, 5035
Schubzahl, 5037
Schuhpappe, 5066
Schuhpappe aus Sekundärfaserstoff, 1642
Schürze, 424
Schussdrähte, 6165
Schussdrähte am Papiersieb, 5085
Schussfaden, 6166
Schussgarn, 6166
Schuss (Gewebe), 6164
Schüttdichte, 946
Schüttelapparatur, 5028
Schüttelmaschine, 6006
schütteln, 3313, 5027
Schütteln, 5031
Schüttelsieb, 5029
Schüttelsortierer, 6004
Schüttelsortierung, 5029
Schüttelvorrichtung, 5028
Schüttelzeit, 5030
Schüttgewicht, 414, 946
Schutz, 1511, 5058
Schutzbrett, 522
schützend, 4481
Schutzgas, 1544
Schutzhaube, 1654
Schutzhülle, 6296
Schutzkleidung, 4557
Schutzmaterial, 586
Schutzpolster, 6041
Schutzschicht, 4558
Schutzüberzug, 4558
Schutzumschlag, 833, 3306
Schutzumschlagpapier, 832, 840
Schutzvorrichtung, 2930
schwach, 3470
schwächer werden, 1954
schwache Säure, 6137
schwaches Rundholz, 5191
schwach geleimt, 5143, 5253
Schwachholz, 5191
Schwaden, 5957, 5966
Schwämme, 2752
schwammig, 5333

schwarzes Albenpapier, 717
Schwarzesche, 718
Schwarzfichte, 722
Schwarzlauge, 720, 776
Schwarzpappel, 721
Schwärzung durch zu hohe Feuchtigkeit
    beim Kalandrieren, 719
Schwärzungsmesser, 1903
Schweben, 2583
Schwebetrockner, 265
Schwefel, 5541, 5547
Schwefeldioxide, 5544
schwefelige Säure, 5546
Schwefelkies, 4598
Schwefelkohlenstoff, 1098
Schwefelsäure, 5545
schwefelsaures Ammoniak, 351
schwefelsaures Natrium, 5235
schwefelsaure Tonerde, 335
Schwefelverbindungen, 5543
Schwefelwasserstoff, 3145
Schwefelziehen, 5542
Schweissdraht, 5329
Schweissen, 6175
Schweissnaht, 6174
Schwemme, 2611
schwenkbar, 5152
schwer arbeiten, 3385, 3387
Schwere, 6170
schwergewichtige Papiere, 3025
Schweröl, 2744
Schwerspat, 593, 4236
Schwiele, 1420
Schwielen (durch den Rollenscheider
    verursacht), 6236
Schwimmen, 2583
Schwimmkraft, 961
Schwimmtrockner, 2568
Schwimmwalze, 5604
Schwingung, 4089, 4585, 6005
Schwingungserzeuger, 4090, 6006
Schwingungsmesser, 4091
Schwingungszahl, 2725
Schwitzen, 5821
Schwund, 2379, 4016, 5082
Sealing (Kraftpapiersorte), 4962
Sediment, 4979
sedimentieren, 5096
Sedimentieren, 5016
See, 3404
Seetang, 294, 1132
Seidenpapier, 5756
Seidensiebdruck, 5095
Seife, 5206
Seifeneinwickelpapier, 5208
Seifenentschäumer, 5207
Seil, 1005, 4835
Seilaufführung, 4837
Seilführung, 4836
Seilmarkierung, 4838
Seiltrommel, 2110
Seitenfalte einer Tragetasche, 2950
Sekundärfaser, 4674
Sekundärstoff, 4973
Sekundärstoffauflauf, 4972
Sekundärwald, 4970
Sekundärzellwand, 4975
Selbstabnahme-Langsiebmaschine, 2706
Selbstabnahmemaschine, 3600, 6312
Selbstabnahmetrockenzylinder, 6311
Selbstregelung, 484
Selbststeuerung, 484
selbsttätiger Ausschalter, 1328
selbsttätiger Schaber, 489
Selektion, 4986
Selektivität, 4987
Senkwaage, 3149
Sensibilisator, 5000
Sensibilisierung, 4998
Sensibilität, 4997
Sensitivierung, 4998
Sensitivität, 4997
Separatstreichen, 1552
Separatstreicher, 4020
separat verarbeiten, 4022
Serviettenpapier, 3939
Serviettentissue, 3940
Servomechanismus, 5011
Servo-Regler, 2212

Setzbord, 6050
Sheffield-Prüfgerät, 5055
Shore-Härte, 5068
Shore-Härtemesser, 5067
sich Abheben (einer Schicht), 3467
sich abmühen, 3385, 3387
sich bauschen, 944
sich durchbiegen, 4884
Sicherheit, 4880
Sicherheitsanlagen, 4882
Sicherheitsausrüstung, 4882
Sicherheitshahn, 4881
Sicherheitspapier, 394, 1236, 4883
sich Erstrecken, 5468
Sicherung, 4880
sich klumpen, 233, 3591
sich kräuseln, 4565
sich rollen, 1751
sich schief ziehen (vom Filz), 2041
sichten, 5279
Sichten, 4947
sich überlappen, 3419
sich verjüngen, 5646
sich verschieben, 5164
sich verziehen, 934
sich wölbend, 429
sich zusammenballen, 1354
Sickerkühlung, 5837
Sieb, 5088, 5445, 6242
Siebabwasser, 6225
Siebanalyse, 5089
Siebbelastung, 6249
sieben, 4945, 5088
Siebende, 6244
Siebfalte, 4796
Siebflecken, 6256
Siebführung, 2936
Sieblaufregler, 486, 6245
Sieblaufzeit, 6247
Siebleder, 424
Siebleiste (zur Regulierung der
    Entwässerung am Sieb), 425
Siebleitlinie, 6248
Siebleitwalze, 2933, 6253
Sieblinien, 6248
Siebloch, 6246
Siebmarkierung, 6248, 6250
Siebmaschenweite, 6246
Sieböffnung, 3756
Siebpartie, 6244, 6254
Siebrückstand, 4948
Siebsaugerbeläger, 5499
Siebsaugermarkierung, 5500
Siebsaugkasten, 5498
Siebschiff, 3630, 6252
Siebschüttelung, 5027
Siebseite, 6255
Siebspannwalze, 6253, 6257
Siebspritzblech, 5621
Siebtisch, 2690
Siebtischentwässerungsmaschine, 6191
Siebtrommel, 2115
Siebtuch, 6243
Siebtuchpresse, 2372
Siebumkehrwalze, 6258
Siebwalze, 1820
Siebwasser, 6225
Siebwasserkasten, 6252
Siebwasser (z.B. in der Rundsiebpartie),
    5635
Siebweite, 3756
Siebwendewalze, 6258
Siebwinkel (nur an der Streichmaschine),
    4946
Siedepunkt, 815
Siederohrkessel, 6114
siegelfähiges Papier, 4963
Siegelmaschine, 541
Siegelpapier, 4963
Siekeneisen, 1672
Signallampe, 3208
Signieren, 3699
Silbereinpackpapier, 412
Silberpapier, 333
Silberschutzseidenpapier, 413
Silbertanne, 5098
Silberverbindungen, 5097
Silikate, 5090
Silikon, 5091

Silikonharz, 5094
Silizium, 5091
Silizium-Flächengleichrichter, 5093
Siliziumverbindungen, 5092
Silo, 689
sinngemäss, 360
Sinterung, 6027
Siphon, 5115
Sisal, 5117
Sisalhanf, 5118
Sitkafichte, 5120
Skala, 4922
Skalenscheibe, 1939
Skidder, 5133
Sockel, 597
Soda, 5209, 5223
Sojaproteine, 5288
Sol, 5257
Sole, 4889
solide, 5260
Solvation, 5273
Solvatisierung, 5273
Sommerholz, 2898, 5548
Sonderzellstoff, 5298
Sorte, 2857, 5279
sortieren, 4945, 5279
Sortieren, 2859, 4947, 5282
Sortierer, 4945, 5280
Sortierquerschneider, 1773
Sortiersaal, 4949
Sortierstoff, 4948
Sortier-und Prüfmethode von Papier,
    2386
Sortierung, 1342
Sortiment, 5012, 5971
Spachtelmasse, 4519
Spalt, 3281, 3974
Spaltbarkeit, 1894
Spalten der Kartonbahn, 773
Spaltfestigkeit, 821, 4390
Spaltmaschine, 5331
Spaltprüfung, 781
Spaltspur (Fehler im Karton), 774
Spalttest, 781
Spaltung, 5332
spanabhebend bearbeiten, 3606
spanabhebende Bearbeitung, 3637
spanabhebendes Werkzeug, 1776
Spanabmessungen, 1285
Spänefang, 1280
Spanen, 1294
Spänesieb, 1295
Spänetransportanlage, 1283
Spangrösse, 4214
Spangutklassierung, 1281
Spanmesser, 1291
Spanner, 5467
Spannkraft, 2201
Spannrolle, 3177
Spannung, 5444, 5685
Spannungs-Dehnung, 3541
Spannungs-Dehnungs-Eigenschaften,
    5464
Spannungsmesser, 5684, 6033
Spannungsprüfer, 5446
Spannungsteiler, 4457
Spannvorrichtung, 1337
Spannwalze, 5463
Spanplatten (aus Stückspänen), 4213
Spanplatte (US), 1278
Spansieb, 1295
Spant, 2687
Spätholz, 3429, 5548
Speckglanz, 5312
Speicher, 3549, 4748
speichern, 5430
Speicherraum, 4379
Speicherung, 141
speisen, 2403, 5565
Speisevorwärmer, 2180
Speisewasser, 2412
Speisung, 2411
spektrale Reflexion, 5307
spektrales Reflexionsvermögen, 5308
Spektralphotometer, 5309
Spektroskop, 5310
Spektroskopie, 5311
Sperre, 585
Sperrschichtmaterial, 586

Sperrschichtpapier, 587, 588
Spezialität, 5294
Spezialkarton, 5295
Spezialebensmittelkarton, 5293
Spezialpapiere, 5297
Spezialwellpappe, 5296
Spezialzellstoff, 5298
spezifiches Gewicht, 1905, 5300
spezifische Leimart zur Erzielung erhöhter
    Wasserfestigkeit in Papier und Karton,
    2695, 2696
spezifische Oberfläche, 5302
spezifischer Widerstand, 4761
spezifisches Volumen, 415, 949, 5303
spezifische Wärme, 5301
Spezifizierung, 5299
Spiegelreflexion, 5313
Spindelpresse, 4952
Spinnen, 5319
Spinnfaser, 5701
Spinngewebe, 5320
spiralförmiger Egoutteur, 5324
spiralförmiger Holzspaner, 5321
spiralförmig mit Draht umwickelter
    Rotationsstabstreicher, 6259
Spiralripp-Egoutteur, 3397
spiralverlegt, 5322
Spitze, 4396
Spitzenpapier, 3388
Spitzverschluss, 2768
Splintholz, 4905
Splitter, 5062, 5174
Splitterfang, 3363
Splitterfänger, 958, 3361
Sprenggerät, 5292
Sprengpulverpapier, 732
sprenkeln, 5305
spriessen, 5358
Spriessen, 2810
springen, 1659
Springen, 1662
Sprinkler, 5356
Spritzauftrag, 5343
Spritzauftragsmaschine, 5342
Spritzbefilmen, 2365
Spritzen, 5339
Spritzgussverfahren, 3231
Spritzkopf, 2367
Spritznarben, 4238
Spritzpistole, 2948
Spritzrohr, 5077
Sprödigkeit, 902
sprudeln, 1659
Sprühanlage, 5356
Sprühauftrag, 5343
Sprühauftragsmaschine, 5342
sprühen, 5341
Sprühtrocknung, 5345
Sprung, 1659
Spuckstoff, 5634
Spule, 4699, 5334
Spulenpappe, 1586
spun-bonded, 5360
Spurenelemente, 5796
Stab, 568, 4813
stabiler Zustand, 5401
Stabilisator, 5371
stabilisieren, 5370
Stabilisierung, 5369
Stabilität, 5367
Stabmühle, 4815
Stabstreicher, 569, 4814
Stadtmüll, 3933
Stahl, 5415
Stahlbeton, 4721
stahlplattiert, 1336
Stahlstichpapier, 4371
Stamm, 5418
Stammförderer, 3555
Stammspaltmaschine, 3561
Stampfen, 5380
Stand, 3452
Standard, 5382
Standard-Buchseitenformat, 829
Stand der Farbe, 3071, 3235
Ständer, 4317
Standfestigkeitsprüfung, 1679
ständig messen, 3887
Standregelung, 3453

Stange, 568, 4813
stanzen, 1955
Stanzen, 1957, 4488
Stanzer, 1956
Stanzform, 2580
Stanzlinge, 731
Stanzpappe, 4484
Stanzwerkzeug, 1956
Stapel, 4322, 5372
Stapelfaser, 5384
Stapeln, 5374
Stapelvorrichtung, 5373
Stapler, 3469, 5373
stark, 5260
stark auftragend, 3039
stark ausmahlen, 2972
stark dehnbare Kartonagenpappe (für
    Verpackungszwecke und Kartons),
    2344
stark dehnbares Papier (für
    Verpackungszwecke), 2346
Stärke, 1943, 2664, 3266, 4464,
    5386, 5459, 5722
Stärkederivat, 5389
Stärke der Mahlung, 642, 1880
Stärkekocher, 5388
starke Säure, 5476
Stärkeumwandlung, 5387
starke Wolkigkeit, 6231
Stärkexanthat, 5390
Stärkexanthid, 5391
Stärkexanthogenat, 5390
stark geleimtes Vorsatzpapier, 835
stark satiniert, 5555
Statiktests, 5397
stationäre Hülsenstange, 4002
statische Elektrizität, 5394
statischer Druck, 5396
statischer Inhibitor, 5395
statische Versuche, 5397
Statistiken, 5399
statistische Analyse, 5398
Stator, 5400
Stau, 4433
Staub, 2164, 3513, 4462
Staubabsaugung an den Rollenenden
    oder Seiten eines Papierstapels, 5503
Staubabscheider, 1790, 2165
Staubbildung, 2168
stauben, 2762
Stauben, 3515
Staubentferner, 1789
Stauben (von Papier), 2764
Staubfaden, 2481
Staubfänger, 2165
Staubfilter, 2167
Staubfliessverfahren, 2608
Staubkontrolle, 2166
Staublech, 521
Staubsammler, 2165
Staubungsneigung, 2600, 2763
Stauchanlage (für non-wovens), 1473
Stauchen (Wellpappe), 1723
Stauchversuch, 1480, 1725
Staudruck, 285
Stauhöhe, 3452
Staukörper, 521
Staulatte, 1804, 5154
Staurohr, 4346
Stearate, 5413
Stearinsäure, 5414
Stecheiche, 3179
Stecker, 2211, 4382
Steg, 5289
Stegkarton, 4216
Stegschlitzmaschine, 4218
stehender Zellstoffkocher, 6000
stehenlassen, 1735
Steichrohmaterial, 1414
Steifigkeit, 4800, 5425
Steifigkeitsprüfer, 5426
Steifigkeitsprüfung, 5427
Steilsieb, 3202, 3203, 5933
Steilsiebmaschine, 3204
Steinberg-Streicher, 1759
Steindruck, 3532
Steindruckpapier, 3531, 3533
Steinöl, 2697
Steinwalze, 2873

Stele, 5417
Stellschraube, 3458
Stellung der Spindel am Stoffauflaug,
    5157
stellvertretender Betriebsleiter, 474
Stengel, 5418
Stereoskop, 5423
Stetigaufschluss, 1536
Stetigholländer, 885
Stetigkocher, 1532
Steuergerät, 1542
steuern, 2932
Steuerung, 1541
Sticheinreissfestigkeit, 4592
Sticheinreissprüfgerät, 4593
Sticheinreissprüfung, 4594
Stickstoff, 3989
Stickstoffverbindungen, 3990
Stiel, 2961, 5418
Stillegung, 5083
Stillsetzen, 5083
Stillstand, 5083
Stillstandszeit, 2084, 3176
Stint, 5192
Stippe, 5304
Stippen (im Bindemittel), 5306
Stirnblatt (als Rollenschutz), 4823
Stirndeckel, 2271
Stirnfläche, 2373
Stirnseite, 2373, 2995
Stockpunkt, 4461
Stoff, 2370, 3713, 5430, 5480
Stoffänger, 4914
Stoffängerauffangwanne, 4916
Stoffaufbereitung, 5435
Stoffauflauf, 2590, 2994, 5434
Stoffauflaufkasten, 893
Stoffauflauflippe, 3516
Stoffauflöser, 4571
Stoffauflösung, 4574
Stoffbatzen, 3591
Stoffbrei, 4187
Stoffbütte, 3609, 5431, 5482
Stoffdichte, 1512, 5483
Stoffdichteanzeige, 1515
Stoffdichteregler, 1514, 1516
Stoffdichteregulierung, 1513
Stoffeintrag, 2761
Stoffeintrag im Holländer, 631
Stoff für die Filterschicht des Stoffängers,
    5600
Stoff für Zeitschriftenpapier, 3642
Stoffgewicht, 5300
Stoff im Absetzkasten, 2092
Stoffleimung, 5486
Stofffluss, 5433
Stoffmühle, 3074
Stoffprüfung, 3717
Stoffpumpe, 5436
Stoffregler, 5437
Stoffregulierung, 5437
stoffreicher Siebwasser, 4794
Stoffstand, 2993
Stoffstandsanzeige, 3455
Stoffstandsregler, 3454
Stoffstauhöhe, 2993
Stoffübergang, 3708
Stoffüberlaufkasten, 5481
Stoffverbrauchen, 3715
Stoffverteiler, 422
Stonit, 5440
Stopfbüchse, 5484
Stopfbüchsenbrille, 5485
Stöpsel, 4382
Störung, 2383
Stoss, 771, 3312, 4322, 5372
Stossschutz, 6041
Stossdämpfung, 936
stossen, 3312
Stossfestigkeit, 5065
Stossmühle, 3187
straff gewickelte Rolle, 5741
strahlen, 2813
Strahlen, 2195
Strahlentrinder, 3120, 3125
Strahlenzelle, 4651
Strahlung, 4613
Strahlungsheizung, 1550, 4612
Strähne, 4387

Strangpressform, 2366
Strangpresswerkzeug, 2366
Strassenkanal, 5022
streckbar, 5675
Strecken, 5468
Streckgrenze, 6320
Streckvorrichtung, 5467
Streckwalze, 5349, 6158
Streichaggregat, 1398
Streichanlage, 1398, 1410
Streichanlage für beidseitiges Streichen,
    5893
Streichanlage mit Kontaktauftrag im
    Gegenlauf, 1537
Streichclay, 1404
Streichdefekt, 1407
Streichen, 1399
Streichen ausserhalb der Maschine,
    1552
Streichen ausserhalb der Papiermaschine,
    4021
Streichen indem die Auftragswalze die
    Bahn berührt, 3349
Streichen in der Maschine, 3614, 4051
Streicher, 1398
Streicher mit überfluteter Einlaufzone,
    2581
Streichfähigkeit, 1383
Streichfarbe, 1405
Streichfarbenstreifen, 1401
Streichfarbenüberschuss, 2337
Streichfarbenzusammensetzung, 1406
Streichhilfsstoffe, 1400
Streichklumpen, 1409
Streichmaschine, 1410
Streichmaschine mit Abquetschwalze,
    5366
Streichmaschine mit Luftrakel, 2569
Streichmasse, 1411
Streichnapf, 1412
Streichpigment, 1404
Streichrohpapier, 1403, 5495
Streichschaber, 2049
Streichschaberbalken, 2052
Streichschabermarkierung, 2056
Streichtrog, 1412
Streichunterlage, 1402
Streichvorrichtung, 419
Streifen, 5455, 5643
Streifenbildung, 5455
Streifenbildung im Stoff (auf dem Sieb)
    hervogerufen durch Fehler im
    Stoffauflauf, 5131
Streifenfehler, 5456
Streifenlackiermaschine, 5473
Streifenlackierung, 5474
Streifenstreichen, 5474
Streifenstreichmaschine, 5473
streifig, 5469
strenge Farbe, 5630
Streueffekt, 4927
Streuung, 1974, 4927
Streuungsanalyse, 365
Strich, 1382, 1399
Strichabhebung, 1413
Strichgewicht, 1417
Strichträgermaterial, 1402
Strichzusatz, 1400
Stroh, 5452
Strohpappe, 5453
Strohpappe von 140-150 g/cm2, 3973
Strohzellstoff, 3737, 5454
Strom, 4464, 4809
Stromabnehmer, 915
Stromdichte, 1758
Strömen, 2588
Stromerzeuger, 2808
Stromkreis, 2209
Stromleistung, 2221
Strommesser, 344
Strommessung, 1639
Strömungslehre, 2602
Strömungsmechanik, 3131
Strömungsmesser, 2594
Strömungsmessung, 2593
Strömungstechnik, 2592
Strömungsverhältnisse, 3119
Stromunterbrecher, 1328
Stromverteilung, 2222

Strudel, 6035
Struktur, 2684, 2862, 5478, 5702
stückiger Kontaktkörper, 4245
stufenloser Antrieb, 5969
Stufenriemenkegel, 5422
Stufenscheibe, 1506
Stuhlung (am Kalander), 2714, 2715
stumpf aneinanderfügen, 987
stumpf gegeneinanderstossende
    Anklebestelle ohne Überlappung, 988
Sturzkocher, 5744
Stütze, 5567
Stutzen, 2995
Stützsieb, 511
Styrol, 5489
Sublimierung, 5491
substantive Farbstoffe, 2007
Substanz, 5493
Substitutionsgrad, 1884
Substrat, 5495
Sucrose, 5496
Sulfamate, 5519
Sulfamidsäure, 5520
sulfamidsaure Salze, 5519
Sulfatablauge, 5523
Sulfataufschluss, 3376, 5525
Sulfatdeckschicht (eines Kartons), 5522
Sulfate, 5528
Sulfatlauge, 5523
Sulfatpapier, 5524
Sulfatpappe, 5521
Sulfatterpentin, 4326, 5527
Sulfatverfahren, 5525
Sulfatzellstoff, 3377, 5526
Sulfatzellstoffpapier von hoher
    Zugfestigkeit für Webzwecke, 1130
Sulfide, 5530
Sulfidierung, 5529
Sulfidität (%-Verhältnis von Na-sulfid zu
    den wirksamen Alkalien), 5531
Sulfit, 5536
Sulfitablauge, 5318
Sulfitanlage, 169
Sulfitaufschluss, 5535
Sulfitlauge, 5532
Sulfitpapier, 5533
Sulfitzellstoff, 5534
Sulfogruppen, 5537
Sulfonate, 5538
Sulfonylgruppen, 5539
Sulfoxide, 5540
Sumpf, 4433, 5469
Sumpfkiefer, 3567, 5287
Sumpfschleppprakelstreicher, 4434
Sumpfzypresse, 544
Superkalander, 5550
Superkalandrieren, 5553
superkalandriertes Papier, 5552
Suspension, 5184
Süsswasser, 2727
Sykomore, 5606
Symmetrie, 5607
synchron, 5609
Synchronismus, 5608
Synchronmotor, 5610
synergetischer Effekt, 5611
Synergismus, 5611
Synthese, 5612
synthetische Faser, 5613
synthetische Polymere, 5614
synthetischer Gummi, 5616
Syringylgruppen, 5617
System, 5618

T

Tabellarisierung, 5626
Tabelle, 1231
Tablette, 4245
Tablettieren, 4246
Tachometer, 5627
Tafel, 789, 4144, 5141
Täfelung, 4144
Tageslicht, 1830
Talkerde, 3643

Talkum, 5636
Tallöl, 5637
Tallölseife, 5638
Tamariske, 5640
Tambour, 6239
Tank, 5641
Tanne, 105, 2520
Tannenholz, 105
Tannenzapfen, 5831
Tantalverbindungen, 5642
Tänzerwalze, 1814
Tapete, 2970
Tapeten, 6045
Tapetenrohpapier, 2971
Tapiokastärke, 5648
Tara, 5651
Tasche, 4393
Tastelement, 5002
Tasterdruck bei der Dickenmessung, 953
tato-Ausstoss einer in Betrieb
   befindlichen Maschine, 1080
tatsächliches Gewicht, 199
Tau, 1005
Tauchauftrag, 2004
Tauchstreichen, 2004
Tauchstreicher, 2003
Tauchwalze, 2701
Tauchwalzenstreicher, 2700
Tauenpapier, 3331, 4839
Taupunkt, 1930
Tauwerk, 4840
Technik, 5665
Techniker, 2282
Technikumsanlage, 4323
technische Anlage, 4353
technisches Natriumkarbonat, 5210
Teebeutelpapier, 5654
Teer, 5650
teeren, 5650
Teerpapier, 5652
Teersatz, 2100
Teich, 4433
Teilchen, 4212
Teilchengrösse, 4214
Teilchengrössenverteilung, 4215
Teiler, 2047
Telefonbuchpapier, 2011
Teleskopbehälter, 5668
Teleskopfutteral, 5668
Teller, 2026
Temperatur, 5670
Temperaturregelung, 5671
Temperaturregler, 5671
Temperaturspannung, 5710
Teppichboden, 1831
Teppich-Tissue, 1130
Terpen, 5689
Terpentine, 5882
Terpentin-Test (auf Fettdichtigkeit), 5883
Terpinen, 5690
Tertiärbehandlung, 5692
Tertiärlamelle, 5691
Tertiärzellwand, 5693
Test-Deckenbahn (Wellpappe), 5697
Tetrachlorkohlenstoff, 1105
Textilfaser, 5701
textilverstärktes Papier, 4722
thermische Behandlung, 3024
thermischer Wirkungsgrad, 2192
Thermocolorstift, 5711
thermodynamische Wärmelehre, 5713
Thermoelement, 5712
thermomechanischer Aufschluss, 5715
Thermometer, 5716
Thermopaar, 5712
Thermoplasten, 5717
thermoplastischer Stoff, 5717
Thermorefiner-Aufschluss, 5715
Thiocarbamate, 5727
Thiocarbonate, 5728
Thiolalkohole, 3749
Thiolignine, 5729
Thixotropie, 5733
Thuja, 5739
Tiefdruck, 2882, 4858
Tiefdruckpapier, 2881, 4857
Tiefe, 1917
Tiefkühlkarton, 2740
Tieftemperatur, 3581

Tiegeldruckpresse, 4375
Tierleim, 376, 377
Tinte, 3232
Tintenwischprüfung, 3242
Tissuepapier, 5756
Tissuepapiermaschine, 5755
Tissues, 3217
Titan, 5757
Titandioxid, 4877, 5759
Titanverbindungen, 5758
Titration, 5762
titrierbares Alkali, 5761
titrierbare Säure, 5760
Titrierung, 5762
Toilettenpapier, 5763
Toilettenpapierrolle, 5764
Toilettenpapierschneide-und
   Perforiermaschine, 5765
Toiletten-Tissue, 5766
Toilettentissue, 4899
Toleranz, 5767
Toluol, 5768
Ton, 1344
Tonband, 3660
tönen, 3103, 5750
Tönen, 5753
Toner, 5770
Tonerde, 1344
Tonne, 584, 5769
Tönung, 3103
Tönungsmittel, 5770
Topfscheibenzerspaner, 2112
Torsionsspannung, 5039
Totaldurchgang, 5787
Tour, 4782
Tourenzähler, 5627
Toxizität, 5795
Tracheide, 5797
Tragant, 2947
Tragantgummi, 2947, 5799
Träger, 5567
Trägerpapier, 5495
Trägerpapier für Asphaltimprägnierung, 808
Trägerstoff, 794
Trägfähigkeit, 961
Tragfläche, 267, 3136
Tragflügel, 267
Tragtrommelaufwickler, 2116
Tragtrommelroller, 4438
Tragtrommelwickler, 2116, 4438
Tragwalze, 3347
Tränken, 5205
Tränkung, 3199
Transfereinrichtung mit glatter
   Abnahmewalze, 4349
Transformator, 1554
Transistor, 5808
transparent, 5818
transparente Flecken im Papier, 6135
Transparenz, 3570, 5815, 5816
Transpiration, 5821
Transplantation, 5822
Transport, 5823
Transportband, 1563
Transporteur, 1562
Transportgerät, 2406
transportieren, 5823
Transportkette, 1221
Transportschnecke, 1564
Transversalflusspresse, 5824
Traubenzucker, 1936
treiben, 2103
Treibmittel, 2631
Treibstoff, 2743
trennen, 1764
Trennmittel, 4730
Trennschärfe, 4987
Trennung, 5003
Trennvorrichtung, 5004
Trichter, 1665, 3086
Tricyanverbindungen, 1782
Triebkraft, 3898
triften, 2103
Trinatriumphosphat, 5853
Trinkbecherpapier, 1741
Triplex, 5851
Triplex-Karton, 5852
Triplexkarton, 3327
trocken, 2117

Trockenabriebfestigkeit, 2142
Trockenausschuss, 2118
Trockenbatteriekarton, 612
Trockenboden, 2139
Trockendestillation, 1100
trockener Zustand, 2141
trockenes Wachspapier, 2143
Trockenfestigkeit, 2144
Trockenfestigkeitsmittel, 2145
Trockenfilz, 2123, 2131
Trockenfilzmarkierung, 2124
Trockenformierung (Rundsieb), 2146
Trockenformierungsmethode, 2133
Trockengehalt, 2141
Trockenglättwerk, 1052
Trockengradeinteilung, 4923
Trockenhaube, 2125
Trockenheit, 2141
Trockenkammer, 2140, 4379
Trockenkreppung, 2120
Trockenleistung, 2192
trocken mahlen, 2972
Trockenmittel, 2135
Trockenofen, 4097
Trockenpartie, 2121, 2126, 2129
Trockenspeicher, 2139, 3552
Trockenstange, 2130
Trockenstreichen (ohne Lösungsmittel),
   2119
Trockentasche, 2128
Trockenteil, 2126
trocken werden, 2117
Trockenzylinder, 1791, 2122, 2138
trocknen, 2117
Trocknen, 2134
trocknend, 2134
Trockner, 2122
Trocknungsanlage, 3345
Trocknungsgeschwindigkeit, 4643
Trocknungsleistung, 1921, 2135
Trocknungsmittel, 2136
Trocknungsvermögen, 2136
Trog, 5829
trogloses Schleifen, 4345
Trogschleifen, 4347
Trommel, 584, 2110
Trompetenbaum, 1155
Tröpfchen, 929
Tropfen, 4985
Tropffilter, 5836
tropfsicherer Behälter, 3444
Tropfstreicher, 615
tropische Pflanze, 5854
trübe, 1368, 5872
Trübung, 788, 2992, 4059, 5874
Trübungsmesser, 4058, 5873
Trübungsmittel, 4057
Tsuga, 5858
Tuch, 2370
Tulpenbaum, 3525, 5868, 6318
Tungöl, 5869
Tupfen, 5305
Turbine, 5875
Turbogebläse, 1208, 1212
Turbogebläsepumpe, 2387
Turbogenerator, 5876
Turbulenz, 5877
Turm, 5794
Tüte, 526
Tütenfutter, 537
Tütenpapier, 540
Twin-wire Maschine, 5886
Tylose, 5898

Übergabepresse, 5806
Übergang, 1991, 5804
Übergangspunkt, 5809
Übergewicht, 4112
übergewichtige Palette, 4114
Übergrösse, 4111
überheizt, 5557
überhitzt, 5557
überhitzter Zellstoff, 968
Überhitzung, 5558
Überholtsein, 4016
überkrusten, 2270
überlagern, 4107
Überlandleitung, 5813
Überlauf, 4105
überlaufen, 4105
Überlaufkasten, 4718
Übermass, 4111
übermässiges Kochen, 4101
Übernahme der Papierbahn durch den
   Oberfilz ohne Anwendung von
   Vakuum, 3464
Überprüfen, 1234
überschreiten, 4110
überschwemmen, 2582, 4110
Überstand, 5560
überströmen, 4105
übertragen, 1561
übertragen (Druck), 5823
Übertragung, 1548, 5804
Übertragungsleitung, 5813
Übertreibgas, 4736
übertrocken, 4104
übertrocknet, 4103
übertrocknetes Papier, 968
überwachen, 3887
Überwachung, 3888, 5563
Überziehen, 4100
überziehen, 4107
überzogen, 1384
Überzug, 1382, 1654
Überzugspapier, 2376
Ulme, 2246, 5902
Ulmenholz, 2246
ultraviolette Strahlung, 5903
Umbau, 4671
umbiegen, 1669
Umbiegen, 616
umbörteln, 1695
umbrechen (Druck), 4110
Umdrehung, 4782, 4856
Umdrehungen pro Minute, 4783
Umdrehungsanzeiger, 5314
Umdruck, 5804
Umdruckpapier, 5805
Umfangs-, 4265
umfangsbedingte, 1331
Umfangsgeschwindigkeit, 1332
Umfangspasser, 1331
Umformer, 1554, 2213, 5807
Umgebung, 2293
umgehen mit, 2961
umgeklappt, 5246
umgeknickte Kante, 5879
umgelegte Kante, 5879
umgewandelte Stärke, 1553
umgewandelte Stärken, 3851
umgürten, 2812
umhüllen, 1654
Umhüllung, 1656, 3306
Umhüllung des Abwasserrohrs, 1427
Umkehrung, 4781
Umkehrwalzenbeschichter, 4779
Umkehrwalzenbeschichtung, 4780
Umkleidung, 1656, 3259
Umkleidungspapier, 5041
umkreisen, 4802
Umlauf, 4782, 4856
umlaufender Filz, 2273
Umlaufholländer, 1784
Umlaufpumpe, 4851
Umlufttrockner, 5738
ummanteln, 1654
Ummantelung, 3259
Ummantelungspapier, 5041
Umpflanzung, 5822
umringen, 4802
umrollen, 4786
Umrollen, 4788

U

Überchlorsäure, 4254
überchlorsaure Salze, 4253
überdimensionieren, 4111
Überdrucken, 4109
übereinanderlegen, 3419
überfliessen, 2582, 4105
überfluten, 2582, 4110
überführen, 1561
Überführung, 1551

Umrühren, 5429
Umsatz, 5881
umschalten, 5605
Umschlag, 1654
Umschlagpapier, 1658
Umschlagpapier (das durch ein
　Trockenglättwerk gelaufen ist), 3621
Umsetzung, 1551
Umspanner, 5807
Umsteuergrösse, 3852
Umsteuerung, 4781
Umwandler, 1554, 3852, 5803
Umwandlung, 1551, 1555, 3935
Umwandlung von Cellulose in
　Alkalicellulose, 5416
Umwelt, 2293
Umweltschutz, 4405
umwenden, 5881
umwickeln, 3419, 6234
Umwickeln, 5319
Umwicklung mit Band, 5647
unausgefüllt, 6028
unbearbeiteter Stoff, 6014
unbedrucktes Buchdruckpapier, 727
unbelastet, 5922
unbeschnitten, 3822
unbeschnittener Karton, 1851
unbeschnittenes Papier, 5927
unbeschwert, 5922
unbeständig, 5968
unbrauchbar, 6056
Undichtheit, 3443
undurchdringlich, 3194
Undurchdringlichkeit, 3193
Undurchlässigkeit, 3193
Unebenheit, 4861
unecht, 795
unechte Akazie, 3546
Unempfindlichkeit, 2977
Unempfindlichmachen, 1920
unfertig, 5916
ungebleicht, 5904
ungebleichter Zellstoff, 5906
ungebleichtes Papier, 5905
ungefalztes Papier, 2747
ungeleimt, 5925
ungeleimtes Papier, 5926, 6080
ungenügend aufgeschlossen, 5913
ungeschlechtliche Fortpflanzung, 455
ungestrichen, 5908
ungestrichener Karton, 4222
ungestrichenes Buchpapier, 5909
ungestrichenes Druckpapier, 5911
ungestrichenes Plakatpapier, 729
Ungleichförmigkeit, 3035
ungleichmässige Durchsicht, 1367
unklar, 1368
Unkosten, 1622
unreines Soda, 718
unsatiniert, 5916, 5917
unsortiertes Altpapier, 3839
unterbrechen, 882
Unterbrechung, 2084
Unterbringung, 3102
unterchlorige Säure, 3173
unterchlorigsaure Salze, 3172
Unterdruck, 3579
unter Druck setzen, 4508
untere Gautschwalze, 851
untere Presse, 854
untere Presswalze, 855
unteres Stammende (Baum), 987
untere Walze, 856
unterfahren, 5915
Unterfilz, 852
Unterfilzpresse, 853
Untergewicht, 5915
Unterklebepapier, 507
Unterlage, 505, 4141, 5567
Unterlagen, 1821
Unterlegbohlen, 2152
Unterlegpapier, 510
Untermesser eines Querschneiders, 646
Unternehmensforschung, 4070
unterstützen, 503
Unterstützung, 5567
Untersuchung, 3253, 5696
Untertauchen, 5492
Unterwalze (eines Kalanders), 3347

ununterbrochen, 1531
unveränderliche Geschwindigkeit, 1517
unvermischbar, 3186
unvollständiges Ries, 908
Urethan, 5937

V

Vakuum, 5940
Vakuumabnahme, 5507, 5947
Vakuumbahnüberführung, 5950
Vakuumfilter, 5943
Vakuumfiltrierung, 5944
Vakuumflachsauger, 5941
Vakuum-Flachsaugkasten, 5941
Vakuumformierung, 5506
Vakuumpumpe, 5512
Vakuumumlader, 5942
Vakuumverformung, 5945
Vakuumzufuhr, 5504
Valenz, 817
Vanillin, 5955
Vanillinderivate, 5956
variabel, 5968
Variable, 5968
Varianzanalyse, 365
Velin, 5983
Velinausrüstung, 5985
Velinausrüstung mit besonders glatter
　Oberfläche, 1056
Velin-Egoutteur, 6292
Velinkarton, 5984
Velinoberfläche, 5985
Velinpapier, 5986, 6291, 6295
Velinpapier-Schöpfform, 6293
Velourspapier, 2578, 5990
Venta-Nip Presse, 4437
Ventil, 5952
Ventilator, 772, 2384, 5997
Ventilatorflügel, 2385
Ventilatorrad, 3191
ventilieren, 2384, 5995
Ventilsack, 5953
Venturi-Düse, 5999
Verabeiten in der Maschine, 4052
Verabeitung des Holzes am Wegerand,
　4811
Veralten, 4016
veränderlich, 5968
Veränderlichkeit, 5967
verankern, 1650
verarbeiten, 4532
verarbeiten, 4534
verarbeitende Industrie, 1556
Verarbeiten in einem Arbeitsgang, 4052
Verarbeitung, 1555
Verarbeitungsbetrieb, 1559
verarbeitungsfähig, 3605
verarbeitungsfähiges Papier, 1558
Verarbeitungsmaschine, 1557
Verarbeitungsmethode, 1560, 4532
Veräusserung, 2031
Verbeulung, 944
verbiegen, 5444
verbinden, 817
Verbinden, 819
Verbindung mit Kohlensäure, 1095
Verbindung mit Wasser, 3117
Verbindungsdraht, 3402
Verbindungsgewicht, 2301
Verbindungshilfsmittel, 1652
Verbindungsklammer, 873
Verbindungsstelle, 3314
Verbindungsstück, 1651
verblassen, 2379
verbleien, 3437
Verblendung, 2377
Verbot, 3227
verbrannt, 968
Verbrauch, 1520
Verbreitung, 2042, 5352
Verbrennen, 1466
Verbrennung, 1466
Verbrennungsprodukte, 1468
Verbrennungsraum, 1467

Verbrennungswärme, 3012
verbuchen, 826
verbundene Rückgewinnung (z.B. Sulfat
　und NSSC), 1710
Verbundkarton, 1461
verdampfen, 2331
Verdampfer, 1495, 1887, 2333
Verdampfung, 2332, 5407, 5959
verdichten, 4959, 4962
Verdichter, 1473
Verdichtung, 1472, 1478, 1902
Verdickungsmittel, 5719
Verdickungsmittel für Papierverleimung,
　2931
verdrückt, 1718
verdünnen, 1990
Verdünnung, 481, 1991, 5725
Verdünnungsmittel, 1989
verdunsten, 2331
Veredeln, 4534
Veredlung, 1555
Veredlungsanlage, 1559
Veredlungsindustrie, 1556
Veredlungsmethode, 1560
Vereinbarkeit, 1474
Vererbbarkeit, 3033
Vererbungslehre, 2809
Veresterung, 2309
Verfahren, 4532, 5618, 5665
Verfahrensgrösse, 365
Verfahrenstechnik, 3778, 5619
Verfall, 1841
Verfärbung, 2018
verfaulen, 1841
verfaultes Holz, 1843
Verfestigung, 5262
Verfilzen, 2425
Verflüchtigung, 6032
verflüssigen, 1990
Verformbarkeit, 3872
Verformung, 1876, 2041
Verfrachten, 3543
Verfrachtung, 5823
Vergasung, 2787
vergiessen, 2925
Vergiessen, 1152
Vergiftung, 1530
Vergilben, 6317
Verglasung, 6027
Vergleichskolorimeter, 1451
Verhalten in der Wärme, 5707
Verhältnis, 4644
Verhältnis von Längs-zu Querrichtung in
　Zerreissfestigkeit, 5661
Verhältnis von Stoffen zueinander, 3714
Verhältnis zwischen Holz und
　Chemikalienzugabe (bei der
　Sulfatkochung), 1259
Verhältnis zwischen Poren/Luftblasenvolu-
　men zum Gesamtvolumen des Papiers,
　6029
Verhaltung, 4769
Verholzung, 3479
Verholzungsgrad, 3481
Verjüngung, 5646
Verkauf, 4885
verkleben, 817
Verkleben, 819
Verklebung, 817
Verklebung eines Rollenendes während
　der laufenden Produktion, 5327
Verkleidung, 1148, 2377, 3306, 4144
Verknittern, 1675
Verkohlung, 1100
Verladen, 3543
Verlader, 3542
Verlagerung, 5059
Verlangsamung, 4768
Verlauf, 4532
verlaufen, 751
Verlaufmittel, 3457
Verlegen des Filzes, 2484
verlegter Filz, 2488
verlegt (Filz), 2485
verlorengehen, 6056
Verlust, 3574
Verlust an Stoffstandshöhe, 2998
vermengen, 756, 3837
Vermessung, 3748

Verminderung, 481
vermodern, 1841
Vermögen, 4552
Vernebeln, 2637
Verneblung, 479
Vernetzbarkeit, 6207
Vernetzung, 1708
Vernichtung, 2031
Vernietung von Wellpappenlagen, 4324
Verpacken, 4132, 4137
verpackter Papierstapel, 1145
Verpackung, 4130, 6297
Verpackung aus Fasergussmasse, 3876
Verpackung mit Stahlband, 5448
Verpackungsform, 4131
Verpackungsgewicht, 5651
Verpackungsindustrie, 4133
Verpackungsmaschine, 4134, 6298
Verpackungsmaterial, 4135
Verpackungsmuster, 4131
Verpackungspapier, 4136, 6299
Verpackungspapier aus
　Sekundärfaserstoff, 811
Verpflanzung, 5822
Verpressbarkeit, 3872
verpuffen, 1663
Verrichtung, 4260
verrussen, 5276
Versagen, 2383, 3670
Versagen des Leims, 2845
Versandbehälter, 5061
versanden, 5096
versandfertige Lagen Nasszellstoff, 3419
Versandschachtel, 1134
Verschalung, 5049
Verschiebung, 5059
Verschiebung des geschnittenen Papiers
　beim Randbeschnitt, 2096
Verschiedenheit, 5971
verschiessen, 2379
Verschleiss, 6138
Verschleissfestigkeit, 4760
Verschleissprüfgerät, 6139
Verschleissprüfung, 6140
Verschluss, 4959
Verschlussmaschine, 4964
verschmutzen, 1529
Verschmutzung, 1530, 4404
Verschmutzung durch Hitze, 5706
Verschmutzung durch Wärme, 5706
Verschnittmittel, 2353
verschroten, 914
verschrotten, 4940
Verschweissbarkeit, 4960
verschweissen, 4962
verschwenden, 6056
Verseifung, 4902
Verseifungszahl, 4903
versetzt, 4024
Versiegelbarkeit, 4960
versiegeln, 4959, 4962
versorgen, 5565
versprühen, 5341
verstärken, 873
verstärkte Kunststoffe, 4723
verstärktes Papier, 4722
verstärktes Tissue, 5328
Verstäuben, 2168
verstäuben, 5341
versteifen, 873
verstellbarer Schraubenschlüssel, 1691
verstopfen, 1354, 4382, 4383
verstopfter Filz, 2488
Verstopfung, 1354
Verstreichwalze, 5199
Versuchspapiermaschine, 2349
Verteiler, 422, 5349
Verteiler (im Stoffauflauf), 2596
Verteilerrohr, 2996
Verteilerstück, 3676
Verteilerwalze, 2043
Verteilung, 2042, 5352
Vertikalformierungseinrichtung, 6002
Vertikall-Zellstoffkocher, 6000
Vertikalpresse, 6001
Verträglichkeit, 1474
Vertrieb, 3696
Vertrieb (von Waren), 2042
verunreinigen, 1529, 5256
Verunreinigung, 1530, 2012, 3200,

4404
Vervielfältigungspapier, 2161, 3677, 4744
Verwalter, 5559
verwaschen, 751
Verweilzeit, 2169, 3070
Verwerfen, 6047
Verwerfen (von Papier), 1754, 6051
Verwertung, 5939
verzahnen, 3313
Verzerrung, 2041
Verzierung, 616
Verzögerung, 4768
Verzögerungsmittel, 4767
Vibration, 4089, 5027, 6005
Vibrationssortierer, 6004
Vickery-Filzinstandhalter, 6007
vielfach, 3676
Vielfalt, 5971
Vielkammersaugwalze, 3910
Vielsiebmaschine, 3932
vierundzwanzig Bogen (aus einem 480er Ries - Grobpapier), 4610
Vinylacetat, 6008
Vinylcellulose, 6009
Vinylchlorid, 6010
Vinylidenchlorid, 6012
Vinylverbindungen, 6011
Viskoelastizität, 6015
Viskometer, 6016, 6022
Viskose, 6018
Viskosefaser, 6019, 6021
Viskosezellstoff, 6020
Viskosimetrie, 6017
Viskosität, 6023
Viskosität (bei bestimmtem Geschwindigkeitsgefälle), 416
Viskositäts-Index, 6025
Viskositäts-Reguliermittel, 6024
Vliesstoff, 4007
voll, 2485
vollgeleimt, 2984
vollgeleimtes Papier, 2985
Vollmacht, 4464
Vollpappe, 1123, 1524, 5261
Vollwellpappe, 1610
Vollzellstoff, 2745
Vollzellstoffkarton, 5264
Voltmessung, 1639
Voltmeter, 6033
Volumen, 6034
Volumeneinheit für Faserholz, 1738
Volumenkarton (nicht kalandriert), 950
Volumenvergrösserung, 1987
Volumen (von Papier oder Karton), 945
voluminös, 957
voluminöse Pappe, 950
voluminöser Karton, 950
voluminöses Buchdruckpapier, 3038
voluminöses Einschlagpapier, 805
voluminöses Papier, 952, 3040
von der Kante aus in das Papier (die Bahn) hineinverlaufend, 2867
Vorarbeiter, 2665
Vorbehandlung, 4511, 4517
Vorbereitung, 4479
Vorbereitung der Baustelle, 5119
Vorbereitung des Geländes, 5119
vorbeugende Wartung, 4512
Vorderkante, 2736
Vorderseite, 2737
Vorfertigung, 4476
Vorfluter, 4094
vorformen, 4477
Vorgautschwalze, 5365
Vorhangstreichen, 1760
Vorhydrolyse, 4478
Vorlage, 3848, 4230
Vorpresse, 2533, 4515
Vorratsbehälter, 5442
Vorratsbunker, 689
Vorsatzpapier, 830
Vorschrift, 2009
Vorschub, 2411
Vorsortiersystem, 133
Vorsortierung, 4516
Vorstellung, 4260
Vorstreichen, 4474
Vorteil, 4546

Vortrockner, 2531, 4475
Vorwärtsregelung, 2408
Vulkanfiber, 6036
Vulkanfiberstoff, 6037
Vulkanisieren, 6038
vulkanisierfähiges Papier, 6039

W

wabenförmige Mittellage, 3082
Wabenkern, 3082
Wacholderbaum, 3322, 3323
Wachs, 6121
Wachsauftragsmaschine, 6130
Wachsbeschichtung, 6122
Wachsbeschichtungspapier, 6131
Wachsemulsion, 6128
Wachsen, 6129
wachsgeleimtes Papier, 6134
Wachsimprägnierpapier, 6131
Wachsimprägniertissue, 6132
Wachskarton, 6123
Wachskaschiermaschine, 6130
Wachskraftpapier, 6125
Wachsleim, 6133
Wachspapier, 6126
Wachspergamin, 6124
Wachstest, 6136
Wachstissue, 6127
Wachstum, 2926
Wachstumsrate, 2928
Wachstumszone, 2929
Wächter, 2930
Wägen, 6167
Wagenladung, 1127
Wagenplane, 1125
Wagenplanierung, 1126
Waggonladung, 1127
Wahl, 4986
wählen, 1939
Wald, 2666, 6260
Waldarbeiter, 2415, 3587
Waldbau, 5099
Waldbewirtschaftung, 2673
Waldbrand, 2669
Waldgebiet, 6260
Waldkrone, 2668
Waldstreu, 2672
Waldwirtschaft, 2671
Walnuss, 3320
Walze, 4816
Walze mit glatter Oberfläche, 4350
walzen, 3816
Walzen-Absenkvorrichtung, 4826
Walzenabstreifer, 4821
Walzenauftragsmaschine, 4825
Walzenbombierung, 4820
Walzenlager, 4824
Walzenmarkierung, 509
Walzenstreichen, 4819
Walzenstreicher, 4818
Walzentrockner, 2113
Walzwerk, 1025
Wandeffekt, 6044
wandern, 1678
Wanderung (z.B. des Bindemittels), 3811
Warenzeichen, 3694
Wärme, 3000
Wärmeabführung, 3023
Wärmeanzeiger, 5711
wärmeaufnehmend, 2275
wärmeaufzehrend, 2275
Wärmeausdehnung, 5704
Wärmeausgleich, 3002
Wärmeausnutzung, 3005
Wärmeaustauscher, 3006
Wärmebehandlung, 3024
Wärmebeständigkeit, 5709
Wärmebilanz, 3002
Wärmeempfindlichkeit, 5708
Wärmefestigkeit, 3014
Wärmegefälle, 3007
Wärmehaushalt, 3002
Wärmeisolierung, 3010, 5705
Wärmeleitfähigkeit, 3003, 5703

Wärmeleitzahl, 3022
Wärmemesser, 1062
Wärmemessung, 1063
Wärmemitführung, 1548
Wärmerückgewinnung, 3013
Wärmeschrank, 2140
Wärmeschutz, 3261
Wärmespannung, 5710
Wärmespeicher, 3001
Wärmeübergang, 3021
Wärmeübergangszahl, 3022
Wärmeübertragung, 1550, 3021, 3023
Wärmeverbrauch, 3004
Wärmevergütung, 3024
Wärmeverlust, 3011
warmfest, 3054
warmwalzen, 3099
Warngerät, 3887
Wartung, 3662
Wartungs- und Instandsetzungswerkstatt, 3663
Waschanlage, 6052
Waschen, 6053
Wäscher, 4954
Waschgrube, 778, 1972
Waschmittel, 1928
Waschtrommel, 2115, 6054
Waschturm, 4955
Waschvorgang, 1973
Wasser, 6063
Wasserabdichtung, 6106
wasserabstossend, 3151, 4742
wasserabstossender Stoff, 4742
Wasserabstreifer, 3137, 6069
wasserabweisender Effekt, 6101
Wasseramt, 6083
Wasseranalyse, 6066
wasseranziehend, 3150
Wasseraufbereitung, 6113
Wasseraufnahme, 6065
wasseraufnahmefähig, 6064
Wasseraufnahmefähigkeit, 3170
Wasserbecken zur Aufnahme von Schleiferholz, 3560
Wasserbeseitigung, 6100
Wasserbeständigkeit, 6102
wasserbindend, 3150
Wasserdampf, 5402, 6116
Wasserdampfdurchlässigkeit, 6117
wasserdampfundurchlässig, 3865
Wasserdampfundurchlässigkeit, 3866
wasserdicht, 6093
wasserdichter Abschluss, 6106
wasserdichter Karton, 6094
wasserdichtes Papier, 6097
Wasserdichtigkeit, 6096
Wasserdichtmachen, 6095
Wasserdruck, 3121
Wasserdurchlässigkeit, 6091
Wassereinzugsgebiet, 6107
Wasserentfernung, 6100
Wasser enthärten, 5245
Wasserenthärtung, 6108
wasserentziehendes Reagens, 1887
Wasserfestigkeit, 6102
Wasserflecken, 6110
Wasserführung, 6078
wassergekühlte Federwalze (am Glättwerk), 6118
Wassergüte, 6098
Wasserhahn, 1419
wasserhaltig, 426, 3114
wasserhaltiger Stoff, 3115
wässerig, 426
Wasserkasten, 6068
Wasserkraft, 3132
Wasserkraftwerk, 3133
Wasserlagerung von Holz, 3394
wasserlöslich, 6109
wassermeidend, 3151
wässern, 6063
Wasserqualität, 6098
Wasserquelle, 6103
Wasserrad, 6118
Wasserretention, 6104
Wassersäule, 6081
Wasserscheide, 6107
Wasserschlauch, 3091
Wasserstoffanlagerung, 3139

Wasserstoffbindung, 3140
Wasserstoffionenkonzentration, 3143
Wasserstoffperoxid, 3144
Wasserstoffverbindung, 3141
Wasserstoffverbindungen, 3142
Wasserstoffzahl, 3143
Wasserstrahl, 6079
Wasserstrahlentrindung, 3120, 3125
Wasserstreifen, 6111
Wasserturbine, 6115, 6118
Wasseruntersuchung, 6066
Wasserverbrauch, 6071
Wasserverschmutzung, 6092
Wasserverseuchung, 6092
Wasserversorgung, 6112
Wasserverwertung, 6099
Wasserzeichen, 6084
Wasserzeichen-Egoutteur, 3702, 6085, 6090
Wasserzeichenpapier, 6089
Wasserzeichenprägewalze, 6087
Wasserzeichenwebegoutteur, 6088
wässriges Calciumsulfat, 1714
Wattebausch, 6040
Watteeinlagestoff, 6042
wattieren, 4141
Wattierung, 6041
Webart, 6144
Webeffekt, 1363
Weben, 6145
Webstoff, 6294
Webstruktur, 6144
Webstruktur im Papier, 1363
Wechsel, 5059
Wechselbeanspruchung, 1786
Wechselgetriebe, 2800
Wechselpapier, 688
Wechselstrom, 100, 326
Wegschlagen (einer Druckfarbe), 5470
Wehr, 1804, 6173
weich, 5241
weicher Stoffbatzen, 5247
weicher Stoff (Zellstoff), 5251
weiches Papier, 5250
Weichharze, 4050
Weichheit, 5249
weichmachen, 4364, 5243
Weichmacher, 4363, 5242
Weide, 4887, 6233
Weissbirke, 6218
Weissbuche, 3088
Weisse, 6222
weisse Birke, 6218
weisser einseitig kaschierter Schrenzkarton, 5114
weisses Papier, 6223
Weissfichte, 6224
Weissgehalt, 6222
Weissgrad, 896, 6222
Weissgradabfall, 897
Weissholz, 5688
Weisslaugenschlamm, 6220
Weisspigment, 5752
Weisstanne, 6219
weiterleiten, 1561
Weiterreissfestigkeit, 5659
weitlumiges Holz, 5355
Weizenstärke, 6216
Weizenstroh, 6217
Wellblech, 1612
Welle, 5026, 6240
Wellen, 1614, 4808
Wellenhöhe (Wellpappe), 3026
Wellenmaschine, 1619
Wellenmaterial, 2622
Wellenpapier, 2622
Wellenstoff, 2621
wellig, 536
wellige Kante, 6120
wellige Oberfläche, 1423
welliges Papier, 1422, 1756
Welligkeit, 6051, 6119
Wellpackpapier (für zerbrechliches Gut), 1613
Wellpappe, 1607
Wellpappenmaschine, 1616
Wellpappenrohstoff, 1615
Wellpappenwelle, 2619
Wellpappenwelle aus Halbzellstoff, 4992

Wellpappenwelle aus Sekundärfaserstoff, 799
Wellpappe von 140-150 g/m2 Flächengewicht, 3971
Wendeantrieb, 4775
Wendepresse, 4777
Wendepressenfilz, 4778
Wenzelwalze (Stoffauflauf), 3078
Werk, 3816, 4353
Werkdruckpapier, 837, 951
Werksdirektor, 3827
Werkseinschlagpapier, 3829
Werksleiter, 3827
Werksmechaniker, 3830
Werkstoff, 1518, 3713
Werkzeug, 1954, 5771
Werkzeug zur Oberflächenbehandlung von Schleifsteinen, 969
Wertigkeit, 5951
Wertminderung, 1916
Wertpapier, 822
Wertverlust, 1916
Wertzeichenpapier, 567
wetterbeständig, 6141
Wetterbeständigkeit, 6143
wetterfest, 6141
wetterfest ausrüsten, 6142
Wichte, 1905
Wickelhülsenpappe, 1586
wickeln, 6234
Wickeln, 6238
Wickelpappe, 5866, 6192
Wicklung, 6234
Widerstand, 4758
Widerstandsfähigkeit, 2977, 4758
wiederanfeuchten, 4785
Wiederanfeuchten und Kochen von Zellstoff, 4746
Wiederaufbau, 4671
Wiederaufforstung, 4711
Wiedergewinnung, 4675
wiederverwenden, 4774
Wiederverwendung des Wassers, 6105
wiederverwerten, 4890
wiegen, 4922
Wiegen, 6167
Wiegetrichter, 6168
wild, 6229
wilde Blattformierung (am Stoffauflauf), 1369
wilde Formierung (in der Blattbildung), 1366
Wildwuchsbäume, 6163
Winden, 5888
Windkessel, 251
Windmessgerät, 369
Windung, 5785, 6234
Winkelblech, 2950
Wirbel, 6035
Wirbelschicht, 2609
Wirbelschichter, 2610
Wirbelschichttechnik, 2608
Wirbelschichtverfahren, 2608
Wirbelsichter, 1350, 1789, 3130
Wirbelströmung, 5878
Wirken, 6145
Wirkstoff, 6294
Wirrfaserlage, 4640
Wirrfaservlies, 4641
wirrverlegt, 277
Wirtschaftsanalyse, 2179
Wischtücher, 3218
Wischwasser (beim Offsetdruck), 2702
Witherit, 572
witterungsbeständig, 6141
Wölbung, 934, 1712
Wolframverbindungen, 5870
wolkig, 1368, 6229
wolkige Blattbildung, 6230
Wolkigkeit (Pap.), 6232
Wolkigkeit (vom Papier), 1367
Wolle, 6285
Wollfilz, 6288
Wollflocke, 2575
Wollpapier, 6287
Wulst, 616
Wurzel, 4834
Wurzel (math.), 4614
Wurzelziehen (math.), 2357

X

Xanthatgruppen, 6303
Xerographie-Papiere, 2238
Xylan, 6304
Xylole, 6305
Xylolsulfonate, 6306
Xylose, 6307

Z

Zacken, 5772
zäher Fluss, 6026
zähflüssig Fliessen, 6026
Zähflüssigkeit, 5788, 6023
Zähigkeit, 5788
zählen, 1640, 4931
Zählen, 1649
Zahlensteuerung, 1984
Zähler, 1641, 3208, 3770
Zählung, 1649
Zählung einer Lieferung durch das Werk, 3821
Zählwerk, 1641, 3770
Zahn, 5772
Zahnabdruckpapier, 437
Zapfen, 3319
Zapfen (bot.), 1505
Zapfenegoutteur, 1360
Zeder, 1174
Zedernholz, 1174
Zedernöl, 1174
Zeichenkohle, 1228
Zeichenpapier, 2095, 2099
Zeilengussetzmaschine, 3512
(Zeilen) hinübernehemen, 4110
Zeilensetzmaschine, 3512
Zeit, 5746
Zeitdauer, 3448, 5746
Zeitmesser, 5747
Zeitschriftenpapier, 3641
Zeitung, 3965
Zeitungsausschnitt, 1775
Zeitungsdruckpapier, 3960, 3966
Zeitungsdruckpapierbogen, 3967
Zeitungsdruckpapier mit glatter Oberfläche, 5561
Zeitungsdruckpapierqualität, 3963
Zeitungsdruckpapierrollen (aus einer kontinuierlichen Bahn), 6153
Zelle, 1177
zellenartiger Kern, 3082
zellenartige Struktur, 3083
Zellenaufbau, 1179
Zellenstruktur, 1179
Zellglas, 1178, 1192
Zellpech (im Zellstoff), 4340
Zellstoff, 1261, 4568
Zellstofffabrik, 4577
Zellstoffabsatz, 4580
Zellstoffaufschluss mit hoher Ausbeute, 3061
Zellstoff aus Kurzfasern, 5073
Zellstoffbeständigkeit (gegen verschiedenste Einflüsse), 5368
Zellstoffblatt, 4582
Zellstoffbogen, 4582
Zellstoffgehalt, 4570
Zellstoffgeschäft, 4580
Zellstoffharz, 4340
Zellstoffharzkontrolle, 4341
Zellstoffindustrie, 4573
Zellstoffkarton, 4569
Zellstoffkocher, 1976
Zellstoffflecken, 2534
Zellstoffpappe, 4569
Zellstoffprobe, 4582
Zellstoffqualität, 4579
Zellstoffverkauf, 4580
Zellstoffwatte, 1197

Zellstruktur, 1181
Zellstrukturpappe, 1180
Zellulose, 1182
Zellulose-Lösungsmittel, 1195
Zellwand, 1199
Zellwandstärke, 1200
Zementsackpapier, 1202
Zentipoise, 1206
zentriertes Wasserzeichen, 3545
zentrifugal, 1207
Zentrifugalfiltrierung, 1210
Zentrifugalgebläse, 1208
Zentrifugalsortierer, 1209, 1211
Zentrifuge, 1214
zentrifugieren, 1214
Zentrifugierung, 1213
zentripetal, 1215
Zeolite, 6323
zerbrechen, 2711
zerdrücken, 1717
Zerdrücken, 1723
zerdrückte Hülse, 1719
zerdrückte Papierrolle, 1721
zerdrücktes Papier, 1032
Zerfall, 1858, 2024
zerfallen, 1841
Zerfaserer, 2025, 4062, 4571
zerfasern, 620, 1862, 2463, 2475
Zerfaserung, 1863, 1866
zerfliessen, 1897
zerfressend, 1604
zergehen (chem.), 1897
Zerkleinerer, 4586
zerkleinern, 1717, 5080
Zerkleinern, 1723
zerkleinert, 1718
Zerkleinerungsmaschine zur Probeentnahme, 4893
Zerlegen in Spaltfäserchen, 2476
Zerlegung, 364, 5332
Zerlegung in Spaltfäserchen, 2477
zermahlen, 2475
Zermahlen, 2476
Zermahlung, 2477
Zermalmen, 1723
zerquetschen, 914
zerreiben, 914, 1717, 2878
Zerreissdehnung, 2248, 5681
zerreissen, 5080
Zerreissen, 889, 5657
Zerreissfaktor, 5658
Zerreissfestigkeit, 5658
Zerreissmühle, 5079
Zerreissprüfgerät, 5682
zerrieben, 1718
zersägen, 4917
Zersägen eines Baumes in kurze Längen, 5076
zerschlagen, 882
zerschnitzeln, 5080
zersetzen, 1841
Zersetzung, 1858, 2024
Zerspanungsmesser, 1292
Zersplitterung, 2712
zerspringen, 972
zerstäuben, 5341
Zerstäuben, 3834, 5348
Zerstäuber, 4586, 5292, 5347
zerstäubte Flüssigkeit, 5341
Zerstäubung, 479
Zerstäubungstrocknung, 5345
Zerstörung (durch Schimmelbildung), 3219
zerstörungsfreie Materialprüfung, 4000
Zerstreuen, 4927
zerstreutporiges Holz, 1970
Zerstreuung, 2029
Zetapotential, 6325
Zettel, 5164
Zickzackfaltung, 1489
Zickzackfalz, 140
Zickzackfalzung, 140
Zickzackwatte, 1683
Ziehpappe, 3873
Ziffernblatt, 1939
Zigarettenpapier, 1326, 1465
Zigarettenseidenpapier, 1327
Zink, 6326
Zinkasche, 6329

Zinke, 5772
Zinkhydrosulfit, 6328
Zinkoxid, 6329
Zinkstearat, 6330
Zinksulfat, 6331
Zinksulfid, 6332
Zinksulfidweiss, 3534
Zinkverbindungen, 6327
Zinn, 5748
Zinnverbindungen, 5749
Zirkonverbindungen, 6333
Zopfwinde beim Altpapier-Pulper, 1918
Zubehör, 139
Zubringer, 1562
Zucht, 1737
Zuckerahorn, 5518
Zuckerrohr, 5516
zu einer Suspension verflüssigen, 5184, 5185
Zufliessystem, 422
Zufluss, 421, 2588
Zufuhr, 2403
zuführen, 2403, 5565
Zufuhr mittels Vakuum, 5504
Zuführung, 2411
Zufuhr unter Hochdruck, 3053
Zufuhr unter Pressluft, 3053
Zug, 5685
Zug-, 5675
Zugabnahme, 2386
Zugänglichkeit, 138
Zugbeanspruchung, 5680
Zugeigenschaften, 5677
zugeschnittene Klebestelle, 5842
zugeschnittenes Papier, 5840
zugespitzte Kante, 2397
Zugfestigkeitsprüfer, 5679
Zugholz, 5688
Zugkraft pro Flächeneinheit um eine Papier- oder Kartonprobe zu zerreissen (Kraft wirkt senkrecht auf Probe), 6322
Zugregelung, 5686
zugrundegelegte Geschwindigkeit, 1924
Zugspannung, 2096, 5680
Zugwalze, 3069, 4567
zu lange gekocht, 4102
zulässige Abweichung, 5767
Zulieferer, 5564
Zunahme, 3205
zünden, 3470
Zündholz, 3711
Zündung, 3178
zu Pulver zerkleinern, 4462
Zurichtebogen, 4107
zurichten, 4107, 5121
Zurichtung, 2517, 3664, 4107
Zurückfedern, 5353
Zurückführung, 4769
zurückgebliebene Papierspannung während der Trocknung, 2101
zurückgebliebene Spannung während der Trocknung, 941
Zurückhalten, 4769
Zurückhalten des Wassers, 6104
Zurückhaltung, 4766
Zurückweisung, 4726
zur Verfügungstellung, 2031
zusammenbacken, 233
Zusammenbacken von Papierbogen, 764
zusammenballen, 233, 3591
zusammenbündeln, 959
Zusammenfassung einzelner Ladungen in Containereinheiten, 1526
Zusammenkleben von Papierbogen, 764
Zusammenlegen, 2646
zusammenpassen, 3711
zusammenpassend, 3712
Zusammenschluss, 3754
zusammenstellbare Containersätze, 3953
zusammenziehen, 4565
Zusatz, 1025
Zusätze, 202
Zusatzeinrichtung, 139
zusätzlich, 139
Zusatzwasser, 3666
Zuschlagstoff, 235
zuschnüren, 1584

zu stark gekocht, 4102
zustopfen, 4382, 4383
Zutat, 3226
zuteilen, 3770
Zuteilsystem, 3773
Zuteilung, 423, 4555
Zu- und Abgang von Barmitteln, 1146
Zuverlässigkeit, 4734
Zuwachs, 2926, 3205
Zuwachsrate, 2928
Zuwachsring, 2929
Zuwachszone, 2927
Zuweisung, 423
Zwangskonvektionstrockner, 3058

zweifach gestrichen, 2065
Zweifachstreicher, 2156
Zweig, 5884
zweikeimblättrig, 1952
Zweikomponentenfaser, 685
zweilagig, 5891
zweilagiger Karton, 5892
zweilagiges Kraftpapier (meist verstärkt
    mit Asphalt oder Glasfaser oder Jute),
    4720
zwei- oder beidseitiges Streichen, 5895
zweischichtig, 5891
zweiseitig, 5896
Zweiseitigkeit, 5897

Zweisiebmaschine, 5886
zweistöckig angeordnete Trockenzylinder,
    2067
zweite Presse, 4976
zweite Qualität, 4977
zweitunterste Walze im Glättwerk, 4608
zweiwertige Phenyle, 706
Zwergkiefer, 3307
Zwiebelschaleneffektpapier, 4056
Zwinge, 2437
Zwischenbehälter, 5594
Zwischenfaserbindung, 3268
Zwischenlage, 1204, 4381
Zwischenlagenkarton, 4216

Zwischenlagenpapier, 3270
Zwischenpapier, 588
Zwischenraum, 5290
Zwischenschichtmaterial, 586
Zwischenwalze, 3272
Zwischenwand, 521, 948, 3810
Zwischenwandkarton, 4216
Zyclodextrin, 1787
Zyklon, 1789
Zyklus, 1783
Zylinder, 1791
zylinderförmiger Behälter (aus Karton-
    Faserplatte-Papier-oder Metallfolie), 1068
Zypresse, 1744, 1802

english

svenska

deutsch

français

español

FRANÇAIS

A

à base de matières de récupération, 795
abattage, 2417, 5834
abattage d'arbres entiers, 2748
abattage des arbres, 5432
abattage du bois en petites dimensions, 5076
abatteuse débardeuse, 2416
abatteuse empileuse, 2415
abattre (un arbre), 2414
abrasifs, 112
abrasion, 108
absorbant, 116, 118, 125
absorber, 114
absorption, 120, 127, 3754
absorption de l'eau, 6065
absorption de l'encre, 3233
absorption de l'énergie de rupture par traction, 5676
absorption de l'huile, 4034
acacia, 128
accélérer, 5316
accessibilité, 138
accessoire, 139
accouplement, 1651
accoupler, 1650
accrocheuse sécheuse, 2442
accroissement, 2926, 3205
accumulateur, 142
accumulateur de chaleur, 3001
accumulation, 141, 5441
acétaldéhyde, 145
acétate, 150
acétate de butyl, 994
acétate de cellulose, 1183
acétate de méthyl, 3779
acétate de polyvinyl, 4426
acétate d'éthyl, 2317
acétate de vinyl, 6008
acétone, 153
acétonitrile, 154
acétyl, 146
acétylation, 155
achat, 4595
acide, 157, 5284
acide acétique, 151
acide acrylique, 189
acide ascorbique, 454
acide benzoïque, 671
acide borique, 845
acide bromhydrique, 3126
acide butyrique, 998
acide chlorhydrique, 3129
acide de cuisson, 2759
acide fluorhydrique, 3135
acide formique, 2688
acide fumarique, 2749
acide hypochloreux, 3173
acide lactique, 3392
acide levulinique, 3462
acide linoléique, 3511
acide maléique, 3668
acide méthacrylique, 3775
acide nitrique, 3981
acide oléique, 4049
acide oxalique, 4116
acide péracétique, 4252
acide perchlorique, 4254
acide phtalique, 4297
acides abiétiques, 106
acides aminés, 340
acides aromatiques, 431
acides carboxyliques, 1118
acides chloroacétiques, 1313
acides gras, 2395
acides inorganiques, 3247
acides lignosulfoniques, 3486
acides organiques, 4083
acides phénoliques, 4280

acides résiniques, 4754
acide stéarique, 5414
acide sulfamique, 5520
acide sulfureux, 5546
acide sulfurique, 5545
acide titrable, 5760
acidification, 166
acidique, 165
acidité, 167, 5286
acier, 5415
acier inoxydable, 5378
à côté de bobine, 5087, 5479
acrilate d'éthyl, 2318
acrylamide, 187
acrylate de méthyl, 3780
acrylates, 188
acrylonitrile, 192
action de brunir, 967
action de rainer, 4937
action (finances), 5430
adaptabilité, 1507
adaptation à l'humidité, 3868
additif de couchage, 1400
additifs, 202
adhérence, 204, 817
adhérence au rouleau de calandre, 765
adhérence en surface, 764
adhésif, 4961
adhésif à base de matière thermoplastique, 3095
adhésifs, 206
adhésion de la cannelure à la couverture (essai à l'aiguille), 4324
adhésion des couches, 4388, 4389
adjuvant de filtration, 2502
adjuvant de rétention, 4770
adjuvant introduit dans la masse, 6185
admissibilité, 132
adoucissement, 5243
adoucisseur, 5242, 5600
adsorbant, 216
adsorber, 214
adsorption, 217, 219
aérateur, 223, 5997
aération, 221
aérer, 220
aéroporté, 246
affaiblissement d'une teinte, 2379
affinité, 225
affleurage, 923
affleurer, 923
affutage, 5032
à fibres courtes, 5071
agalite (poudre d'amiante), 227
agar, 228
agent, 232, 4071
agent d'amélioration de la résistance à l'état sec, 2145
agent de contrôle de la viscosité, 6024
agent d'écoulement, 2589
agent de transmission, 1652
agent épaississeur, 5721
agent hydrotropique, 3154
agent oxydant, 4121
agent réducteur, 4694
agent tampon, 937
agent tensio-actif, 5593
agglomération, 234
aggloméré, 233
agitateur, 241, 3063
agitateur à secousses, 5028
agitation, 240, 5031, 5429
agiter, 239, 5429
agrafage, 5385
aide-conducteur, 517, 3627, 5731
aigre, 5284
ailette, 5954
aimant, 3654
air, 242
air comprimé, 1476
à jeter, 2030
ajouture, 988, 5325

ajouture rognée, 5842
ajutage, 4009
albumines, 286
alcali, 299
alcali actif, 196, 2189
alcali cellulose, 300
alcalinité, 313
alcali titrable, 5761
alcali total, 5786
alcool de furfuryl, 2759
alcool de polyvinyl, 4427
alcool isobutyl, 3299
alcool isopropyl, 3302
alcool méthylique, 3781
alcool propylique, 4556
alcools, 290
aldéhydes, 292
alfa, 2305
algicides, 295
alginate de sodium, 5216
alginates, 296
alignement, 297, 3510
alimentation, 2403, 2411
alimentation de la bande, 6150
alimentation en feuilles, 5045
alimentation par aspiration, 5504
alkylation, 318
alliage, 320
allongement, 1812, 2354, 5465, 5681
allongement avant rupture, 2248
allongement de rupture, 2247
allonger, 5468
alpha cellulose, 322
alpha protéine, 324
altération, 2379
altération du bois, 1841
aluminate de sodium, 5217
aluminates, 328
aluminum, 329, 330
alun, 327, 4171
alun basique, 600
amaigrir, 3602
amélioration, 667
amélioration de la production, 4538
améliorer, 5932
amiante, 446
amides, 337
amidon, 1593, 5386, 6216
amidon cationique, 1164
amidon chloruré, 1300
amidon de riz, 4792
amidon de tapioca, 5648
amidon dialdéhyde, 1941
amidon oxydé, 4127
amidon perle, 4235
amidons modifiés, 3851
amidon transformé, 1553
amines, 339
aminoéthyl cellulose, 341
aminopropyl cellulose, 343
ammoniaque, 345
amorce, 4519
amorphe, 353
amortir, 1763
ampèremètre, 344
amphotère, 356
ampoule, 243, 758, 4303
ampoule de protection, 762
amylase, 357
amylose, 358
analogue, 360
analyse, 364
analyse aux rayons X, 6308
analyse de la tension de retrait, 5462
analyse de l'eau, 6066
analyse des fibres, 2446
analyse de variation, 365
analyse économique, 2179
analyse gazeuse, 2781
analyse granulométrique par tamisage, 5089
analyse pris au hasard, 2856

analyse qualitative, 4603
analyse quantitative, 4606
analyses chimiques, 1258
analyse statistique, 5398
anatase, 367
anémomètre, 369
angiosperme, 1846
angiospermes, 370
angle de contact, 1521
angle de coupe, 5157
angle de racle d'égouttage, 2639
anhydres, 371
anhydride acétique, 152
anhydride maléique, 3669
anhydride phtalique, 4298
anhydride sulfureux, 5544
anhydrite, 372
aniline, 373
anionique, 380
anions, 382
anneau, 4802
anneau d'accroissement annuel, 383
anneau de croissance, 2929
anneaux de Raschig, 4642
annelation circulaire, 2812
annulaire, 384
anode, 386
anti-acide, 387
anti-adhérent, 4004
anti-adhésif, 4730
anti-buée, 5958
antichlores, 390
anti-corrosion, 1598
anti-dérapant, 4003
antiflexion, 392
anti-friction (métal blanc), 502
anti-mousse, 395, 2629, 2633, 2634
anti-oxydant, 4875
antioxydants, 397
anti-rouille, 405
anti-statique, 5395
anti-ternissure, 410
aplat du papier, 2553
appareil, 3255
appareil d'alimentation à basse pression, 3580
appareil d'alimentation à haute pression, 3053
appareil de contrôle, 3887
appareil de contrôle électrique, 2212
appareil de levage, 3066
appareil de mesure, 3208
appareil de mesure de la cohésion interne, 3276
appareil de mesure de la résistance de la coloration à la lumière,
appareil de mesure de la résistance des joints de collage, 2378, 2842
appareil de mesure de l'égouttage en continu, 3246
appareil de mesure de l'épair, 2686
appareil de préhension, 2910
appareil d'essai de résistance à l'éclatement, 976
appareil d'essais à l'usure, 6139
appareil d'essais d'absorption, 123
appareil d'exploration optique, 4076
appareil enregistreur, 4681
appareil Gurley de mesure de la perméabilité à l'air, 2949
appareil mesureur, 3771
appareil "MIT" de mesure de la résistance au pliage, 3835
appareil pour contrôler la résistance au gondolage, 1676
appareil pour déterminer l'état de surface d'un papier, 5202
appareil pour éliminer l'air contenu dans la pâte, 1836, 1860
appareil pour essais d'arrachage, 4312
appareil pour l'étude des fibres par fractionnement,

Français

appareil pour l'étude des fibres par la méthode du fractionnement, 614, 1343, 2453, 4572
appareil pour mesurer la résistance à la déchirure, 5663
appareil pour mesurer le bruit, 3996
appareil pour mesurer le rayonnement bêta, 677
appareil pour mesurer l'humidité, 3863
appareil Sheffield de mesure de la perméabilité à l'air, 5055
apparition de l'encre d'imprimerie à travers le papier, 3241
appel d'offres, 5672
application de contrainte, 5463
application de la colle, 2848
apprêt, 2516
apprêt de papier, 5576
apprêté, 3596, 3598, 3622, 5917
apprêté machine, 3826
appropriation, 423
approvisionnement, 5565
approvisionnement en énergie, 4468
approvisionnement en eau, 6112
appui, 5567
aptitude à la conversion en sulfure, 5531
aptitude à l'adhésion, 4960
aptitude à la moisissure, 3872
aptitude à la reproduction, 4743
aptitude à l'impression, 4527
aptitude à l'ondulation, 6119
aptitude au blanchiment, 734, 748
aptitude au blanchiment au chlore, 1311
aptitude au couchage, 1383
aptitude au drapement, 2094
aptitude au façonnage, 2679
aptitude au passage sur machine, 3605
aptitude au pelurage, 4240
aptitude au pliage, 664, 2656, 4380
aptitude au raffinage, 621
aptitude au vernissage, 5973
aqueux, 426
arabinose, 428
arbre, 5830
arbre feuillu, 1846
arbres surrégénérés, 4386
arbre (transmission), 5026
archives, 4676
archiviste, 4678
aréomètre, 3149
arête, 4796
argile, 1344
armé, 4722
arpentage, 6271
arqué, 429
arrachage, 2537, 4301, 4307, 4310
arrachage causé par la calandre, 1043, 4304
arrachage des fibres, 4306
arrachage de souches, 5488
arrachage dû au rouleau égoutteur, 1819
arrachage du couchage, 1413, 4305
arrachage en surface, 3467, 5578
arracher, 5660
arrangement, 2031
arrêt, 3176
arrêt (machine), 5083
arrivée, 3244
arrivée de pâte sur machine, 5433, 5434
arrondir, 2023
article en carton, 4156
articles en fibrociment, 2451
article tissé, 3356
arts graphiques, 2875
ascenseur, 3466
ascension, 1678
ascension capillaire, 115, 6227
aspect, 417
aspect crayeux, 1223
aspect du papier vu par transparence, 5078
aspect nuageux, 1367
aspersion, 5357
asphalte, 462
aspiration, 472, 5497
aspiration de poussières, 5503

assemblage, 235
assiettes en papier, 4180
assimilation, 473
assortiment, 5012
assortisseur, 4945, 5280
asymétrie, 475
atelier de finissage, 2519
atelier d'entretien, 3633, 3663
atelier de transformation, 1559
atmosphère, 476
atmosphère contrôlée, 1544
atomisation, 479
atomiser, 5341
attacher, 5628
attapulgite, 480
atténuation, 481
attirer, 2096
aubier, 4905
augmentation, 3205
aulne, 321
aune, 293
autoclave, 482
auto-copiant, 3936
automatisation, 487
avec barbes sur les deux bords, 2068
axial, 493
azote, 3989
azurage optique, 4078
azuré, 783

B

bacholle, 4916
bactéricides, 520
bactéries, 518
bagasse, 527
bague, 4802
bague fendue, 5330
baguette, 4813
bain de couchage, 1405, 1411, 1456
balai, 915
balance, 543, 6171
balance à papier, 4183
balle, 545, 959
balle de cassés, 906
balle de pâte humide, 4576
balle de pâte humide en feuilles repliées, 3419
bambou, 559
banc d'essai, 5694
banc de trieuse de chiffons, 5281
bande, 5447, 5472, 5643
bande adhésive, 208, 696
bande de papier pour séparer les rames, 2047
bande gommée, 208
bande gommée sur toile pour emballage, 3688
bandelettre, 5472
bande magnétique, 3660
bande perforée, 4590
bandes de papier pour séparer les rames, 4666
barrage, 1804, 6173
barre, 568, 4813
barre déplisseuse, 5350
barrer, 589
baryte, 590
barytes, 593
baryum, 571
bascule, 6171
base, 595
basicité, 602
basse pression, 3579
basse température, 3581
bassin, 4433
bassin à rondins, 3560
bassin de décantation, 5017, 5018
bassin de floculation, 2574
bassin de réserve, 5442
bassine de mouillage, 2699
bâti, 2714, 2715
bâtiment des lessiveurs, 1978
bâton, 5424
bavure (encre), 2401

bécher, 617
bénéfice, 4546
benne, 5132
bentonite, 668
benzène, 669
benzoates, 670
benzyl cellulose, 672
béril, 674
bêta cellulose, 675
béton armé, 4721
bétula, 679
biais, 681
bicarbonate de soude, 5219
bicarbonates, 684
bidon, 1068
bilan calorifique, 3002
bilan énergétique, 2278
billot, 987
biocides, 698
biodégradable, 699
biodégradation, 700
bioxyde de soufre, 1463
bioxyde de titane, 5759
biphényles, 706
bisulfates, 709
bisulfite, 180
bisulfite d'ammoniaque, 347
bisulfite de calcium, 1013, 1014
bisulfite de magnésium, 3646
bisulfite de sodium, 5220, 5229
bisulfites, 713
bisulfure de diméthyl, 1996
bitume, 462, 714
blanc fixe, 590, 726
blanc fluorescent, 2616
blanchet, 4523
blancheur, 896, 2765, 6222
blanchiment, 733, 741
blanchiment à haute densité, 3043
blanchiment en plusieurs stades, 3924
blanchiment optique, 4072
blanchir, 6226
blanchissable, 735
blanchisserie, 740
blanc neutre, 1832
blanc nuancé, 5752
blancophore, 4079
blanc satin, 4908
bleu, 783
bloc, 1370
blocage, 3703
blutoir, 6233
blutoir à chiffons, 4628, 4634
bobinage sur double tambour, 2073
bobinage sur mandrin, 1205
bobinage sur tambour porteur, 2116
bobine, 4699, 4816
bobineau, 5086
bobine brute (non rebobinée), 3321, 3828
bobine de carton, 4701
bobine de papier, 4702
bobine de papier carbone, 1101
bobine détériorée aux extrémités, 5392
bobine écrasée, 1721, 2555
bobine foirée, 5667
bobine mère, 3321, 4210
bobine molle, 3573, 4435
bobine peu serrée, 3573, 4435, 5252
bobine (pour fil), 5334
bobine qui a glissé sur son mandrin, 5166
bobiner, 6234
bobine serrée, 5741
bobine terminée, 6290
bobineuse, 6235
bobineuse à double tambour, 2071
bobineuse-refendeuse, 5171
bobineuse trancheuse, 5173
B.O.D., 703
B.O.D. (consommation d'oxygène), 499, 697
bois, 6260
bois à pores diffus, 1970
bois à pores en anneaux, 4805
bois d'automne, 3429
bois (de charpente), 5745
bois de coeur, 2999
bois de compression, 1481

bois de conifères, 1510, 5254
bois de fer, 3297
bois de feuillu, 2987
bois de papeterie, 4584
bois de petite dimension, 5191
bois de printemps, 2176, 3334, 5355
bois de réaction, 4659
bois de résineux, 1510, 4757, 5254
bois d'été, 5548
bois de tension, 5688
bois écorcé, 4241
bois en grume, 3586, 5745
bois feuillu, 1847
bois gelé, 2741
bois humide, 6214
bois mort, 1833
bois mûr, 3722
bois poreux, 4448
bois pourri, 1843
bois vert, 2898
boîte, 862, 1140
boîte à conserves, 1068
boîte à eau de calandre, 1028
boîte d'engrenages, 2800
boîte de vapeur, 2361
boîte en fer-blanc, 1071
boîte extensible, 2344
boîte pliante, 2649
boîte pliante en carton, 1134
boîtes montées en carton, 5020
bombage, 1715
bombé, 1712
bombé concave, 3951
boom, 841
borate de sodium, 5221
borates, 842
borax, 843
bord, 2183, 6179
bord à la cuve (barbé), 1852
bord coupé au couteau, 3355
bord craquelé, 1660
bord d'attaque, 3438
bord de la caisse à pâte en tête de machine, 3516
bord détendu, 5142
bordeur, 5849
bordeuse, 2186
bord gondolé, 6120
bord grainé, 2867
bord imitant celui d'un papier à la main, 5782
bord irrégulier, 907
bord non rogné, 3823
bord retourné, 5879
bord roulé, 4822
bords barbés, 3309, 3822
bords barbés amincis, 2399
bords barbés obtenus sur machine par jet d'air, 256
bore, 848
borohydrure de sodium, 5222
bosselé, 2254
botte (de paille), 6040
bouchon (centre de bobine), 1588
bouchon en bois, 4382, 6264
boucle d'accrocheuse, 2440
boue, 3906, 5160, 5181
boue activée, 194
boue de chaux, 3490
boue de lessive neuve, 6220
boue liquide, 5186
boue résiduaire de chaux, 3492
bouffant, 945, 957
bouleau, 707
bouleau à papier, 4151
bouleau blanc, 6218
boulette, 4245
boulon de scellement, 368
bourgeonner, 5358
bourre, 2598
bourreur de copeaux, 1289
bouton, 935
boutons (dans le papier), 993, 4186
brai, 5650
branche (arbre), 5424
branlement, 5027, 5031
bras de levier, 3460
bride (tuyau), 2543
brillant, 896, 2813, 2826, 3595, 4877

brillant à faible angle, 3575
brindille, 5884
brique réfractaire, 2522
brique résistant aux acides, 171
bristol à base de vieux papiers, 798
bristol couché, 1391
bristol duplex, 2155
bristol multijet fabriqué sur forme ronde, 2487
bristol non apprêté, 399
bristol pliant, 2650
bristols, 901
bristols fabriqués sur forme ronde, 1793
broche d'enroulage, 1589, 6240
brome, 910
bronze, 912, 2948
bronzer, 913
brosse, 915
brosses du milieu (couchage), 757
broyage, 889, 2960
broyer, 1717
broyeur, 1722, 2025, 3350, 4571
broyeur à barres, 4815
broyeur à copeaux, 4673
broyeur à marteaux, 2959
broyeur de cassés, 909
broyeur de copeaux, 1284
bruit, 3994
brûlé, 968
bruleur, 964
brunisseur, 966
brut, 1377, 5907, 5916
brut de machine, 3620
bûcheron, 3587
bûchette, 5062, 5141, 5174, 5634
buée, 5957, 5966
bulle, 509, 929
bulles d'air, 249
burette, 962
buse, 4009
but, 681
butadiène, 982
butanes, 983
butée d'engagement de la feuille, 2409
butée de presse, 4494
butylcaoutchouc, 996
butyrate de cellulose, 1184
butyrates, 997

C

câblage, 5888
câble, 1005, 4835
câble de montée, 5933
câble pour passer la feuille (dans la sècherie), 4836
cacheter, 4959
cachet (pour sceller ou clore), 4962
cadmium, 1008
cadran, 1939
cadre, 2714, 2715
caisse, 862, 1140, 1266
caisse à doubles parois, 2082
caisse à eau, 6068
caisse à eau de rouleau de calandre, 1057
caisse à pâte ouverte, 4064
caisse aspirante, 2549, 5498, 5941
caisse aspirante Uhle, 5901
caisse carton, 864
caisse d'alimentation en circuit fermé, 1361
caisse d'arrivée de pâte, 2590, 2994
caisse d'arrivée de pâtes, 893
caisse d'arrivée de pâte sous pression, 4501
caisse d'égouttage, 2091
caisse de mélange, 3843
caisse de tête, 2590
caisse de tête intermédiaire, 3273
caisse de tête primaire, 4514
caisse en carton ondulé, 1609
caisse plate aspirante, 5505
caisse pour transport, 5061
calandrage, 860, 1038, 2823

calandrage humide, 6074, 6077
calandré, 1035
calandre, 1025
calandre à bâtis ouverts, 4063
calandre à friction, 2731
calandré avec docteur à eau, 6075
calandre de bout de machine, 1052
calandre de machine, 3608
calandre de satinage, 2828
calandré deux fois, 2064
calandre de bout de machine, 1052
calandré en continu, 6148
calandre gaufreuse, 2258
calandre humide, 6070
calandre intermédiaire, 3271
calandres combinées, 1464
calandreur, 1039, 1042
calcaire, 3494
calcination, 1010
calcination de pyrite, 4599
calcium, 1011
calculateur analogique, 361
calculateur numérique, 1983
calculatrice, 1024
calibre, 1061, 2687
calibrer, 1058
calorimètre, 1062
calorimétrie, 1063
cambium, 1066
came, 1064
camion, 5855
canal d'arrivée d'eau, 2611
canalisation, 4336
canaux résinifères, 4756
canne à sucre, 5516
cannelé, 2620
canneler, 2915
cannelure, 1614, 2619, 2621
cannelure d'entraînement, 5329
cannelure kraft, 3370
cannelure mi-chimique, 4992
cannelure moyenne, 1004
cannelure vieux papiers, 799
caoutchouc, 2940, 4864
caoutchouc synthétique, 5616
capabilité d'hériter, 3033
capacité, 1076, 1080
capacité de blanchiment, 737
capacité de pointe, 3724
capacité de séchage, 2136
capillaire, 1082
capillarité, 1081, 1678
capital, 1083, 5430
capot, 1148
caprolactam, 1086
capsule, 1087
capsule d'air, 250
caractéristique, 1227
caractéristique de la molette de rhabillage des meules de défibreur, 971
carbamates, 1089
carbamides, 1090
carbonate de baryum, 572
carbonate de calcium, 1015
carbonate de chaux pur, 4236
carbonate de magnésie, 3647
carbonate de sodium, 5209
carbonate de soude, 5210, 5223
carbonates, 1094
carbone, 1093
carbone activé, 193
carbonisation, 1095, 1100
carbonyls, 1107
carboxyalkylation, 1110
carboxyalkyl cellulose, 1111
carboxyéthylation, 1113
carboxyéthyl cellulose, 1114
carboxylation, 1116
carboxyméthylation, 1119
carboxyméthyl cellulose, 1120
carbures, 1091
carneau, 2148
carré, 5361
carreau, 5742
carrosserie, 4145
carte de voeux, 2899
carte mince, 1122
carte perforée, 4589
carter, 1148, 1654, 3102
carteux, 4645

carton, 789
carton à alvéoles, 1180
carton à base de jute, 3330
carton à base de vieux papiers, 797
carton à bords barbés, 1853
carton à doubler, 506
carton à fermeture pignon, 2768
carton à l'enrouleuse, 6192
carton assez fort, 1123
carton au sulfate(kraft), 5521
carton baryté, 591
carton bituminé, 715
carton bois, 6277
carton bristol, 900
carton brun de pâte mi-chimique, 3679
carton compact, 5261
carton compact à base de pâtes, 5264
carton compact pour caisses, 1524
carton composé de plusieurs couches de papier, 3915
carton comprimé, 4484
carton contrecollé, 731, 3411, 4220, 4223
carton couché, 1387
carton couché au kaolin, 1347
carton couché chrome, 1323
carton couché machine, 3612
carton couché pour affiches, 443
carton couché pour boîtes, 1390
carton cuir (pour chaussures), 2447
carton cuir pour chaussures, 5066
carton d'amiante, 447
carton de construction, 939, 6043
carton de forme ronde, 1792
carton d'emballage, 4138
carton de pâte mécanique et de vieux papiers, 4569
carton de pâtes mi-chimiques, 4991
carton de pâtes mi-chimiques, 3972
carton deux côtés blancs, 3817
carton de vieux papiers doublé manille sur les deux faces, 2079
carton de vieux papiers pour mandrins, 5067
carton de vieux papiers pour boîtes plaintes ordinaires, 4700, 4988
carton de vieux papiers recouvert d'une feuille de papier d'emballage, 5110
carton double, 3506
carton doublé d'une feuille faite sur forme ronde, 5977, 5978
carton duplex, 2154, 5109, 5892
carton duplex à base de vieux papiers, 801
carton dur, 2973
carton en plusieurs jets, 3921
carton épais pour protéger les fonds des bobines emballées, 2187
carton fabriqué sur table plate, 2703
carton feutre pour toitures, 2429, 4832
carton goudronné, 469
carton goudronné pour toitures, 467
carton gris, 4152
carton gris à base de vieux papiers fabriqué sur machine multiforme, 3968
carton gris à cloisons intérieures, 4216
carton gris compact, 4348
carton gris contrecollé, 4224
carton gris de vieux journaux, 3961
carton gris (de vieux papiers), 1278
carton gris doublé manille sur les deux faces, 2078
carton gris multijet, 1459
carton gris recouvert d'une feuille de papier kraft, 5107, 5114
carton hydrochine, 1274
carton imperméable à l'eau, 6094
carton ingraissable, 2885, 2891
carton isolant, 3257
carton isolant (de paille de lin), 2560
carton kraft, 3369
carton kraft liner, 3372
carton laminé, 2820
carton laminé à la plaque, 4372, 4377
carton laminé dur pour poulies, 2730
carton léger, 950
carton liner au sulfate (kraft), 5522

carton liner léger, 3476
carton lissé, 2820
carton lustré, 4484
carton moulé, 3873
carton multiforme, 2486
carton multijet, 1458, 2491
carton multijet de vieux papiers pour boîtes montées, 2489
carton multijets, 3914
cartonnerie, 793, 4155
carton ondulé, 1607, 1615
carton ondulé à base de microcannelure, 5104
carton ondulé double-double, 2074
carton ondulé en plusieurs épaisseurs, 3931
carton ondulé microcannelé, 3796
carton ondulé pour caisses, 1610
carton ondulé simple face, 5103
carton ordinaire pour boîtes pliantes, 660
carton paille, 3973, 4700, 5453
carton paraffiné, 6123
carton plâtre, 2952, 4358
carton pliant, 2647, 2651
carton pour accumulateurs, 612
carton pour affiches, 4449
carton pour billets de chemin de fer, 5740
carton pour boîtes, 863
carton pour boîtes doublé de papier écru, 3682
carton pour boîtes montées, 5019
carton pour boîtes pliantes, 2648, 3327
carton pour buvards, 769
carton pour caisses à savon résistant à l'alcali, 316
carton pour capsules de bouteilles, 850
carton pour contreforts de chaussures, 1642
carton pour coutellerie, 411
carton pour dessous de bocks, 652, 1381
carton pour dessous de verres, 3710
carton pour étiquettes, 5632
carton pour gobelets, 1739
carton pour isolation acoustique, 183
carton pour joints, 2789
carton pour la partie cylindrique de récipients, 1072
carton pour l'emballage du lait, 3814
carton pour livres, 831
carton pour mandrins, 1586
carton pour pâtissiers, 1661
carton pour patrons de découpe, 4231
carton pour pochettes d'allumettes, 836
carton pour pochoir, 5420
carton pour produits alimentaires, 2661, 2740
carton pour produits alimentaires congelés, 2661, 2740
carton pour pyrotechnie, 2398
carton pour reliure, 692
carton pour toitures, 1175
carton pour tubes, 5866, 5867
carton pour usages spéciaux, 5295
carton pour visières, 645
carton recouvert de papier à base de papiers recyclés, 3964
carton recouvert sur les deux faces de papier fabriqué sur forme ronde, 2081
carton redoublé, 1461
cartons affichés blanchis, 3499
carton satiné, 5196
cartons huilés, 4037
carton silicaté, 4358
carton simili cuir, 3184, 3445
carton simili kraft, 5065
carton spécial pour produits alimentaires, 5293
cartons pour l'emballage des oeufs, 2196
carton support pour goudronnage, 470
carton triples, 5852
carton un jet, 5112
carton vélin, 5984
caséine, 1142
cash flow, 1146
cassé à la calandre, 1029

cassé de fabrication, 905
cassée de massicot, 1771
cassés de fabrication, 3820
cassés de finissage, 2518
cassés de machine, 3607
cassés de papier résistant à l'état humide, 6203
cassés humides, 6182
cassés s'accumulant sur le docteur, 2050 -
cassés secs, 2118
cassure, 882, 2383
cassure au pli rainé, 4932
Catalpa, 1155
catalyseur, 1156
cathode, 1157
cation, 1159
cationique, 1162
causticité, 1167
caustificateur, 1169
caustification, 1170
caustifier, 1168
caustique, 1166
cavitation, 1172
cèdre, 1173, 1174, 1224, 3463
ceinturage, 2812
célite (terre de diatomée), 1176
celloderme à l'enrouleuse, 3818
celloderme pour carrosserie, 488, 1125
celloderme pour valises, 4145
cellophane, 1178, 1192
cellule, 1177
cellule de Denver (désencrage), 1909
cellule d'électrolyse pour la production de chlore, 1304
cellule d'électrolyseur, 2230
cellule de pesage, 3539
cellule photoélectrique, 4289
cellulose, 1182
cellulose amorphe, 354
cellulose cristallisée, 1729
cellulose de coton, 1625
cellulose engraissée, 3113
cellulose mésomorphe, 3757
cellulose microcristalline, 3797
cellulose modifiée, 3850
cellulose régénérée, 4715
cellulose résistant à l'alcali, 304
cellulose transparente, 5819
cémentation, 1141
cendre, 456
cendres volatiles, 2623
centipoise, 1206
centrale de force motrice, 4467
centrale électrique, 4467
centrale hydroélectrique, 3133
centrifugation, 1213
centrifuge, 1207, 1214
centripète, 1215
céramique, 1218
cerclage, 5448
cerclage de bobines avec du feuillard, 5449
cercler avec du feuillard, 562
cerisier, 1265
cetènes, 3343
cétones, 3344
chaîne (d'une toile métallique), 6046
chaland, 570
chaleur, 3000
chaleur de combustion, 3012
chaleur progressive, 3007
chaleur spécifique, 5301
chambre à air, 251
chambre de combustion, 1467
chambre d'extraction, 2358
chambre plénière, 4379
champ électromagnétique, 2231
champignons, 2752
champ magnétique, 3656
changement de couverte, 1226
changement de direction de feuille, 6162
chanvre, 3031
chanvre de Manille, 103, 3678, 3681
chanvre de sisal, 5118
chapeau de presse-étoupe, 5485
charbon de bois, 1228
charge, 2493, 3538, 3544

charge alcaline, 308
charge-allongement, 3541
charge de base, 647
charge du lessiveur, 1977
charge électrostatique, 2239
charge fibreuse, 2480
chargement, 2411, 3543
chargement de la pile, 632
chargement d'une machine, 3628
charge minérale, 2494
charges cycliques, 1786
chargeur, 3542, 5439
chargeur de copeux, 1279
chariot, 5855
chariot élévateur, 3469
chariot élévateur à fourche, 2678
charme, 1131, 3088
charpie, 3513
châssis, 2714, 2715
châssis de caisse aspirante, 5499
chataignier, 1267
chaudière, 812
chaudière à eau multitubulaire, 6114
chauffage, 3008
chauffage à la vapeur, 5406
chauffage diélectrique, 1963
chauffage direct, 2008
chauffage indirect, 3211
chauffage par induction, 3214
chauffage par infra-rouges, 3224
chauffage par rayonnement, 4612
chaufferie, 814
chauffeur, 5439
chaux, 3487
chaux éteinte, 5144
chaux vive, 1021
chef de fabrication, 5559
chef de fabrication des presses, 4492
chelates, 1239
chelateur, 1240
chelation, 1241
chemin de fer, 4635
cheminée, 1273
chemisage du lessiveur, 1979
chemise, 3306
chêne, 4015, 4609
cheval-vapeur, 3090
chicane, 521, 523
chiffon, 4621
chiffonnerie, 4631
chiffons coupés, 4627
chiffons pour essuyer, 3218
chiffres, 1821
chimiste en chef, 1269
chiné, 1365, 3833, 4321
chiner, 3834
chlorate de sodium, 5224
chlorates, 1297
chlore, 1303
chlore actif, 197
chlorite de sodium, 5226
chlorites, 1312
chlorolignine, 1299
chlorolignines, 1314
chlorose, 1315
chloruration, 1301
chloruration en plusieurs stades, 1302
chlorure de calcium, 1016
chlorure de chaux, 744
chlorure de lithium, 3528
chlorure de magnésie, 3648
chlorure de polyvinyl, 4428
chlorure de polyvinylidène, 4431
chlorure de sodium, 5225
chlorure de vinyl, 6010
chlorure de vinyldiène, 6012
chlorures, 1298
chlorures acides, 158
choc, 5063
choix, 4986
chromates, 1316
chromaticité, 1317
chromatographie des gaz, 2782
chromatographie sur papier, 4159
chrome, 1320
chute d'eau, 4094
chute de pression, 4497
chute des copeaux, 1280
chutes, 5034

chutes de mandrins, 1591
circuit électrique, 2209
circuit fermé, 1356
cire, 6121
cire de carnauba, 1129
cire de paraffine, 4197
cire microcristalline, 3798
cisaillement, 5035
citerne, 5641
clapet à bille, 556
clarificateur, 1340
clarification, 1339
clarifier, 1341
classage, 2859, 4947, 5282
classement, 1342
classement (d'après la granulation), 4555
classer, 5279
classeur, 2513, 4945
classeur centrifuge, 1209
classeur mécanique pour l'étude des fibres par la méthode du fractionnement, 3728
classification, 5626
cliver, 1893
cloison, 948
cloison centrale (épi d'une pile), 3810
cloison mobile de la fosse du défibreur, 1804
cloisons de caisses, 872
coagulant, 1372, 1374
coagulation, 1376
coaguler, 1373
C.O.D., 1003, 1250
coéfficient d'absorption, 121
coefficient d'absorption, 1426
coefficient d'échange thermique, 3022
coefficient de réfraction, 4712
coefficient de transparence, 5817
cohésion, 1428
cohésion interne, 3275
coin, 5132
collage, 5128
collage à l'alcali, 312
collage à la paraffine, 6133
collage à la résine, 178, 4842, 4844
collage au bobinage, 5327
collage dans la masse, 5486
collage en cuve, 5864
collage en cuve à la gélatine, 377
collage en pâte, 5486
collage en pile, 638, 2284
collage en surface, 5585, 5780
collage interne, 3278
collage neutre, 3956
collage renforcé, 2695
collages sur bobine, 2625
collagimètre, 5129
collant, 5630
collé, 4221
colle, 4219
collé à la cuve, 5861
collé à la résine, 4843
colle animale, 376
colle (collage du papier), 5121
collecteur, 2995
collecteur à poussières, 2165
collecteur (de gaz), 3676
collé en pile, 637
collé et surcalandré, 5122
colle forte, 2840
colle stable en milieu acide, 179
collet, 1435
colleuse, 4227
collier, 1337
colliers de rouleau, 4817
colloïdal, 1436
colloïdes, 1438
collure, 5325
collure rognée, 5842
colonne barométrique, 2108, 6081
colorant, 2170

colorant acide, 160
colorant pour aviver, 5770
colorants, 1447
colorants acides, 159
colorants à l'aniline, 374
colorants à l'indigo, 3210
colorants basiques, 601
colorants de cuve, 5976
colorants directs, 2007
colorants fluorescents, 2614
coloration, 1446, 2172, 5375, 5377, 5753
coloration au plongé, 5859
coloration au pulvérisateur, 5346
coloration dans la pile, 629
coloration en surface, 5573
coloration sur calandre, 1054
coloré, 1440, 2171
coloré à la pile, 628
coloré en surface, 5574
coloré en surface sur calandre, 1030
coloré naturellement, 3942
coloré sur calandre, 1034
colorimètre, 1443
colorimètre à cellule photoélectrique, 4290
colorimètre à comparaison, 1451
colorimètre, 1445, 1639
combustible, 2743
combustion, 1466
combustion de soufre, 5542
commande, 2104, 2106
commande auxiliaire, 3028
commande à vitesses variables, 5969
commande de la machine à papier, 4169
commande différentielle, 1967
commande électrique, 2216
commande inversée, 4775
commande mécanique, 3731
commande par action directe, 2408
commande par courroie, 655
commande par friction, 2732
commande par vis sans fin, 6289
commande périphérique, 4266
commande sectionnelle, 4978
commande sur fabrication, 3667
commercialisation, 3696
compacité, 1472
compacteur, 1473
comparaison, 3712
comparaison des couleurs, 1452
comparer, 3711
compatibilité, 1474
complexe metal-carton, 1475
complexe plastique, 3413
comportement, 4870
composés acryliques, 190
composés aliphatiques, 298
composés ammoniacaux quaternaires, 4607
composés anioniques, 381
composés aromatiques, 432
composés cationiques, 1163
composés cycliques, 1785
composés d'alumine, 331
composés d'ammoniaque, 348
composés d'azote, 496
composés de bore, 849
composés de brome, 911
composés de cadmium, 1009
composés de calcium, 1017
composés de chlore, 1305
composés de chrome, 1321
composés de cyanure, 1782
composés de l'argent, 5097
composés de l'azote, 3990
composés de l'hydrogène, 3142
composés de l'iode, 3285
composés de magnésium, 3649
composés de potassium, 4453
composés de silicium, 5092
composés d'étain, 5749
composés de titane, 5758
composés d'halogène, 2956
composés d'oxygène, 4125
composés du fer, 3295
composés du fluor, 2618
composés du lithium, 3529

composés du manganèse, 3675
composés du mercure, 3753
composés du molybdène, 3886
composés du nickel, 3970
composés du phosphore, 4288
composés du sodium, 5227
composés du soufre, 5543
composés du tantale, 5642
composés du tunstène, 5870
composés du vinyl, 6011
composés du zinc, 6327
composés du zirconium, 6333
composés hétérocycliques, 3034
composés inorganiques, 3248
composés métallifères, 3760
composés organiques, 4084
composition, 5900
composition de fabrication, 2761
composition de la lessive de cuisson, 1571
composition d'un papier, 4188
composition fibreuse, 2454
compressibilité, 1477
compresseur, 1482
compresseur aspirant centrifuge, 1212
compressibilité, 955
compression, 1478, 4488
comptage, 1649
comptage à l'usine, 3821
compter, 1640
compte-tours, 1641
compteur, 1641
concasseur, 1722, 2025
concentration, 1487, 1512, 1849, 1905, 1932
concentration hydroxyle-ion, 3166
concentration ionique, 3290
concentration très élevée, 3041
conception, 1923
conception d'un ensemble, 4131
condensateur, 1079, 1495
condensateur à ruissellement, 5837
condensation, 1492
condensation polymérisation, 1493
condensats, 1491
conditionné, 1498, 5370
conditionnement, 1501, 1749
conditionnement du feutre, 2420
conditionneur de feutre, 2419
conditionneur de feutre type vickery, 6007
conditions adiabatiques, 209
conducteur de machine, 3634
conducteur électrique, 2210
conductivité, 1504
conductivité calorifique, 3003
conductivité électrique, 2204
conductivité thermique, 5703
conduit, 2148
conduite, 4334
conduite d'air, 280
conduite de feuille, 3437
cône, 1505, 4859
cône à étages, 5422
cône (arbre), 5831
cône de raffineur cônique, 3317, 4382
conifère, 1509
cônique, 5646
connaissance, 1655
conservation, 1511, 4480, 4769, 5441
conservation de l'eau, 6104
consistance, 1512
consommation, 1520
consommation chimique, 1244
consommation d'eau, 6071
consommation de chaleur, 3004
consommation de matières premières, 3715
consommation d'énergie, 2279
consommation d'oxygène, 4126
constante diélectrique, 1960
constitution de la feuille, 2685
containerisation, 1526
contamination, 1530
contaminer, 1529
conteneur, 1523
continu, 1531
"contracoater", 1537
contracter (se,) 5081

contraction, 5082
contrainte, 4766
contrecollage, 4226
contre-collage sous pression, 3414
contre-courant, 1538, 1643, 1646
contremaître, 2665
contremaître de faction, 5790
contreplaqué, 5992
contrepression, 514
contrôle, 1234, 1541, 3888
contrôle automatique, 484
contrôle biologique, 701
contrôle de consistance, 1513
contrôle de la charge, 3540
contrôle de la fabrication, 4533
contrôle de la poix, 4341
contrôle de la pollution, 4405
contrôle de la pression, 4496
contrôle de la production, 4540
contrôle de la qualité, 4605
contrôle de la réalimentation, 2405
contrôle de la résine, 4341
contrôle de la température, 5671
contrôle de la tension, 5686
contrôle de la vitesse, 5988
contrôle de l'humidité, 3108, 3862
contrôle des odeurs, 4018
contrôle des poussières, 2166
contrôle des stocks, 3283
contrôle du bruit, 3995
contrôle du débit, 2592
contrôle du niveau, 3453
contrôle du pH, 4275
contrôle électronique, 2232
contrôle numérique, 1984
contrôle (ou commande) à distance, 4740
contrôle par câble, 4837
contrôles des microorganismes, 3804
contrôleur, 1545
contrôleur de consistance, 1514
contrôleur de niveau, 3454
convection, 1548
conversion, 1551
conversion de l'amidon, 5387
convertisseur analogique à numérique, 363
convertisseur de fréquence, 2726
convoyeur à pesage, 6169
copalme d'Amérique, 3518, 4690, 5601
copeau, 1276
copeaux acceptés, 135
copeaux classés, 135
copeaux de bois, 6261
copeaux pour pâte mécanique, 1287
copier, 1578
copolymères, 1579
copolymères de cellulose, 1186
copolymères greffés, 2861
coquille, 5057
cordage, 4835
corde, 1584
cordons au bobinage, 6236, 6237
corriger la teinte, 211, 2058
corroder, 1596
corrosif, 1604
corrosion, 1597
corrosion galvanique, 2774
corrosivité, 1605
cortex, 1621
côté commande, 2107
côté commande, 515
côté conducteur, 2737, 5673, 5674
côté fabrication, 4068
côté feutre, 2432, 5779
côté feutre de la feuille, 5773
côté supérieur, 5779
côté toile, 6255
côté transmission, 503
coton, 1624
couchage, 1399
couchage à la brosse, 921
couchage à la lame d'air, 248
couchage à lame d'air, 275
couchage à la lame traînante, 5801
couchage à la pyroxiline (nitrate de cellulose), 4602
couchage anti-adhésif, 4732

couchage anti-dérapant, 407
couchage à plusieurs composants, 3911
couchage au glacis, 1151
couchage au plongé, 2004
couchage avec séchage sur tambour poli, 4401
couchage cloqué, 930
couchage de protection, 4558
couchage électrostatique, 2240
couchage etc.), 598
couchage floconneux, 2576
couchage hors machine, 1552, 4021
couchage imperméable à la graisse, 2886
couchage irrégulier, 1415
couchage métallique (sur caséine ou laque), 3765
couchage multiple, 3917
couchage par bandes, 5474
couchage par brosses, 917
couchage par collage à chaud, 3096
couchage par extrusion, 2365
couchage par gravure, 2880
couchage par impression, 4529
couchage par lame, 725
couchage par pulvérisation, 5343
couchage par râcle, 3354
couchage par rouleaux, 4819
couchage par séchage, 3349
couchage rougeâtre, 787
couchage sur coucheuse à nappe, 1760
couchage sur coucheuse à rouleaux marchant en sens inverse, 4780
couchage sur deux faces, 5895
couchage sur le principe d'une presse typo, 4531
couchage sur machine, 3614, 4051
couchage sur par préalablement gommé, 2119
couchage sur presse, 4486
couchage translucide, 5812
couché, 1384, 2266, 2267
couche, 1382, 1416, 3434, 4387
couché à la cuve, 4229
couché au kaolin, 1345
couche centrale, 1203
couche de croissance, 2927
couche d'encre, 3237
couché deux faces, 1002, 2065
couche gélatineuse, 2804
couche inférieure, 5495
couche monomoléculaire, 3891
coucher, 1630, 1632
couches, 4381
couche supérieure de la doublure recouvrant un carton,
couche supérieure de la doublure recouvrant une boîte, 5776
couché une face, 1001, 5100
couché une ou deux faces, 5572
coucheuse, 1398, 1410
coucheuse à barre, 4814
coucheuse à barre filetée, 6259
coucheuse à bourrelet, 615
coucheuse à lame, 569, 724
coucheuse à lame à "fontaine", 2700
coucheuse à lame d'air, 247, 274
coucheuse à lame traînante, 5800
coucheuse à lame traînante immergée, 4434
coucheuse à nappe, 1759
coucheuse à pince submergée, 2581
coucheuse à râcles, 3353
coucheuse à rouleau barreur, 2796
coucheuse à rouleau essoreur, 5366
coucheuse à rouleaux, 4818, 5200
coucheuse à rouleaux inversés, 1537
coucheuse à rouleaux marchant en sens inverse, 4779
coucheuse au glacis, 1150
coucheuse au plongé, 2003
coucheuse avec râcle entre deux rouleaux, 2569
coucheuse deux faces, 2066, 5893
coucheuse duplex, 2156
coucheuse hors machine, 4020
coucheuse par bandes, 5473
coucheuse par extrusion, 2364

coucheuse par léchage, 4825
coucheuse par pulvérisation, 5342
coucheuse par séchage (au rouleau), 3348
coucheuse sur le principe d'une presse typo, 4530
coucheuse sur presse, 4485
coucheuse sur rouleau égoutteur, 1817
coucheuse type Champion, 1225
coucheuse type Massey, 3706
coudre, 4968
coulée, 2603
couleur, 1439, 1456
couleur laque, 1449
couleur marbrée, 3901
couleur solide à la lumière, 2388
coupe, 1711, 1775
coupé à l'emporte-pièce, 1955
coupe-circuit, 1328
coupe en feuilles rotative, 4853
coupe-feuille hydraulique, 1767
coupe par molettes sur cylindre en acier trempé, 4933
coupe par mollettes sur cylindre en acier trempé (bobineuse-trancheuse), 1768
coupe rase, 1352
coupeur, 1774
coupeur (ouvrier), 5170
coupeuse, 1770, 3065, 4163
coupeuse à bagasse, 528
coupeuse à bois, 1291
coupeuse à disques, 2019
coupeuse à molettes, 4934
coupeuse de chiffons, 4626
coupeuse en bout de machine, 3245
coupeuse en feuilles, 5043
coupeuse en long, 5167
coupeuse en spirale, 5321
coupeuse rotative, 4848, 4854
coupeuse simplex, 5113
coupeuse-trieuse, 1773
couple (mouvement de rotation), 5784
coupure, 1764, 5147
coupure d'ampoule, 760
coupure de fibre, 2456
coupure de lame, 1765
coupure de poil, 1766
courant, 2588
courant alternatif, 100, 326
courant continu, 1803, 2006
courant d'air, 266
courant d'arrivée d'eau, 4211
courant de foucault, 2214
courant de pâte en avant de la machine à papier, 421
courant de retour, 504
courant électrique, 2214
courant visqueux, 6026
courbe, 1761, 1938
courbe de débit, 2591
courbure, 659, 2563
couronne, 1594, 1712
courroie, 653, 5447
courroie de transporteur, 1563
courroie-guide, 1856
cours d'eau, 6078
coût, 1622
couteau, 3351
couteau circulaire, 1330, 2021
couteau fixe, 646
couteau rotatif, 2626, 4784
couteaux, 3357
coûts de fabrication, 4067
couture, 4966
couture rognée, 1353
couvercle, 1654
couvercle en carton ondulé pour les fûts en carton, 1608
couverte, 1850
couverte fixe, 1851
couverture, 1654, 1655, 1656, 6296
couverture bico-kraft, 3333
couverture couchée, 436
couverture forte, 5697
couverture forte kraft, 3378
couverture morte, 2672
crampon, 1337
craquelure, 1659, 1663, 1667, 5204
craquelures de déchage, 2137

cratère, 1665
crêpage, 1687
crêpage à l'état humide, 6073
crêpage à sec, 2120
crêpé, 1680, 1681
crêpé croisé, 1705
crêpé sur machine, 3615
crépine, 5445
crésols, 1692
crible, 5088
crispage, 1420, 1424
cristal, 1728, 2816
cristallinité, 1731
cristallisation, 1732
crochet, 2968
croissance, 2926, 3205
croissance primaire, 4513
croissance secondaire, 4971
croûte de calandre, 1053
cryogénie, 1727
cubage, 6271, 6313
cuiseur d'amidon, 5388
cuisson, 1565, 1567, 1981, 1982
cuisson au chlorite, 3077
cuisson brute, 4646
cuisson en phase liqueur, 3523
cuisson hydrotropique, 3156
cuisson poussée, 2974
cuivre, 1580
culture, 1737
Cupressus, 1744
cuve, 5975
cuvette, 5829
cuvette d'une coucheuse, 1412
cuvier, 1266
cuvier à dissoudre, 2037
cuvier à pâte, 5431, 5482
cuvier de machine, 3609
cuvier de mélange, 3844
cuvier de pile, 627
cyanamides, 1777
cyanates, 1778
cyanoéthylation, 1780
cyanoéthyl cellulose, 1781
cyanures, 1779
cycle, 1783
cyclodextrines, 1787
cyclohexane, 1788
cyclone, 1789
cylindre, 1791, 4816
cylindre à brosses, 928
cylindre aspirant de la toile, 5502
cylindre chauffé à la vapeur, 3099
cylindre de format, 5126
cylindre de la pile, 635
cylindre de presse aspirante, 5510
cylindre de presse coucheuse, 1637
cylindre de presse inférieure, 855
cylindre en lave, 594
cylindre frictionneur, 2824
cylindre perforé, 4255
cylindre refroidisseur, 1575, 5597, 5598
cylindre sécheur, 2122, 2138
cylindre sécheur à grande vitesse, 3058
cymènes, 1801
cyprès, 1802
cyprès chauve, 544

D

dalle à eaux blanches, 4916
dalle pour recueillir les eaux sous toile, 5829
débardage, 1351, 5134
débardage par câble (système "high-lead"),
débardage par câble (système skyline), 3049, 5140
débardeur, 5133
débarrasser de cire, 1933
débit, 2588, 4096
débit à contre-courant, 1644
débit constant, 1698
débitmètre, 2594

débobinage, 5930
débobiner, 5928
débobineuse, 5929
débouché orifice d'évacuation, 4095
débusquage, 5134
débusqueur, 5133
débusqueur à pinces, 2877
décalque, 5804
décantation, 2252, 5016
décanter, 2251, 3394
décanteur, 5004
décapage, 1351
décélération, 1844
déchargé, 5922
décharge, 2015, 4735
décharge corona, 1595
décharge d'impuretés, 5828
décharge électrique, 2215
déchargement, 5924
déchargeur, 5923
déchet, 4724, 5634, 6056
déchets, 809, 4948
déchets de bois, 6281
déchets de bois écrasé (combustible), 3064
déchets de contreplaqué, 5994
déchets de rayonne, 4655
déchets de scierie, 4921
déchets de tronçonnage, 3558
déchiquetage, 5080
déchiqueteur, 3350, 5079
déchiqueteur à tambour, 2112
déchiqueteuse, 3065
déchirure, 5655, 5657
déclecher, 5013
décollement de la couche, 1662
décoloration, 1857, 2018
décomposition, 1858
découpage, 1775
découpe à l'emporte-pièce, 1957
découpeuse de fente des séparateurs, 4218
découverte, 1927
décrasse-meule, 969
défaut, 1861, 2557
défaut de collage, 2845
défaut de couchage, 1407
défaut de la bobine provenant d'une tension non uniforme de la feuille et provoquant la formation de plis, 533
défectueux, 536
défibrage, 1863, 2906
défibrage à chaud, 3093
défibrage à froid, 1431
défibrage avec fosse, 4347
défibrage sans fosse, 4345
défibration, 1866
défibrer, 1862, 1864
défibreur, 2900, 6269
défibreur à chambres, 2903, 4394
défibreur à magazine, 3640
défibreur à trois chambres (presses), 5736
défilage, 889, 5188
déflecteur, 521, 1872
déflecteur d'eau, 2640
déflecteur de pontuseau, 5622
défloculation, 1874
déformation, 1876, 2563
déformation incorporée, 941
dégager, 4729
dégât, 1805
dégazage, 771, 1878, 4735, 4739
dégazeur, 1877
dégradation, 1879
dégradation chimique, 1246
dégradation de la cellulose, 1187
dégradation mécanique, 3730
dégraissage d'une feutre, 4939
degré, 2857
degré de blanchiment, 1881
degré de cuisson, 1570
degré d'engraissement, 6194
degré de polymérisation, 1883
degré de raffinage, 642, 1880
degré de substitution, 1884
degré d'orientation, 1882
degré hygrométrique, 4727
degré Kelvin, 3338

délayer, 1990
délignification, 1895
deliquescence, 1897
démarrage, 5393
demi-brillant, 2197
demi-pâte, 2954
dendrologie, 1899
denier, 1900
densimètre, 1903, 1907
densimétrie, 1904
densité, 1512, 1905
densité courante, 1758
densité de la pâte, 5483
densité de masse, 946
densité très élevée, 3042
dent (d'engrenage), 5772
département des presses, 4493
déperdition de chaleur, 3011
dépastiller, 2463
dépastilleur, 1869
dépolarisation, 1913
dépôt, 1914, 1915, 5016, 5160, 5181
dépôt minéral, 3831
dépoussiérage, 2168
dépréciation, 1916
dérivés de la cellulose, 1188
dérivés de l'amidon, 5389
dérivés de la vanilline, 5956
dérivés du glucose, 2834
déroulement, 5930
désagrégation, 2024
description, 5299
désencrage, 1891
désencrage par flottation, 2585
désencrer, 1889
désensibilisation, 1920
deshydratation, 1886
deshydrater, 1885
deshydrateur, 1887
déshydratation, 1932
déshydrogénation, 1888
désintégrateur, 2025
désintégration, 2024, 5188
désintoxication, 1929
desionisation, 1892
désodorisation, 1910
désorption, 1925
désoxygénation, 1911
desserrer, 4729
dessin, 1923
désurchauffeur, 1926
détartrage, 1919
détecteur, 5002
détecteur à infra-rouges, 3221
détecteur de défauts, 2558
détecteur de mâtons, 3594
détecteur d'incendie, 2523
détection, 1927
détendeur, 4695
détersif, 1928
deux faces, 5896
deuxième choix, 4977
déversoir, 6173
déviation, 1871
dévidoir, 5931
dévidoir pour bobines, 516
dévidoir support de bobines, 4831
dévier, 1870
dextrines, 1935
dextrose, 1936
diagnostic, 1937
diagramme, 1231, 1938
diagramme circulaire, 1329
dialdéhydrate cellulose, 1940
dialyse, 1942
diamètre, 1943
diamètre d'une fibre, 2457
diamètre intérieur, 3252
diaphragme, 1944
diaphragme de débit, 4087
diatomes, 1947
diatomite, 1946
dichloroéthane, 1950
dichlorométhane, 1951
dicotylédons, 1952
dicyandiamide, 1953
diélectrique, 1959
diffraction, 1969

diffuseur, 1971
diffusion, 1974
diisocyanates, 1986
dilatabilité, 1987
dilatant, 1988
dilatation thermique, 5704
dilation, 2348
diluant, 1989
diluer, 1990
dilution, 1991
dilution (de la pâte), 5725
dimension, 1992
dimension des particules, 4214
dimension d'un orifice, 3073
dimension du pore, 4442
dimensions d'une fibre, 2458
dimère, 1995
diméthyl formamide, 1997
diminuer, 2109
diminution de la pression, 4503
diode, 2002
dipôle, 2005
directeur, 3672
directeur de la production, 2807, 4541
directeur d'usine, 3827
directeur général, 2806
direction, 2009
disolvant, 5274
dispersant, 2027
dispersion, 2029
dispersion de la lumière, 3472
dispersion optique, 4077
dispositif d'alimentation, 2407
dispositif d'amenée du courant de pâte, 422
dispositif de descente, 3582
dispositif de fluidification, 2610
dispositif d'encrage, 3236
dispositif de tension pneumatique, 278
dispositif d'homogénéisation, 3081
dispositif marqueur, 3700
dispositif mesureur, 3773
dispositif pour protéger les bobines expédiées, 4829
disposition, 2031
disposition des presses, 4483
disque, 2014, 2026
disque de raffinage, 4705
dissipation de chaleur, 2028
dissolvant, 2035
dissolvants de la cellulose, 1195
dissolveur, 2034
dissolveur à chaux, 5145
dissoudre, 2032
distillat, 2038
distillation, 2039
distorsion, 2041
distributeur, 4176
distribution, 2042
distribution d'énergie électrique, 2222
disulfure de carbone, 1098
disulfures, 2044
diversification, 2045
diversité, 5971
diviseur, 2047
docteur, 2048
docteur à eau, 6069
docteur crêpeur, 1688
document, 4676
dolomite, 2060
domaine traité, 1655
dommage, 1805
donnée, 4676
données, 1821
dosage, 2062
doseur, 4554
doublage, 505
doublage de cartons et caisses, 790
doublage de livres, 835
doublé, 3498
double encollage, 2080
double liaison, 2063
doublé sur une seule face, 5108
doublure, 3505, 3509
doublure à base de vieux papiers, 796
doublure arrière, 512
doublure pour boîtes en carton, 1527
doublures, 2377
doublures de sacs, 537

douceur, 5249
doux, 5241
duplex, 2153, 5891
durcir, 913
durcissement, 2976
durcissement en surface, 5577
durcisseur, 2975
durée, 3465, 5746
durée de cuisson, 1569
durée de séjour, 2169
durée de stockage, 5056
durée d'une réaction, 4658
durée d'une toile, 6247
durée d'utilisation, 5010
dureté, 2977, 2979
dureté brinell, 898
dureté en degrés Rockwell, 4812
dureté shore, 5068
duromètre, 2163
duromètre Bekk, 2981
duromètre brinell, 899
duromètre Shore, 5067
duvet, 2598, 2762
dynamique des fluides, 2602
dynamomètre, 5679, 5682

E

eau, 6063
eau additionnelle, 3666
eau blanche, 6225
eau brute (ni filtrée ni épurée), 4649
eau collée, 4794, 6225
eau d'alimentation, 2412
eau d'alimentation de la chaudière, 813
eau de fabrication, 4536
eau de refroidissement, 1577
eau de surface, 5590
eau distillée, 2040
eau douce, 5245
eau fraîche, 2727
eau liée, 857
eau oxygénée, 3144
eau résiduaire, 5635, 6062
eau salée, 4889
ébranchage, 1896
écaille, 4922
écaille de sécheur, 4923
écailles de calandre, 1047
écartement, 5290
échange d'anions, 378
échange de cations, 1160
échange d'ions, 3286
échange thermique, 3021
échangeur d'anions, 379
échangeur de cations, 1161
échangeur de chaleur, 3006
échangeur d'ions, 3289
échantillon, 4891
échantillonnage, 4892, 4894
échantillonner, 3711
échantillon pour essai, 2856
échantillon pris sur la bobine à la
    machine, 4703
échappement, 2339, 4735
échéancier, 5747
échec, 2383
échelle, 4922
échelle de dureté, 2980
éclaircie (forêt), 5725
éclaircissant, 895
éclaircissant fluorescent, 2613
éclaircisseur optique, 4073
éclat, 896, 2813, 3595
éclatement, 974
éclater, 972
éclatomètre, 979
éclatomètre Mullen, 3908
écologie, 2178
économiseur, 2180
écope, 4930
écorçage, 577, 1839
écorçage chimique, 1242, 1245
écorçage de jeunes arbres, 4904
écorce, 574

écorce caryocostine, 1621
écorcer, 1837
écorceuse, 576, 1838, 4242
écorceuse à anneau, 4803
écorceuse à chaîne, 1220
écorceuse à couteaux, 3352
écorceuse à friction, 2729
écorceuse à tambour, 2111
écorceuse hydraulique, 3120, 3125
écotype, 2181
écoulement, 2588, 4871
écoulement à courant transversal, 1706
écoulement de matières solides, 5268
écoulement laminaire, 3408
écoulement plastique, 4361
écoulement turbulent, 5878
écran, 5058
écrasé, 1718
écrasé au passage dans la calandre,
    1031, 1032
écrasement, 1723
écrasement des fibres, 914
écraser, 1717
écrin, 1140
écru, 5904
écumage, 5137
écume, 2628, 2738, 4958
écumoire, 5136
écumoire à savon, 5207
édifice, 5478
effervescence, 2190
effet de paroi, 6044
effet pelure d'orange, 4082
efficience, 2191
effilocher, 3515
effilocheur de chiffons, 1918
effilochure, 3513, 5169
effluent, 2193
effluent de blanchiment, 742
effluent d'usine, 3825
effort, 3385, 5444, 5461, 5685
effort de tension, 5680
égout, 5022
égouttage, 2086, 2093, 4985
égoutter, 2085
égratignures, 4944
éjecteur, 2198
élasticité, 961, 2200, 4751
élastomères, 2202
électricité statique, 5394
électrode, 2227
électrolyse, 2228
électrolyte, 2229
électron, 2236
électronique, 2235
électrophorèse, 2237
électrostatique, 2244
élément, 3226
éléments traceurs, 5796
élévateur à godets, 3324
élimination de l'air contenu dans la pâte,
    1834, 1835
élimination des fines, 2515
éliminer, 5013
éluat, 2249
élution, 2250
émaillé, 2267
emballage, 4132, 6297
emballage en pâte moulée, 3876, 4578
emballage en sachets, 4459
emballage ondulé, 1613
emballages à base de vieux papiers, 811
emballages kraft, 3380
emballage sous macule, 4669
emballages pour boucherie, 986
emballages pour savons, 5208
emballages qui s'emboîtent les uns dans
    les autres, 3953
emballeur, 6296
embarqueur, 2531
emboîter, 2253
emboutisseuse pour gobelets, 1740
embrayage, 1371
émission, 2260
emmagasinage de données, 1827
empaquetage, 4137
empêchement, 3227
empêchement de la formation de
    mousse, 2632

empilage, 5374
empileur de bois, 5373
emploi, 418
emporte-pièce, 1956
émulsion, 2265
émulsion aqueuse de paraffine, 6128
émulsion d'asphalte, 463
émulsion de bitume, 716
émulsionnant, 2263
émulsionnement, 2262
émulsion, 2264
encapsulation, 2269
encastrer, 2253
encastrer (s'), 5666
enchaînement par liaison hydrogénée,
    3141
enchaînement par liaison transversale,
    1708
enchevêtrement des fibres, 2460
encoller, 2849
encolleuse, 2841, 2847, 2945, 4225,
    5633
encre, 3232
encre d'imprimerie, 4524
encre fixable par la chaleur, 3019
encre magnétique, 3657
encre métallisée, 3766
encre pour impression offset, 4025
encre pour typographie, 3449
endosperme, 2274
endothermique, 2275
enducteur, 419
enduction, 5570
enduction du papier carbone, 1097
enduit, 1382, 4107
énergie, 2277, 4464
énergie cinétique, 3346
énergie électrique, 2221
énergie électrique thermique, 5404
énergie hydroélectrique, 3132
en forme de croissant, 1691
engagement du papier dans la machine,
    5734
engrais, 2439
engraissement, 3117
engrenage, 2799
engrenage cônique, 680
engrenage d'angle, 680
engrenage différentiel, 1968
en phase vapeur, 5961
en plusieurs jets, 3922
enregistrement, 4680
enregistrement de données, 1825
enregistreur, 4678
enregistreur de consistance, 1515
enregistreur de degré de raffinage,
    2719
enregistreur de données, 1822, 1824
enroulage, 6238
enroulement très serré des bords
    (transformation), 616
enrouler, 6234
enrouleuse à carton, 6181
enrouleuse en bout de machine, 4699
enrouleuse Pope, 4438
ensemble, 4130
ensemble des programmes constructeurs
    pour mettre en oeuvre l'ordinateur,
    1985
entaille, 3341
enthalpie, 2286
entomologie, 2287
entraînement, 2288, 4871
entraînement à la vapeur, 5410
entraînement par air, 262
entraîner, 2096
entrée, 3244
entre-lame, 2047
entre-lames d'un cylindre de pile, 5289
entre-lames d'un rotor, 5289
entretien, 3662
entretien préventif, 4512
entropie, 2289
enveloppe, 5057, 6296
enveloppes, 2292
envers, 2432, 6255
envers du papier (coloration), 5897
environnement, 2293
enzymes, 2294

épair, 2684, 3570
épair fondu, 1362
épair irrégulier, 6230
épair nuageux, 1369, 6231
épair régulier, 5918
épaisseur, 2769, 5722
épaisseur de la paroi de cellule, 1200
épaisseur du film d'encre, 3234
épaisseur d'un carton, 1060
épaisseur d'une rame sous une pression
    donnée, 954
épaisseur d'un papier, 1060
épaissir, 1931
épaississement, 1849, 1902, 1932,
    5720
épaississeur, 1488, 1848, 5719
épandage, 5021, 5352
éparpiller, 4927
epibromhydrine, 2295
épicéa, 4300, 5359
épicéa d'Engelmann, 2281
épicéa de Sitka, 5120
épicéa du Canada, 6224
épicéa noir, 722
epichlorhydrine, 2296
épissure, 2788, 5325
épreuve, 5696
épuisement, 2357
épur, 1938
épurateur, 1350, 4945, 5445
épurateur à vibrations, 6004
épurateur centrifuge, 1211
épurateur de pâte, 4581
épurateur incliné, 3202
épurateur plat, 2556
épurateur rotatif, 4852
épurateur tourbillonnaire, 3130
épuration, 1351, 4947, 5475
épuration de l'eau, 6108
épuration des fines, 2514
épuration par l'air, 270
épuration par sablier, 4799
équation, 2298
équerré, 5361
équilibre, 543, 2299
équilibre pondéral, 3714
équipe, 1693, 5059
équipement, 2300
équipement de guidage électronique du
    papier, 4925
équipement pneumatique, 4392
équipement radioactif, 4615
équipement transistorisé, 5265
érable, 144
érable blanc, 3689
érable rouge, 5248, 5518
érosion, 2304
essai, 2350, 5696
essai aux cires Dennison, 1901
essai de jaunissement, 238
essai de paraffine, 6136
essai de pâtes au verre bleu, 784
essai de perméabilité aux liquides, 3520
essai de réception, 134
essai de résistance d'un récipient, 1528
essai de roulage des bords, 1755
essai d'oxydation accélérée, 130
essai préliminaire, 5835
essais à la tache d'encre, 3242
essais à l'essence de térébenthine, 5883
essais biologiques, 704
essais Cobb, 1418
essais Concora, 1490
essais de capillarité, 1679
essais d'éclatement, 980
essais de collage, 5130
essais de compression, 1480
essais de corrosion, 1603
essais d'écrasement à plat, 2551
essais d'écrasement en anneau, 4804
essais de dureté, 2982
essais de fatigue, 2394
essais de flexion, 666
essais de fonctionnement, 4261
essais de matières premières, 3717
essais de mouillabilité en surface, 5592

essais de papiers, 4190
essais de pelurage, 4244
essais de perforation, 4594
essais de pile, 640
essais de résistance, 5460, 5664
essais de résistance à l'affaiblissement, 2381
essais de résistance au chiffonage, 1725
essais de résistance au pliage, 2654
essais de résistance aux acides, 175
essais de résistance aux chocs, 3189
essais de résistance des bords à l'écrasement, 2184
essais de rigidité, 5427
essais de rupture, 5040
essais des cartons, 4158
essais de soufflage, 781
essais de tension, 5683
essais de tirage, 2098
essais de travail, 2174
essais d'humidité, 3869
essais d'imperméabilité à l'huile, 4041
essais d'imprimabilité, 4528
essais d'usure, 6140
essais non destructifs, 4000
essais statiques, 5397
essayeur de degré de raffinage, 623, 2720
essayeur de raffinage en continu, 1534
essence de térébenthine provenant de liqueur kraft, 5527
essuie-mains, 5792
estérification, 2309
esters, 2310
esters de cellulose, 1189
esters de polyvinyl, 4429
esters du glucose, 2835
estimation, 2330
établissement d'une formule, 2694
étachéité à l'eau, 6096
étain, 5748
étalonnage, 1059
étanche à la vapeur, 5964
état de régime, 5401
état des stocks, 3282
état de surface, 5575
état de surface d'un papier lisse et uni, 5201
état de surface grossier obtenu par laminage, 4807
étendeur, 2139
éthane, 2311
éthanol, 2312
éthanolamines, 2313
ether diéthyl, 1966
éthérification, 2315
éthers, 2316
éthers de cellulose, 1190
éthers de polyvinyl, 4430
éthers du glucose, 2836
éthylamine, 2319
éthylation, 2320
éthyl cellulose, 2321
éthylènediamine, 2323
éthylèneinine, 2324
éthyl mercaptan, 2326
étiquetage, 3382
étiquette, 3381, 5631
étiquette couchée, 1397
étiquettes à base de vieux papiers, 810
étiquettes pour boîtes, 1143
étiquettes pour boîtes pliantes, 1136
étiquettes pour caisses, 1143
étiquettes pour rames, 4665
étirage du feutre, 2434
étoffe croisée, 5885
étoffe non tissée, 4007
étoupe, 2472
être rejeté en arrière, 5353
étude, 1923
étude de marché, 3698
étude des fibres par la méthode du fractionnement, 2452
études des pâtes par la méthode du fractionnement, 2709
étuvage, 5407
étuvage direct, 2010
étuvage indirect, 3212
étuve, 2140, 4097

eucalyptus, 2327
eutrophication, 2328
évacuation de l'eau, 6100
évacuation des boues, 5182
évacuation des déchets, 6057
évacuation des gaz, 2783
évacuer, 2329
évacuer par soufflage, 777
évaluation, 2330
évaporateur, 2333
évaporation, 2332
évaporer, 2331
évolution, 2335
exactitude, 143, 4734
examen, 5696
excédent fabriqué sur une commande, 4110
excès de couchage, 2337
excitateur, 2338
exécution, 4260
expansion, 2348
expérience, 2350
expertise, 2330
exploitation forestière, 2673, 2674
expulser, 4729
exsiccateur, 1922
exsudation, 2368
extenseur, 2353
extension, 2354, 5465
extensomètre, 2356
extincteur de la chaux, 3491, 5146
extracteur, 2361
extraction, 2357
extraction de l'alcali, 301
extraction de l'alcali à chaud, 3092
extraction de la moelle, 1912
extrait gélatineux de Canagheen, 1132
extrudeuse, 2362
extrusion, 2363

F

fabricant de papier, 4175
fabrication, 3687, 4534
fabrication d'acide, 168
fabrication de pâtes à haut rendement, 3061
fabrication de pâtes à l'acide nitrique, 3983
fabrication de pâtes à la soude, 5214
fabrication de pâtes au bisulfite, 712, 5535
fabrication de pâtes chimiques, 1253
fabrication de pâtes en continu, 1536
fabrication de pâtes en phase vapeur, 5962
fabrication de pâtes kraft, 3376
fabrication de pâtes mécaniques, 2924, 3735
fabrication de pâtes mi-chimiques au monosulfite,
fabrication de pâtes mi-chimiques, 3959, 4994
fabrication de pâtes par le procédé de la soude à froid, 1434
fabrication de pâtes thermomécaniques, 5715
fabrication de pâtes au polysulfure, 4423
fabrication du papier, 4173
fabrication en discontinu, 609
fabrication hors machine, 4022
fabrication sur machine, 4052
fabrique, 3687
fabriqué sur forme ronde, 2491
facilité, 2375
facilité de solubilité, 2177
façonnage, 1555
façonnage en bord de route, 4811
façonnier, 1554
facteur de déchirure, 5656
facteur de frottement, 2733
facteur d'égouttage, 2087
facteur de puissance, 4465
faction, 5059, 5789

"fadeomètre", 2378
Fagus, 2382
faible consistance, 3576
faible densité, 3577
faire passer dans une qualité supérieure, 5932
faisceau, 959
faisceau de fibres, 2449
fardage, 2152
farine, 2587
fatigue, 2391
faux-acacia, 3546
faux-pli, 1669, 6301
faux-pli de calandre, 1670
fécondation, 2438
fécule, 5386
fécule d'arrow root, 434
fécule de pomme de terre, 4456
fêlure, 2557
fendeur, 5331
fendeuse à bois, 3561
fendillement, 1667
fendre, 5332
fente, 2557, 5176
fer, 3294
fermer avec une bande gommée, 5647
fermeture (usine), 5083
feu, 2521
feuillard, 564
feuille, 3440, 5042
feuille continue humide, 6212
feuille de doublage, 510
feuille défectueuse, 3192, 5783
feuille de papier, 4184, 5052
feuille de papier d'emballage, 1648
feuille de papier en continu, 4194, 6146
feuille de papier offset, 4030
feuille de pâte, 4582
feuille de pâte humide, 5141
feuille de placage, 5992
feuille de protection, 588
feuille d'essai, 5700
feuille dont la résistance est la même dans les deux sens, 5364
feuille en continu orientée au hasard, 4641
feuille en plusieurs jets, 3919
feuille faite à la main, 2966
feuille mince d'aluminum, 332
feuille mince de métal, 2638, 3763
feuille ondulée, 1612
feuilleter (un livre), 5881
feuille volante, 2723
feuillu, 903, 1845
feutrage, 2425
feutre, 2418
feutre aiguilleté, 3950
feutre amortisseur, 1831, 2430
feutre à toile nappée, 1460
feutre bituminé, 464
feutre bituminés, 2431
feutre combiné, 1460
feutre coucheur, 1631, 4487, 6187
feutre d'amiante, 449
feutre d'amiante imperméable, 453
feutre d'amiante pour toitures, 452
feutre de laine, 6288
feutre de papeterie, 4172
feutre de presse inférieure, 853
feutre de rechange, 5291
feutre duplex, 2157
feutre fait à la main, 2962
feutre inférieur, 852
feutre leveur, 4315
feutre marqueur, 3701
feutre montant, 4778
feutre pour forme ronde, 3878
feutre preneur, 4315
feutre preneur supérieur, 5775
feutres à imprégner, 4910
feutre saturé, 2488
feutre sec, 2131
feutre sécheur, 2123
feutre tissé sans fin, 2273
fiabilité, 4734
fibre, 2445, 2473
fibre acrylique, 191
fibre à deux composants, 685

fibre d'acétate, 147
fibre d'amiante, 450
fibre de bagasse, 530, 5517
fibre de bois, 6265
fibre de carbone, 1099
fibre de cellulose, 1191
fibre de céramique, 1217
fibre de chanvre de Manille, 3441, 3680
fibre (de coton ou de textiles artificiels), 5384
fibre de jute, 3328
fibre de lin, 3502
fibre de nylon, 4013
fibre de ramie, 4639
fibre de rayonne, 4653
fibre de récupération, 4674
fibre de verre, 2815
fibre de viscose, 6019
fibre inorganique, 3249
fibre libérienne, 606
fibre métallique, 3761
fibre minérale, 3832
fibre naturelle, 3943
fibres de récupération, 4973
fibres disposées au hasard, 4640
fibres séparées par explosion, 2351
fibre synthétique, 5613
fibre textile, 5701
fibre transversale, 5825
fibreux, 2479
fibre végétale (autre que bois), 4355
fibre vulcanisée, 6036, 6037
fibrillation, 923, 2476, 2477
fibrille, 2474
fibrille élémentaire, 2245
fibriller, 2475
fibrilles, 2478
fibro-ciment, 448
ficelle en papier, 4193
fiche, 5631
fiche d'enregistrement, 4677
figer, 1373
filament, 2481
filature, 5319
fil de fer de cerclage, 552
fil de trame, 6166
fil de trame d'une toile métallique, 6165
fil (électrique ou métallique), 6242
fil en papier, 4195
filetage, 5734
filière, 1954
filière d'une extrudeuse, 2366
filigrane, 6084
filigrane localisé, 3545
filigraneur, 1820
film, 2498
film mince, 5724
fils de chaîne, 6048, 6049
fils vergeurs, 6315
fil textile, 2482, 6315
filtrabilité, 2501
filtrant, 2508
filtration, 2505, 2509
filtration centrifuge, 1210
filtration sous vide, 5944
filtre, 2500
filtre à air, 263
filtre à huile, 4039
filtre à poussière, 2167
filtre à rondelles, 2020
filtre à ruissellement, 5836
filtre à sable, 4895
filtre à tambour, 2114
filtre à vide, 5943
filtrer, 2085
filtre rotatif, 4850
filtre sous pression, 4498
fin de bobine, 4823
fines, 1694, 2512
finesse (pâte mécanique), 2510
fini, 2516
fini antique, 401
fini brosse, 920
fini brut, 1378
fini calandré, 1037
fini crêpé, 1684
fini écrasé, 1720
fini entoilé, 963, 1067

fini gondolé, 1423
fini intermédiaire, 3742
fini lisse et uni, 5159
fini marbré, 3902
fini mat, 2552, 3721
fini obtenu par pression entre deux
    cartons épais, 1124
fini parchemin, 4202
finissage, 2517
fini surglacé, 5556
fini sur une face, 4053
fini vélin, 5985
fini vélin en fin de machine, 6186
fini vélin sur calandre, 1056
fini vélin sur machine, 6188
fini velours, 5991
fini vergé, 3398
fixateur de l'humidité, 3864
fixation des faux-plis, 1674
flan de clicherie, 2580
flèche, 3026
flèche de direction de marche (sur le
    feutre), 2421
fléchir, 4884
fléchisseur, 658
fleuve, 4809
flexibilité, 2562, 5788
flexiomètre, 661, 5426
flexion, 659, 1871
flexographie, 2564
flocon, 2575, 2579
floconneux, 2570
floculat, 2571
floculation, 2573, 2577
flottage des bois, 4620, 4810
flottation, 2583
fluide, 2601
fluidification, 2608
fluidité, 2607
fluor, 2617
fluorescence, 2612
fonceuse, 1398, 1410
fonceuse à brosses, 916, 927
fonctionnement, 4066, 4069
fonctionnement de l'épuration primaire,
    4516
fonctionnement d'une machine, 3629
fonctionnement intermittent, 3274
fondation, 2698
fond d'un récipient, 1525
fond mobile, 3536
fongicides, 2753
fonte, 1153
force, 2103, 2277, 2664, 4464, 5459
force à l'état sec, 2144
force du courant, 5458
force motrice, 4466
force vive, 3346
forestation, 2667
forêt, 2666
forêt secondaire, 4970
formaldéhyde, 2680
formaline, 2681
formamide, 2682
format, 1769, 4185, 5121
format de base, 603
format de la feuille, 5054
format fini, 5841
formation, 2684, 2689, 5702
formation de cratères, 1666
formation de la feuille, 5047
formation d'entaille, 5178
formation de traînée, 5131
formation sous vide, 5945
formats en stock, 5438
forme à main, 2965
forme pour la formation de feuille, 5051
forme pour papiers destinés aux essais de
    pâtes, 2967
forme ronde, 1791, 1797
forme ronde non noyée, 2146
forme sous vide, 5506
formette tiroir (de caisse aspirante),
    1850
formeur multijet, 3918
forme vélin, 6293
forme vergée, 3400
formiates, 2683
formulaires commerciaux, 981

formulaires en continu (LFC), 1533
formule, 2693
fortement collé, 2984
fosse, 4339
fosse couverte, 4965
fosse d'aération, 222
fosse de décantation, 3393
fosse de décharge, 780
fosse de décharge par soufflage, 778
fosse pour eau condensée, 3101
fosse sous la presse humide, 1635
fosse sous le défibreur, 2902
fosse sous machine, 3630
fosse sous toile, 6252
foulage, 4500
four, 2760, 3345, 4097
four à chaux, 3488
four à pyrite, 4600
fourchette de débrayage, 656
four de fusion, 5193
four de récupération, 4685
fourneau, 2760
fournisseur, 5564
fourniture, 1898, 5565
fraction, 2707
fraction de vide, 6029
fractionnement, 1342, 2708
fractionnement des fibres, 2461
fracture, 2711
fragilité, 902
fragment, 4940
fragmentation, 2712
fragments de fibres, 2713
frais fixes, 2536
franco à bord, 2369, 2721
freinage, 877
freins, 876
frêne, 460, 2716
fréquence, 2725
friction, 2728, 4869, 4956
frictionné, 3624
frictionné sur machine, 3626
frictionneur (sécheur), 3599
frisons de papier, 2336
fronteau, 948
frottement, 2728, 4869, 4956
frottement humide, 6198
fuel, 2744
fuite, 1764, 3443
fuite d'eau, 6067
fumée, 5195
fumées, 2750
furans, 2756
furfurane, 2758
furfurol, 2757
fusible électrique, 2217
fusion, 1152, 3745, 3754
fût, 584

G

gabarit, 2687
galactans, 2770
galactomannan, 2771
galactose, 2772
galet tendeur, 3177
galvaniser, 2775, 2776
galvanomètre, 2777
gamma cellulose, 2778
gangue, 691
garde-mains, 3975
garni, 2485
garnissage, 2535
garniture, 3509
gâteau de filtre presse, 2504
gauchissement, 6047
gaufrage, 1696, 2257
gaufrage en continu, 6149
gaufrage toile, 1363, 3503
gaufré, 2254
gaufreuse, 2256
gaz, 2780
gaz de carneau, 2597
gaz d'échappement, 2341, 4736
gaz de chlore, 1307

gazéification, 2787
gazeux, 2784
gaz naturel, 3944
gélatiné, 5582
gélatine, 376, 2802, 2840
gélatinisé, 2803
gélification, 2805
gel (suspension colloïdale), 2801
générateur, 2808
générateur de gaz, 2785
génétique, 2809
génétique forestière, 2670
genévrier, 3322, 3323
gerbage, 5374
gerbeur, 5373
germe, 4983
germer, 5358
germination, 2810
gestion de l'eau, 6083
gicleur, 5356
glaçage, 2823
glaçage sur brosseuse, 922
glacé, 2819
glacé sur calandre à friction, 2734
glaise, 1344
glarimètre, 2830
glissement, 5165
glucomannan, 2832
glucose, 2833
glucosides, 2838
glucosides de méthyl, 3786
glycérine, 2850
glycérol, 2851
glycols, 2853
glyoxal, 2854
gobelets, 1747
gode, 1420, 1751
godets, 1747
gommage, 2944
gomme, 2940
gomme adragante, 2947, 5799
gomme arabique, 129, 427, 2941
gomme de dextrine, 1934
gomme de Guar, 2931
gomme de robinier, 3547
gondolage, 934, 1424, 1675, 1754,
    6047
gonflement, 5602
gorge, 2912
goudron, 5650
gouverneur de pile, 633
graduation, 4922
grain de sable, 2911
grain du bois, 6268
grain (du papier), 2862
grain du papier, 5772
graine, 4981
grains d'adhésif, 5306
graissage, 2894, 3585
graisse, 2883
grammage, 414, 604, 5493
gramme, 2869
grammes/m2, 2767
grande cannelure, 102
grandeur nature, 2747
grande vitesse, 3056
granit, 2871
granité, 3833
graniter, 3834
granulation, 2865
granule, 2874
graphique, 1231
gravier, 2911
greffe, 2860
grenu, 2866
grillage, 2878
grille, 2878
grille de renforcement (pour non tissé),
    4953
gros bout, 987, 989
gros sécheur frictionneur, 6311
gros sécheur frictionneur nerveux
    intérieurement, 3280
grossier, 1377
groupe, 1370
groupement acétyl, 156
groupement acides, 163
groupes alcooliques, 288
groupes aldéhydiques, 291

groupes alkyl, 319
groupes aminés, 342
groupes aromatiques, 433
groupes benzyl, 673
groupes butyl, 995
groupes butyryl, 999
groupes carbonyl, 1106
groupes carboxyalkyl, 1112
groupes carboxyéthyl, 1115
groupes carboxyl, 1117
groupes carboxymérhyl, 1121
groupes des nitrates, 3978
groupes du phénol, 4279
groupes esters, 2308
groupes éther, 2314
groupes éthyl, 2325
groupes éthylène, 2322
groupes fonctionnels, 2751
groupes hydroxyalkyl, 3159
groupes hydroxyéthyl, 3162
groupes hydroxyle, 3165
groupes méthyl, 3787
groupes phényl, 4284
groupes sulfo, 5537
groupes sulfonyl, 5539
groupes syringyl, 5617
groupes xanthate, 6303
grue, 1664
guidage de la toile, 2936
guidage du feutre, 2934
guidage du papier, 2935
guide, 2932
guide feuille, 5048
guide-feuille, 6152
guide-toile, 6245
guillotine, 2938
guirlande, 2779
gymnospermes, 2951

H

habillage, 1364, 2535
habillage de machine, 3610
halogènes, 2957
halogénures, 2955
haut, 5773
haute énergie, 3045
haute pression, 3052
hauteur, 1917, 3026
hauteur de charge, 2993
hauteur de refoulement, 2016
haut rendement, 3059
hélice, 4551
hélicoïdal, 3027
hémicelluloses, 3029
hétérogènéité, 3035
hêtre, 649, 2382
holocellulose, 3076
homogénéisation, 3080
homogénéité, 3079
hotte, 2342, 3084
hotte fermée, 1357
hotte sécheuse, 2125
huilé, 4036
huile, 4032
huile d'abrasin, 5869
huile de graissage, 3584
huile de pin, 5637
huile de tung, 5869
humectage, 1810
humectage à la vapeur, 5405
humectant, 3104
humecteur, 1807
humecteuse, 1809, 3106, 3858
humide, 1806, 3857, 6180
humidification, 1808, 3105
humidité, 3107, 3860
humidité absolue, 113
humus, 3905
hybride, 3111
hydracides, 164
hydrafiner (raffineur cônique), 3112
hydraté, 3114
hydrate de calcium, 1018
hydrate de cuprammonium, 1742

hydrate de cuproéthylènediamine, 1745
hydrate de magnésium, 3650
hydrate de sodium, 5230
hydrates, 3116, 3157
hydrates de carbone, 1092
hydration, 3117
hydraulique, 3118, 3119
hydrazine, 3123
hydrocarbures, 3127
hydrocellulose, 3128
hydrodilatabilité, 3134
hydrodynamique, 3131
hydrogénation, 3139
hydrogène sulfuré, 3145
hydrolysats, 3148
hydrolyse, 3146
hydrophile, 3150
hydrophobie, 6101
hydrophobique, 3151
hydrostatique, 3152
hydrosulfite de zinc, 6328
hydrosulfure de sodium, 5228
hydroxyalkyl cellulose, 3158
hydroxyde d'ammoniaque, 349
hydroxyéthylation, 3160
hydroxyéthyl cellulose, 3161
hydroxyle, 3163
hydroxypropyl cellulose, 3167
hydroxypropylméthyl cellulose, 3168
hydrures, 3124
hydrures de bore, 847
hygrodilatabilité, 3169
hygroscopique, 3170
hypobromites, 3171
hypochlorite de calcium, 1019
hypochlorite de soude, 5231
hypochlorites, 3172
hysteresis, 3174

I

identification, 3175
if, 5653
ignifugé, 2540
ignition, 1466, 3178
ilex, 3179
image latente, 3427
imides, 3180
imines, 3181
imitation de papier à la main, 3182
immersion, 5492
impédance, 3190
impédance électrique, 2218
imperméabilisation, 6095
imperméabilité, 6096
imperméabilité à la graisse, 2888
imperméabilité à l'huile, 4040
imperméabilitée, 3193
imperméable, 3194
imperméable à l'eau, 6093
imprégnation, 3199, 4913
imprégné, 3196, 4909
impression, 4522, 5380
impression à l'écran de soie, 5095
impression à l'aniline, 375
impression en héliogravure, 4858
impression en offset, 4029
impression gaufrée, 1958
impression hélio, 2882
imprimabilité, 3238, 4520
imprimerie, 4522
imprimeur, 4521
impureté, 2012, 3200
impuretés, 1539
incendie, 2521
incendie de forêt, 2669
incruster, 2270
incuit, 5174
inculte, 6232
indicateur, 2798, 3208
indicateur de niveau, 3455
indice d'acidité ionique, 3143, 4274,
　　4276
indice de chlore, 1309
indice d'éclatement, 973, 977

indice de cuivre, 1581
indice d'égouttage Schopper, 4929
indice d'égouttge Canadien Standard,
　　1069
indice de permanganate, 4269
indice de résistance à l'état humide,
　　6206
indice de saponification, 4903
indice de volume, 949
indice NU, 4010
industrie de la transformation, 1556
industrie de l'emballage, 4133
industrie des pâtes, 4573
industrie forestière, 2671
industrie papetière, 4165
inertie du papier, 1993
infalsifiable, 4883
infestation, 3219
inflammabilité, 2542, 3178
infra-rouge, 3220
ingénierie, 2283
ingénierie industrielle, 3215
ingénieur, 2282
ingénieur en chef, 1270
ingraissable, 2884
ingrédient, 3226
inhibiteur, 3228
inhibition, 3227
injection, 3230
inondation, 2582
inspection, 3253, 5563
installation, 3254, 4353
installation de blanchiment, 747
installation de chauffage, 3009
installation de raffinage par soufflage,
　　775
installation de refroidissement, 1576
installation de soufflage, 774
installation d'évaporation des lessives,
　　3522
installation thermique, 5408
instrument, 3255
instrumentation, 3256
instrument de mesure, 3727
insuffisance, 5069
intégration, 3265
intensité, 3266, 5033
intercalage d'une feuille entre deux
　　autres, 3269
interchangeabilité, 3267
intérieur de carton à base de vieux
　　papiers, 4005
intérieur vieux papiers, 2491
interrupteur, 3495, 5605
interstice, 3281
intervalle, 3281
intervention, 4069
inventaire, 3282
inverser, 5881
inversion, 4781
inversion de l'éclat, 897
inversion des couleurs, 1453
iode, 3284
iodure de potassium, 4454
ionisation, 3291
ionomères, 3292
ions, 3293
irradier, 3298
irrégulier, 6229
isocyanates, 3300
isolation acoustique, 184
isolation calorifique, 3010
isolation électrique, 2219
isolation thermique, 3261
isolation thermique calorifugeage, 5705
isomères, 3301
isothermes, 3303
isotopes, 3304
isotropie, 3305

J

jaillissement, 5339
jante, 4801
jaquette, 833

jaspage, 3693
jauge, 2769, 2798
jauge alpha, 323
jauge bêta, 676
jauge de contrainte, 5446
jauge de copeaux, 1288
jauge mécanique de tension, 3736
jaunissement, 231, 2018
jaunissement (du papier), 6317
jet, 3308, 4387
jet d'air, 272
jet d'eau, 6079
jet de couchage, 1638
jets, 4381
jeu, 5012
jeu (ensemble de rouleaux), 4830
jeune arbre, 4901
joint, 2788, 3314
joint hydraulique, 6106
joints de collage, 2844
jonction, 4967
joue (poulie), 2543
journal, 3965
journal amélioré, 5561
journal offset, 4027
journaux à plat invendus bouillons, 4106
Juglans, 3320
jute, 3325

K

kaolin, 1275, 1404, 1408, 3336
kénaf, 3339
kératine, 3340
kérosène, 3342
kraft, 3366
kraft liner, 3371
kraftliner, 3055
kraftliner fabriqué sur forme ronde,
　　1795
kraft neutre, 3955

L

laboratoire, 3386
laine, 6285
laine de bois, 6282
laineux, 6286
lait de chaux, 3489, 3815
lait de colle, 5124
laiton, 878
lamage d'une pile, 639, 5629
lame, 723, 3351
lame d'air, 244, 273
lame d'air pour égaliser la couche sur une
　　fonceuse, 269
lame de coupeuse à bois, 1292
lame de docteur, 2049
lame de massicot, 2939, 5847
lame (de pile), 568
lame (de pile), 626
lame de racle d'égouttage, 3137
lame de rotor, 4860
lamelle, 3406
lamelle moyenne, 3809
lamelle tertiaire, 5691
lames, 3357
lames de pile, 884
laminage à la plaque, 4378
laminé, 3410
laminé à la plaque, 4367, 4368, 4369
laminé en parallèle, 4198
laminer, 3409
laminoir, 4376
lampe, 5865
lampe à rayons cathodiques, 1158
lance de séchoir, 2130
langage de programmation, 4548
laque, 3389, 3404
largeur, 6228
largeur de table (machine), 2373

largeur de table utilisée, 3616
largeur de toile utile, 3725
largeur rognée, 754, 5843
largeur rognée d'une machine à papier,
　　5838
largeur utile maximum d'une machine à
　　papier, 2484
laser, 3425
latex, 3430, 3431, 4866
lavage, 6053
lavage à l'eau acidulée, 182
lavage des gaz, 2793
lavage du feutre, 2436
lavage par diffusion, 1973
laver, 6055
laver à l'eau acidulée, 5284
laveur de feutre, 1500, 2419
laveur de gaz, 2792, 4954
laveur par diffusion, 1972
lecteur, 4662
léger, 3475
le solvant étant récupéré par évaporation,
　　5275
lessivage, 1565, 1567, 1982
lessive, 157, 3435, 3521, 5211
lessive à l'ammoniaque, 346
lessive au bisulfite, 710
lessive au sulfate (kraft), 5523
lessive au sulfite neutre, 3957
lessive composée, 3838
lessive de bisulfite, 5532
lessive de complément, 3665
lessive de cuisson, 1572
lessive faible (bisulfite), 6137
lessive forte (bisulfite), 5476
lessive kraft, 3373
lessive neuve, 6221
lessive pour la fabrication de pâtes,
　　4575
lessiver, 1975
lessive résiduaire, 6058
lessive résiduaire de bisulfite, 5318
lessive résiduaire (noire), 5317
lessiveur, 812, 1566, 1976
lessiveur à chiffons, 4622
lessiveur en continu, 1532
lessiveur en métal plaqué, 1333
lessiveur incliné, 5744
lessiveur rotatif, 4849
lessiveur sphérique, 2825
lessiveur vertical, 6000
lessive verte, 2897
levage à vide, 5947
levier, 3459
lèvre de la caisse de tête, 5155
levure de lessive, 6316
liaison, 817
liaison atomique, 478
liaison chimque, 1243
liaison covalente, 1653
liaison des fibres, 2448
liaison entre deux feuilles de multijets,
　　819
liaison glucosique, 2839
liaison hydrogénée, 3140
liaison hydroxyle, 3164
liaison interfibre, 3268
liant, 691, 820
liard, 1629
libérer (gaz), 4729
lies, 2100
ligne d'arbres, 3508
ligne d'axe, 495
ligne de collage, 2846
ligne de contact entre deux presse,
　　3974
ligne de contact entre deux rouleaux,
　　3974
ligne de partage des eaux, 6107
ligne de transmission, 5813
lignes de toile, 6248
ligneux, 3478
lignification, 3479
lignine hydrate de carbone, 3480
lignines, 3482
lignines alcalines, 302
lignines alcooliques, 289
lignines au périodate, 4263
lignines de bois broyé, 3824

lignines d'hydrolyse, 3147
lignines hydrotropiques, 3155
lignines organo-solubles, 4085
lignines phénoliques, 4282
lignocellulose, 3483
lignols, 3484
lignosulfonate de calcium, 1020
lignosulfonates, 3485
limoneux, 5163
lin, 2559, 3500
liner, 3505
linotype, 3512
linters, 1627, 3514
liqueur à base de calcium, 1012
liqueur à base de magnésium, 3645
liqueur à base de sodium, 5218
liqueur à base soluble, 5271
liqueur de soufflage, 776
liqueur (lessive) noire, 720
liquide, 3517
liquides, 2606
lissage, 2823
lissé, 4907, 5201
lisse, 886
lissé à la calandre de friction, 965
lissé à la pierre, 965
liste de contrôle, 1235
lit de filtration, 2503
lit fluidisé, 2609
litharge, 3526
lithium, 3527
litho en offset, 4026
lithographie, 3532
litho offset, 4293
lithopone, 3534
litière de la forêt, 2672
livraison, 1898
livre, 826
lixiviation, 3436
logement de palier, 619
longévité, 2162, 3563
longueur, 3448
longueur de la fibre, 2465
longueur de rupture, 890
lot de papier soldé, 3310
lot refusé à la livraison, 3310
lubrifiant, 3583
lumen, 3588
lumière, 3470
lumière du jour, 1830
luminosité, 3589
lustrage, 1038
lustrage sur brosseuse, 926
lustre, 2195, 2818, 2826, 4400

M

macération, 3604, 5416
macéré, 3603
macérer, 3602
machine, 3606
machine à carton, 792
machine à carton multiforme, 3927
machine à cercler avec du feuillard, 563
machine à coller les étiquettes, 3383
machine à couper et à perforer le papier
    hygiénique, 5765
machine à courber, 662
machine à crêper, 1689
machine à doubler le carton, 791
machine à doubler par collage, 3417
machine à écorcer, 579
machine à emballer, 4134
machine à emballer, 6298
machine à enrouler les mandrins, 1592
machine à ensacher, 535
machine à enveloppes, 2290
machine à fabriquer les boîtes pliantes,
    1135
machine à fabriquer les caisses, 871
machine à fabriquer les sacs, 539, 541
machine à formation verticale, 6002
machine à forme ronde, 1796, 1800,
    3880, 5979
machine à forme ronde multicylindres,

    3912
machine à mettre en balles la pâte
    humide en feuilles repliées, 3420
machine à onduler, 1619, 1620
machine à onduler double face, 2076
machine à onduler simple face, 5105
machine à papier, 4168
machine à papier à table inversée, 2989
machine à papier avec gros cylindre
    frictionneur, 6312
machine à papier expérimentale, 2349
machine à papier mousseline, 5755
machine à plusieurs toiles, 3932
machine à rainer, 1672
machine à sachets, 4458
machine à sacs, 538
machine à sacs en papier, 4148
machine à sceller, 4964
machine à table plate, 2704, 3909
machine à table plate à trois toiles,
    5850
machine à table plate avec gros sécheur
    frictionneur, 2706
machine à tables multiples, 5886
machine à thermocoller, 3017
machine à toile inclinée, 3204
machine de formation de feuille, 5046
machine de formation de film, 2499
machine de transformation, 1557
machine d'impression, 4525
machine frictionneuse, 3600
machine monocylindrique, 5102
machine monotype, 5938
machine multiforme, 3928
machine pour ôter la moelle de la
    bagasse, 529
machines, 3632
macro-instruction, 4547
macromolécules, 3638
macule à base de vieux papiers, 805
macule pour balles, 550
macules pour emballage, 4670
macules pour rames, 3829
M.A.E., 822
magazine amélioré, 5562
magnésie, 3643
magnésium, 3644
maille (toile de machine), 3756
main, 945
main de papier, 4610
main d'oeuvre, 3387
maintenance, 3662
maintien, 4480
malaxeur, 3816
mal équerré, 4031
maltose, 3671
manchon, 3306, 5149
manchon de presse humide, 1633
mandrin, 1585, 3673
mandrin (de tour), 1325
mandrin écrasé, 1719
mandrin en bois, 6263
mandrin en nid d'abeilles, 3082
mandrin en papier, 4162
mandrin extensible, 3206
mandrin extensible (compensateur),
    2345
mandrin perdu, 4002
manette, 2961
manganèse, 3674
manipulation, 4069
mannans, 3684
mannogalactan, 3685
mannose, 3686
manoeuvre, 6314
manoeuvres, 5938
manomètre, 4499
manque, 5069
manufacture, 3687
manutention de charges unitaires, 5921
manutention des bois, 6270
manutention des matières premières,
    3716
manutention en vrac, 947
marbe en tête de machine à papier,
    2690
marbré, 3900
marbrer, 3904
marbrure, 3899

mare, 4433
margeur, 2407
marketing, 3696
marquage, 3699
marquage au feutre, 2426
marquage sur la calandre, 1041
marque, 3694
marque de fabrique, 3819
marque de la lame d'air, 276
marque du feutre, 728, 2427
marque du rouleau aspirant, 5514
marque du rouleau coucheur, 1634
marque du rouleau de soutien, 509
marque du rouleau égoutteur, 1818
marques à la molette, 4489
marques d'agrafe, 1338
marques d'arrachage du sécheur, 2127
marques de brosse, 925
marques de câble, 4838
marques de calandre, 1046
marques de la caisse aspirante, 5500
marques de la lèvre, 5156
marques de la toile, 6250
marques de planche à laver, 6051
marques de plaque, 4228
marques du feutre sécheur, 2124
marqué sur la calandre, 1040
marqueur, 3695, 4936
marronier d'Inde, 933
marronnier d'Inde, 3089
marteau, 2958
masse, 945, 3705
masse filtrante, 2506
massicot, 874, 2937, 4163, 5844,
    5845, 5848
matelas (de fibres), 3709
matelas de fibres, 2467
matériaux de construction, 1518
matériaux d'emballage, 4135
matériaux de protection, 586
matériaux isolants, 3258
matériaux poreux, 4447
matériel, 2300, 3713
matériel d'apprentissage, 5802
matériel de contrôle, 1542
matériel de coupe, 5050
matériel de levage, 3468
matériel de mesure, 3772
matériel électronique, 2233, 2234
matériel magnétique, 3655
matériel pour la fabrication du papier,
    4174
matériel sécurité, 4882
matériel sous pression, 4509
matière d'arrachage, 4308
matière de protection, 585
matières colorantes, 2173
matière solide (restant dans une solution),
    5260
matières premières, 4647, 4648, 5430
matières premières pour couche centrale,
    1204
matières premières pour mandrins, 1590
matières premières pour support tenture,
    2971
matière thermoplastique, 3094
mâton, 3591
matrice, 1954
maturité, 3723
mauvais fonctionnement, 3670
mécanicien d'usine, 3830
mécanique, 3739
mécanique de la réaction, 4656
mécaniques des fluides, 2604
mécanisation, 3740
mécanisme, 3632
mécanisme pour retirer un rouleau,
    4826
mélangeur à disques, 3187
mélamine, 3743
mélange, 756, 3842, 3846
mélanger, 3837
mélangeur, 3841
mélangeur du milieu, 757
mélangeur Kady, 3335
mélèze, 3423, 3424
mélèze d'Amérique, 5639
membrane, 3747
membrane semi-imperméable, 4996

mercaptans, 3749
mercapto lignines, 3750
mercerisation, 3751
mercure, 3752
méristème, 3755
mesurage, 3748
mesure, 3726
mesure américaine de volume pour bois
    non écorcés, 1738
mesure de capacité = gallons par jour,
    2766
mesure de densité, 1906
mesure de l'absorption, 122
mesure de la pression, 4502
mesure de la vitesse, 5989
mesure de l'humidité, 3109
mesure du débit, 2593
mesure mécanique, 3732
mesure optique, 4074
mesure sans contact, 1522
mesures de précaution contre l'incendie,
    2524
mesures de protection contre l'incendie,
    2528
métacellulose, 3758
métal, 3759
métallisation, 3762, 3764
métallisation sous vide, 5946
métal plaqué, 1335
météorologie, 3769
méthacrylate de méthyl, 3789
méthacrylates, 3774
méthane, 3776
méthanol, 3777
méthode de production, 4542
méthode d'essai, 5698
méthodes du chemin critique, 1699
méthylation, 3782
méthyl cellulose, 3783
méthylène cellulose, 3784
méthyl ethyl ketone, 3785
méthyl mercaptan, 3788
méthyl orange, 3790
métrage (en pieds anglais), 2663
mètre, 3770
meule, 2908
meule concave, 1486
meule de défibreur, 2905, 4583
meuleton, 3365
mi-blanchi, 4989
mica, 3793
micro-analyse, 3795
microcannelure, 2175, 3971
microfibrille, 3799
micro fiche, 3800
micro film, 3801
micromètre d'épaisseur, 3802, 5723
microorganisme, 3803
micro-porosité, 3805
microscopie, 3806
microtome, 3807
migration, 3811
migration de la colle, 207
millième de pouce, 3812
minérale, 3249
mise à niveau, 3456
mise en balle de pâte humide en feuilles
    repliées, 3422
mise en balles, 549, 960
mise en boulettes, 4246
mise en copeaux, 1294
mise en éventail, 2386
mise en feuille, 5049
mise en marche, 5393
mise en place d'une toile, 6249
mise en tableaux, 5626
mise en train, 3664
mis en balles, 546
mixture, 213
mode d'action, 3847
modèle, 3848, 4230, 4893
modèle de rupture, 4873
modèle mathématique, 3718
modernisation, 3849
modificateur, 3852
modification, 5970
module, 3854
module d'élasticité, 3855
module de rupture, 3856

moelle, 4344
moelleux, 5249
moisissure, 3813, 3871, 3905
moisissure de la toile, 6251
moisson, 2991
moissoneur à rondins, 5075
moissonneur, 2990
molécules, 3884
mollette de marquage, 3702
mollette pour rhabiller les meules, 2909
molybdène, 3885
monocotylédons, 3889
monofibre, 5106
monofilament, 3890
monomères, 3892
montage, 5014
montage en porte-à-faux, 1073, 1074
monte-charge, 3466
montmorillonite, 3894
monture, 3067
monture de docteur, 2052, 2053
morceau, 4940
mordant, 3895
mortaiseuse, 5177
mortier, 3896
mortier liquide, 2925
moteur, 3898
moteur électrique, 2220
moteur synchrone, 5610
mou, 5241
moucheture, 5305
mouillabilité, 6207
mouillabilité à l'huile, 4046
mouillabilité en surface, 5591
mouillage, 3859
mouillant, 6209
mouilleuse, 1809
mouilleuse à brosses, 918
mouilleuse à pulvarisation, 5344
moulage, 3879
moulage par injection, 3231
moulin colloïdal, 1437
mousse, 2628, 2738, 4958
mousseline, 1084, 5756
mousseline à fruits, 2742
mousseline pour coutellerie, 413
mousseline pour emballage, 4140
mouvement, 3897
multicouche, 3913
multijet (trois couches), 5735
mutation, 3935

N

nacelle, 4930
naphte, 3938
nappes en papier, 4178
nébulisation, 5959
négoce de papiers, 2042
négociant en papiers, 4176
négociant en vieux papiers, 6060
néoprène, 3952
ne pouvant être rogné au format prévu, 4111
nettoyage par le vide, 5942
neutralisation, 3954
neutralisation de la corrosion, 1600
neutre, 387
nickel, 3969
nitrate, 3979
nitrate de cellulose, 1193
nitration, 3980
nitrification, 3985
nitriles, 3986
nitrites, 3987
nitrocellulose, 3988
nitrolignines, 3991
nitrures, 3984
niveau, 3452
nodule, 3993
noeuds, 3360
noeuds de fibres, 2464
noircissement, 719
noir de fumée, 1096
nombre de bactéries sur une surface

donnée, 519
nombre d'impuretés sur une surface donnée de papier, 2013
nom de l'inventeur du raffineur cônique, 3315
non apprêté, 3620
non collé, 5925
non couché, 5908
non-miscible, 3186
non tissés, 4008
normal, 5382
normale, 1758
normalisé, 5382
norme, 1454
notice, 3442
nourrissant, 4012
noyau, 1585
N.S.S.C., 3937
nuageuses, 1365, 1366, 1368
nuance, 3103
nutritif, 4012

O

obscurité, 3426
obstruction par accumulation de matière pompée, 4383
obstruer, 1354
odeur, 4017
oeil de poisson, 2534
offset, 4024
oléates, 4047
oléfines, 4048
oléorésines, 4050
ombres, 5025
onde sonore (lumineuse), 5283
ondulation, 1420, 1618, 1751
ondulation pour usages spéciaux, 5296
ondulé, 1606
onduler, 4808
onduleuse, 1616
onglet, 2930
opacificateur, 4057
opacimètre, 4058, 5873
opacité, 4059, 5874
opacité de contraste, 1540
opaque, 4060
opérateur, 4071
opération, 4069
opération de contrôle de routine, 4863
optimisation, 4080
ordinateur, 1483
ordinateur analogique, 362
ordures municipales, 3933
orientation, 4086
orientation de trame, 4946
orifice, 3072, 4062
orme, 2246, 5902
oscillant, 4088
oscillateur, 4090
oscillation, 4089
oscillographe, 4091
osier, 4887
osmose, 4092
osmose inverse, 4776
ouate de cellulose, 1197, 6041, 6042
ouate de cellulose crêpée, 1683
outil, 5771
outil de découpage, 1776
outillage, 2300
ouverture, 5995
ouvrier défibreur, 2901
ouvrier des lessiveurs, 1980
overlay, 4107
oxalates, 4115
oxycellulose, 4122
oxydants, 4117
oxydation, 231, 4118
oxyde de chrome, 1322
oxyde de magnésium, 3651
oxyde de zinc, 6329
oxydes, 4120
oxydes de fer, 3296
oxygénation, 4124
oxygène, 4123

oxygène dissous, 2033
ozone, 4128
ozonisation, 4129

P

paille, 5452
paille de blé, 6217
paille de riz, 4793
paille de seigle, 4878
pale, 5954
pale de ventilateur, 2385
palette, 4142
palettisation, 4143
palier, 618
palier à manchon, 5150
palier à rouleaux côniques, 5645
pallette qui dépasse le poids de celle qui a été commandée, 4114
panneau, 4144
panneau de particules, 4213
par raffinage, 6122
papeterie, 4177
papetier, 4170, 4175
papier, 4146
papier à articuler (pour dentistes), 437
papier à base de bagasse, 531
papier à base de bambou, 560
papier à base de chanvre, 3032
papier à base de chiffons, 4625
papier à base de vieux journaux, 3962
papier à beurre, 991
papier à bords barbés, 1854
papier à bords frangés imitant le papier à la main, 3729
papier à bords parallèles après perforation, 4257
papier abrasif, 1109
papier absorbant, 683
papier à cigarettes, 1326, 1327
papier à cigarettes combustible, 1465
papier à couvrir les boîtes en carton, 867
papier à dessin, 2095
papier à dessin à base de vieux papier, 800
papier à doubler, 2376, 3507
papier à doubler les boîtes pliantes, 1137
papier à doubler les livres, 828
papier à fibres longues, 3564
papier à fibres longues pour support de couche, 2658
papier à filer, 4191, 5889
papier à la main, 2964
papier à la main sec, 6080
papier antiseptique, 3741
papier à paraffiner (ou à filer), 5890
papier à plat, 2554
papier à plat livré emballé, 1145
papier apprêté, 3597, 3623
papier à reports, 5805
papier à sacs, 4879
papier au ferrocyanure, 785
papier au sulfate (kraft), 5524
papier auto-adhésif, 4506
papier auto-copiant, 3992
papier autographique, 5805
papier avec bois, 6262
papier avec lignes d'eau, 6082
papier avion, 279
papier à vulcaniser, 6039
papier baryté, 592
papier bible, 682, 3207
papier bicoloré, 2159
papier bien collé, 6178
papier bien garni, 6177
papier blanc, 6223
papier bouffant, 952, 2402
papier braille (pour aveugles), 875
papier brocart, 904
papier brut (emballage), 1379
papier brut pour photo, 4291
papier bulle, 938, 5905
papier buvard, 770

papier calandré, 1036
papier calandré par docteur à eau, 6076
papier calque, 5798
papier calque au ferrocyanure, 786
papier carbone, 1103
papier carbone une fois, 4055
papier chargé, 2490
papier collant, 2942
papier collé, 5123
papier collé à la cuve, 5862
papier collé à la colle, 2843
papier collé à la paraffine, 6134
papier conditionné, 1499
papier contenant des fibres de coton, 1626
papier contrecollé, 3412
papier contrecollé alu, 2642
papier couché, 441
papier couché à couvrir les boîtes, 869
papier couché à la colle, 2843
papier couché anti-adhésif, 4731
papier couché au polyéthylène, 4413
papier couché chrome, 1324
papier couché chrome superbrillant, 1149
papier couché classique, 1385
papier couché deux faces, 5894
papier couché machine, 3613
papier couché mat, 2149
papier couché pour lithographie, 3530
papier couché pour magazines (LWC), 1393
papier couché une face, 4054, 5101
papier couché une face à la brosse, 919
papier couverture à base de vieux papiers, 803
papier couverture non apprêté, 3621
papier crêpé, 1682, 1685
papier crêpé incombustible, 2525
papier cristal, 2817
papier cristal couché, 1392
papier cristal paraffiné, 6124
papier d'alfa, 2306
papier d'aluminum, 333
papier d'amiante, 451
papier d'arsenic, 435
papier de cellulose à la soude, 4963
papier de chanvre de Manille, 4839
papier de chiffons, 3504, 4629
papier de China, 4791
papier de construction, 940, 5041
papier de construction isolant, 3259
papier de construction renforcé, 4720
papier d'édition, 837
papier d'édition avec bois, 2917
papier d'édition couché, 1389, 2821
papier d'édition non couché, 5909
papier d'édition très bouffant, 3038
papier de fibres de verre, 2814
papier de liber, 607
papier d'emballage, 4136, 4139, 6299
papier d'emballage à base de vieux papiers, 804
papier d'emballage pour boulangeries, 881
papier d'emballage pour cadeaux, 2811
papier d'emballage pour magasins de détail, 1647
papier d'emballage pour produits alimentaires, 2662
papier dentelle, 3388
papier de pâte chimique écrue, 3683
papier de riz, 4791
papier de sécurité, 394, 4737
papier destiné à être transformé, 1558
papier de sûreté, 4883
papier de tenture, 2970
papier diélectrique, 2206
papier d'impression, 4526
papier d'impression couché, 1396
papier d'impression non couché, 5911
papier d'impression offset en continu, 6155
papier dont le fini sur chaque face est différent, 2158
papier double, 2075
papier draft goudronné, 3368
papier draft pour sacs, 3367
papier duplex, 2159
papier échangeur d'ions, 3287

papier écriture, 6302
papier écriture coquille, 822
papier écriture couché, 1388
papier écriture emballé en caisse, 868
papier écriture parcheminé, 4208
papier écriture simili parcheminé, 4201
papier écriture vergé, 3403
papier écriture vergé bleuté, 497
papier écru, 5905
papier édition, 829
papier électrophotographique, 2238
papier en bandes, 5644
papier en continu, 6156
papier enduit de matières plastiques, 4360
papier en plusieurs jets, 3923
papier en-tête, 2997
papier entoilé, 3501, 4722
papier épais pour protéger les bords des bobines emballées, 2271
papier étanche à la vapeur, 5965
papier exempt d'acide, 162
papier extensible, 2346, 2355, 5466
papier fabriqué sur forme ronde, 1798
papier fabriqué sur machine à tables multiples, 5887
papier fantaisie pour couvrir les boîtes en carton, 1268
papier feutre, 2428
papier fiduciaire, 1757
papier filigrané, 6089
papier filtre, 2507
papier filtre analytique, 366
papier filtre pour brasseries, 651
papier filtre pour laboratoires, 1248
papier filtre sans cendres, 459
papier fini à la main, 2963
papier fluorescent, 2615
papier fortement apprêté, 2285
papier fortement collé, 2985
papier frictionné, 3601, 3625
papier frisé, 1756
papier gaufré, 2255, 2864
papier gaufré imitation toile, 2371
papier gélatiné, 5583
papier genre velours, 2578
papier glacé, 2268
papier gondolé, 1422
papier goudronné, 5652
papier goudronné pour construction, 468
papier goudron stratifié, 465
papier grainé, 404, 2864, 2868
papier granité, 2872
papier hélio, 2881, 4857
papier hygiénique, 4898, 4899, 5763, 5766
papier imitant le bois, 5993
papier imperméable à l'eau, 6097
papier imperméable à l'huile, 4042
papier imperméable pour emballer les caisses carton contenant des produits alimentaires, 1138
papier impression bouffant, 951
papier impression non apprêté, 398
papier incombustible, 2540
papier inflammable, 2527
papier ingraissable, 2892
papier isolant, 1962
papier journal, 3966
papier journal en continu, 6153
papier journal en feuilles, 3967
papier journal standard, 5383
papier jute, 3331
papier jute pour couverture de caisses, 3329
papier jute pour sacs, 3326
papier kraft, 3375
papier kraft en plusieurs épaisseurs pour sacs G.C., 3930
papier kraft paraffiné, 6125
papier kraft pour câbles, 3379
papier laineux pour calandres, 6287
papier lissé à l'agathe, 2567
papier litho, 3531, 3533
papier marbré, 1318, 3691, 3692, 3903
papier marbré à l'agate, 229
papier mat, 3719

papier métallisé, 3767, 3768
papier micacé, 3794
papier millimétré, 4924
papier mince, 3477, 5726, 5756
papier mince à filer, 5320
papier mince crêpé, 1690
papier mince et doux, 5250
papier mince isolant pour câbles, 3260
papier mince neutre, 388
papier mince paraffiné, 6127
papier mince pour boucherie, 984
papier mince pour collage au bobinage, 5328
papier mince pour condensateurs, 1078, 1497
papier mince pour mettre sous les tapis, 1130
papier mince pour nappes, 3940
papier mince pour paraffinage, 6132
papier mince pour usages industriels, 3217
papier mis au rebut, 1736
papier mousseline pour patrons, 4233
papier multijet, 2490
papier ne correspondant pas à la commande mais néanmoins vendable, 1469
papier noir pour albums, 717
papier non apprêté couleur coquille d'oeuf, 400
papier non collé, 5926
papier noncollé, 6080
papier non collé pour affiches, 729
papier non couché, 5910
papier non rogné, 5927
papier offset, 4028
papier offset couché, 1394
papier offset en continu, 6154
papier opaque, 4061
papier ouaté, 599
papier overlay, 4108
papier paraffiné, 6126
papier paraffiné avec saturation, 2143
papier paraffiné en surface, 6211
papier paraffiné pour confiserie, 5889
papier paraffiné sur calandre, 1049
papier parcheminé grattable, 2303
papier peint, 5376
papier pelure pour doubles, 3677
papier photo sensibilisé, 4999
papier pour accumulateurs, 613
papier pour affiches, 686, 4450
papier pour albums photo, 287
papier pour annuaires, 2011
papier pour billets de banque, 565, 1757
papier pour boulangerie, 542
papier pour câbles, 1007
papier pour caisses enregistreuses, 200
papier pour cannelure, 2622
papier pour capsules de bouteilles, 1085
papier pour cartes géographiques, 3690
papier pour cartes statistiques, 5625
papier pour cartouches, 352, 1139
papier pour catalogues, 1154
papier pour chèques, 1236, 1264
papier pour chromatographie, 1319
papier pour condensateurs, 1496
papier pour copies de lettres, 1582
papier pour coutellerie, 412
papier pour couverture de livres, 1658
papier pour couvrir les livres, 832
papier pour décalcomanie, 1840
papier pour dessin au fusain, 1229, 1230
papier pour diagrammes, 1232
papier pour diagrammes d'appareil enregistreur, 4679
papier pour dossiers, 2483
papier pour doubler les boîtes, 1144
papier pour doubler des caisses, 1144
papier pour doublure arrière, 513
papier pour doublure de livres, 827
papier pour duplicateurs, 2161
papier pour effets de commerce, 688
papier pour emballage de livres, 840
papier pour emballage du pain, 880

papier pour emballer les pneus, 490
papier pour enregistrements, 4682
papier pour enveloppes, 2291
papier pour essuie-mains, 5793
papier pour étiquettes, 203, 3384
papier pour feuillets mobiles, 3572
papier pour filtres à air, 264
papier pour flans, 3720
papier pour gobelets, 1741
papier pour impression en relief, 5714
papier pour intercalaires, 3270
papier pour intérieur de cartons, 2495
papier pour isolation électrique, 1503
papier pour isolation thermique, 3262
papier pour joints, 2790
papier pour la détermination du pH, 3209
papier pour liasses et formulaires en continu, 491
papier pour machine à écrire, 566
papier pour machines comptables, 687, 834
papier pour magazines, 3641, 3642
papier pour mandrins, 1587
papier pour matrices, 507
papier pour mise en train (impr.), 5899
papier pour munitions, 732
papier pour nappes, 3939
papier pour ondulé, 1617
papier pour ordinateur, 1485
papier pour patrons, 4232
papier pour photocopies, 4295
papier pour pyrotechnie, 2400
papier pour registres, 4717
papier pour reliure, 830
papier pour reproductions, 4744
papier pour rouleaux de caissses enregistreuses, 1147
papier pour rouleaux de calandres, 861, 1045
papier pour sachets de thé, 5654
papier pour sacs, 540
papier pour sacs à pain, 879
papier pour sacs ciment, 1202
papier pour salle de bains, 611
papier pour stratifiés, 5451
papier pour taille-douce, 4371
papier pour timbres-poste, 5381
papier pour toitures, 4833
papier pour typographie, 3450
papier pour usages électriques, 430, 2207
papier pour usages spéciaux, 5297
papier "procédé", 4941, 4943
papier protecteur, 587
papier pour condensateurs, 1077
papier quadrillé, 5363
papier qui ne roule pas, 3999
papier registre, 727, 3447
papier résistant à l'alcali, 315
papier résistant à l'état humide, 6204
papier résistant aux acides, 173, 177
papier résistant aux taches de sang, 767
papier rogné, 5840
papiers à base de matières de récupération, 806
papiers abrasifs (emeri), 111
papiers absorbants, 117
papiers à démaquiller, 2374
papiers à dessin, 2099
papiers adhésifs, 205
papiers à doubler à base de matières de récupération, 807
papiers à imprégner à base de matières de récupération, 808
papiers à la main, 5980
papier sans bois, 1247, 6267
papiers anti-adhésifs, 4733
papiers anti-rouille, 406
papier satiné, 2822, 4368
papier satiné fini antique, 402
papier satiné sur calandre, 2831
papiers avec bois, 2921
papiers colorés, 1441
papiers contenus dans la caisse d'égouttage, 2092
papiers couchés, 1395
papiers culturels, 1471

papiers d'emballage écrus, 3941
papiers diazo, 1949
papiers d'impression avec bois, 2922
papier séché à l'air, 258, 3551
papier sensible à la lumière, 3473
papiers et cartons coucheé au kaolin, 1346
papiers et cartons gaufrés, 2259
papiers et cartons résistant aux moisissures, 3881
papiers goudronnés, 466
papiers huilés, 4038
papier simili cuir, 439, 3446
papiers imprégnés, 3197
papiers inorganiques, 3250
papiers mêlés, 3839
papiers minces autres que pour emballage, 1470
papier soldé, 3311
papiers pour boucherie, 985
papiers pour décoration, 1859
papiers pour usages industriels, 3216
papiers réactifs, 3198, 4911
papiers résistant au blanchiment, 749
papiers sans bois, 2511
papiers support pour imprégnation, 3198, 4911
papier stable, 4268
papier stencil, 5421
papiers traités au latex, 3432
papier sulfite, 5533
papier sulfurisé, 4206
papier sulfurisé véritable, 5981
papier support, 596, 794
papier support carbone, 1102
papier support de couche, 1403
papier support (enduction, 598
papier support photo, 5001
papier support pour contrecollage, 4167
papier support pour gommage, 2946
papier support pour goudronnage, 471
papier support pour paraffinage, 6131
papier support pour parcheminage, 4205
papier support pour plastification, 4150
papier support pour stratifiés, 3415, 4149, 4167
papier support tenture, 2969
papier surcalandré, 5552
papier surcalandré dont le fini sur chaque face est différent, 2160
papier surglacé, 5552
papier tenture, 6045
papier thermocollable, 3018
papier tournesol, 3535
papier très bouffant, 3040
papier très chargé à haute teneur en cendres, 1348
papier très glacé, 3047
papier vélin, 5986, 6295
papier velours, 5990
papier vergé, 3401
papier vergé fini antique, 403
papier verni, 3390
papyrus, 4196
paquet (colis), 4130
paraffinage, 6129
paraffinage sur calandre, 1050
paraffine, 6121
paraffineuse, 6130
paramètre, 4199
parc à bois, 6283
parcelle, 5304
parcelles, 2512
parcheminage, 4203
parchemin couché, 442
parcheminer, 4204
parchemin (peau), 4200
parchemin végétal couché, 444
parenchyme, 4209, 4344
paroi de cellule, 1199
paroi primaire, 4518
paroi secondaire, 4975
paroi tertiaire, 5693
partage, 2042
particule, 4212
particule de charge transparente après calandrage, 5060
particules de bois, 6272

partie humide, 6184
parties annulaires, 385
passage, 5995
passerelle, 1165
passer la toile à l'acide, 5285
pâte, 4568, 5480
pâte acceptée, 137
pâte à haut rendement, 3060
pâte à l'acétate, 148
pâte à l'acide nitrique, 3982
pâte à la soude, 5213
pâte à la soude à froid, 1433
pâte alcaline, 310
pâte à papier, 4187
pâte au bisulfite, 711, 5534
pâte au polysulfure, 4422
pâte au sulfate (kraft), 5526
pâte blanchie, 738
pâte blanchissable, 736
pâte bucheteuse, 3364
pâte chimicomécanique, 1262
pâte chimique, 1252
pâte commercialisée, 3697
pâte courte, 5073
pâte d'alfa, 325, 2307
pâte de bagasse, 532
pâte de bambou, 561
pâte de bois, 6276
pâte de bois chimique, 1261
pâte de bois mécanique, 3738
pâte de bouleau, 708
pâte de chanvre de Manille, 104
pâte de chiffons, 2986, 4630
pâte de confières, 5255
pâte de cordages, 2986
pâte de feuillus, 2988
pâte défibrée, 1865
pâte défibrée à froid, 1430
pâte de hêtre, 650
pâte de jute, 2986, 3332
pâte de linters, 1628
pâte de noeuds, 3362
pâte de paille, 5454
pâte de pin au sulfate, 4328
pâte de pin au sulfite, 4329
pâte de résineux, 5255
pâte de vieux papiers désencrés, 1890
pâte dure, 2983
pâte écrue, 5906
pâte en bourre, 2599
pâte engraissée, 2895, 3115
pâte entièrement chimique, 2745
pâte épurée, 136, 137
pâte exportée, 2352
pâte grasse, 5180, 6197
pâte humide en feuilles repliées, 3421
pâte incuite, 5913
pâte kraft, 3377
pâte liquide, 5184, 5186, 5190, 5430
pâte liquide en fabrication, 5185
pâte longue, 3568
pâte maigre, 2390, 2722, 2724
pâte mécanique, 2916, 2923, 3734
pâte mécanique de copeaux, 2919
pâte mécanique de paille, 3737
pâte mi-chimique, 4993
pâte michimique, 1263
pâte mi-chimique au sulfite neutre
   (monosulfite), 3958
pâte molle, 5251
pâte moulée, 3875
pâte neuve, 6014
pâte pour rayonne, 4654
pâte pour usages spéciaux, 5298
pâte pour usages textiles à dissoudre,
   2036
pâte raffinée maigre, 2717
pâtes contenus dans la caisse
   d'égouttage, 2092
pâte séchée à la vapeur, 2544
pâtes mi-blanchies, 4990
pâte vendue dans une filiale, 1088
pâte vierge, 6014
pâte viscose, 6020
pathologie forestière, 2675
patin, 5132
pâton, 3591
pâton de couchage, 1409
pâton de couleur, 1450

pâton de glaise, 1349
pâton dur, 2978
pâton mou, 5247
patron, 4230, 4893, 5419
pauvre, 6229
pavillon, 1641, 3695
pavillon pour indiquer l'emplacement
   d'une collure dans une bobine, 5326
Pavillons, 4666
pectine, 4239
pellicule, 2498
peluchage, 2168, 2764
peluche, 2762
pelurage, 4243
pelurage en surface, 5579
pelure, 5756
pelure d'orange, 4081
pelure pour doubles de lettres, 1583
pelure surglacée, 4056
pénétrant, 4248
pénétration, 4249
pénétration capillaire, 6227
péniche, 570
pentachlorophénol, 4250
pente, 2858, 5175
pentosans, 4251
pénurie, 5069
pépinière, 4011, 4982, 5832
percer, 5471
perchlorates, 4253
perforamètre, 4593
perforateur, 4259
perforation, 4258, 4591
perforer, 4256
perforeuse, 4259
périderme, 4262
périodates, 4264
périphérique, 4265
permanence, 2162, 4267
permanganate de potassium, 4455
permanganates, 4270
perméabilité, 4271
perméabilité à l'air, 281
perméabilité à la vapeur, 5960
perméabilité à la vapeur d'eau, 6117
perméabilité à l'eau, 6091
perméabilité aux gaz, 2791
perméabilité aux liquides, 3519
peroxyde de chlore, 1306
peroxyde de sodium, 5232
peroxydes, 4272
persulfates, 4273
perte, 3574
perte à l'égout, 5023
perte chimique, 1249
perte de charge, 2998
perte diélectrique, 1961
perte par frottement, 2735
pesage, 6167
pesage à la bascule, 3461
pesée, 6167
petite branche, 5884
petite cannelure, 501
pétrole brut, 1716, 2697
pétrole lampant, 3342
peu collé, 5143, 5253
peuplements d'arbres de même âge,
   2334
peuplements hétérogènes, 3840
peuplements vierges, 6013
peuplier, 4439, 4440
peuplier baumier, 558
peuplier du Canada, 1629
peuplier jaune, 6318
peuplier noir, 721
pH, 4274
phase solide, 5263
phénol, 4283
phénolphthaléine, 4283
phloème, 4285
phosphate de trisodium, 5853
phosphates, 4286
phosphore, 4287
photomètre, 4292
photosensibilité, 4294
photosynthèse, 4296
pick up, 4314
pick up de pauvre, 4436

pied à coulisse, 1061
pigment, 4319
pigment colorant, 1448
pigmenté, 4321
pigment laqué, 3405
pile, 4322, 5372
pile blanchisseuse, 739
pile de bois, 6273
pile défileuse, 883, 3074, 5187
pile défileuse (à chiffons), 4623
pile laveuse, 6052
pile raffineuse, 624, 3075
pile sans platine, 885
pin, 4325, 4333
pin à feuilles courtes, 5072
pin d'Ecosse, 4938
pin de murray, 3548
pin de Murray, 3934
pin du sud, 5287
pinènes, 4326
pin gris, 3307
pin jaune, 4342
pin jaune à feuilles longues, 3567
pinorésinol, 4332
pin sylvestre, 4938
piquage (d'une meule), 5032
piqûre au filigrane, 1818
pivoter, 5152
planche à laver, 6050
planche de garde, 522
plan d'une machine, 3617
plantation, 4354, 4356
plante tropicale, 5854
plaquages de calandre, 1051
plaque, 4366
plaque d'assise, 648, 5259
plaque de fondation, 597, 5259
plaque d'extraction, 2359
plaque revêtue d'acier, 1336
plaquer un métal sur un autre, 1334
plasticité, 4362
plastifiant, 4363
plastifier, 4364
plastique, 4359
plastique renforcé, 4723
plastiques à base de cellulose, 1194
plastomètre, 4365
platane, 4352
plateau, 4373
platine, 4373
platine de pile, 634, 648
plein, 1362
pli, 1695, 2643, 4565
pliage, 2646
pliage à la main de plusieurs feuilles à la
   fois, 5246
plié, 2644
plié dans le sens longitudinal, 3566
plié en accordéon, 140
plié en long, 3565
plié en trois à l'état humide, 6189
plieuse, 2655
plieuse colleuse, 2645
pli humide, 6215
pliographe, 2653, 2660
pli rainé, 4931
plis au bobinage, 6236, 6237
plis de calandre, 1033, 1421
plissé, 1697
pliure en accordéon, 1489
plomb, 3437
plombage, 719
plombage (à la calandre), 1027
plombé (à la calandre), 1026
plongeur, 4385
plume, 4247, 4384
poche, 2128
pochoir, 5419
poids, 6170
poids brut, 4650
poids de la couche, 1417
poids d'une rame, 4668
poids d'une rame par rapport à une autre
   d'un poids différent, 2301
poids du support de couche, 5912
poids humide, 6213
poids lourds, 3025
poids moléculaire, 3883
poids nominal, 3997

poids réel, 199
poids sec absolu, 825, 4099
poids spécifique, 5300
poids spécifique apparent, 415
poignée, 2961
poils de chien, 2059
poils de feutre, 2424
point, 4396
point d'ébullition, 815
point de fusion, 3746
point de passage, 5809
point de ramollissement, 5244
point de rosée, 1930
point de saturation des fibres, 2470
point de transition, 5809
point éclair (huiles), 2547
pointillé, 5428
point lustré, 5060
points de mousse, 2636
poivre, 5304
poix, 4340
polarisation, 4398
polarité, 4397
poli spéculaire, 5312
polissage, 4402
poli subjectif, 5490
pollinisation, 4403
pollution, 4404
pollution de l'air, 282
pollution de l'eau, 6092
pollution thermique, 5706
polyacrylamide, 4406
polyacrylate, 4407
polyamides, 4409
polyélectrolytes, 4410
polyesters, 4411
polyéthylène, 4412
polyéthylenimine, 4414
polymères, 4418
polymères amines, 338
polymères de complément, 201
polymères synthétiques, 5614
polymérisation, 4416
polymériser, 4417
polyméthacrylate, 4419
polymides, 4415
polypropylène, 4420
polystyrène, 4421
polysulfures, 4424
polyuréthanes, 4425
polyvinyls, 4432
pompage, 4588
pompe, 4587
pompe à hélice, 494
pompe à piston, 5436
pompe aspirante, 5512
pompe à vide, 5948
pompe de mélange, 2387, 3845
pompe multicellulaire, 3926
pompe rotative, 4851
ponctuation aréolée, 844
pontuseau, 5620, 5621
pore, 4441
poreux, 4446
porosimètre, 1908, 4443, 4445
porosité, 4444
porosité horizontale, 3087
porosité latérale, 3428
porosité transversale, 5826
porte, 2795
portée, 2103
porter à haute température, 483
possibilité, 2396
post-raffinage, 4451
post-sécheur, 226
potassium, 4452
potentiel électrocinétique, 6325
potentiel redox, 4691
potentiomètre, 4457
pouce par heure, 3201
pouche de défibreur, 4393
poudrage, 1222
poudre, 4462
poudre d'amiante, 445
poulie, 4566
poulie cônique, 1506
pour cartes géographiques, 1232
pourriture, 1841
poussière, 2164

poussière aux couteaux circulaires, 5168
poussière de docteur, 2051
poussière de massicot, 1772
pouvant emboîter (s'), 5666
pouvoir abrasif, 110
pouvoir absorbant, 115, 126
pouvoir absorbant à l'huile, 4033
pouvoir adsorbant, 218
pouvoir calorifique, 3005
pouvoir couvrant, 1657, 3037
pouvoir diélectrique, 1965
pouvoir liant, 695
pouvoir lustrant, 2827
pouvoir réfléchissant, 3590, 4709
pouvoir réfléchissant spectral, 5307
précipitateur, 4472
précipitateur électrostatique, 2243
précipitation, 1915, 4471
précipitation électrostatique, 2242
précipité, 4469
précipiter, 4470
précision, 143, 4473
préfabrication, 4476
préformation, 4477
préhydrolyse, 4478
prélèvement de rondelles, 846
première couche d'un couchage, 2497, 4474
première presse coucheuse, 2533
première presse coucheuse aspirante, 5511
premier épurateur dégraisseur, 1380
préparation, 4479
préparation du bois, 6274
préparation du site, 5119
pré-sécheur, 4475
pressage humide, 6196
presse, 4482
presse à balles, 548
presse à cylindre, 1799
presse à décalquer, 5806
presse à double feutre, 2077
presse à écorcer, 580
presse à écoulement transversal, 5824
presse à écoulement vertical, 6001
presse à emballer, 551
presse à haute intensité, 3048
presse à plateaux, 4375
presse à report, 5806
presse à tissu, 2372
presse à vis, 4952
presse combinée, 1462
presse coucheuse, 1636, 6195
presse coucheuse aspirante, 5509
presse coucheuse principale, 2532
presse divisée, 2046
presse double, 2147
presse double divisée, 2070
presse en blanc, 2548
presse encolleuse, 5125, 5584
presse encolleuse au plongé, 5863
presse en pierre, 2873
presse-étoupe, 5481, 5484
presse feuille à feuille, 5044
presse inférieure, 854
presse manchonnée, 5151
presse montante, 4777
presse offset, 5198
presse-pâte, 6191
presse pick up, 4317
presse primaire, 4515
presse principale, 3661
presse rainurée, 2913
pression, 4488, 4495, 5461
pression atmosphérique, 477
pression barométrique, 583
pression de la vapeur, 5963
pression en livres par point, 4460
pression exacte, 1700
pression hydraulique, 3121
pression linéaire, 3496, 3977
pression osmotique, 4093
pression sous laquelle l'épaisseur d'une rame est déterminée, 953
pression statique, 5396
pressurisation, 4508
pré-traitement, 4511
prise automatique, 3464, 4314

prise d'air, 271
prise d'encre, 3235
procédé, 4532
procédé à contre-courant, 1645
procédé aérobique, 224
procédé à la soude, 5212
procédé à la soude à froid, 1432
procédé alcalin, 309
procédé alcalin de fabrication des pâtes, 311
procédé anaérobique, 359
procédé au chlore, 1310
procédé au sulfate, 5525
procédé Ben Day (écran), 657
procédé de formation à sec, 2133
procédé d'épuration par boues activées, 195
procédé discontinu, 608
procédé en continu, 1535
procédé en plusieurs stades, 3925
procédé humide, 6190
procédé mi-chimique au sulfite neutre, 3937
procédé Mitscherlich (bisulfite), 3836
procédé pour améliorer la qualité des pâtes, 1752
processus, 4532
processus de corrosion, 1599
production, 4096, 4539
production de cordons sur une bobine, 4840
production de mousse, 2630
productivité, 4543
produit, 232, 4537
produit accélérant, 131
produit adsorbé, 215
produit agissant en surface, 5569
produit anti-buée, 396
produit anti-dérapant, 391
produit anti-mousse, 1875
produit antistatique, 409
produit coagulant, 1375
produit correcteur, 1750
produit défloculateur, 1873
produit de flottation, 2584
produit de nettoyage, 1350
produit de remplacement, 5494
produit d'imprégnation, 3195
produit égalisateur, 3457
produit empêchant l'adhérence en surface, 389
produit en formation, 2691
produit floculant, 2572
produit gonflant, 5603
produit mouillant, 6210
produit permettant d'améliorer qualité des pâtes, 1753
produit provoquant la formation de mousse, 2631
produit provoquant un blocage, 3704
produit remplaçant le papier, 4189
produit rendant résistant à l'état humide, 6202
produits amorphes, 355
produits chimiques, 1257
produits corrosifs, 1601
produits de combustion, 1468
produits en pâte moulée, 3874, 3877
produits forestiers, 2676, 6275
produits papetiers, 4181
produits siccatifs, 2135
profil, 4544
profil transversal, 1709
profit, 4546
profondeur, 1917
programmateur, 3887
programmation, 3888, 4928
programmation non linéaire, 4001
programme d'ordinateur, 1484
programme linéaire, 3497
propagation, 4549
propane, 4550
propionates, 4553
propriété, 4552
propriétés acoustiques, 185
propriétés chimiques, 1251
propriétés des cartons, 4157
propriétés de surface, 5580
propriétés de traction, 5677

propriétés d'extension, 5464
propriétés diélectriques, 1964
propriétés d'imprégnation, 4912
propriétés du papier, 4182
propriétés électriques, 2208
propriétés magnétiques, 3658
propriétés mécaniques, 3733
propriétés optiques, 4075
propriétés physiques, 4299
propriétés rhéologiques, 4789
propriétés thermiques, 5707
propulseur, 3191, 4551
prospectus, 3442
protection, 4480
protection contre les champignons, 2754
protection contre les faux-plis, 1671
protection du plancher d'un camion avec du celloderme, 1126
protéines, 4559
protéines de soja, 5288
protolignines, 4560
protons, 4561
protoxyde de chlore, 1308
pruche du Canada, 5858
Prunus, 4562
pseudoplasticité, 4563
puces, 3036
puisard, 5549
puissance, 4464
puits, 6176
pulsation, 4585
pulvérisateur, 4586, 5347
pulvérisation, 5348
pulvérisé, 4463
pulvériser, 5341
pureté, 4597
pureté colorimétrique, 1444
purge (d'air ...), 5998
purgeur à vapeur, 5411
purgeur de pile, 887
purgeur (vapeur), 5827
purification, 4596, 5475
pyrite, 4598
pyrolyse, 4601
pyrotechnie, 2397

Q

qualité, 2857, 4552, 4604
qualité de l'air, 283
qualité de l'eau, 6098
qualité d'une pâte, 4579
qualité d'une pâte grasse, 5179
qualité d'une pâte maigre, 2718
qualité journal, 3963
quantité chargée, 2761
quantité chargée dans la pile, 631
quantité fabriquée en moins sur une commande, 5915
qui absorbe l'eau, 6064
qui dépasse le poids commandé, 4112
qui déteint, 1703
qui est extrait, 2360
qui n'a pas la teinte voulue, 4019
qui ne répond pas à la qualité voulue, 4023
qui ne se courbe pas, 3998

R

raccord, 3976
raccordement, 4967
raccordement électrique, 2211
raccorder, 4968
raccord soudé, 6174
raccourcissement des fibres, 5070
racine, 4834
râcle, 2048
râcle de couchage par léchage, 2054
racle d'égouttage, 2088, 3136

radeau, 4619
radiation, 4613
radical, 4614
radioactivité, 4616
raffinage, 641, 4706
raffinage à chaud, 3098
raffinage à haute densité, 3044
raffinage au raffineur cônique, 3316
raffinage des refus, 4725
raffinage poussé, 2972
raffinage sous pression, 4510
raffiné, 622
raffiner, 620, 4618
raffiner dans un raffineur à boulets, 4238
raffineur, 4704
raffineur à boulets, 554, 4237
raffineur à disques, 2022
raffineur à double disque, 2069
raffineur associé à un cuvier, 1784
raffineur cônique, 1508, 3318
raideur, 5443
raies de docteur, 2055, 2056
raies en long sur papier calandré, 1055
rail de pontuseau, 5623
rainé, 4935
rainer, 2915
rainurage, 5138, 5178
rainure, 2912, 5176
rainure de la caisse aspirante, 5501
ramasse-feuilles, 3433
ramasse-pâte, 1488, 4914, 4915
ramasse-pâte à flottation, 2586
rame, 4664
rame cachetée, 4667
rame incomplète, 908
ramie, 4638
ramollissement, 5243
râperie, 2920
rapport, 4644, 4676
rapport de vide, 6030
rapport liqueur/bois, 3524
rapport produits chimiques/bois, 1259
raser, 107
rayon (d'une cercle), 4617
rayon médullaire, 4651
rayonne, 4652
rayonne à haute ténacité, 3051
rayonne à l'acétate, 149
rayonne cupro-ammoniacale, 1734
rayonnement, 4613
rayonnement infra-rouge, 3225
rayonnement ultraviolet, 5903
rayonne par pneus, 5754
rayonne viscose, 6021
rayons bêta, 678
rayons X, 6309
réacteur, 4661
réactif, 4663
réaction chimique, 1254
réaction d'oxydo-réduction, 4692
réaction endothermique, 2276
réaction exothermique, 2343
réactivité, 4660
réalimentation, 2404
réalisation, 4260
rebobinage, 4788
rebobiner, 4786
rebobineuse, 4787
rebobineuse à double tambour, 2072
reboisement, 4711
reboisement naturel, 3945
rebondissement, 4751, 4752, 5360
rebord antérieur, 2736
rebut, 1735, 6056
rebut (arbre), 6163
réceptivité, 4672
réchauffeur à infra-rouges, 3223
réchauffeur d'air, 268
réchauffeur de gaz, 2786
recherche, 4747
recherche opérationnelle, 4070
récipient, 1523, 6003
récipient à cloisons intérieurs, 4217
récipient à dessus ouvrant, 4065
récipient avec couvercle à charnière, 3062
récipient cylindrique en carton, 1475
récipient en carton, 1070, 2450

récipient en carton gris, 4153
récipient étanche, 3444
récipient s'emboîtant dans un autre, 5668
reconstruction, 4671
recouvrement, 3419
rectificateur, 4687
recul, 4772
récupération, 4675, 4683, 4684
récupération chimique, 1255
récupération de chaleur, 3013
récupération de l'eau, 6099
récupération des faux-plis, 1673
récupération des fibres, 2468, 2469
récupération transversale, 1710
recyclage, 4689
recyclage de l'eau, 6105
redreseur au silicium, 5093
redresseur, 4687
réducteur de vitesse, 5315
réduction, 4696
réduire en pâte, 4574
réemploi, 4774
refendeuse, 5167
reflectomètre, 4710
reflet, 2813
réflexion, 4708
réflexion (optique), 4707
réflexion spectrale, 5308
réflexion spéculaire, 5313
reforestation, 4711
réfractaire, 4713
réfrigérant, 1573
refroidi à l'air, 253
refroidissement, 1574
refroidissement à l'air, 254
refroidisseur, 1573
refus d'épuration, 4948
région cristallinisée, 1730
réglage, 1541
réglage de la chaleur, 3020
règle automatique, 489
règle de caisse d'arrivée de pâte, 524
règle (table de fabrication), 5154
réglette, 1851
régularité, 5919
régulateur, 2855, 4719
régulateur de densité, 1516, 4718
régulateur de pâte, 5437
régulateur de pression, 4504
régulation en boucle fermée, 1358
régulier, 1362
réhumidifier, 4785
rejet, 4726, 5634
relâchement, 4728
relais électrique, 2224
relentissant, 4767
relèvement de données, 1826
reliure, 694
remise en pâte, 4746
rempli, 2485
remplissage, 2497
remplissage de la pile, 630
remuer, 3313, 5429
rendement, 2191, 6319
rendement constant, 5596
rendement de l'actif, 4773
rendement du séchage, 2192
rendre imperméable à la graisse, 2887
rendre incombustible, 2526
rendre résistant aux intempéries, 6142
rendre transparent, 5820
renflement, 944
renseignements, 1821
rentabilité, 6319
rentabilité de capital, 4773
rentrer, 5470
répartisseur de pâte, 5349
répartiteur de pâte, 2596
répartition de la taille des particules, 4215
répartition des longueurs de fibres, 2466
répartition et bourrage des copeaux, 1290
repère, 4716
report, 5804
repos, 2061, 4728
reproduction cryptogame, 455

reprographie, 4745
répulsif, 4742
réseau public de distribution d'électricité, 2226
réservior, 5641
réservoir, 1266, 4748
réservoir à salin, 5194
réservoir couvert, 4966
réservoir de compensation, 5594
résidu, 4750
résidu de la combustion de la lessive noire, 718
résidus, 2100
résidus solides, 5266
résidu végétal, 4357
résilience, 4752, 5360
résinates, 4755
résine, 4327, 6278
résine (colophane), 4841
résine de condensation, 1494
résine hydrogénée, 3138
résine mélamine-formaldéhyde, 3744
résine phénol-formaldéhyde, 4278
résine renforcée, 2696
résines, 4753
résines de silicium, 5094
résines échangeuses d'ions, 3288
résines époxy, 2297
résines phénoliques, 4281
résines polyamides, 4408
résines résistant à l'état humide, 6205
résines urée, 5936
résines urée-formaldéhyde, 5935
résine synthétique, 1748, 5615
résineux, 1509
résistance, 3337, 4758, 4761, 5459
résistance à l'abrasion, 109
résistance à la chaleur, 3014
résistance à la compression, 1479
résistance à la corrosion, 1602
résistance à la déchirure, 5658, 5659, 5662
résistance à l'adhérence, 766
résistance à la fatigue, 2393
résistance à l'affaiblissement, 2380
résistance à la flexion, 663, 2565
résistance à la graisse, 2890
résistance à l'alcali, 303
résistance à l'allongement, 4611
résistance à la pénétration par un liquide, 4759
résistance à la perforation, 4592
résistance à la pourriture, 1842
résistance à l'arrachage, 821, 823, 4309, 4311
résistance à l'arrachage en surface, 5571
résistance à la rupture, 891, 892, 5038
résistance à la rupture par traction, 5678
résistance à la traction à l'état humide, 6208
résistance à la traction à mâchoires jointes, 6324
résistance à l'eau, 6102
résistance à l'éclatement, 975, 978
résistance à l'éclatement à l'état humide, 6193
résistance à l'éclatement d'après Mullen, 3907
résistance à l'écrasement, 1726
résistance à l'écrasement à plat, 2550
résistance à l'égouttage, 2089
résistance à l'encre, 3239
résistance à l'état humide, 6201
résistance à l'huile, 4043
résistance à l'humidité, 3866, 3867
résistance à l'oxydation, 4119
résistance à l'usure, 4760
résistance au bombement, 943
résistance au chiffonage, 1724
résistance au choc, 3188
résistance au cisaillement, 5036
résistance au déchirement amorcé, 3277, 3279
résistance au feu, 2529
résistance au frottement, 4957
résistance au frottement à l'huile, 4044
résistance au frottement à sec, 2142

résistance au frottement humide, 6199
résistance au gondolage, 1668, 1677
résistance au grattage, 2302
résistance au jaunissement, 237
résistance au pliage, 665, 2652, 2657, 2659
résistance au sang, 768
résistance au thermocollage, 3015
résistance aux acides, 174
résistance aux champignons, 2755
résistance aux chocs, 5065
résistance aux intempéries, 6143
résistance aux taches, 5379
résistance aux taches d'alcali, 306
résistance chimique, 1256
résistance dans le sens Z, 6322
résistance de l'encre au frottement, 3240
résistance des bords du papier à la déchirure, 2188
résistance des couches à l'arrachage, 4390
résistance d'une caisse à la compression, 866
résistance d'un papier goudronné à la transudation du goudron, 753
résistance élastique, 2201
résistance électrique, 2225, 4762
résistance en colonne, 1457
résistance en surface, 5581
résistance initiale à la déchirure, 3229
résistance superficielle, 5586
résistant à l'acide, 170
résistant à l'alcali, 314
résistant à l'humidité, 3865
résistant aux acides, 176
résistant aux intempéries, 6141
résistibilité, 4761
résonnance, 4763
resorcinol, 4764
ressources en eau, 6103
ressources naturelles, 3946
restriction, 4766
retardement, 4768
retardeur de flamme, 2541
retardeur d'incendie, 2530
rétention, 4769
rétention de charge, 2496
retenue d'encre (résistance du papier à la pénétration de l'encre), 3071
réticulé, 4771
retour, 4781
retourner, 5881
retrait, 5082
retrait de la feuille au cours du séchage, 2101
rétrécir, 5081
rétrogradation, 4772
réutilisation, 4774
revêtement, 1399, 3509, 5058
revêtement de boîtes, 870
revêtement de céramique, 1216
revêtement d'une cuve, 5860
revêtement en céramique de la face supérieure d'une caisse aspirante, 1219
revêtement résistant aux acides, 172
révolution, 4782
rhabillage des meules d'un défibreur, 970
rhéologie, 4790
rider, 4808
rigidité, 2566, 4800, 5425
rigidité en anneau, 4806
rigole à copeaux, 1286
rinceur, 5077
rinceur de choc de cassé, 3358
ripage, 5165
rivière, 4809
robinet, 1419, 5952
robinet à pointeau, 3949
robinet de sécurité, 4881
robinier, 3546
rogné, 5839
rogner, 751
rogneuse, 1770
rognure, 5846
rognures, 5034

rognures blanches, 730
rognures collées, 4222
rognures de caisses, 865
rognures de livres, 838
rognures de papier couché, 1386
rognures de relieur, 693
rondeur, 4862
rondin, 3554
rondin en longueur d'arbre, 5833
rondins en désordre, 3559
ronger, 1596
roseau, 4698
rotation, 4856
rotomètre (indicateur de débit), 4847
rotor, 3191, 4859
roue hydraulique, 6118
rougeur, 788
rouille, 4874
roulage des bords, 2185
roulage des bords du papier, 1754
rouleau, 4699, 4816
rouleau anti-déflecteur, 393
rouleau aspirant, 5513, 5949
rouleau aspirant multicellulaire, 3910
rouleau bombé, 1713, 4820
rouleau bouché, 1065
rouleau briseur de mâtons, 3592
rouleau cannelé, 1611
rouleau chauffé à la vapeur, 5409
rouleau cintré, 858, 1762
rouleau danseur, 1814
rouleau de calandre, 859, 1044
rouleau de commande de toile, 6258
rouleau de formation, 2692
rouleau de lisse, 5199
rouleau de papier, 4828
rouleau de papier d'une calandre, 2492
rouleau de papier hygiénique, 5764
rouleau déplisseur, 2347, 5351, 6158
rouleau de presse coucheuse, 4490
rouleau de presse supérieure, 5777
rouleau de renvoi de la feuille, 2627
rouleau de retour de toile, 6258
rouleau de soutien, 508
rouleau de tension refroidi à l'eau, 6072
rouleau de tête, 894
rouleau de traction, 4567
rouleau de transmission de sauce, 2701
rouleau de ventilation des pouches de sécherie, 4395
rouleau distributeur, 2043
rouleau écraseur de mâtons, 3593
rouleau égalisateur, 2057
rouleau égaliseur, 4821
rouleau égoutteur, 1816, 1820
rouleau égoutteur à tourillon fermé, 1360
rouleau égoutteur vergé, 6086
rouleau élastique de calandre, 2199
rouleau en caoutchouc, 4867
rouleau enducteur, 420
rouleau en lame, 1272
rouleau entraîné, 2105
rouleau essoreur, 5365, 6300
rouleau filigraneur, 6085, 6087, 6090
rouleau filigraneur vergeur, 6088
rouleau garne de caoutchouc, 4865
rouleau-guide automatique, 485
rouleau guide rouleau conducteur, 3439
rouleau guide-toile, 6253
rouleau guide-toile automatique, 486
rouleau guideur, 2933
rouleau inférieur, 856, 3347
rouleau inférieur de presse humide, 851
rouleau intermédiaire, 3272
rouleau leveur aspirant, 4318, 5507
rouleau monté sur ressorts, 5354
rouleau mouilleur, 1811
rouleau perforé, 4255, 4688
rouleau perforé avec déflecteur, 525
rouleau perforé (caisse d'arrivée), 3078
rouleau pivotant, 5153
rouleau plongeur, 5604
rouleau plus lourd que celui qui a été commandé, 4113
rouleau porteur, 1133
rouleau presseur, 3069, 4795
rouleau presseur sur bobine, 4797
rouleau rainuré, 2914

rouleau refroidisseur, 1271, 5599
rouleau souffleur, 284, 779
rouleau supérieur, 4608, 5778
rouleau supérieur de presse humide, 5774
rouleau tendeur, 5687
rouleau tendeur de feutre, 2435
rouleau tendeur oscillant, 1815
rouleau transporteur, 2406
rouleau vergeur, 3397, 6292
rouleau vergeur en spirale, 5324
rouleaux porteurs de la feuille, 5053
roulement, 618, 4856
roulement à aiguilles, 3948
roulement à billes, 553
roulement à rouleaux, 4824
ruban, 5472, 5643
ruban adhésif, 2943
rugosité, 4861
rupture, 882, 2383, 2711, 4310
rupture de la feuille sur machine, 6147
rupture dûe à la fatigue, 2392

S

sablier, 4798
sablier (de pile), 4896
sac, 526
sac à fond croisé, 1704
sac à ordures, 4714
saccharose, 5496
sac en papier, 4147
sac en plusieurs épaisseurs de papier, 3929
sac G.C. à valve, 5953
saleté, 2012
salin, 5192
salinité, 4886
salle conditionnée, 3110
salle de fabrication, 636
salle de préparation de la lessive au bisulfite, 169
salle de préparation du bois, 6279
salle d'épuration, 4949
salle des défibreurs, 2904
salle des fournitures, 5566
salle des machines, 3631
salle des piles, 636
salle des presses, 4491
sans acide, 161
sans apprêt, 5907, 5916
sans bois, 2918, 6266
sans cendres, 458
sans piqûres, 4331
sapin, 105, 2520
sapin argenté, 5098
sapin baumier, 557
sapin concolore, 6219
sapin de Douglas, 2083, 4564
sapin du Canada, 3030
sapin élancé de Vancouver, 2870
saponification, 4902
sassafrass, 4906
satinage, 5197
satinage à chaud, 3097
satinage en continu, 6151
satinage humide, 6157
satinage sur calandre, 2829
satiné, 2132, 2819, 4367
saturation, 4913
saturé, 4909
sauce (de couchage), 1416, 5164
sauce de couchage, 1406
saule, 4887, 6233
saut, 5139
sauvetage, 4890
savon, 5206
savon à l'huile de pin, 5638
savon de chaux, 3493
scarification, 4926
sceller, 4959
sciage, 4919
scie, 4917
scierie, 4920
sciure, 4918

sec, 2117
sec absolu, 498, 824, 4014, 4098
sec à l'air, 101, 259
séchage, 2134
séchage à fond, 5738
séchage à grande vitesse, 3057
séchage à la barre, 4399
séchage à la flamme, 2538
séchage à l'air, 261, 3553
séchage à la perche, 4399
séchage à la vapeur, 2545, 2546
séchage excessif de la feuille, 5064
séchage par convection, 1550
séchage par infra-rouges, 3222
séchage par pulvérisation, 5345
séchage par accrocheuse, 2443, 3571
séché à l'air, 257, 3550
sécherie, 2121, 2126, 2129
sécher sur accrocheuse, 2444
séché sur accrocheuse, 2441
séché sur cylindre, 1794
séché sur machine, 3619
sécheur, 2122, 2126
sécheur à air, 260
sécheur à air à grande vitesse, 2953
sécheur à deux rangées, 2067
sécheur à feuille aéroportée, 265
sécheur à tambour, 2113
sécheur de feutre, 2422
sécheur flottant, 2568
sécheur par convection, 1549
sécheur type Madeleine, 3639
séchoir, 2139
séchoir à écorcés, 575
séchoir à l'air, 3549, 3552
séchoir à platines, 4374
séchoir à tunnel, 5871
secondaire, 139
seconde caisse d'arrivée de pâte, 4972
seconde presse, 4976
secousse, 5027, 5031
section des calandres, 1048
section toile, 6254
sécurité, 4880
sécurité (de fonctionnement) certitude, 4734
sédiment, 4979
sédimentation, 4980
sélection, 4986
sélectivité, 4987
sels inorganiques, 3251
semence, 4981
semi-conducteur, 4995
semis, 4984
sens, 2009
sensibilisateur, 5000
sensibilisation, 4998
sensibilité, 4997
sensibilité à la pression, 4507
sensibilité thermique, 5708
sens machine, 2863, 3562, 3618
sens travers, 186, 1707
séparateur, 2710, 5004
séparateur à cyclone, 1790
séparateur magnétique, 3659
séparation, 5003
séparation des couches, 4391
séparation des plis, 1894
séparations en papier (emballage), 4179
sequoia, 5007
sequoia toujours vert, 4697
série, 5005, 5012
sérier, 5006
serpentins de chauffage, 5403
serré, 5008
service, 5009
serviette, 5791
serviettes en papier, 4192
serviettes hygiéniques, 4897
servomécanisme, 5011
seuil de plasticité, 6320
seuil de rentabilité, 888
sève, 4900
siccatif, 1921
siccité, 2141
siège d'une réaction, 4657
signet, 3695
silicate de soude, 5233
silicates, 5090

silicium, 5091
silo, 689
silo à copeaux, 1277, 1296
silo à fond mobile (avec vis transporteuse), 3537
simili cuir, 438, 3183
simili kraft, 802
simili sulfurisé, 440, 3185
simili-sulfurisé ingraissable, 2889
siphon, 5115
siphon à eau, 582
sisal, 5117
size press, 5125
slotter, 5177
sodium, 5215
sol, 5256, 5257
solde, 543
solidification, 5262
solidité à la lumière, 2389
solidité de la couleur à la lumière, 1442
solidité à la lumière, 3471
solubilisation, 5270
solubilité, 5269
soluble dans l'alcali, 305
soluble dans l'eau, 6109
solution, 3521, 5272
solution de blanchiment, 743, 746
solution de mouillage, 2702
solution de résine dissoute dans un solvant appliqué sur un papier non collé, 5275
solution de salin récupérée, 2897
solution filtrée, 2508
solvant, 5274
solvation, 5273
sommet, 5773
sondeur, 5002
sonnant, 4645
sorbent, 5277
sorption, 5278
sorte, 2857
sorte de carton, 4154
sorte de papier, 4164
sorte de pince pour saisir les bobines, 2876
souche (d'arbre), 987, 5487
soude, 5209
soude à froid, 1429
soude caustique, 1171
souder, 4968
soudure, 5258, 6175
soufflage, 771
soufflet (sac), 2950
soufflette, 243
soufre, 5541, 5547
soufre actif, 198
soufre réductible, 4693
souiller, 1529
soupape, 5952
soupape à boulet, 556
source de lumière, 3474
sous-couche, 5914
sous-directeur, 474
sous-produit, 1000
sous-produit rejeté, 4871
soutien, 505, 5567
soutirage, 752
soutirer, 751
spectrophotomètre, 5309
spécialités, 5294
spécification, 5299
spectromètre à rayons X, 6310
spectroscope, 5310
spectroscopie, 5311
spectroscopie aux microondes, 3808
spectroscopie de masse, 3707
spectroscopie d'émission, 2261
spongieux, 5333
stabilisant, 4481, 5371
stabilisateur, 5371
stabilisation, 5369
stabilisation de l'inertie, 1994
stabilisé, 5370
stabilité, 4267, 5367
stabilité d'une pâte, 5368
stabilité thermique, 5709
standard, 5382
statistiques, 5399
stator, 5400

stéarate d'alumine, 334
stéarate d'ammoniaque, 350
stéarate de soude, 5234
stéarate de zinc, 6330
stéarates, 5413
stéarates de glycérol, 2852
stencil, 5419
stéréoscope, 5423
stériliser, 483
stock, 5430
stockage, 5441
stockage en vrac, 956
stock de chiffons, 4633
stock de livres, 839
stock de papier M.A.E., 567
stonite, 5440
stratification, 3414, 3416, 5450
stratifié, 3410
strié, 5469
strie de lame, 4302, 4942
strier, 5457
stries, 4944
structure, 4188, 5478
structure cellulaire, 1181
structure cristalline, 1733
structure de la cellulose, 1196
structure des cellules, 1179
structure des fibres, 2471
structure du bois, 6280
structure en nid d'abeilles, 3083
structure lamellaire, 3407
structure moléculaire, 3882
styrolène, 5489
sublimation, 5491
substituant, 5494
succédané, 5494
succession, 5005
suie, 5276
suintement, 4985
suite, 5005
sulfamates, 5519
sulfate d'alumine, 335, 4171
sulfate d'ammoniaque, 351
sulfate de baryum, 573
sulfate de calcium, 1022
sulfate de calcium brut, 4888
sulfate de calcium hydraté, 1714
sulfate de calcium précipité (charge), 4234
sulfate de diméthyl, 1998
sulfate de magnésium, 3652
sulfate de soude, 5235
sulfate de zinc, 6331
sulfates, 5528
sulfite de calcium, 1023
sulfite de magnésium, 3653
sulfite de sodium, 5237
sulfites, 5536
sulfonate de xylène, 6306
sulfonates, 5528
sulfonates d'alkylaryle, 317
sulfone de diméthyl, 2000
sulfoxyde de diméthyl, 2001
sulfoxydes, 5540
sulfuration, 5529
sulfure de carbonyl, 1108
sulfure de diméthyl, 1999
sulfure de sodium, 5236
sulfure de zinc, 6332
sulfures, 5530
sulfurisation, 4203
sulphydrates, 3153
supercalandre, 5550
supervision, 5563
support, 595, 3067, 5567
support de couche, 1402, 1414
support pour contrecollé, 2641
surcalandré, 3050, 5551, 5554
surchauffé, 5557
surchauffe, 5558
surcouchage, 4100
surcuisson, 4101
sureau, 2203
surface, 5568
surface de défibrage, 2907
surface mate, 2150
surface rugueuse, 3578
surface spécifique, 5302
surglaçage, 5553

surglacé, 5554, 5555
surimpression, 4109
surnageant, 5560
surséché, 4103, 4104
surveillance, 5563
suspension de pâte, 5595
sycomore, 5606
sylviculture, 2677, 5099
symétrie, 5607
synchrone, 5609
synchronisme, 5608
synergisme, 5611
synthèse, 5612
système, 5618
système binaire, 690
système de branlement, 5030
système de collage automatique, 2624
système de collage bout-à-bout de deux
  feuilles de papier, 992
système de contrôle, 1547
système d'élimination des condensats du
  type siphon, 5116
système d'engagement de la feuille,
  6160
système de séchage par feuille
  aéroportée, 267
système de transport par gravité, 2879
système en circuit fermé, 1359
système hydraulique, 3122
système métrique, 3791
système pneumatique de formation de la
  feuille, 277

T

tableau de contrôle, 1546
tableaux, 5624
tableaux de données, 1828
tableaux de rendement, 6321
table de fabrication, 6244
table de fabrication amovible, 4741
table plate sortante, 4827
table plate type Cantilever, 1075
tablier, 4864
tablier de machine à papier, 424
tache, 5304, 5375
tache d'eau, 6110
tache de colorant de couchage, 5337
tache de couleur, 1455
tache de liant, 5336
tache de paraffine, 6135
tache de rouleau de soutien, 5335
tache de toile, 6256
tacher, 755
taches d'alun, 336
taches de boue, 5162
taches de caoutchouc, 4868
taches de carbone, 1104
taches d'écorce, 581
taches de graisse, 2893
taches de mousse, 2635, 2739
taches de résine, 4343, 4845, 4846,
  5127
taches de rouille, 4876
taches d'huile, 4045
taches provoquées par des résidus de
  blanchiment, 750
taché sur les bords, 1855
tacheter, 5338
tachymètre, 5314, 5627
taille des copeaux, 1285
taille-douce, 4370
talc, 5636
tall oil, 5637
tamaris, 5640
tambour, 584, 2110
tambour à lattes, 5292
tambour de bobinage, 6239
tambour écorceur, 578
tambour en fibre, 2459
tambour laveur, 2115, 6054
tamis, 4945, 5088
tamis à buchettes, 958
tamis à copeaux, 1295
tampon, 936, 4141

tampon absorbant, 119
taquer (un paquet de feuilles), 3312
taraudage, 5649
tare, 5651
tas, 4322
tas de copeaux, 1293
tassement, 1472
taux, 4644
taux de crêpage, 1686
taux de croissance, 2928
taux de déchirure, 5661
taux de raffinage, 643
taux de séchage, 4643
taux de viscosité, 6025
technique, 4532, 5665
technique de l'électricité, 2205
technique des couts, 1623
technique des méthodes, 3778
technique des ordinateurs, 5619
technique des structures, 5477
teinté, 2171, 5751
teinte, 3103, 5024, 5750
teinté en surface, 5588
teinture, 2172
télévision, 5669
température, 5670
température de coulée, 4461
température élevée, 3054
température exacte, 1701
temps, 5746
temps d'arrêt, 2084
temps d'égouttage, 2090
temps de montage, 5015
temps de prise, 3070
temps de raffinage, 644
temps de réponse, 4765
temps machine, 3635
temps perdu, 3176
tendance, 681, 2103
tendance à se mettre en bourre, 2600
tendance à tomber en désuétude, 4016
tendance au peluchage, 2763
tendeur, 5467
tendeur de feutre, 2433
tendeur de toile, 6257
tendre, 5241, 5468
teneur en air, 252
teneur en cellulose, 1185
teneur en cendres, 457
teneur en chiffons, 4624
teneur en chlore actif, 492
teneur en eau, 3861
teneur en fibres, 2455, 2462
teneur en lignine, 3481
teneur en lignine résiduelle, 4749
teneur en matières solides, 5267
teneur en pâte, 4570
tensiomètre, 5684
tension, 5444, 5675, 5685
tension de la feuille au cours du séchage,
  2102
tension de la feuille en continu, 6159
tension d'enroulage, 6241
tension de rupture, 5039
tension incorporée, 942
tension superficielle, 5587
tension thermique, 5710
térébenthine, 5882
ternir, 755
terpènes, 5689
terpinènes, 5690
terrain, 5256
terre à chaux, 3494
terre à foulon, 2746
terre de remblai, 3418
terre d'infusoires, 1945
tête, 2993
tête d'extrusion, 2367
tétrachlorure de carbone, 1105
texture, 5702
thermocollable, 3016
thermocouple, 5712
thermodurcissement, 5718
thermodynamique, 5713
thermomètre, 5716
thermomètre celsius, 1201
thermoplastiques, 5717
thiocarbamates, 5727
thiocarbonates, 5728

thiocyanate de sodium, 5238
thiolignines, 5729
thiosulfate de sodium, 5239
thiosulfates, 5730
thixotropie, 5733
thuya, 5739
tige, 4813, 5418
tige (veget.), 5417
tilleul, 5743
tilleul d'Amérique, 605
timbrage, 5380
tirage, 2097
tirage dans la partie humide, 6183
tirage électrostatique, 2241
tirage sur papier diazo, 1948
tirant, 873
tissage, 6145
tisser, 6144
tissu, 2370
tissu tissé, 6294
titane, 5757
titrage, 5762
toile de la machine, 3636
toile de la machine à table plate, 2705
toile de soutien, 511
toile inclinée, 3203
toile métallique, 6242, 6243
toile sans fin, 2272
toile unie, 4351
tolérance, 5767
tolérance de poids, 6172
toluène, 5768
tonnage habituellement livré dans un
  wagon complet, 1128
tonne, 5769
tonne américaine, 5074
tonne anglaise, 3569
tonneau, 584
tonne métrique, 3792
torsion, 5785, 5888
toucher, 2413
tour, 4782, 5789, 5794
tourbillon, 6035
tourbillonnaire, 2182
tour d'absorption, 124
tour de blanchiment, 745
tour de lavage, 4955
tour de récupération, 4686
tourillon, 3319, 5857
tours par minute, 4783
tous les produits extraits du bois de pin,
  3947
tout produit non fibreux introduit dans la
  composition de fabrication, 625
toxicité, 5795
traces d'humidité, 3870
trachéides, 5797
traction, 2370
traînée, 5455
traînée de couchage, 1401, 5456
traînées d'eau sur la feuille humide,
  6111
traînées humides, 6200
traînées humides dans la feuille, 1813
trait, 3341
traitement à l'acide, 181
traitement à l'alcali, 307
traitement anti-dérapant, 408
traitement biologique, 705
traitement biologique des effluents, 702
traitement chimique, 1260
traitement de la pâte, 5435
traitement de la pâte à chaud, 3100
traitement de l'eau, 6113
traitement des effluents, 2194
traitement en surface, 5589
traitement modificateur, 3853
traitement primaire, 4517
traitement secondaire, 4974
traitement tertiaire, 5692
traitement thermique, 3024
trame, 6164
trame de toile métallique, 6284
trame d'une toile métallique, 5085
trancher, 5172
transducteur (de pression), 5803
transfert, 5804
transfert de la feuille, 6161
transfert de l'encre, 3243

transfert de masse, 3708
transfert d'énergie, 2280
transformateur, 1554, 5807
transformateur de papiers, 4161
transformateur électrique, 2213
transformation, 1555, 1560
transformation des papiers, 4160
transistor, 5808
translucidité, 5811
transmission, 515, 1502, 1651, 3263
transmission de chaleur, 3023
transmission de données, 1829
transmission (de la lumière à travers un
  papier), 5814
transmission d'énergie électrique, 2223
transmission totale (de la lumière à
  travers le papier), 5787
transparence, 3570, 5810, 5815, 5816
transparent, 5818
transpiration, 5821
transplantation, 5822
transport, 1561, 5823
transport des arbres abattus par câble,
  1006
transport des bois par ballon, 555
transporteur, 1562
transporteur à chaîne, 1221
transporteur à courroie, 654
transporteur à godets, 932
transporteur à vis, 1564
transporteur à vis sans fin, 4951
transporteur de copeaux, 1283
transporteur-élévateur de rondins, 3555
transvaser sous vide, 5950
trappe, 5827
travail, 3385
travail manoeuvre, 4069
traverse de fixation du tablier, 425
traverser, 5471
tremble, 461
trembleur, 5028
trémie, 3086
trémie de chargement, 2410
trémie de pesage, 6168
trempage, 5205
trempe, 2976
très lustré, 3046
tresser, 6144
triage, 5282
tri des chiffons, 4632
tri des copeaux, 1281
trier, 5279
trieur, 5280
trieur de copeaux, 1282
trieur de noeuds, 3359, 3363
trieur de noeuds à secousses, 5029
trieur de noeuds et buchettes, 3361
triplex, 5851
triturateur, 1867, 4571, 5189
triturateur en discontinu, 610
triturateur (pour balles entières), 547
trituration, 1868
troisième presse, 5732
trompe, 582
tronc d'arbe, 816
tronçonnage du bois en rondins, 3556
tronçonneuse, 3557
tronçonneuse à scies multiples, 5148
tronquer, 5856
trop cuite (pâte), 4102
trop-plein, 4105
trou, 3072, 5995
trouble, 5872, 5874, 5877
trou d'épingle, 4330
trou de toile, 6246
trou dû à la présence de boues, 5161
tube, 5865
tube de Pitot, 4346
tube en hélice, 5323
tube venturi, 5999
tuile, 5742
tulipier, 3525, 5868
turbine, 5875
turbine à gaz, 2794
turbine à vapeur, 5412
turbine hydraulique, 6115
turbogénérateur, 5876
turbulence, 5877
tuyau, 4334

tuyau collecteur, 2996
tuyau flexible, 3091
tuyau souterrain, 4338
tuyauterie, 4335
tuyauterie cordon (défaut du papier),
4337
tuyaux aspirants, 5508
tyloses, 5898
type étalonné, 5382
typographie, 3451

U

uni, 4907
uniformité, 5919
unité, 5920
unité de chaleur anglaise, 500
unité de glucose, 2837
unité de mesure pour liquides, 2773
un jet, 5111
urée, 5934
uréthanes, 5937
usage, 418, 5009
user en frottant, 107
usinage, 3637
usine, 3816, 4353
usine clés en main, 5880
usine de pâtes, 4577
usine intégrée, 3264
usine kraft, 3374
usine pilote, 4323
usure, 6138
utilisation, 5939
utilisation des données, 1823
utilisations multiples, 3920
utilité, 5009

V

valence, 5951
valve, 5952

vanilline, 5955
vanne, 5952
vanne à papillon, 990
vanne à passage direct, 2797
vanne d'admission dans une machine à
vapeur, 5737
vanne d'arrêt, 1237, 5084
vanne de contrôle, 1543
vanne de décharge, 782, 2017, 4738
vannette, 2795
vannette à pâte, 5183
vapeau, 5957
vapeur, 2992, 5402, 5966
vapeur d'eau, 6116
vaporisation, 5959
varech, 294
variable, 5968
variable de traitement, 4535
variation, 5967, 5970
variation saisonnière, 4969
variété, 5971
vase, 5096, 5160, 6003
vaseux, 5163, 5872
végétal autre que de bois, 4006
véhicule, 5982
vélin, 5983, 6291
vélin parcheminé, 4207
venta nip du pauvre, 4437
vente, 4885
ventes de pâtes, 4580
ventilateur, 245, 772, 2384, 5997
ventilateur aspirant, 2340
ventilateur à tirage indirect, 3213
ventilateur centrifuge, 1208
ventilation, 773, 5996
vergé, 3395
vergé au feutre, 2423
vergé non apprêté, 3396
vergeur, 1820
vergeur en spirale, 5322
vergeures, 3399, 3402
vérificateur de papiers, 4166
vérification, 1234, 3253
vernis, 3389, 5972
vernissage, 3391, 5974
vert, 2896

vêtements de protection, 4557
vibration, 6005
vibration des rouleaux, 1233
vibreur, 6006
vidange, 2015, 2151
vidange de l'eau, 6100
vide, 5940, 6028
vieillesse, 230
vieillissement, 231, 236
vieux journaux, 3960
vieux papiers, 6059, 6061
vieux sacs de jute classés, 534
vinyl cellulose, 6009
virole, 2437
virole de sécheur, 5057
vis, 4950
viscoélasticité, 6015
viscose, 6018
viscosimètre, 6016
viscosimètre, 6022
viscosimètre rotatif, 4855
viscosimétrie, 6017
viscosité, 6023
viscosité apparente, 416
viscosité du cuprammonium, 1743
viscosité du cuproéthylènediamine,
1746
vis d'ajustement, 3458
vis réglable, 212
vitesse, 5987
vitesse circonférentielle, 1332
vitesse constante, 1517
vitesse d'écoulement, 2595
vitesse de coupe, 5158
vitesse de jaillissement, 5340
vitesse de pression de l'air, 285
vitesse de rupture, 5037
vitesse exacte, 1702
vitesse prévue, 1924
vitesse variable, 210
vitrification, 6027
voie de chemin de fer, 4637
voile en offset, 2637
voiture, 5982
volatilité, 6032
voltmètre, 6033

volume, 945, 6034
volume de vide, 6031
volume spécifique, 5303
voûte des arbres, 2668
vrac, 945
vulcaniser, 6038

W

wagon complet, 1127
wagon de chemin de fer, 4636
wagonnet, 5855

X

xanthate d'amidon, 5390
xanthate de cellulose, 1198
xanthure d'amidon, 5391
xylem, 6304
xylènes, 6305
xylènesulfonate de sodium, 5240
xylose, 6307

Z

zéolites, 6323
zinc, 6326
zone d'adhérence,

english

svenska

deutsch

français

español

ESPAÑOL

A

ábaco, 1231
abarquillamiento, 934, 1424, 1751
abastecedor, 5564
abatidor de rodillo, 4826
abedul, 707, 6218
abedul para pasta, 4151
abertura de enlejiado, 933
abeto, 105, 2520
abeto balsámico, 557
abeto blanco, 6219
abeto del Canadá, 3030
abeto Douglas, 2083
abeto grandis, 2870
abeto plateado, 5098
ablandador, 5242
ablandamiento, 5243
ablandar, 5672
abrasión, 108
abrasivos, 112
abridor, 4062
abrillantado, 4402
abrillantado de la hoja continua, 6151
abrillantado por calandra, 860
abrillantador, 895, 5060
abrillantador fluorescente, 2613
abrillantamiento, 2823
abrillantamiento de ángulo reducido, 3575
absorbedor, 118
absorbencia, 127
absorbente, 116, 125, 5277
absorber, 114
absorción, 120, 5278
absorción del aceite, 4034
absorción de la energía tensora, 5676
absorción del agua, 6065
absorción de la tinta, 3233
acabado, 2517
acabado abrasivo, 1720
acabado a la plancha, 4367
acabado al cepillo, 920
acabado al vapor, 5405
acabado a mano, 2963
acabado antiguo, 401
acabado basto, 1378
acabado bruñido, 965
acabado cartón, 1124
acabado con arrugas, 1423
acabado crepé, 1684
acabado China, 2197
acabado débil, 3578
acabado de desfibrado en calandra, 1032
acabado duplex, 2158
acabado en calandra, 1037
acabado en fábrica, 3826
acabado en la sección húmeda, 6186
acabado en máquina, 3620
acabado en rizo, 4807
acabado en superficie, 5576
acabado especial en máquina, 3050
acabado fieltro, 2423
acabado filigrana, 3398
acabado glaseado, 5556
acabado húmedo, 6188
acabado jaspeado, 3902
acabado liso, 5159
acabado mate, 2150, 2552, 3721
acabado medio, 3742
acabado metálico, 3762
acabado pergamino, 4202
acabado satinado, 4907
acabado tela, 963, 1067
acabado terciopelo, 5991
acabado una cara, 4053
acabado vitela, 5985
acabado vitela en calandra, 1056

acacia, 128
acacina, 129
acanalado, 1675, 2915
acanaladura, 2621, 2912
acaracolado, 5203
accesibilidad, 138
accesorio, 139
accionamiento de velocidad variable, 5969
accionamiento eléctrico, 2216
accionamiento por fricción, 2732
acción de marcar, 3699
acebo, 3179
aceite, 4032
aceite de palo, 5869
aceite lubrificador, 3584
acelerar, 5316
acero, 5415
acero inoxidable, 5378
acetales, 146
acetato de butilo, 994
acetato de celulosa, 1183
acetato de vinilo, 6008
acetatos, 150
acetilación, 155
acetona, 153
acetonas, 3344
acetonitrilo, 154
acidez, 167, 5286
acidificador, 165
acidificar, 168
ácido, 157, 5284
ácido acético, 151
ácido acrílico, 189
ácido ascórbico, 454
ácido benzoico, 671
ácido bórico, 845
ácido bromhídrico, 3126
ácido butírico, 998
ácido clorhídrico, 3129
ácido concentrado, 5476
ácido débil, 6137
ácido de cocción, 1568
ácido esteárico, 5414
ácido fluorhídrico, 3135
ácido fórmico, 2688
ácido ftálico, 4297
ácido fumárico, 2749
ácido hipocloroso, 3173
ácido láctico, 3392
ácido levulínico, 3462
ácido linoleico, 3511
ácido málico, 3668
ácido metacrílico, 3775
ácido nítrico, 3981
ácido oléico, 4049
ácido oxálico, 4116
ácido peracético, 4252
ácido perclórico, 4254
ácidos abiéticos, 106
ácidos borhídricos, 847
ácidos carboxílicos, 1118
ácidos cloracéticos, 1313
ácidos de resina, 4754
ácidos fenólicos, 4280
ácidos grasos, 2395
ácidos inorgánicos, 3247
ácidos lignosulfónicos, 3486
ácidos odoríferos, 431
ácidos orgánicos, 4083
ácido sulfámico, 5520
ácido sulfúrico, 5545, 5546
ácido titulable, 5760
acidulación, 166, 5285
aclarado, 5725
acodar, 3313
acondicionado, 1498
acondicionador, 1500
acondicionamiento, 1501, 1749
acondicionar, 1748
acoplamiento, 1651
acoplar, 1650

acortado de las fibras, 5070
acrilamida, 187
acrilatos, 188
acrilonitrilo, 192
acumulación, 141
acumulador, 142
acumulador térmico, 3001
acuoso, 426
adelgazado, 5646
adherencia, 819
adherencia al rodillo de calandra, 765
adherencia de clavija, 4324
adherencia de las capas, 4388, 4389
adherencia en superficie, 764
adhesión, 204
aditivo, 213
aditivo de la sección húmeda, 6185
aditivo del estucado, 1400
aditivo de refino, 625
aditivos, 202
administración, 5563
administración forestal, 2673
administración hidráulica, 6083
administrador, 3672
admisibilidad, 132
adragante, 2947, 5799
adsorbente, 216
adsorber, 214
adsorbido, 215
adsorción, 217
afieltrado, 2425
afilado, 5032
afilar, 5032
afinidad, 225
afloramiento, 924
agar, 228
agente, 232
agente abrillantador, 2827
agente acelerador, 131
agente acondicionador, 1750
agente activo de superficie, 5569
agente amortiguador, 937
agente antiadherente, 389
agente antiarrastre, 391
agente anticorrosivo, 1598
agente antiestático, 409
agente antinebuloso, 396
agente blanqueante, 733
agente coagulante, 1375
agente de acoplamiento, 1652
agente de azulamiento óptico, 4079
agente de control de la viscosidad, 6024
agente de floculación, 2572
agente de flotación, 2584
agente de flujo, 2589
agente de obturación, 3704
agente de resistencia en estado húmedo, 6202
agente desfloculador, 1873
agente espesador, 5721
agente esponjador, 5603
agente espumante, 2631
agente hidrotrópico, 3154
agente humectante, 6210
agente nivelador, 3457
agente oxidante, 4121
agente quelato, 1240
agente resistente en seco, 2145
agente soltador, 4730
agentes secadores, 2135
agitación, 240
agitado, 5429
agitador, 241
agitar, 239
aglomeración, 234
aglomerar, 233
agregar, 235
agua, 6063
agua colada, 6225

agua colada enriquecida, 4794
agua cruda, 4649
agua de alimentación, 2412
agua de alimentación para caldera, 813
agua de descarga, 5635
agua de elaboración, 4536
agua de enfriamiento, 1577
agua destilada, 2040
agua dulce, 2727
agua fijada, 857
aguante, 3071
agua oxigenada, 3144
aguarrás, 5882
agua salobre, 4889
aguas residuales, 6062
agua superficial, 5590
aguas usadas, 5021
agudeza, 5033
agujetas, 1421
ahilamiento, 4840
aire, 242
aireador, 223
aire comprimido, 1476
aislamiento, 3261
aislamiento acústico, 184
aislamiento eléctrico, 2219
aislamiento térmico, 3010, 5705
ajuste de tubos, 4336
álamo, 4439
álamo amarillo, 6318
álamo negro, 721
álamo tremblón, 461
alargamiento, 2247, 5465
alargamiento a la tracción, 2248
albúminas, 286
albura, 4905
álcali, 299
álcali activo, 196
álcali celulosa, 300
álcali efectivo, 2189
álcali ligninas, 302
alcalinidad, 313
álcali soluble, 305
álcali titulable, 5761
álcali total, 5786
alcohol de polivinilo, 4427
alcoholes, 290
alcohol furfuril, 2759
alcohol isobutil, 3299
alcohol metílico, 3781
alcohol propílico, 4556
aldehida acética, 145
aldehidos, 292
aleación, 320
alerce, 3423, 3424, 5639
alfa celulosa, 322
alfa proteína, 324
algas, 294
algicidas, 295
alginatos, 296
alginato sódico, 5216
algodón, 1624
alimentación, 2403, 2411, 3543
alimentación de hojas, 5045
alimentación del rollo de hoja continua, 6150
alimentado de hoja, 5044
alimentado por aspiración, 5504
alimentador, 2407
alimentador de alta presión, 3053
alimentador de baja presión, 3580
alimentador de virutas, 1286
alineación, 297, 5701
alisado, 2132, 5201
alisadura, 5197
aliso, 293, 321
almacén, 3549
almacenaje, 5441
almacenamiento de bultos, 956
almacenamiento de datos, 1827
almacén de fichas de tabulación, 5625
almiar, 5372

almidón, 5386
almidón catiónico, 1164
almidón clorado, 1300
almidón convertido, 1553
almidón de arroz, 4792
almidón de arrurruz, 434
almidón de maíz, 1593
almidón de patata, 4456
almidón de tapioca, 5648
almidón de trigo, 6216
almidón dialdehído, 1941
almidones modificados, 3851
almidón perlado, 4235
almohadilla absorbente, 119
almohadillas de papel, 4179
alquilaril sulfonatos, 317
alquilización, 318
alquitrán, 5650
alta densidad, 3042
alta presión, 3052
alta velocidad, 3056
altura, 3026
altura de descarga, 2016
alumbre, 327
alumbre básico, 600
alumbre del fabricante de papel, 4171
aluminatos, 328
aluminato sódico, 5217
aluminio, 329, 330
alvéolos, 1747
alzamiento, 3467
amarilleamiento, 6317
amianto, 446
amidas, 337
amilasa, 357
amilosa, 358
aminas, 339
aminas polimeras, 338
aminoácidos, 340
amoniaco, 345
amorfo, 353
amperímetro, 344
ampolla, 758, 762
ampolla (estucado), 759
ampollamiento, 761, 931
ampolla (separación de capas), 763
análisis, 364
análisis cualitativo, 4603
análisis cuantitativo, 4606
análisis de fibras, 2446
análisis de gas, 2781
análisis del agua, 6066
análisis de la tensión, 5462
análisis de tamiz, 5089
análisis de variación, 365
análisis económico, 2179
análisis estadísticos, 5398
análisis volumétrico, 5762
análogo, 360
anaranjado de metilo, 3790
anatasa, 367
ancho, 6228
ancho del recorte, 5843
anchura de la tela utilizada, 3616
anchura útil de la tela, 3725
anemómetro, 369
anfótero, 356
angioesperma, 370
ángulo de contacto, 1521
ángulo de la criba, 4946
ángulo de lámina, 2639
anhídrido acético, 152
anhídrido ftálico, 4298
anhídrido málico, 3669
anhídridos, 371
anhídrido de manosa, 3684
anhídrido sulfuroso, 5544
anhidrita, 372
anilina, 373
anillo, 4802
anillo anual, 2929
anillo de crecimiento, 383
anillo partido, 5330
anillos, 385
anillos de Raschig, 4642
aniones, 382
aniónico, 380
ánodo, 386
antiácido, 176, 387

antiadherente, 4004
anticloros, 390
anticorrosivo, 4875
antideflexión, 392
antideslizante, 4003
antideslustrante, 410
antiespuma, 1875
antiespumante, 395, 2634
antioxidante, 405
antioxidantes, 397
anublado, 2637
anular, 384
apagado de la cal, 3491
apagador, 5145
aparato de cronometraje, 5747
aparato de medición, 3771
apergaminar, 4204
apertura de inspección, 5078
apilado por aire, 277
apilador, 5373
apilamiento, 5374
apisonador, 1473
aplanado, 2553
aplicación de la cola, 2848
aplicador de cola, 2841
apoyo, 505
apoyo filtrante, 2502
aprestador, 4519
apresto, 2516
apretado en el bobinado, 6241
apropiación, 423
aprovechamiento forestal completo, 2748
a prueba de humedad, 3865
a prueba de intemperie, 6141
aptitud al acanalado, 1668
aptitud a la humectación, 6207
aptitud a la impresión, 4520
aptitud a la pelusa, 2763
aptitud al barnizado, 5973
aptitud al descortezado, 4240
aptitud al engrasamiento, 4046
aptitud al moldeado, 3872
aptitud al plegado, 2094
aptitud al refino, 621
aptitud al sellado, 4960
aptitud de la superficie a la humectación, 5591
aqua de rellenar, 3666
arabinosa, 428
árbol, 5653, 5830
árboles positivos, 4386
árboles renacidos, 4971
árbol frondoso, 1846
arborización, 2667
arca, 689
arca de virutas, 1277
arce blanco, 3689
arce de azúcar, 5518
arce rojo, 5248
arcilla, 1344
arcilla de batán, 2746
arcilla de estucado, 1408
arcilla para estucar, 1404
arcilla para rellenar, 2494
área unida, 818
arenero, 4798, 4896, 5017
arenilla, 2911
armadura, 3067
armadura de pila, 639
armadura de rasqueta, 2052, 2053
arpillera, 534
arrancamiento, 2537, 4301, 4307
arrancamiento del estucado, 1413
arrancamiento de ruptura, 4310
arrancamiento en el estucado, 4305
arrancamiento en el rodillo desgotador, 1819
arrancamiento en la calandra, 4304
arrancamiento en la fibra, 4306
arrancamiento por las ampollas, 4303
arrancar, 5660
arranque de calandra, 1043
arrastrador, 2910
arrastrador (de troncos), 5133
arrastramiento, 1678
arrastrar (troncos), 5132
arrastre, 2096
arrastre de maderas, 3555

arrastre (de troncos), 5134
arrastre de troncos por cable aéreo, 3049
arrastre de troncos por globo aerostático, 555
arrastre por aire, 262
arrastre por cable, 1006
arreglo en series, 5006
arrollamiento del borde, 2185
arrollamiento de los bordes, 1754
arrugar, 1420
arrugas de rasqueta, 2056
artesa tomadora, 4316
artes gráficas, 2875
asbestina, 445
aserradero, 4920
aserradura, 4919
asfalto, 462
asfalto emulsificado, 463
asimetría, 475
asimilación, 473
asir, 2876
aspecto, 417
aspecto gredoso, 1223
aspereza, 4861
aspersión, 5357
aspiración, 472, 5497
astilla, 5062
atapulguita, 480
atascamiento de maderas, 3559
atenuación, 481
atmósfera, 476
atmósfera controlada, 1544
atomización, 479
a través de las fibras, 186
autoclave, 482, 483
automatización, 487
auto-rebanador, 489
avivamiento de muela, 970
axial, 493
ayudante de conductor, 517
ayuda retentiva, 4770
azufre, 5541, 5547
azufre activo, 198
azulamiento óptico, 4078
azular, 783

B

bacteria, 518
bactericidas, 520
bagazo, 527
baja consistencia, 3576
baja densidad, 3577
baja presión, 3579
bala de recortes de fabricación, 906
balance de materiales, 3714
baldosa, 5742
balsa, 4619
bambú, 559
banco de clasificadora de trapos, 5281
banco de pruebas, 5694
bandeja, 4376, 4958, 5829
bandeja de estucado, 1412
baño de enlucido, 4107
baño de estucado, 1405, 1411, 1416, 5164
barcaza, 570
bario, 571
barita, 590
baritina, 593
barniz, 5972
barnizado, 3391, 5974
barra, 568
barra de bobinadora, 1589, 6240
barra de desplisar, 5350
barras del molino, 884
barrera de vapor, 5958
barrido, 5152
barril, 584
base, 595
basicidad, 602
bastidor, 2714
bastidor (en calandra), 2715
basto, 1377
basura municipal, 3933

basuras, 2103
batán, 6300
benceno, 669
beneficio, 667, 4546
bentonita, 668
benzoatos, 670
berilio, 674
betacelulosa, 675
Betula, 679
betún, 714
bias, 681
bicarbonato de soda, 5219
bicarbonatos, 684
bien encolado, 6178
bifenilos, 706
biócidos, 698
biodegradable, 699
biodegradación, 700
biselado, 2401
bisulfatos, 709
bisulfito, 180
bisulfito a base de calcio, 1014
bisulfito de amonio, 347
bisulfito de magnesio, 3646
bisulfito de sodio, 5220
bisulfitos, 713
bisulfuro de carbono, 1098
bisulfuros, 2044
blanco de perla, 4236
blanco de yeso, 6226
blanco fijo, 726
blanco flúor, 2616
blanco neutro, 1832
blanco satino, 4908
blancura, 896, 2765, 6222
blando, 5241
blandura, 5249
blanqueabilidad, 734
blanqueable, 735
blanqueado a la tina, 5977
blanqueado de alta densidad, 3043
blanqueado de varios pasos, 3924
blanqueador, 739
blanqueador óptico, 4073
blanqueado una cara, 5108
blanqueamiento, 741
blanqueo óptico, 4072
blanquería, 740
bobina abombada, 1713
bobina aplastada, 1721
bobina bruta, 3321
bobina corrida, 5667
bobina de cartón, 4701
bobina de papel, 4702
bobinado, 6238
bobinado a tambor doble, 2073
bobinadora, 6235
bobinadora cortadora, 5171
bobinadora de mandril, 1592
bobinado sobre mandril, 1205
bobinado sobre tambor, 2116
bobina dura, 5741
bobina estropeada en los bordes, 5392
bobina floja, 3573, 4435
bobina madre, 4210
bobinar, 6234
bobina terminada, 6290
bolita, 4245
bolsa de desperdicios, 4714
bolsa de papel, 4147
bolsa de papel a válvula, 5953
bolsillo, 4393
bomba, 4587
bomba aspirante, 5512
bomba axial, 494
bomba de aletas, 2387
bomba de mezcla, 3845
bomba de vacío, 5436
bomba multicelular, 3926
bomba para pasta, 5436
bomba rotatoria, 4851
bombeado, 1712, 1715
bombeado del cilindro, 4820
bombeo, 4588
bond registro, 4717
boquilla, 4009
borato de sodio, 5221
boratos, 842
bórax, 843

borbor, 5339
borde, 2183
borde aflojado, 5142
borde agrietado, 1660
borde conductor, 3438
borde de caja de entrada, 3516
borde de la hendedora, 5169
borde delantero, 2736
borde granoso, 2867
borde laminado, 3823, 4822
borde ondeado, 6120
borde quebrado, 907
bordes con barbas, 1852
borde torneado, 5879
boro, 848
borra de algodón, 1627, 3514
bosque, 2666
bosque de misma edad, 2334
bosque secundario, 4970
bote, 1071
brazo de palanca, 3460
brida, 2543
brillantez, 2195, 3589
brillo, 2813, 2826, 3595, 4400
brillo subjetivo, 5490
brístol, 900
Brístol antiguo, 399
bristol duplex, 2155
bristol estucado, 1391
bristol plegable, 2650
bristol rellenado, 2487
bristol simili, 798
bromo, 910
bronce, 912
bronceado, 913
brote, 935
bruñido, 967
bruñidor, 966
burbuja, 243, 929
burbujas de aire, 249
bureta, 962
butadieno, 982
butanos, 983
butirato de celulosa, 1184
butiratos, 997

C

caballo de fuerza, 3090
cable, 1005
cables de acero, 6248
C.A. (corriente alterna), 100
cadena de troncos, 841
cadena transportadora, 1221
cadmio, 1008
caducar, 2109
caduco, 1845
caída de presión, 4497
caja, 862, 1140, 1266
caja aspirante, 2549, 4965, 5498, 5941
caja de agua, 6068
caja de agua para calandra, 1057
caja de aguas residuales, 4915
caja de calandra, 1028
caja de cartón, 864
caja de cartón con paredes dobles, 2082
caja de cartón ondulado, 1609
caja de cerco, 564
caja de engranajes, 2800
caja de entrada, 893, 2590, 2994
caja de entrada abierta, 4064
caja de entrada a presión, 4501
caja de entrada cerrada, 1361
caja de entrada primaria, 4514
caja de entrada secundaria, 4972
caja de escurrido, 2091
caja de mezcla, 3843, 3844
caja de pila, 627
caja de vapor, 2361
caja extensible, 2344
caja intermedia, 3273
caja plegable, 2649
caja reguladora, 4718

cajas de cartón para montar, 5020
cajas para huevos, 2196
caja Uhle, 5901
cal, 3487
calandra, 1025
calandra de ancho total, 4063
calandra de combinación, 1464
calandra de fricción, 2731
calandra de máquina, 3608
calandra de satinado, 2828
calandrado, 1038
calandrado de hoja continua, 6148
calandrado de satinado, 2829
calandrado húmedo, 6075, 6077
calandrado por doctor de agua, 6074
calandra húmeda, 6070
calandra intermedia, 3271
cal apagada, 5144
calar, 5471
calcinación, 1010
calcio, 1011
calculadora, 1024, 1483
calculadora analógica, 361
caldeo dieléctrico, 1963
caldera, 812
caldera de trapos, 4622
calefacción, 3008
calefacción de convección, 1550
calefacción directa, 2008
calefacción radiante, 4612
calentador de gas, 2786
calentador de vapor, 5403
calentador infrarrojo, 3223
calentamiento indirecto, 3211
calentamiento infrarrojo, 3224
calentamiento por inducción, 3214
calentamiento por vapor, 5406
calibrador, 2769, 2798
calibrador de chorro, 3309
calibrar, 1058
calibre, 1061
calibre para vasos, 1740
calidad, 4604
calidad de curvatura, 664
calidad del agua, 6098
calidad del aire, 283
calidad de la pasta, 4579
calidad del plegado, 2656
calidad inferior, 4023
calor, 3000
calor de combustión, 3012
calor específico, 5301
calorimetría, 1063
calorímetro, 1062
cal viva, 1021
cámara de aire, 251
cámara de combustión, 1467
cámara de pleno, 4379
cámara de secadero, 2128
cambiadora de rollos de hoja continua, 6161
cambiador aniónico, 379
cambio de cubierta, 1226
cambium, 1066
camisa, 3306
campana, 3084
campana de aspiración, 2342
campana de extracción, 2358
campana de la secadora, 2125
campo electromagnético, 2231
campo magnético, 3656
caña, 4698
caña de azúcar, 5516
canal, 2611, 5183
canaladura, 1614, 2619
canaladuras A, 102
canaladuras B, 501
canaladuras C, 1004
canaladuras E, 2175
canales resiníferos, 4756
canal medidor de Parshall, 4211
cáñamo, 3031
cáñamo de Manila, 103, 3681
cáñamo sisal, 5118
canjilón, 4930
canto vivo, 2397
caolin, 1275, 3336
capa, 3434, 4387
capa central, 1203

capacidad, 1080
capacidad de adsorción, 218, 219
capacidad de cubrimiento, 1657
capacidad de secado, 2136
capacidad encubridora, 3037
capacidad máxima, 3724
capacitancia, 1076
capa de aire, 244, 273
capa de fibras, 2467
capa de impregnación, 1382
capa gelatinosa, 2804
capa líquida en sistema de extracción solvente, 4618
capas, 4381
capa superior, 5776
capataz, 2665
capataz de la sala de prensas, 4492
capataz de servicio, 5790
capilar, 1082
capilaridad, 1081
capital, 1083
caprolactam, 1086
cápsula, 1087
cápsula corrugada, 1608
cápsula de aire, 250
captador de polvo, 2165
cara, 2373
característico, 1227
cara del fieltro, 2432
cara fieltro (de un papel), 5779
cara inferior, 6255
carbamatos, 1089
carbamidas, 1090
carbón activado, 193
carbonatación, 1095
carbonato de bario, 572
carbonato de calcio, 1015
carbonato de sosa, 5210, 5223
carbonato magnésico, 3647
carbonatos, 1094
carbonilos, 1107
carbonización, 1100
carbono, 1093
carbón vegetal, 1228
carboxialquilación, 1110
carboxietilación, 1113
carboxilación, 1116
carboximetilación, 1119
carburos, 1091
carga, 2493, 2993, 3538
carga alcalina, 308
carga de la máquina, 3628
carga de lejiadora, 1977
carga de pila, 632
cargador, 3542, 5439
cargador de estucadora, 1415
cargador de virutas, 1279
carga electroestática, 2239
carga en la pila, 647
carga fibrosa, 2480
carga integrada, 942
cargamento de la tela, 6249
cargas cíclicas, 1786
carpeta, 2658
Carpinus, 1131
carretilla elevadora, 3469
carretilla elevadora de horquilla, 2678
carro, 5855
carro completo, 1128
carteo, 1663, 4645
cárter, 1654, 3102
cartivana, 2930
cartón, 789, 4152
cartón acolchado, 425
cartón acústico, 183
cartón a encolar, 4220
cartón aislante, 3257
cartonaje, 1134
cartonaje con tapas de faldón, 2768
cartonaje estucado caolina, 1347
cartonaje plegable, 2651
cartonajes estucados, 1390
cartón a la plancha, 4372
cartón antideslustrante, 411
cartón a ondular, 1615, 1617
cartón a ondular semiquímico, 4992
cartón asbestos, 447
cartón asfaltado, 469
cartón bituminoso, 715

cartón blanqueado una cara, 5109
cartón calandrado, 5196
cartón celuloso, 1180
cartón cemento de asbesto, 448
cartón combinado, 1458, 1461
cartón comprimido, 4484
cartón con barbas, 1853
cartón consistente, 5261
cartón con suspensión de barita, 591
cartón contrafuerte, 1642
cartón cromo, 1323
cartón cuero, 3445
cartón cuero para calzado, 5066
cartón de apoyo, 506
cartón de canto vivo, 2398
cartón de China, 1274
cartón de fibra para paredes, 645
cartón deflector, 522
cartón de fricción, 2730
cartón de Manila, 3679
cartón de moldear, 3873
cartón de pasta, 4569
cartón de pasta de madera, 6277
cartón de pasta real, 5264
cartón de pasta rellenado, 2491
cartón de prueba, 5695
cartón de yeso, 2952
cartón de yute, 3327
cartón duplex, 2154, 5892
cartón duplex y triplex, 2648
cartón duro, 2447
cartón encartelado a la tina, 5978
cartón encerado, 6123
cartonería, 793
cartones aceitados, 4037
cartones encartelados, 3499
cartones grises, 3818
cartones multiplex, 3817
cartón esponjoso, 950
cartón estucado, 1387
cartón estucado en máquina, 3612
cartón forrado de papel periódico, 3964
cartón gris blanqueado en tina, 5114
cartón gris de recortes forrado en tina, 3968
cartón gris de recortes para divisiones, 4216
cartón gris forrado una cara kraft, 5107
cartón gris pegado, 4224
cartón imitación cuero, 3184
cartón impermeable, 6094
cartón kraft, 3369
cartón kraft a ondular, 3370
cartón laminado, 3411
cartón libro, 831
cartón lino, 2560
cartón lustrado, 2820
cartón mate, 3710
cartón micro-ondulado, 3796
cartón moldeado, 4578
cartón multicapa, 3914, 3915, 3921
cartón ondulado, 1607
cartón ondulado de varias hojas, 3931
cartón ondulado dos caras, 2074
cartón ondulado para cajas, 1610
cartón ondulado simili, 799
cartón ondulado una cara, 5103
cartón paja, 5453
cartón paja del nueve, 3973
cartón para alimentos especiales, 5293
cartón para asfaltar, 470
cartón para automóviles, 488
cartón para bandejas, 4377
cartón para cajas, 1524
cartón para cajas forrado de manila, 3682
cartón para cápsulas de botella, 850
cartón para carteles, 4449
cartón para cartonajes, 863
cartón para cartuchos, 1661
cartón para cilindros, 1792
cartón para contenedores de leche, 3814
cartón para embalaje del jabón resistente al álcali, 316
cartón para empaquetaduras, 2789
cartón para empaquetar, 4138
cartón para encuadernaciones, 692
cartón para fósforos, 836

cartón para mandriles, 1586
cartón para matrices, 2580
cartón para modelos, 4231, 5139
cartón para montar, 5019
cartón para paneles, 4145
cartón para paneles de construcción, 939
cartón para pilas, 612
cartón para posavasos, 652
cartón para productos alimenticios, 2661
cartón para recipientes, 1070
cartón para stencil, 5420
cartón para techos, 1175
cartón para usos especiales, 5295
cartón para vasos, 1739
cartón pegado, 4223
cartón piedra, 2973
cartón plegable, 2647
carton prensa, 3961
cartón prensa manila encartelado dos caras, 2079
cartón prensa rellenado, 2489
cartón rellenado, 2486
cartón resistente a la grasa, 2891
cartón semiquímico, 4991
cartón semiquímico del nueve, 3972
cartón simili, 797
cartón simili sulfurizado, 2885
cartón sulfato, 5521
cartón tríplex, 5852
cartón tubo, 5866
cartón una capa, 5112
cartón vitela, 5984
cartón-yeso, 4358
cartulina, 1123
cartulina de estraza, 1278, 1459
cartulina de estraza encartelado de manila dos caras, 2078
cartulina estucada para carteles, 443
cartulina para billetaje, 5740
cartulina para cilindros, 1793
cartulina para etiquetas, 5632
cartulinas brístol, 901
cartulina secante, 769
cartulina simple de estraza, 4348
caseína, 1142
castaño, 1267
castaño de indias, 3089
catalizador, 1156
Catalpa, 1155
catión, 1159
catiónico, 1162
cátodo, 1157
caucho butilo, 996
caudal, 1898, 5458
caudal crítico, 1698
causticador, 1169
causticar, 1168
causticidad, 1167
cáustico, 1166
cáustico frío, 1429
caustificación, 1170
cavitación, 1172
cedazo para trapos, 4628
cedro, 1173
Cedrus, 1174
celita, 1176
celofán, 1178, 1192
célula, 1177
célula de Denver, 1909
célula de radiación, 4651
célula fotoeléctrica, 4289
celulosa, 1182
celulosa al benzilo, 672
celulosa aminoetílica, 341
celulosa aminopropilo, 343
celulosa amorfa, 354
celulosa cianoetil, 1781
celulosa cristalina, 1729
celulosa de algodón, 1625
celulosa de carboxialquilo, 1111
celulosa de carboxietilo, 1114
celulosa de carboximetilo, 1120
celulosa dialdehida, 1940
celulosa mesoforma, 3757
celulosa microcristalina, 3797
celulosa modificada, 3850
celulosa regenerada, 4715

celulosa resistente a los álcalis, 304
celulosa transparente, 5819
cementación, 1141
cemento armado, 4721
ceniza, 456
ceniza fina, 2623
ceniza negra, 718
centelleo, 2546
centipoise, 1206
central eléctrica, 4467
centrifugación, 1213
centrífugo, 1207, 1214
centrípeto, 1215
centro de bobina, 1588
cepa, 5487
cepillo, 915
cepillo rotativo, 928
cera, 6121
cera de carnauba, 1129
cera de parafina, 4197
cerámica, 1218
cera microcristalina, 3798
cerezo, 1265
cianamidas, 1777
cianatos, 1778
cianoetilación, 1780
cianuros, 1779
ciclo, 1783
ciclo de lejiación, 1569
ciclodextrinas, 1787
ciclohexano, 1788
ciclón, 1789
cien pies cúbicos de madera compacta, 1738
cierre hidráulico, 6106
cilindro, 1791
cilindro aspirante, 5502
cilindro de holandesa, 635
cilindro de satinado, 2824
cilindro distribuidor, 2043
cilindro enfriador, 1575, 5597, 5599
cilindro formato, 5126
cilindro satinador, 3599
cilindro secador, 2138
cimenos, 1801
cinc, 6326
cinta, 5472, 5643
cinta adhesiva, 208
cinta de bobina, 4700
cinta engomada, 2943
cinta estucada, 1401
cinta magnética, 3660
cinta perforada, 4590
cintas de remate, 2271
cinta transportadora, 1563
ciprés, 1744, 1802
ciprés pelado, 544
circuito cerrado, 1356
circuito eléctrico, 2209
ciruelo, 4562
clarificación, 1339
clarificador, 1340
clarificar, 1341
clase de cartón, 4154
clase de papel, 2857, 4164
clasificación, 1342, 2859
clasificación de fibras, 5385
clasificación de la fibra, 2452
clasificación de los trapos, 4632
clasificación de virutas, 1281
clasificador, 2483, 5280
clasificadora, 1343
clasificador Bauer-McNett, 614
clasificador centrífugo, 1209
clasificador de fibras, 2453
clasificador de la fibra de pasta, 4572
clasificador de virutas, 1282
clasificador mecánico, 3728
clasificar, 5279
clónico, 1355
clorato de sodio, 5224
cloratos, 1297
clorito de sodio, 5226
cloritos, 1312
cloro, 1303
cloro activo, 197, 492
cloroligninas, 1299, 1314
clorosis, 1315
cloruración, 1301

cloruro de cal, 744
cloruro de calcio, 1016
cloruro de litio, 3528
cloruro de polivinilideno, 4431
cloruro de polivinilo, 4428
cloruro de sodio, 5225
cloruro de viniliden, 6012
cloruro de vinilo, 6010
cloruro magnésico, 3648
cloruros, 1298
cloruros ácidos, 158
coagulación, 1374, 1376
coagulante, 1372
coagular, 1373
cobre, 1580
cocción, 1565, 1567
cocción a fondo, 2974
cocción fase lejía, 3523
cocedora de almidón, 5388
cocido incompleto, 5913
coeficiente de absorción, 121, 1426
coeficiente de fricción, 2733
coeficiente de refinado, 643
coeficiente de refracción, 4712
coeficiente de transferencia térmica, 3022
cohesión, 1428
cojinete, 618
cojinete de manguito, 5150
cojinete de rodillos ahusados, 5645
cojinete de rulemán, 4824
cola, 2840, 5121
cola ácida, 178
cola a la cera, 6133
cola animal, 376, 377
cola básica, 603
cola de resina, 4842
colagímetro, 5129
cola neutra, 3956
colar, 5444
cola reforzada, 2695
cola resistente a los ácidos, 179
colector, 2995, 3676
colector de agua, 6107
colector de condensamiento, 3101
colgamiento, 2969
coloidal, 1436
coloides, 1438
color, 1439, 1456
coloración, 1446, 2172, 5753
coloración en calandra, 1034
coloración en pila, 629
coloración en superficie, 5573
coloración en tina, 5859
coloración por pulverización, 5346
colorante, 2170
colorante ácido, 160
colorante (para avivar un matiz), 5770
colorantes, 2173
colorantes ácidos, 159
colorantes a la tina, 5976
colorantes básicos, 601
colorantes de anilina, 374
colorantes directos, 2007
colorantes fluorescentes, 2614
coloreado, 1440, 2171, 5751
coloreado en calandra, 1030
coloreado en el marco, 1855
coloreado en pila, 628
coloreado en superficie, 5574, 5588
coloreado natural, 3942
color fijo, 2388
colorimetría, 1445
colorimetría de comparación, 1452
colorímetro, 1443
colorímetro de comparación, 1451
colorímetro fotoeléctrico, 4290
color jaspeado, 3901
columna barométrica, 6081
columna de agua, 2108
collar, 1435
combadura, 1761
combinador eléctrico, 2212
combustible de leña, 3064
combustible fósil, 2697
combustión, 1466
combustión de azufre, 5542
comerciante en papel, 4176
compactación, 1472

comparación, 3712
compartimento de carga, 3539
compatibilidad, 1474
complejos lignina-carbohidrato, 3480
componentes de potasio, 4453
comportamiento, 4870
composición, 631, 2761, 5900
composición de la fibra, 2454
composición de la lejía de cocción, 1571
composición del estucado, 1406
composición fibrosa de la pasta, 2462
comprar, 4595
compresibilidad, 1477
compresión, 1478
compresión de caja, 866
compresor, 1482
compresor centrífugo de vacío, 1212
compuerta, 948
compuestos acrílicos, 190
compuestos alifáticos, 298
compuestos aniónicos, 381
compuestos aromáticos, 432
compuestos azoicos, 496
compuestos catiónicos, 1163
compuestos cianúricos, 1782
compuestos cíclicos, 1785
compuestos cuaternarios del amoniaco, 4607
compuestos de alumino, 331
compuestos de amonio, 348
compuestos de azufre, 5543
compuestos de boro, 849
compuestos de bromo, 911
compuestos de cadmio, 1009
compuestos de calcio, 1017
compuestos de cinc, 6327
compuestos de circonio, 6333
compuestos de cloro, 1305
compuestos de cromo, 1321
compuestos de estaño, 5749
compuestos de flúor, 2618
compuestos de fósforo, 4288
compuestos de hidrógeno, 3142
compuestos de hierro, 3295
compuestos de litio, 3529
compuestos de magnesio, 3649
compuestos de manganeso, 3675
compuestos de mercurio, 3753
compuestos de metal, 3760
compuestos de molibdeno, 3886
compuestos de níquel, 3970
compuestos de nitrógeno, 3990
compuestos de oxígeno, 4125
compuestos de plata, 5097
compuestos de silicio, 5092
compuestos de sodio, 5227
compuestos de tántalo, 5642
compuestos de titanio, 5758
compuestos de tungsteno, 5870
compuestos de vinilo, 6011
compuestos de yodo, 3285
compuestos halógenos, 2956
compuestos heterocíclicos, 3034
compuestos inorgánicos, 3248
compuestos orgánicos, 4084
cóncavo, 2023
concentración, 1487
concentración de hidroxilión, 3166
concentración hidrogeniónica, 3143
concentrador, 1488
condensación, 1492
condensador, 1079, 1495
condensador del escurrimiento, 5837
condensados, 1491
condiciones adiabáticas, 209
conducción, 1502, 4338
conducción de la tela metálica, 2936
conducción del fieltro, 2934
conducción del papel, 2935
conductividad, 1504
conductividad eléctrica, 2204
conductividad térmica, 3003, 5703
conducto, 2148
conducto de desagüe, 5022
conductor de calandra, 1039, 1042
conductor de guillotina, 5043
conductor de máquina de papel, 3634
conductor eléctrico, 2210

conector eléctrico, 2211
conexión transversal, 1708
conformidad, 1507
conífera, 1509
conmutador, 5605
cono, 1505
cono de Jordan, 3317
cono de panal, 3082
cono sin retorno, 4002
conservación, 1511
consistencia, 1512
consistencia de la pasta, 5483
consistómetro, 1514
constante dieléctrica, 1960
consumo, 1520
consumo de agua, 6071
consumo de energía, 2279
consumo de materiales, 3715
consumo de oxígeno, 499, 697, 703
consumo químico, 1244
consumo térmico, 3004
contado, 1649
contaminación, 1530, 4404
contaminación atmosférica, 282
contaminación del agua, 6092
contaminación térmica, 5706
contaminar, 1529
contar, 1640, 3770
contenedor, 1523
contenedor a prueba de fugas, 3444
contenedor con tapas articuladas, 3062
contenedor de cartón, 4153
contenedor de transporte, 5061
contenedor dividido, 4217
contenedor enchufador, 5668
contenedores encajables, 3953
contenedor sin tapa, 4065
contenido de aire, 252
contenido de celulosa, 1185
contenido de fibra, 2455
contenido de humedad, 3861
contenido de lignina, 3481
contenido de pasta, 4570
contenido de sólidos, 5267
contenido de trapos, 4624
contenido en cenizas, 457
contenido en sulfitos, 5531
contenido residual de lignina, 4749
contextura del papel, 2684
continuo, 1531
contracolado, 4226
contracorriente, 1646
contraestucadora, 1537
contraflujo, 504, 1538, 1643, 1644
contraforro para libros, 828
contrahoja, 1648
contrapresión, 514
contrarodillo, 1647
contraste, 1059
control, 1234, 1541
controlador, 1545
controlador de nivel, 3454
control automático, 484
control biológico, 701
control de bucle cerrado, 1358
control de calidad, 4605
control de carga, 3540
control de consistencia, 1513
control de humedad, 3108, 3862
control de la contaminación, 4405
control de la resina, 4341
control del flujo, 2592
control del inventario, 3283
control del olor, 4018
control de microorganismo, 3804
control de nivel, 3453
control de pH, 4275
control de polvo, 2166
control de presión, 4496
control de procedimientos, 4533
control de producción, 4540
control de realimentación, 2405
control de reenvío de alimentación, 2408
control de ruido, 3995
control de rutina, 4863
control de temperatura, 5671
control de tensión, 5686
control de velocidad, 5988

control digital, 1984
control electrónico, 2232
control remoto, 4740
convección, 1548
conversión del almidón, 5387
convertidor de frecuencia, 2726
convertidor de sistema analógico en numérico, 363
convertidor eléctrico, 2213
copal, 4690, 5601
copal rojo, 3518
copiado electroestático, 2241
copiadora, 1578
copias diazo, 1948
copo, 2575
copolímeros, 1579
copolímeros de celulosa, 1186
cordel de papel, 4193
coriente alterna (C.A.), 326
corona, 1594
corona negativa, 3951
corpulento, 957
correa, 653, 5447
correa guía, 1856
corrección de coloración, 211
corrector de pruebas, 4662
correr por debajo, 5915
corriente continua, 1803, 2006
corriente de aire, 266
corriente de pasta antes de la máquina, 421
corriente eléctrica, 2214
corriente parásita, 2182
corrientes de agua, 6111
corroer, 1596
corrosión, 1597
corrosión galvánica, 2774
corrosividad, 1605
corrosivo, 1604
cortadas, 1033
cortadas de calandra, 1670
cortado de árboles bajos, 5076
cortado limpio, 1352
cortador, 1774, 5148, 5170
cortadora, 1770
cortadora clasificadora, 1773
cortadora de árboles bajos, 5075
cortadora de bagazo, 528
cortadora de discos, 2019
cortadora de madera, 3561
cortadora de moleta, 4934
cortadora de papel, 4163
cortadora de trapos, 4626
cortadora de una hoja, 5113
cortadora longitudinal en línea, 3245
cortadora perforadora para papel higiénico, 5765
cortadora rotativa, 4848
cortadura, 5035
cortar un círculo alrededor del tronco, 2812
corte, 1764, 1775, 3341
corte de ampollas, 760
corte de árboles, 5834
corte de fábrica, 3822
corte de fibra, 2456
corte de pasta, 5432
corte de sangría, 754
corte hidráulico, 2605
corte por cuchilla, 1765
corte por moleta, 4933
corteza, 574, 1621
cortina de aire, 255
cosido, 4968
coste, 1622
costero, 5141
costo fijo, 2536
costura, 4967
costura de abrazadera, 1353
cráter, 1665
crecimiento, 2926
crecimiento primario, 4513
crepé, 1680
crepé incombustible, 2525
cresoles, 1692
criba de incocidos, 3363
criba de nudos, 3359
criba de troncos, 958
cribado, 4947

cribado fino, 2514
criba fina, 2513
criba inclinada, 3202
criba para astillas, 1295
criba plana, 2556
cribar, 5088
criba rotativa, 4852
criba vibrante, 5029
criógenos, 1727
cristal, 1728, 2816
cristalinidad, 1731
cristalización, 1732
cromaticidad, 1317
cromatografía de gases, 2782
cromatografía del papel, 4159
cromatos, 1316
cromo, 1320
cuadrante, 1939
cuadro de flujo, 2591
cuadro de mando, 1546
cuarteamiento, 1667
cuarto de calderas, 814
cuarto de lejiación, 1978
cuarto de pila, 636
cuba de aireación, 222
cubierta, 1655, 5057
cubierta cerrada, 1357
cubierta doble, 2068
cubierta en bisel, 2399
cubierta fija, 1851
cubiertas de la caja aspirante, 5499
cubilete, 617
cubrejuntas de calandra, 1046
cubre libro, 833
cubrimiento, 1656
cubrimiento de la caperuza, 3085
cuchilla, 723, 3351
cuchilla circular, 1330
cuchilla chata, 3065
cuchilla de disco, 2021
cuchilla de guillotina, 2939
cuchilla del rascador, 2049
cuchilla del rotor, 4860
cuchilla de pila, 626
cuchilla de troceadora, 1292
cuchilla fija, 646
cuchilla hidrodinámica, 3137
cuchilla móvil, 2626, 4784
cuchilla recortadora, 5847
cuchillas, 3357
cuenta, 1640
cuenta en fábrica, 3821
cuerda, 1584, 4835
cuerda para cubiertas, 5754
cuerda (para pasar la hoja en la sequería), 4836
cuero artificial, 438
cuerpo de la lámina, 2641
culombimetría, 1639
cultivo, 1737
cumbre, 5773
curvatura, 429

CH

Chamaecyparis, 1224
chapeado, 1334
cheviot, 1268
chimenea, 1273
chirrido de los rodillos, 1233
chopo, 4440
chopo balsámico, 558
chopo del Canadá, 1629
chorreo, 4985
chorro, 3308
chorro cortador de orillo, 2186
chorro de agua, 6079
chorro de aire, 272
chorro de orillo, 5849

D

daño, 1805
dar la vuelta, 5881
datos, 1821
debilitación, 2379
decantación, 5016
decantación por arenero, 4799
deceleración, 1844
de ciento ochenta grados, 2534
decoloración, 1857, 2018
deculador, 1860
dechirómetro, 5663
defecto, 1861, 2557
defecto del estucado, 1407
defensas, 3975
de fibras cortas, 5071
deflector, 521, 1872
deflector de lámina, 2640
deflector del rodillo desgotador, 5621, 5622
deflector de rodillo, 4821
deflector múltiple, 3916
deflexión, 1871
deformación, 1876, 2041
deformación plástica, 4361
degradación, 1879
degradación de la celulosa, 1187
degradación mecánica, 3730
degradación química, 1246
de hoja ancha, 903
delantal, 424
deleznamiento, 5146
demanda de cloro, 1311
demanda de oxígeno, 4126
demanda química de oxígeno, 1003, 1250
demasiado cocido, 4102
dendrología, 1899
densidad, 1905
densidad aparente, 414
densidad de corriente, 1758
densidad en masa, 946
densificación, 1902
densímetro, 1907, 1908, 3149
densitometría, 1904
densitómetro, 1903
dentado, 5008
deposición, 1914
depósito, 1915
depósito mineral, 3831
depósitos del vacío, 5942
depreciación, 1916
depuración, 1351, 4596
depuración de gas, 2793
depurador, 1350, 4945, 5445
depuradora Ben Day, 657
depuradora de gas, 2792
depurador centrífugo, 1211
depurador de desbaste, 1380
depurador de pasta, 4581
depurador vibrante, 6004
derechura, 5443
derivados de glucosa, 2834
derivados de la celulosa, 1188
derivados del almidón, 5389
derivados de vanillina, 5956
desabsorción, 1925
desaereación, 1835
desaereador, 1836
desaerear, 1834
desajustado, 4031
desarrollo del producto, 4538
descarga, 2015, 2151, 5998
descarga de residuos, 5828
descargado, 5922
descargador, 5923
descarga eléctrica, 2215
descarga gaseosa, 2783
descarga por efecto corona, 1595
descargar, 5924
descargar a presión, 777
descenso del nivel, 2097
descomposición, 1858

descortezado, 577, 1839, 4243
descortezado químico, 1242, 1245
descortezadora, 576, 1838, 4242
descortezadora de anillos, 4803
descortezadora de cuchilla, 3352
descortezadora de fricción, 2729
descortezadora hidráulica, 3120
descortezador de cadena, 1220
descortezador de tambor, 2111
descortezado superficial, 5579
descortezar, 1837
desecador, 1922
desecante, 1921
desecar, 1931
desechos, 4940
desechos del enchapado, 5994
desembocadura, 4094
desempolvado, 2168
desempolvado por succión, 5503
desengrase, 4939
desenrollado, 5930
desenrolladora, 5929
desenrollar, 5928
desensibilización, 1920
desestucado, 5474
desestucadora, 5473
desfibrado, 1863, 1866, 2906, 5080,
  5188
desfibrado con foso, 4347
desfibrado de algodón, 3515
desfibrado en calandra, 1031
desfibrado en caliente, 3093
desfibrado en continuo, 1536
desfibrado en frío, 1431
desfibrador, 1722, 4571
desfibradora, 1867, 2900, 5079, 6269
desfibradora de almacén, 3640
desfibradora de prensas, 4394
desfibradora de tres cámaras, 5736
desfibrador de recortes, 909
desfibrador discontinuo, 610
desfibrado sin foso, 4345
desfibrar, 1862, 1864, 5186
desfilochadora, 1918
desfloculación, 1874
desgasificación, 4735, 4739
desgasificado, 1878
desgasificador, 1877
desgastar, 107
desgaste, 6138
desgotador, 1816
desgotador de vitela, 6292
desgote standard canadiense, 1069
deshidratación, 1886
deshidratador, 1887
deshidratar, 1885
deshidrogenación, 1888
desigual, 6229
desincrustación, 1919
desintegración, 2024
desintegrador, 2025
desintoxicación, 1929
desionización, 1892
deslaminación, 1894
deslaminar, 1893
desleimiento, 1991
deslignificación, 1895
deslizamiento, 5165
desmedular, 1912
desmenuzadora, 3350
desmonte, 5475
desodorización, 1910
desoxigenación, 1911
desparafinaje, 1933
desparramiento, 4927
despastillado, 2463
despastillador, 1869
desperdicio, 6056
desperdicios de encuadernación, 693
desperdicios de la madera, 6281
desperdicios sólidos, 6281
desplisador de la hoja continua, 6158
despolarización, 1913
desprendimiento, 1222
desprendimiento de polvo de una tinta,
  1222
despumación, 5137
desrecalentador, 1926
desteñido, 4019

destilación, 2039
destilado, 2038
destintado, 1891
destintado de flotación, 2585
destintar, 1889
desviar, 1870
detección, 1927
detector de defectos, 2558
detector de fuego, 2523
detector de grumos, 3594
detector de infrarrojos, 3221
de tensión, 5675
detergente, 1928
deterioro, 1841
de tres capas, 5735
de una capa, 3891
de una cara, 5105
devanadera, 5334, 5931
dextrinacaucho, 1934
dextrinas, 1935
dextrosa, 1936
diafragma, 1944
diagnosis, 1937
diagrama, 1938
diagrama circular, 1329
diálisis, 1942
diámetro, 1943
diámetro de la fibra, 2457
diámetro interior, 3252
diatomas, 1947
diciandiamida, 1953
dicloroetano, 1950
diclorometano, 1951
dicotiledones, 1952
dieléctrico, 1959
diente, 5772
dietil éter, 1966
difracción, 1969
difusión, 1974
difusor, 1971
diisocianatos, 1986
dilatador, 2353
dilatancia, 1987
dilatante, 1988
diluente, 1989
diluir, 1990
dimensión, 1992
dimensión de la partícula, 4214
dimensión de las virutas, 1285
dimensión del poro, 4442
dimensiones de la fibra, 2458
dímero, 1995
dimetil bisulfuro, 1996
dimetil formamida, 1997
dimetil sulfato, 1998
dimetil sulfona, 2000
dimetil sulfóxido, 2001
dimetil sulfuro, 1999
dinámicas de los flúidos, 2602
dinamómetro de tracción, 5679
diodo, 2002
dióxido combinado de azufre, 1463
dióxido de cloro, 1306
dióxido de titanio, 5759
dipolo, 2005
dique provisorio, 1427
dirección, 2009
dirección de la granulación superficial,
  2863
dirección de máquina, 3618
dirección larga, 3562
director de fábrica, 3827
director de producción, 2807, 4541
director gerente, 2806
disco, 2014, 2026
disco de refino, 4705
diseño del embalaje, 4131
diseño, 1923
diseño de máquina, 3617
diseño de voladizo, 1074
disolutivo, 2035
disolvente, 2034, 5274
disolventes celulósicos, 1195
disolver, 2027
dispersador, 2027
dispersión, 2028, 2029
dispersión de luz, 3472
dispersión óptica, 4077
disponible, 2030

disposición, 2031
disposición del lodo, 5182
dispositivo de descenso, 3582
dispositivo de repartición de pasta,
  5349
distribución, 2042
distribución de partículas por tamaño,
  4215
distribución de prensas, 4483
distribución longitudinal de la fibra,
  2466
distribuidor de fuerza eléctrica, 2222
disyuntor, 1328
disyuntor automático, 3495
diversificación, 2045
división de caja, 872
doblado, 6047
dobladora, 658
doctor, 2048
doctor de agua, 6069
dolomita, 2060
dos caras, 5896
dosificación, 2062, 4555
dosificador, 4554
dos o más capas, 3922
duplex, 2153, 5891
duplex simili, 801
duración, 3465
duración de la tela, 6247
duración de servicio, 5010
dureza, 2979
dureza de Brinell, 898
dureza Rockwell, 4812
dureza Shore, 5068
durómetro, 2163, 2981
durómetro Shore, 5067

E

ecología, 2178
economizador, 2180
ecotipo, 2181
ecuación, 2298
efecto anublado, 1366
efecto de pared, 6044
efecto mondo de naranja, 4082
efervescencia, 2190
eficiencia, 2191
eficiencia del secado, 2192
efluente, 2193
efluente de blanqueo, 742
efluentes de fábrica, 3825
eje, 495, 5026
eje de línea, 3508
elaboración, 4534
elaboración de pasta, 4574
elaboración de pasta alcalina,
elaboración de pasta al ácido nítrico,
elaboración de pasta al polisulfuro,
elaboración de pasta al bisulfito, 311,
  3983, 4423, 5535
elaboración de pasta con fase de vapor,
  5962
elaboración de pasta hidrotrópica, 3156
elaboración de pasta kraft, 3376
elaboración de pasta mecánica, 2924,
  3735
elaboración de pasta química, 1253
elaboración de pasta semiquímica al
  sulfito neutro,
elaboración de pasta semiquímica,
  3959, 4994
elaboración de pasta sódica, 5214
elaboración termomecánica de pasta,
  5715
elasticidad, 2200
elastomeros, 2202
electricidad estática, 5394
electrodo, 2227
electroestática, 2244
electrofóresis, 2237
electrólisis, 2228
electrólito, 2229
electrón, 2236
electrónica, 2235

elementos de calco, 5796
elementos de madera separando las
  cuchillas, 2047
elevada consistencia, 3041
elevador, 3466
elevado rendimiento, 3059
eliminación de desperdicios, 6057
elución, 2250
eluir, 2249
elutriación, 2252
elutriar, 2251
embalado en resma, 4669
embaladoras en fábrica, 3829
embalaje, 4132, 6297
embalaje corrugado, 1613
embalaje de carnicerías, 986
embalaje de pan, 880
embalaje de ruedas de coche, 490
embalaje en bolsitas, 4459
embalaje en cajas de cartón, 1526
embalaje kraft, 3380
embalaje para mantequilla, 991
embalajes de resma, 4670
embalajes de tahona, 542
embalaje simili, 811
embarcador, 2531
embarcadora de la hoja continua, 6160
embarcamiento, 5734
embarque en tren, 2288
embrague, 1371
empacado, 546, 549
empacadora, 548
empalmador, 992
empalme, 5327
empalme recortado, 5842
empañado, 3834
empapado, 5205
empaque, 1763
empaquetado, 960
empaquetadura de algodón, 6227
empaquetamiento, 4137
empergaminamiento, 4203
emulgente, 2263
emulsificación, 2262
emulsión, 2265
emulsionar, 2264
emulsión bituminosa, 716
emulsión de cera, 6128
encaje de papel, 3388
encapsulación, 2269
encargado de refinos, 633
encendido, 3178
encintado, 5647
encoger, 5081
encogido, 5082
encolado, 2849, 4221, 5128
encolado al álcali, 312
encolado a la resina, 4843, 4844
encolado de calandra, 1050
encolado disolutivo, 5275
encolado doble, 2080
encolado en pasta, 5486
encolado en pila, 637, 638, 2284
encolado en rollo, 6157
encolado en superficie, 5582, 5585
encolado en tina, 5861, 5864
encolado fuerte, 2984
encolado interno, 3278
encolado ligero, 5143, 5253
encoladora, 2847
encoladora plegadora, 2645
encolador de cola, 5633
encolado superior, 5780
encolado y satinado, 5122
encolar, 4219
encuadernación, 694
encuadernación estucada, 436
encuadrado, 5361
encharcar, 3394
enchufado, 5666
endospermo, 2274
endotérmico, 2275
endurecedor, 2975
endurecimiento, 2976
endurecimiento en superficie, 5577
enebro, 3322, 3323
energía, 2277
energía cinética, 3346
energía eléctrica, 2221

enfardado grande, 3039
enfardadora de pasta húmeda, 3420
enfardador de virutas, 1289
enfardeladura de virutas, 1290
enfriado por aire, 253
enfriador, 1573
enfriamiento, 1574
enfriamiento por aire, 254
enganchador secador, 2442
engomado, 2944
engomadora, 2945
engranaje, 2799
engranaje cónico, 680
engranaje diferencial, 1968
engrasamiento fuerte, 2972
engrase, 3585
enguirnaldar, 2444
enlace atómico, 478
ennegrecido en calandra, 1026
ennegrecimiento, 719
ennegrecimiento en calandra, 1027
enrolladora, 4699, 6181
enrolladora de cartón, 6192
enrolladora de tambor doble, 2071
enrolladora Pope, 4438
enrosque de hembra, 5649
ensayo por máquina Brinell, 899
ensayos a la tracción, 5683
ensortijado, 5888
entalpía, 2286
entalladura, 4931, 4937
entomología, 2287
entrada de extrusión, 2367
entrada de pasta, 5434
entropía, 2289
envases para alimentos congelados, 2740
envejecimiento, 236
envejecimiento (del papel), 231
envoltura, 1148, 6296
envoltura de libros, 840
envoltura para jabón, 5208
envolturas para pan, 881
enzimas, 2294
epibromohidrin, 2295
epícea, 4300, 5359
epícea de Engelmann, 2281
epícea del Canadá, 6224
epícea negro, 722
epícea Sitka, 5120
epiclorohidrin, 2296
equilibrio, 543, 2299
equilibrio calorífico, 3002
equilibrio de energía, 2278
equipo, 1693, 2300
equipo calefactor, 3009
equipo de elevación, 3468
equipo de medición, 3772
equipo magnético, 3655
equipo para la puesta en hojas, 5050
equipos de control, 1542
equipos de primera calidad, 5265
equipos de seguridad, 4882
equipos didácticos, 5802
equipos neumáticos, 4392
equipos para fabricar papel, 4174
equipos presurizados, 4509
erosión, 2304
escala, 4922
escala Celsius, 1201
escala de blanqueo, 750
escamas de calandra, 1047
escape, 2339
escarificación, 4926
escasez, 5069
escoger, 1735
escogido, 5282
escurrido, 2093
escurrimiento, 2086
escurrir, 2085
esfuerzo de corte, 5039
esfuerzo de tensión, 5680
esmaltado, 2267
esmalte, 2266
esparto, 2305
especialidades, 5294
especificación, 5299
especificación de color, 1454
espectrofotómetro, 5309

espectrógrafo de rayos X, 6310
espectrorreflectancia, 5307
espectrorreflexividad, 5308
espectroscopia, 5311
espectroscopia de emisión, 2261
espectroscopia de la masa, 3707
espectroscopia de microondas, 3808
espectroscopio, 5310
espesador, 1848, 5719
espesamiento, 1849, 1932, 5720
espesor, 1060, 5722
espesor de cuerpo, 954
espesor de la pared de célula, 1200
espesor de la película de tinta, 3234
espiga, 3810
espitado de la savia, 4904
esponjoso, 5333
espuma, 2628, 2738, 4958
espumadera, 5136
espumadera del jabón, 5207
espumante, 2630
estabilidad, 5367
estabilidad a la luz, 3471
estabilidad de la pasta, 5368
estabilidad térmica, 5709
estabilización, 5369
estabilización del papel, 1994
estabilizado, 5370
estabilizador, 5371
estadística, 5399
estado de superficie, 5575
estado latente, 3426
estado permanente, 5401
estampación con cuño, 1958
estampado, 5380
estampado de hoja continua, 6149
estanco al vapor, 5964
estaño, 5748
estanque para rollos (de madera), 3560
estarcido, 5419
estátor, 5400
estearato de aluminio, 334
estearato de amonio, 350
estearato de cinc, 6330
estearato de sodio, 5234
estearatos, 5413
estela, 5417
estereoscopio, 5423
ésteres, 2310
ésteres de celulosa, 1189
ésteres de glucosa, 2835
ésteres de polivinilo, 4429
esterificación, 2309
estiraje, 5468
estiramiento tensor, 5681
estireno, 5489
estonita, 5440
estratificación, 3414, 5450
estría de cuchilla, 4942
estriado, 5469
estriados, 4489
estrías de calandra, 1055
estructura, 5478
estructura celular, 1181
estructura de celulosa, 1196
estructura de cristal, 1733
estructura de la célula, 1179
estructura del papel, 4188
estructura de madera, 6280
estructura de panal, 3083
estructura fibrosa, 2471
estructura laminar, 3407
estructura molecular, 3882
estucado, 1384, 1399, 1632
estucado a copos, 2576
estucado a la piroxilina, 4602
estucado antiadhesivo, 4732
estucado antideslizante, 407
estucado cera, 6122
estucado cerámico, 1216
estucado compuesto, 3911
estucado con acabado al cepillo, 921
estucado con ampollas, 930
estucado de arcilla, 1345
estucado de aspersión, 5343
estucado de carbono, 1097
estucado de cuchilla, 3354
estucado de labio soplador, 275
estucado de lamedura, 3349

estucado de manufactura, 1552
estucado de prensa, 4486
estucado de rodillo estampador, 4531
estucado de rodillos, 4814, 4819
estucado dos caras, 1002, 2065, 5895
estucado electroestático, 2240
estucado en capas, 1760
estucado en máquina, 3611, 3614
estucado en seco, 2119
estucado en superficie, 5572
estucado fuera de máquina, 4021
estucado metálico, 3765
estucado múltiple, 3917
estucado para grabado, 2880
estucado patentado, 4229
estucado pergamino, 442
estucado por cuchilla, 725
estucado por cuchillas de arrastre, 5801
estucado por escobillas, 917
estucado por extrusión, 1151, 2365
estucado por fundido en caliente, 3096
estucado por inmersión, 2004
estucado por labio soplador, 248
estucado por rodillos invertidos, 4780
estucado por tambor lustrado, 4401
estucado protector, 4558
estucador, 419
estucadora, 1398, 1410
estucadora bordeadora, 615
estucadora con cuchillas de arrastre de bandeja, 4434
estucadora con rodillo desgotador, 1817
estucadora con rodillo escurridor, 5366
estucadora con rodillo regulador, 2796
estucadora con rodillos invertidos, 4779
estucadora de ambos lados, 5893
estucadora de cepillos, 916
estucadora de cuchilla, 724
estucadora de cuchillas,
estucadora de cuchillas flotantes,
estucadora de cuchillas de arrastre, 569, 2569, 3353, 5800
estucadora de fuente, 2700
estucadora de inundación entre rodillos, 2581
estucadora de labio soplador, 247, 274
estucadora de lamedura, 3348
estucadora de lamedura por cuchilla, 2054
estucadora de prensa, 4485
estucadora de rodillo alisador, 5200
estucadora de rodillo estampador, 4530
estucadora de rodillo lamedor, 4825
estucadora de rodillos, 4818
estucadora de rodillos armados, 6259
estucadora dos caras, 2066
estucadora duplex, 2156
estucadora en capas, 1759
estucadora fuera de máquina, 4020
estucadora Massey, 3706
estucadora por aspersión, 5342
estucadora por extrusión, 1150, 2364
estucadora por inmersión, 2003
estucadora principal, 1225
estucado resistente a las grasas, 2886
estucado sobre máquina, 4051
estucado soflama, 787
estucado translúcido, 5812
estucado una cara, 1001, 5100
estucar, 1630
estudio de mercados, 3698
estufa, 2140
etano, 2311
etanol, 2312
etanolaminas, 2313
etapa de cloruración, 1302
éteres, 2316
éteres de celulosa, 1190
éteres de glucosa, 2836
éteres de polivinilo, 4430
etificación, 2315
etil acetato, 2317
etilación, 2320
etil acrilato, 2318
etilamina, 2319
etil celulosa, 2321
etilenamina, 2324
etilenodiamina, 2323
etil mercaptan, 2326

etiqueta, 3381, 5631
etiqueta de pegar, 5326
etiquetado, 3382
etiquetas de resmilla, 4665
etiquetas estucadas, 1397
etiqueta simili para equipaje, 810
etiquetas para cajas, 1143
etiquetas para cartonajes, 1136
eucalipto, 2327
eutrofisación, 2328
evacuación del agua, 6100
evacuar, 2329
evacuar por soplado, 771
evaluación, 2330
evaluación del flujo, 2595
evaporación, 2332
evaporador, 2333
evaporar, 2331
evolución, 2335
excedente de fabricación, 4110
exceso de estucado, 2337
excitatriz, 2338
expansión, 2348
expansión térmica, 5704
experimentación, 2350
exploración, 4925
explorador optoelectrónico, 4076
explotación forestal, 3556
extensibilidad, 2354
extensión de la carga, 3541
extensómetro, 2356
extracción, 2357
extracción de álcali, 301
extracción de condensados por sifón, 5116
extracción del álcali en caliente, 3092
extractivo, 2360
extremidad del rodillo, 4823
extremidad seca, 2121
extrusión, 2363
extrusionadora, 2362
exudación, 2368
exudación acuosa, 6067
eyector, 2198

F

fábrica, 3816
fabricación, 2689, 3687
fabricación al vacío, 5945
fabricación de papel, 4173
fabricación de pasta al bisulfito,
fabricación de pasta a la sosa fría, 712, 1434
fabricación de pasta de gran rendimiento, 3061
fabricación de planchas, 4370
fábrica de coloide, 1437
fábrica de papel, 4177
fábrica de pasta, 4577
fábrica de pasta mecánica, 2920
fábrica integrada, 3264
fábrica Kady, 3335
fábrica kraft, 3374
fabricante de papel, 4170, 4175
facilidad, 2375
facilitación de solubilidad, 2177
factor de desgarre, 5656
factor de potencia, 4465
fadómetro, 2378
faja húmeda en una hoja, 1812
falsa acacia, 3546
falso pliegue, 1669, 6301
falla, 2383
fallo de la cola, 2845
fanfo de cal, 3492
fardo, 545, 959
fardo de pasta húmeda, 3419, 4576
fase de vapor, 5961
fase fija, 5263
fatiga, 2391
fenol, 4277
fenolftaleína, 4283
ferrocarril, 4635
fertilización, 2438

fertilizante, 2439
férula, 2437
festón, 2440
fibra, 2445, 2473
fibra acrílica, 191
fibra bicomponente, 685
fibra cerámica, 1217
fibra de acetato, 147
fibra de amianto, 450
fibra de bagazo, 530
fibra de caña de azúcar, 5517
fibra de cáñamo de Manila, 3680
fibra de carbono, 1099
fibra de celulosa, 1191
fibra de liber, 606
fibra de lienzo, 3502
fibra de madera, 6265
fibra de ramio, 4639
fibra de rayón, 4653
fibra de recuperación, 4674
fibra de vidrio, 2815
fibra de yute, 3328
fibra-estopa, 2472
fibra inorgánica, 3249
fibra metálica, 3761
fibra mineral, 3832
fibra natural, 3943
fibra nilón, 4013
fibras corrientes, 5384
fibra sencilla, 5106
fibra sintética, 5613
fibras orientadas al azar, 4640
fibras reventadas, 2351
fibra textil, 5701
fibra transversal, 5825
fibra vegetal, 4355
fibra viscosa, 6019
fibra vulcanizada, 6036
fibrilación, 2477
fibrilado, 2475
fibrilla, 2474
fibrilla elemental, 2245
fibrillas, 2478
fibrocementos, 2451
fibroso, 2479
ficha de registro, 4677
fieltro, 2418
fieltro a la aguja, 3950
fieltro a la forma, 2962
fieltro asbestino impermeabilizante, 453
fieltro asbestino para tejados, 452
fieltro asfaltado, 464
fieltro de combinación, 1460
fieltro de forma, 3878
fieltro de lana, 6288
fieltro del fabricante de papel, 4172
fieltro de prensa, 4487
fieltro duplex, 2157
fieltro estucador, 1631
fieltro húmedo, 6187
fieltro inferior, 852
fieltro marcador, 3701
fieltro montante, 5458
fieltro para tejados, 4832
fieltros de impregnación, 4910
fieltros de saturación, 2431
fieltro secador, 2123
fieltro seco, 2131
fieltros insonorizantes, 2430
fieltro tejido sin fin, 2273
fijación, 4766
filamento, 2481
filamento textil, 2482
filatura, 5319
fileteador, 1672
filigrana, 6084
filigrana localizada, 3545
filo, 4796
filo de cuchilla, 3355
filtrabilidad, 2501
filtración, 2509
filtración al vacío, 5944
filtración centrífuga, 1210
filtrado, 2508
filtrar, 2505
filtro, 2500, 4850
filtro al vacío, 5943
filtro a presión, 4498
filtro de aceite, 4039

filtro de aire, 263
filtro de arena, 4895
filtro de asbesto, 449
filtro de discos, 2020
filtro de escurrimiento, 5836
filtro de hoja, 3441
filtro de polvo, 2167
filtro de recambio, 5291
filtro de tambor, 2114
filtro insonorizante, 1831
filtro rellenado, 2488
fineza, 2510
finos, 2512
flecha de dirección, 2421
flexibilidad, 2562, 4380
flexión, 659, 2563
flexografía, 2564
floculación, 2573, 2577
floculador, 2574
floculante, 2570
flocular, 1370, 2571
floema, 4285
flotación, 961, 2583, 4620, 4810
fluidez, 2607
fluídica, 2606
fluidificación, 2608
fluidificador, 2610
flúido, 2601
flujo, 2588
flujo de abertura, 4087
flujo de agua, 6078
flujo de corriente transversal, 1706
flujo de la pasta, 5433
flujo de sólidos, 5268
flujo flúido, 2603
flujo laminar, 3408
flujo turbulento, 5878
flujo viscoso, 6026
flúor, 2617
fluorescencia, 2612
F.O.B. (franco a bordo), 2369, 2721
folio verso, 503
fondo de cajas de cartón, 1525
fondo de cárter, 5549
fondo vivo, 3536
forma, 3871, 3905
forma a mano, 2965
formabilidad, 2679
formación compacta, 6177
formación de bolitas de papel, 4246
formación de bolsas, 533
formación de cráteres, 1666
formación de fibrilla, 2476
formación de hojas, 5047
formación de la hoja, 2685
formación uniforme, 5918
forma de saco, 536
forma de tela, 6251
formaldehido, 2680
formamida, 2682
forma redonda, 1797
formato acabado, 5841
formato de corte, 1769
formiatos, 2683
formol, 2681
fórmula, 2693
formulación, 2694
formulación del pigmento, 4320
formularios estucados caolina, 1346
formularios pegados, 4222
forrado para coches, 1126
forrar, 4141
forro, 3505, 3509
forro a toda prueba, 3055
forro de cajas de cartón, 1527
forro de cartón y cajas, 790
forro de yute, 3329
forro interior, 512
forro kraft del cilindro, 1795
forro para coches, 1125
forro para libros, 835
forro posterior simili, 796
forro resistente a los ácidos, 172
forros de caja, 870
forros para sacos, 537
forro sulfático, 5522
fosa bajo la prensa húmeda, 1635
fosa bajo la tela, 3630, 6252
fosfatos, 4286

fosfato trisódico, 5853
fósforo, 3711, 4287
fósforo comercial, 1469
foso, 4339
fotómetro, 4292
foto-offset, 4293
fotosensibilidad, 4294
fotosíntesis, 4296
fracción, 2707
fraccionador, 2710
fraccionamiento, 2708, 2709
fracción de huecos, 6029
fracción fibrosa, 2461
fragilidad, 902
fragmentación, 2712
fraguado al fuego, 3020
fraguado térmico, 5718
frecuencia, 2725
frenaje, 877
frenos, 876
fresno, 460, 2716
fricción, 2728, 4956
frotación, 4869
frotación húmeda, 6198
fuego, 2521
fuel, 2743
fuelle, 2950
fuente, 2699
fuente luminosa, 3474
fuerza, 2664, 4464, 5459
fuerza cortante, 5038
fuerza de adherencia, 821
fuerza de flexión, 665
fuerza de reventamiento, 975
fuerza de unión, 823
fuerza eléctrica generada por vapor, 5404
fuerza flexional, 2566
fuerza hidroeléctrica, 3132
fuga, 3443
funcionamiento, 4066
funcionamiento defectuoso, 3670
funcionamiento de la máquina, 3629
funcionamiento de prueba, 5835
fundación, 2698
fundición, 1152
fundición en caliente, 3094
fungicidas, 2753
furfurano, 2758
furfuranos, 2756
furfurol, 2757
fusible eléctrico, 2217
fusión, 3745, 3754

G

galactanos, 2770
galactómano, 2771
galactosa, 2772
galga aferradora, 2877
galón, 2773
galones por día, 2766
galvanización, 2776
galvanizar, 2775
galvanómetro, 2777
gamma celulosa, 2778
gas, 2780
gas de cloro, 1307
gas de combustión, 2597
gas de escape, 2341
gaseoso, 2784
gases de escape, 4736
gasificación, 2787
gas natural, 3944
gastos de funcionamiento, 4067
gel, 2801
gelación, 2805
gelatina, 2802
gelatinizado, 2803
generación de fuerza, 4466
generador, 2808
generador de gases, 2785
genética, 2809
genética forestal, 2670
germinación, 2810

gimnospermo, 2951
glicerina, 2850
glicerol, 2851
glicerol estearatos, 2852
glicoles, 2853
glioxal, 2854
glucómano, 2832
glucosa, 2833
gofrado, 2254
gofrador, 2256
gofradora, 2258
gofrado tela, 1363, 3503
goframiento, 2257
goma, 2940, 4864
goma arábiga, 427, 2941
goma de acacia, 129
goma de algarroba, 3547
goma de guar, 2931
goma látex, 4866
goma sintética, 5616
gradiente de presión, 4500
grado de blanqueo, 1881
grado de engrase, 6194
grado de lejiación, 1570
grado del secado, 4643
grado de orientación, 1882
grado de polimerización, 1883
grado de refinado, 642
grado de refino, 1880, 2718
grado de refino de Schopper-Riegler, 4929
grado de substitución, 1884
grado Kelvin, 3338
graduación del papel, 4183
gramo, 2869
gramos por metro cuadrado, 2767
graneo, 5428
granito, 2871
grano, 4981, 5304
grano de madera, 6268
granos de arena, 993
granoso, 2866
gran potencia, 3045
granulación, 2862, 2865
granulación en molino de piedras, 4238
gránulo, 2874
grapa, 1337
grapas de bobina, 4817
grasa, 2883
grasiento, 2895
grieta, 1659
grietas del secado, 2137
grifo, 1419
grifo de seguridad, 4881
grúa, 1664
grumo, 2579, 3591
grumo blando, 5247
grumo de arcilla, 1349
grumo del color, 1450
grumo duro, 2978
grumos en el estucado, 1409
grupos acetilos, 156
grupos ácidos, 163
grupos alcohilo, 319
grupos aldehidos, 291
grupos amino, 342
grupos aromáticos, 433
grupos benzilo, 673
grupos butiril, 999
grupos carbonilo, 1106
grupos carboxialquilo, 1112
grupos carboxietilo, 1115
grupos carboxilo, 1117
grupos carboximetilo, 1121
grupos celindos, 5617
grupos de alcohol, 288
grupos de butilo, 995
grupos de fenilo, 4284
grupos de fenol, 4279
grupos de metilo, 3787
grupos de oxidrilo, 3165
grupos de xantato, 6303
grupos éster, 2308
grupos éter, 2314
grupos etilenos, 2322
grupos etilos, 2325
grupos funcionales, 2751
grupos hidroxialquil, 3159
grupos hidroxietil, 3162

grupos nitrato, 3978
grupos sulfonil, 5539
guardacantar, 562
guarnición, 2497, 2535, 5629
guarnición de pila, 630
guarniciones, 1364, 3610
guata, 6040, 6041
guata de celulosa, 1197
guata rizada, 1683
guía, 2932
guía de la hoja, 5048
guía del cable, 4837
guías de la hoja continua, 6152
guías-tela, 6245
guillotina, 2937, 2938, 5844
guillotina de cartela, 874
guirnalda, 2779

H

hacer a prueba de intemperie, 6142
hacer pasta de nuevo, 4746
hacer transparente, 5820
halógenos, 2957
haluros ácidos, 164
harina, 2587
haya, 649, 2382
haya blanca, 3088
haz de fibras, 2449
helicoidal, 3027
hemicelulosas, 3029
hendedora, 4854, 5167
hendedura, 5332
henedura, 5172
heredable, 3033
herramienta, 5771
herramienta para avivar la muela, 969
herramienta para cortar, 1776
herrumbre, 4874
heterogeneidad, 3035
hez, 2100
híbrido, 3111
hidracina, 3123
hidratación, 3117
hidratado, 3114
hidratos, 3116
hidratos de carbono, 1092
hidráulica, 3119
hidráulico, 3118
hidrocarburos, 3127
hidrocelulosa, 3113, 3128
hidrociclón, 3130
hidrodescortezadora, 3125
hidrodesfibrador, 547
hidrodinámica, 3131
hidroexpansividad, 3134
hidrófilo, 3150, 6064
hidrófobo, 3151
hidrogenación, 3139
hidrólisis, 3146
hidrolizados, 3148
hidrostático, 3152
hidrosulfito de cinc, 6328
hidrosulfito de sodio, 5229
hidrosulfuros, 3153
hidrotecnia, 3119
hidroxialquil celulosas, 3158
hidróxido de amonio, 349
hidróxido de calcio, 1018
hidróxido de cupramonio, 1742
hidróxido de cuprietilenamina, 1745
hidróxido de magnesio, 3650
hidróxidos, 3157
hidroxietilación, 3160
hidroxietil celulosa, 3161
hidroxipropil celulosa, 3167
hidroxipropil-metil celulosa, 3168
hidruro de boro sódico, 5222
hidruros, 3124
hierro, 3294
hierro colado, 1153
higroexpansividad, 3169
higroscopicidad, 3170
hilas, 3513
hilera de una extrusionadora, 2366

hilo de papel, 4195
hilo de trama, 6166
hilos de urdimbre, 6048
hilo textil, 6315
hinchamiento, 5602
hipobromitos, 3171
hipoclorito de calcio, 1019
hipoclorito de sodio, 5231
hipocloritos, 3172
histéresis, 3174
hoja, 3440, 5042
hoja continua, 6146
hoja continua de papel, 4194
hoja continua húmeda, 6212
hoja cuadrada, 5364
hoja de apoyo, 510
hoja defectuosa, 3192
hoja de madera, 5992
hoja de papel, 4184, 5052
hoja de pasta, 4582
hoja de protección, 588
hoja de prueba, 5700
hoja de registro, 2966
hoja desgarrada, 5783
hoja libre, 2723
hoja multicapa, 3919
hoja offset, 4030
hojas de fibra prensada, 6043
hojas de papel prensa, 3967
hojilla, 3442
holocelulosa, 3076
holodesfibrado, 3077
hollín, 5276
hollinado, 1703
homogeneidad, 3079
homogeneización, 3080
homogeneizador, 3081
hongos, 2752
hornada de cal, 3488
horno, 2760, 4097
horno de fundición, 5193
horno de recuperación, 4685
horno de secar, 3345
horquilla de correa, 656
hueco, 6028
huellas del estucado, 5456
huellas húmedas, 1813
humectación, 1808, 1810
humectadora, 1807, 3106, 3858
humectadora de escobillas, 918
humectadora por pulverización, 5344
humectante, 3104
humectar, 3859
humedad, 3107, 3860
humedad absoluta, 113
humedad relativa, 4727
húmedo, 1806, 3857, 6180
húmedo Mullen, 6193
humidificación, 3105
humo, 5195

I

identificación, 3175
idioma de programación, 4548
ignífugo, 2526
igualar, 3312
igualdad en la impresión, 4527
imagen latente, 3427
imán, 3654
imidas, 3180
iminas, 3181
imitación cuero, 3183
imitación pergamino, 3185
impactadora de aire, 269
impacto de aire, 270
impedancia, 3190
impedancia eléctrico, 2218
impedimento de la espuma, 2632
imperfección, 755
impermeabilidad, 3193, 4759, 6096, 6101
impermeabilidad a la grasa, 2888
impermeabilización, 6095
impermeabilización a la grasa,

impermeabilización a los líquidos, 2887, 3520
impermeable, 3194, 6093
impermeable a las grasas, 2884
impregnabilidad, 1383
impregnación, 418, 3199, 4913, 5570
impregnado, 3196
impresión, 4522
impresión de anilina, 375
impresión de heliograbado, 2882
impresión en huecograbado, 4858
impresión offset sobre rollo, 6155
impresión sobre estucado, 4529
impresión sobre pantalla de seda, 5095
impresión tipográfica, 3451
impresor, 4521
impresos de negocios, 981
impresos en blanco, 731
impresos en continuo, 1533
impresos estucados, 1386
imprimir, 4871
impulso mecánico, 3731
impureza, 2012, 3200
impureza aparente, 1539
incendio forestal, 2669
inclinación, 2858
incocido, 5174
incombustible, 2539
incremento, 3205
incrustar, 2270
indicador, 2798, 3208
indicador alfa, 323
indicador beta, 676
indicador de consistencia, 1515
indicador de dureza, 2980
indicador de humedad, 3863
indicador de nivel, 3455
indicador de opacidad, 4058
indicador de peso, 6171
indicador de presión, 4499
indicador de rayos beta, 677
indicador de ruido, 3996
indicador de velocidad, 5314
índice de acidez real, 4276
índice de bicromato, 1425
índice de cloro, 1309
índice de cobre, 1581
índice de crecimiento, 2928
índice de desgote, 2087
índice del rasgado, 5661
índice de mano, 949
índice de permanganato, 4269
índice de rendimiento, 6321
índice de resistencia en estado húmedo, 6206
índice de reventamiento, 973, 977
índice de saponificación, 4903
índice de viscosidad, 6025
índice Kappa, 3337
índice Nu, 4010
índices de datos, 1828
industria cartonera, 4155
industria de la pasta, 4573
industria de la transformación, 1556
industria del embalaje, 4133
industria del papel, 4165
industria forestal, 2671
inercia del papel, 1993
infestación, 3219
inflamabilidad, 2542
infrarrojo, 3220
ingeniería, 2283
ingeniería de estructuras, 5477
ingeniería de sistemas, 5619
ingeniería eléctrica, 2205
ingeniería industrial, 3215
ingeniero, 2282
ingeniero de costos, 1623
ingeniero jefe, 1270
ingrediente, 3226
inhibición, 3227
inhibidor, 3228
inhibidor de espuma, 2633
inhibidor estático, 5395
injertar, 2860
injertos copolimeros, 2861
inspección, 3253
inspector del papel, 4166
instalación, 3254

instrucción programada, 4547
instrumentación, 3256
instrumento, 3255
instrumento electrónico, 2234
instrumento gráfico, 4681
instrumento marcador, 3700
instrumentos de medida, 3727
integración, 3265
intensidad, 3266
intensidad iónica, 3290
intercambiabilidad, 3267
intercambiador de calor, 3006
intercambiador de cationes, 1161
intercambiador iónico, 3289
intercambio aniónico, 378
intercambio de cationes, 1160
intercambio iónico, 3286
interponer una hoja entre otras dos, 3269
intersticio, 3281
inundación, 2582
inventario, 3282
inversión, 4781
inversión de blancura, 897
investigación, 4747
investigación operacional, 4070
inyección, 3230
iones, 3293
ionización, 3291
ionomeros, 3292
irradiar, 3298
isocianatos, 3300
isomeros, 3301
isopropil alcohol, 3302
isotermos, 3303
isótopos, 3304
isotropía, 3305

J

jabón, 5206
jabón de tall oil, 5638
jaspeado, 3900, 3904
jaspear, 3693, 3899
jefe de fabricación, 5559
jefe químico, 1269
Jordan, 3315
juego, 5012
juego de laminadores, 4831
juego de rodillos, 4830
junta, 2788, 3314
juntarse, 5470
juntas encoladas, 2844

K

kenaf, 3339
keroseno, 3342
kraft, 3366
kraft encerado, 6125
kraft liner, 3371
kraft linerboard, 3372
kraft neutro, 3955
kraft para hilados, 3379
kraft simili, 802

L

labio de la caja de alimentación, 5155
labio superior de la caja de alimentación, 5781
laboratorio, 3386
labrabilidad, 3605
labrado, 3637
laca, 3389, 3404
laca de color, 1449
lado conductor, 2737, 5674

lado de mando, 4068
lado de transmisión, 515, 2107
lado de transmisión de la máquina, 3627
ladrillo refractario, 2522
ladrillo resistente a los ácidos, 171
laguna, 3393
lámina, 2638
lámina acanalada, 1612
laminación, 3416
lámina de desgote, 2088
laminado, 3410
laminado con base de papel, 4149
laminado de papel, 4167
laminado entre chapas, 4369
laminado paralelo, 4198
laminador, 3417
laminar, 3409
laminilla, 3406
laminilla de aluminio, 332
laminilla media, 3809
laminilla metalizada, 3763
laminilla terciaria, 5691
lana, 6285
lanoso, 6286
lápiz de color térmico, 5711
láser, 3425
lastrado de palanca, 3461
látex, 3430
látexes, 3431
latón, 878
lavado, 6053, 6055
lavado con agua acidulada, 182
lavado de fieltros, 2436
lavado por difusión, 1973
lavador de fieltros, 2419
lavador de fieltro Vickery, 6007
lavador de gas, 4954
lechada de cal, 3489, 3493, 3815
lechada de cola, 5124
lecho de crecimiento, 2927
lecho del bosque, 2672
lecho de simiente, 4982
lecho filtrante, 2503
lecho fluidizado, 2609
lejía, 3521
lejía a base de calcio, 1012
lejía agotada, 5317
lejía al bisulfito, 710
lejía al sulfito, 5532
lejía al sulfito agotada, 5318
lejiación, 1981, 1982
lejiación cruda, 4646
lejía con base de sodio, 5218
lejía con base soluble, 5271
lejía con mezcla de bases, 3838
lejía de base amónica, 346
lejía de blanqueo, 743, 746
lejía de cocción, 1572
lejía de complemento, 3665
lejía de magnesio, 3645
lejía de sulfito neutro, 3957
lejiadora, 1566, 1976, 4849
lejiadora basculante, 5744
lejiadora chapada, 1333
lejiadora en continuo, 1532
lejiadora esférica, 2825
lejiadora vertical, 6000
lejía fresca, 6221
lejía para fabricar pasta, 4575
lejiar, 1975
lejía residual, 6058
lejía sódica, 5211
lejía sopladora, 776
lejía sulfática, 5523
lejía verde, 2897
leñador, 3587
leña muerta, 1833
lengüeta postiza, 5329
letargo, 2061
letras en cajetín, 868
leva, 1064
levadura, 6316
levantamiento superficial, 5578
Libocedrus (lat.), 3463
libras por punto, 4460
libro, 826
licor negro, 720
licuarse, 1897

lienzo, 3500
ligante, 691, 820
ligero, 3475
lignificación, 3479
ligninas, 3482
ligninas de alcohol, 289
ligninas de fenol, 4282
ligninas de madera trituradas, 3824
ligninas hidrolíticas, 3147
ligninas hidrotrópicas, 3155
ligninas organo-solubles, 4085
lignocelulosa, 3483
lignoles, 3484
lignoso, 3478
lignosulfonato de calcio, 1020
lignosulfonatos, 3485
limpieza con mazqueta, 2058
limpieza de fieltros, 2420
línea de cola, 2846
línea de soplado, 774
línea de tangencia, 3974
línea de transmisión, 5813
linerboard, 3506
linerboard de yute, 3330
linerboard fino, 3476
liners de tina, 5860
lino, 2559
linotipia, 3512
líquido, 3517
líquido impregnante, 3195
Liriodendron, 3525
lisa, 886, 1052
lista de comprobación, 1235
litargirio, 3526
litio, 3527
litografía, 3532
litografía offset, 4026
litopono, 3534
lixiviación, 3436
lixiviar, 3435
lodo, 3906, 5181
lodo activado, 194
lodo de cal, 3490
lodo de lejía fresca, 6220
longevidad, 3563
longitud, 3448
longitud de la fibra, 2465
longitud de ruptura, 890
longitud en pies, 2663
lubricación, 2894
lubricado, 4036
lubricante, 3583
lumen, 3588
lustrado con cepillo, 922, 926
lustre, 2818
luz, 3470
luz diurna, 1830

LL

llanta, 4801
llave de punzón, 3949

M

maceración, 3604, 5416
macerado, 3603
macerar, 3602
macromoléculas, 3638
maculación, 5338
maculado, 5013
machacadura, 914
madera, 6260
madera a comprimir, 1481
madera de coníferas, 1510, 5254
madera de corazón, 2999
madera de corazón triturada, 1719
madera de frondosas, 1847, 2987
madera de otoño, 3429
madera de porosidad decreciente hacia el extremo de los anillos, 4805

madera de porosidad uniforme, 1970
madera de primavera, 2176, 5355
madera de reacción, 4659
madera descortezada, 4241
madera de verano, 5548
madera en pie, 5488
madera en troncos, 3586, 5745
madera helada, 2741
madera húmeda, 6214
madera joven, 3334
madera madura, 3722
madera para estibar, 2152
madera para pasta, 4584
madera porosa, 4448
madera putrefacta, 1843
madera rámea, 5688
madera recia tropical, 3297
madera resinosa, 4757
madera verde, 2898
madera vírgen, 6013
maderería, 6283
madero, 3554
madero corto, 5073
maderos pequeños, 5191
madurez, 3723
magnesia, 3643, 3651
magnesio, 3644
maleza, 6163
maltosa, 3671
malla, 3756
mancha, 5375
mancha del rodillo de sostén, 5335
mancha de tela, 6256
manchado de calandra, 1054
manchas de aceite, 4045
manchas de agua, 6110
manchas de alumbre, 336
manchas de calandra, 1051
manchas de carbón, 1104
manchas de cera, 6135
manchas de color, 1455
manchas de espuma, 2636
manchas de grasa, 2893
manchas de herrumbre, 4876
manchas de humedad, 6200
manchas de la goma, 4868
manchas de posos, 5162
manchas en el color de estucado, 5337
manchón de prensa, 1633
mando, 2104
mando auxiliar, 3028
mando de la máquina de papel, 4169
mando de retroceso, 4775
mando periférico, 4266
mando seccional, 4978
mandril, 1325, 1585, 3673
mandril de incremento, 3206
mandril de madera, 4263
mandril de papel, 4162
mandril extensible, 2345
manganeso, 3674
manguera, 3091
manguera de corte, 1767
manguito, 5149
manguito roscado, 3976
manguito sujetador, 3036
manila, 3678
manila para carnicerías, 984
manila simili, 804
manilla, 2961
manipulación de la madera, 6270
manipulación de los bultos, 947
manipulación de los rechazos de depuración en el refino, 4725
manipulación de materiales, 3716
manipulación junto al camino, 4811
manipulación unificada, 5921
mano, 945
mano de obra, 3385, 3387
mano de papel, 4610
manoglactan, 3685
manosa, 3686
mantenimiento, 3662
mantenimiento de prevención, 4512
mantilla, 4523
máquina, 3606
máquina cercadora, 563
máquina con varias telas, 3932
máquina de cartón, 792

máquina de cartón multiforma, 3927
máquina de cilindro único, 5102
máquina de formas, 3880
máquina de formas redondas, 1796
máquina de imprimir, 4525
máquina de mesa plana para desenrollar, 4827
máquina de ondular, 1616, 1619
máquina de papel, 4168
máquina de papel de mesa invertida, 2989
máquina de papel de mesa plana, 2704
máquina de papel de seda, 5755
máquina de papel de telas gemelas, 5886
máquina de papel triple de mesa plana, 5850
máquina de papel yankee de mesa plana, 2706
máquina de parafinar, 6130
máquina de sacos, 538
máquina de saquitos, 4458
máquina de satinar al cepillo, 927
máquina descortezadora, 579
máquina desmeduladora de bagazo, 529
máquina de tela metálica inclinada, 3204
máquina de varios cilindros, 3912
máquina dobladora, 662
máquina embaladora, 6298
máquina empaquetadora, 4134
máquina etiquetadora, 3383
máquina experimental de papel, 2349
máquina friccionadora, 3600, 6312
máquina hendedora, 5173
máquina humectadora, 1809
máquina multiforma, 3928
máquina múltiple de mesa plana, 3909
máquina para bolsas de papel, 4148
máquina para cajas, 871
máquina para cartonajes, 1135
máquina para fabricar sacos, 539
máquina para hacer sacos, 535
máquina para sobres, 2290
máquina pegadora, 4227
máquina plegadora, 2655
máquina recortadora, 5848
maquinaria, 3632
máquina rizadora, 1689
máquina run-out de mesa plana, 4872
máquina selladora, 4964
máquina selladora de sacos, 541
máquina taladora, 3557
máquina transformadora, 1557
maraña de fibra, 2460
marca, 3694
marca de bayeta, 1723
marca de fábrica, 3819
marca del estucado, 1634
marca del pegamento, 4228
marca del rodillo aspirante, 5514
marca del rodillo desgotador, 1818
marca del rodillo de sostén, 509
marca del labio soplador, 276
marca de papel, 1641, 3695
marcado de calandra, 1041
marcado del fieltro, 2426
marcado en calandra, 1040
marcador, 4936
marcadores de resma, 4666
marcaje de salto, 5138
marcas de arranque en la secadora, 2127
marcas de espuma, 2635, 2739
marcas de grapa, 1338
marcas de la cuerda, 4838
marcas de la rasqueta, 2055
marcas de la tela, 6250
marcas del fieltro, 2427
marcas del fieltro secador, 2124
marcas del sifón, 5500
marcas de regla, 5156
marcas de resina, 4343, 4846
marcas de rodapié, 6051
marcas de sombra, 5025
marco, 1850
marco de aire, 256
marco rasgado, 5782
marcha lateral, 5087

marchitamiento, 4326
marketing, 3696
mármol de mesa de fabricación, 2690
martillo, 2958
masa, 3705
masa filtrante, 2506
mate, 3709
materia de protección, 585
material, 3713
material aislante, 3258
material amorfo, 355
material a ondular del nueve, 3971
material de carga, 3544
material de coloración, 1447
material de embalaje, 4135
material de protección, 586
material electrónico, 2233
materiales para la construcción, 1518
material poroso, 4447
material radioactivo, 4615
materia prima, 4647
matiz, 3103
matriz vitela, 6293
mecánica, 3739
mecánica de los flúidos, 2604
mecánico, 3830
mecanismo de corrosión, 1599
mecanismo de reacción, 4656
mecanización, 3740
mecano-química, 1262
medición, 3726, 3748
medición de la densidad, 1906
medición de la madera, 6271
medición de la presión, 4502
medición de la velocidad, 5989
medición de la viscosidad, 6017
medición del flujo, 2593
medición mecánica, 3732
medición óptica, 4074
medida de la absorción, 122
medida de la humedad, 3109
medida sin contacto, 1522
medidor de deformación, 5446
medidor de flujo, 2594
medidor de formación, 3736
medidor de virutas, 1288
medio ambiente, 2293
médula, 4344
mejora de la calidad, 5932
melamina, 3743
membrana, 3747
membrana semipermeable, 4996
mercaptanos, 3749
mercapto ligninos, 3750
mercerización, 3751
mercurio, 3752
meristemos, 3755
mesa de fabricación, 6244
mesa de formación vertical, 6002
mesa plana, 2703, 2705
mesa plana amovible, 4741
mesa plana de tipo voladizo, 1075
metacelulosa, 3758
metacrilatos, 3774
metal, 3759
metal antifricción, 502
metal chapado, 1335
metalización, 3764
metalización al vacío, 5946
metano, 3776
metanol, 3777
meteorología, 3769
metil acetato, 3779
metilación, 3782
metil acrilato, 3780
metil celulosa, 3783
metileno celulosa, 3784
metil etil cetona, 3785
metil glucósidos, 3786
metil mercaptan, 3788
metil metacrilato, 3789
método del cristal azul, 784
método de producción, 4542
método de prueba, 5698
métodos de dirección, 3778
métodos de trayectorias críticas de rutas
   de proceso, 1699
mezcla, 756, 3842, 3846
mezclado pastoso, 5185

mezclador, 757, 3841
mezclar, 3837
mica, 3793
microanálisis, 3795
microfibrilla, 3799
microficha, 3800
microfilme, 3801
micrómetro, 3802, 5723
microorganismo, 3803
microporosidad, 3805
microscopio, 3806
microtomo, 3807
migas de lisa, 1053
migración, 3811
migración del adhesivo, 207
milipulgada, 3812
modelo, 4230
modelo matemático, 3718
modernización, 3849
modificador, 3852
módulo, 3854
módulo de elasticidad, 3855
módulo de ruptura, 3856, 4873
módulos de cuerpo, 955
moho, 3813
moldear, 616, 3848, 3879
molde de hoja, 5051
molde verjurado, 3400
moléculas, 3884
moleta, 2909, 3702
molibdeno, 3885
molino de bolas, 554
molino de martillos, 2959
molino de piedras, 4237
molino de rodillos, 4815
mondar, 1766
mondo de naranja, 4081
monitor, 3887
monitoraje, 3888
monocotiledones, 3889
monofilamento, 3890
monomeros, 3892
monotipia, 3893
monóxido de cloro, 1308
montmorillonita, 3894
mordedura, 5204
mordidas de la cuchilla, 4302
mordientes, 3895
mortero, 3896
mortero líquido, 2925
mota, 5305
motas de cola, 5127
motas de corteza, 581
motas del ligante, 5306
motas de resina, 4845
motas en el papel, 4186
motor, 3898
motor eléctrico, 2220
motor sincrónico, 5610
movimiento, 3897
mucilagocidas, 5163
muela de desfibradora, 2905, 2908
muela desfibradora, 4583
muela móvil, 3365
muestra, 4891
muestra fortuita, 2856
muestra tomada sobre bobina, 4703
multicapa, 3913
Mullen, 3907
muñón, 5857
muselina aisladora, 5328
muselina aislante, 3260
muselina anticorrosiva, 413
muselina de embalaje, 4140
muselina de rizo, 2336
muselina higiénica, 4899
muselina industrial, 3217
muselina para condensadores, 1078,
   1497
muselina para embalaje de frutas, 2742
muselina para encerar, 6132
muselina parafinada, 6127
muselina para hilar, 5320
muselina para modelos, 4233
muselina para parafinar, 5890
muselina para rizar, 1690
muselina para servilletas, 3940
muselinas comerciales, 1470
muselinas para el baño, 611

musgo perlado, 1132
mutación, 3935
muy brillante, 3046

N

nafta, 3938
neblina, 3833
necesidad de blanqueo, 737
negador, 1900
negociante en papeles viejos, 6060
negro de humo, 1096
neopreno, 3952
neutralización, 3954
niebla, 2992
níquel, 3969
nitración, 3980
nitrato de celulosa, 1193
nitratos, 3979
nitrificación, 3985
nitrilos, 3986
nitritos, 3987
nitrocelulosa, 3988
nitrógeno, 3989
nitroligninas, 3991
nitruros, 3984
nivel, 3452
nivelación, 3456
no calandrado, 5907
no doblador, 3998
nódulo, 3993
no estucado, 5908
no mezclable, 3186
norma, 5382
no tejidos, 4008
nubosidad, 1367
nuboso, 1368
núcleo, 1585
núcleo de pasta, 1204
nudos, 3360
nudos de la fibra, 2464
nuez, 3320
número de bacterias, 519
número de impurezas sobre el papel,
   2013
número de rodana, 971
nutritivo, 4012

O

obrero de patio, 6314
obsolescencia, 4016
obstrucción, 589
obstruir, 1354
obturación, 3703
offset, 4024, 4029
oleatos, 4047
olefinas, 4048
oleoresinas, 4050
olmo, 2246, 5902
olor, 4017
onda sonora, 5283
ondulación, 1618, 6119
ondulado, 1606, 1697, 2620, 4808
ondulado para uso especial, 5296
opacidad, 4059
opaco, 4060
operación, 4069
operación de cribado primario, 4516
operación forestal, 2674
operación intermitente, 3274
operador de desfibradora, 2901
operador de lejiadora, 1980
operario, 4071
optimización, 4080
ordenador digital, 1983
orden de fabricación, 3667
orientación, 4086
orientación de las fibras al azar, 4641
orificación, 4383
orificio, 3072

orificio de salida, 4095
orificio en la tela, 6246
orificio para posos, 5161
orillos, 1238
oscilación, 4089
oscilador, 4090
oscilante, 4088
oscilógrafo, 4091
ósmosis, 4092
ósmosis inversa, 4776
oxalatos, 4115
oxialmidón, 4127
oxicelulosa, 4122
oxidación, 4118
oxidantes, 4117
óxido de cinc, 6329
óxido de cromo, 1322
óxidos, 4120
óxidos de azufre, 5540
óxidos de hierro, 3296
oxidrilo, 3163
oxigenación, 4124
oxígeno, 4123
oxígeno disuelto, 2033
ozonización, 4129
ozono, 4128

P

paja, 5452
paja de arroz, 4793
paja de centeno, 4878
paja de trigo, 6217
pala de ventilador, 2385
palanca, 3459
paleta, 4142, 5954
paletización, 4143
palo, 5424
palpar, 2413
pandeo, 4884
panel, 4144
panel de partículas, 4213
paño tomador, 4315
pantalla de protección, 5058
papel, 4146
papel absorbente, 683, 6080
papel absorbente para platillos, 1381
papel acabado a la plancha, 4368
papel acabado tela, 2371
papel aceitado, 4038
papel acondicionado, 1499
papel a estratificar, 3415
papel ágata jaspeado, 229
papel a la plancha, 4371
papel a la tina, 2964
papel alquitranado, 5652
papel aluminio, 333
papel anticorrosivo, 412
papel antiguo, 404
papel a perforar, 4257
papel arrugado, 1422
papel arsenical, 435
papel asbestino, 451
papel asfáltico para puesta en hojas,
   468
papel a troquelar, 729
papel autográfico, 5805
papel autotipia, 491
papel avión, 279
papel bambú, 560
papel bancario, 566
papel barba, 1854
papel barnizado, 3390
papel biblia, 682, 3207
papel blanco, 6223
papel bond, 822
papel bond estucado, 1388
papel braille, 875
papel brillo, 2831
papel brochado, 904
papel calandrado húmedo, 6076
papel calco, 5798
papel calcomanía, 1840
papel calco para planos, 786
papel Cap para envolver, 1084

papel carbón, 1103
papel carbón de una sola vez, 4055
papel carbón vegetal, 1230
papel carborundo, 1109
papel cartográfico, 3690
papel cebolla, 1102, 4056
papel combustible, 1465
papel con asiento de guata, 599
papel conductivo, 1503
papel con fibra de algodón, 1626
papel con manchas, 5376
papel con pasta mecánica, 6262
papel con suspensión de barita, 592
papel copia, 1582
papel crepé, 1685
papel cristal, 2817
papel cristal encerado, 6124
papel cromático, 1318
papel cromatográfico, 1319
papel cromo, 1324
papel cuadriculado, 4924, 5363
papel cuero, 3446
papel charol, 919
papel de álbum, 287
papel de apoyo, 507
papel de avance, 2997
papel de bagazo, 531
papel de canto vivo, 2400
papel de cilindro, 1798
papel de condensadores, 1496
papel de coronación, 1085
papel de China antiguo, 400
papel de dibujo simili, 800
papel de embalaje para alimentos, 2662
papel de embalar, 4136
papel de encuadernar satinado, 3621
papel de escritura, 6302
papel de escritura azulado, 497
papel de esparto, 2306
papel de fibra de vidrio, 2814
papel de fotograbado, 2881
papel de fumar, 1327
papel de girasol, 3535
papel de imprimir, 4526
papel de intercambio iónico, 3287
papel delgado, 5726
papel de liber, 607
papel de manila, 3032
papel de manila antiácido, 388
papel de mica, 3794
papel de municiones, 352
papel de pasta mecánica para libros,
papel de pasta mecánica,
papel de pasta mecánica para imprimir,
   2917, 2921, 2922
papel de protección, 587
papel de revestimiento, 2376, 5993
papel de rizar, 2622
papel de seda para copias, 1583
papel de trapos, 4629
papel de voladura, 732
papel de yute, 3331
papel de yute para bolsas, 3326
papel dieléctrico, 1962, 2206, 2207
papel dos caras, 2075
papel dos telas, 5887
papel duplex, 2159
papel duradero, 4268
papel electrofotográfico, 2238
papel embreado para tejados, 4833
papel encartelado al agua, 6082
papel encerado, 6126
papel encerado húmedo, 6211
papel encolado, 5123
papel encolado a la cera, 6134
papel encolado en calandra, 1049
papel encolado en superficie, 5583
papel encolado en tina, 5862
papel en fardos grandes, 3040
papel engomado, 2942
papel en hoja continua, 6156
papel en mandril, 1587
papel en rollo, 4828
papel entelado, 3501
papel entresacado, 1736
papel envoltura para regalos, 2811
papeles abrasivos, 111
papeles absorbentes, 117

papeles adhesivos, 205
papeles a la tina, 5980
papeles antiadhesivos, 4733
papeles antioxidantes, 406
papeles asfálticos, 466
papeles bastos, 1379
papeles coloreados, 1441
papeles con contenido de trapos, 4625
papeles decorativos, 1859
papeles de correspondencia, 1471
papeles de dibujo, 2099
papeles de impregnación, 4911
papeles diazo, 1949
papeles estucados, 1395
papeles finos, 2511
papeles impregnados, 3197
papeles industriales, 3216
papeles inorgánicos, 3250
papel esmaltado, 2268
papel esmaltado para cajas, 869
papel esmeril, 2567
papeles mezclados, 3839
papeles para impregnar, 3198
papeles para servilletas, 3939
papel especial, 5297
papel especial de periódicos, 5561
papeles pintados, 6045
papel esponjoso, 952, 2402
papel esponjoso para libros, 951
papeles reactivos, 5699
papeles resistentes al blanqueo, 749
papeles simili, 806
papeles simili de impregnación, 808
papeles simili vitro contracolado, 807
papel estanco al vapor, 5965
papeles tratados al látex, 3432
papel estratificado, 5451
papel estucado, 441
papel estucado al polietileno, 4413
papel estucado antiadhesivo, 4731
papel estucado dos caras, 5894
papel estucado encolado, 2843
papel estucado en máquina, 3613
papel estucado lito, 3530
papel estucado mate, 2149
papel estucado offset, 1394
papel estucado para impresión artística,
   1385
papel estucado para imprimir, 1396
papel estucado para libros, 1389
papel estucado para revistas, 1393
papel estucado plástico, 4360
papel estucado por extrusión, 1149
papel estucado una cara, 4054, 5101
papel estucado y satinado para libros,
   2821
papeles y cartones resistentes al moldeo,
   3881
papel extensible, 2346, 2355, 5466
papel ferroprusiato, 785
papel fibra química, 1247
papel fieltro, 2428
papel filigrana, 3403
papel filigranado, 6089
papel filtro, 2507
papel filtro analítico, 366
papel filtro para cerveza, 651
papel filtro químico, 1248
papel filtro sin cenizas, 459
papel fino, 3477
papel fluorescente, 2615
papel forrado con tela, 4722
papel forrado con tela para la
   construcción, 4720
papel forro para cajas, 1144
papel forro para cajas de cartón, 1137
papel forro simili, 803
papel fotográfico, 4291
papel fotosensible, 3473
papel fuertemente alisado, 2285
papel fuertemente encolado, 2985
papel glaseado, 5554, 5555
papel gofrado, 2255
papel gráfico para registrador, 4679
papel granito, 2872
papel granulado, 2864, 2868
papel higiénico, 4898, 5763
papel higiénico de muselina, 5766
papel imitación barba, 3182

papel impermeable, 6097
papel impermeable al aceite, 4042
papel impregnado, 4108
papel incombustible, 2527, 2540
papel indicador, 3209
papel infalsificable, 394, 4883
papel jaspeado, 3692, 3903, 4791
papel kraft, 3375
papel kraft embreado, 3368
papel kraft para sacos, 3367
papel kraft para sacos de varias hojas,
   3930
papel Manila, 3683
papel mate, 3719
papel mecánico de barba, 3729
papel medicinal, 3741
papel metálico, 3767
papel metalizado, 2642, 3768
papel moneda, 565, 1757
papel multicapa, 3923
papel muselina, 5756
papel muy abrillantado, 3047
papel negro para álbum, 717
papel no abarquillable, 3999
papel no blanqueado, 5905
papel offset en rollo, 6154
papel opaco, 4061
papel ordinario de periódicos, 3960
papelote, 6059
papel para aislamiento, 3262
papel para anuarios, 2011
papel para apergaminar, 4205
papel para articular, 437
papel para asfaltar, 471
papel parabólico, 861
papel para bolsas de pan, 879
papel para bositas de té, 5654
papel para cables, 1007
papel para cajas registradoras, 1147
papel para carboncillo, 1229
papel para carnicerías, 985
papel para carteles, 4450
papel para cartuchos, 1139
papel para catálogos, 1154
papel para cigarrillos, 1326
papel para cintas, 1544
papel para condensadores, 1077
papel para copias, 3677
papel para cordel, 4839
papel para cubiertas, 1658
papel para cubiertas de libro, 832
papel para cheques, 1236, 1264
papel para desmaquillado, 2374
papel para dibujar, 2095
papel para dibujo al raspado, 4941,
   4943
papel para duplicador, 2161
papel para empacado, 550
papel para empaquetaduras, 2790
papel para empaquetar, 4139
papel para encerar, 6131
papel para encuadernadores, 830
papel para engomar, 2946
papel para envolver, 2490
papel para etiquetas, 3384
papel para etiquetas de direcciones,
   203
papel para filtración de aire, 264
papel parafinado en seco, 2143
papel para forrado interior, 513
papel para forrar, 3507
papel para hilados, 5889
papel para hojas reemplazable, 3572
papel para huecograbado, 4857
papel para inducidos, 430
papel para interponer, 3270
papel para jaspear, 3691
papel para la construcción, 1519
papel para las letras de cambio, 688
papel para libros, 837
papel para libros en blanco, 917
papel para libros en fardos grandes,
   3038
papel para litografía, 3450, 3531
papel para mapas, 1232
papel para máquinas computadoras,
   1485
papel para máquinas contables, 834
papel para máquinas facturadoras, 687

papel para máquinas sumadoras, 200
papel para matrices, 3720
papel para modelos, 4232, 5139
papel para offset, 4028
papel para paneles de construcción, 940
papel para pilas, 613
papel para pulir, 938
papel para registro, 4682
papel para rellenar, 2495
papel para reproducciones, 4295
papel para reproducir, 4744
papel para revestir, 5041
papel para revistas, 3641
papel para sacos, 540, 4879
papel para sacos de cemento, 1202
papel para sellado de cajas de cartón,
   1138
papel para sellado térmico, 3018
papel para sensibilizar, 5001
papel para sobres, 2291
papel para tableros de avisos, 686
papel para tapas de caja, 867
papel para timbres, 5381
papel para toallas, 5793
papel para transformación, 1558
papel para vasos, 1741
papel para vulcanizar, 6039
papel pergamino, 4206
papel periódico en blanco, 730
papel periódico forrado manila una cara,
   5110
papel periódico offset, 4027
papel pintado, 2970
papel prensa, 3966
papel prensa especial, 5562
papel prensa standard, 5383
papel recortado, 5840
papel registro, 3447
papel relieve, 4737
papel rellenado, 2490
papel relleno de arcilla, 1348
papel resistente a la grasa, 2892
papel resistente a la sangre, 767
papel resistente a los ácidos, 173, 177
papel resistente a los álcalis, 315
papel resistente en estado húmedo,
   6204
papel rizado, 1682, 1756
papel rugoso para empapelar, 2578
papel satinado, 1036, 2822, 3597,
   3623, 5552
papel satinado antiguo, 402
papel satinado en máquina, 3625
papel satinado una cara, 3601
papel secado al aire, 258, 3551
papel secante, 770
papel sellador, 4963
papel sensibilizado, 4999
papel sensible a la presión, 4506
papel simil cuero, 439
papel simili para periódicos, 3962
papel simili sulfurizado, 2889
papel sin ácido, 162
papel sin cola, 5926
papel sin desbarbar, 5927
papel sin estucar, 5910
papel sin estucar para imprimir, 5911
papel sin estucar para libros, 5909
papel sin necesidad de carbono, 3992
papel sin pasta mecánica, 6267
papel soporte, 596
papel soporte de estucado, 1403
papel soporte para libros, 827
papel stencil, 5421
papel suave, 5250
papel sulfato, 5524
papel sulfito, 5533
papel tela, 3504
papel terciopelo, 5990
papel termográfico, 5714
papel tímpano, 5899
papel tipo periódico, 3963
papel transparente estucado, 1392
papel verjurado, 3401
papel vitela, 5986, 6295
papel y cartón para gofrar, 2259
papel laminado, 3412
papel laminado al asfalto, 465
papel lanoso, 6287

papel liso, 2554
papel lito, 3533
papel litos antiguo, 398
papiro, 4196
paquete, 4130
paquete de pasta de moldear, 3876
parafinado, 6129
parámetro, 4199
páramo, 6232
parar (una máquina), 5083
pardos naturales, 3941
pared de célula, 1199
pared primaria, 4518
pared secundaria, 4975
pared terciaria, 5693
parénquima, 4209
par motor, 5784
partícula, 4212
partícula de madera, 6272
partida de cajas, 1145
partida de papel con rebaja de precio, 3310, 3311
partidor, 5331
paryodato ligninas, 4263
paryodatos, 4264
pasarela, 1165
pasillo conductor, 5673
pasta, 4568
pasta al ácido nítrico, 3982
pasta a la sosa fría, 1433
pasta al bisulfito, 711, 5534
pasta alcalina, 310
pasta alfa, 325
pasta al polisulfuro, 4422
pasta al sulfato, 5526
pasta blanda, 5251
pasta blanqueable, 736
pasta blanqueada, 738
pasta cáustica fría, 1430
pasta cautiva, 1088
pasta central, 1590
pasta con astillas, 3364
pasta con pelusa, 2599
pasta cruda, 4648, 5906
pasta de abacá, 104
pasta de abedul, 708
pasta de acetato, 148
pasta de algodón, 1628
pasta de bagazo, 532
pasta de bambú, 561
pasta de desgote, 2092
pasta de elevado rendimiento, 3060
pasta de esparto, 2307
pasta de exportación, 2352
pasta de fibras largas, 3564
pasta de fibra vulcanizada, 6037
pasta de haya, 650
pasta de madera, 6276
pasta de madera de coníferas, 5255
pasta de madera de frondosas, 2988
pasta de madera química, 1261
pasta de mandriles, 5867
pasta de moldear, 3875
pasta de nudos, 3362
pasta de paja, 5454
pasta de periódicos, 3642
pasta de pino al bisulfito, 4329
pasta de pino al sulfato, 4328
pasta depurada, 136, 137
pasta de rayón, 4654
pasta de recortes, 6061
pasta de secado instantáneo, 2544
pasta desfibrada, 1865
pasta destintada, 1890
pasta de trapos, 4630, 4633
pasta de yute, 3332
pasta disoluble, 2036
pasta dura, 2983, 2986
pasta en cartones, 3421
pasta engrasada, 5180, 5190, 6197
pasta enteramente química, 2745
pasta hidratada, 3115
pasta kraft, 3377
pasta larga, 3568
pasta lechada, 5184
pasta magra, 2390, 2722, 2724
pasta mecánica, 2916, 2923, 3734, 3738
pasta mecánica de paja, 3737

pasta mecánica de virutas, 1287, 2919
pasta mecano-química, 1263
pasta para guata, 6042
pasta para la venta, 3697
pasta para usos especiales, 5298
pasta química, 1252
pasta secundaria, 4973
pasta semiblanqueada, 4990
pasta semiquímica, 4993
pasta semiquímica de sulfito neutro, 3958
pasta sódica, 5213
pasta soporte, 598
pasta (tratada), 5430, 5480
pasta virgen, 6014
pasta viscosa, 6020
patinaje, 5131
patinaje por sobrecarga, 4114
patología forestal, 2675
pectina, 4239
pegadura, 4225, 5325
pegadura asteada, 988
pegadura rápida, 2624
pegaduras de corrido, 2625
pegajoso, 5630
película, 2498
película fina, 5724
pelos del fieltro, 2424
pelusa, 2598, 2762, 2764
pendiente, 5175
pendiente calórica, 3007
penetración, 4249
penetración del aceite, 4040
penetrante, 4248
pentaclorofenol, 4250
pentosanas, 4251
percloratos, 4253
pérdida, 3574
pérdida de calor, 3011
pérdida de carga, 2998
pérdida de carga por fricción, 2735
pérdida del vertedero, 5023
pérdida dieléctrica, 1961
pérdida química, 1249
perfil, 4544
perfilador, 4545
perfil transversal, 1709
perforación, 4256, 4258, 4591
perforadora, 4259
perforámetro, 4593
pergamino, 4200
pergamino artificial, 440
pergamino bond, 4201
pergamino bond borrable, 4203
pergamino para escribir, 4208
pergamino vegetal, 5981
pergamino vegetal estucado, 444
peridermis, 4262
periférico, 4265
periódico, 3965
periódicos viejos, 4106
período, 230
período de paralización de trabajo, 2084
permanencia, 2162, 4267
permanencia del colorante, 1442
permanganato de potasio, 4455
permanganatos, 4270
permeabilidad, 4271, 6091
permeabilidad al aire, 281
permeabilidad al gas, 2791
permeabilidad a los líquidos, 3519
permeabilidad al vapor, 5960
permeabilidad al vapor de agua, 6117
perno de anclaje, 368
peróxido de sodio, 5232
peróxidos, 4272
persulfatos, 4273
pertrechos navales, 3947
pesada, 6167
peso, 6170
peso bruto, 4650
peso de la resma, 4668
peso de la resma de papel, 604
peso del estucado, 1417
peso equivalente, 2301
peso específico, 5300
peso húmedo, 6213
peso molecular, 3883
peso nominal, 3997

peso real, 199
peso seco absoluto, 4099
peso seco absoluto a cien grados, 498, 825
peso sin estucado, 5912
pesos pesados, 3025
petróleo, 2744
petróleo bruto, 1716
pH, 4274
piedra caliza, 3494
pigmentado, 4321
pigmento, 4319
pigmento de coloración, 1448
pigmento de la laca, 3405
pila, 4322
pila de cloro, 1304
pila de leña, 6273
pila de papel, 5135
pila desfibradora, 883, 5187
pila de virutas, 1293
pila electrolítica, 2230
pila filochadora, 4623
pila holandesa, 3074, 3075
pila lavadora, 6052
pila lavadora de difusión, 1972
pila refinadora, 624
pila refinadora de rotación, 1784
pila refinadora holandesa, 885
pila termoeléctrica, 5712
piña de árbol, 5831
pino, 4325, 4333
pino alerce, 3548
pino amarillo, 3567, 4342
pino austral, 5287
pino de hojas cortas, 5072
pino del Canadá, 3307
pino de Murray, 3934
pinoresinol, 4332
pino silvestre, 4938
pirita, 4598
pirólisis, 4601
pistola, 2948
pistón inmergible, 4385
placa de acero chapeado, 1336
placa de asiento, 597
placa de extracción, 2359
placa deflectora, 523
placa de soporte, 5259
plancha, 4366
plancheado, 4378
planta, 4353
plantación, 4354, 4356
planta criada de semilla, 4984
planta de blanqueo, 747
planta de evaporación de lejías, 3522
planta de reacción, 4657
planta de transformación, 1559
planta generadora de vapor, 5408
planta hidroeléctrica, 3133
planta llave en mano, 5880
planta piloto, 4323
planta sin pasta mecánica, 4006
planta tropical, 5854
plantilla, 2687
plantilla aspiradora, 5506
plantilla de cuba en seco, 2146
plantilla de dos o más capas, 3918
plantilla de hojas, 5046
plantilla de hojas-registro, 2967
plantilla de lúnula, 1691
plantilla de película, 2499
plasticidad, 4362
plástico, 4359
plástico laminado, 3413
plástico reforzado, 4723
plásticos con base de papel, 4150
plásticos de celulosa, 1194
plastificación, 4364
plastificante, 4363
plastómetro, 4365
plátano, 4352
platina, 4373
platina de holandesa, 634, 648
platos de papel, 4180
plegado, 1696, 2644, 2646
plegado en acordeón, 140, 1489
plegado en tres, 3422
plegado en tres húmedo, 6189
plegámetro, 2653

pliegue, 1695, 2643, 4565
pliegue falso húmedo, 6215
pliegue largo, 3565
pliegue longitudinal, 3566
pliegues de humedad, 3870
pliegues falsos de bobinadora, 6237
pliegue suave, 5246
plomo, 3437
pluma, 4247, 4384
poda de ramas, 1896
poder abrasivo, 110
poder absorbente, 115
poder absorbente de aceite, 4033
poder ligante, 695
polaridad, 4397
polarización, 4398
polea, 4566
polea de cono, 1506
polea escalonada, 5422
poliacrilamida, 4406
poliacrilato, 4407
poliamidas, 4409
polielectrólitos, 4410
poliésteros, 4411
poliestireno, 4421
polietilenimina, 4414
polietileno, 4412
polimenoros de adición, 201
polimeración, 4416
polimeración de condensación, 1493
polimerizar, 4417
polimeros, 4418
polimeros sintéticos, 5614
polimetacrilato, 4419
polimidas, 4415
polinación, 4403
polipropileno, 4420
polisulfuros, 4424
poliuretanos, 4425
polivinil-acetato, 4426
polivinilos, 4432
polvo, 2164, 4462
polvo de amianto, 227
polvo de cortadora, 1772
polvo de la hendedora, 5168
polvo de la rasqueta, 2051
poro, 4441
poros, 4330
porosidad, 126, 4444
porosidad horizontal, 3087
porosidad lateral, 3428
porosidad transversal, 5826
porosímetro, 4443
poroso, 4446
posibilidad, 2396
posición de la regla, 5157
poso, 5160
postrefinado, 4451
post-secador, 226
potasio, 4452
potencial redox, 4691
potencial zeta, 6325
potenciómetro, 4457
pozo, 6176
precipitación, 4471
precipitación electroestática, 2242
precipitado, 4470
precipitante, 4472
precipitante electroestático, 2243
precipitar, 4469
precisión, 143, 4473
pre-estucado, 4474
prefabricación, 4476
preformación, 4477
prehidrólisis, 4478
prendas de protección, 4557
prensa, 4482
prensa afieltrada doble, 2077
prensa aspirante, 5509, 5515
prensa auxiliar, 5732
prensa combinada, 1462
prensa con tela de plástico, 2372
prensa de cilindros, 1799
prensa de corteza, 580
prensa de desfibradora, 2903
prensa de enfardar, 551
prensa de fieltro inferior, 853
prensa de fieltro superior, 5775
prensa de flujo transversal, 5824

prensa de flujo vertical, 6001
prensa de gran intensidad, 3048
prensa de manguito, 5151
prensa de platinas, 4375
prensa de reporte, 5806
prensa de toma automática, 4317
prensa de tornillo, 4952
prensa dividida, 2046
prensado, 4488
prensa doble, 2147
prensa doble dividida, 2070
prensado en caliente, 3097
prensado húmedo, 6196
prensa encoladora, 5125
prensa encoladora de superficie, 5584
prensa encoladora de tina, 5863
prensaestopas, 5484
prensa húmeda, 1636, 6195
prensa inferior, 854
prensa montante, 4777
prensa offset, 5199
prensa offset (sobre máquina de papel),
    5198
prensa-pastas, 6191
prensa plana, 2548
prensa primaria, 4515
prensa principal, 3661
prensa ranurada, 2913
prensa recortadora, 5845
preparación, 4479
preparación de la madera, 6274
preparación de la pasta, 5435
preparación del emplazamiento, 5119
preparación de muestrario, 4894
preparación de muestras, 4892
presa, 1804
presecador, 4475
preservación, 4480
presión, 4495
presión atmosférica, 477
presión barométrica, 583
presión de cuerpo, 953
presión del vapor, 5963
presión dinámica, 285
presión estática, 5396
presión hidráulica, 3121
presión lineal, 3496, 3977
presión osmótica, 4093
presurización, 4508
pretratamiento, 4511
prevención de corrosión, 1600
prevención de incendios, 2524
primera prensa, 2533
primera prensa principal, 2532
probador de absorción, 123
probador de adherencia de la cola,
    2842
probador de adherencia interna, 3276
probador de arrancamiento, 4312
probador de contextura, 2686
probador de desgaste, 6139
probador de fatiga de flexión, 661
probador del alisado, 5202
probador de porosidad, 4445
probador de resistencia al reventamiento,
probador de resistencia a los pliegues,
    976, 1676
probador de reventamiento, 979
probador de rigidez, 5426
probador de tracción, 5682
probador Gurley, 2949
probador MIT de pliegues, 3835
probador Mullen de reventamiento,
    3908
probador Sheffield, 5055
procedimiento, 4532
procedimiento de sulfato, 5525
procedimiento discontinuo, 609
proceso aeróbico, 224
proceso alcalino, 309
proceso anaeróbico, 359
proceso continuo, 1535
proceso de cloro, 1310
proceso de contraflujo, 1645
proceso de datos, 1823
proceso de fabricación en seco, 2133
proceso del desfibrado, 1868
proceso de sosa fría, 1432
proceso de transformación, 1560

proceso de varios pasos, 3925
proceso discontinuo, 608
proceso Mitscherlich, 3836
proceso sódico, 5212
producción, 4539
productividad, 4543
producto, 4537
producto de cartón, 4156
productos corrosivos, 1601
productos de combustión, 1468
productos de papel, 4181
productos de pasta moldeada, 3877
productos forestales, 2676
productos madereros, 6275
productos moldeados, 3874
productos químicos, 1257
profundidad, 1917
programación, 4928
programación lineal, 3497
programa de calculadora, 1484
programado no lineal, 4001
propagación, 4549
propano, 4550
propiedad, 4552
propiedades acústicas, 185
propiedades de la superficie, 5580
propiedades del cartón, 4157
propiedades del papel, 4182
propiedades de saturación, 4912
propiedades dieléctricas, 1964
propiedades eléctricas, 2208
propiedades físicas, 4299
propiedades magnéticas, 3658
propiedades mecánicas, 3733
propiedades ópticas, 4075
propiedades químicas, 1251
propiedades reológicas, 4789
propiedades tenso - deformadoras, 5464
propiedades tensoras, 5677
propiedades térmicas, 5707
propionatos, 4553
proporción lejía-madera, 3524
propulsor, 3191, 4551
protección contra el fuego, 2528
protector, 4481
protector del rodillo, 4829
protector de orillo, 2187
proteínas, 4559
proteínas de soya, 5288
protoligninas, 4560
protones, 4561
protuberante, 944
prueba, 5696
prueba de aceptación, 134
prueba de cera Dennison, 1901
prueba de envejecimiento acelerado,
prueba de envejecimiento, 130, 238
prueba de espito, 4244
prueba del cristal azul, 784
pruebas biológicas, 704
pruebas Cóncora, 1490
pruebas de absorción de aceite, 4035
pruebas de aplastamiento de los bordes,
pruebas de aplastamiento, 2184, 2551
pruebas de arrancamiento, 4313
pruebas de arrollamiento de los bordes
    de una hoja de papel, 1755
pruebas de cajas de cartón, 1528
pruebas de cera, 6136
pruebas de Cobb, 1418
pruebas de compresión, 1480
pruebas de corrosión, 1603
pruebas de corte, 5040
pruebas de debilitación, 2381
pruebas de descenso del nivel, 2098
pruebas de desgaste, 6140
pruebas de deslizamiento, 1679
pruebas de dureza, 2982
pruebas de encolado, 5130
pruebas de fatiga, 2394
pruebas de flexión, 666
pruebas de humectación superficial,
    5592
pruebas de humedad, 3869
pruebas de impacto, 3189
pruebas de impresión, 4528
pruebas del aguarrás, 5883
pruebas del cartón, 4158
pruebas del papel, 4190

pruebas de manchas de tinta, 3242
pruebas de materiales, 3717
pruebas de penetración del aceite, 4041
pruebas de perforación, 4594
pruebas de pliegue, 2660
pruebas de rasgado, 5664
pruebas de rendimiento, 4261
pruebas de resistencia, 5460
pruebas de resistencia a los ácidos,
pruebas de resistencia al plegado, 175,
    2654
pruebas de reventamiento, 980
pruebas de rigidez, 5427
pruebas de soplado, 781
pruebas de trituración, 1725
pruebas de trituración por anillos, 4804
pruebas dinámicas, 2174
pruebas en pila, 640
pruebas estáticas, 5397
pruebas indestructivas, 4000
pruebas por rayos X, 6308
pruebas químicas, 1258
Pseudotsuga, 4564
puerta, 2795
puesta a punto, 3664
puesta de tinta, 3237
puesta en abanico, 2386
puesta en espiral, 5322
puesta en hojas, 5049
puesta en hojas en rotativa, 4853
puesta en marcha, 5393
pulgada/hora, 3201
pulido especular, 5312
pulidora, 5593
pulsación, 4585
pulverización, 5348
pulverizado, 4463
pulverizador, 4586, 5347
pulverizar, 5341
puntal, 873
punta secadora, 2130
puntiagudo, 144
punto, 4396
punto de colada, 4461
punto de deformación, 6320
punto de fusión, 3746
punto de reblandecimiento, 5244
punto de rocío, 1930
punto de rotura uniforme, 888
punto de saturación de la fibra, 2470
punto de transición, 5809
punto inflamador, 2547
pureza, 4597
pureza colorimétrica, 1444
purgador, 5411, 5827
purgador de pilas desfibradoras, 887

Q

que hace opaco, 4057
quelación, 1241
quelatos, 1239
quema de piritas, 4599
quemado, 968
quemador, 964
quemador de piritas, 4600
que puede reproducirse, 4743
queratina, 3340
quetenos, 3343

R

radiación, 4613
radiación infrarroja, 3225
radiación ultravioleta, 5903
radical, 4614
radio, 4617
radioactividad, 4616
ragado, 5657
raíz, 4834
ramio, 4638

ramita, 5884
ranura, 5176
ranurado, 5178
ranurado de corte, 1768
ranurador, 5177
ranuradora divisoria, 4218
ranuras del sifón, 5501
rascador de rizado, 1688
rasgar, 5655
raspado, 5457
rayado, 4935
rayador, 5914
rayas, 4944
rayón, 4652
rayón cuam, 1734
rayón de acetato, 149
rayón de módulo grande, 3051
rayos beta, 678
rayos X, 6309
reacción exotérmica, 2343
reacción química, 1254
reacción redox, 4692
reación endotérmica, 2276
reactividad, 4660
reactivo, 4663
reactor, 4661
realimentación, 2404
rebobinado, 4788
rebobinadora, 4787
rebobinar, 4786
reborde, 6179
rebordes de bobinadora, 6236
rebosante, 4105
recalentado, 5557
recalentador de aire, 268
recalentamiento, 5558
receptividad, 4672
receptividad de la tinta, 3238
recipiente, 1068, 6003
recipiente de fibra, 2450
recipiente mixto, 1475
reclamación, 4675
recogedor de desperdicios, 3324
recoge hojas, 3433
recogepasta, 4914
recogepasta por flotación, 2586
reconstrucción, 4671
recortado, 5839
recortadora a troquel, 1956
recortadrua, 5846
recortar, 5838
recortar a troquel, 1957
recorte a troquel, 1955
recorte de máquina, 3607
recortes, 5034
recortes de cajas, 865
recortes de cortadora, 1771
recortes de fabricación,
recortes de fabricación en calandra,
    905, 1029, 3820
recortes de libros, 838
recortes de rasqueta, 2050
recortes de trapos, 4627
recortes en el acabado, 2518
recortes húmedos, 6182
recortes resistentes en estado húmedo,
    6203
recortes secos, 2118
rectificador, 4687
rectificador de silicio, 5093
recubrimiento, 2253
recuperación, 4683, 4684, 4890
recuperación de datos, 1826
recuperación de fibras, 2468, 2469
recuperación del calor, 3013
recuperación química, 1255
recuperación transversal, 1710
recursos generados, 1146
recursos hidráulicos, 6103
recursos naturales, 3946
rechazo, 4696
rechazo de depuración, 4724, 4948,
    5634
rechazos de núcleo, 1591
rechazos de rayón, 4655
rechazos simili de depuración, 809
redondez, 4862
reducción, 4696
reducción de presión, 4503

reductor, 4694
reductor de velocidad, 5315
reenrolladora de tambor doble, 2072
refinado, 622, 641, 4706
refinado cónico, 3316
refinado de alta densidad, 3044
refinado en caliente, 3098
refinado en línea de soplado, 775
refinado ligero, 923
refinado magro, 2717
refinador hidráulico, 3112
refinador Jordan, 3318
refinar, 620
refino, 4704
refino bajo presión, 4510
refino cónico, 1508
refino de discos, 2022
refino de doble disco, 2069
refinómetro, 623, 2720
refinómetro en continuo, 1534
reflectancia, 4707
reflectómetro, 4710
reflexión, 4708
reflexión especular, 5313
reflexividad, 4709
reflexividad luminosa, 3590
refractario, 4713
regadera, 5077, 5356
regadera de descontinuo, 3358
región cristalina, 1730
registrador, 4678
registrador de datos, 1824
registrador de grado de refino, 2719
registrador del grado de refino en línea, 3246
registro, 4676, 4680, 4716
registro circunferencial, 1331
registro de datos, 1825
regla, 5154
regla deflectora, 524
reguero, 5455
regulación, 5014
regulación de la humedad, 3868
regulador, 1543, 2855, 4719
regulador de la consistencia, 1516
regulador de pasta, 5437
regulador de presión, 4504
rehacimiento de pliegues, 1673
rehumectar, 4785
reinversión, 4773
rejilla, 2878
relación, 4644
relación de contraste, 1540
relación de rizado, 1686
relación de transparencia, 5817
relación de vacío, 6030
relación largo/ancho de una pista, 5362
relación químico-maderera, 1259
relajación, 4728
relé eléctrico, 2224
rellenar, 2484
relleno, 2485
relleno de tierras, 3418
remojo, 6209
rendimiento, 4096, 4260, 6319
rendimiento calórico, 3005
rendimiento estable, 5596
renuevo, 4901
reología, 4790
repartición, 5352
repartidor de pasta, 2596
repaso, 4689
repelente, 4742
repoblación forestal, 4711
repoblación forestal natural, 3945
reporte litográfico, 5804
reproducción asexual, 455
reprografía, 4745
requisitos de blanqueo, 748
resecado, 4103
resecar, 4104
reserva, 4748
reserva de libros, 839
reserva de papel, 4187
reservas bancarias, 567
residuo, 4750
residuos de aserradero, 4921
residuos de fabricación, 4357
residuos de la tala maderera, 3558

resiliencia, 4751, 4752
resina, 4340, 4841
resina de condensación, 1494
resina de fenol-formaldehido, 4278
resina de la madera, 6278
resina de melamina-formaldehido, 3744
resina de pino, 4327
resina hidrogenada, 3138
resina reforzada, 2696
resinas, 4753
resinas de intercambio iónico, 3288
resinas de poliamida, 4408
resinas de silicio, 5094
resinas de urea, 5936
resinas de urea-formaldehido, 5935
resinas epoxi, 2297
resinas fenólicas, 4281
resina sintética, 5615
resinas resistentes en estado húmedo, 6205
resinatos, 4755
resistencia, 4758
resistencia a la abrasión, 109
resistencia a la adherencia, 766
resistencia al aceite, 4043
resistencia a la coloración por álcali, 306
resistencia a la compresión, 1479
resistencia a la corrosión, 1602
resistencia a la debilitación, 2380
resistencia a la deformación transversal, 4611
resistencia a la fatiga, 2393
resistencia a la flexión, 663, 2565
resistencia a la fricción, 4957
resistencia a la fricción en seco, 2142
resistencia a la frotación húmeda, 6199
resistencia a la grasa, 2890
resistencia al agua, 6102
resistencia a la humedad, 3866, 3867
resistencia a la oxidación, 4119
resistencia a la perforación, 4592
resistencia al aplastado, 2550
resistencia al aplastamiento, 1726
resistencia a la rotura, 891, 892
resistencia al arrancamiento, 4309, 4311, 5571
resistencia al arrugado, 1724
resistencia a la sangre, 768
resistencia a las intemperies, 6143
resistencia a la tensión de espacio nulo, 6324
resistencia a la tracción, 5678, 6208
resistencia a la tracción en estado húmedo, 5678, 6208
resistencia al calor, 3014
resistencia al combado, 943
resistencia al corte, 5036
resistencia al desgaste, 4760
resistencia al deterioro, 1842
resistencia al envejecimiento, 237
resistencia al frotamiento con aceite, 4044
resistencia al fuego, 2529
resistencia al impacto, 3188
resistencia al manchado, 5379
resistencia a los ácidos, 174
resistencia a los álcalis, 303
resistencia a los golpes, 5065
resistencia a los hongos, 2755
resistencia al pandeo, 1457
resistencia al plegado, 1677, 2652, 2657, 2659
resistencia al rasgado, 5658, 5659, 5662
resistencia al rasgado del borde, 2188
resistencia al rasgado interno, 3277
resistencia al reventamiento, 978
resistencia al rozamiento, 2302
resistencia a sangrar, 753
resistencia de desgote, 2089
resistencia de dirección Z, 6322
resistencia de la adherencia interna, resistencia de la adherencia de capas, 3275, 4390
resistencia de la tinta, 4611
resistencia de la tinta al rozamiento, 3239, 3240
resistencia del sellado térmico, 3015

resistencia de un colorante, 2389
resistencia dieléctrica, 1965
resistencia elástica, 2201
resistencia eléctrica, 2225
resistencia en estado húmedo, 6201
resistencia en seco, 2144
resistencia inicial al rasgado, 3229
resistencia interna al rasgado, 3279
resistencia química, 1256
resistencia superficial, 5581, 5586
resistencia térmica, 5710
resistente a los ácidos, 170
resistente a los álcalis, 314
resistividad, 4761
resistividad eléctrica, 4762
resma, 4664
resma incompleta, 908
resonancia, 4763
resorcina, 4764
respiradero, 5995
restauración del agua, 6099
retardación, 4768
retardador, 4767
retardador de incendios, 2530
retardador de llamas, 2541
retención, 4769
retención de agua, 6104
retención de la carga, 2496
retención de los pliegues, 1674
reticular, 4771
retraso, 5179
retroceadora, 4673
retroceder, 5353
retrogradación, 4772
reutilización del agua, 6105
reutilizar, 4774
reventamiento, 972, 974
reversión de color, 1453
revés (de un papel), 5897
revés doble, 2076
revestido, 3498
revestimiento aislante, 3259
revestimiento de basalto, 594
revestimiento de lejiadora, 1979
revestimientos, 2377
revoluciones por minuto, 4783
ribete de encuadernación, 696
riel del rodillo desgotador, 5623
rigidez, 4800, 5425
rigidez del anillo, 4806
río, 4809
rizado, 1681, 1687, 1752
rizado en máquina, 3615
rizado en seco, 2120
rizado húmedo, 6073
rizador, 1753
rizado transversal, 1705
roble, 4015, 4609
robustez, 2977
rociador, 5292
rodamiento a bolas, 553
rodamiento de agujas, 3948
rodapié, 6050
rodillo, 4816
rodillo antideflector, 393
rodillo aplanado, 2555
rodillo a presión, 4505
rodillo aspirante, 5513, 5949
rodillo aspirante multicelular, 3910
rodillo a vapor, 5409
rodillo blando, 5252
rodillo bombeado, 1065
rodillo cabecero, 894
rodillo caliente, 3099
rodillo carbonizado, 1101
rodillo cóncavo de base, 1486
rodillo conducido, 2105
rodillo conductor, 3439
rodillo curvo, 1762
rodillo de barrido, 5153
rodillo de calandra, 859, 1044
rodillo de caucho, 4867
rodillo de enfriamiento, 1271
rodillo de formación, 2692
rodillo de granito, 2873
rodillo de hojas, 5053
rodillo de máquina de ondular, 1620
rodillo de prensa, 4490

rodillo de prensa aspirante, 5510
rodillo de prensa húmeda, 1637
rodillo de prensa húmeda inferior, 851
rodillo de prensa inferior, 855
rodillo de prensa superior, 5777
rodillo de rasqueta, 2057
rodillo de resorte, 5354
rodillo de resorte refrigerado por agua, 6072
rodillo de retorno de tela, 6258
rodillo desgotador, 1820, 5620, 6087
rodillo desgotador bobinado en espiral, 5324
rodillo desgotador para soporte de cojinete cerrado, 1360
rodillo desgrumador, 3593
rodillo de sostén, 508
rodillo desplazado, 5166
rodillo desplisador, 5351
rodillo de superficie lisa, 4350
rodillo de tela, 6253
rodillo de tope, 5479
rodillo de tracción, 4567
rodillo elástico de calandra, 2199
rodillo enfriado, 1272
rodillo escurridor, 5365
rodillo estucador, 420
rodillo expansor, 2347
rodillo filigranador, 3397, 6085, 6086, 6090
rodillo filigranador de vitela, 6088
rodillo flotante, 2627
rodillo fuente, 2701
rodillo guía, 2933, 3177
rodillo humectador, 1811
rodillo igualador, 1814, 1815
rodillo inclinado, 858
rodillo inferior, 856
rodillo inferior de calandra, 3347
rodillo intermedio, 3272
rodillo lateral, 5086
rodillo lleno, 2492
rodillo natatorio, 5604
rodillo ondulado, 1611
rodillo para una cara, 5104
rodillo perforado, 3078, 4688
rodillo perforado de deflección, 525
rodillo perforado desgotador, 4255
rodillo portador, 1133
rodillo prensor, 4795, 4797
rodillo ranurado, 2914
rodillo revestido de caucho, 4865
rodillo soplador, 284, 779
rodillo sujetador, 3069
rodillo superior, 4608, 5778
rodillo superior de prensa húmeda, 5774
rodillo tensador de fieltros, 2435
rodillo tensor, 5687
rodillo tomador, 4318
rodillo tomador aspirante, 5511
rodillo tomador de succión, 5507
rodillo ventilador del bolsillo, 4395
rollo de máquina, 3828
rollo de papel higiénico, 5764
rollo de papel para periódico, 6153
rollo de sobrepeso, 4113
rollo-guía automático, 485
rollo-guía de tela automático, 486
rompedor de espuma, 2629
rotación, 4856
rotámetro, 4847
rotor, 4859
rotura, 1662, 2711, 6147
rueda hidráulica, 6118
ruido, 3994
ruptura, 882
ruptura por entalladuras, 4932
ruptura por fatiga, 2392
rutilo, 4877

S

sacabocados, 1954
sacamuestras, 4893

saco, 526
saco de fondo transversal, 1704
saco de varias hojas, 3929
sacudidor, 5028
sacudimiento, 5031, 5063
sala de acabado, 2519
sala de depuración, 4949
sala de desfibradoras, 2904
sala de máquinas, 3631
sala de prensas, 4491
sala de preparación de lejía, 169
sala de suministros, 5566
sala húmeda, 3110
sales fundidas, 5192
sales haloideas, 2955
sales inorgánicas, 3251
salinidad, 4886
sangrar, 751
sangría, 752
saponificación, 4902
sasafrás, 4906
satinación, 5551, 5553
satinado, 1035, 2819, 3596, 3598, 3622
satinado dos caras, 2064
satinado en máquina, 3624
satinado sobre calandra por fricción, 2734
satinado una cara, 3626
satinómetro, 2830
saturado, 4909
sauce, 4887, 6233
saúco, 2203
savia, 4900
secadero, 2126, 2139
secado, 2134
secado al aire, 257, 261, 3550
secado al chorro de llama, 2538
secado con enganchadora, 3571
secado de pulverización, 5345
secado directo, 5738
secado en cilindro, 1794
secado en enganchadera, 2441, 2443
secado en máquina, 3619
secado en varas, 4399
secado excéntrico, 5064
secado infrarrojo, 3222
secado instantáneo, 2545
secado por aire, 3553
secador, 2122
secadora de alta velocidad, 3058
secadora de convección, 1549
secadora de corteza, 575
secadora de cubierta doble, 2067
secadora de platinas, 4374
secadora de tambor, 2113
secadora de túnel, 5871
secadora enfriadora, 5598
secadora flotadora, 2568
secador aireador, 3552
secadora Madeleine, 3639
secado rápido, 3057
secadora yankee con estriado interno, 3280
secador de fieltros, 2422
secador friccionador, 6311
seccionar, 1711
sección de calandra, 1048
sección de prensas, 4493
sección de telas, 6254
sección húmeda, 6184
seco, 2117
seco absoluto, 824
seco absoluto a cien grados, 4014, 4098
seco al aire, 101, 259
secoya, 4697, 5007
seda artificial viscosa, 6021
sedimentación, 4980
sedimento, 4979
sedimento en la secadora, 4923
sedimentos, 5096
segadora, 2990
segunda calidad, 4977
segunda capa (cartón), 791
segunda prensa, 4976
seguridad, 4734, 4880
selección, 4986

selectividad, 4987
sellado, 4962
sellado en resma, 4667
sellador, 4961
selladora térmica, 3017
sellado térmico, 3016
sellar, 4959
sembrar, 4983
semiblanqueado, 4989
semiconductor, 4995
semilla, 4981
semillero, 4011
semipasta, 2954
señal de mantilla, 728
señales del cepillo, 925
sensibilidad, 4997
sensibilidad a la presión, 4507
sensibilidad térmica, 5708
sensibilización, 4998
sensibilizador, 5000
sensor, 5002
sentido transversal, 1707
separación, 5003, 5290
separación de finos, 2515
separación de hojas por vapor, 5410
separación de las capas, 4391
separador, 5004
separador centrífugo, 1790
separador de cuchillas, 5289
separador de nudos, 3361
separador magnético, 3659
separar, 4308
sequedad, 2141
sequería, 2129
sequería de hoja llevada por aire, 265
sequería por aire de alta velocidad, 2953
serie, 5005
serrín, 4918
servicio, 5009
servicio público, 5938
servicio público de electricidad, 2226
servilleta, 5791
servilletas de papel, 4178
servilletas higiénicas, 4897
servomecanismo, 5011
seudoplasticidad, 4563
sicamoro, 5606
siega, 2991
sierra, 4917
sifón, 582, 5115
sifón aspirador, 5505
sifón aspirante de prensa, 3068
silicato de sodio, 5233
silicatos, 5090
sílice diatomácea, 1946
silicio, 5091
silo de fondo vivo, 3537
silo de virutas, 1296
silvicultura, 2677, 5099
simetría, 5607
simili, 795
sin abrillantar, 5917
sin ácido, 161
sin blanquear, 5904
sin cenizas, 458
sin cola, 5925
sincrónico, 5609
sincronismo, 5608
sinergismo, 5611
sin necesidad de carbono, 3936
sin pasta mecánica, 2918, 6266
sin poros, 4331
sin satinar, 5916
síntesis, 5612
sisal, 5117
sistema, 5618
sistema analógico, 362
sistema binario, 690
sistema cerrado, 1359
sistema de admisión, 3263
sistema de clasificación previa, 133
sistema de control, 1547
sistema de corriente de pasta antes de la máquina, 422
sistema de entintado, 3236
sistema de estucado húmedo, 6190
sistema de medición, 3773
sistema digital, 1985

sistema enfriador, 1576
sistema hidráulico, 3122
sistema métrico, 3773, 3791
sistema operacional, 3847
sobrecielo del bosque, 2668
sobrecocción, 4101
sobreestucado, 4100
sobreimpresión, 4109
sobrenadante, 5560
sobrepeso, 4112
sobres, 2292
sobretamaño, 4111
sodio, 5215
soflamado, 788
sol, 5257
soldadura, 5258, 6175
soldaduras industriales, 3218
solidificación, 5262
sólido, 5260
soltar, 4729
solubilidad, 5269
solubilización, 5270
soluble en agua, 6109
solución, 5272
solución de origen, 2702
solución kraft, 3373
solvatación, 5273
someter a tensión, 5463
sopladura, 773
soporte, 794, 2968, 5567
soporte de cojinete, 619, 3319
soporte de estucado, 1414
soporte del estucado, 1402
soporte para libros, 829
soporte para papel pintado, 2971
soporte posterior, 516
soportes mezclados, 3840
sosa, 5209
sosa cáustica, 1171, 5230
stock de recipientes, 1072
suavizador, 5600
suavizador del agua, 6108
suavizar el agua, 5245
sublimación, 5491
subproducto, 1000
substancia, 5493
substituto, 5494
substituto del papel, 4189
substrato, 5495
sucrosa, 5496
sulfamatos, 5519
sulfato de aluminio, 335
sulfato de amonio, 351
sulfato de bario, 573
sulfato de calcio, 1022
sulfato de calcio precipitado, 1714, 4234
sulfato de cinc, 6331
sulfato de magnesio, 3652
sulfato de sodio, 5235
sulfato de sodio sin refinar, 4888
sulfatos, 5528
sulfhidrato sódico, 5228
sulfito a base de calcio, 1013
sulfito de calcio, 1023
sulfito de magnesio, 3653
sulfito de sodio, 5237
sulfito neutro semiquímico, 3937
sulfitos, 5536
sulfogrupos, 5537
sulfonatos, 5538
sulfonatos de xileno, 6306
sulfuración, 5529
sulfuro carbonilo, 1108
sulfuro de cinc, 6332
sulfuro de hidrógeno, 3145
sulfuro de sodio, 5236
sulfuro reducible, 4693
sulfuros, 5530
sumersión, 5492
suministro, 5565
suministro de agua, 6112
suministro de energía, 4468
supercalandra, 5550
super duplex, 2160
superficie, 5568
superficie aerodinámica, 267
superficie de desfibrado, 2907
superficie de reacción hidráulica, 3136

superficie específica, 5302
superintendente auxiliar, 474
suspensión de la pasta, 5595

T

tablas, 5624
tabulación, 5626
tabulador automático de datos, 1822
taco, 4382, 6264
tacómetro, 5627
tachuela, 5628
tala, 2417
tala de árboles por cable, 5140
talador agrupador de troncos, 2415
talador arrastrador de troncos, 2416
talar, 2414
talco, 5636
taller de maderas, 6279
taller mecánico, 3633, 3663
tallo, 5358, 5418
tall oil, 5637
tamaño de la hoja, 5054
tamaño del orificio, 3073
tamaño del papel, 4185
tamaño natural, 2747
tamaños corrientes, 5438
tamarisco, 5640
tambor, 2110
tambor de enrollar, 6239
tambor de fibra, 2459
tambor descortezador, 578
tambor lavador, 2115, 6054
tamiz, 5088
tampón, 936
tanque, 5641
tanque almacenador, 5442
tanque de decantación, 5018
tanque de fundición, 5194
tanque de soplado, 780
tanque igualador, 5594
tanques de destensado, 2561
tapa cerámica de sifón, 1219
tapa de prensaestopas, 5485
tapadero de prensa, 4494
tara, 5651
tarjeta, 1122
tarjeta de felicitación, 2899
tarjeta perforada, 4589
técnica, 5665
techado asfáltico, 467
techado de fieltro, 2429
tejer, 6144
tejido, 6145, 6294
tejido de alfombras, 1130
tejido esponja, 5792
tejido leñoso, 6304
tela, 2370
tela ascendente, 5933
tela de enfardar, 552
tela de formación, 2691
tela de forro, 4953
tela de punto, 3356
tela de trama, 6165
tela inclinada, 3203
tela lisa, 4351
tela metálica, 6242, 6243
tela metálica con trama de seda, 5085
tela metálica de la máquina de papel, 3636
tela metálica sin fin, 2272
tela no tejida, 4007
tela sargada, 5885
tela soporte, 511
telas verjuradas, 3402
televisión, 5669
temperatura, 5670
temperatura baja, 3581
temperatura crítica, 1701
temperatura de ebullición, 815
temperatura elevada, 3054
tenacidad, 5788
tendencia a la pelusa, 2600
teñido blanco, 5752
tensado de fieltros, 2434

tensiómetro, 5684
tensión, 5461, 5685
tensión crítica, 1700
tensión de la hoja continua, 6159
tensión en seco, 2101, 2102
tensión integrada, 941
tensión superficial, 5587
tensor, 5467
tensor cargado de aire, 278
tensor del fieltro, 2433
tensor de tela, 6257
tercera mano, 5731
termodinámica, 5713
termómetro, 5716
termoplásticos, 5717
terpenos, 5689
terpinenos, 5690
terreno, 5256
testlinerboard, 5697
test liner de yute, 3333
test liner kraft, 3378
tetracloruro de carbón, 1105
textiles de papel, 4191
textura, 5702
tiempo, 5746
tiempo de ajuste, 5015
tiempo de detención, 2169
tiempo de escurrimiento, 2090
tiempo de máquina, 3635
tiempo de ocupación, 3070
tiempo de reacción, 4658
tiempo de refinado, 644
tiempo de repuesta, 4765
tiempo muerto, 3176
tierra diatomácea, 1945
tilo, 5743
tilo de América, 605
tilosis, 5898
tina, 4433, 5975, 5979
tina de aspiración, 4966
tina de cilindro, 1800
tina de descarga por soplado, 778
tina de desfibradora, 2902
tina de máquina, 3609, 5481, 5482
tina de pasta, 5431
tina orillada, 844
tina para disolver, 2037
tinta, 3232
tintado, 5377
tinta fijada por calor, 3019
tinta magnética, 3657
tinta para imprimir, 4524
tinta para metales, 3766
tinta para offset, 4025
tinta para tipografía, 3449
tinte, 5024, 5750
tintes de añilina, 3210
tiocarbamatos, 5727
tiocarbonatos, 5728
tiocianato de sodio, 5238
tiolígninas, 5729
tiosulfato de sodio, 5239
tiosulfatos, 5730
tiro húmedo, 6183
titanio, 5757
títulos verjurados antiguos, 403
tixotropía, 5733
toallas de papel, 4192
tobera de aire, 280
tobera de estucado, 1638
tolerancia, 5767
tolerancia de peso, 6172
tolueno, 5768
tolva, 3086
tolva de alimentación, 2410
tolva de pesadas, 6168
toma, 3244
toma automática, 3464, 4314
toma automática de la humedad, 3864
toma automática de vacío, 5947
toma de aire, 271
toma de muestras circulares, 846
toma de tinta, 3235
tonelada, 5769
tonelada americana, 5074
tonelada inglesa, 3569
tonelada métrica, 3792
tope, 987, 989
tope de guía de alimentación, 2409

tornillo, 4950
tornillo de ajuste, 212
tornillo de transportadora, 1564
tornillo nivelador, 3458
torno elevador, 3066
torre, 5794
torre de absorción, 124
torre de blanqueo, 745
torre de lavado, 4955
torre de recuperación, 4686
torsión, 5785
torta de filtro prensa, 2504
toxicidad, 5795
trama, 6164, 6284
transductor, 5803
transferencia con rodillo tomador liso, 4349
transferencia de energía, 2280
transferencia de la masa, 3708
transferencia de tinta, 3243
transferencia de vacío, 5950
transformación, 1551, 1555
transformación del papel, 4160
transformador, 1554, 5807
transformador de papel, 4161
transistor, 5808
translucidez, 5810, 5811
transmisión, 2106
transmisión de calor, 3023
transmisión de datos, 1829
transmisión de fuerza eléctrica, 2223
transmisión diferencial, 1967
transmisión por correa, 655
transmisión por tornillo sin fin, 6289
transmitencia, 5814
transmitencia total, 5787
transparencia, 5815, 5816
transparencia de la tinta, 3241
transparencia desigual, 6230, 6231
transparencia nubosa, 1369
transparencia uniforme, 1362, 3570
transparente, 5818
transpiración, 5821
transportado por aire, 246
transportador, 1562
transportador alimentador, 2406
transportador de cangilones, 932
transportador de correa, 654
transportador de gravedad, 2879
transportador de pesos, 6169
transportador de tornillo, 4951
transportador de virutas, 1283
transporte, 1561, 5823
trapería, 4631
trapo, 4621
traqueidas, 5797
traqueo, 5027
traspaso de calor, 3021
trasplante, 5822
tratameinto de superficie, 5589
tratamiento ácido, 181
tratamiento al álcali, 307
tratamiento antideslizante, 408
tratamiento antipliegues, 1671
tratamiento a prueba de hongos, 2754
tratamiento biológico, 702, 705
tratamiento del agua, 6113
tratamiento de la pasta en caliente, 3100
tratamiento de los efluentes, 2194
tratamiento del lodo activado, 195
tratamiento de modificación, 3853
tratamiento directo en estufa de vapor, 2010
tratamiento en estufa de vapor, 5407
tratamiento fuera de máquina, 4022
tratamiento indirecto en estufa de vapor, 3212
tratamiento primario, 4517
tratamiento químico, 1260
tratamiento secundario, 4974
tratamiento sobre máquina, 4052
tratamiento terciario, 5692
tratamiento térmico, 3024
trementina sulfática, 5527
trillador de trapos, 4634
tríplex, 5851
trituración, 889
trituración por martillos, 2960

triturado, 1718
triturador, 5189
trituradora de madera, 3063
trituradora de martillos, 3187
triturador de grumos, 3592
triturador de virutas, 1284
triturar, 1717
troceadora, 1291
troceadora de tambor, 2112
trocha, 5147
troncar, 5856
tronco, 816
tronco entero, 5833
tronzadora helicoidal, 5321
trozos de fibra, 2713
Tsuga, 5858
tubería, 4337
tubero, 4335
tubo, 4334, 5865
tubo colector, 2996
tubo de Pitot, 4346
tubo de rayos catódicos, 1158
tubo de Venturi, 5999
tubo helicoidal, 5323
tubo hervidor, 6114
tubos de succión, 5508
tulipero de Virginia, 5868
turbidímetro, 5873
turbieza, 5874
turbina, 5875
turbina a gas, 2794
turbina de vapor, 5412
turbina hidráulica, 6115
turbio, 5872
turbogenerador, 5876
turbulencia, 5877
turno, 5059, 5789
tuya, 5739

U

una capa, 5111
unidad, 5920
unidad de glucosa, 2837
unidad sacudidora, 5030
unidad térmica británica, 500
uniformidad, 5919
unión, 817
unión con hilado, 5360
unión covalente, 1653
unión de fibras, 2448
unión de hidrógeno, 3140, 3141
unión doble, 2063
unión entre fibras, 3268
unión glucósida, 2839
unión oxidrilo, 3164
unión por puntos, 5336
unión química, 1243
unión soldada, 6174
urdimbre, 6046, 6049
urea, 5934
uretanos, 5937
uso múltiple, 3920
utilización, 5939

V

vaciado por inyección, 3231
vacío, 5940
vagonada, 1127
vagón de tren, 4636
valencia, 5951
válvula, 5952
válvula de bola, 556
válvula de cierre, 5084
válvula de compuerta, 2797
válvula de desahogo, 4738
válvula de descarga, 2017
válvula de estrangulación, 5737
válvula de mariposa, 990
válvula de retención, 1237

válvula de soplado, 782
válvula reductora, 4695
vanillina, 5955
vapor, 5402, 5957, 5966
vapor de agua, 6116
vapores, 2750
vaporización, 5959
variabilidad, 5967
variable, 5968
variable de elaboración, 4535
variación, 5970
variación estacional, 4969
variedad, 5971
varilla, 4813
vehículo, 5982
velocidad, 5987
velocidad ajustable, 210
velocidad circunferencial, 1332
velocidad constante, 1517
velocidad crítica, 1702
velocidad de borbor, 5340
velocidad de corte, 5037
velocidad de la regla, 5158
velocidad de régimen, 1924
venta, 4885
ventas de pasta, 4580
ventilación, 221, 5996
ventilador, 2384, 5997
ventilador aspirador, 3213
ventilador de aspiración, 2340
ventilador soplador, 245, 772
ventilador soplador centrífugo, 1208
ventilar, 220
verde, 2896
verjurado, 3395
verjurado antiguo, 3396
verjurados, 3399
vertedero, 6173
vertedor de virutas, 1280
vía ferroviaria, 4637
vibración, 6005
vibrador, 6006
vida del entrepaño, 5056
vinil celulosa, 6009
viruta, 1276
viruta de dobladura, 660
viruta de semidoblamiento, 4988
virutado, 1294
virutas aceptadas, 135
virutas de madera, 6261
virutas finas de madera, 6282
viruta sin probar, 4005
viscoelasticidad, 6015
viscómetro, 6016
viscosa, 6018
viscosidad, 6023
viscosidad aparente, 416
viscosidad de la cuprietilenamina, 1746
viscosidad del cupramonio, 1743
viscosímetro, 6022
viscosímetro rotativo, 4855
vitela, 5983, 6291
vitela pergamino, 4207
vitrificación, 6027
vivero de árboles, 5832
voladizo, 1073
volatilidad, 6032
volteadora de rollos de hoja continua, 6162
voltímetro, 6033
volumen, 6034
volumen de vacío, 6031
volumen específico, 5303
volumen específico aparente, 415
vórtice, 6035
vuelta, 4782
vulcanizado, 6038

xantato celulósico, 1198

xantato de almidón, 5390
xanturo de almidón, 5391
xilenos, 6305
xilenosulfonato de sodio, 5240

xilosa, 6307
yardaje, 6313
yodo, 3284
yoduro de potasio, 4454

yute, 3325
zeolitas, 6323
zunchamiento, 5448
zunchamiento de los rollos,

INDEX TO ADVERTISERS

Bird Machine Co., Inc. 113

Black Clawson Co. 73

Gebr. Böhler/Vereinigte Edelstahlwerke 131

Cameron Machine S.A. 119

Cellier S.A. 53, 99

Dravo Corp., Water & Waste Treatment Div. 137

English China Clays Sales Co. Ltd. 35

Enso-Gutzeit 8

Hooker Chemicals & Plastics Corp. 33

Miller Freeman Publications, Inc. 235

Nash Engineering Co. .... Opp. Table of Contents

Norden Automation Systems, AG, Zurich 41

Ross Pulp & Paper Div., Midland-Ross of Canada Ltd., Ross Air Systems Div. 103

Testing Machines, Inc. 128

Valmet Oy 6

Vickery's Ltd. 51

Westvaco 69

the *PULP & PAPER* bookshelf

TRANSPORT & HANDLING IN THE PULP & PAPER INDUSTRY

Brings together all material presented and discussed at the first international Symposium, Rotterdam, Holland, April, 1974. All papers, all discussions and all illustrative materials shown to delegates have been carefully edited for presentation in this book. Its 25 chapters are arranged in five major groups: raw material storage and handling; finished products handling in mill and terminal; integrated transport systems; and future trends. 360 pages, 200 illustrations, LC 74-20162, ISBN 0-87930-033-7, $27.50.

FIBER CONSERVATION AND UTILIZATION

Covers the complete proceedings, including discussion periods and all visual material presented at the seminar, Chicago, May 1974. The first section of 12 chapters, Wastepaper as a Fiber Resource, discusses the impact of recycling on fiber supply, improving recycling efficiency, processing techniques, competitive demands, transport costs, government regulations. The 10 chapters in the second section, Wood Resources for Present and Future Needs, discusses methods for meeting growing wood fiber needs, residuals, forest wastes, harvesting techniques, genetics, pulping, energy generation. 288 pages, 100 illustrations, LC 74-20163, ISBN 0-87930-032-9, $27.50.

PRACTICAL COMPUTER APPLICATIONS FOR THE PULP AND PAPER INDUSTRY

Brings together in one source the latest developments in computer control in all areas of pulp/and paper mill operations. The 40 chapters in this manual, selected from recent issue of PULP & PAPER and PULP & PAPER INTERNATIONAL, written by experienced industry experts, are arranged by sections: fundamentals of computer control; evaluating control systems; papermaking applications; pulp mill; bleaching; coating/coloring; shipping/finishing; stores/inventory control. 186 pages, 206 illustrations, LC ·74-20166, ISBN 0-87930-036-1, $12.50.

1976 Post's PULP & PAPER DIRECTORY

The premier directory of the pulp, paper and paperboard industry. Gives complete detail on every mill, all major converting plants, all headquarters offices in the USA and Canada. Lists over 11,000 management, production, technical and engineering people. Complete information on process equipment, machinery, production. Contains classified mill buyers' guide to 2,700 items of mill equipment, supplies, chemicals, services. 668 pages, $35.

ATLAS OF WESTERN EUROPEAN PULP & PAPER INDUSTRY

On nine separate large-scale maps, each 20" x·14" (50 x 35 cm), this new atlas shows all pulp, paper and paperboard mills in Western Europe over 5,000 tpy. Symbols show size ranges and type of mill at each location. A separate index page lists companies and their mills by country with map coordinates for locating each. Prepared

METRICATION FOR THE PULP AND PAPER INDUSTRY

By Kenneth E. Lowe, this new book was written specifically to help the industry get ready for and carry out metric conversion. It provides an overall understanding of the system for those unfamiliar with it. It is a ready reference guide to the metric system, the new SI system and to the ISO paper standards. It is also a valuable text for training paper industry employees in the actual use of metric. Useful metric procedures, based on current industry practices, are detailed. Over 70 separate tables; specific conversion tables; illustrated definitions of technical terms; appendixes; index; 192 pages, 8½ x 11" (21.6 27.9 cm), LC 74-20165, ISBN O-87930-034-5, $24.50.

X

PULP AND PAPER MILL MAP—1976

The new 6th edition of a colorful map of the USA and Canadian industry, featuring exact location of *every* pulp, paper and paperboard mill, each shown with a symbol identifying type. A complete index lists all mills by city, state/province. The six major North American forest types, with a legend describing species in each, are shown in colors: West Coast forest, Western, Central hardwood, Northern, Southern, Tropical. Size 42" x 29" (approx. 107 x 74 cm), rolled in container. $12.50 (professional mounting/framing services available; write for brochure).

INTERNATIONAL PULP & PAPER DIRECTORY

Prepared by the editors of PULP & PAPER INTERNATIONAL, this is the most comprehensive, complete source of information about the major pulp, paper and paperboard mills of the world (excluding USA & Canada). Lists: mills (over 10,000 tpy) and headquarters by country; over 5,300 officers, management, production, technical, engineering personnel; grades, capacity; production facilities, equipment, machine sizes/speeds; coating; converting; power. Separate worldwide alphabetical indexes: mills; companies; personnel. International classified equipment, supplies buyers' guide. 452 pages. $30.

AN INTRODUCTION TO PAPER INDUSTRY INSTRUMENTATION

By John R. Lavigne, this comprehensive book is the only basic reference source of all aspects of instrumentation and control. Valuable to those experienced in the field, it is especially useful as a self-training course for those with limited experience. The four major areas of automatic control systems are covered: primary measurements; signal transmission; automatic controllers; and final control elements. Examples of actual installations are shown. 456 pages, 300 illustrations, LC 72-78929, ISBN 0-87930-013-2, $15.

by the editors of PULP & PAPER INTERNATIONAL, drawn by a noted European cartographer and printed by map specialists, it also features in color the major forest areas, main roads, rivers and cities. Flexible binding allows removal of pages for assembly into a large wall map. $25.

PULP & PAPER, Book Department, 500 Howard Street, San Francisco, Cal. 94105, U.S.A.

Please send the following publications for 10-day examination. After reviewing we will have your invoice paid or return any not wanted. (☐ Check here if you enclose payment with order, *including sales tax in Calif. & N.Y.* Publisher will pay postage and shipping expense. Prompt refund if returned.)

☐ Transport & Handling in the Pulp & Paper Industry, US $27.50
☐ Fiber Conservation & Utilization, US $27.50
☐ Practical Computer Applications for the Pulp & Paper Industry, US $12.50
☐ An Introduction to Paper Industry Instrumentation, US $15
☐ Metrication for the Pulp & Paper Industry, US $24.50
☐ 1976 Post's Pulp & Paper Directory, US $35
  Pulp & Paper Mill Map (USA & Canada), 1976, 6th Edition, US $12.50
☐ International Pulp & Paper Directory, US $30
☐ Atlas of Western European Pulp & Paper Industry, US $25

name _____

company or organization name   title or position

address _____

city _____

state/province   zip